T0350195

CAMBRIDGE STUDIES IN ADVANCED MATHEMATICS 212

Editorial Board
J. BERTOIN, B. BOLLOBÁS, W. FULTON, B. KRA, I. MOERDIJK, C. PRAEGER, P. SARNAK, B. SIMON, B. TOTARO

ANALYTIC COMBINATORICS IN SEVERAL VARIABLES

Discrete structures model a vast array of objects ranging from DNA sequences to internet networks. The theory of generating functions provides an algebraic framework for discrete structures to be enumerated using mathematical tools. This book is the result of 25 years of work developing analytic machinery to recover asymptotics of multivariate sequences from their generating functions, using multivariate methods that rely on a combination of analytic, algebraic, and topological tools. The resulting theory of analytic combinatorics in several variables is put to use in diverse applications from mathematics, combinatorics, computer science, and the natural sciences. This new edition is even more accessible to graduate students, with many more exercises, computational examples with Sage worksheets to illustrate the main results, updated background material, additional illustrations, and a new chapter providing a conceptual overview.

Robin Pemantle is Merriam Term Professor of Mathematics at the University of Pennsylvania, working in the fields of probability theory and combinatorics. He received his bachelor's degree from Berkeley and his PhD from MIT. He is a Fellow of the AMS and IMS and a winner of the Rollo Davidson Prize.

Mark C. Wilson is Senior Teaching Faculty at the College of Information and Computer Sciences at the University of Massachusetts, Amherst. He received his PhD in mathematics from the University of Wisconsin–Madison. He is Editor-in-Chief of *Notices of the American Mathematical Society* and life member of the Combinatorial Mathematics Society of Australasia.

Stephen Melczer is Assistant Professor in the Department of Combinatorics and Optimization at the University of Waterloo. He received doctorates from the École normale supérieure de Lyon and the University of Waterloo. He is a recipient of a Governor General Silver Academic Medal and previously published the textbook *An Invitation to Analytic Combinatorics*.

CAMBRIDGE STUDIES IN ADVANCED MATHEMATICS

Editorial Board

J. Bertoin, B. Bollobás, W. Fulton, B. Kra, I. Moerdijk, C. Praeger, P. Sarnak, B. Simon, B. Totaro

All the titles listed below can be obtained from good booksellers or from Cambridge University Press. For a complete series listing, visit www.cambridge.org/mathematics.

Already Published

172. T. Ceccherini-Silberstein, F. Scarabotti & F. Tolli *Discrete Harmonic Analysis*
173. Garrett *Modern Analysis of Automorphic Forms by Example, I*
174. Garrett *Modern Analysis of Automorphic Forms by Example, II*
175. G. Navarro *Character Theory and the McKay Conjecture*
176. P. Fleig, H. P. A. Gustafsson, A. Kleinschmidt & D. Persson *Eisenstein Series and Automorphic Representations*
177. E. Peterson *Formal Geometry and Bordism Operators*
178. A. Ogus *Lectures on Logarithmic Algebraic Geometry*
179. N. Nikolski *Hardy Spaces*
180. D.-C. Cisinski *Higher Categories and Homotopical Algebra*
181. A. Agrachev, D. Barilari & U. Boscain *A Comprehensive Introduction to Sub-Riemannian Geometry*
182. N. Nikolski *Toeplitz Matrices and Operators*
183. A. Yekutieli *Derived Categories*
184. C. Demeter *Fourier Restriction, Decoupling and Applications*
185. D. Barnes & C. Roitzheim *Foundations of Stable Homotopy Theory*
186. V. Vasyunin & A. Volberg *The Bellman Function Technique in Harmonic Analysis*
187. M. Geck & G. Malle *The Character Theory of Finite Groups of Lie Type*
188. B. Richter *Category Theory for Homotopy Theory*
189. R. Willett & G. Yu *Higher Index Theory*
190. A. Bobrowski *Generators of Markov Chains*
191. D. Cao, S. Peng & S. Yan *Singularly Perturbed Methods for Nonlinear Elliptic Problems*
192. E. Kowalski *An Introduction to Probabilistic Number Theory*
193. V. Gorin *Lectures on Random Lozenge Tilings*
194. E. Riehl & D. Verity *Elements of ∞-Category Theory*
195. H. Krause *Homological Theory of Representations*
196. F. Durand & D. Perrin *Dimension Groups and Dynamical Systems*
197. A. Sheffer *Polynomial Methods and Incidence Theory*
198. T. Dobson, A. Malnič & D. Marušič *Symmetry in Graphs*
199. K. S. Kedlaya *p-adic Differential Equations*
200. R. L. Frank, A. Laptev & T. Weidl *Schrödinger Operators: Eigenvalues and Lieb–Thirring Inequalities*
201. J. van Neerven *Functional Analysis*
202. A. Schmeding *An Introduction to Infinite-Dimensional Differential Geometry*
203. F. Cabello Sánchez & J. M. F. Castillo *Homological Methods in Banach Space Theory*
204. G. P. Paternain, M. Salo & G. Uhlmann *Geometric Inverse Problems*
205. V. Platonov, A. Rapinchuk & I. Rapinchuk *Algebraic Groups and Number Theory, I (2nd Edition)*
206. D. Huybrechts *The Geometry of Cubic Hypersurfaces*
207. F. Maggi *Optimal Mass Transport on Euclidean Spaces*
208. R. P. Stanley *Enumerative Combinatorics, II (2nd Edition)*
209. M. Kawakita *Complex Algebraic Threefolds*
210. D. Anderson & W. Fulton *Equivariant Cohomology in Algebraic Geometry*
211. G. Pineda Villavicencio *Polytopes and Graphs*

Analytic Combinatorics in Several Variables

Second Edition

ROBIN PEMANTLE
University of Pennsylvania

MARK C. WILSON
University of Massachusetts, Amherst

STEPHEN MELCZER
University of Waterloo, Ontario

CAMBRIDGE
UNIVERSITY PRESS

CAMBRIDGE
UNIVERSITY PRESS

Shaftesbury Road, Cambridge CB2 8EA, United Kingdom

One Liberty Plaza, 20th Floor, New York, NY 10006, USA

477 Williamstown Road, Port Melbourne, VIC 3207, Australia

314–321, 3rd Floor, Plot 3, Splendor Forum, Jasola District Centre, New Delhi – 110025, India

103 Penang Road, #05–06/07, Visioncrest Commercial, Singapore 238467

Cambridge University Press is part of Cambridge University Press & Assessment,
a department of the University of Cambridge.

We share the University's mission to contribute to society through the pursuit of
education, learning and research at the highest international levels of excellence.

www.cambridge.org
Information on this title: www.cambridge.org/9781108836623

DOI: 10.1017/9781108874144

First edition © Robin Pemantle and Mark C. Wilson 2013
Second edition © Robin Pemantle, Mark C. Wilson, and Stephen Melczer 2024

This publication is in copyright. Subject to statutory exception and to the provisions
of relevant collective licensing agreements, no reproduction of any part may take
place without the written permission of Cambridge University Press & Assessment.

First published 2013
Second edition 2024

A catalogue record for this publication is available from the British Library

A Cataloging-in-Publication data record for this book is available from the Library of Congress

ISBN 978-1-108-83662-3 Hardback

Cambridge University Press & Assessment has no responsibility for the persistence
or accuracy of URLs for external or third-party internet websites referred to in this
publication and does not guarantee that any content on such websites is, or will
remain, accurate or appropriate.

To the memory of Philippe Flajolet, on whose shoulders stands all of the work herein.

Contents

Preface to the second edition

There is no perfect time to write a textbook for a field in its infancy. Act too early and the theory might not have the cohesive structure it will eventually develop, but act too late and you might miss an opportunity to encourage new collaborators to enter and shape the area. The first edition of this book was timed to strike a balance between these two extremes: after a sufficient framework for analytic combinatorics in several variables had been developed, but during a time when fundamental results were still being discovered and incorporated.

As a consequence of this choice, the first edition of the text, while influential and put to use by many others in enumerative combinatorics, was presented in a way that many end users found difficult to follow. Having been given the opportunity to create a second edition of this text, after a decade of further development, we are now able to improve both the content and presentation of the field. We have been conscious in this rewriting of making the book more useful for a variety of readers having different motivations, including making it easier to look up and cite desired asymptotic results.

For the second edition, the original authors welcome our active collaborator Stephen Melczer, whose own introductory book on this topic [Mel21] was published recently, and who has rejuvenated the entire enterprise. In contrast to [Mel21], which skips much of the advanced topological and geometric approach to analytic combinatorics in several variables (ACSV) to focus more on elementary arguments and explicit computation, this text remains dedicated to developing the theory in its most general, and most powerful, form. The field of ACSV has flourished since the publication of the first edition, including numerous workshops, seminars, summer school courses, and many publications exploring applications of the theory [Wil15; dALN15; MM16; Pan17; Vid17; Kov19; Mis19; MW19; RWZ20; GE20; Geo21; GFS21; KLM21; GWW21; Len+23].

Over the last ten years, we have gained an improved understanding of the technical parts of the theory. Perhaps the largest change is to give ACSV a rigorous foundation using stratified Morse theory, whereas in the first edition Morse-theoretic arguments were used to motivate constructions that were then verified with other techniques. In addition to fixing numerous typographical and other errors in the first edition, some the fault of the authors and some of the publisher, the following content changes have been made to improve the book.

- The chapters in Parts I (Combinatorial Enumeration) and II (Mathematical Background) have mainly kept their general structure, however much of their discussion has been rewritten. Of particular note, Section 2.4 has been revised to better explain how ACSV for rational functions extends to algebraic functions via diagonal embeddings, and Section 5.4 has been revised to better explain the proof of Theorem 5.3 (formerly Theorem 5.4.8). Chapter 6 in the first edition has also been moved after the former Chapters 7 and 8 so that Part II now ends in Chapter 6 (which was Chapter 7 in the first edition).
- Part III (Multivariate Enumeration) begins with Chapter 7, which has been completely overhauled and is almost entirely new. The second edition is constructed to put the large majority of the topological and homological arguments in this chapter and the appendices. The main output of the chapter is an expression for coefficient asymptotics as a finite integer sum of saddle-point-like integrals, and those wanting to skip the homological material can simply assume this decomposition in later chapters.
- Chapter 8, which is a complete re-imagining of Chapter 6 in the first edition, discusses how to compute the quantities needed for an asymptotic analysis in a computer algebra system. In contrast to the first edition, we now put additional focus on computing the quantities needed for ACSV – this explains its postponement until Part III.
- Chapters 9–11 have been reworked to begin from the decomposition described in Chapter 7, streamlining their presentation, and to have more explicit results that can be easily cited. Section 11.4 has also been expanded to include a worked example of solving a connection problem via creative telescoping.
- The appendices have been revised and enlarged to be more self-contained, and to give readers a more complete explanation of the constructions they will need for ACSV.
- We have greatly increased the number of exercises and examples, and added many more signposts and guides so that readers with different motivations

can find what they are looking for. Exercises have been split into in-text (shorter and more straightforward, meant to help the reader think over the material) and end-of-chapter (more challenging) problems. We have also listed open problems and ongoing research in Chapter 13.
- Finally, we have created Sage worksheets that cover most of the examples in the book, and some of the exercises.

Supplementary material, including Sage worksheets and a maintained list of errata, are available from the book website

`http://acsvproject.org/acsvbook`

More general resources for the ACSV project are available at

`http://acsvproject.org`

The authors thank our colleagues who helped with proofreading and otherwise checking the manuscript, including Nick Beaton, Jeremy Chizewer, Jacob Cordeiro, William Dugan, Stephen Gillen, Kaitian Jin, Alexander Kroitor, Geoffrey Pritchard, Stephan Ramon Garcia, and Josip Smolčić. We thank the anonymous reviewers consulted by the publisher, and in particular one reviewer who set us straight on the inner workings of Thom's Isotopy Theorem. We thank Herman Gluck for help with topology and Frank Sottile for considerable help with computational algebra. A special word of thanks is due to Yuliy Baryshnikov. Not only did we learn most of the recent material from him or with him, but he has remained available for consultation during the entire production of the second edition. The authors each thank their families for their patience and support.

Preface to the first edition

The term "Analytic Combinatorics" refers to the use of complex analytic methods to solve problems in combinatorial enumeration. Its chief objects of study are generating functions [FS09, page vii]. Generating functions have been used for enumeration for over a hundred years, going back to Hardy and, arguably, to Euler. Their systematic study began in the 1950s [Hay56]. Much of the impetus for analytic combinatorics comes from the theory of algorithms, arising for example in the work of Knuth [Knu06]. The recent, seminal work [FS09] describes the rich univariate theory with literally hundreds of applications.

The multivariate theory, as recently as the mid-1990s, was still in its infancy. Techniques for *deriving* multivariate generating functions have been well understood, sometimes paralleling the univariate theory and sometimes achieving surprising depth [FIM99]. Analytic methods for recovering coefficients of generating functions once the functions have been derived have, however, been sorely lacking. A small body of analytic work goes back to the early 1980s [BR83]; however, even by 1995, of 100+ pages in the Handbook of Combinatorics devoted to asymptotic enumeration [Odl95], multivariate asymptotics received fewer than six.

This book is the result of work spanning nearly 15 years. Our aim has been to develop analytic machinery to recover, as effectively as possible, asymptotics of the coefficients of a multivariate generating function. Both authors feel drawn to this area of study because it combines so many areas of modern mathematics. Functions of one or more complex variables are essential, but also algebraic topology in the Russian style, stratified Morse theory, computational algebraic methods, saddle point integration, and of course the basics of combinatorial enumeration. The many applications of this work in areas such as bioinformatics, queueing theory, and statistical mechanics are not surprising when we realize how widespread is the use of generating functions in applied combinatorics and probability.

The purpose of this book is to pass on what we have learned, so that others may learn it and use it before we forget it. The present form of the book grew out of graduate-level mathematics courses that developed, along with the theory, at the University of Wisconsin, Ohio State University, and the University of Pennsylvania. The course was intended to be accessible to students in their second year of graduate study. Because of the eclectic nature of the required background, this presents something of a challenge. One may count on students having seen calculus on manifolds by the end of a year of graduate studies, and some complex variable theory. One may also assume some willingness to do some outside reading. However, some of the more specialized areas on which multivariate analytic combinatorics must draw are not easy to get from books. This includes topics such as the theory of amoebas [GKZ08] and the Leray–Petrovsky–Gårding theory of inverse Fourier transforms. Other topics such as saddle point integration and stratified Morse theory exist in books but require being summarized in order not to cause a semester-long detour.

We have dealt with these problems by summarizing a great amount of background material. Part I contains the combinatorial background and will be known to students who have taken a graduate-level course in combinatorial enumeration. Part II contains mathematical background from outside of combinatorics. The topics in Part II are central to the understanding and execution of the techniques of analytic combinatorics in several variables. Part III contains the theory, all of which is new since the turn of the millennium and only parts of which exist in published form. Finally, there are appendices, almost equal in total size to Part II, which include necessary results from algebraic and differential topology. Some students will have seen these but for the rest, the inclusion of these topics will make the present book self-contained rather than one that can only be read in a library.

We hope to recruit further researchers into this field, which still has many interesting challenges to offer, and this explains the rather comprehensive nature of the book. However, we are aware that some readers will be more focused on applications and seek the solution of a given problem. The book is structured so that after reading Chapter 1, it should be possible to skip to Part III, and pick up supporting material as required from previous chapters. A list of papers using the multivariate methods described in this book can be found on our website: `http://acsvproject.org`.

The mathematical development of the theory belongs mostly to the two authors, but there are a number of individuals whose help was greatly instrumental in moving the theory forward. The complex analysts at the University of Wisconsin-Madison, Steve Wainger, Jean-Pierre Rosay, and Andreas Seeger, helped the authors (then rather junior researchers) to grapple with the problem in its earliest incarnation. A similar role was played several years later by Jeff McNeal. Perhaps the greatest thanks are due to Yuliy Baryshnikov, who translated the Leray–Petrovsky theory and the work of Atiyah–Bott–Gårding into terms the authors could understand, and coauthored several papers. Frank Sottile provided help with algebra on many occasions; Persi Diaconis arranged for a graduate course while the first author visited Stanford in 2000; Richard Stanley answered our numerous miscellaneous queries. Thanks are also due to our other coauthors on papers related to this project, listed on the project website linked from the book website. Alex Raichev and Torin Greenwood helped substantially with proofreading and with computer algebra implementations of some parts of the book. All software can be located via the book website.

On a more personal level, the first author would like to thank his wife, Diana Mutz, for encouraging him to follow this unusual project wherever it took him, even if it meant abandoning a still productive vein of problems in probability theory. The sentiment in the probability theory community may be otherwise, but the many connections of this work to other areas of mathematics have been a source of satisfaction to the authors. The first author would also like to thank his children, Walden, Maria, and Simi, for their participation in the project via the Make-A-Plate company (see Figure 0.1).

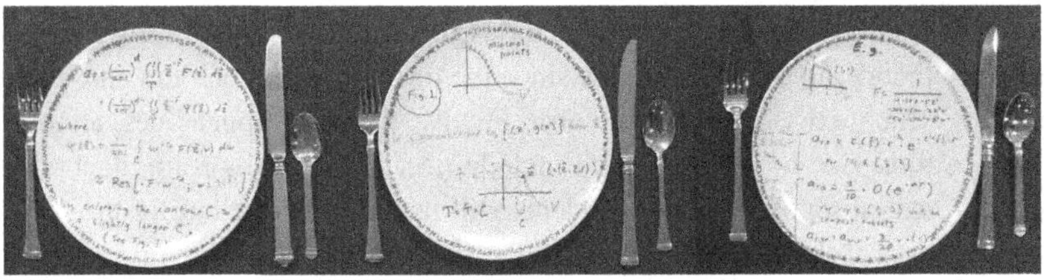

Figure 0.1 Customized "asymptotics of a multivariable generating function" dinner plates.

The second author thanks his wife Golbon Zakeri, children Yusef and Yahya, and mother-in-law Shahin Sabetghadam for their help in carving out time for him to work on this project, sometimes at substantial inconvenience to themselves. He hopes they will agree that the result is worth it.

List of Symbols

$\boldsymbol{r}$	generic multi-index for multivariate array, page 3
$\boldsymbol{z}$	generic element of $\mathbb{C}^d$, page 4
$\lvert\boldsymbol{r}\rvert$	1-norm of multi-index $\boldsymbol{r}$, page 7
$\hat{\boldsymbol{r}}$	unitized vector representing direction determined by $\boldsymbol{r}$, page 7
$\mathbf{1}$	vector with all components equal to 1, page 10
F	generic meromorphic multivariate generating function, page 11
P	numerator of generating function F, page 11
Q	denominator of generating function F, page 11
$\texttt{critical}(\boldsymbol{r})$	set of critical points in direction $\hat{\boldsymbol{r}}$, page 11
$\texttt{contrib}(\boldsymbol{r})$	set of contributing points, page 11
$\Phi_{\boldsymbol{w}}(\boldsymbol{r})$	formula for the contribution from the point $\boldsymbol{w}$ to the asymptotic series for $a_{\boldsymbol{r}}$, page 12
δ_j	vector of length d with a 1 in its jth coordinate and a 0 elsewhere, page 17
$\mathbb{C}[[z_1, \ldots, z_d]]$	ring of formal power series in $\boldsymbol{z} = (z_1, \ldots, z_d)$ with complex coefficients, page 17
$\mathbf{0}$	vector all of whose components equal 0, page 18
∂_k	partial derivative operator with respect to kth variable, page 18
$\mathbb{C}\{z_1, \ldots, z_d\}$	ring of germs of analytic functions, page 19
$\sqcup$	disjoint union, page 22
$\mathbb{P}$	probability measure, page 29
$\mathbb{C}((z))$	field of formal Laurent series, page 42
Re	real part of a complex number, page 77
Im	imaginary part of a complex number, page 77
ϕ	generic phase function of Fourier–Laplace integral, page 89
A	generic amplitude function of Fourier–Laplace integral, page 89
$C(k, \ell)$	constants defined in terms of Gamma function, appearing in Fourier–Laplace integral formulae, page 90

C^M	class of functions whose derivatives to order M are continuous, page 95
C^∞	class of functions whose derivatives of all orders exist, page 95
$\mathfrak{p}$	principal value of kth root, page 98
$\mathcal{I}(\lambda)$	two-sided Fourier–Laplace integral, page 105
$\mathcal{I}_+(\lambda)$	one-sided univariate Fourier–Laplace integral, page 110
Ai	Airy function, page 113
$\mathcal{H}$	Hessian matrix of second partial derivatives, page 114
∇	gradient map, page 114
Relog	coordinatewise log modulus map, page 135
$\mathbb{C}\left[z, z^{-1}\right]$	ring of Laurent polynomials, page 135
$\mathbb{L}(z)$	space of formal Laurent expressions, page 135
amoeba	polynomial amoeba, page 141
hull	convex hull, page 142
$\mathcal{N}$	Newton polytope, page 143
ν	order map, page 143
$\mathtt{tan}_x(B)$	geometric tangent cone to B at x, page 150
$\mathtt{normal}_x(B)$	outward normal cone, dual to $\mathtt{tan}_x(B)$, page 150
K^*	dual cone of K, page 150
$\mathcal{V}$	singular variety of generating function, page 153
$\nabla_{\log}$	logarithmic gradient, page 159
C	amoeba contour, page 160
deg	order of vanishing of power series, page 162
$\mathtt{algtan}_x(f)$	algebraic tangent cone of f at x, page 162
hom	homogeneous part of power series, page 162
$\tilde{Q}$	square-free part of Q, page 182
LT	leading term with respect to monomial order, page 223
$\mathbf{L}(z)$	logarithmic normal space to the stratum containing z, page 229
$\mathcal{G}$	Gauss map, page 281
$\mathcal{K}$	Gaussian curvature, page 281
$M(\mathcal{A})$	matroid of a hyperplane arrangement, page 311
$\mathcal{O}_p$	local ring of analytic germs at p, page 320
Δ	standard (embedded) simplex, page 330
$\mathbf{t}$	generic variable for simplex, page 330
$\pi\Delta$	shadow simplex, page 330
$\mathcal{S}$	set of critical points for a multiple point Fourier–Laplace integral, page 333
$\mathbf{K}^z(A)$	cone of hyperbolicity for the homogeneous polynomial A, page 348
$\mathbf{K}^{A,C}(x)$	family of cones for a homogeneous polynomial A, page 349

$\mathbf{K}^{q,B}(\boldsymbol{z})$	family of cones when q is log-Laurent polynomial, page 350
A^*	algebraic dual to the homogeneous polynomial A, page 369
$\mathfrak{n}$	density of standard normal distribution, page 413
δ	coboundary map, page 475
$\boldsymbol{\Gamma}_\Psi$	augmented lognormal matrix, page 492

PART I

COMBINATORIAL ENUMERATION

1

Introduction

Consider an array[1] of complex numbers

$$\left\{a_{\boldsymbol{r}} : \boldsymbol{r} \in \mathbb{N}^d\right\} := \{a_{r_1,\dots,r_d} : r_1, \dots, r_d \in \mathbb{N}\}$$

where, as in the rest of this book, we include zero in the set $\mathbb{N} = \{0, 1, 2, \dots\}$. The numbers $a_{\boldsymbol{r}}$ usually come with a story – a reason they are interesting. Often, they count a class of objects parametrized by $\boldsymbol{r}$. For example, it could be that $a_{\boldsymbol{r}}$ is the multinomial coefficient $a_{\boldsymbol{r}} = \binom{|\boldsymbol{r}|}{r_1 \cdots r_d}$, in which case $a_{\boldsymbol{r}}$ counts sequences of elements in $\{1, \dots, d\}$ with r_1 occurrences of 1, r_2 occurrences of 2, and so forth up to r_d occurrences of the symbol d. Another frequent source of these arrays is probability theory, where the numbers $a_{\boldsymbol{r}} \in [0, 1]$ are probabilities of events parametrized by $\boldsymbol{r}$. For example, a_{rs} might be the probability that a simple random walk of r steps in $\{-1, 1\}$ ends at the integer point s.

Definition 1.1 (running notation). Throughout this text we use d to denote the dimension of an arbitrary array, and often employ r, s, and t as synonyms for r_1, r_2, and r_3, respectively, so as to avoid subscripts in low-dimensional examples. We also use the notation $|\boldsymbol{r}| := \sum_{j=1}^{d} |r_j|$ for any vector $\boldsymbol{r}$, which helps us normalize in a way convenient for combinatorial examples.

How might one *understand* an array of numbers? In some cases there may be a simple explicit formula, for instance the multinomial coefficients are given by a ratio of factorials. When a formula of such brevity exists, we don't need fancy techniques to describe the array. Unfortunately, this rarely happens. Often, if a formula exists at all, it will not be in closed form but will include indefinite summation. As Stanley [Sta97, Ex.1.1.4] notes in his foundational text on enumeration, "There are actually formulas in the literature (nameless here

[1] To simplify our presentation in this introduction we consider arrays indexed by vectors of natural numbers, while later in the text we generalize to arrays indexed by integer vectors.

forevermore) for certain counting functions whose evaluation requires listing all of the objects being counted! Such a 'formula' is completely worthless." Less egregious are the formulae containing functions that are rare or complicated and whose properties are not immediately familiar to us. It is not clear how much good comes from this kind of formula.

Another way of describing arrays of numbers is via recursions. The simplest examples are finite linear recurrences, such as the recurrence $a_{r,s} = a_{r-1,s} + a_{r,s-1}$ for the binomial coefficients $a_{r,s} = \binom{r+s}{r}$. A recursion for $a_{\boldsymbol{r}}$ in terms of values $\{a_{\boldsymbol{s}} : \boldsymbol{s} < \boldsymbol{r}\}$ whose indices precede $\boldsymbol{r}$ in the coordinatewise partial order may be unwieldy, perhaps requiring evaluation of a complicated function of all $a_{\boldsymbol{s}}$ with $\boldsymbol{s} < \boldsymbol{r}$, but if the recursion is of bounded complexity then it can give an efficient algorithm for computing $a_{\boldsymbol{r}}$. Still, we will see that even in the case of simple recursions the estimation of $a_{\boldsymbol{r}}$ may not be straightforward. Thus, while we look for recursions to help us understand number arrays, and for efficient methods of computation, they rarely provide definitive descriptions.

A third way of understanding an array of numbers is via an estimate. For instance, Stirling's formula, which approximates

$$n! \approx \frac{n^n}{e^n} \sqrt{2\pi n}$$

for large n, yields an approximation

$$\binom{r+s}{r} \approx \left(\frac{r+s}{r}\right)^r \left(\frac{r+s}{s}\right)^s \sqrt{\frac{r+s}{2\pi rs}} \tag{1.1}$$

for the binomial coefficients when r and s are large. If number-theoretic properties of the binomial coefficients are required then we are better off sticking with a ratio of factorials; when their approximate size is paramount, the estimate (1.1) is better.

A fourth way to understand an array of numbers is to encode it algebraically. The *generating function* (often abbreviated GF) of the array $\{a_{\boldsymbol{r}}\}$ is the formal series $F(\boldsymbol{z}) := \sum_{\boldsymbol{r}\in\mathbb{N}^d} a_{\boldsymbol{r}} \boldsymbol{z}^{\boldsymbol{r}}$. Here $\boldsymbol{z}$ is a d-dimensional vector of indeterminates $(z_1, \ldots, z_d)$ and we use the notation $\boldsymbol{z}^{\boldsymbol{r}} := z_1^{r_1} \cdots z_d^{r_d}$. In our running example of multinomial coefficients, we have the generating function

$$F(\boldsymbol{z}) = \sum_{\boldsymbol{r}\in\mathbb{N}^d} \binom{|\boldsymbol{r}|}{r_1 \ \cdots \ r_d} z_1^{r_1} \cdots z_d^{r_d} = \frac{1}{1 - z_1 - \cdots - z_d},$$

where the final expression can be viewed either as a multiplicative inverse in a formal power series ring, or as an analytic function over an appropriate domain of $\mathbb{C}^d$. Stanley calls the generating function "the most useful but the most difficult to understand" method for describing a sequence or array.

The algebraic form of a generating function is intimately related to recursions – and exact formulae – for its coefficient sequence $a_{\mathbf{r}}$, as well as combinatorial decompositions for the objects enumerated by $a_{\mathbf{r}}$. In a complementary manner, the analytic properties of a generating function correspond to estimates of $a_{\mathbf{r}}$.

1.1 Generating functions and asymptotics

In this text we are chiefly concerned with the asymptotic behavior of $a_{\mathbf{r}}$ as $\mathbf{r} \to \infty$ in certain *directions*. To discuss the behavior of sequences as their indices go off to infinity, we introduce some standard asymptotic notation.

Definition 1.2 (asymptotic notation). If f and g are real-valued functions then we write

- $f = O(g)$ if and only if $\limsup_{x \to x_0} |f(x)/g(x)| < \infty$,
- $f = o(g)$ if and only if $\lim_{x \to x_0} f(x)/g(x) = 0$,
- $f \sim g$ if and only if $\lim_{x \to x_0} f(x)/g(x) = 1$,
- $f = \Omega(g)$ when $g = O(f)$, and
- $f = \Theta(g)$ when both $f = O(g)$ and $g = O(f)$,

for some value x_0 understood in context, typically 0 or $+\infty$.

As $n \to \infty$ the function $f(n)$ is said to be ***rapidly decreasing*** if $f(n) = O\left(n^{-K}\right)$ for every $K > 0$, ***exponentially decaying*** if $f(n) = O(e^{-cn})$ for some $c > 0$, and ***super-exponentially decaying*** if $f(n) = O(e^{-cn})$ for every $c > 0$.

Remark. An alternative definition is that $f = O(g)$ when there exists $C > 0$ and an open neighborhood N of x such that $f(x) \le Cg(x)$ for all $x \in N$. In this case C is called an ***implied constant***. One may increase C and decrease N and still maintain the inequality, so implied constants are not unique, even if they are chosen to give a tight inequality.

Example 1.3. As $n \to \infty$ the function $f(n) = 1/n!$ decays super-exponentially, while 2^{-n} decays exponentially and $e^{-\sqrt{n}}$ approaches zero but does not decay exponentially. ◁

An ***asymptotic scale*** is a sequence $\{g_j : j \in \mathbb{N}\}$ of functions satisfying $g_{j+1} = o(g_j)$ for all $j \ge 0$. An ***asymptotic expansion*** (also called ***asymptotic series*** or ***asymptotic development***)

$$f \approx \sum_{j=0}^{\infty} c_j g_j$$

for a function f in terms of an asymptotic scale $\{g_j : j \in \mathbb{N}\}$ and constants $c_j \in \mathbb{C}$ is said to hold if

$$f - \sum_{j=0}^{M-1} c_j g_j = O(g_M) \tag{1.2}$$

for every $M \geq 1$.

Remark. It is possible that $c_j = 0$ for all j. For example, this will happen if $g_j(n) = n^{-j}$ and f is exponentially decaying. In this case there is no leading term in the expansion. Otherwise, the ***leading term of an asymptotic expansion*** is the first non-zero term $c_k g_k$ in the expansion.

Example 1.4. Stirling's famous approximation to the factorial can be refined to give an asymptotic series

$$n! \approx \left(\frac{n}{e}\right)^n \sqrt{2\pi n} \sum_{\ell \geq 0} c_\ell n^{-\ell}$$

with coefficient sequence $\{c_\ell\}$ beginning $1, 1/12, 1/288, -139/51840, \ldots$. ◁

Example 1.5. Let $f \in C^\infty(\mathbb{R})$ be a smooth real function defined on a neighborhood of zero, so that $c_n = f^{(n)}(0)/n!$ is the nth term in its Taylor expansion. If f is not analytic then this expansion may not converge to f, and may even diverge for all non-zero x, but Taylor's Theorem with remainder always implies

$$f(x) = \sum_{n=0}^{M-1} c_n x^n + c_M \xi^M$$

for some $\xi > 0$ bounded close to the origin. This proves that

$$f \approx \sum_{n \geq 0} c_n x^n$$

is always an asymptotic expansion for f near zero. ◁

Remark. Following Poincaré, many authors use the symbol $\sim$ to denote both asymptotic equivalence of functions and asymptotic series expansions. However, this overloading of notation can lead to inconsistencies. We thus follow texts such as [dBru81] in using $\approx$ for asymptotic expansions.

Exercise 1.1. Let $f(x) = e^x$. Prove that $f(x) \sim 1$ as $x \to 0$ but $f(x) \not\approx 1$ as an asymptotic expansion in powers of x at $x = 0$.

All these notations hold in the multivariate case as well, except that if the limit value $\mathbf{z}_0$ is infinity then a statement such as $f(\mathbf{z}) = O(g(\mathbf{z}))$ must also specify how $\mathbf{z}$ approaches the limit. A ***direction*** is a ray in $\mathbb{R}^d$ defined by all positive multiples of a fixed non-zero vector, which can also be viewed as an element of $(d-1)$-dimensional real projective space $\mathbb{RP}^{d-1}$. Often we will parametrize directions of interest by taking $\boldsymbol{r} \to \infty$ while fixing or bounding the normalized vector $\hat{\boldsymbol{r}} := \boldsymbol{r}/|\boldsymbol{r}|$, where, as introduced above,

$$|\boldsymbol{r}| = |r_1| + |r_2| + \cdots + |r_d|.$$

Sometimes we shall loosely refer to "the direction $\boldsymbol{r}$", by which we mean the direction parametrized by $\hat{\boldsymbol{r}}$, or the ray determined by $\boldsymbol{r}$.

Definition 1.6. A ***multivariate asymptotic expansion***

$$f_{\boldsymbol{r}} \approx \sum_{j=0}^{\infty} c_j g_j(\boldsymbol{r})$$

holds on a compact set of directions $D \subseteq \mathbb{RP}^{d-1}$ if each $c_j \in \mathbb{C}$, each $g_j = o(g_{j+1})$, and $f_{\boldsymbol{r}} - \sum_{j=0}^{M-1} g_j(\boldsymbol{r}) = O(g_M)$ for each M as $\boldsymbol{r} \to \infty$ with $\hat{\boldsymbol{r}} \in D$. This asymptotic expansion is a ***uniform asymptotic expansion*** on D if the implied constants can be chosen independently of the sequence $\boldsymbol{r}$ as long as $\hat{\boldsymbol{r}} \in D$.

Example 1.7. In Chapter 9 we shall derive the result

$$\binom{r+s}{s} \sim \frac{(r+s)^{(r+s)}}{r^r s^s} \sqrt{\frac{r+s}{2\pi rs}}$$

for all $r, s > 0$ as $(r, s) \to \infty$ with $r/(r+s)$ and $s/(r+s)$ remaining bounded and away from 0. This gives the first term of an asymptotic series which is uniform provided r/s and s/r are bounded away from 0, with all terms in the series varying smoothly with direction. Because of our restrictions on r/s, this asymptotic series can be expressed in terms of the asymptotic scale

$$g_j(r, s) = \frac{(r+s)^{(r+s)}}{r^r s^s} \sqrt{\frac{r+s}{rs}} (r+s)^{-j},$$

an asymptotic scale involving decreasing powers s^{-j} of s, or an asymptotic scale involving decreasing powers r^{-j} of r. Note that this multivariate asymptotic approximation is not uniform for all real directions: for instance, if $r = 0$ then $\binom{r+s}{s} = 1$ for all s. ◁

Remark. Throughout this book, we typically use $f(z)$ and a_n instead of $F(z)$ and $a_{\boldsymbol{r}}$ when dealing with the univariate case.

As we will see in Chapter 3, the generating function $f(z)$ for a univariate sequence $\{a_n : n \in \mathbb{N}\}$ leads, almost automatically, to asymptotic estimates for

a_n as $n \to \infty$. To estimate a_n when its generating function f is known, we begin with Cauchy's integral formula

$$a_n = \frac{1}{2\pi i} \int_C z^{-n-1} f(z)\, dz\,. \tag{1.3}$$

Equation (1.3) represents a_n by a complex contour integral on a sufficiently small circle C around the origin, and one may apply complex analytic methods to obtain an asymptotic estimate. The necessary knowledge of residues and contour shifting may be found in an introductory complex variables text such as [Con78b; BG91], with a particularly nice treatment of univariate saddle point integration found in [Hen88; Hen91]. In particular, the singularities of $f(z)$ play a large role in characterizing asymptotic behavior.

The situation for multivariate arrays is nothing like the situation for univariate arrays. In 1974, when Bender published his review article [Ben74] on asymptotic enumeration, the literature on asymptotics of multivariate generating functions was in its infancy. Bender's concluding section urges research in this area:

> Practically nothing is known about asymptotics for recursions in two variables even when a generating function is available. Techniques for obtaining asymptotics from bivariate generating functions would be quite useful.

In the 1980s and 1990s, a small body of results was developed by Bender, Richmond, Gao, and others, giving the first partial answers to asymptotic questions for multivariate generating functions. The first paper to concentrate on extracting asymptotics from multivariate generating functions was [Ben73], already published at the time of Bender's survey, but the seminal paper is [BR83]. The authors work under the hypothesis that F has a singularity of the form $A/(z_d - g(\boldsymbol{x}))^q$ on the graph of a smooth function g, for some real exponent q, where $\boldsymbol{x}$ denotes $(z_1, \ldots, z_{d-1})$. They show, under appropriate further hypotheses on F, that the probability measure μ_n one obtains by renormalizing $\{a_{\boldsymbol{r}} : r_d = n\}$ to sum to 1 converges to a multivariate normal distribution when appropriately rescaled. Their method, which we call the ***GF-sequence method***, is to break the d-dimensional array $\{a_{\boldsymbol{r}}\}$ into a sequence of $(d-1)$-dimensional slices and consider the sequence of $(d-1)$-variate generating functions

$$f_n(\boldsymbol{x}) = \sum_{\boldsymbol{r}:r_d=n} a_{\boldsymbol{r}} \boldsymbol{x}^{\boldsymbol{r}}\,.$$

They show that, asymptotically as $n \to \infty$,

$$f_n(\boldsymbol{x}) \sim C_n g(\boldsymbol{x}) h(\boldsymbol{x})^n \tag{1.4}$$

and that sequences of generating functions obeying (1.4) satisfy a central limit theorem and a local central limit theorem.

The GF-sequence method is limited to the single, though important, case where the coefficients $a_{\mathbf{r}}$ are nonnegative and possess Gaussian (central limit) behavior. The work of [BR83] has been greatly expanded upon, but always in a similar framework. For example, it has been extended to matrix recursions [BRW83] and, in [GR92; BR99], from algebraic to algebraico-logarithmic singularities of the form $F \sim (z_d - g(\boldsymbol{x}))^q \log^\alpha(1/(z_d - g(\boldsymbol{x})))$. The difficult step under these hypotheses is deducing asymptotics from the *quasi-power* hypothesis (1.4).

1.2 New multivariate methods

The research presented in this book grew out of several problems encountered by the first author, concerning bivariate and trivariate arrays of probabilities. One might have thought, based on the situation for univariate generating functions, that there would be well-known, neatly packaged results yielding asymptotic estimates for the probabilities in question. At that time, the most recent and complete reference on asymptotic enumeration was a 1995 survey of Odlyzko [Odl95]. As mentioned in the preface, only six of the over 100 pages of the survey are devoted to multivariate asymptotics, mainly to the GF-sequence results of Bender et al., and its section on multivariate methods closes with a call for further work in this area. Evidently, a general asymptotic method was not known in the multivariate case, even for the simplest non-trivial class of rational functions.

This stands in stark contrast to the univariate theory of rational functions, which is trivial in combinatorial applications (see Chapter 3). The relative difficulty of the problem in higher dimensions is perhaps unexpected, but connections to other areas of mathematics such as Morse theory are quite intriguing. These connections, as much as anything else, have caused us to pursue this line of research long after the urgency of the original motivating problems had faded.

Odlyzko [Odl95] describes why he believes multivariate coefficient estimation to be difficult. First, generating function singularities are no longer isolated, but generally form $(d-1)$-dimensional hypersurfaces, so even multivariate rational functions have an infinite set of singularities. Second, the multivariate analogue of the one-dimensional residue theorem is the considerably more difficult theory of Leray residues [Ler59]. This theory is fleshed out in the text of Aizenberg and Yuzhakov [AY83], who also spend a few pages [AY83, Section 23] on generating functions and combinatorial sums. Further progress in using multivariate residues to evaluate coefficients of generating functions was made by Bertozzi and McKenna [BM93], though at the time of Odlyzko's survey none of the papers based on multivariate residues such as [Lic91; BM93] had resulted in any kind of systematic application of these methods to enumeration. It is interesting to note that several of these early works, such as [BM93; KY96], are centered on queueing theory applications.

The focus of this book is a more recent vein of research, begun in [PW02], continued in its infancy in [PW04; Lla03; Wil05; Lla06; RW08; RW11; PW08; DeV10; PW10], and now comprising a stable and ever-growing component of enumerative combinatorics. This research extends ideas that are present to some degree in [Lic91; BM93; KY96], using complex methods that are genuinely multivariate to evaluate coefficients via the multivariate Cauchy formula

$$a_{\mathbf{r}} = \left(\frac{1}{2\pi i}\right)^d \int_T \mathbf{z}^{-\mathbf{r}-\mathbf{1}} F(\mathbf{z})\, d\mathbf{z}\,, \tag{1.5}$$

where T is a suitable product of circles in each variable. We hope that by avoiding the symmetry-breaking decompositions of the GF-sequence method we will obtain methods that are more universally applicable. In particular, much of this past work can be viewed as instances of a more general result estimating the Cauchy integral via topological reductions of the cycle T of integration. These topological reductions, while not fully automatic, are algorithmically decidable in many cases. The ultimate goal, now well on its way to fruition [Mel21, Chapter 7], is to develop software to automate all of the computation.

We can by no means say that the majority of multivariate generating functions fall prey to these new techniques. Nevertheless, as illustrated in this text and a steadily increasing number of papers, we can treat a large number of combinatorially interesting examples. The class of functions to which the methods described in this book may be applied is larger than the class of rational functions, but similar in spirit: the function must have singularities, and the singularities dictating asymptotics must be poles. This translates to the requirement that the function be meromorphic in a neighborhood of a certain polydisk, which means that it has a representation, at least locally, as a quotient of analytic functions.

Throughout this book, we reserve the symbols F, P, and Q for a meromorphic function F expressed as the quotient P/Q of analytic functions with a convergent series expansion

$$F(\boldsymbol{z}) = \frac{P(\boldsymbol{z})}{Q(\boldsymbol{z})} = \sum_{\boldsymbol{r}} a_{\boldsymbol{r}} \boldsymbol{z}^{\boldsymbol{r}} .$$

Although this introduction has focused on power series expansions, we will develop the theory for convergent Laurent expansions, allowing the index $\boldsymbol{r}$ to range over $\mathbb{Z}^d$. The set $\mathcal{V}$ of singularities of F, which is crucial to the asymptotic analysis, is known as its *singular variety*. For instance, if P and Q are coprime polynomials then the singular variety is the algebraic set $\mathcal{V} = \{\boldsymbol{z} \in \mathbb{C}^d : Q(\boldsymbol{z}) = 0\}$.

We now briefly describe the ACSV approach to computing multivariate asymptotics. A more detailed overview is provided in Chapter 7.

(i) Use the multidimensional Cauchy integral (1.5) to express $a_{\boldsymbol{r}}$ as an integral over a d-dimensional torus (product of circles) T in $\mathbb{C}^d$.
(ii) Observe that T may be replaced by any cycle homologous to $[T]$ in $H_d(\mathcal{M})$, where $\mathcal{M}$ is the domain of analyticity of the integrand.
(iii) Deform the cycle T to lower the modulus of the integrand as much as possible. Morse-theoretic arguments imply that local maxima are characterized by the set $\texttt{critical}(\boldsymbol{r})$ of *critical points* of $\mathcal{V}$, which depend only on the direction $\hat{\boldsymbol{r}}$ of $\boldsymbol{r}$ as $\boldsymbol{r} \to \infty$ and are saddle points for the magnitude of the integrand.
(iv) Use algebraic methods to encode the elements of $\texttt{critical}(\boldsymbol{r})$ by a finite collection of equalities and inequalities (defined by polynomials when F is rational).
(v) Use topological methods to find certain minimax cycles $C(\boldsymbol{w})$ near each critical point $\boldsymbol{w}$, termed *quasi-local cycles*, such that the homology class $[T]$ can be represented by a sum $\sum_{\boldsymbol{w}} n_{\boldsymbol{w}} C(\boldsymbol{w})$ with each $n_{\boldsymbol{w}} \in \mathbb{Z}$.
(vi) Refine the set of critical points to the set $\texttt{contrib}(\boldsymbol{r})$ of *contributing points* that maximize the modulus of the Cauchy integrand among the critical points $\boldsymbol{w}$ with $n_{\boldsymbol{w}} \neq 0$. In the vast majority of cases for which we have explicit asymptotic results, it is the case that $n_{\boldsymbol{w}} \in \{0, \pm 1\}$.
(vii) Asymptotically approximate integrals over the $C(\boldsymbol{w})$ as $\boldsymbol{w}$ ranges over the set of contributing points, using a combination of residue and saddle point techniques.

When successful, this approach leads to an asymptotic representation of the coefficients $a_{\boldsymbol{r}}$ that is uniform as $\boldsymbol{r}$ varies on the interior of finitely many cones that partition $\mathbb{R}^d$. As $\hat{\boldsymbol{r}}$ varies over compact subsets in the interior of such cones, the elements of $\texttt{contrib}(\boldsymbol{r}) \subseteq \mathcal{V}$ vary smoothly with $\hat{\boldsymbol{r}}$ and there exist asymptotic series $\{\Phi_{\boldsymbol{w}}(\boldsymbol{r}) : \boldsymbol{w} \in \texttt{contrib}(\boldsymbol{r})\}$ whose terms can be computed explicitly such that

$$\begin{aligned} a_{\boldsymbol{r}} &= \frac{1}{(2\pi i)^d} \int_{[T]} \boldsymbol{z}^{\boldsymbol{r}-\boldsymbol{1}} F(\boldsymbol{z})\, d\boldsymbol{z} \\ &= \sum_{\boldsymbol{w} \in \texttt{critical}(\boldsymbol{r})} \frac{n_{\boldsymbol{w}}}{(2\pi i)^d} \int_{C(\boldsymbol{w})} \boldsymbol{z}^{\boldsymbol{r}-\boldsymbol{1}} F(\boldsymbol{z})\, d\boldsymbol{z} \\ &= \sum_{\boldsymbol{w} \in \texttt{contrib}(\boldsymbol{r})} n_{\boldsymbol{w}} \Phi_{\boldsymbol{w}}(\boldsymbol{r})\,. \end{aligned} \tag{1.6}$$

The first line in this chain of equalities reflects steps (i) and (ii) of our program above, while the second is the result of steps (iii)–(v), and the final line comes from steps (vi) and (vii). The set $\texttt{critical}(\boldsymbol{r})$ is algorithmically computable in reasonable time, while determining membership in the subset $\texttt{contrib}(\boldsymbol{r})$ can be extremely challenging. The explicit formulae $\Phi_{\boldsymbol{w}}(\boldsymbol{r})$ in the last line are sometimes relatively easily to compute (see Chapter 9) and sometimes more difficult (see Chapter 10 and especially Chapter 11).

1.3 Outline of the remaining chapters

This book is divided into three parts, of which the third part is the heart of the subject: deriving asymptotics in the multivariate setting once a meromorphic generating function is known. Nevertheless, some discussion is required on how generating functions are obtained, how to interpret them, what the chief motivating examples and applications are, and what we knew how to do before the line of research described in Part III. These topics also make the book into a self-contained reference, and allow one to obtain asymptotics by deriving new forms of a generating function, turning an intractable analysis into a tractable one by changing variables, re-indexing, aggregating, and so forth. Consequently, the first three chapters comprising Part I form a crash course in univariate analytic combinatorics. Chapter 2 explains generating functions and their uses, introducing formal power series, their relation to combinatorial enumeration, and the combinatorial interpretation of rational, algebraic, and transcendental operations on power series. Chapter 3 is a review of univariate asymptotics, much of which serves as mathematical background for the multivariate case. While some excellent sources are available in the univariate case, for example [dBru81; Wil06; FS09], none of these is concerned with providing the brief yet reasonably complete summary of analytic techniques that we provide here. It seems almost certain that someone trying to understand the main subject of this text will profit from a review of the essentials of univariate asymptotics.

Carrying out the multivariate analyses described in Part III requires a fair amount of mathematical background. Most of this is at the level of graduate coursework, ideally already known by practicing mathematicians but in reality often forgotten, never learned, or not learned in sufficient depth. The required background is composed of small to medium-sized chunks taken from many areas: undergraduate complex analysis, calculus on manifolds, saddle point integration (both univariate and multivariate), algebraic topology, computational algebra, and Morse theory. Many of these background topics would require a full semester's course to learn from scratch. That is too much material to include here, but we also want to avoid the scenario where a reference library is required each time a reader picks up this book. Accordingly, we have included substantial background material.

This background material is separated into two pieces. The first piece is the three chapters that comprise Part II, which contains material that we feel should be read or skimmed before the central topics are tackled. The topics in Part II have been sufficiently pared down that it is possible to learn them from scratch if necessary. Chapters 4 and 5 describe how to asymptotically evaluate saddle point integrals in one and several variables, respectively. Familiarity with these results is needed for the final steps in the analyses in Part III to make sense. Most of the results in these chapters can be found in a reference such as [BH86]; the treatment here differs from the usual sources in that Fourier and Laplace type integrals are treated as instances of a single complex-phase case. Working in the holomorphic setting, analytic techniques (contour deformation) are used whenever possible, after which comparisons are given to the corresponding C^∞ approach (which uses integration by parts in place of contour deformation). Chapter 6 covers domains of convergence of multivariate power series and Laurent series, the notion of polynomial amoebas, and results relating amoebas to domains of convergence of Laurent series. We also note that much of Chapter 8, which recalls several tools from polynomial system solving such as Gröbner bases, morally belongs with the background material in Part II; we have placed it in Part III so that we can compute quantities appearing in our multivariate analyses that are introduced in Chapter 7. It is possible to skip Chapter 8, if one wants to understand the theory and does not care about computation; however, few users of analytic combinatorics live in a world where computation does not matter.

The remaining background material is relegated to the four appendices, each of which contains a reduction of a semester's worth of material. It is not expected that the reader will go through these in advance. Rather, they serve as references so that frequent library visits will not be necessary. Appendix A presents for beginners all relevant knowledge about calculus on manifolds and algebraic topology. Manifolds and tangent and cotangent vectors are defined, differential forms in $\mathbb{R}^n$ are constructed from scratch, and integration of forms is developed. The appendix ends with a short treatment of complex differential forms. Appendix B reviews the essentials of algebraic topology: chain complexes, homology and cohomology, relative homology, Stokes's Theorem, and some important exact sequences. Appendix C summarizes classical Morse theory – roughly the first few chapters of Milnor's classic text [Mil63] – after which Appendix D introduces the notion of stratified spaces and describes stratified Morse theory as developed by Goresky and MacPherson [GM88]. Part II and the appendices also have a second function: some of the results used in Part III are often quoted in the literature from sources that do not provide a proof. On more than one occasion, when organizing the material in this book, we found that a purported reference to a proof ultimately led to nothing. Beyond serving as a mini-reference library, therefore, the background sections provide some key proofs and corrected citations to eliminate ghost references and the misquoting of existing results.

The heart of this book, Part III, is devoted to new results in the asymptotic analysis of multivariate generating functions. Chapter 7 sets out the theory by which multivariate asymptotics are derived, greatly expanding the outline given in Section 1.2. The internal structure of Chapter 7 is described at length in the beginning of the chapter. Because some of the material in this long chapter relies on specialized topological knowledge, it is possible to take a conceptual off-ramp after most sections, which get progressively more general as the chapter proceeds. We begin with extended examples in Section 7.1, before describing the argument in the simpler case when $\mathcal{V}$ is a smooth manifold in Section 7.2. Section 7.3 covers the general case, ultimately deriving the fundamental result of the chapter: a decomposition (7.2) for $a_{\boldsymbol{r}}$ as an integer sum of quasi-local cycles near critical points, without any specification of the set $\texttt{contrib}(\boldsymbol{r})$ or the asymptotic series $\Phi_{\boldsymbol{z}}$.

Having reduced the computation of $a_{\boldsymbol{r}}$ to saddle point integrals with computable parameters, plugging in results on saddle point integration yields theorems for the end user. These break into several types. Chapter 9 discusses the case when the singular variety $\mathcal{V}$ is smooth near the contributing points. This case is simpler than the general case in several respects: the residues are more straightforward, so multivariate residue theory is not always needed, and only classical Morse theory is required. Chapter 10 discusses the case where $\mathcal{V}$ is locally the union of smooth hypersurfaces near contributing points, which is also a case that is reasonably well understood. Finally, we discuss rational functions with singularities having non-trivial monodromy. In this case our knowledge is limited, but some known results are derived in Chapter 11. This chapter is not quite as self-contained as the preceding ones; in particular, some results from [BP11] are quoted without proof. This is because the technical background for these analyses exceeds even the relatively large space we have allotted for background. The paper [BP11], which is self-contained, already reduces by a significant factor the body of work presented in the celebrated paper [ABG70], and further reduction is only possible by quoting key results. Chapter 12 works out a large number of examples following the theory in Chapters 9–11. Finally, Chapter 13 is devoted to further topics, including higher order asymptotics, algebraic generating functions, diagonals, and a number of open problems.

Notes

The overall viewpoint on enumeration discussed here is heavily influenced by [Sta97] and [FS09]. The two, very different, motivating problems alluded to in Section 1.2 were the hitting time generating function from [LL99] and the Aztec Diamond placement probability generating function from [JPS98]. The first versions of the seven step program at the end of Section 1.2 that were used to obtain multivariate asymptotics involved expanding a torus of integration until it was near a critical point on the boundary of the domain of convergence of the series under consideration, and then doing some surgery to isolate the main asymptotic contribution as the integral of a univariate residue over a complementary $(d-1)$-dimensional chain. This was carried out in [PW02; PW04] and was brought to the attention of the authors by several analysts at Wisconsin, among them S. Wainger, J.-P. Rosay, and A. Seeger. Although their names do not appear in any bibliographic citations associated with this project, they are acknowledged in these early publications and should be credited with useful contributions to this enterprise.

Additional exercises

Exercise 1.2. (asymptotic expansions need not converge) Find an asymptotic expansion $f \approx \sum_{j=0}^{\infty} g_j$ for a function f as $x \downarrow 0$ such that $\sum_{j=0}^{\infty} g_j(x)$ is not convergent for any $x > 0$. Conversely, suppose that $f(x) = \sum_{j=0}^{\infty} g_j(x)$ for $x > 0$ and $g_{j+1} = o(g_j)$ as $x \downarrow 0$ — does it follow that $\sum_{j=0}^{\infty} g_j$ is an asymptotic expansion of f at the origin?

Exercise 1.3. Prove or give a counterexample: if g is a continuous function and for each λ we have $a_{rs} = g(\lambda) + O\left((r+s)^{-1}\right)$ as $r, s \to \infty$ with $r/s \to \lambda$, then $a_{rs} \sim g(r/s)$ when $r, s \to \infty$ as λ varies over a compact interval in $\mathbb{R}^+$.

Exercise 1.4. (Laplace transform asymptotics) Let A be a smooth real function in a neighborhood of zero and define its Laplace transform by

$$\hat{A}(\tau) := \int_0^\infty e^{-\tau x} A(x)\, dx .$$

Writing $A(x) = \sum_{n\geq 0} c_n x^n$ with $c_n = A^{(n)}(0)/n!$ and integrating term by term using

$$\int_0^\infty x^n e^{-\tau x}\, dx = n!\tau^{-n-1}$$

suggests the series

$$\sum_{n\geq 0} A^{(n)}(0)\tau^{-n-1} \tag{1.7}$$

as a possible asymptotic expansion for $\hat{A}$. Although the term-by-term integration is completely unjustified, show that the series (1.7) is a valid asymptotic expansion of $\hat{A}$ in decreasing powers of τ as $\tau \to \infty$.

Exercise 1.5. Recall Stirling's approximation from Example 1.4. Use a computer algebra system to experiment, for $1 \leq m \leq 5$, with the mth order approximation for $n = 1, \ldots, 50$. For each such value of n, find the best value m at which to truncate the asymptotic series. For each n, what is the best relative error in the approximation to $n!$ that we can obtain in this way?

Exercise 1.6. Use a computer algebra system to experiment for $1 \leq m \leq 20$ with the error in the mth order Stirling approximation to $n!$ when $n = 1$. After which value of m does the error become noticeably bad?

2

Generating functions

This chapter provides a crash course on generating functions and their uses in enumeration. For a more lengthy introduction, we recommend [Wil06] – other standard references include [FS09, Chapters I–III], [vLW01, Chapter 14], [Sta97, Section 1.1], and the comprehensive treatment in [GJ04].

Throughout this book, we use the notation $[n] := \{1, \dots, n\}$ and write $\boldsymbol{\delta}_j$ for the vector of length d with a 1 in its jth coordinate and a 0 elsewhere.

2.1 Power series

Generating functions impose an algebraic structure on arrays of complex numbers, analogous to (and thus most easily expressed using the same notation as) convergent power series. To that end, let $\boldsymbol{z} = (z_1, \dots, z_d)$ be a vector of indeterminates and consider the set of formal expressions

$$\mathbb{C}[[\boldsymbol{z}]] = \mathbb{C}[[z_1, \dots, z_d]] := \left\{ \sum_{\boldsymbol{r} \in \mathbb{N}^d} a_{\boldsymbol{r}} \boldsymbol{z}^{\boldsymbol{r}} : a_{\boldsymbol{r}} \in \mathbb{C} \text{ for all } \boldsymbol{r} \in \mathbb{N}^d \right\}.$$

If $F(\boldsymbol{z}) = \sum_{\boldsymbol{r}} f_{\boldsymbol{r}} \boldsymbol{z}^{\boldsymbol{r}} \in \mathbb{C}[\boldsymbol{z}]$ then f is called a ***formal power series***, the ***coefficient of*** $\boldsymbol{z}^{\boldsymbol{r}}$ ***in*** F is $[\boldsymbol{z}^{\boldsymbol{r}}]F(\boldsymbol{z}) := f_{\boldsymbol{r}}$, and we also single out the ***constant coefficient*** $F(\mathbf{0}) := [\boldsymbol{z}^{\mathbf{0}}]F(\boldsymbol{z}) = f_{\mathbf{0}}$. The set $\mathbb{C}[[\boldsymbol{z}]]$ becomes the ***ring of formal power series*** by defining addition term-wise

$$[\boldsymbol{z}^{\boldsymbol{r}}]\big(F(\boldsymbol{z}) + G(\boldsymbol{z})\big) := [\boldsymbol{z}^{\boldsymbol{r}}]F(\boldsymbol{z}) + [\boldsymbol{z}^{\boldsymbol{r}}]G(\boldsymbol{z})$$

and multiplication by ***convolution***

$$[\boldsymbol{z}^{\boldsymbol{r}}]F(\boldsymbol{z})G(\boldsymbol{z}) := \sum_{\boldsymbol{s} \in \mathbb{N}^d} [\boldsymbol{z}^{\boldsymbol{s}}]F(\boldsymbol{z}) \cdot [\boldsymbol{z}^{\boldsymbol{r}-\boldsymbol{s}}]G(\boldsymbol{z}),$$

where $[z^{r-s}]G(z)$ is defined to be zero if $r - s$ has a negative coordinate (so the sum in the convolution is always finite). The ***generating function*** of the array $\{a_r : r \in \mathbb{N}^d\}$ is the formal power series $\sum_{r\in\mathbb{N}^d} a_r z^r \in \mathbb{C}[[z]]$.

The additive identity in $\mathbb{C}[[z]]$ is the series 0 (where every coefficient is zero) and the multiplicative identity is the series 1 (which has a coefficient of zero for every index vector $r \neq \mathbf{0}$).

Exercise 2.1. Prove that $F(z) \in \mathbb{C}[[z]]$ has a multiplicative inverse if and only if $F(\mathbf{0}) \neq 0$.

Exercise 2.1 implies $\mathbb{C}[[z]]$ is a *local ring*, meaning that it has a unique maximal ideal $\mathfrak{m}$ that, in this case, consists of all non-units. Local rings come equipped with a notion of convergence: a sequence of elements $F_n(z) \in \mathbb{C}[[z]]$ converges to $F(z) \in \mathbb{C}[[z]]$ if and only if for every $k \in \mathbb{N}$ the difference $F_n(z) - F(z)$ lies in $\mathfrak{m}^k$ for all sufficiently large n. An easier way to say this is that $F_n(z) \to F(z)$ if and only if for any $r \in \mathbb{N}^d$ there exists $N_r \in \mathbb{N}$ such that $[z^r]F_n(z) = [z^r]F(z)$ for all $n \geq N_r$. In other words, each coefficient eventually stabilizes.

For $1 \leq k \leq d$ we write ∂_k for the ***formal partial derivative operator*** which takes $f(z) = \sum_{r\in\mathbb{N}^d} a_r z^r$ and returns

$$(\partial_k f)(z) = \frac{\partial}{\partial z_k} f(z) := \sum_{r\in\mathbb{N}^d} r_k a_r z^{r-\delta_k}.$$

Powers ∂_k^r of ∂_k denote repeated differentiation, with $(\partial_k^0 f)(z) = f(z)$, and a monomial $\boldsymbol{\partial}^r = \partial_1^{r_1} \cdots \partial_d^{r_d}$ represents repeated differentiation with respect to each variable. We also use the shorthand $f_{z^r}(z)$ and $f^{(r)}(z)$ for $(\boldsymbol{\partial}^r f)(z) = (\partial^{|r|} f / \partial_{z_1}^{r_1} \cdots \partial_{z_d}^{r_d})(z)$.

Although formal power series are simply algebraic objects, we often consider them as representing convergent series expansions, and thus analytic functions, in certain subsets of $\mathbb{C}^d$. The (open) ***polydisk*** centered at $p \in \mathbb{C}^d$ with ***polyradius*** $b \in \mathbb{R}^d_{>0}$ is the set

$$D_b(p) := \{z \in \mathbb{C}^d : |z_j - p_j| < b_j \text{ for } 1 \leq j \leq d\},$$

and a ***neighborhood*** of a point $p \in \mathbb{C}^d$ is any polydisk containing p. The ***torus*** centered at $p \in \mathbb{C}^d$ with ***polyradius*** $r \in \mathbb{R}^d_{>0}$ is the set

$$T_p(r) := \{z \in \mathbb{C}^d : |z_j - p_j| = r_j \text{ for } 1 \leq j \leq d\}.$$

To ease notation we write $T(r)$ for a torus centered at the origin $p = \mathbf{0}$, and for every $w \in \mathbb{C}^d_*$ (here $\mathbb{C}_* = \mathbb{C} \setminus \{0\}$ is the set of non-zero complex numbers) we define

$$\mathbf{T}(w) = T(|w_1|, \ldots, |w_d|) = \{z \in \mathbb{C}^d : |z_j| = |w_j| \text{ for } 1 \leq j \leq d\}.$$

Let $\mathcal{N}$ be a neighborhood of the origin and suppose that $F, G \in \mathbb{C}[[\boldsymbol{z}]]$ are the generating functions of the sequences $\{f_{\boldsymbol{r}}\}$ and $\{g_{\boldsymbol{r}}\}$. If $\sum_{\boldsymbol{r}\in\mathbb{N}^d} |f_{\boldsymbol{r}} \boldsymbol{w}^{\boldsymbol{r}}|$ and $\sum_{\boldsymbol{r}\in\mathbb{N}^d} |g_{\boldsymbol{r}} \boldsymbol{w}^{\boldsymbol{r}}|$ are finite for all $\boldsymbol{w} \in \mathcal{N}$ then F and G are absolutely convergent series on $\mathcal{N}$, meaning they represent analytic functions, and taking the sum $F(\boldsymbol{z}) + G(\boldsymbol{z})$ or product $F(\boldsymbol{z})G(\boldsymbol{z})$ as formal series or analytic functions gives the same result. If f represents an analytic function on a neighborhood of $\mathbb{C}^d$ then every partial derivative of f does as well.

Since a finite intersection of neighborhoods of the origin is a neighborhood of the origin, the collection of series that converge in some neighborhood of the origin is a proper subring $\mathbb{C}\{\boldsymbol{z}\}$ of $\mathbb{C}[[\boldsymbol{z}]]$, called the ***ring of germs of analytic functions***. The interior $\mathcal{D} \subset \mathbb{C}^d$ of the domain on which the formal power series F converges is the union of open polydisks, and is hence characterized by its intersection $\mathcal{D}_{\mathbb{R}}$ with $\mathbb{R}^d$. The domain $\mathcal{D}$ is in fact *log-convex*, meaning that the set $\{\boldsymbol{x} \in \mathbb{R}^d : (e^{x_1}, \ldots, e^{x_d}) \in \mathcal{D}_{\mathbb{R}}\}$ is convex; see Chapter 6 for further details and generalizations.

Because there are formal power series that fail to converge in any open polydisk centered at the origin, to which we cannot apply analytic methods, it can be convenient to work with a rescaled series $\sum_{\boldsymbol{r}\in\mathbb{N}^d} \frac{a_{\boldsymbol{r}}}{g(\boldsymbol{r})} \boldsymbol{z}^{\boldsymbol{r}}$ for a judiciously chosen function g. In combinatorial contexts, it is often useful to let $g(\boldsymbol{r})$ be a product of factorials $r_i!$, and a generating function normalized in this way is called an *exponential generating function*. In addition to facilitating convergence of power series, exponential generating functions also have important combinatorial interpretations (see Section 2.5 below).

Remark. In examples we often write (x, y, z) for (z_1, z_2, z_3), and (r, s, t) for (r_1, r_2, r_3), to remove the need for subscripts.

Exercise 2.2. Which of the following formal power series lie in $\mathbb{C}\{x\}$?

(a) $\displaystyle\sum_{n=0}^{\infty} n!x^n$

(b) $\displaystyle\sum_{n=0}^{\infty} n^2 x^n$

(c) $\displaystyle\sum_{n=1}^{\infty} \frac{x^n}{n}$

2.2 Rational operations on generating functions

A ***combinatorial class*** (in d variables) is a set $\mathcal{A}$ together with a partition of $\mathcal{A}$ into finite sets $\{\mathcal{A}_{\boldsymbol{r}} : \boldsymbol{r} \in \mathbb{N}^d\}$ encoded by the ***weight map*** ϕ of the class, which takes $\boldsymbol{x} \in \mathcal{A}$ to the unique index $\phi(\boldsymbol{x}) = \boldsymbol{r} \in \mathbb{N}^d$ with $\boldsymbol{x} \in \mathcal{A}_{\boldsymbol{r}}$. In one variable, we have a disjoint union $\mathcal{A} = \cup_{n=0}^{\infty}\mathcal{A}_n$ and call the weight function $\phi(x)$ the ***size*** of $x \in \mathcal{A}$, so that $\mathcal{A}_n$ contains the number of elements in $\mathcal{A}$ of size n.

Exercise 2.3. Prove that the objects in any combinatorial class form a (possibly finite) countable set.

The ***generating function of the combinatorial class*** $\mathcal{A}$ is the formal power series

$$A(\boldsymbol{z}) = \sum_{\boldsymbol{x}\in\mathcal{A}} \boldsymbol{z}^{\phi(\boldsymbol{x})} = \sum_{\boldsymbol{r}\in\mathbb{N}^d} |\mathcal{A}_{\boldsymbol{r}}|\boldsymbol{z}^{\boldsymbol{r}} \in \mathbb{C}[[\boldsymbol{z}]],$$

and we say that A ***enumerates*** $\mathcal{A}$ under the weighting ϕ.

Arithmetic operations in the ring of formal power series were defined so as to correspond to existing operations on analytic power series. However, it is instructive to interpret these operations combinatorially. We begin with a list of set-theoretic interpretations for rational operations; the combinatorial wealth of these interpretations helps explain why there are so many rational generating functions in combinatorics.

Equality corresponds to bijection

Equality between two generating functions A and B corresponds to bijective correspondence between the classes $\mathcal{A}$ and $\mathcal{B}$ that they enumerate, because $A(\boldsymbol{z}) = B(\boldsymbol{z})$ as formal series if and only if $|\mathcal{A}_{\boldsymbol{r}}| = |\mathcal{B}_{\boldsymbol{r}}|$ for all $\boldsymbol{r} \in \mathbb{N}^d$.

Multiplying by z_j corresponds to re-indexing

In the univariate case, the product $zF(z)$ enumerates the right-shifted sequence $0, a_0, a_1, a_2, \ldots$. More generally, the product $z_jF(\boldsymbol{z})$ enumerates the right-shifted array $\{a_{\boldsymbol{r}-\boldsymbol{\delta}_j}\}$, where $a_{\boldsymbol{r}-\boldsymbol{\delta}_j}$ is defined to be zero if any coordinate of $\boldsymbol{r} - \boldsymbol{\delta}_j$ is negative. Conversely, the left-shifted sequence $\{a_{\boldsymbol{r}+\boldsymbol{\delta}_j}\}$ is enumerated by $(F(\boldsymbol{z}) - F_j(\boldsymbol{z}))/z_j$, where $F_j(\boldsymbol{z}) = [z_j^0]F(\boldsymbol{z})$ is obtained from F by taking only the terms that are free of z_j.

Sum corresponds to disjoint union

If $A(\boldsymbol{z})$ and $B(\boldsymbol{z})$ enumerate classes $\mathcal{A}$ and $\mathcal{B}$, respectively, then $A(\boldsymbol{z}) + B(\boldsymbol{z})$ enumerates the ***disjoint union class*** $C = \mathcal{A} \sqcup \mathcal{B}$ where $C_{\boldsymbol{r}}$ is the *disjoint* union of $\mathcal{A}_r$ and $\mathcal{B}_r$ (so that objects from $\mathcal{A}$ are always considered distinct from objects in $\mathcal{B}$).

The combinatorial interpretations of equality, multiplication by variables and disjoint union are fairly simple, but already yield interesting examples.

Example 2.1 (binary sequences with no repeated 1s). Let $\mathcal{A}_n$ be the set of sequences of 0s and 1s of length n that do not begin with 1 and have no two consecutive 1s. Each such sequence ends either in 0 or in 01, and the previous terms can be any sequence in $\mathcal{A}_{n-1}$ or $\mathcal{A}_{n-2}$, respectively. Thus, stripping off the last one or two symbols yields a bijective correspondence between $\mathcal{A}_n$ and the disjoint union $\mathcal{A}_{n-1} \sqcup \mathcal{A}_{n-2}$. At the generating function level, this would translate to $A(z) = zA(z) + z^2A(z)$, except the correspondence fails for $n = 0$ (it works for $n = 1$ if we take $\mathcal{A}_n$ to be empty for $n < 0$). If we consider $\mathcal{A}$ to have a single object of size zero (the empty string) then correcting for this base case gives the equation

$$A(z) = 1 + zA(z) + z^2A(z).$$

Formal power series manipulations then allow us to rearrange and divide by $1 - z - z^2$ to obtain the generating function

$$A(z) = \frac{1}{1 - z - z^2} = 1 + z + 2z^2 + 3z^3 + 5z^4 + \cdots$$

enumerating binary sequences with no repeated 1s. One may recognize $A(z)$ as the (shifted) generating function for the Fibonacci numbers. ◁

Example 2.2 (binomial coefficients). Let $\mathcal{A}_{r,s}$ be the set of colorings of the set $[r + s] = \{1, \ldots, r + s\}$ for which r elements are red and s are green. Decomposing according to the color of the last element, $\mathcal{A}_{r+s}$ is in bijective correspondence with the disjoint union of $\mathcal{A}_{r-1,s}$ and $\mathcal{A}_{r,s-1}$. This is a combinatorial interpretation of the identity $\binom{r+s}{r} = \binom{r+s-1}{r} + \binom{r+s-1}{r-1}$ and holds as long as $r + s > 0$. It follows that $F(x, y) - 1 = xF(x, y) + yF(x, y)$, and solving for F gives the generating function

$$F(x, y) = \frac{1}{1 - x - y} = \sum_{r,s \geq 0} \binom{r+s}{r} x^r y^s$$

of the binomial coefficients. ◁

Product corresponds to convolution

Let $A(z)$ and $B(z)$ enumerate classes $\mathcal{A}$ and $\mathcal{B}$, respectively. The definition of multiplication of power series shows that $A(z)B(z)$ enumerates the ***product***

class $C = \mathcal{A} \times \mathcal{B}$ defined by letting $C_{\boldsymbol{r}}$ be the disjoint union of Cartesian products $\mathcal{A}_{\boldsymbol{s}} \times \mathcal{B}_{\boldsymbol{r}-\boldsymbol{s}}$ over all $\boldsymbol{s} \in \mathbb{N}^d$ that are coordinatewise less than or equal to $\boldsymbol{r}$. This is the canonical definition of product in any category of graded objects.

Students of probability theory will recognize the product class as a convolution. Suppose that F and G are any power series with nonnegative coefficients whose sets of coefficients both sum to 1. Then F is the ***probability generating function*** for a probability distribution on $\mathbb{N}^d$ that gives mass $[z^{\boldsymbol{r}}]F(\boldsymbol{z})$ to the points $\boldsymbol{r} \in \mathbb{N}^d$, and the analogous result holds for G. The series coefficients in the product FG characterize the convolution of these distributions, which is the distribution of the sum of independent picks from the two given distributions. Thus the study of sums of independent, identically distributed random variables taking values in $\mathbb{N}^d$ is subsumed by the study of powers of generating functions. The laws of large numbers in probability theory may be derived via generating function analyses, and the central limit theorem is usually proved this way. In Chapter 12, versions of these laws are proved for coefficients of generating functions beyond powers of probability generating functions.

Example 2.3 (enumerating partial sums)**.** Let $A(z)$ enumerate a class $\mathcal{A}$ and let $B(z) = 1/(1-z)$ enumerate the class $\mathcal{B}$ with $|\mathcal{B}_n| = 1$ for all n. Then $A(z)B(z)$ enumerates the class C where C_n is the disjoint union $\sqcup_{j=0}^{n} \mathcal{A}_j$ of objects in $\mathcal{A}$ with size at most n, and the coefficients of $C(z) = A(z)/(1-z)$ form the partial sum sequence $c_n = \sum_{j=0}^{n} a_j$. ◁

Quasi-inverse corresponds to finite tuples

Let $\mathcal{B}$ be a combinatorial class with $\mathcal{B}_{\boldsymbol{0}} = \emptyset$. Then we can construct the ***sequence class*** $\mathcal{A} = \mathrm{SEQ}(\mathcal{B})$ containing all finite tuples of elements of $\mathcal{B}$, where $\mathcal{A}_{\boldsymbol{r}}$ consists of all tuples $(x_1, \ldots, x_k) \in \mathcal{B}^k$ (of any length k) with $\phi(x_1) + \cdots + \phi(x_k) = \boldsymbol{r}$. For instance, in the univariate case $\mathcal{A}_n$ contains all finite tuples of elements in $\mathcal{B}$ whose sizes sum to n. Since $\mathcal{A}$ is the disjoint union of the empty sequence, the class of singleton sequences, the class of sequences of length 2, and so forth, we have the generating function equation $A(\boldsymbol{z}) = 1 + B(\boldsymbol{z}) + B(\boldsymbol{z})^2 + \cdots = 1/(1 - B(\boldsymbol{z}))$, where convergence in the ring of formal power series follows from the fact that $B(\boldsymbol{0}) = |\mathcal{B}_{\boldsymbol{0}}| = 0$.

Example 2.4 (binary strings by zeros and ones)**.** Consider the class $\mathcal{A}$ of binary strings in Example 2.1 enumerated by the number of zeros and the number of ones they contain, rather than by total length. Any such sequence can be uniquely decomposed into a finite sequence of the blocks 0 and 01. If $\mathcal{B}$ is the combinatorial class containing these blocks, with weight function $\phi(0) = (1, 0)$

and $\phi(01) = (1, 1)$ encoding the blocks by the number of zeros and ones they contain, then $\mathcal{B}$ has generating function $x + xy$. The generating function for $\mathcal{A}$ is therefore $A(x, y) = 1/(1 - x - xy)$. Note that setting $x = y = z$ recovers the univariate generating function $A(z, z) = 1/(1 - z - z^2)$ counting these strings by total length. ◁

Example 2.5 (prefix codes). This example illustrates the use of *prefix codes* to encode messages over an arbitrary alphabet using binary strings. Fix a finite rooted planar binary tree T with at least two vertices such that every vertex has exactly zero or exactly two children. We label each non-root vertex v of T with the binary string obtained by taking the unique path from the root of T to v and recording a 0 whenever we move to the left and a 1 whenever we move to the right. Thus, the elements in T of distance d from the root are labeled with binary strings of length d (we can consider the root itself to have the empty string as its label).

Any binary string may be uniquely decomposed into blocks by repeatedly stripping off the left-most substring that corresponds to a label of a leaf in T, where the decomposition may end with a partial block corresponding to an internal node of T. Let $\mathcal{B}$ be the class of binary sequences enumerated by their length and the number of blocks they have under this decomposition. If $L(x)$ is the univariate generating function enumerating the number of leaves in T by depth (i.e., distance to the root of T) then the subclass of $\mathcal{B}$ containing strings consisting of a single block under this decomposition has generating function $yL(x)$. If $I(x)$ is the univariate generating function enumerating the internal vertices in T by depth then $1 + y(I(x) - 1)$ enumerates the possible partial blocks that a binary string can end with under our decomposition, including having no partial block. As an arbitrary binary string can be uniquely split into a sequence of blocks followed by a (possibly empty) partial block, the generating function $B(x, y)$ enumerating $\mathcal{B}$ is

$$B(x, y) = \frac{1 + y(I(x) - 1)}{1 - yL(x)}.$$

To use a prefix code over the alphabet A one creates a binary tree T whose leaves correspond to the symbols in A and parses a binary string into a word over A using the block decomposition described above. The generating function $B(x, y)$ encodes how many different messages can be transmitted through binary strings of a fixed length (in applications one would like to pick the tree T to maximize this quantity, perhaps under some constraints). ◁

The field of lattice path enumeration also yields a large and well-studied class of examples.

Example 2.6. Let S be a finite subset of $\mathbb{N}^d \setminus \{\mathbf{0}\}$, let $\boldsymbol{x}_0 = \mathbf{0}$ and let $\mathcal{A}$ be the class of finite tuples $\sigma = (\boldsymbol{x}_0, \boldsymbol{x}_1, \ldots, \boldsymbol{x}_\ell)$ with elements in $\mathbb{N}^d$ such that $\boldsymbol{x}_k - \boldsymbol{x}_{k-1} \in S$ for all $1 \leq k \leq \ell$. The elements of $\mathcal{A}$ are the ***lattice paths with steps in*** S (starting at the origin). The name *lattice path* comes from concatenating the vectors $\boldsymbol{x}_k$ of σ in the plane, giving a path of steps from the origin to the endpoint $\phi(\sigma) = \sum_{k=0}^{\ell} \boldsymbol{x}_k$.

The polynomial $S(\boldsymbol{z}) = \sum_{\boldsymbol{r} \in S} \boldsymbol{z}^{\boldsymbol{r}}$ enumerates S by the endpoint of its steps, so the generating function of $\mathcal{A}$ with respect to the endpoint is $A(\boldsymbol{z}) = 1/(1 - S(\boldsymbol{z}))$. Many other combinatorial classes can be studied in the context of lattice path enumeration, for instance the multinomial coefficient $\binom{|\boldsymbol{r}|}{r_1,\ldots,r_d} = \frac{(r_1+\cdots+r_d)!}{r_1!\cdots r_d!}$ counts the number of paths ending at $\boldsymbol{r}$ with the step set $S = \{\mathbf{e}_1, \ldots, \mathbf{e}_d\}$ of standard basis vectors, giving the generating function $1/(1 - \sum_{j=1}^{d} z_j)$ for the multinomial coefficients. ◁

Example 2.7 (Delannoy numbers)**.** If $\mathcal{A}$ is the class of lattice paths in $\mathbb{N}^2$ using the steps North $(0, 1)$, East $(1, 0)$, and Northeast $(1, 1)$ enumerated by endpoint then the counting sequence $a_{\boldsymbol{r}} = |\mathcal{A}_{\boldsymbol{r}}|$ defines the ***Delannoy numbers*** [Com74, Exercise I.21]. The generating function $x + y + xy$ for the set of steps leads to the Delannoy generating function

$$F(x, y) = \frac{1}{1 - x - y - xy}.$$

◁

Example 2.8 (no gaps of size 2)**.** For any $n \in \mathbb{N}$ let $\mathcal{B}_n$ be the collection of all subsets of $[n]$ where no two consecutive members are absent. One can enumerate $\mathcal{B}_n$ by mapping bijectively to Example 2.1. In [CLP04] an estimate was required on the number of such sets that were mapped into other sets of the same form by a random permutation. That paper showed that in order to compute the second moment of this random variable it suffices to count the pairs $(S, T) \in \mathcal{B}_n^2$ where the parameters $n, |S|, |T|$, and $|S \cap T|$ are fixed. To that end, let $F(x, y, z, w)$ be the 4-variable generating function for the class $\mathcal{F}$ of all pairs $(S, T) \in \mathcal{B}_n^2$ for any $n \in \mathbb{N}$, using the weight function $\phi(S, T) = (n, |S|, |T|, |S \cap T|)$.

We may derive F by investigating what happens between consecutive elements of $S \cap T$. Identify $(S, T) \in \mathcal{B}_n^2$ with a sequence

$$\alpha \in \{(0, 0), (0, 1), (1, 0), (1, 1)\}^n$$

where

$$\alpha_j = \begin{cases} (1,1) & \text{if } j \in S \cap T, \\ (1,0) & \text{if } j \in S \setminus T, \\ (0,1) & \text{if } j \in T \setminus S, \\ (0,0) & \text{if } j \notin S \cup T. \end{cases}$$

If j and $j + r$ are positions of consecutive occurrences of $(1, 1)$ in α then the possibilities for the string $\alpha_{j+1} \cdots \alpha_{j+r}$ are as follows.

(i) If $\alpha_{j+1} = (1, 1)$ then the only possibility is $r = 1$.
(ii) If $\alpha_{j+1} = (0, 0)$ then the only possibility is $r = 2$ and $\alpha_{j+1} \cdots \alpha_{j+r} = ((0, 0), (1, 1))$.
(iii) If $\alpha_{j+1} = (1, 0)$ then $r \geq 2$ may be arbitrary and $\alpha_{j+1} \cdots \alpha_{j+r}$ alternates between $(1, 0)$ and $(0, 1)$ until the final $(1, 1)$.
(iv) If $\alpha_{j+1} = (0, 1)$ then $r \geq 2$ may be arbitrary and $\alpha_{j+1} \cdots \alpha_{j+r}$ alternates between $(0, 1)$ and $(1, 0)$ until the final $(1, 1)$.

We will build an element of $\mathcal{F}$ uniquely from blocks of one of these four types. In the first case, the generating function $G_1(x, y, z, w)$ for a single block is simply $xyzw$, while in the second case we obtain $G_2(x, y, z, w) = x^2yzw$. In the third case, we can write the block as either $((1, 0))$ or $((1, 0), (0, 1))$, followed by zero or more alternations of length two; decomposing this way shows the generating function to be $G_3(x, y, z, w) = xyzw\frac{xy+x^2yz}{1-x^2yz}$. Similarly, in the fourth case we obtain $G_4(x, y, z, w) = xyzw\frac{xz+x^2yz}{1-x^2yz}$. Summing these four cases gives the generating function

$$G(x, y, z, w) = xyzw\frac{(1 + x)(1 - x^2yz) + xy + xz + 2x^2yz}{1 - x^2yz}$$

that enumerates all possible configurations of one block.

Stringing together blocks of the four types gives all elements of $\mathcal{F}$ that end in $(1, 1)$, and this subclass thus has generating function $1/(1 - G(x, y, z, w))$. An element $(S, T) \in \mathcal{B}_n^2$ of this subclass must have $n \in S \cap T$, and by removing n we obtain a bijection of such pairs to $\mathcal{B}_{n-1}^2$, except that when $n = 0$ it is not possible to delete n. Applying this bijection reduces the weight of each pair in $\mathcal{F}$ by $(1, 1, 1, 1)$, ultimately giving

$$\begin{aligned} F(x, y, z, w) &= \left(\frac{1}{1 - G(x, y, z, w)} - 1\right) \Big/ (xyzw) \\ &= \frac{(1 + x)(1 - x^2yz) + xy + xz + 2x^2yz}{1 - x^2yz - xyzw[(1 + x)(1 - x^2yz) + xy + xz + 2x^2yz]}. \end{aligned}$$

◁

Exercise 2.4. Fix a natural number n.

(a) What is the generating polynomial for the number of strictly increasing sequences of length k chosen from $[n]$?
(b) Let S be the set of sequences of symbols from $\{0\} \cup [n]$ starting with a 0 and such that the subsequence between any two consecutive zeros is strictly increasing. Find the bivariate generating function counting elements of S by their length and number of zeros.

Transfer matrices and restricted transitions

Suppose we want to count words (finite sequences) on an alphabet V such that consecutive pairs of letters must come from a fixed set E. Allowed words of length n are equivalent to paths of length n in the directed graph whose vertices are identified by the elements of V and whose edges are identified by the elements of E. To count such paths by number of steps, let $\boldsymbol{M}$ be the *adjacency matrix* of (V, E): the square matrix indexed by V with $M_{vw} = 1$ if $(v, w) \in E$ and $M_{vw} = 0$ otherwise. The number of allowed paths of length n from v to w is $(\boldsymbol{M}^n)_{vw}$. If we wish to enumerate paths by length we must sum $(z\boldsymbol{M})^n$ over n, so the generating function counting finite paths from v to w weighted by their length is

$$F_{vw}(z) = \sum_{n=0}^{\infty} ((z\boldsymbol{M})^n)_{vw} = \left[(\boldsymbol{I} - z\boldsymbol{M})^{-1}\right]_{vw} .$$

This approach, known as the ***transfer matrix method***, is quite versatile. For instance, to count all allowed paths by length we may sum in v and w, giving the generating function $F(z) = \text{trace}((\boldsymbol{I} - z\boldsymbol{M})^{-1}\boldsymbol{J})$ where $\boldsymbol{J}$ is the $|V| \times |V|$ square matrix of 1s. More generally, we can enumerate by the number of each type of transition: if $E = \{\boldsymbol{p}_1, \ldots, \boldsymbol{p}_r\}$ and $\tilde{M}_{vw} = z_k$ if $\boldsymbol{p}_k = (v, w)$ and $\tilde{M}_{vw} = 0$ if $(v, w) \notin E$ then $\tilde{F}_{vw}(\boldsymbol{z}) = [(\boldsymbol{I} - \tilde{\boldsymbol{M}})^{-1}]_{vw}$ enumerates paths from v to w by the number of each transition, and $\tilde{F}(\boldsymbol{z}) = \text{trace}((\boldsymbol{I} - \tilde{\boldsymbol{M}})^{-1}\boldsymbol{J})$ enumerates all paths by the number of each transition.

Example 2.9 (binary strings revisited)**.** The transfer matrix method may be used to re-derive the generating function in Example 2.1. Let $V = \{0, 1\}$ and $E = \{(0, 0), (0, 1), (1, 0)\}$ contain all directed edges except $(1, 1)$. Then

$$\boldsymbol{M} = \begin{bmatrix} 1 & 1 \\ 1 & 0 \end{bmatrix}$$

so the entries of

$$\mathbf{Q} = (\boldsymbol{I} - z\boldsymbol{M})^{-1} = \frac{1}{1 - z - z^2} \begin{bmatrix} 1 & z \\ z & 1 - z \end{bmatrix}$$

enumerate binary strings without consecutive ones, depending on their starting and ending symbols. The paths from 0 to 0 having $n \geq 0$ transitions are in one-to-one correspondence, via stripping off the last 0, to the words in Example 2.1 of length n. Thus, the generating function for the class in Example 2.1 is the $(0, 0)$-entry of $\mathbf{Q}$, namely $1/(1 - z - z^2)$. ◁

Composition corresponds to block substitution

Let $F(\boldsymbol{z})$ be a d-variate formal power series and $G_1, \ldots, G_d$ be d formal power series in any number of variables, all with vanishing constant terms. We define the formal composition $F(G_1, \ldots, G_d)$ as a limit in the formal power series ring,

$$F(G_1, \ldots, G_d) := \lim_{n\to\infty} \sum_{|\boldsymbol{r}|\leq n} a_{\boldsymbol{r}} \mathbf{G}(\boldsymbol{z})^{\boldsymbol{r}} . \tag{2.1}$$

The degree of the term $\mathbf{G}(\boldsymbol{z})^{\boldsymbol{r}} := G_1(\boldsymbol{z})^{r_1} \cdots G_d(\boldsymbol{z})^{r_d}$ is at least $|\boldsymbol{r}| = \sum_{j=1}^{d} r_j$ by the assumption that $G_j(\mathbf{0}) = 0$ for all j, hence the coefficients of degree at most δ in the sum in (2.1) do not change once $n > \delta$, and the limit exists in the formal power series ring.

Remark. Even if some G_j has a non-zero constant term, it may still happen that the sum converges in the ring of analytic functions, meaning that the infinitely many contributions to all coefficients are absolutely summable. The composition $\exp(1 - x)$ is not formally allowed in $\mathbb{C}[[x]]$ because, for instance, the constant term cannot be computed in a finite number of operations, but $\exp(1 - x) = e \sum_{n=0}^{\infty} \frac{(-x)^n}{n!}$ represents the composition as a convergent series for all $x \in \mathbb{C}$, meaning it belongs to $\mathbb{C}\{x\}$.

A slightly unwieldy combinatorial interpretation of composition is given in [GJ04, Section 2.2.20]. Suppose $\mathcal{A} = \Phi(\mathcal{S})$ where Φ is a specification built from the combinatorial sum, product and sequence operations described above and $\mathcal{S} = \{\boldsymbol{s}_1, \ldots, \boldsymbol{s}_d\}$ is a class with d distinct elements such that $\phi(\boldsymbol{s}_k)$ is the elementary basis vector $\mathbf{e}_k$. If $\mathcal{B}_1, \ldots, \mathcal{B}_d$ are combinatorial classes with no objects of weight zero, then the ***composition class*** $\mathcal{A} \circ (\mathcal{B}_1, \ldots, \mathcal{B}_d)$ enumerated by $A(B_1, \ldots, B_d)$ is the class obtained from $\mathcal{A}$ by replacing each occurrence of $\boldsymbol{s}_k$ by an element of $\mathcal{B}_k$ in all possible ways.

Example 2.10 (queries). Queries from a database have integer computation times associated with them. Suppose we have a database such that for each $k \geq 1$ there are b_k different queries that take k time units, and the protocol in use does not allow two large queries in a row, where a large query is one of size greater than some fixed number M. How many query sequences are there of total time n?

The sequences of queries are bijectively equivalent to the composition $\mathcal{A} \circ (\mathcal{B}_1, \mathcal{B}_2)$, where $\mathcal{A}$ is the class from Example 2.9, counted by numbers of 0s and 1s, and $\mathcal{B}_1$ and $\mathcal{B}_2$ are respectively the short queries and the long queries, counted by computation time. Thus,

$$A(B_1, B_2) = \frac{1}{1 - B_1(z) - B_1(z)B_2(z)}$$

enumerates query sequences in this model by length, where $B_1(z) = \sum_{k=1}^{M} b_k z^k$ and $B_2(z) = \sum_{k>M} b_k z^k$. ◁

Our next example may seem a natural candidate for the transfer matrix method, but is simpler to analyze from the viewpoint of compositions.

Example 2.11 (Smirnov words). Let $\mathcal{A}$ be the class of ***Smirnov words*** on the alphabet $[d]$, which are words where no consecutive repetition of any symbol is allowed. The definition immediately implies that there are $d \cdot (d-1)^{n-1}$ words of length n, and we now perform a more refined enumeration tracking each symbol.

Let $A(\boldsymbol{z})$ enumerate Smirnov words and $B(\boldsymbol{z})$ enumerate the class $\mathcal{B}$ of all words on the alphabet $[d]$, both weighted by the number of occurrences of each symbol. Starting with $\boldsymbol{x} \in \mathcal{A}$ and substituting an arbitrary nonempty string of the symbol j for every occurrence of j in $\boldsymbol{x}$ produces each element of $\mathcal{B}$ in a unique way. The generating function for a nonempty string of js is $\frac{z_j}{1-z_j}$, whence

$$B(\boldsymbol{z}) = A\left(\frac{z_1}{1-z_1}, \ldots, \frac{z_d}{1-z_d}\right).$$

We can solve for A by setting $y_j = z_j/(1-z_j)$ in this equation, since inverting the substitution gives $z_j = y_j/(1+y_j)$ and thus

$$A(\boldsymbol{y}) = B\left(\frac{y_1}{1+y_1}, \ldots, \frac{y_d}{1+y_d}\right).$$

Since $B(\boldsymbol{z}) = 1/(1 - \sum_{j=1}^{d} z_j)$ we ultimately obtain

$$A(\boldsymbol{z}) = \frac{1}{1 - \sum_{j=1}^{d} \frac{z_j}{1+z_j}}.$$

◁

In probability theory, the study of branching processes is almost always dealt with by means of analytic generating functions.

Example 2.12 (*Galton–Watson process*). Let $f(z)$ be a *probability generating function* supported on $\mathbb{N}$, meaning that $f(z) = \sum_{n=0}^{\infty} p_n z^n$ with $p_n \geq 0$ and $\sum_{n=0}^{\infty} p_n = 1$. A *branching process* with offspring distribution f is a random tree with one vertex in generation 0 where each individual in each generation has a random number of children, and the numbers of children born to the individuals in a generation are independent and equal to n with probability p_n. If Z_n is the random variable tracking the number of individuals in generation n then we can compute the probability generating function $g_n(z) = \sum_{k \geq 0} p_{n,k} z^k$ for Z_n inductively, where $p_{n,k}$ is the probability that $Z_n = k$.

Indeed, the probability generating function for Z_1 is simply $f(z)$. In a configuration with $Z_n = k$, the next generation is composed of a sequence of k families, each independently having size j with probability p_j. The probability generating function for such a sequence is $f(z)^k$, whence $g_{n+1} = g_n \circ f$. Inductively then, $g_n = f \circ \cdots \circ f$ is the n-fold composition of f with itself. Observe that, unless $p_0 = 0$ (no extinction), this composition is not defined in the formal power series ring, but since all functions involved are convergent on the unit disk, the compositions are well defined analytically. ◁

Example 2.13 (branching random walk). Associate to each particle in a branching process a real number, which we interpret as the displacement in one dimension between its position and that of its parent. If these are independent of each other and of the branching, and are identically distributed, then one has the classical branching random walk. A question that has been asked several times in the literature, for instance in [Kes78], is how to determine when there exists a line of descent from a single particle at position 1 that remains to the right of the origin for all time. Here we consider the simplest non-trivial case, where the branching process is deterministic binary splitting ($p_2 = 1$) and the displacement distribution is a random walk that moves one unit to the right with probability $p < 1/2$ and one unit to the left with probability $1 - p$.

If we modify the process so that particles stop moving or reproducing when they hit the origin, then $p < 1/2$ implies an infinite line of descent to the right of the origin is equivalent to infinitely many particles reaching the origin. To analyze the process we thus let X be the number of particles ever to hit the origin, beginning with a single particle at 1, and let ϕ be the probability generating function

$$\phi(z) = \sum_{n=0}^{\infty} a_n z^n \quad \text{where } a_n = \mathbb{P}(X = n).$$

If the initial condition is changed to a single particle at position 2, then the number of particles ever to reach the origin will have probability generating function $\phi \circ \phi$. To see this, apply the analysis of Example 2.12, noting that the number of particles ever to reach 1 before any ancestor has reached 1, together with their collections of descendants who ever reach 0, form two generations of a branching process with offspring distribution the same as X. If the initial condition is changed to a single particle at position 0 then the generating function is z.

Each of the two children in the first generation is located at 0 with probability $1 - p$ and at 2 with probability p, so the probability generating function for the contribution to X of each child is $(1-p)z + p\phi(\phi(z))$. The two contributions are independent so their sum is a convolution, whose probability generating function is therefore the square of this. Thus, we have the identity

$$\phi(z) = [(1 - p)z + p\phi(\phi(z))]^2 . \tag{2.2}$$

While this does not produce an explicit formula for ϕ, it is possible from this to derive asymptotics for $\phi(t)$ as $t \uparrow 1$, allowing us to use so-called *Tauberian theorems* to recover asymptotic information about a_n. See Example 3.17 for more information. ◁

Derivation corresponds to marking an atom

If $A(z) = \sum_{n \geq 0} a_n z^n$ enumerates the class $\mathcal{A}$ then the definition of the formal derivative implies that $zA'(z) = \sum_{n \geq 1} n a_n z^n$. Thus, $zA'(z)$ can be viewed combinatorially as the generating function of a new class obtained from $\mathcal{A}$ by taking each object and marking, in all possible ways, one of the atomic pieces it is composed of. Similarly, if $A(\mathbf{z})$ is a multivariate generating function then $(z_k \partial_k A)(\mathbf{z})$ can be interpreted as the generating function of a combinatorial class obtained by marking pieces defining the parameter of $\mathcal{A}$ tracked by the variable z_k.

Exercise 2.5. Let $f(z) = \sum_{n=0}^{\infty} p_n z^n$ be the probability generating function for the probability that a randomly chosen household in some town consists of n people. Let $g(z) = zf'(z)/f'(1)$. Show that g is a probability generating function and determine what sampling probability it represents.

2.3 Algebraic generating functions

A formal power series $F(\mathbf{z})$ is called an ***algebraic power series*** if there exists a polynomial $P(\mathbf{z}, y) \in \mathbb{C}[\mathbf{z}, y]$ such that $P(\mathbf{z}, F(\mathbf{z})) = 0$. Algebraic series are often considered to be the second simplest class of generating functions, after rational series, to arise frequently in combinatorics. The main reason algebraic series arise often is that a decomposition of the elements in a combinatorial class into products or disjoint unions involving smaller elements from the same class yields an algebraic equation satisfied by the generating function of the class. Perhaps the most famous example of a (non-rational) algebraic generating function is the following.

Example 2.14 (binary trees and Catalan numbers)**.** Let C be the class of planar rooted binary trees, defined recursively by saying that the empty tree lies in C and every element of C with $n \geq 1$ vertices is formed by a root vertex together with a left subtree $L \in C$ and a right subtree $R \in C$ such that the number of vertices in L and R sum to $n - 1$. The counting sequence $c_n = |C_n|$ defines the ***Catalan numbers***; Stanley [Sta15] lists 214 combinatorial classes enumerated by the Catalan numbers.

The recursive definition for the elements of C gives a bijection of combinatorial classes between C and $\Upsilon \sqcup \mathcal{R} \times C \times C$, where Υ is the combinatorial class containing only the empty tree (to account for the case $n = 0$) and $\mathcal{R}$ is the class containing only the tree with one node (corresponding to a root vertex). At the generating function level, this bijection yields the algebraic equation

$$F(z) = 1 + zF(z)^2, \tag{2.3}$$

which has a formal power series solution corresponding to the generating function $C(z)$ of C. To solve (2.3) in the ring of formal power series we first find solutions in the ring of germs of analytic functions, since we can perform our usual algebraic operations. The quadratic formula yields two solutions

$$F_{\pm}(z) = \frac{1 \pm \sqrt{1-4z}}{2z}.$$

The solution $F_+(z)$ is not analytic at the origin, as its denominator vanishes when $z = 0$ but its numerator does not, so the generating function of C is

$$C(z) = F_-(z) = \frac{1 - \sqrt{1-4z}}{2z} = 1 + z + 2z^2 + 5z^3 + \cdots.$$

From this explicit expression, the generalized binomial theorem yields the closed form $c_n = \frac{1}{n+1}\binom{2n}{n}$. ◁

Exercise 2.6. Define the class of *d-ary until the end* trees by altering the definition in Example 2.14 so that each vertex has at most d children *unless* all the children are leaves, in which case an arbitrary number is permitted. Adapt the argument from Example 2.14 to compute the generating function enumerating these trees by their number of vertices.

The kernel method

We now give an account of the ***kernel method***, one of the most prolific sources of algebraic generating functions in combinatorics. The kernel method is a means of producing a generating function for an array $\{a_{\boldsymbol{r}}\}$ satisfying a linear recurrence

$$a_{\boldsymbol{r}} = \sum_{\boldsymbol{s} \in E} c_{\boldsymbol{s}} a_{\boldsymbol{r}-\boldsymbol{s}} \tag{2.4}$$

for some constants $\{c_{\boldsymbol{s}} : \boldsymbol{s} \in E\}$ defined over a finite set $E \subset \mathbb{Z}^d$, except when $\boldsymbol{r}$ lies in a set of ***boundary conditions*** to be made precise. We will see in Lemma 2.16 below that this recursion is well founded whenever the convex hull of E does not intersect the non-positive orthant $\mathbb{R}^d_{\leq 0}$.

If the index set $E \subseteq \mathbb{N}^d$ then Example 2.6 generalizes to show that $F(\boldsymbol{z}) = \sum_{\boldsymbol{r} \in \mathbb{N}^d} a_{\boldsymbol{r}} \boldsymbol{z}^{\boldsymbol{r}}$ is rational. The kernel method is of interest to this text because it often produces generating functions which, even though they are not rational, satisfy the meromorphicity assumptions that allow us to compute their asymptotics. It is shown in [BP00] that the complexity of F increases with the number of coordinates in which points of E are allowed to take negative values: just as allowing no negative coordinates in E causes $F(\boldsymbol{z})$ to be rational, it turns out that allowing only one negative coordinate in E causes $F(\boldsymbol{z})$ to be algebraic whenever the generating function encoding its initial conditions is algebraic. When E contains points with two different negative coordinates it is possible to have very pathological behavior (including sequences whose generating functions do not satisfy polynomial differential equations). Our presentation of the kernel method draws heavily on [BP00].

Example 2.15 (a random walk game). Suppose two players play a game, moving their respective tokens along a track of squares by flipping a fair coin at each time step to see who moves forward one square. The second player starts behind the first player and the game ends as follows: if the second player passes the first player, the second player wins; if the first player reaches a fixed goal square, the first player wins; if both players are on the square immediately preceding the goal, then it is a draw.

Let a_{rs} be the probability of a draw, when the players start at respective distances $1 + r$ and $1 + r + s$ from the goal. For convenience we set $a_{00} = 1$ and extend our sequence to have integer indices by defining $a_{rs} = 0$ if at least one of r or s is negative. Conditioning on which player moves first yields the recursion

$$a_{rs} = \frac{a_{r,s-1} + a_{r-1,s+1}}{2},$$

which is valid for all (r, s) with nonnegative coordinates except for $(0, 0)$. Because shifting indices corresponds to multiplication of the generating function by monomials, this recurrence suggests that we multiply the generating function $F(x, y) = \sum a_{rs} x^r y^s$ for $\{a_{rs}\}$ by $1 - (1/2)y - (1/2)(x/y)$. Clearing denominators, we let $Q(x, y) = 2y - y^2 - x$ and note from the recurrence that all coefficients of $Q(x, y)F(x, y)$ vanish with two exceptions: the coefficient $[x^0 y^1]Q(x, y)F(x, y) = 2a_{0,0} - a_{0,-1} - a_{-1,1} = 2$, because the recursion does not hold at $(0, 0)$, and the coefficients $[x^j y^0]Q(x, y)F(x, y) = 2a_{j,-1} - a_{j,-2} - a_{j-1,0}$ with $j \geq 1$ do not vanish because only the third term is non-zero. In other words,

$$Q(x, y)F(x, y) = 2y - h(x), \tag{2.5}$$

where $h(x) = \sum_{j \geq 1} a_{j-1,0} x^j = xF(x, 0)$ will not be known until we solve for F.

This generating function is in fact a simpler variant of the one derived in [LL99] for the waiting time until the two players collide, which is needed in the analysis of a sorting algorithm. Their solution is to observe that there is an analytic curve in a neighborhood of the origin on which Q vanishes. Solving $Q = 0$ for y yields two solutions, with the solution $y = \xi(x) = 1 - \sqrt{1 - x}$ vanishing at the origin. Since $\xi(x)$ is analytic at the origin we also have $Q(x, \xi(x)) = 0$ at the level of formal power series, and substituting $y = \xi(x)$ in (2.5) gives

$$0 = Q(x, \xi(x))F(x, \xi(x)) = 2\xi(x) - h(x)\,.$$

Thus, $h(x) = 2\xi(x)$ and

$$F(x, y) = 2\frac{y - \xi(x)}{Q(x, y)} = \frac{2}{1 + \sqrt{1 - x} - y}\,.$$

◁

An explanation of the kernel method

Let $\mathbf{p}$ be the ***valley*** of E, defined as the coordinatewise minimum of the points in $E \cup \{\mathbf{0}\}$ and define

$$Q(\mathbf{z}) = \mathbf{z}^{-\mathbf{p}} \left(1 - \sum_{\mathbf{s} \in E} c_{\mathbf{s}} \mathbf{z}^{\mathbf{s}} \right),$$

where the normalization by $\mathbf{z}^{-\mathbf{p}}$ guarantees that Q is a polynomial but not divisible by any z_j. We assume $\mathbf{p} \neq \mathbf{0}$, since we already understand how when $\mathbf{p} = \mathbf{0}$ the recursion leads to a rational generating function. A set of ***boundary locations*** is any set $B \subseteq \mathbb{N}^d$ closed under the coordinatewise partial-order $\leq$, with a corresponding set of ***boundary values*** $\{b_{\boldsymbol{r}} : \boldsymbol{r} \in B\}$. We study the initial value problem with initial conditions

$$a_{\boldsymbol{r}} = b_{\boldsymbol{r}} \text{ for all } \boldsymbol{r} \in B, \tag{2.6}$$

where the recursion (2.4) holds for all $\boldsymbol{r} \in \mathbb{N}^d \backslash B$ (summands with $\boldsymbol{r}-\boldsymbol{s} \notin \mathbb{N}^d$ are defined to be zero). Our goal here is to take the set of shifts E, the polynomial Q, and the sets of boundary locations and boundary values, and determine the generating function enumerating $\{a_{\boldsymbol{r}}\}$. Figure 2.1 shows an example setup where $E = \{(2,-1),(-1,2)\}$ and B is the y-axis. We encode the terms in each polynomial that appears by their ***Newton diagrams***, the set of vectors defined by the exponents of their monomials.

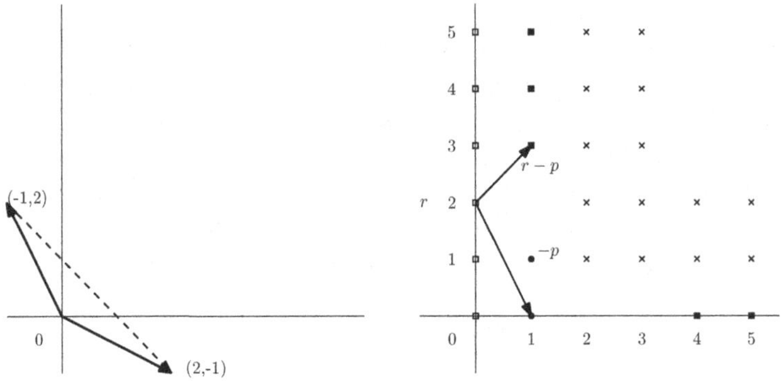

Figure 2.1 *Left:* The set E and its convex hull (dashed). *Right:* A Newton diagram of K (marked by ×), U (marked by ■), and B (marked by □). The quantities $K, U, B, \boldsymbol{r}, \boldsymbol{s}$, and $\boldsymbol{p}$ are defined in the text.

Let $Z = \mathbb{N}^d \setminus B$ and decompose $F(\boldsymbol{z}) = F_Z(\boldsymbol{z}) + F_B(\boldsymbol{z})$ where

$$F_Z(\boldsymbol{z}) = \sum_{\boldsymbol{r} \in Z} a_{\boldsymbol{r}} \boldsymbol{z}^{\boldsymbol{r}}$$

is the generating function for the indices where the recursion (2.4) holds and

$$F_B(\boldsymbol{z}) = \sum_{\boldsymbol{r} \in B} b_{\boldsymbol{r}} \boldsymbol{z}^{\boldsymbol{r}}$$

enumerates the boundary conditions. Following the method of Example 2.15, to apply the kernel method we study the product $Q(\boldsymbol{z})F_Z(\boldsymbol{z})$. Examining this product shows that there are two types of terms with non-zero coefficients.

- First, for every pair $(\boldsymbol{r}, \boldsymbol{s})$ with $\boldsymbol{r} \in Z$, $\boldsymbol{s} \in E$, and $\boldsymbol{r} - \boldsymbol{s} \in B$ there is a coefficient

$$\underbrace{[z^{r-p}]Q(z)F(z)}_{0} - \underbrace{[z^{r-p}]Q(z)F_B(z)}_{-c_s b_{r-s}} = c_s b_{r-s}\,,$$

where the first extraction is zero because $r \in Z$ implies the recurrence (2.4) holds, and the second extraction is computed directly. Let

$$K(z) = \sum_{\substack{r \in Z, s \in E \\ r-s \in B}} c_s b_{r-s} z^{r-p}$$

enumerate such terms. The K stands for *known*, because the coefficients of K are determined by the known boundary conditions, which are specified in the problem. The example in Figure 2.1 marks the terms of K in the first two rows and columns of Z.

- Second, for every pair (r, s) with $r \notin Z, s \in E$, and $r - s \in Z$, direct expansion shows that

$$[z^{r-p}]Q(z)F_Z(z) = -c_s a_{r-s},$$

reflecting the fact that the recursion does not hold at r. Let

$$U(z) = \sum_{\substack{r \notin Z, s \in E \\ r-s \in Z}} c_s a_{r-s} z^{r-p}$$

enumerate such terms. The U stands for *unknown*, because these coefficients are not explicitly determined from the boundary conditions. In the example from Figure 2.1, U has one row and one column of terms. The value of r leading to the xy^3-term of U is pictured.

We now give a sufficient condition for the recurrence (2.4) to be well founded, and prove that the series F_Z and U are unique.

Lemma 2.16 ([BP00, Theorem 5]). *Let E be a finite subset of $\mathbb{N}^d$ whose convex hull does not intersect the non-positive orthant, let $\{c_s : s \in E\}$ be constants, let p be the valley of E, let $B \subseteq \mathbb{N}^d$ be closed under the coordinatewise partial order $\leq$ with complement $Z = \mathbb{N}^d \setminus B$, let $\{b_s : s \in B\}$ be constants, and let*

$$K(z) = \sum_{\substack{r \in Z, s \in E \\ r-s \in B}} c_s b_{r-s} z^{r-p}.$$

Then there is a unique set of values $\{a_r : r \in Z\}$ such that (2.4) holds for all $r \in Z$ and $a_r = b_r$ for all $r \in B$. Consequently, there is a unique pair of formal power series F_Z and U such that

$$Q(z)F_Z(z) = K(z) - U(z)\,.$$

Furthermore, if K is analytic in a neighborhood of the origin, then so are F_Z and U.

Proof The convex hull of E and the closed non-positive orthant are disjoint convex polyhedra so there is a hyperplane that separates them and meets neither. The normal vector may be perturbed slightly to obtain a rational vector $\boldsymbol{v}$ such that $\boldsymbol{v} \cdot \boldsymbol{s} > 0$ for all $\boldsymbol{s} \in E$ and $\boldsymbol{v} \cdot \boldsymbol{s} < 0$ for all $\boldsymbol{s} \neq \boldsymbol{0}$ in the non-positive orthant. The vector $\boldsymbol{v}$ must have positive coordinates and, clearing denominators, we may assume $\boldsymbol{v}$ has integer coordinates. Linearly order $\mathbb{N}^d$ by the value of the dot product with $\boldsymbol{v}$, breaking ties arbitrarily, to produce a well ordering $\preceq$ of $\mathbb{N}^d$.

If $\boldsymbol{m} \in Z$ is fixed then $\boldsymbol{s} \in E$ and $\boldsymbol{r} \prec \boldsymbol{m}$ implies $\boldsymbol{r} - \boldsymbol{s} \prec \boldsymbol{m}$. Consequently, the validity of (2.4) for all $\boldsymbol{r} \prec \boldsymbol{m}$ depends only on values $a_{\boldsymbol{r}}$ with $\boldsymbol{r} \prec \boldsymbol{m}$. In particular, if there is a unique set of values of $\{a_{\boldsymbol{r}} : \boldsymbol{r} \prec \boldsymbol{m}\}$ such that (2.4) holds for $\boldsymbol{r} \prec \boldsymbol{m}$ then setting $\boldsymbol{r} = \boldsymbol{m}$ in (2.4) uniquely specifies $a_{\boldsymbol{m}}$. Existence and uniqueness thus follow by induction, with the base case the minimal vector $\boldsymbol{v} \in Z$ under $\preceq$.

To show analyticity of U and F_Z, let $\gamma' = \log \sum_{\boldsymbol{s} \in E} |c_{\boldsymbol{s}}|$. By analyticity of K we may choose $\gamma' \geq \gamma$ such that $|a_{\boldsymbol{r}}| = |b_{\boldsymbol{r}}| \leq \exp(\gamma \boldsymbol{r} \cdot \boldsymbol{v})$ for any $\boldsymbol{r} \in B$. Furthermore, for $\boldsymbol{r} \in Z$ we have by induction that

$$\begin{aligned}
|a_{\boldsymbol{r}}| &\leq \left(\sum_{\boldsymbol{s} \in E} |c_{\boldsymbol{s}}| \right) \sup_{\boldsymbol{s} \in E} |a_{\boldsymbol{r}-\boldsymbol{s}}| \\
&\leq e^{\gamma} \sup_{\boldsymbol{m} \cdot \boldsymbol{v} < \boldsymbol{r} \cdot \boldsymbol{v}} |a_{\boldsymbol{m}}| \\
&\leq e^{\gamma} e^{\gamma(\boldsymbol{r} \cdot \boldsymbol{v} - 1)} \\
&= e^{\gamma \boldsymbol{r} \cdot \boldsymbol{v}},
\end{aligned}$$

establishing an exponential bound on $|a_{\boldsymbol{r}}|$ for all $\boldsymbol{r} \in \mathbb{N}^d$. Analyticity of $F_Z(\boldsymbol{z})$ follows directly, which then implies analyticity of $U(\boldsymbol{z}) = K(\boldsymbol{z}) - Q(\boldsymbol{z})F_Z(\boldsymbol{z})$. □

The previous lemma is based on a formal power series approach. Another way of thinking about this is that F_Z is trying to be the power series K/Q but, since Q vanishes at the origin, one must subtract some terms from K to cancel whatever factor of Q vanishes at the origin. The kernel method turns this intuition into a precise statement, when the boundary locations are specified using the coordinatewise partial order $\geq$ on $\mathbb{N}^d$.

Theorem 2.17 ([BP00, Theorem 13]). *Let $d \geq 2$ be arbitrary and suppose the boundary locations B are of the form $\{\boldsymbol{r} : \boldsymbol{r} \not\geq \boldsymbol{s}\}$ for some $\boldsymbol{s} \in \mathbb{N}^d$. If the* *coordinates of the valley* $\mathbf{p}$ *satisfy* $p_1, \ldots, p_{d-1} \geq 0 > p_d$ *and the boundary generating function* $K(\boldsymbol{z})$ *is algebraic then* F *is algebraic.*

Proof Suppose $\boldsymbol{r} \notin Z$ and $\boldsymbol{r} - \boldsymbol{s} \in Z$ with $\boldsymbol{s} \in E$. If $\boldsymbol{s}'$ is the vector whose first $d-1$ coordinates match $\boldsymbol{s}$ but $\boldsymbol{s}'_d = 0$ then $\boldsymbol{r} \notin Z$ implies $\boldsymbol{r} - \boldsymbol{s}' \notin Z$ because the complement of Z is closed under coordinatewise $\leq$ and the hypothesis on $\boldsymbol{p}$ implies the first $d-1$ coordinates of any point in E are nonnegative. Thus, $s_d - p_d < r_d \leq s_d$ and it follows that U is a polynomial of degree at most $p_d - 1$ in z_d.

The polynomial Q is equal to $z_d^{p_d} - \sum_{\boldsymbol{s} \in E} c_{\boldsymbol{s}} z_d^{p_d} \boldsymbol{z}^{\boldsymbol{s}}$. It is convenient to regard this as a polynomial in z_d of degree at least p_d, over the field of algebraic functions in $z_1, \ldots, z_{d-1}$. We denote the roots of this polynomial by $\xi_i(z_1, \ldots, z_{d-1})$ and note that p_d of the roots, counted with multiplicities, vanish at the origin since $(0, \ldots, 0, j) \notin E$ for any $j < 0$ implies $Q(0, \ldots, 0, z_d)$ has multiplicity p_d at $z_d = 0$.

If the p_d roots of Q vanishing at the origin are distinct then the equation $QF_Z = K - U$ evaluated at each ξ_i leads to p_d equations $U(\xi_i) = K(\xi_i)$. The Lagrange interpolation formula [PS98, Section V1.9] produces a polynomial P given its values $y_1, \ldots, y_k$ at any k points $x_1, \ldots, x_k$,

$$P(x) = \sum_{j=1}^{n} y_j \prod_{i \neq j} \frac{x - x_i}{x_j - x_i}. \tag{2.7}$$

Over any field of characteristic zero, and in particular over the field of algebraic functions in $x_1, \ldots, x_d$, this is the unique polynomial of degree at most $k - 1$ passing through the k points. Applying (2.7) to the p_d equations $U(\xi_i) = K(\xi_i)$ thus uniquely determines the polynomial U. In particular, U is a rational function of algebraic quantities, and is thus also algebraic.

If the ξ_i are not distinct, one has instead the p_d equations

$$U(\xi_i) = K(\xi_i),\ U'(\xi_i) = K'(\xi_i),\ \ldots\ , U^{(m_i-1)}(\xi_i) = k^{(m_i-1)}(\xi_i),$$

where m_i is the multiplicity of the root ξ_i. One may replace the Lagrange interpolation formula by the Hermite interpolation formula [IK94, Section 6.1, Problem 10], which again gives U as a rational function of each $K(\xi_i)$ and its derivatives. □

Restricting to $d = 2$ and $B = \{\mathbf{0}\}$ gives a more explicit formula for F.

Corollary 2.18. *Further specialize the hypotheses of Theorem 2.17 to assume that* $d = 2$, *the point* $\mathbf{p} = (0, -p)$ *for some* $p > 0$, *and* $B = \{\mathbf{0}\}$ *with boundary value* $b_{\mathbf{0}} = 1$. *Then there are exactly* p *formal power series* $\xi_1(x), \ldots, \xi_p(x)$

such that $\xi_j(0) = 0$ *and* $Q(x, \xi_j(x)) = 0$. *Furthermore,*

$$Q(x,y) = -C(x)\prod_{j=1}^{p}(y-\xi_j(x))\prod_{j=1}^{r}(y-\rho_j(x))$$

for some algebraic functions $C(x), \rho_1(x), \dots, \rho_P(x)$ *not vanishing at the origin, and*

$$F_Z(x,y) = \frac{K(x,y)-U(x,y)}{Q(x,y)} = \frac{\prod_{j=1}^{p}(y-\xi_j(x))}{Q(x,y)} = \frac{1}{-C(x)\prod_{j=1}^{r}(y-\rho_j(x))}.$$

Proof We work in the ring $\mathbb{C}\{x\}[y]$ of polynomials in y with coefficients in the ring of power series in x converging in a neighborhood of zero. The asserted factorization of Q follows from its vanishing to order p at $y = 0$ and having some degree $p + r$ as a polynomial in y. By definition $K(x, y) = y^p$ and, recalling that the degree of $U(x, y)$ in y is at most $p - 1$, it follows that the degree of $K(x, y) - U(x, y)$ in y is exactly p. Since $K(x, y) - U(x, y)$ vanishes in a neighborhood of zero when $y = \xi_j(x)$ for any j, it is divisible by $\prod_{j=1}^{p}(y - \xi_j(x))$. The leading coefficient of $K - U$ is the same as the leading coefficient of K, namely 1, so $K(x, y) - U(x, y) = \prod_{j=1}^{p}(y - \xi_j(x))$ as claimed. □

Dyck, Motzkin, Schröder, and generalized Dyck paths

Let $S = \{(r_1, s_1), \dots, (r_k, s_k)\}$ be a set of integer vectors with $r_j > 0$ for all j and $\min_j s_j = -p < 0 < \max_j s_j = P$. The class $\mathcal{A}$ of ***generalized Dyck paths*** starting at $(0, 0)$ and taking steps in S consists of all finite tuples $(\boldsymbol{\sigma}_1, \dots, \boldsymbol{\sigma}_\ell)$ of any length $\ell \in \mathbb{N}$ with elements in S whose partial sums $\boldsymbol{\sigma}_1 + \cdots + \boldsymbol{\sigma}_j$ for all $1 \le j \le \ell$ have nonnegative coordinates. The elements of $\mathcal{A}$ can be viewed as lattice paths starting at the origin which never go below the x-axis. The weight of an element in $(\boldsymbol{\sigma}_1, \dots, \boldsymbol{\sigma}_\ell) \in \mathcal{A}$ is the endpoint $(r, s) = \boldsymbol{\sigma}_1 + \cdots + \boldsymbol{\sigma}_\ell$ of the path.

Let $F(x, y) = \sum_{r,s\ge0} a_{rs}x^r y^s$ enumerate this class. Because a generalized Dyck path is either empty, or is a smaller generalized Dyck path with a single step added we obtain a recurrence of the form treated by the kernel method. In fact, we are precisely in the situation covered by Corollary 2.18 where $Q(x, y) = y^p(1-\sum_{i=1}^{k} x^{r_i}y^{s_i})$ and $F(x, y) = F_Z(x, y)$. The special case $p = P = 1$, that is, vertical displacement of at most 1 per step, occurs often in classical examples.

Proposition 2.19. *Assume the above setup. If* $p = P = 1$ *then the generating*

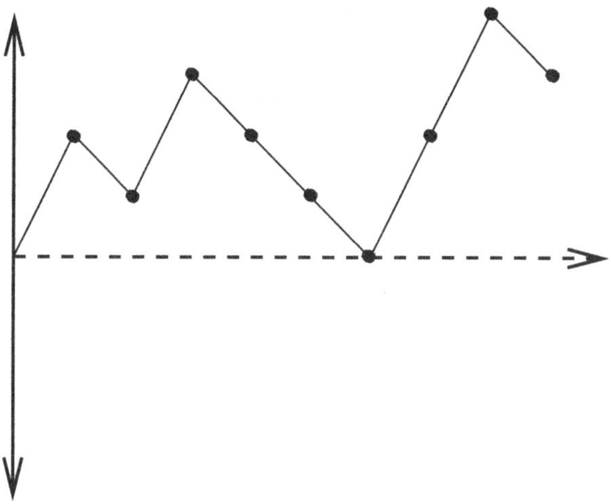

Figure 2.2 A generalized Dyck path of length nine with $E = \{(1,2),(1,-1)\}$.

function for generalized Dyck paths with steps from E is given by

$$F(x,y) = \frac{\xi(x)}{a(x) - C(x)\xi(x)y},$$

where $a(x) = \sum_{i:s_i=-1} x^{r_i}$, *the algebraic function* $\xi(x)$ *is the unique root of* $Q(x,y)$ *in* y *that vanishes at the origin, and* $C(x) = \sum_{i:s_i=P} x^{r_i}$.

Proof Under these hypotheses, Q is a quadratic polynomial in y with leading coefficient $C(x)$ stated in the proposition. If $\xi(x)$ and $\rho(x)$ are the roots of $Q(x,y)$ in y then the product $\xi\rho$ equals $a(x)/C(x)$ where $a(x)$ is stated in the proposition. The formula in Corollary 2.18 thus simplifies to

$$F(x,y) = \frac{\xi(x)/a(x)}{1 - [C(x)\xi(x)/a(x)]y}.$$

□

We end this section with three of the most famous examples of lattice path classes.

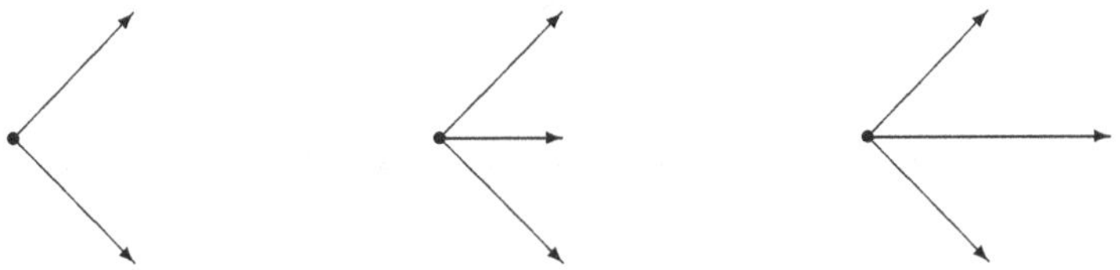

Figure 2.3 Legal steps for three types of paths.

- *Classical Dyck paths* occur when $S = \{(1,1),(1,-1)\}$. Computing the quantities in Proposition 2.19 implies

$$F(x,y) = \frac{1}{-x(y-\rho(x))} = \frac{\xi(x)/x}{1-y\xi(x)}$$

where $\xi(x) = (1-\sqrt{1-4x^2})/(2x)$. Setting $y = 0$ proves the well-known fact that the Dyck paths coming back to the x-axis at $(2n, 0)$ are counted by the Catalan number c_n.

- ***Motzkin paths*** occur when $S = \{(1,1),(1,0),(1,-1)\}$, ultimately giving the generating function

$$F(x,y) = \frac{2}{1-x+\sqrt{1-2x-3x^2}-2xy}.$$

- ***Schröder paths*** occur when $S = \{(1,1),(2,0),(1,-1)\}$, ultimately giving the generating function

$$F(x,y) = \frac{2}{1+\sqrt{1-6x^2+x^4}-x^2-2xy}.$$

2.4 D-finite generating functions and diagonals

The more explicitly a generating function is described, the better the prospects are for extracting information from it. This includes not only exact formulae and asymptotic estimation, but also additional properties like bijections to other classes and random sampling algorithms for objects in the class. Rational generating functions (with coefficients in the integers or rational numbers) are easy to work with because they are specified by the finite data of their numerator and denominator polynomials.

We have seen that some common and very natural combinatorial operations take us from the class of rational functions to the larger class of algebraic generating functions. Algebraic functions also have canonical representations: if f is algebraic then there is a minimal polynomial P for which $P(f(z), z) = 0$ and the series f may be specified by writing down the coefficients of P, which are themselves polynomials and therefore finitely specified, and enough initial terms to uniquely determine f among all roots of P. In Chapter 8 we discuss techniques in computational algebra that allow one to manipulate algebraic quantities implicitly using symbolic methods, making the class of algebraic generating functions reasonably nice to work with. There are, however, common combinatorial operations that take us out of the class of algebraic functions, and this drives us to consider a more general class of generating functions involving differential equations. A more complete discussion of the hierarchy for univariate generating functions may be found in [Mel21, Chapter 2] and [Sta99, Chapter 6].

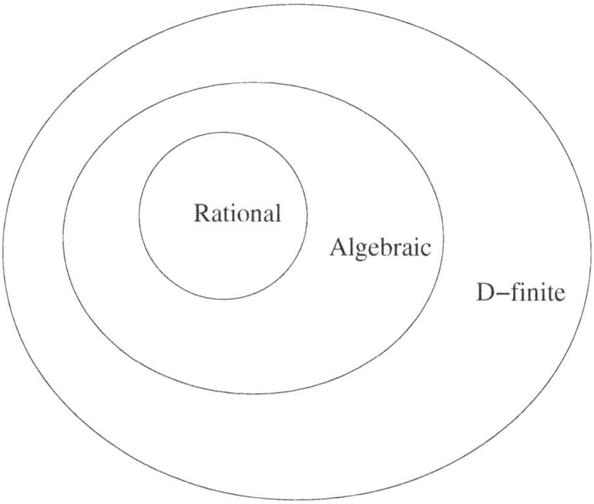

Figure 2.4 Common classes of generating functions.

Univariate D-finite functions

Because any formal power series with a non-zero constant is invertible, the field of fractions of $\mathbb{C}[[z]]$ is the ***field of formal Laurent series***

$$\mathbb{C}((z)) = \left\{ \sum_{n \geq \kappa} a_n z^n : \kappa \in \mathbb{Z} \text{ and } a_n \in \mathbb{C} \text{ for all } n \right\}$$

containing series with a finite number of terms with negative exponents, where addition is still defined termwise and multiplication is still defined by convolution. Section 6.1 gives further information on Laurent series.

Definition 2.20. A formal power series $f(z) \in \mathbb{C}[[z]]$ is a ***D-finite series*** if and only if there is an integer m and polynomials $P_0(z), \ldots, P_m(z) \in \mathbb{C}[z]$ with $P_m(z) \neq 0$ such that

$$P_0(z)f(z) + P_1(z)f'(z) + \cdots + P_m(z)f^{(m)}(z) = 0 . \tag{2.8}$$

Equivalently, f is D-finite if and only if f and its derivatives span a finite dimensional vector space in $\mathbb{C}((z))$ over the field $\mathbb{C}(z)$ of rational functions.

Remark. The more natural definition of a D-finite series is one whose derivatives span a finite dimensional space over the polynomial ring $\mathbb{C}[z]$. Vector spaces over fields are simpler than modules over rings, so we phrase this in terms of vector space dimension over the rational functions instead. This requires that the rational functions act on $\mathbb{C}[[z]]$, which requires us to extend $\mathbb{C}[[z]]$ to $\mathbb{C}((z))$.

Remark. If $f(z)$ is a formal power series that satisfies a *non-homogeneous* linear equation

$$P_0(z)f(z) + P_1(z)f'(z) + \cdots + P_m(z)f^{(m)}(z) = P(z)$$

with $P(z), P_0(z), \ldots, P_m(z) \in \mathbb{C}[z]$ then repeated differentiation proves that $f(z)$ is D-finite.

Variants of this definition are discussed in [Sta99, Proposition 6.4.1]. The Venn diagram in Figure 2.4 depicting the hierarchy of generating function class is justified by the following classical proposition.

Proposition 2.21. *If $f(z) \in \mathbb{C}[[z]]$ is algebraic then $f(z)$ is D-finite.*

Proof Since $f(z)$ is algebraic there exists $P(z, y) = \sum_{j=0}^{m} P_j(z)y^j \in \mathbb{C}[z, y]$ such that $P(z, f(z)) = 0$. Assume P is selected among all such polynomials to have minimal degree in y, so that the partial derivative $P_y(z, y)$, which has smaller degree in y than $P(z, y)$, is non-zero when $y = f(z)$. Implicit differentiation of the equation $P(z, f(z)) = 0$ shows that $P_z(z, f(z)) + f'(z)P_y(z, f(z)) = 0$, so $P_y(z, f(z)) \neq 0$ implies

$$f'(z) = -\frac{P_z(z, f(z))}{P_y(z, f(z))}. \tag{2.9}$$

We have now shown that $f'(z)$ lies in the field of fractions of the ring $\mathbb{C}[z][f]$. The quotient rule thus implies that the derivative of any element of the field extension $\mathbb{C}(z, f)$ is again in $\mathbb{C}(z, f)$, so induction implies all derivatives of f are in $\mathbb{C}(z, f)$. Because f is algebraic, $\mathbb{C}(z, f)$ is a finite extension of $\mathbb{C}(z)$. Thus, f and its derivatives span a finite vector space over $\mathbb{C}(z)$, and f is D-finite. □

Recall that a series $f = \sum_{n=0}^{\infty} a_n z^n$ is rational if and only if the sequence $\{a_n : n \geq 0\}$ satisfies a linear recurrence with constant coefficients. There is no equivalent linear characterization for the coefficients of algebraic generating functions, but there is for D-finite functions.

Definition 2.22 (P-recursiveness). A univariate sequence $\{a_n : n \geq 0\}$ is said to be a ***P-recursive sequence*** (or *polynomially recursive*) if there exist polynomials $P_0(n), \ldots, P_r(n) \in \mathbb{C}[n]$ with $P_m(n) \neq 0$ such that

$$P_r(n)a_{n+r} + P_{r-1}(n)a_{n+r-1} + \cdots + P_0(n)a_n = 0 \tag{2.10}$$

for all $n \geq 0$.

Example 2.23. If $a_n = 1/n!$ then $\{a_n\}$ is P-recursive as $(n+1)a_{n+1} - a_n = 0$ for all $n \geq 0$. Furthermore, if $f(z)$ enumerates $\{a_n\}$ then $\sum_{n\geq 0}(n+1)a_{n+1}z^n = f'(z)$, so $f'(z) - f(z) = 0$ and f is D-finite. We can compute f by solving this linear differential equation with the initial condition $a_0 = 1$, obtaining the expected solution $f(z) = e^z = \sum_{n\geq 0} z^n/n!$. ◁

The correspondence between P-recursive sequences and D-finite series is a classical result, known at least as far back as Frobenius's work on computing series solutions to ordinary differential equations.

Theorem 2.24 (P-recursive corresponds to D-finite). *A sequence $\{a_n : n \geq 0\}$ is P-recursive if and only if its generating function $f(z) = \sum_{n=0}^{\infty} a_n z^n$ is D-finite.*

Proof First, suppose f is D-finite and satisfies (2.8). Differentiating the power series for f shows that for any $n \in \mathbb{N}$ and $k \in [n]$ the coefficient $[z^{n-k}]f^{(j)}(z)$ equals $(n-k+j)_j a_{n-k+j}$, where $(u)_j := u(u-1)\cdots(u-j+1)$ denotes the falling factorial. Let $b_{k,j} = [z^k]P_j(z)$ and let D denote the maximum degree of the coefficient polynomials $P_0(z), \ldots, P_m(z)$. Then for any $n \geq 0$, extracting the coefficient of z^n in (2.8) gives the linear equation

$$\sum_{j=0}^{m}\sum_{k=0}^{D} b_{k,j}\,(n-k+j)_j\,a_{n-k+j} = 0$$

having polynomial coefficients in n of degrees at most m. This is a non-trivial linear recurrence for $\{a_n\}$, because for any k such that $b_{k,m} = [z^k]P_m(z) \neq 0$ the coefficient of a_{n-k+m} is $b_{m,k}n^m + O(n^{m-1}) \neq 0$.

Conversely, suppose that $\{a_n\}$ is P-recursive and satisfies (2.10). The set of falling-factorial polynomials $\{(n+j)_j : j \geq 0\}$ forms a basis for $\mathbb{C}[n]$ which is triangular with respect to the basis $\{n^j : j \geq 0\}$, hence each $P_k(n)$ is a finite linear combination $P_k(n) = \sum_j c_{k,j}(n+j)_j$ and substitution into (2.10) yields

$$\sum_{k=0}^{m}\sum_{j=0}^{\deg P_k} c_{k,j}\,(n+j)_j a_{n+k} = 0\,.$$

The rules for differentiating formal power series extend to the differentiation of Laurent series, and imply that the coefficient of z^n in the repeated derivative $(z^{j-k}f(z))^{(j)}$ is $(n+j)_j\,a_{n+k}$. Multiplying our last equality by z^n and summing over all values of n thus gives the relation

$$\sum_{k=0}^{m}\sum_{j=0}^{\deg P_k} c_{k,j}\left(z^{j-k}f(z)\right)^{(j)} = 0\,.$$

Using the product rule, this becomes a non-trivial linear ODE in $f(z)$ with coefficients in $\mathbb{C}[z, z^{-1}]$, and multiplying through by a sufficiently high power of z proves that $f(z)$ is D-finite. □

Multivariate D-finite functions

Generalizing the univariate definition, a multivariate formal power series $f(\boldsymbol{z}) \in \mathbb{C}[[\boldsymbol{z}]]$ is called a ***D-finite sequence*** when f and all its iterated partial derivatives generate a finite dimensional vector space over $\mathbb{C}(\boldsymbol{z})$. The correct analogue of P-recursiveness in the multivariate case is not as obvious: the following definition, recursive in the dimension d, comes from [Lip89, Definition 3.2].

Definition 2.25 (P-recursiveness in arbitrary dimension)**.** Suppose $d > 1$ and P-recursiveness has been defined for arrays of dimension $d-1$. Then the multivariate sequence $\{a_{\boldsymbol{r}} : \boldsymbol{r} \in \mathbb{N}^d\}$ is said to be a ***P-recursive sequence*** if there is some positive integer k such that the following two conditions hold.

(i) For each $j \in [d]$ there are polynomials $\left\{P^j_{\boldsymbol{\tau}}(n) : \boldsymbol{\tau} \in \{0, \dots, k\}^d\right\}$, not all zero, such that

$$\sum_{\boldsymbol{\tau} \in \{0,\dots,k\}^d} P^j_{\boldsymbol{\tau}}(r_j) a_{\boldsymbol{r}-\boldsymbol{\tau}} = 0 \quad \text{for all } r_1, \dots, r_d \geq k.$$

(ii) All the $(d-1)$-variate arrays obtained from $\{a_{\boldsymbol{r}}\}$ by holding one of the d indices fixed at a value less than k are P-recursive.

The following result, proved in [Lip89, Theorem 31], shows that this definition of P-recursiveness is the correct one.

Theorem 2.26. *The array $\{a_{\boldsymbol{r}} : \boldsymbol{r} \in \mathbb{N}^d\}$ is P-recursive if and only if its generating function $f(\boldsymbol{z}) = \sum_{\boldsymbol{r}} a_{\boldsymbol{r}} \boldsymbol{z}^{\boldsymbol{r}}$ is D-finite.* □

2.4.1 Diagonals

D-finite generating functions are finitely specifiable, the arrays they enumerate satisfy nice recursions, and they appear often in combinatorics. In addition, D-finite functions form an interesting family to study because they satisfy nice closure properties. First, we note that D-finite functions are closed under addition and multiplication, just like rational and algebraic functions.

Theorem 2.27. *If $f(\boldsymbol{z})$ and $g(\boldsymbol{z})$ are D-finite then so are $f(\boldsymbol{z}) + g(\boldsymbol{z})$ and $f(\boldsymbol{z})g(\boldsymbol{z})$.*

Proof Let V be the $\mathbb{C}(\boldsymbol{z})$-subspace of $\mathbb{C}((\boldsymbol{z}))$ spanned by $f(\boldsymbol{z})+g(\boldsymbol{z})$ and all its derivatives. Then V is contained in the sum of subspaces $V_f + V_g$ spanned by the derivatives of f and g. Hence, V is a finite-dimensional space with dimension at most $\dim(V_f) + \dim(V_g)$ and $f(\boldsymbol{z}) + g(\boldsymbol{z})$ is D-finite.

Similarly, the products $f^{(\boldsymbol{r})}(\boldsymbol{z})g^{(\boldsymbol{s})}(\boldsymbol{z})$ of the partial derivatives of f and g span a finite dimensional space W of dimension at most $\dim(V_f) \cdot \dim(V_g)$. By the product rule, every derivative of $f(\boldsymbol{z})g(\boldsymbol{z})$ lies in W, so $f(\boldsymbol{z})g(\boldsymbol{z})$ is D-finite. □

A more interesting closure property has to do with *diagonals* of multivariate series.

Definition 2.28 (diagonal of a formal power series). Let $F(\boldsymbol{z}) = \sum_{\boldsymbol{r} \in \mathbb{N}^d} a_{\boldsymbol{r}} \boldsymbol{z}^{\boldsymbol{r}}$ be a formal power series and $\boldsymbol{s} \in \mathbb{R}^d$ be a fixed vector. The ***s*-diagonal** of F is the univariate series

$$(\Delta_{\boldsymbol{s}} F)(z_1) := \sum_{n \geq 0} a_{n\boldsymbol{s}} z_1^n$$

consisting of the series coefficients of F whose indices are multiples of $\boldsymbol{s}$, where $a_{n\mathbf{s}} = 0$ if $n\mathbf{s} \notin \mathbb{N}^d$ (so the $\mathbf{s}$-diagonal is only non-trivial when $\mathbf{s}$ has nonnegative rational coordinates). When $\mathbf{s} = \mathbf{1}$ the $\mathbf{s}$-diagonal of F is called the ***main diagonal*** and denoted diag F or ΔF. For $1 \leq i < j \leq d$ the (z_i, z_j)-***elementary diagonal*** of F is the formal power series

$$(\Delta_{z_i=z_j} F)(z_1, \ldots, z_{j-1}, z_{j+1}, \ldots, z_d)$$

obtained by taking the terms of F where the exponent of z_i equals the exponent of z_j and setting $z_j = 1$. Note that the main diagonal of F can be computed by taking $d - 1$ elementary diagonals.

Exercise 2.7. Show that if $f(x)$ is the diagonal of $F(x, y)$ then it is also the diagonal of $F(-x, -y)$.

Example 2.29. Consider the generating function

$$F(x, y, z) = \frac{1}{1 - x - y - z} = \sum_{r,s,t \geq 0} \frac{(r + s + t)!}{r!s!t!} x^r y^s z^t$$

enumerating the trinomial coefficients. Then the (x, y)-elementary diagonal of F is the series

$$(\Delta_{x=y} F)(x, z) = \sum_{r,t \geq 0} \frac{(2r + t)!}{(r!)^2\, t!} x^r z^t ,$$

while the main diagonal is

$$(\Delta F)(x) = \sum_{n \geq 0} \frac{(3n)!}{(n!)^3} x^n = 1 + 6x + 90x^2 + \cdots .$$

◁

Exercise 2.8. Let $F(x, y) = \sum_{r,s \geq 0} a_{rs} x^r y^s$ be a bivariate formal power series. What is the difference between diag F and $F(x, x)$?

It is interesting to note that even though the generating function in Example 2.29 is rational, the main diagonal ΔF is transcendental (i.e., not algebraic). Perhaps the easiest way to see this is to compute asymptotics of the diagonal coefficients using the methods of Chapter 9 and verify that they do not have asymptotic growth compatible with being an algebraic series [Mel21, Corollary 2.1].

Although the classes of rational and algebraic series are not closed under taking diagonals, the class of D-finite series is.

Theorem 2.30 (Lipshitz [Lip88]). *Any elementary diagonal of a D-finite series in d variables is a D-finite series in d − 1 variables. In particular, the main diagonal of a D-finite series is a univariate D-finite function.* □

Because many other operations on power series can be phrased as diagonals, Theorem 2.30 finds a wide variety of applications.

Example 2.31. The ***Hadamard product*** of two formal power series $F(\mathbf{z}) = \sum_{\mathbf{r}\in\mathbb{N}^d} a_{\mathbf{r}}\mathbf{z}^{\mathbf{r}}$ and $G(\mathbf{z}) = \sum_{\mathbf{r}\in\mathbb{N}^d} b_{\mathbf{r}}\mathbf{z}^{\mathbf{r}}$ is the series $(F \odot G)(\mathbf{z}) = \sum_{\mathbf{r}\in\mathbb{N}^d} a_{\mathbf{r}}b_{\mathbf{r}}\mathbf{z}^{\mathbf{r}}$. The repeated diagonal expression

$$(F \odot G)(\mathbf{z}) = \Delta_{z_1=y_1} \cdots \Delta_{z_d=y_d} F(\mathbf{z})G(\mathbf{y})$$

implies that $F \odot G$ is D-finite whenever F and G are D-finite. See [CS98] for an algorithm to compute an annihilating D-finite equation for $F \odot G$ from annihilating D-finite equations for F and G. ◁

Exercise 2.9. The ***z-constant term*** of an analytic series $F(\mathbf{z}, t) \in \mathbb{C}[[\mathbf{z}, t]]$ is the series $F(\mathbf{0}, t) \in \mathbb{C}[[t]]$. Prove that the $\mathbf{z}$-constant term of F is the main diagonal of $F(\mathbf{z}, z_1 \cdots z_d t)$.

Finally, we note that the class of D-finite series is not closed under composition or division.

Exercise 2.10. Let $f(x) = \sin x$ and $g(x) = 1/x$. Prove that f and g are D-finite but $g(f(x))$ is not.

Algebraic series and diagonals

The family of series obtained as main diagonals of rational functions appears in many combinatorial applications, and sits naturally between the classes of algebraic and D-finite series. Indeed, it turns out that the diagonals of bivariate rational series form precisely the set of algebraic series. This means that computation of asymptotics for the main diagonal of a bivariate rational series can be reduced to computing asymptotics of a univariate algebraic series (see the discussion in Section 13.1) or vice versa (see Section 9.3.1).

The proof that the diagonal of a bivariate rational series is algebraic was sketched by Furstenberg [Fur67] and given in more detail by Hautus and Klarner [HK71]. Because we work with series over the complex numbers, we give a straightforward analytic proof that helps compute an annihilating polynomial for the diagonal. Stanley [Sta99, Theorem 6.3.3] gives an algebraic proof holding over arbitrary fields.

Theorem 2.32 (bivariate diagonal extraction). *If $F(x, y)$ is a rational power series in two variables then ΔF is algebraic.*

Proof Write $F(x, y) = P(x, y)/Q(x, y)$ for coprime polynomials P and Q with $Q(0, 0) \neq 0$, and let $h(y) = (\Delta F)(y)$. Since F converges in a neighborhood of the origin, when $|y|$ is sufficiently small the function $F(z, y/z)$ is absolutely convergent for z in some annulus $A(y)$. Treating y as a constant, we view $F(z, y/z)$ as a convergent Laurent series in z inside the annulus $A(y)$ with constant term $h(y)$. The Cauchy Integral Formula then gives the integral expression

$$h(y) = \frac{1}{2\pi i} \int_C \frac{P(z, y/z)}{zQ(z, y/z)} \, dz,$$

where C is any positively oriented circle in the annulus of convergence $A(y)$. By the Residue Theorem,

$$h(y) = \sum_{\alpha} \operatorname*{Res}_{z=\alpha(y)} \left(\frac{P(z, y/z)}{zQ(z, y/z)} \right), \tag{2.11}$$

where the sum is over all poles $\alpha(y)$ inside the circle C. These residues are all rational expressions of the poles, which are roots of $Q(z, y/z)$, so we have represented the diagonal as a sum of algebraic functions, which is also algebraic. □

Remark. Taking $|y|$ sufficiently small in the proof of Theorem 2.32 allows one to pick a circle C such that the poles of the Cauchy integrand inside the circle are precisely the roots $z = \alpha(y)$ of $Q(z, y/z)$ that approach zero as $y \to 0$. An annihilating polynomial for the diagonal can be computed from (2.11) using algebraic tools like the resultant, although in low-degree cases it can usually be computed directly.

Example 2.33 (Delannoy numbers, continued)**.** Recall from above that the generating function of the Delannoy numbers is $F(x, y) = 1/(1 - x - y - xy)$, so

$$x^{-1}F(x, y/x) = \frac{1}{x - x^2 - y - xy}$$

has poles at

$$x_{\pm}(y) = \frac{1 - y \pm \sqrt{1 - 6y + y^2}}{2}$$

with only $x_-(y) \to 0$ as $y \to 0$. Since $x^{-1}F(x, y/x) = -1/[(x - x_-(y))(x - x_+(y))]$, the sequence of Delannoy numbers where both indices are equal has generating function

$$(\Delta F)(y) = \operatorname*{Res}_{x=x_-(y)} x^{-1}F(x, y/x) = \frac{-1}{x_-(y) - x_+(y)} = \frac{1}{\sqrt{1 - 6y + y^2}}.$$

◁

The converse of Theorem 2.32 also holds. First, we give an explicit formula of Furstenberg [Fur67] that holds in a special case.

Proposition 2.34. *Let $f(x)$ be an algebraic univariate power series satisfying $P(x, f(x)) = 0$ for some $P(x,y) \in \mathbb{C}[x,y]$. Suppose further that $f(0) = 0$ and $P_y(0,0) \neq 0$. Then*

$$f(x) = \Delta\left(\frac{y^2 P_y(xy,y)}{P(xy,y)}\right).$$

The proof of Proposition 2.34 follows directly from factoring $P(x,y) = (y - f(x))u(x,y)$ for some $u \in \mathbb{C}[[x]][y]$ with $u(0,0) \neq 0$ and then examining which terms lie on the diagonal of $y^2 P_y(xy,y)/P(xy,y)$.

Example 2.35. The generating function for binary trees counted by external nodes is the shifted Catalan generating function $f(x) = (1 - \sqrt{1-4x})/2$ with minimal polynomial $P(x,y) = y^2 - y + x$. Proposition 2.34 yields the rational function $F(y,z) = y(1-2y)/(1-x-y)$ whose diagonal is $f(x)$. ◁

The hypothesis that $P_y(0,0) \neq 0$ in Proposition 2.34 restricts it to the case where $P(x,y)$ has a single root that passes through the origin. One natural generalization to the multivariate setting is the following, which can be found in [Saf00, Lemma 2].

Proposition 2.36. *Let $f(\mathbf{z})$ be an algebraic power series with $P(\mathbf{z}, f(\mathbf{z})) = 0$ for some $P(\mathbf{z},y) \in \mathbb{C}[\mathbf{z},y]$. Suppose further that f is divisible by z_d and that in some neighborhood of $\mathbf{0}$ there is a positive integer k and factorization $P(\mathbf{z},y) = (y - f(\mathbf{z}))^k u(\mathbf{z},y)$ such that $u(\mathbf{0},0) \neq 0$. Then $f(\mathbf{z})$ is the elementary diagonal $\Delta_{z_d=y}F(\mathbf{z},y)$ of the rational function*

$$F(\mathbf{z},y) = \frac{y^2 P_y(z_1,\dots,z_{d-1},yz_d,y)}{kP(z_1,\dots,z_{d-1},yz_d,y)}.$$

Example 2.37. The ***Narayana numbers*** [FS09, Example III.13], defined by the explicit formula

$$a_{rs} = \frac{1}{r}\binom{r}{s}\binom{r}{s-1},$$

are a refinement of the Catalan numbers which enumerate, for example, rooted ordered trees by edges and leaves. The Narayana generating function

$$f(x,y) = \frac{1 + x(y-1) - \sqrt{1 - 2x(y+1) + x^2(y-1)^2}}{2}$$

is a root of the polynomial

$$P(x,y,w) = w^2 - w\left[1 + x(y-1)\right] + xy = \left[w - f(x,y)\right]\left[w - \overline{f}(x,y)\right],$$

where $\overline{f}$ denotes the algebraic conjugate of f, obtained by changing the sign in front of the square root in f. Proposition 2.36 gives the diagonal expression

$$f(x, y) = \Delta_{x=w}\left(\frac{w(1 - 2w - wx(1 - y))}{1 - w - xy - wx(1 - y)}\right).$$

Note that we cannot use Proposition 2.36 with respect to the variable y. ◁

The following example from [Saf00] shows that the hypotheses of Proposition 2.36 are necessary for its conclusion.

Example 2.38. Let $f(x, y) = y\sqrt{1 - x - y}$. If $f(x, y)$ could be expressed as the (y, w)-elementary diagonal of a trivariate rational series $F(x, y, w)$ then

$$F_{yw}(x, 0, 0) = [yw]F(x, y, w) = [y]f(x, y) = f_y(x, 0) = \sqrt{1 - x},$$

contradicting the fact that $\sqrt{1 - x}$ is irrational while $F_{yw}(x, 0, 0)$ is rational. ◁

To extend Proposition 2.36 to larger classes of algebraic series we can either increase the dimension of the rational function in question, or relax our definition of a diagonal. In the first direction, Denef and Lipshitz proved that every algebraic function has a rational diagonal expression in at most double the number of variables.

Proposition 2.39 (Denef and Lipshitz [DL87, Theorem 6.2]). *If $f(\mathbf{z})$ is an algebraic power series in d variables then there exists a rational series $R(\mathbf{z}, \mathbf{y})$ in $2d$ variables such that $f(\mathbf{z}) = \Delta_{z_1=y_1} \cdots \Delta_{z_d=y_d} R(\mathbf{z}, \mathbf{y})$.* □

In the second direction, Safonov gave an algorithm to express any algebraic function in d variables as a generalized diagonal of a rational function in $d + 1$ variables.

Definition 2.40. A ***unimodular matrix*** is a matrix having integer entries and determinant ± 1. Let $F(\mathbf{z}, y) = \sum_{(\mathbf{r},s)\in\mathbb{N}^{d+1}} a_{\mathbf{r},s}\mathbf{z}^{\mathbf{r}} y^s$ be a formal power series in $d + 1$ variables and let M be a unimodular $d \times d$ matrix with nonnegative entries. The ***M-diagonal*** or ***generalized diagonal*** (with respect to M) of F is the formal power series in d variables given by $\sum_{\mathbf{r}\in\mathbb{N}^d} b_{\mathbf{r}}\mathbf{z}^{\mathbf{r}}$ with $b_{\mathbf{r}} = a_{\mathbf{s},s_d}$, where $\mathbf{s} = M\mathbf{r}$.

Exercise 2.11. Let $M = \left(\begin{smallmatrix}1 & 0\\ 1 & 1\end{smallmatrix}\right)$ and suppose that $F(x, y)$ is the M-diagonal of $H(x, y, z)$. Express ΔF in terms of the coefficients $a_{r,s,t}$ of H.

Theorem 2.41 (Safonov [Saf00]). *Let $f(\mathbf{z})$ be an algebraic function in d variables. Then there is a unimodular matrix $M \in \mathbb{N}^{d\times d}$ and a rational function $F(\mathbf{z}, y)$ in $d + 1$ variables such that f is the M-diagonal of F.*

Proof idea The basic idea is to apply a sequence of *blowups* to a polynomial annihilating $f(z)$ to resolve any singularities it has at the origin, until we arrive at a case where Proposition 2.36 applies. These changes of variable are monomial substitutions of the form $z_i \mapsto z_i z_j$, which yields the unimodular matrix. Full details can be found in [Saf00, Theorem 1]. □

Example 2.42. If $f(x, y) = y\sqrt{1 - x - y}$ is the function from Example 2.38 then Safonov's algorithm implies that f is the M-diagonal of

$$F(x, y, w) = wx + \frac{2wx\left(w^2 + w\right)}{2 + w + x + xy} = \sum_{r,s,t \geq 0} a_{r,s,t} x^s y^t w^r,$$

where $M = \left(\begin{smallmatrix} 1 & 0 \\ 1 & 1 \end{smallmatrix}\right)$. In particular, $f(x, y) = \sum_{s,t \geq 0} a_{s+t,s+t,t} x^s y^t$. ◁

2.5 Labeled classes

We end this chapter with additional constructions that produce a large variety of combinatorial classes. Each combinatorial class that we have seen so far is ultimately constructed from *atomic pieces* via constructions such as the disjoint union and product, and we can enrich the objects in a class by adding labels to their atoms. More formally, let the ***atomic class*** $\mathcal{Z}$ be the class containing a single object of size 1 and no objects of other sizes, and let the ***neutral class*** $\mathcal{E}$ be the class containing a single object of size 0 and no objects of other sizes. A ***labeled combinatorial class*** is a univariate combinatorial class $\mathcal{A}$ constructed from copies of the atomic and neutral classes, where the size of an object is the number of atoms it contains and for any $n \in \mathbb{N}$ and $\alpha \in \mathcal{A}_n$ the atoms in α are given unique labels from the set $[n]$. The ***exponential generating function*** $A(z)$ of the labeled class $\mathcal{A}$ is the formal power series $A(z) = \sum_{n \geq 0} \frac{|\mathcal{A}_n|}{n!} z^n \in \mathbb{C}[[z]]$.

Example 2.43 (Basic labeled classes)**.** The ***labeled atomic class*** $\mathcal{Z} = \{①\}$ is obtained from the atomic class by giving its single element of size one the label 1. The ***labeled neutral class*** is the same as the neutral class, as an object of size zero has no atoms. ◁

Example 2.44. By viewing a permutation of $[n]$ in one-line notation, the class of permutations can be constructed as the labeled class consisting of all finite sequences of labeled atoms. ◁

We can construct labeled classes iteratively using labeled versions of the constructions discussed above. If $\mathcal{A}$ and $\mathcal{B}$ are two labeled combinatorial classes then, just like in the unlabeled case, the ***labeled disjoint union*** class $C = \mathcal{A} \sqcup \mathcal{B}$

is simply the class whose set of objects is the disjoint union of the elements in $\mathcal{A}$ and $\mathcal{B}$. The exponential generating function of the labeled disjoint union is

$$C(z) = \sum_{n\geq 0} \frac{|\mathcal{A}_n| + |\mathcal{B}_n|}{n!} z^n = A(z) + B(z),$$

where A and B are the exponential generating functions of $\mathcal{A}$ and $\mathcal{B}$.

Creating a product for labeled classes is harder: when building tuples of objects, relabeling must occur in order to achieve uniqueness of labels. The ***labeled product*** $\mathcal{A} \star \mathcal{B}$ of labeled classes $\mathcal{A}$ and $\mathcal{B}$ is formed by first taking the product of the underlying unlabeled classes, and then labeling in all possible ways that are consistent with the order of labels on the component pieces. More formally, if $\alpha \in \mathcal{A}$ and $\beta \in \mathcal{B}$ then we say that the pair (α,β) is ***pair labeled*** if the sizes of α and β sum to n and the atoms appearing in both elements of the pair are together given distinct labels from $[n]$. If ρ is the ***reduction map*** that takes an object of size k whose atoms have distinct labels from any subset of $\mathbb{N}$ and reduces the labels to lie in $[k]$, then the labeled product of $\mathcal{A}$ and $\mathcal{B}$ is

$$\mathcal{A}\star\mathcal{B} = \{(\alpha',\beta') : \alpha \in \mathcal{A},\ \beta \in \mathcal{B},\ \rho(\alpha') = \alpha,\ \rho(\beta') = \beta; (\alpha',\beta') \text{ is pair labeled}\}.$$

As in the unlabeled case, the size of a labeled pair is the sum of the sizes of its elements.

Example 2.45 (Labeled product)**.** The labeled product of the atomic class with itself is the labeled class

$$\mathcal{Z}^2 = \mathcal{Z} \star \mathcal{Z} = \{(\text{①},\text{②}),\ (\text{②},\text{①})\} = \{\text{①} - \text{②},\ \text{②} - \text{①}\}$$

containing two elements of size two, where we draw elements of a tuple separated by dashes to simplify notation. If $\mathcal{A} = \{\sigma\}$ is the labeled class containing a single object, which is the set $\sigma = \{\text{①},\text{②}\}$, then the labeled class $\mathcal{Z}^2 \times \mathcal{A}$ contains the six elements

$$\begin{gathered}(\text{①} - \text{②}, \{\text{③},\text{④}\}), (\text{①} - \text{③}, \{\text{②},\text{④}\}), (\text{①} - \text{④}, \{\text{②},\text{③}\}),\\ (\text{②} - \text{③}, \{\text{①},\text{④}\}), (\text{②} - \text{④}, \{\text{①},\text{③}\}), (\text{③} - \text{④}, \{\text{①},\text{②}\}),\end{gathered}$$

coming from pair labelings (α',β') whose components reduce to $\rho(\alpha') = \text{①} - \text{②}$ and $\rho(\beta') = \sigma$, together with another six elements whose first component reduces to ② – ①. ◁

If $\alpha \in \mathcal{A}$ has size k and $\beta \in \mathcal{B}$ has size $n - k$ then there are $\binom{n}{k}$ elements $(\alpha',\beta') \in \mathcal{A} \star \mathcal{B}$ with $\rho(\alpha') = \alpha$ and $\rho(\beta') = \beta$, since choosing the k labels from $[n]$ to give to the elements of α' uniquely determines (α',β'). Thus, the exponential generating function of $C = \mathcal{A} \star \mathcal{B}$ is

$$\begin{aligned} C(z) &= \sum_{n\geq 0}\sum_{k=0}^{n}\binom{n}{k}|\mathcal{A}_k|\,|\mathcal{B}_{n-k}|\,\frac{z^n}{n!} \\ &= \sum_{n\geq 0}\sum_{k=0}^{n}\frac{|\mathcal{A}_k|}{k!}z^k\cdot\frac{|\mathcal{B}_{n-k}|}{(n-k)!}z^{n-k} \\ &= A(z)B(z)\,, \end{aligned}$$

where A and B are the exponential generating functions of $\mathcal{A}$ and $\mathcal{B}$. This nice behavior is one reason why exponential generating functions are used for labeled classes.

A labeled sequence construction is defined analogously to the unlabeled case. Because sequences can be represented in terms of disjoint union and product, and the exponential generating functions of labeled classes behave the same under these operations as ordinary generating functions do in the unlabeled case, the exponential generating function of the sequence class $\mathrm{SEQ}(\mathcal{A})$ for any labeled class $\mathcal{A}$ with no objects of size zero is still $1/(1-A(z))$.

Example 2.46 (permutations as labeled sequences). As discussed in Example 2.44, the class of permutations has the labeled construction $\mathcal{P} = \mathrm{SEQ}(\mathcal{Z})$, so that $P(z) = 1/(1-z)$. As a sanity check, we note that there are $n![z^n]P(z) = n!$ permutations of size n. ◁

Finally, we introduce two new constructions, which can be used in the unlabeled case but are much easier when dealing with labeled objects. Let $\mathcal{A}$ be a labeled class with no objects of size zero and exponential generating function $A(z)$. The ***set class*** $\mathrm{SET}(\mathcal{A})$ is the labeled class containing all sets of objects in $\mathcal{A}$, which can be viewed as $\mathrm{SEQ}(\mathcal{A})$ after identifying tuples whose elements differ only by a permutation. Because there are $k!$ ways to permute the elements in a tuple of length k, the exponential generating function of $\mathcal{B} = \mathrm{SET}(\mathcal{A})$ is

$$B(z) = \sum_{k\geq 0}\frac{A(z)^k}{k!} = e^{A(z)}\,,$$

where $e^z = \exp(z)$ denotes the formal power series $\sum_{n\geq 1}\frac{z^n}{n!}$. Similarly, the ***cycle class*** $\mathrm{CYC}(\mathcal{A})$ is the labeled class containing the elements of $\mathrm{SEQ}(\mathcal{A})$ after identifying tuples whose elements differ only by a cyclic shift. By convention, we do not add an element of size zero to the cycle class. Because there are k ways to cyclically shift the elements in a tuple of length k, the exponential generating function of $C = \text{CYC}(\mathcal{A})$ is

$$C(z) = \sum_{k\geq 1} \frac{A(z)^k}{k} = \log\left(\frac{1}{1-A(z)}\right),$$

where log is the formal power series defined by $\log\left(\frac{1}{1-z}\right) = \sum_{n\geq 1} \frac{z^n}{n}$.

Example 2.47 (permutations as labeled cycles)**.** Viewing a permutation as a set of disjoint cycles, the class of permutations has the labeled construction $\mathcal{P} = \text{SET}(\text{CYC}(\mathcal{Z}))$, so that $P(z) = \exp\left[\log\left(\frac{1}{1-z}\right)\right]$. Comparing this to Example 2.46 gives a combinatorial proof that

$$\exp\left[\log\left(\frac{1}{1-z}\right)\right] = \frac{1}{1-z}$$

as formal power series. ◁

Exercise 2.12. Prove the following statements for formal power series using algebraic arguments.

- If $A(0) = B(0) = 0$ then $e^{A(z)+B(z)} = e^{A(z)}e^{B(z)}$.
- If $A(0) = B(0) = 1$ then $\log(A(z)B(z)) = \log A(z) + \log B(z)$.
- If $A(0) = 0$ then $\log(\exp(A(z))) = A(z)$.
- If $A(0) = 0$ then $\exp(\log(1 - A(z))) = 1 - A(z)$.

Often, we wish to track the size of the tuple that defines an element in a sequence, set, or cycle class. If $\mathcal{A}$ is a labeled class with no objects of size zero and exponential generating function $A(z)$ then the ***semi-exponential generating function*** tracking the elements of $\text{SEQ}(\mathcal{A})$ by size and tuple length is

$$\sum_{k\geq 0} y^k A(z)^k = \frac{1}{1-yA(z)}.$$

The term *semi-exponential* refers to the fact that the coefficient of $y^k z^n$ is divided by $n!$ but not by $k!$. Similarly, the semi-exponential generating function for $\text{SET}(\mathcal{A})$ is

$$\sum_{k\geq 0} y^k \frac{A(z)^k}{k!} = \exp(yA(z))$$

and the semi-exponential generating function for $\text{CYC}(\mathcal{A})$ is

$$\sum_{k\geq 0} y^k \frac{A(z)^k}{k} = \log\frac{1}{1-yA(z)}.$$

We end by listing several examples of (semi-)exponential generating functions that we return to in Chapter 3 when deriving asymptotics.

Example 2.48 (permutations by number of cycles). Continuing Example 2.47, the semi-exponential generating function for permutations enumerated by size and number of cycles is

$$f(z,y) = \exp\left(y \log \frac{1}{1-z}\right).$$

The entries of the bivariate sequence counting permutations of size n with k cycles are called the ***Stirling numbers of the second kind*** . ◁

Example 2.49 (involutions). An involution is a permutation whose square is the identity. Equivalently, an involution is a permutation with cycles of length one or two. The exponential generating function for the class of cycles of length one or two is $z + z^2/2$, so the exponential generating function for the class of involutions is $\exp(z + z^2/2)$. ◁

Example 2.50 (set partitions). Let $\mathcal{B} = \text{SEQ}_{>0}(\mathcal{Z})$ be the labeled class with a single object of all sizes $n \geq 1$ and no object of size zero, with exponential generating function $g(z) = \exp(z) - 1$. The labeled class $\exp(\mathcal{B})$, with exponential generating function $\exp(e^z - 1)$, can be interpreted as the partitions of $[n]$ into nonempty sets. Tracking both the size and number of sets in these *set partitions* gives the semi-exponential generating function $\exp(yg(z)) = \exp(y(e^z - 1))$. The entries of the bivariate sequence counting partitions of $[n]$ into k nonempty sets are called the ***Stirling numbers of the first kind***. ◁

Example 2.51 (set partitions into tuples). If $\mathcal{P}_{>0}$ is the labeled class of nonempty permutations, then $\text{SET}(\mathcal{P}_{>0})$ is the class of set partitions into tuples, containing sets $\{(x_{11}, \ldots, x_{1n_1}), \ldots, (x_{k1}, \ldots, x_{kn_k})\}$, where each element of $[n]$ appears exactly once. The semi-exponential generating function $F(y,z)$ for partitions into tuples by size and number of sets is thus

$$F(y,z) = \exp\left(y \frac{z}{1-z}\right). \tag{2.12}$$

◁

Example 2.52 (2-regular graphs). A 2-regular graph is a simple graph (with no loops or multiple edges) in which every vertex has degree 2. A labeled 2-regular graph is the union of labeled undirected cycles and, because we consider simple graphs, every cycle has length at least three. The exponential generating function for undirected cycles of length at least three is

$$u(z) = \frac{1}{2}\left(\log \frac{1}{1-z} - z - \frac{z^2}{2}\right),$$

derived by taking the expression for all cycles and removing the terms for cycles of length one and two, then dividing by 2 to account for orientation. The exponential generating function for labeled 2-regular graphs is thus

$$e^{u(z)} = \frac{\exp\left(-\frac{1}{2}z - \frac{1}{4}z^2\right)}{\sqrt{1-z}}.$$

◁

Example 2.53 (surjections). A ***surjection of size*** n is any mapping from $[n]$ to a set $[k]$ for any positive integer k. Because a surjection of size n onto $[k]$ can be viewed as the sequence $(f^{-1}(1), \ldots, f^{-1}(k))$ of nonempty sets, and the exponential generating function of nonempty sets is $e^z - 1$, the exponential generating function of surjections is $1/(2 - e^z)$. ◁

A wide variety of additional combinatorial constructions, for labeled and unlabeled classes, can be found in [FS09].

Notes

The transfer matrix method is classical, and discussed in [Sta97, Section 4.7] and [GJ04, Chapter 2], among many other places. Our discussion of the kernel method borrows liberally from [BP00]. The method itself, which appears to have been re-discovered several times, has been taken much further; a historical account and survey of recent results can be found in [Mel21, Section 4.2.1].

Several of our proofs in Section 2.4 are taken from [Sta99]. An earlier definition of P-recursiveness appeared in the literature but it was discarded, due to its failure to be equivalent to D-finiteness; counterexamples are given in [Lip89]. Lipshitz's Theorem replaced two earlier proofs [Ges81; Zei82] with gaps and solved a problem of Stanley [Sta80, Question 4e]. Algorithms for finding a differential equation satisfied by an algebraic function go back at least to Abel, and work on efficient algorithms for this purpose is still ongoing [Bos+07]. The approach of Exercise 2.14 is due to Comtet [Com64].

The method of Denef and Lipshitz [DL87] for expressing a d-variate algebraic function as the diagonal of a $2d$-variate rational function is not, to our knowledge, computationally effective, as it relies on the existence of a generator for a certain type of ring extension. Further discussion can be found in [AB13], which gives a different, computationally effective, embedding procedure. Denef and Lipshitz also give an embedding as an M-diagonal in dimension $d + 1$, and they allow a wider range of substitutions than do the other methods.

By Theorem 2.30, the main diagonal of every rational function is a D-finite univariate generating function. The converse is not true, but Christol conjectured that every D-finite univariate function that satisfies a mild condition called *global boundedness* is the main diagonal of some rational function (see [Chr15] for a recent survey of progress). Rational diagonals occur very often in applications: see, for instance, [Mel21, Ch. 3.4] for a list of interesting examples.

There are several methods for computing asymptotics of P-recursive sequences, which can be used to compute asymptotics for rational diagonal sequences, including the *method of Frobenius* for D-finite series and methods arising from the work of Birkhoff and Trjitzinsky [WZ85]. The difficulty in such work is typically expressing the sequence of interest as a linear combination of certain known series expansions, which is a type of *connection problem.* In Chapter 8 we discuss how to combine methods for D-finite series and multivariate asymptotics to resolve the connection problem, and how to effectively compute with algebraic and D-finite series.

Moving on from D-finite series, one might consider the even larger class of ***differentially algebraic*** series, defined to be those that satisfy an equation $P(z, f, f', \dots, f^{(m)}) = 0$ for some positive integer m and some polynomial P. The question of possible behaviors of the coefficient sequence of such a function is wide open; some of the few known results are summarized in [Rub83; Rub92]. Some negative results using model theory can be found in [MM08].

In the other direction, the theory of *hypergeometric sequences* is well developed. These are the coefficients of univariate formal power series satisfying a *first order* linear recurrence with polynomial coefficients (equivalently, they are sequences where the ratio of successive coefficients a_{n+1}/a_n is some fixed rational function of n), and hence correspond to a special subclass of D-finite generating functions. A substantial algorithmic theory exists, well described at an elementary level in the book [PWZ96].

Additional exercises

Exercise 2.13. A *domino* or *dimer* is a union of two unit squares along a common edge. Let a_{nk} be the number of ways of placing k non-overlapping dominoes on a $2 \times n$ grid. Find the generating function enumerating $\{a_{nk}\}$.

Exercise 2.14. From the defining algebraic equation for the Catalan number generating function $C(z) = \sum_{n \geq 0} c_n z^n$, derive a first order linear differential equation with polynomial coefficients for $C(z)$, and then a first order linear recurrence for c_n. Use this to deduce an explicit formula for c_n in terms of factorials. *Hint:* After expressing C' as a rational function A/B, use the fact that B and the minimal polynomial P of C are relatively prime, and apply the Euclidean algorithm to eventually express $B^{-1}(z)$ modulo $P(z)$.

Exercise 2.15. Given a generating function $F(x, y)$ in dimension d, consider the monomial change of variables $F(x^a y^b, x^c y^d)$, where $a, b, c, d \in \mathbb{N}$. How does the main diagonal of F relate to the $\boldsymbol{r}$-diagonal for other values of $\boldsymbol{r}$?

Exercise 2.16. Using the method shown in this chapter, obtain the generating function $T(z)$ for the $(2, 1)$-diagonal $\{a_{2n,n}\}$ of the Delannoy numbers and compare the algebraic complexity of T to the diagonal generating function in Example 2.33.

Exercise 2.17. Let $\boldsymbol{\alpha} \in \mathbb{N}^d$ represent a direction (with all coefficients nonzero and relatively prime). Show how to express the $\boldsymbol{\alpha}$-diagonal of the GF $F(\boldsymbol{z}) = \sum_{\boldsymbol{r} \in \mathbb{N}^d} a_{\boldsymbol{r}} \boldsymbol{z}^{\boldsymbol{r}}$ as the main diagonal of a function related to F in a simple way. *Hint:* Consider roots of unity.

Exercise 2.18. (a D-finite generating function) Let $p_0 = 1$ and define $\{p_n : n \geq 1\}$ recursively by

$$p_n = \frac{1}{3n+1}\left(2p_{n-1} + \sum_{j=2}^{n} p_{j-2}p_{n-j}\right).$$

This sequence from [LP04] gives the probability that a genome in a certain model due to Kaufmann and Levin cannot be improved by changing one allele. Find a differential equation satisfied by the generating function $f(z) = \sum_{N=0}^{\infty} p_n z^n$, then use a computer algebra system to solve the resulting Riccati equation explicitly in terms of Bessel functions. Among the solutions, find the only one that is analytic in a neighborhood of the origin.

Exercise 2.19. A *left-to-right maximum* (respectively right-to-left maximum) of a permutation π of size n is a position i for which $\pi_i > \pi_j$ for all $j < i$ (respectively all $j > i$). Derive an explicit formula for the bivariate semi-exponential generating function that enumerates permutations by length and number of left-to-right maxima. Then derive the trivariate generating function that also counts right-to-left maxima.

Exercise 2.20. The Catalan generating function $f(x) = (1 - \sqrt{1-4x})/(2x)$ does not satisfy the hypotheses of Proposition 2.34. We can get around this by multiplying by x, as in Example 2.35. Find another simple change to f that allows Proposition 2.34 to be applied, and compare results.

Exercise 2.21. The Narayana numbers can be further refined by considering different types of leaves in a rooted ordered tree. Following [CDE06], we say that the leaf of a tree is *old* if it is the leftmost child of its parent, and *young* otherwise. The authors of [CDE06] enumerate such trees according to the number of old leaves, number of young leaves and number of edges, finding the algebraic equation

$$G(x, y, z) = 1 + \frac{z(G(x, y, z) - 1 + x)}{1 - z(G(x, y, z) - 1 + y)}.$$

Use the embedding procedure above to express this as a diagonal of a 4-variable rational function R. How exactly do we obtain the Narayana generating function from G? Does the embedding procedure commute with this operation?

3

Univariate asymptotics

In this chapter we review some classical results on the asymptotics of univariate generating functions. Throughout, $f(z) = \sum_{n=0}^{\infty} a_n z^n$ will be a univariate generating function for the sequence $\{a_n\}$, and for any complex function $g(z)$ analytic at the origin we write $[z^n]g(z)$ for the coefficient of z^n in the power series expansion of $g(z)$ at the origin.

3.1 An explicit formula for rational functions

For rational functions in one variable, it is possible to determine an exact formula for a_n when n is sufficiently large. For instance, when $\rho \neq 0$ the equality

$$[z^n]\frac{1}{(1 - z/\rho)^k} = \binom{n+k-1}{k-1}\rho^{-n} \tag{3.1}$$

holds for $k = 1$ as the left-hand side is a geometric series, and repeated differentiation proves inductively that it holds for all $k \in \mathbb{N}$. More generally, suppose $f(z) = p(z)/q(z)$ is any *rational function* that is analytic at $z = 0$. We assume, without loss of generality, that p and q are relatively prime polynomials with q having the distinct roots $\rho_1, \ldots, \rho_t \in \mathbb{C}$, and that $q(0) = 1$. For each $j \in \{1, \ldots, t\}$, let m_j denote the multiplicity of the root ρ_j and let $q_j(z) = q(z)/(1 - z/\rho_j)^{m_j}$.

Because the q_j have no common root, there exist polynomials $p_1, \ldots, p_t \in \mathbb{C}[z]$ such that the numerator p of f can be written as a linear combination $p(z) = \sum_{j=1}^{t} p_j(z)q_j(z)$. This yields a ***partial fraction decomposition***

$$f(z) = \frac{p(z)}{q(z)} = \sum_{j=1}^{t} \frac{p_j(z)q_j(z)}{q(z)} = h_0(z) + \sum_{j=1}^{t} \frac{h_j(z)}{(1 - z/\rho_j)^{m_j}},$$

where $h_0, \ldots, h_t \in \mathbb{C}[z]$ and for every $j \in \{1, \ldots, t\}$ the polynomial $h_j(z)$ has degree at most $m_j - 1$ and does not vanish at ρ_j. Further decomposing each term $h_j(z)/(1 - z/\rho_j)^{m_j}$ as a sum $\sum_{i=0}^{m_j} c_{ij}/(1 - z/\rho_j)^i$ for constants $c_{ij} \in \mathbb{C}$ and applying (3.1) then implies

$$a_n = \sum_{j=1}^{t} \sum_{i=0}^{m_j} c_{ij} \binom{n+i-1}{i-1} \rho_j^{-n} \quad \text{when } n > \deg(p_0). \tag{3.2}$$

In this way, a partial fraction decomposition of f results in an explicit expression for the coefficients in its power series expansion.

Proposition 3.1 (univariate rational coefficients). *Suppose $f(z) = p(z)/q(z)$ is the ratio of coprime polynomials p and q, where q has the distinct roots $\rho_1, \dots, \rho_t \in \mathbb{C}$ and $q(0) \neq 0$. Then there exist $N \in \mathbb{N}$ and polynomials*

$$P_1(n, x), \dots, P_t(n, x) \in \mathbb{Q}[n, x]$$

such that

$$[z^n]f(z) = \sum_{k=1}^{t} P_k(n, \rho_k) \rho_k^{-n}$$

for all $n \geq N$. If the zero ρ_k of $q(z)$ has multiplicity m_k then as a function of n the polynomial $P_k(n, x)$ has degree $m_k - 1$ and leading term expansion

$$P_k(n, x) = n^{m_k - 1} (-1)^{m_k} \frac{m_k\, p(x)}{x^{m_k}\, q^{(m_k)}(x)} + O(n^{m_k - 2}). \tag{3.3}$$

If $q(z)$ has degree d then the polynomials $P_1, \dots, P_t$ have degree at most d in x, and they can all be computed explicitly in polynomial time with respect to d.

Proof The stated decomposition and the degree of $P_k(n, x)$ as a polynomial in n follow from (3.2) after noting that the binomial coefficient $\binom{n+i-1}{i-1}$ is a polynomial of degree $i - 1$ in n. The fact that each $P_k(n, x)$ has rational coefficients, and a method to compute them, follows most easily from an analytic argument given in Lemma 3.6 below. Each ρ_j is an algebraic number of degree at most d, so the degree of $P_k(n, x)$ in x can be taken to be at most d. □

Proposition 3.1 has strong consequences for asymptotics. Most importantly, the roots of q each give a *contribution* to the asymptotics of a_n, with the roots of minimal modulus having the exponentially largest asymptotic contributions.

Remark 3.2. When $q(x)$ has a unique root ρ_1 of minimal modulus then

$$a_n = P_1(n, \rho_1) \rho_1^{-n} + O(\rho_\dagger^{-n}),$$

where $\rho_\dagger$ is any root of $q(z)$ with second smallest modulus. When ρ_1 has multiplicity one then $P_1(n, \rho_1)$ is constant, and the expression

$$a_n = -\frac{p(\rho_1)}{\rho_1\, q'(\rho_1)}\rho_1^{-n} + O(\rho_\dagger^{-n})$$

gives an asymptotic expansion of a_n with exponentially small error term. When $q(x)$ has a unique root ρ_1 of minimal modulus with multiplicity larger than one then, because the coefficients of $P_1(n, x)$ get rather unwieldy to compute, it is common to give only the leading term (or the first few terms) in $P_1(n, x)$ and leave an error of polynomially smaller size. Algorithms to separate the roots of a univariate polynomial by modulus are discussed in [GS96] and [MS21].

If there are several roots of q with minimal modulus then only those with maximum multiplicity contribute to dominant asymptotic behavior. The existence of several roots of minimal modulus and maximum multiplicity means one must compute the terms in Proposition 3.1 coming from each, and potentially deal with cancellation in the powers of these roots. Because it can be very difficult to track algebraic relations between terms involving powers of algebraic numbers with the same modulus, there are (perhaps surprisingly) still open problems related to when such cancellation can occur. Thankfully, the very pathological cases where it is difficult to detect dominant asymptotic behavior do not arise for combinatorial examples; see [Mel21, Section 2.2] for further discussion of these issues.

Exercise 3.1. Let a_0 and a_1 be any real numbers and suppose $a_{n+1} = 10a_n - 25a_{n-1}$ for all integers $n \geq 1$. Explicitly determine the generating function $f(z) = \sum_{n=0}^{\infty} a_n z^n$ and use this to determine the asymptotic behavior of a_n as $n \to \infty$. Split the parameter space determined by a_0 and a_1 into regions depending on the different asymptotic behaviors of a_n.

3.2 Meromorphic asymptotics

Partial fraction decomposition gives a simple algebraic method to determine asymptotics for rational generating functions. In this section we introduce *analytic* techniques, allowing for a vast generalization from rational functions to functions that behave locally like rational functions. Our arguments make use of standard results about meromorphic functions and, although we make our presentation as self-contained as possible, the reader not familiar with this aspect of complex analysis can consult [Con78b] for further background.

Analytic methods require that the series $f(z)$ represents an analytic function at the origin. To that end, we now assume that the sequence $\{a_n\}$ behaves exponentially, meaning there exist $C_1, C_2 > 0$ such that $C_1^n < |a_n| < C_2^n$ for all sufficiently large n. Under this assumption, the open domain of convergence of $f(z)$ is a finite open disk around the origin, and the Cauchy Integral Formula implies

$$a_n = \frac{1}{2\pi i} \int_C f(z) \frac{dz}{z^{n+1}} \tag{3.4}$$

whenever C is a simple closed contour enclosing the origin in this disk. The domain of integration in a complex integral can be deformed without changing the value of the integral, as long as the deformation stays where the integrand is analytic. It is therefore crucial to understand where $f(z)$ is *not* analytic.

Definition 3.3 (singularities). If $\mathcal{D}_1$ and $\mathcal{D}_2$ are domains (connected open subsets) of $\mathbb{C}$ and $g_1(z)$ and $g_2(z)$ are analytic functions that agree on $\mathcal{D}_1 \cap \mathcal{D}_2 \neq \emptyset$ then we say g_2 is a ***direct analytic continuation*** of g_1 to $\mathcal{D}_2$. More generally, we say that $g_2(z)$ is an ***analytic continuation*** of $g_1(z)$ if there exists a sequence of direct analytic continuations on consecutively overlapping domains that begins with g_1 and end with g_2. If $f(z)$ can be analytically continued to the interior of a simple closed curve γ but cannot be analytically continued to a neighborhood of a point $\omega \in \gamma$ then we call ω a ***singularity*** of f.

Example 3.4. If $f(z)$ is a rational function with coprime numerator and denominator then f has singularities at the roots of its denominator. Aside from division by zero, the most common types of singularities encountered in combinatorial applications include substitution of zero into an algebraic root or logarithm (see Section 3.4 below). ◃

One implication of the Cauchy Integral Formula is that the radius of convergence $0 < R < \infty$ of f equals the minimum modulus of a singularity of f, and this correspondence allows us to obtain a rough estimate of the growth of a_n. Since

$$|a_n| = \left| \frac{1}{2\pi i} \int_{|z|=R-\varepsilon} z^{-n-1} f(z)\, dz \right| \leq (R-\varepsilon)^{-n} \sup_{|z|=R-\varepsilon} |f(z)|,$$

we see that $\limsup_{n\to\infty} |a_n|^{1/n} \leq (R-\varepsilon)^{-1}$ for all $0 < \varepsilon < R$. Conversely, because there is a singularity of modulus R the power series for f does not converge for $|z| > R$, so for any $0 < \varepsilon < R$ we have $|a_n|^{1/n} \geq (R-\varepsilon)^{-1}$ infinitely often. Thus, the ***exponential growth rate*** $\limsup_{n\to\infty} |a_n|^{1/n}$ of a_n satisfies

$$\limsup_{n\to\infty} |a_n|^{1/n} = \frac{1}{R} = \frac{1}{|\rho|}, \tag{3.5}$$

where ρ is a singularity of $f(z)$ with minimal modulus. The exponential growth rate of a_n can be viewed as the coarsest measure of its asymptotic behavior.

Just as for rational functions, the singularities of $f(z)$ give contributions to the asymptotic behavior of a_n. For the rest of this section, we restrict to the case where f locally behaves like a rational function near the singularities that determine dominant asymptotics of a_n. The asymptotic contributions of more general types of singularities are discussed in Section 3.4 below.

Definition 3.5 (poles and meromorphic functions). We say that f has a ***pole*** (or polar singularity) of order $\kappa \in \mathbb{Z}_{>0}$ at the point $z = \omega$ if $f(z)$ cannot be analytically continued to $z = \omega$ but $(z - \omega)^\kappa f(z)$ can, and κ is the smallest positive integer with this property. A pole of order one is usually called a ***simple pole***. If f is either analytic or has a pole at every point of a set $\mathcal{D} \subset \mathbb{C}$ then we say $f(z)$ is a ***meromorphic function*** on $\mathcal{D}$.

Suppose now that $f(z)$ is analytic on a closed disk $D = \{z \in \mathbb{C} : |z| \leq S\}$ for some $S > 0$, except at a nonempty collection $\{\rho_1, \ldots, \rho_t\}$ of poles of orders $\kappa_1, \ldots, \kappa_t$ which lie in the interior of D (by our running assumption that a_n grows exponentially, the ρ_j must be non-zero). If C_- is any positively oriented circle around the origin with radius less than $R = \min_j |\rho_j|$, and C_+ is the positively oriented circle around the origin with radius S, then the Cauchy residue theorem implies

$$\frac{1}{2\pi i}\int_{C_+} f(z)\frac{dz}{z^{n+1}} - \frac{1}{2\pi i}\int_{C_-} f(z)\frac{dz}{z^{n+1}} = \sum_{j=1}^{t} \operatorname*{Res}_{z=\rho_j}\left[z^{-n-1}f(z)\right]. \tag{3.6}$$

For readers unfamiliar with complex residues, the residue of a function $g(z)$ at a pole $z = \omega$ of order k can be defined by the explicit formula

$$\operatorname*{Res}_{z=\omega} g(z) = \frac{1}{(k-1)!}\lim_{z\to\omega}\left[\frac{d^{k-1}}{dz^{k-1}}\left((z-\omega)^k g(z)\right)\right], \tag{3.7}$$

which for a simple pole reduces to

$$\operatorname*{Res}_{z=\omega} g(z) = \lim_{z\to\omega}\left[(z-\omega)g(z)\right]. \tag{3.8}$$

It is a classic result in complex analysis that if $z = \omega$ is a pole of $f(z)$ of order k then for z in a neighborhood of ω we can write f as a ratio $f(z) = p(z)/q(z)$ of analytic functions with $p(\omega), q^{(k)}(\omega) \neq 0$ and $q(\omega) = \cdots = q^{(k-1)}(\omega) = 0$ (taking a series expansion of q at $z = \omega$ shows the converse is also true). We can use such a representation to compute the residue of f at $z = \omega$.

Lemma 3.6. *Under the assumptions of the previous paragraph,*

$$\operatorname*{Res}_{z=\omega} f(z)z^{-n-1} = \omega^{-n}P(n),$$

where $P(n)$ is a polynomial in n of degree $k-1$ with leading term expansion

$$P(n) = n^{k-1}\,(-1)^{k-1}\frac{k\,p(\omega)}{\omega^k\,q^{(k)}(\omega)} + O(n^{k-2}).$$

Proof The vanishing of the repeated derivatives of q implies $q(z) = \frac{q^{(k)}(\omega)}{k!}(z-\omega)^k + h(z)(z-\omega)^{k+1}$ for some function $h(z)$ analytic at $z=\omega$, so (3.7) yields

$$\operatorname*{Res}_{z=\omega} f(z)z^{-n-1} = \frac{1}{(k-1)!}\lim_{z\to\omega}\left[\frac{d^{k-1}}{dz^{k-1}}\left(\frac{p(z)}{\frac{q^{(k)}(\omega)}{k!} + (z-\omega)h(z)}z^{-n-1}\right)\right].$$

The product rule gives a finite sum expression for the repeated derivative in this limit, each term of which is a multiple of ω^{-n} times a polynomial in n. The leading term of this polynomial in n is contained only in the summand

$$\lim_{z\to\omega}\left[\frac{p(z)}{\frac{q^{(k)}(\omega)}{k!} + (z-\omega)h(z)}\cdot\frac{d^{k-1}}{dz^{k-1}}\left(z^{-n-1}\right)\right] = \frac{k!\,p(\omega)}{q^{(k)}(\omega)}\,\omega^{-n-k}\,(-n)^{k-1} + O(\omega^{-n-k}n^{k-2}),$$

and algebraic simplification gives the stated result. □

Exercise 3.2. Compute $P(n)$ in Lemma 3.6 when $f(z) = p(z)/q(z) = (2z-1)/(2-z-e^{1-z})$.

Combining Lemma 3.6 with (3.7) gives a generalization of Proposition 3.1 from rational functions to functions whose closest singularities to the origin are poles.

Proposition 3.7 (meromorphic coefficients). *Suppose that $f(z)$ is analytic on a closed disk $D = \{z \in \mathbb{C} : |z| \le S\}$ for some $S > 0$, except at a nonempty collection $\{\rho_1,\dots,\rho_t\}$ of non-zero poles of orders $\kappa_1,\dots,\kappa_t$ lying in the interior of D. Then there exist polynomials $P_1(n),\dots,P_t(n)$ in n such that*

$$[z^n]f(z) = -\sum_{j=1}^{t} P_j(n)\,\rho_j^{-n} + O(S^{-n}),$$

where $P_j(n)$ has degree $\kappa_j - 1$. If $f(z) = p(z)/q(z)$ represents f as a ratio of analytic functions at $z = \rho_j$ with $p(\rho_j) \neq 0$ then the polynomial $P_j(n)$ is the polynomial $P(n)$ in Lemma 3.6 with $\omega = \rho_j$ and $k = \kappa_j$. The terms of $P_j(n)$ can be computed explicitly from the evaluations of the first $\kappa_j - 1$ derivatives of p and the first $2\kappa_j$ derivatives of q at $z = \rho_j$.

Proof Because f is analytic inside and on C_-, the Cauchy integral formula implies $\frac{1}{2\pi}\int_{C_-} f(z)\frac{dz}{z^{n+1}} = a_n$, and (3.6) can be rearranged to give

$$a_n = -\sum_{j=1}^{t} \operatorname*{Res}_{z=\rho_j}\left[z^{-n-1} f(z)\right] + \frac{1}{2\pi i}\int_{C_+} f(z)\frac{dz}{z^{n+1}}\,.$$

Since $f(z)$ is analytic on the circle C_+, which is a compact set, the function $|f(z)|$ is bounded on C_+, and

$$\left|\frac{1}{2\pi i}\int_{C_+} f(z)\frac{dz}{z^{n+1}}\right| \leq \max_{z\in C_+}|f(z)|\cdot S^{-n} = O(S^{-n}).$$

The stated forms for the residues follow from Lemma 3.6. □

As was observed above for rational functions, if $f(z)$ satisfies the conditions of Proposition 3.7 and has a unique singularity closest to the origin then the contribution of this point determines dominant asymptotics of a_n, up to an exponentially smaller error. If $f(z)$ has multiple poles of minimal modulus, and several of them have maximum order, then we must consider cancellation between their asymptotic contributions.

Example 3.8 (surjection asymptotics). As seen in Example 2.53, the number a_n of surjections from a set of size n has exponential generating function

$$f(z) = \frac{1}{2 - e^z}.$$

This function is meromorphic in the entire complex plane, with poles at the solutions $\Xi = \{\log 2 + k2\pi i : k \in \mathbb{Z}\}$ to the equation $2 - e^z = 0$ in the complex plane. Writing $p(z) = 1$ and $q(z) = 2 - e^z$, we see that $\omega \in \Xi$ implies $p(\omega), q'(\omega) \neq 0$, so every element of Ξ is a simple pole. Because $\log 2$ is the unique element of Ξ with minimal modulus, Proposition 3.7 implies

$$\frac{a_n}{n!} \sim \frac{-p(\log 2)}{(\log 2)q'(\log 2)}\left(\frac{1}{\log 2}\right)^n = \frac{1}{2}\left(\frac{1}{\log 2}\right)^{n+1}.$$

In fact, because all singularities of $f(z)$ are poles, Proposition 3.7 allows us to obtain an asymptotic expansion of a_n to arbitrary accuracy. If $S > 0$ is not equal to the modulus of any element in Ξ, and Ξ_S denotes the elements of Ξ with modulus at most S, then

$$\frac{a_n}{n!} = \sum_{\omega\in\Xi_S} \frac{-p(\omega)}{\omega q'(\omega)}\omega^{-n} + O\left(S^{-n}\right) = \frac{1}{2}\sum_{\omega\in\Xi_S} \omega^{-n-1} + O\left(S^{-n}\right).$$

◁

3.3 Darboux's method

The fact that the asymptotic contribution of a pole singularity ω is easy to compute using the theory of residues is partially a reflection of the fact that $f(z)$ is analytic in a *punctured disk* around ω, which is a disk centered at ω with the center removed. Unfortunately, this property no longer holds near singularities where $f(z)$ locally behaves like a complex logarithm or a non-integral power, due to the branch cuts required to define such functions. A singularity where branch cuts are required to discuss local behavior of $f(z)$ is called a *branch point*, and in this section we illustrate a classical method for computing asymptotics in the presence of a branch point coming from a non-integral power.

Our first general asymptotic result (3.5), which bounded the exponential growth of a_n, was achieved by pushing the domain of integration in the Cauchy integral to the boundary of the domain of convergence of $f(z)$. Crucially, even if $f(z)$ admits a branch point on the boundary of its domain of convergence, such a deformation can be performed without needing to cross a branch cut. Integrating a slight modification of $f(z)$ on the boundary of the domain of convergence leads to *Darboux's method* and *Darboux's Theorem*. Before describing Darboux's method we require two preliminary results, the first of which asymptotically bounds integrals of smooth functions.

Lemma 3.9. *Suppose a complex-valued function f is k times continuously differentiable on the circle γ of radius R for some integer $k \geq 0$. Then*

$$\int_\gamma z^{-n-1} f(z)\, dz = O\left(n^{-k} R^{-n}\right)$$

as $n \to \infty$.

Proof Replacing $f(z)$ by $f(z/R)$ we may assume without loss of generality that $R = 1$. Integrating by parts shows

$$\int_\gamma z^{-n-1} f(z)\, dz = \int_\gamma \frac{1}{n} z^{-n} f'(z)\, dz,$$

where the term involving $\frac{z^{-n}}{-n} f(z)$ vanishes because γ has no boundary, and induction on k implies

$$\int_\gamma z^{-n-1} f(z)\, dz = \frac{1}{k!\binom{n}{k}} \int_\gamma z^{k-n-1} f^{(k)}(z)\, dz.$$

Since $f^{(k)}$ is continuous, it is bounded on the unit circle. Thus, the last integral above is bounded independently of n and our result follows from the behavior $k!\binom{n}{k} \sim n^k$ when k is fixed and $n \to \infty$. □

Our second preliminary result concerns expansions of power functions.

Lemma 3.10. *For any $\alpha \in \mathbb{C}$ the series $(1-z)^\alpha$ has the power series expansion*

$$(1-z)^\alpha = \sum_{n\geq 0} (-1)^n \binom{\alpha}{n} z^n,$$

which converges for $|z| < 1$, where

$$\binom{\alpha}{n} = \frac{\prod_{j=1}^{n} (\alpha - j + 1)}{n!}.$$

Furthermore, if $\alpha \notin \mathbb{N}$ then there is a series expansion

$$\binom{\alpha}{n} = \frac{n^{-\alpha-1}}{\Gamma(-\alpha)} \left(1 + \sum_{k=1}^{\infty} \frac{e_k}{n^k} \right)$$

as $n \to \infty$, where each e_k is a polynomial in α of degree $2k$ that can be computed explicitly.

Proof The series expansion of $(1-z)^\alpha$ is Newton's generalized binomial theorem. The series expansion for $\binom{\alpha}{n}$ follows from an asymptotic analysis of the Euler Gamma function, and can be found in [FS09, Theorem VI.1]. □

Darboux's method consists of decomposing a generating function of interest into the sum of an error term that can be bounded by Lemma 3.9 and a finite number of terms that can be asymptotically approximated with Lemma 3.10. The method dates back to nineteenth-century work of Darboux on complex functions with algebraic singularities.

Example 3.11. The techniques of Section 2.5 often give non-rational (or even non-algebraic) generating functions to which Darboux's method can be applied. For instance, if C denotes the class of even length cycles of length at least four, with exponential generating function

$$C(z) = \sum_{n\geq 2} \frac{z^{2n}}{(2n)!} = \frac{1}{2} \log \frac{1}{1-z^2} - \frac{z^2}{2},$$

then the class $\mathcal{P}$ of permutations with disjoint cycles of even length at least four has exponential generating function

$$P(z) = e^{C(z)} = \frac{e^{-z^2/2}}{\sqrt{1-z^2}}.$$

Since P is a function of z^2, we make the substitution $t = z^2$ and analyze $f(t) =$

$P(\sqrt{t}) = e^{-t/2}/\sqrt{1-t}$. The Taylor series expansion $e^{-t/2} = e^{-1/2} + \frac{e^{-1/2}}{2}(1-t) + O((1-t)^2)$ at $t = 1$ proves that we can write

$$f(t) = e^{-1/2}(1-t)^{-1/2} + \frac{e^{-1/2}}{2}\sqrt{1-t} + \psi(t)$$

for some C^1 function $\psi(t)$. Lemma 3.9 then implies

$$[t^n]f(t) = e^{-1/2}[t^n](1-t)^{-1/2} + \frac{e^{-1/2}}{2}[t^n](1-t)^{1/2} + o\left(n^{-1}\right),$$

so Lemma 3.10 shows that the counting sequence p_n of $\mathcal{P}$ satisfies

$$\frac{p_{2n}}{(2n)!} = [t^n]f(t) \sim \frac{e^{-1/2}}{\Gamma(-\alpha)}n^{-1/2} = \frac{e^{-1/2}}{\sqrt{\pi n}}.$$

◁

Exercise 3.3. Find real constants $C \neq 0$ and β such that the generating function $f(t)$ in Example 3.11 satisfies

$$[t^n]\,f(t) = \frac{e^{-1/2}}{\sqrt{\pi n}} + (C + o(1))n^{\beta}.$$

As seen in Example 3.11, it is common that a generating function is an analytic function multiplied by a pure power. Applying Darboux's method in this context gives the following result.

Theorem 3.12 (Darboux's Theorem). *Suppose that $f(z) = (1 - z/R)^{\alpha}\psi(z)$ for some $R > 0$, where $\alpha \notin \mathbb{N}$ and ψ is analytic on the closed disk $|z| \leq R$ and satisfies $\psi(R) \neq 0$. If the expansion of ψ about R is $\psi(z) = \sum_{n=0}^{\infty} b_n(R-z)^n$ then the power series coefficients $\{a_n\}$ of f have an asymptotic expansion*

$$a_n \approx R^{-n}\sum_{k=0}^{\infty} c_k n^{-\alpha-1-k},$$

where the coefficient c_k is an explicit linear combination of $b_0, \ldots, b_k$. In particular,

$$a_n \sim \frac{\psi(R)}{\Gamma(-\alpha)}n^{-\alpha-1}R^{-n}.$$

Proof Again, by rescaling our variable we assume without loss of generality that $R = 1$. Lemma 3.10 implies that we can expand $\binom{\alpha}{n}$ into a series in decreasing powers $n^{-\alpha-1-k}$ with explicit coefficients, making it possible to convert an asymptotic series of the form $a_n \approx \sum_{k=0}^{\infty} c'_k(-1)^n\binom{\alpha+k}{n}$ into a series $a_n \approx \sum_{k=0}^{\infty} c_k n^{-\alpha-1-k}$ with $c_0 = c'_0/\Gamma(-\alpha)$. Thus, to prove our claimed result it is sufficient to show that a_n can be expressed as a series in $b_k(-1)^n\binom{\alpha+k}{n}$.

Let m be a positive integer greater than $\mathrm{Re}\{-\alpha\}$ and let ψ_m be the Taylor series remainder such that

$$\psi(z) - \sum_{k=0}^{m} b_k(1-z)^k = (1-z)^{m+1}\psi_m(z).$$

Multiplying by $(1-z)^\alpha$ yields

$$f(z) - \sum_{k=0}^{m} b_k(1-z)^{\alpha+k} = (1-z)^{\alpha+m+1}\psi_m(z)$$

on the open unit disk, and taking the coefficient of z^n on both sides implies

$$a_n - \sum_{k=0}^{m-1} b_k(-1)^n\binom{\alpha+k}{n} + O(n^{-\alpha-m-1}) = [z^n](1-z)^{\alpha+m+1}\psi_m(z). \tag{3.9}$$

By assumption $\alpha + m + 1 \geq 0$, so the function $(1-z)^{\alpha+m+1}\psi_m$ is $\lfloor \alpha + m + 1 \rfloor$ times continuously differentiable on the unit circle and Lemma 3.9 implies the right-hand side of (3.9) is $O(n^{-\alpha-m})$. Since this argument works for any m sufficiently large, this proves the desired series for a_n exists. □

Example 3.13 (2-regular graphs: an algebraic singularity). Let

$$f(z) = e^{-z/2-z^2/4}/\sqrt{1-z}$$

be the exponential generating function for the number a_n of 2-regular graphs that was derived in Example 2.52 of Chapter 2. Applying Darboux's Theorem with $R = 1$, $\alpha = -1/2$, and $\psi = \exp(-z/2 - z^2/4)$ gives

$$\frac{a_n}{n!} \sim \frac{\psi(1)}{\Gamma(-\alpha)} n^{-1/2} = \frac{e^{-3/4}}{\sqrt{\pi n}}.$$

◁

Exercise 3.4. Use Darboux's Theorem to compute an asymptotic estimate for the coefficients of the generating function $f(z) = \dfrac{1}{1-4z+z^2}$.

3.4 Transfer theorems

Our proof of Darboux's Theorem uses analyticity of $\psi(z) = f(z)/(R-z)^\alpha$ beyond the disk of radius R only to provide a series development of f at $z = R$. By making stronger use of analytic properties, and using a sharper estimate than Lemma 3.9 to bound error terms, it is possible to do better. There are various results along these lines, our favorite being the *transfer theorems* of Flajolet and Odlyzko [FO90]. The idea of this approach is to establish an estimate of the form $a_n = O(n^{-\alpha-1})$ for the coefficients of *any* power series $f(z)$ that is analytic in neighborhood of the unit disk in a *slit plane*, except at $z = 1$ where $f(z) = O((1-z)^\alpha)$.

Remark. To simplify our notation in this section we state our results for functions with singularities at $z = 1$. As noted above, this loses no generality since $[z^n]f(z/R) = R^{-n}[z^n]f(z)$ for any non-zero constant R and analytic function f.

The transfer theorem method is also flexible enough to extend beyond powers to other branch singularities. Let **alg-log** be the class of functions that are a product of a power of $1 - z$, a power of $z^{-1}\log(1/(1-z))$, and a power of $\log\left[z^{-1}\log(1/(1-z))\right]$. We begin with a description of asymptotics for all functions in the class **alg-log**, then discuss asymptotics of functions which locally behave as if they are in **alg-log** near their singularities.

Proposition 3.14. *Let $\alpha, \gamma, \delta \in \mathbb{C} \setminus \mathbb{N}$ and let*

$$f(z) = (1-z)^\alpha \left(\frac{1}{z}\log\frac{1}{1-z}\right)^\gamma \left(\frac{1}{z}\log\left(\frac{1}{z}\log\frac{1}{1-z}\right)\right)^\delta .$$

Then the power series coefficients $\{a_n\}$ of f satisfy

$$a_n \sim \frac{n^{-\alpha-1}}{\Gamma(-\alpha)}(\log n)^\gamma (\log\log n)^\delta .$$

Proof See [FO90, Theorem 3B]. □

Remark. When at least one of α, γ, or δ is a nonnegative integer, different formulae can hold. For example, when $\gamma \notin \mathbb{N}$ but $\delta = 0$ and $\alpha \in \mathbb{N}$ the coincidence of α with a nonnegative integer decreases the exponent of the logarithm by one, giving the estimate

$$a_n \sim C n^{-\alpha-1}(\log n)^{\gamma-1} . \tag{3.10}$$

For any $R > 0$ and $\varepsilon \in (0, \pi/2)$, the **Δ-*domain*** (or ***Camembert-shaped region***) defined by R and ε is

$$\Delta(R,\varepsilon) = \{z \in \mathbb{C} : |z| < R + \varepsilon,\ z \neq R,\ |\arg(z - R)| \geq \pi/2 - \varepsilon\} ,$$

pictured in Figure 3.1.

Theorem 3.15 (Transfer Theorem). *Let $f(z) = \sum_{n=0}^\infty a_n z^n$ be analytic in a Δ-domain $\Delta(1,\varepsilon)$. If $g(z) = \sum_{n=0}^\infty b_n z^n$ is in* **alg-log** *then the following statements hold.*

(i) *If $f(z) = O(g(z))$ as $z \to 1$ then $a_n = O(b_n)$ as $n \to \infty$.*

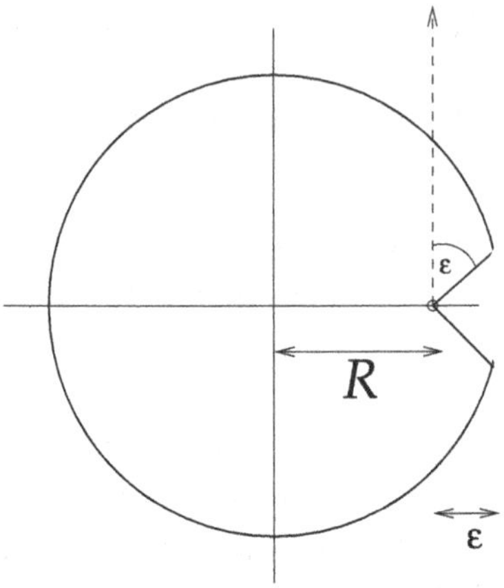

Figure 3.1 A Δ-domain.

(ii) *If* $f(z) = o(g(z))$ *as* $z \to 1$ *then* $a_n = o(b_n)$ *as* $n \to \infty$.
(iii) *If* $f(z) \sim g(z)$ *as* $z \to 1$ *then* $a_n \sim b_n$ *as* $n \to \infty$.

Theorem 3.15 with $g(z) = C(1-z)^\alpha$ strengthens Theorem 3.12. So as not to devote too much space to computation, we only prove Theorem 3.15 for the subset of **alg-log** given by powers $(1-z)^\alpha$.

Proof for $g(z) = (1-z)^\alpha$. We need only prove the first two statements in the theorem, as the third follows as an immediate consequence. Cauchy's integral formula implies a_n can be expressed as a sum of integrals

$$\begin{aligned} a_n = &\frac{1}{2\pi i}\int_{\gamma_1} f(z)z^{-n-1}dz + \frac{1}{2\pi i}\int_{\gamma_2} f(z)z^{-n-1}dz \\ &+ \frac{1}{2\pi i}\int_{\gamma_3} f(z)z^{-n-1}dz + \frac{1}{2\pi i}\int_{\gamma_4} f(z)z^{-n-1}dz \end{aligned}$$

defined by two parameters ξ and η, where

- γ_1 is the circular arc parametrized by $1 + n^{-1}e^{-it}$ for $\xi \le t \le 2\pi - \xi$,
- γ_2 is the line segment between $1 + n^{-1}e^{i\xi}$ and the number β of modulus $1+\eta$ and $\arg(\beta - 1) = \xi$,
- γ_3 is the arc on the circle of radius $1+\eta$ running between β and $\overline{\beta}$ the long way, and
- γ_4 is the conjugate of γ_2, oriented oppositely.

Our argument works with any $0 < \eta < \varepsilon$ and any $0 < \xi < \pi/2$ large enough so that the curves are contained in $\Delta(1,\varepsilon)$; see Figure 3.2 for an illustration.

Suppose first that $f(z) = O((1-z)^\alpha)$ near $z = 1$, so that for some $K > 0$ the inequality $|f(z)| \le K|1-z|^\alpha$ holds everywhere on the curves.

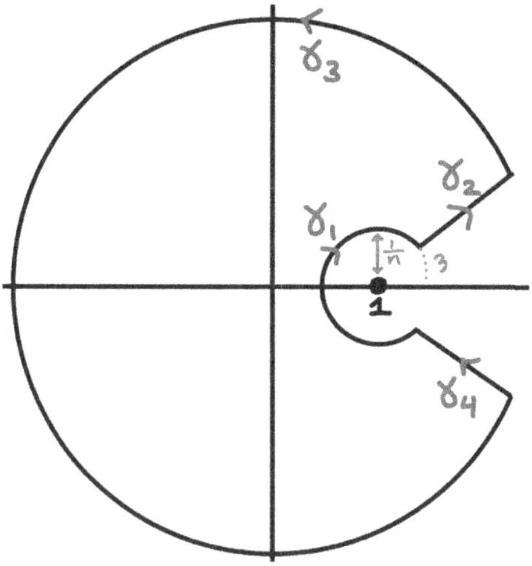

Figure 3.2 The contour γ.

On γ_1 the modulus of f is at most $Kn^{-\alpha}$ and the modulus of z^{-n-1} is at most $(1-n^{-1})^{-n-1} \le 2e \le 6$ so, since the length of the curve is less than $2\pi n^{-1}$, the Cauchy integral over γ_1 has size at most $6Kn^{-\alpha-1}$.

On γ_3 the z^{-n-1} factor reduces the modulus of the integrand to at most $C(\eta)(1+\eta)^{-n}$, where $C(\eta)$ grows at most polynomially with η. Thus, the Cauchy integral over γ_3 is $O(n^{-N})$ for any $N \in \mathbb{N}$.

By symmetry, it remains only to bound the integral over γ_2. Set $\omega = e^{i\xi}$ and parametrize the integral as $z = 1 + (\omega/n)t$ for t from 1 to En, where $E = |\beta - 1|$. We have $|f(z)| \le K|z-1|^\alpha = K(t/n)^\alpha$ and $|z|^{-n-1} = \left|1 + \frac{\omega t}{n}\right|^{-n-1}$, so

$$\left|\int_{\gamma_2} f(z) z^{-n-1} dz\right| \le \int_{\gamma_2} |f(z)||z^{-n-1}||dz \le \int_1^{En} K\left(\frac{t}{n}\right)^\alpha \left|1 + \frac{\omega t}{n}\right|^{-n-1} \frac{dt}{n}$$

$$\le Kn^{-\alpha-1} \int_1^\infty t^\alpha \left|1 + \frac{\omega t}{n}\right|^{-n-1} dt. \tag{3.11}$$

The inequality $|1 + \omega t/n| \ge 1 + \mathrm{Re}\{\omega t/n\} = 1 + (t/n)\cos(\xi)$ implies an upper bound of

$$\int_1^\infty t^\alpha \left(1 + \frac{t\cos(\xi)}{n}\right)^{-n-1} dt$$

for the integral in (3.11), which can be relaxed to

$$J_n = \int_1^\infty t^\alpha \left(1 + \frac{t\cos(\xi)}{n}\right)^{-n} dt$$

because $\cos(\xi) > 0$. The integrand of J_n monotonically decreases as n increases, and is finite for any positive n larger than the real part of α, so the decreasing limit is

$$J = \lim_{n\to\infty} J_n = \int_1^\infty t^\alpha e^{-t\cos(\xi)}\, dt\,,$$

which is finite as $0 < \xi < \pi/2$. We have now bounded all four integrals by multiples of $n^{-\alpha-1}$, so the proof of statement *(i)* in the theorem is complete.

The proof of statement *(ii)* is contained in this argument too. When $|f(z)| \leq Kg(z)$ then the integral over γ_1 is bounded above by $6Kn^{-\alpha-1}$, the integral over γ_3 is $o(n^{-\alpha-1})$, and the integrals over γ_2 and γ_4 are bounded by $JKn^{-\alpha-1}$. Furthermore, the contributions to each of these four integrals from parts of γ at distance greater than any fixed $\delta > 0$ from 1 are $o(n^{-\alpha-1})$. If $f(z) = o(g(z))$ at $z = 1$ then for any $\varepsilon > 0$ there is a δ such that $|f(z)| \leq \varepsilon|g(z)|$ when $|1 - z| \leq \delta$. It follows that $a_n \leq (2J + 6 + o(1))\varepsilon n^{-\alpha-1}$. This is true for every $\varepsilon > 0$, whence $a_n = o(n^{-\alpha-1})$. □

Example 3.16 (Catalan asymptotics). Let $a_n = \frac{1}{n+1}\binom{2n}{n}$ be the nth Catalan number, whose generating function

$$f(z) = \sum_{n=0}^\infty a_n z^n = \frac{1 - \sqrt{1-4z}}{2z} = \frac{1 - 2\sqrt{\frac{1}{4} - z}}{2z}$$

was described in Example 2.14 of Chapter 2. The function $f(z)$ has an algebraic singularity at $z = 1/4$, near which the asymptotic expansion for f begins

$$f(z) = 2 - 4\sqrt{\frac{1}{4} - z} + 8\left(\frac{1}{4} - z\right) - 16\left(\frac{1}{4} - z\right)^{3/2} + O\left(\left(\frac{1}{4} - z\right)^2\right).$$

Note that $f(z)/\sqrt{1/4 - z}$ is not analytic in any disk of radius $1/4 + \varepsilon$, since both integral and half-integral powers appear in f, but f is analytic in a Δ-domain. Since the integral powers of $(1 - z)$ do not contribute to asymptotic behavior as they are polynomials, Theorem 3.15 thus gives an expansion

$$a_n = -4 \cdot \underbrace{4^n \left(-\frac{1}{4n^{\frac{3}{2}}\sqrt{\pi}} - \frac{3}{32n^{\frac{5}{2}}\sqrt{\pi}} + O\left(n^{-\frac{7}{2}}\right) \right)}_{[z^n](1/4-z)^{1/2}} - 16 \cdot \underbrace{4^n \left(\frac{-3}{32n^{\frac{5}{2}}} + O\left(n^{-\frac{7}{2}}\right) \right)}_{[z^n](1/4-z)^{3/2}}$$

$$+ \underbrace{O\left(4^n n^{-\frac{7}{2}}\right)}_{[z^n]O((1/4-z)^2)}$$

$$= 4^n \left(n^{-\frac{3}{2}} \frac{1}{\sqrt{\pi}} - n^{-\frac{5}{2}} \frac{9}{8\sqrt{\pi}} + O\left(n^{-\frac{7}{2}}\right) \right).$$

◁

Exercise 3.5 (common subexpression problem)**.** Flajolet and Odlyzko [FO90] quote the generating function

$$f(z) = \frac{1}{2z} \sum_{p \geq 0} \frac{1}{p+1} \binom{2p}{p} \left[\sqrt{1 - 4z + 4z^{p+1}} - \sqrt{1 - 4z} \right]$$

involved in the representation of trees by directed acyclic graphs.

(a) Show that the minimal modulus singularity occurs at $z = 1/4$, around which

$$f(z) \sim \frac{c}{\sqrt{(1-4z)\log(1-4z)^{-1}}}.$$

(b) Compute the asymptotic behavior of the coefficients of f (you can check your answer against [FO90, (6.7b)]).

Example 3.17 (branching random walk: logarithmic singularity)**.** For an example including a logarithmic term, recall from Example 2.13 the implicit equation

$$\phi(z) = [(1-p)z + p\phi(\phi(z))]^2.$$

This characterizes the probability generating function for the number X of particles to reach the origin in a binary branching nearest-neighbor random walk with absorption at the origin. Aldous (see [AB05, Theorem 29] and [Ald98, Theorem 6]) showed that there is a critical value $p = p_*$ satisfying $16p_*(1 - p_*) = 1$, such that if $p > p_*$ then X is sometimes infinite, while if $p < p_*$ then X is never infinite. At the critical value X is always finite, and it is of interest to know the likelihood of large values of X.

Below, we show that

$$\phi(z) = 1 - \frac{1-z}{4p} - (c + o(1))\frac{1-z}{\log(1/(1-z))}, \tag{3.12}$$

where $c = \log(1/(4p))/(4p)$ and the statement holds for $z \in [0, 1]$ (the interesting situation is when $z \to 1$). If we knew this for all z in a Δ-domain, we could use (3.10) to conclude $a_n \sim cn^{-2}(\log n)^{-2}$, so that X has a first moment but not a "1 + log" moment. Here we establish (3.12) on the unit interval, although it is probably true in a Δ-domain and this is left to the interested reader. Just knowing (3.12), we can deduce information on the partial sums $\sum_{k=0}^{n} a_k$ via a *Tauberian theorem* of Hardy and Littlewood, and perhaps asymptotic information on a_n itself (see [FS09, Sec. VI.11]).

To show (3.12), fix $0 < z_0 < 1$ and consider the iterates $z_n = \phi^{(-n)}(z_0)$ of the inverse of ϕ. The function ϕ is convex on $[0, 1]$ with $\phi(0) > 0$, $\phi(1) = 1$, and one other fixed point k with $p_* < k < 1$. Because $\phi(x) < x$ on $(k, 1)$, if we iterate ϕ on any point in $(c, 1)$ it converges downward to c. Likewise, if we iterate the inverse function ϕ^{-1} starting with any point in $(c, 1)$, it converges upwards to 1, so $z_n \uparrow 1$. The recursion for ϕ gives

$$z_n = ((1-p)z_{n+1} + pz_{n-1})^2,$$

and changing variables to $y_n = 1 - z_n$ implies

$$\begin{aligned} y_n &= 1 - ((1-p)(1-y_{n+1}) + p(1-y_{n-1}))^2 \\ &= 1 - (1 - ((1-p)y_{n+1} - py_{n-1}))^2. \end{aligned}$$

Solving for y_{n+1} gives

$$y_{n+1} = \frac{1 - \sqrt{1-y_n} - py_{n-1}}{1-p}.$$

Setting $x_n = y_n/(4p)^n$ and using $16p(1-p) = 1$ results in

$$x_{n+1} = 2x_n - x_{n-1} + O(y_n)^2.$$

Verifying first that y_n is small, we then approximately solve the linear recurrence for x_n to obtain $x_n \sim An + B$, for some constants A, B, whence $y_n \sim (4p)^n(An + B)$. We may write this as

$$y_{n+1} = 4py_n + (1 + o(1))\frac{y_{n+1}}{n+1} = 4py_n + (1 + o(1))\frac{y_{n+1}}{\log y_{n+1}/\log(4p)}.$$

Let $z = 1 - y_{n+1}$ so $\phi(z) = 1 - y_n$. We then have

$$1 - \phi(z) = \frac{1-z}{4p} - (1 + o(1))\frac{1-z}{4p}\frac{\log(4p)}{\log(1-z)}$$

for all real $z \uparrow 1$, proving (3.12). ◁

3.5 The saddle point method

One of the crowning achievements of complex analysis is the development of techniques to evaluate integrals through clever deformations of their contours of integration. Much of this work can be grouped together under the umbrella of the ***saddle point method***, aimed at discovering the best deformation for an asymptotic analysis. Unlike the techniques discussed above, saddle point methods do not require an integrand to have singularities, and it is common to use a saddle point analysis in situations where transfer theorems cannot be utilized (in fact, the presence of singularities can complicate the saddle point method). In this section we give a short overview of univariate saddle point techniques, with further development of the univariate case covered in Chapter 4 and multivariate generalizations discussed in Chapter 5.

The heart of the saddle point method is the following statement: *when the modulus of an integrand falls steeply on either side of its maximum, most of the contribution to the integral comes from a small interval about the maximum.* If the descent is steep enough, multiplying the integrand by the length of the interval where the modulus is sufficiently near its maximum (or doing something slightly more fancy) gives an accurate estimate. Most contours, however, cannot be used for this purpose: such an estimate cannot hold if the contour can be deformed so as to decrease the maximum modulus of the integrand, since then the integral would be less than the claimed estimate.

Let γ be a contour and let $I = \log f(z) - (n+1)\log z$ be the logarithm of the Cauchy integrand in (3.4). Fixing $z_0 \in \gamma$, we write $\text{Re}\{I'\}$ and $\text{Im}\{I'\}$ for the real and imaginary parts of the derivative at z_0 of I restricted to the curve γ. If z_0 maximizes the modulus of the Cauchy integrand on γ then $\text{Re}\{I'\} = 0$, however it is not usually true that $\text{Im}\{I'\} = 0$. In fact, the Cauchy–Riemann equations imply that $\text{Im}\{I'\}$ equals the real part of the derivative at z_0 of I along any curve *perpendicular* to γ at z_0. Thus, when $\text{Im}\{I'\} \neq 0$ the curve γ may be locally perturbed, fixing the endpoints but pushing the center in the direction of increasing $\text{Re}\{I\}$, thereby decreasing the maximum modulus of the Cauchy integrand on the contour. In other words, if the modulus of the integrand is maximized on γ at z_0, and this maximum cannot be reduced by perturbing γ, then I' must vanish at z_0. The univariate saddle point method thus consists of the following steps:

(i) locate the zeros of I', which form a discrete set of points,
(ii) see whether the contour of integration can be deformed so as to minimize $\text{Re}\{I\}$ at such a point,
(iii) estimate the integral via a Taylor series expansion of the integrand.

In Chapter 4 we will see that for integrals of the form

$$\int A(z) \exp(-\lambda\phi(z))$$

with parameter λ going to infinity, including the Cauchy integral, one can get away with approximating the critical point $z_0(\lambda)$ by the critical point z_0 for ϕ, ignoring A and removing the dependence of z_0 on λ. This approximation is often good enough to provide an asymptotic expansion of the integral, but here we consider cases where we can deal with $z_0(\lambda)$ directly. For the second step above not to fail, either f must be entire or the saddle point where I' vanishes must have smaller modulus than the singularities of f. In practice this is often satisfied, and this classic method is widely applicable. For instance, the seminal paper [Hay56] defines a broad class of functions, called *admissible functions*, for which the saddle point method works and can be automated.

Examples of saddle point integrals

Because we go into great detail on saddle point integrals in Chapter 4, here we simply present two examples illustrating the theory. At their heart, these examples rely on the estimate

$$\int_\gamma A(z) \exp(-\lambda\phi(z))\, dz \sim A(z_0)\sqrt{\frac{2\pi}{\phi''(z_0)\lambda}} \exp(-\lambda\phi(z_0)), \tag{3.13}$$

where A and ϕ are smooth functions with $\text{Re}\{\phi\}$ minimized in the interior of γ at a point z_0 where ϕ'' does not vanish. The approximation (3.13) follows from Theorem 4.1, however we compute it directly in our first example to illustrate why it is true.

Example 3.18 (ordered-set partitions: an isolated essential singularity). Example 2.51 implies that the exponential generating function for the number a_n of ordered-set partitions of $[n]$ is

$$f(z) = \exp\left(\frac{z}{1-z}\right).$$

Our goal here is to prove the estimate

$$a_n \sim n! \sqrt{\frac{1}{4\pi e}} n^{-3/4} \exp\left(2\sqrt{n}\right),$$

starting with the Cauchy integral expression

$$\frac{a_n}{n!} = \frac{1}{2\pi i} \int_{|z|=\varepsilon} \exp\left(\frac{z}{1-z}\right) z^{-n-1}\, dz$$

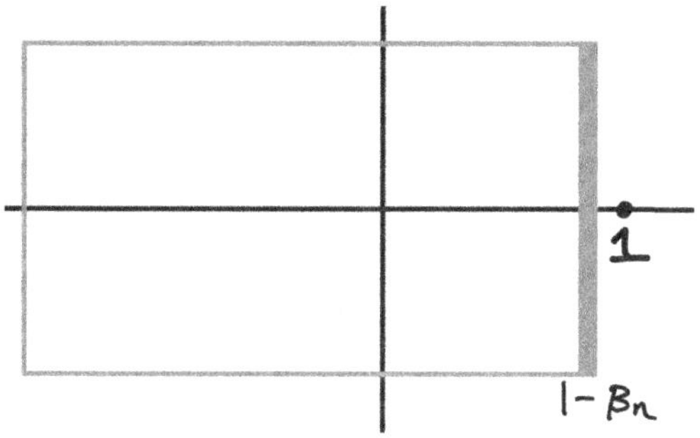

Figure 3.3 The circle $|z| = 1 - \beta_n$ can be deformed to a rectangle with right edge (bold) on the line $x = 1 - \beta_n$ and other edges arbitrarily far from the origin.

that holds for any $0 < \varepsilon < 1$ (in fact, we will select ε to vary with n, with $\varepsilon \to 0$ as $n \to \infty$).

Following the outline of the saddle point method above, we let

$$I(z) = I_n(z) = -(n+1)\log z + \frac{z}{1-z}$$

be the logarithm of the integrand and begin by computing the points where the derivative

$$I'(z) = \frac{-n-1}{z} + \frac{1}{(1-z)^2}$$

vanishes. The closest solution of $I'(z) = 0$ to the origin is $1 - \beta_n$ where

$$\beta_n = n^{-1/2} - \frac{1}{2}n^{-1} + O\left(n^{-3/2}\right), \tag{3.14}$$

and we thus take the Cauchy contour of integration to be the circle of radius $\varepsilon = 1 - \beta_n$ (which is less than one for all n sufficiently large). Because the only singularities of the Cauchy integrand lie at the origin and the point $z = 1$, without changing the value of the Cauchy integral we can deform this circle to a rectangle with right edge on the line $x = 1 - \beta_n$ and all other points arbitrarily far from the origin (see Figure 3.3). When $|z| \geq 2$ the modulus of the Cauchy integrand is upper bounded by $\exp\left(\frac{|z|}{|z|-1}\right)|z|^{-n-1} \leq e^2|z|^{-n-1}$, meaning we can take the left, top, and bottom edges of the rectangle in Figure 3.3 to infinity and use the change of variables $z = 1 - \beta_n + it$ to obtain

$$\frac{a_n}{n!} = \frac{1}{2\pi i}\int_{-\infty}^{\infty} \exp(I(1 - \beta_n + it))\,(i\,dt) \tag{3.15}$$

for n sufficiently large, fulfilling the second step of the saddle point method.

The final step is to prove an approximation of the form

$$\frac{1}{2\pi}\int_{-\infty}^{\infty}\exp(I(1-\beta_n+it))\,dt \sim \frac{1}{2\pi}\int_{-\infty}^{\infty}\exp\left[I(1-\beta_n)+\frac{1}{2}I''(1-\beta_n)(it)^2\right]dt, \tag{3.16}$$

where $I(1-\beta_n+it)$ is replaced by its second-degree Taylor approximation. This is very useful because, as a Gaussian integral, the right-hand side of (3.16) can be computed exactly,

$$\frac{1}{2\pi}\int_{-\infty}^{\infty}\exp\left[I(1-\beta_n)+\frac{1}{2}I''(1-\beta_n)(it)^2\right]dt = \sqrt{\frac{1}{2\pi I''(1-\beta_n)}}\exp(I(1-\beta_n)). \tag{3.17}$$

The estimate (3.16) – comparable to (3.13) with $\lambda = n+1$ and $\phi(t) = -I(1-\beta_n+it)$ – is verified through several integral bounds. The approximation (3.14) for β_n implies

$$I''(1-\beta_n) = \frac{n+1}{(1-\beta_n)^2}+\frac{2}{\beta_n^3} = (2+o(1))n^{3/2}, \tag{3.18}$$

so that the right-hand side of (3.17) is

$$\begin{aligned}
&\sqrt{\frac{1}{2\pi I''(1-\beta_n)}}\exp(I(1-\beta_n))\\
&\sim \sqrt{\frac{1}{4\pi n^{3/2}}}\exp\left(-(n+1)\log(1-\beta_n)-1+\frac{1}{\beta_n}\right)\\
&\sim \sqrt{\frac{1}{4\pi n^{3/2}}}\exp\left(-(n+1)\left(-n^{-1/2}+O\left(n^{-3/2}\right)\right)-1+n^{1/2}+\frac{1}{2}+O\left(n^{-1/2}\right)\right)\\
&\sim \sqrt{\frac{1}{4\pi e}}n^{-3/4}\exp\left(2\sqrt{n}\right).
\end{aligned}$$

Our claimed asymptotic result for a_n thus holds as long as we can show the left and right sides of (3.16) are equal up to an error that is $o\left(n^{-3/4}\exp\left(2\sqrt{n}\right)\right)$. The approximation (3.18) suggests that the main contributions to the integrals in (3.16) come from the region where $t^2n^{3/2}$ is not too small, meaning $|t|$ is roughly $n^{-3/4}$ or smaller. Accordingly, we pick a cutoff $L = 2n^{-3/4}\log n$ a little greater than that and break our integrals into the two parts $|t| \le L$ and $|t| > L$. Up to the cutoff the two integrals are close, and past the cutoff they are both small. More precisely, define

$$M_1 = \int_{|t|\geq L} \left| \exp\left[I(1-\beta_n) + \frac{1}{2} I''(1-\beta_n)(it)^2 \right] \right| dt$$

$$M_2 = \int_{|t|\geq n^{-1/2}} \left| \exp[I(1-\beta_n + it)] \right| dt$$

$$M_3 = \int_{n^{-1/2}>|t|\geq L} \left| \exp[I(1-\beta_n + it)] \right| dt$$

$$M_4 = \int_{|t|<L} \left| \exp\left[I(1-\beta_n) + \frac{1}{2} I''(1-\beta_n)(it)^2 \right] - \exp\left[I(1-\beta_n + it) \right] \right| dt$$

so that

- M_1 is the integral on the right-hand side of (3.16) beyond L,
- the sum of M_2 and M_3 bounds the integral on the left-hand side of (3.16) beyond L,
- M_4 bounds the difference between the left- and right-hand sides of (3.16) on $[-L, L]$,
- and the modulus of the difference between the left and right sides of (3.16) is bounded by the sum $M_1 + M_2 + M_3 + M_4$.

Letting $M = \exp[I(1-\beta_n)]$, we prove M_1, M_2, and M_3 have upper bounds of the form $M \cdot \exp(-c(\log n)^2)$ for some $c > 0$, and that $M_4 = o(Mn^{-3/4})$. These bounds all lie in $o\left(n^{-3/4} \exp\left(2\sqrt{n}\right)\right)$, completing our derivation of asymptotics for this example.

Bound on M_1 We bound M_1 with a standard Gaussian tail estimate. For any $a, C > 0$

$$\begin{aligned}\int_{|t|\geq C} e^{-at^2} dt = 2\int_{t\geq C} e^{-at^2} dt &= 2e^{-aC^2} \int_{t\geq 0} e^{-at^2-2aCt} dt \\ &\leq 2e^{-aC^2} \int_{t\geq 0} e^{-at^2} dt \\ &= \sqrt{\pi/a} e^{-aC^2},\end{aligned}$$

so the growth rates of $I''(1-\beta_n)$ and L give the asserted upper bound on M_1 for any $c < 8$.

Bound on M_2 To bound M_2, observe first that if $|t| \geq n^{-1/2}$ then the exponent $-t^2/(\beta_n^3 + \beta_n t^2)$ decreases to $-\beta_n^{-1} \sim -\sqrt{n}$. This is small, but integrating it over the unbounded region $[n^{-1/2}, \infty]$ requires us to be careful. In particular, we use the upper bound

$$\frac{|\exp(I(1-\beta_n+it))|}{\exp(I(1-\beta_n))} \le \frac{|1-\beta_n|^n}{|1-\beta_n+it|^n} \exp\left(\mathrm{Re}\left\{\frac{1-\beta_n+it}{\beta_n-it} - \frac{1-\beta_n}{\beta_n}\right\}\right)$$
$$\le (1+t^2)^{-n/2} \exp\left(-(1+o(1))n^{1/2}\right),$$

where we can bound the power of n in the first line by $(1+t^2)^{-n/2}$ since $1-\beta_n < 1$ and $|x/(x+it)|$ is increasing in $x \ge 0$. Integrating the factor $(1+t^2)^{-n/2}$ as t ranges from $n^{-1/2}$ to infinity gives a term of size $o(1)$, so

$$\frac{M_2}{M} \le \exp\left(-(1+o(1))n^{1/2}\right) = o\left(\exp\left[-c(\log n)^2\right]\right)$$

for any $c > 0$.

Bound on M_3 To bound M_3 we pull out the factor of M, obtaining

$$M_3 \le M \int_{L<|t|<n^{-1/2}} \exp\left[\mathrm{Re}\left\{I(1-\beta_n+it) - I(1-\beta_n)\right\}\right] dt.$$

The real part of $-(n+1)\log(1-\beta_n+it)$ is maximized at $t = 0$, whence

$$\mathrm{Re}\left\{I(1-\beta_n+it) - I(1-\beta_n)\right\} \le \mathrm{Re}\left\{\frac{1-\beta_n+it}{\beta_n-it} - \frac{1-\beta_n}{\beta_n}\right\},$$

and

$$M_3 \le M \int_{L<|t|<n^{-1/2}} \exp\left(\mathrm{Re}\left\{\frac{1-\beta_n+it}{\beta_n-it} - \frac{1-\beta_n}{\beta_n}\right\}\right) dt$$
$$= M \int_{L<|t|<n^{-1/2}} \exp\left(\frac{-t^2}{\beta_n^3+\beta_n t^2}\right)$$
$$\le M \int_{L<|t|<n^{-1/2}} \exp\left(-\frac{t^2}{2\beta_n^3}\right) dt$$

because the β_n^3 term is the greatest term in the denominator when $t < n^{-1/2}$. The behavior $\beta_n \sim n^{-1/2}$ and $L = 2n^{-3/4}\log n$ proves the desired upper bound on M_3 for any constant $c < 2$.

Bound on M_4 Finally, for M_4 we use the Taylor approximation

$$\left|I(1-\beta_n+it) - I(1-\beta_n) + \frac{1}{2}t^2 I''(1-\beta_n)\right| \le \frac{t^3}{6} \sup_{|s|\le L} |I'''(1-\beta_n+s)|.$$

Differentiating $I'(z) = -(n+1)/z + 1/(1-z)^2$ twice we find that $I'''(z) \sim 6/(1-z)^4$ near $z = 1$, and hence that the right-hand side is bounded by $(k+o(1))t^3n^2 =$

$(k + o(1))n^{-1/4} \log^3 n$ for some $k > 0$. Because the integrand on the right-hand side of (3.16) is everywhere positive, this implies the existence of $c > 0$ such that $M_4 \leq cn^{-1/4} \log^3 n$ times the value of M, as desired. ◁

Remark. The approach of Example 3.18 yields a full asymptotic development of a_n with minor modifications.

Our second example simply assumes the approximation (3.13), greatly reducing the amount of work.

Example 3.19 (involutions: an entire function)**.** Let $f(z) = \exp(z+z^2/2)$ be the exponential generating function for the number a_n of involutions in the permutations group S_n, as discussed in Example 2.49. This is an entire function, and we apply a saddle point analysis. Let

$$I(z) = \log(f(z)z^{-n-1}) = z + \frac{z^2}{2} - (n+1)\log z\,.$$

Setting the derivative of I equal to zero gives the quadratic $z^2 + z - (n+1) = 0$ with roots $-\frac{1}{2} \pm \sqrt{n + \frac{5}{4}}$. The series coefficients a_n of f are positive, whereas $\exp(I(z))$ alternates in sign near the negative root, meaning a_n cannot be approximated by the integrand near the negative root.

We thus let γ be the positively oriented circle around the origin through $z_0 = \sqrt{n + \frac{5}{4}} - \frac{1}{2}$. The real part of I on γ is maximized at z_0, so the estimate (3.13) implies

$$[z^n]f(z) = \frac{1}{2\pi i}\int_\gamma \exp(I(z_n))\,dz \sim \exp(I(z_0))\sqrt{\frac{1}{2\pi I''(z_0)}}\,.$$

From the approximations

$$\begin{aligned} z_0 &= n^{1/2} - \frac{1}{2} + \frac{5}{8}n^{-1/2} + O(n^{-3/2}) \\ \frac{z_0^2}{2} &= \frac{1}{2}n - \frac{1}{2}n^{1/2} + \frac{3}{4} + O(n^{-1/2}) \\ \log(z_0) &= \frac{1}{2}\log n - \frac{1}{2}n^{-1/2} + \frac{1}{2}n^{-1} + O(n^{-3/2}) \end{aligned}$$

it follows that

$$I(z_0) = -\frac{1}{2}n\log n + \frac{1}{2}n + n^{1/2} - \frac{1}{2}\log n - \frac{1}{4} + O(n^{-1/2})$$

and $I''(z_0) = 2 + o(1)$. Thus,

$$\begin{aligned} a_n &\sim n!\ \exp(I(z_0))\sqrt{\frac{1}{2\pi I''(z_0)}} \\ &= n!\ \exp\left(-\frac{1}{2}n\log n + \frac{1}{2}n + n^{1/2} - \frac{1}{2}\log n - \frac{1}{4} + O(n^{-1/2})\right) 2^{-1/2} \\ &\sim n^{n/2}e^{\sqrt{n}-n/2}\frac{1}{\sqrt{2\sqrt{e}}}, \end{aligned}$$

where the final line follows from Stirling's approximation for $n!$. ◁

Notes

One of the earliest and most well-known uses of a modern generating function analysis to obtain asymptotics was Hardy and Ramanujan's derivation of asymptotics for the number of partitions of an integer [HR00a]. Their original argument used a *Tauberian theorem* and the behavior of the generating function $f(s)$ as $s \uparrow 1$ through real values, though later work such as [HR00b] used a *circle method* obtained by integrating over a circle near the boundary of the domain of convergence. Saddle point methods are even more classical, dating back centuries. As mentioned in the chapter, [Hay56] was an influential work in developing the modern general theory.

The exposition in this chapter does not follow any one source, though it owes a debt to Chapter 11 of [Hen91] and to the beautiful paper [FO90]. A nice reference book for univariate asymptotics is the exemplary text [FS09].

Additional exercises

Exercise 3.6. The explicit leading term formulae in Lemmas 3.6 and Proposition 3.7 are only useful when the numerator of the meromorphic generating function is non-zero at the pole in question. Extend these two results to capture vanishing numerators and find the leading asymptotic term for the series coefficients of $f(z) = (1 - z)/(2 - z - e^{1-z})$ as $n \to \infty$.

Exercise 3.7. (set partition asymptotics) Use the exponential generating function $f(z) = \exp(e^z - 1)$ for the number a_n of set partitions of $[n]$ from Example 2.51 to derive the estimate

$$a_n = (\log n + O(1))^n .$$

Exercise 3.8. (Exercise 2.18 continued) Using the fact that the series coefficients a_n of the generating function f in Exercise 2.18 are positive, prove that its smallest positive singularity has the least modulus of any singularity of f. Approximate this singularity and then estimate the logarithmic exponential growth rate $\limsup_{n\to\infty} n^{-1} \log a_n$. Prove that this limsup is equal to the liminf, so the limit exists.

Exercise 3.9. Sometimes, even when f is given explicitly, it can be tricky to compute the minimal modulus of the singularities of f in order to obtain the exponential coefficient behavior using (3.5). The power series coefficients of the function

$$f(z) = \frac{\arctan \sqrt{2e^{-z} - 1}}{\sqrt{2e^{-z} - 1}}$$

were shown by H. Wilf to yield rational approximations to π. An asymptotic analysis was provided by [War10]; do the first step by finding the radius of convergence of the power series for f at zero.

Exercise 3.10. Suppose $P(x)$ is a polynomial of degree k with leading coefficient $a_k \neq 0$. What does the saddle point method tell you about the asymptotics of the Maclaurin coefficients a_n of $e^{P(x)}$? Specifically, can you identify an exponent β such that $\lim_{n\to\infty} n^{-\beta} \log |a_n|$ is finite?

Exercise 3.11. (Open Problem) Is the generating function ϕ from Example 3.17 analytic in a Δ-domain?

PART II

MATHEMATICAL BACKGROUND

4

Fourier–Laplace integrals in one variable

In this chapter we perform a systematic asymptotic study of ***Fourier–Laplace integrals*** having the form

$$\int_\gamma A(z) \exp(-\lambda \phi(z))\, dz\,,$$

as the parameter $\lambda \to \infty$. The functions A and ϕ are called the ***amplitude*** and ***phase*** functions, respectively (note that when $\phi(z) = -i\rho(z)$ is purely imaginary some authors use the term *phase* to denote ρ rather than $i\rho$). The univariate setting covered in this chapter gives a basis for the multivariate case handled in Chapter 5, which in turn underlies the asymptotic results of analytic combinatorics in several variables. Our main result is the following theorem, which is proved in Section 4.2.

Theorem 4.1 (univariate Fourier–Laplace asymptotics). *Let A and ϕ be analytic functions in a neighborhood $\mathcal{N} \subseteq \mathbb{C}$ of the origin. Let*

$$A(z) = \sum_{j=0}^{\infty} b_j z^j$$

$$\phi(z) = \sum_{j=0}^{\infty} c_j z^j$$

be the power series for A and ϕ at the origin, and let $\ell \geq 0$ and $k \geq 2$ be the indices of the least nonvanishing terms in these series, so that $b_\ell, c_k \neq 0$ and $b_j = c_i = 0$ for any $j < \ell$ and $i < k$. Let $\gamma : [-\varepsilon, \varepsilon] \to \mathbb{C}$ be any smooth curve with $\gamma(0) = 0 \neq \gamma'(0)$ and assume that $\text{Re}\{\phi(\gamma(t))\} \geq 0$ with equality only at

$t = 0$. *Denote*

$$\mathcal{I}_+(\lambda) := \int_{\gamma|_{[0,\varepsilon]}} A(z)\exp(-\lambda\phi(z))\,dz$$

$$\mathcal{I}(\lambda) := \int_{\gamma} A(z)\exp(-\lambda\phi(z))\,dz$$

$$C(k,\ell) := \frac{\Gamma((1+\ell)/k)}{k},$$

where Γ *is the Euler gamma function. Then there are asymptotic expansions*

$$\mathcal{I}_+(\lambda) \approx \sum_{j=\ell}^{\infty} a_j C(k,j)(c_k\lambda)^{-(1+j)/k} \tag{4.1}$$

$$\mathcal{I}(\lambda) \approx \sum_{j=\ell}^{\infty} \alpha_j C(k,j)(c_k\lambda)^{-(1+j)/k} \tag{4.2}$$

with the following explicit description.

(i) a_j *is a polynomial expression, explicitly constructed in our proof, in the values* $b_\ell, \ldots, b_j, c_k^{-1}, c_{k+1}, \ldots, c_{k+j-\ell}$ *whose first two values are* $a_\ell = b_\ell$ *and* $a_{\ell+1} = b_{\ell+1} - b_\ell \dfrac{2+\ell}{k}\dfrac{c_{k+1}}{c_k}$,

(ii) the choice of kth root in the expression $(c_k\lambda)^{-(1+j)/k}$ *is made by taking the principal root in* $x^{-1}(c_k\lambda x^k)^{1/k}$ *where* $x = \gamma'(0)$,

(iii) the numbers α_j *are related to the numbers* a_j *by*

$$\alpha_j = \begin{cases} 2a_j & \text{if } k \text{ is even and } j \text{ is even} \\ 0 & \text{if } k \text{ is even and } j \text{ is odd} \\ \left(1-\zeta^{j+1}\right)a_j & \text{if } k \text{ is odd,} \end{cases}$$

where

$$\zeta = -\exp\left(\frac{i\pi}{k}\,\mathrm{sgn}\,\mathrm{Im}\,\{\phi(\gamma'(0))\}\right).$$

Remarks. (i) If $\phi(0) = \nu \neq 0$ but $\mathrm{Re}\{\phi(x)\}$ is still minimized at $x = 0$, then one may apply this result by replacing $\phi(x)$ with $\phi(x) - \nu$ and multiplying the asymptotic behavior by $\exp(\lambda\nu)$.

(ii) The hypothesis that the minimum of $\mathrm{Re}\{\phi\}$ occurs only at 0 will be removed when we reach the multivariate setting. In one variable, due to analyticity, either the minimum occurs only at zero in some neighborhood of the origin or the real part of ϕ is identically zero in that neighborhood. The analysis of a purely imaginary phase function takes place more naturally with C^∞ methods, which are discussed in Section 4.3 below.

Exercise 4.1. Explain why the term "stationary phase integral" is usually reserved for $k \geq 2$.

Exercise 4.2. Although Theorem 4.1 assumes $k \geq 2$, the expansion for $\mathcal{I}_+$ given in part (*i*) by (4.1) holds when $k = 1$. What is ζ in the case $k = 1$ and what expansion for $\mathcal{I}$ holds as a consequence?

To those unfamiliar with saddle point methods, this result may seem difficult to decipher, but both the statement and proof are actually quite intuitive. When A and ϕ are real, their orders of vanishing dictate the order of magnitude of such an integral after direct integration. Changing variables to simplify the exponent produces a full asymptotic development of the integral. When the phase is complex, one can use integration by parts in order to cancel oscillations, or one can reduce to the real case by a contour shift. The latter approach requires stronger hypotheses (analyticity rather than smoothness) but gives stronger results (exponentially small remainders rather than rapidly decreasing remainders). In order to give all of the intuition, we take a route to the derivation that is longer than necessary. We begin with a stripped down special case, in which direct integration suffices, then give the arguments that hold in greater generality.

4.1 Real integrands

For univariate Fourier–Laplace integrals, Theorem 4.1 states the existence of an asymptotic expansion

$$\int_0^{\varepsilon} A(x) \exp(-\lambda\phi(x))\, dx \approx \sum_{j=\ell}^{\infty} a_j C(k, j)(c_k \lambda)^{-(1+j)/k}$$

with explicitly computable constants a_j. The main result of this section, Theorem 4.6, yields this expansion for real integrands, along with further information about the constants.

Complex analytic techniques are not needed when working on the real line, and consequently we need to assume only differentiability and not analyticity of A and ϕ. We build the argument in three steps: first taking A and ϕ to be monomials, then taking ϕ to be a monomial but allowing A to be free, and finally handling the general case. The first step is accomplished via an exact computation, the second via a remainder estimate, and the third with a change of variables.

A and ϕ are monomials

On the positive half-line, we can get away with a change of variables involving a fractional power. This allows us to handle the special case of monomial phase and amplitude by an exact integral, holding for any positive real α and nonnegative real β. The change of variables $y = \lambda x^\alpha$ gives

$$\int_0^\infty x^\beta \exp(-\lambda x^\alpha)\, dx = \int_0^\infty \left(\frac{y}{\lambda}\right)^{\beta/\alpha} e^{-y} \frac{1}{\alpha} \frac{y^{1/\alpha-1}}{\lambda^{1/\alpha}}\, dy$$

$$= \frac{1}{\alpha} \lambda^{-(1+\beta)/\alpha} \int_0^\infty y^{\frac{1+\beta}{\alpha}-1} e^{-y}\, dy\,,$$

so, by the definition of the Gamma function, we have the exact evaluation

$$\int_0^\infty x^\beta \exp(-\lambda x^\alpha)\, dx = C(\alpha,\beta)\, \lambda^{-(1+\beta)/\alpha} \tag{4.3}$$

with $C(\alpha,\beta) = \frac{1}{\alpha}\Gamma\left(\frac{1+\beta}{\alpha}\right)$ as above.

Remark 4.2. All of the contribution to (4.3) comes from a neighborhood of zero: for any $\varepsilon > 0$ the substitution $w = \lambda(x^\alpha - \varepsilon^\alpha)$ proves that the difference

$$\left| \int_0^\varepsilon x^\beta \exp(-\lambda x^\alpha)\, dx - C(\alpha,\beta)\, \lambda^{-(1+\beta)/\alpha} \right| = \int_\varepsilon^\infty x^\beta \exp(-\lambda x^\alpha)\, dx$$

$$= \frac{1}{\alpha} \lambda^{-(1+\beta)/\alpha} e^{-\lambda\varepsilon^\alpha} \int_0^\infty (w + \lambda\varepsilon^\alpha)^{\frac{1+\beta}{\alpha}-1} e^{-w}\, dw$$

$$= O\left(\lambda^{-(1+\beta)/\alpha} e^{-\lambda\varepsilon^\alpha}\right)$$

decays exponentially in λ. The fact that a Fourier–Laplace integral decays exponentially when integrated over a region where its phase does not vanish will be used multiple times in this chapter.

Exercise 4.3. Suppose x has units of time (or length, or any other physical unit). Explain the right-hand side of (4.3), except for the constant, via a soft analysis: what are the free variables on the left, what units must λ have in order to abide by the principle of unitless exponentiation, what units must the left-hand side of (4.3) have (remembering to include the units of dx), and what power must λ therefore be given on the right-hand side?

When $\beta = \ell$ is an integer and $\alpha = 2k$ is an even integer the corresponding two-sided integrals make sense as well, giving

$$\int_{-\infty}^\infty x^\ell \exp\left(-\lambda x^{2k}\right) dx = \begin{cases} 2C(2k,\ell)\, \lambda^{-(1+\ell)/(2k)} & \text{if } \ell \text{ is even} \\ 0 & \text{if } \ell \text{ is odd} \end{cases}. \tag{4.4}$$

ϕ is a monomial and A is anything

Generalizing our argument to the case of general amplitude requires the following estimate.

Lemma 4.3 (Big-O Lemma). *Let $k, \ell > 0$ with k an integer. If A and ϕ are real-valued functions such that $A(x) = O\left(x^\ell\right)$ and $\phi(x) \sim x^k$ at $x = 0$, and $\phi(x)$ vanishes in $[0, \varepsilon]$ only at 0, then*

$$\int_0^\varepsilon A(x) \exp(-\lambda\phi(x))\, dx = O\left(\lambda^{-(1+\ell)/k}\right)$$

as $\lambda \to \infty$.

Proof Pick any $K > 0$ such that $|A(x)| \le K|x|^\ell$ on $[0, \varepsilon]$. Because $\phi(x) \sim x^k$ at $x = 0$, for any $\delta \in (0, 1)$ there is some interval $[0, \varepsilon']$ with $\varepsilon' \le \varepsilon$ such that $\phi(x) \ge (1-\delta)x^k$ on $[0, \varepsilon']$. Fixing any such δ and ε', nonvanishing of ϕ on $(0, \varepsilon]$ implies that $\inf_{\varepsilon' \le x \le \varepsilon} \phi(x)$ is positive, hence the portion of the integral coming from $[\varepsilon', \varepsilon]$ decays exponentially in λ. On $[0, \varepsilon']$,

$$\begin{aligned}
\left|\int_0^{\varepsilon'} A(x) \exp(-\lambda\phi(x))\, dx\right| &\le K \int_0^{\varepsilon'} x^\ell \exp(-\lambda(1-\delta)x^k)\, dx \\
&\le K \int_0^{\infty} x^\ell \exp(-\lambda(1-\delta)x^k)\, dx \\
&= O\left((\lambda(1-\delta))^{-(1+\ell)/k}\right) \qquad \text{by (4.3)} \\
&= O\left(\lambda^{-(1+\ell)/k}\right),
\end{aligned}$$

as desired. □

For monomial phase functions and general amplitude functions, we now have the following result.

Lemma 4.4. *Suppose that A is a real function with*

$$A(x) = \sum_{j=\ell}^{M-1} b_j x^j + O(x^M)$$

as $x \to 0$. Then

$$\int_0^\varepsilon A(x) \exp(-\lambda x^k)\, dx = \sum_{j=\ell}^{M-1} b_j C(k, j) \lambda^{-(1+j)/k} + O\left(\lambda^{-(1+M)/k}\right),$$

where $C(k, j) = \Gamma((1+j)/k)/k$.

Remark. The hypothesis on A in Lemma 4.4 is quite weak. In particular, A need not even be continuously differentiable, so it is useful for examples such as $A(x) = x\sin(x^{-1})$. If A is represented by an infinite asymptotic series then an asymptotic expansion for the integral follows by applying the lemma for each M.

Proof Multiply the estimate

$$A(x) - \sum_{j=0}^{M-1} b_j x^j = O(x^M)$$

by $\exp(-\lambda\phi(x))$ and integrate. Using Lemma 4.3 to bound the integral of the right-hand side gives

$$\left| I - \sum_{j=0}^{M-1} \int_0^{\varepsilon} b_j x^j \exp(-\lambda x^k)\, dx \right| = O\left(\lambda^{-(1+M)/k}\right),$$

and applying (4.3) to each integral with monomial amplitude gives the desired conclusion. □

Exercise 4.4. Apply Lemma 4.4 with $M = 1$ to give an asymptotic estimate with remainder for

$$\int_0^1 (1 + x\sin(x^{-1}))e^{-\lambda x^2/2}\, dx\,.$$

General A and ϕ

A change of variables reduces the general case to Lemma 4.4, although a bit of care is required to ensure we understand the asymptotic series for the functions involved in the change of variables.

Lemma 4.5. *Let $M \geq 2$ be an integer and suppose*

$$y(x) = c_1 x + \cdots + c_{M-1}x^{M-1} + O(x^M) \tag{4.5}$$

in a neighborhood of zero, where $c_1 \neq 0$. Then there is a neighborhood of zero on which y has a compositional inverse. The inverse function $x(y)$ has an expansion

$$x(y) = a_1 y + \cdots + a_{M-1}y^{M-1} + O(y^M),$$

where each a_j is a polynomial in $c_1, \ldots, c_j$ and c_1^{-1}.

Proof Suppose that $c_1 = 1$. From $y = x + O(x^2)$ we see that $y \sim x$ at zero, hence

$$x = y + O(x^2) = y + O(y^2).$$

Now let $2 \leq n < M$ and suppose inductively that $x = y + a_2 y^2 + \cdots + a_{n-1}y^{n-1} +$

$O(y^n)$, where each coefficient a_j is a polynomial in $c_2, \ldots, c_j$. Let a be an indeterminate, and substitute the value of y in (4.5) into the quantity

$$\Phi(x, y) = x - (y + a_2 y^2 + \cdots + a_{n-1} y^{n-1} + a y^n) .$$

The result $\Phi(x, y(x))$ is a polynomial in x, whose coefficients in degrees up to $n-1$ vanish due to the induction hypothesis, plus a remainder of $O(x^M)$ coming from our starting assumption. The coefficient of the x^n term may be written as $a - P(a_2, \ldots, a_{n-1}, c_2, \ldots, c_n)$, where P is a polynomial. By induction, this is a polynomial in $c_2, \ldots, c_n$. Setting $a_n = P(a_2, \ldots, a_{n-1}, c_2, \ldots, c_n)$, we see that

$$x - y - \sum_{j=2}^{n} a_j y^j = O(x^{n+1}) ,$$

completing the induction. When $n = M - 1$, observing that $O(x^M) = O(y^M)$ completes the proof of the lemma for $c_1 = 1$. For general $c_1 \neq 1$ we can use this argument to represent x as a function of y/c_1, which shows that $x = \sum_{j=1}^{M-1} a_j y^j + O(y^M)$ with each $c_1^j a_j$ a polynomial in $c_2, \ldots, c_j$. □

Exercise 4.5. Write down the quadratic Taylor expansion (with remainder term $O(x^3)$) near zero for $y = 1 - \sqrt{1-x}$ and find its compositional inverse. Check your work afterward by computing x as a function of y explicitly.

We can now establish Theorem 4.1 for real integrands. Recall that for each natural number M, a real function f is said to be of ***class*** C^M or C^M***-smooth*** if all derivatives of f up to and including order M are continuous; if this holds for all M we say f belongs to C^∞ and is ***smooth***.

Theorem 4.6. *Let M be a positive integer and let k and ℓ be integers with $0 \leq k, \ell \leq M$, and $k \geq 2$. Suppose that A is a real function, and ϕ is a C^M-smooth real function, with series expansions*

$$A(x) = \sum_{j=\ell}^{M-1} b_j x^j + O\left(x^M\right)$$

$$\phi(x) = \sum_{j=k}^{M+k-1} c_j x^j + O\left(x^{M+k}\right)$$

as $x \to 0$, where $b_\ell, c_k \neq 0$. Then $\mathcal{I}_+(\lambda) = \int_0^\varepsilon A(x) \exp(-\lambda\phi(x))\, dx$ has the asymptotic expansion

$$\mathcal{I}_+(\lambda) = \sum_{j=\ell}^{M-1} a_j C(k, j)(c_k \lambda)^{-(1+j)/k} + O\left(\lambda^{-(1+M)/k}\right) \tag{4.6}$$

as $\lambda \to \infty$, *with* $C(k,j) = k^{-1}\Gamma\left(\frac{1+j}{k}\right)$ *as above and the coefficient* a_j *given by a polynomial in* $b_\ell, \ldots, b_j$ *and* $c_k^{-1}, c_{k+1}, \ldots, c_{k+j-\ell}$. *The first two coefficients in this expansion are*

$$\begin{aligned} a_\ell &= b_\ell \\ a_{\ell+1} &= b_{\ell+1} - b_\ell \frac{2+\ell}{k} \frac{c_{k+1}}{c_k} . \end{aligned} \tag{4.7}$$

Proof Applying the change of variables $y = \phi(x)^{1/k}$ to

$$\phi(x) = c_k x^k \left(1 + \frac{c_{k+1}}{c_k} x + \cdots + \frac{c_{M+k-1}}{c_k} x^{M-1} + O\left(x^M\right)\right)$$

shows that

$$y = c_k^{1/k} x \left(1 + \frac{c_{k+1}}{c_k} x + \cdots + \frac{c_{M+k-1}}{c_k} x^{M-1} + O\left(x^M\right)\right)^{1/k}, \tag{4.8}$$

and the binomial expansion for $(1+u)^{1/k}$ gives

$$y = c_k^{1/k} \sum_{j=1}^{M} d_j x^j + O\left(x^{M+1}\right),$$

where each d_j is a polynomial in $c_{k+1}, \ldots, c_{k+j-1}$ and c_k^{-1}.

By Lemma 4.5, the inverse function $x = x(y)$ defined by this equation satisfies

$$x = \sum_{j=1}^{M} e_j \left(\frac{y}{c_k^{1/k}}\right)^j + O\left(y^{M+1}\right), \tag{4.9}$$

where e_j is a polynomial in $c_{k+1}, \ldots, c_{k+j-1}$. A function of class C^M with nowhere vanishing derivative has an inverse of class C^M, which justifies term-by-term differentiation and yields

$$x'(y) = c_k^{-1/k} \sum_{j=1}^{M} j e_j \left(\frac{y}{c_k^{1/k}}\right)^{j-1} + O\left(y^M\right).$$

The change of variables formula gives

$$\mathcal{I}_+(\lambda) = \int_0^{y(\varepsilon)} \tilde{A}(y) \exp(-\lambda y^k)\, dy,$$

where $\tilde{A}(y) = A(x(y))x'(y)$. Plugging the series for x and x' into the definition of $\tilde{A}$ implies

$$\tilde{A}(y) = c_k^{-1/k} \sum_{j=\ell}^{M-1} \tilde{b}_j \left(\frac{y}{c_k^{1/k}}\right)^j + O\left(y^M\right),$$

where $\tilde{b}_j$ is a polynomial in $b_\ell, \ldots, b_j, c_k^{-1}, c_{k+1}, \ldots, c_{k+j-\ell}$.

The existence of the expansion (4.6), with $a_j = \tilde{b}_j$ for all j, now follows from the monomial exponent case in Lemma 4.4, and we compute the leading terms (4.7) by finding $\tilde{b}_\ell$ and $\tilde{b}_{\ell+1}$. Inverting the expression

$$
\begin{aligned}
y &= c_k^{1/k} x \left(1 + \frac{c_{k+1}}{c_k} x + O\left(x^2\right)\right)^{1/k} \\
&= c_k^{1/k} x \left(1 + \frac{c_{k+1}}{k\, c_k} x + O\left(x^2\right)\right) \qquad (4.10)
\end{aligned}
$$

to expand x as a series in y is a matter of plugging $x = c_k^{-1/k} y + a y^2 + O(y^3)$ into (4.10), setting the result equal to x, and solving for a to obtain

$$
\begin{aligned}
x &= \frac{y}{c_k^{1/k}} - \frac{1}{k} \frac{c_{k+1}}{c_k} \left(\frac{y}{c_k^{1/k}}\right)^2 + O(y^3) \\
x'(y) &= \frac{1}{c_k^{1/k}} - \frac{2}{c_k^{2/k}} \frac{c_{k+1}}{k\, c_k} y + O(y^2). \qquad (4.11)
\end{aligned}
$$

Composing A with $x(y)$ gives

$$
A(x(y)) = b_\ell \frac{1}{c_k^{\ell/k}} y^\ell + \left(b_\ell \frac{\ell}{c_k^{(\ell-1)/k}} \frac{-c_{k+1}}{k c_k^{1+2/k}} + b_{\ell+1} \frac{1}{c_k^{(\ell+1)/k}}\right) y^{\ell+1} + O(y^{\ell+2}),
$$

and multiplying this expansion by the expression for $x'(y)$ in (4.11) shows that the leading series coefficients of $\tilde{A}(y) = A(x(y))x'(y)$ are

$$
\begin{aligned}
\tilde{b}_\ell &= b_\ell \\
\tilde{b}_{\ell+1} &= b_{\ell+1} - b_\ell \frac{\ell+2}{k} \frac{c_{k+1}}{c_k},
\end{aligned}
$$

giving (4.7). □

Exercise 4.6. Give an example to show that the hypotheses do not imply that ϕ is C^{M+1}-smooth.

4.2 Complex phase

Extending the results of the previous section to complex amplitudes is trivial: by linearity of the integral, the result holds separately for $\mathrm{Im}\{A\}$ and $\mathrm{Re}\{A\}$, and these may be recombined to give the result for complex A. When it comes to complex phases, we are faced with a choice. If we assume A and ϕ are analytic in a neighborhood of zero then we are entitled to move the contour, which is the quickest justification for extending the conclusion to complex phases and is the approach taken in this section. Later, in Section 4.3, we discuss an alternative using only smoothness instead of analyticity.

We now prove Theorem 4.1, starting with the one-sided integral $\mathcal{I}_+$ before addressing the double-sided integral $\mathcal{I}$.

Step 1: evaluation of the one-sided integral $\mathcal{I}_+$

Let $\gamma_+ : [0, \varepsilon] \to \mathbb{C}$ denote the restriction of γ to $[0, \varepsilon]$ so that

$$\mathcal{I}_+ = \int_{\gamma_+} A(z) \exp(-\lambda\phi(z))\, dz.$$

To evaluate $\mathcal{I}_+$ we employ the same change of variables $y = \phi(z)^{1/k}$ as in the proof of Theorem 4.6, only we need to be careful in choosing a branch of the kth root. Formula (4.8) defines k different functions, one for each choice of kth root for $c_k^{1/k}$. It follows from Lemma 4.5 that each of these k functions and their inverses are analytic in a neighborhood of the origin. To discuss the ***principal kth root*** – the analytic function from the ***slit plane*** $\mathbb{C} \setminus \mathbb{R}_{<0}$ to the cone $K = \{z : -\pi/k < \arg(z) < \pi/k\}$ – we write $\mathfrak{p}(u^{1/k}) = z$ for the unique $z \in K$ such that $z^k = u$, and let $v = \gamma'(0)$. Near the origin $\phi(z) \sim c_k z^k$, so the requirement that $\operatorname{Re}\{\phi(\gamma(t))\} \geq 0$ forces the quantity v to be in the windmill-shaped set of preimages of the right half-plane under $c_k z^k$, shown in Figure 4.1.

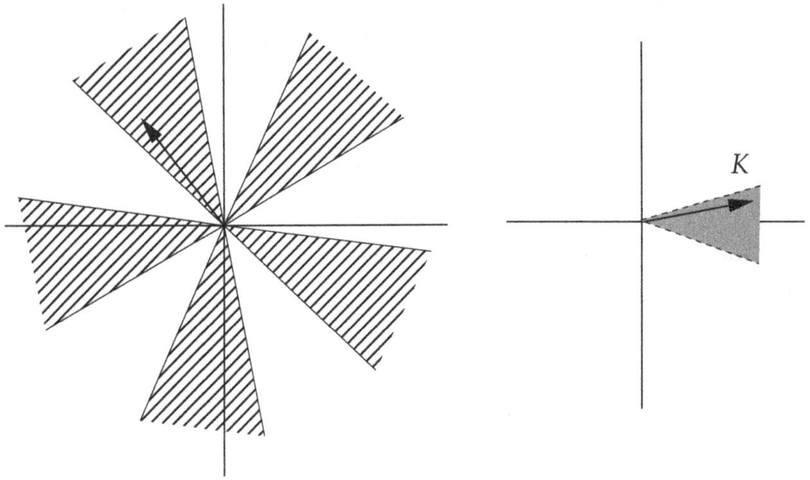

Figure 4.1 Arrows represent v and $(d/dt)|_{t=0} f(\gamma^+(t))$.

Define $f(x) = \mathfrak{p}(\phi(x)^{1/k})$. Since the path $\phi(\gamma_+(t))$ remains in the positive real half-plane for $0 < t \leq \varepsilon$, it also remains in the slit plane, and hence maps the image of γ_+ bi-analytically into the cone K. With this choice of kth root, the change of variables (4.8) becomes

$$y = f(x) = \eta x \left(1 + \cdots + \frac{c_M}{c_k} x^{M-k} + O\left(x^{M-k+1}\right)\right)^{1/k}, \tag{4.12}$$

where $\eta = v^{-1}\,\mathfrak{p}(c_k v^k)^{1/k}$ and the branch of the kth root of the series in parentheses fixes 1. Thus $f'(0) = \eta$ and the inverse function $x(y)$ is defined as in the proof of Theorem 4.6, with $c_k^{1/k} = \eta$ in (4.9). Analogously to the proof of Theorem 4.6, we then have

$$\mathcal{I}_+ = \int_{\tilde{\gamma}} \tilde{A}(y) \exp(-\lambda y^k)\, dy, \tag{4.13}$$

where $\tilde{\gamma} = f \circ \gamma_+$ is the image of γ_+ under our change of variables.

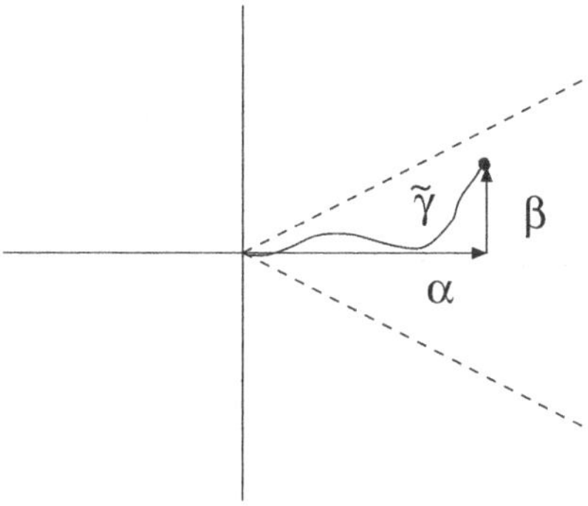

Figure 4.2 The path $\tilde{\gamma}$ in the cone K and the line segments α and β.

Let $p = f(\gamma(\varepsilon))$ denote the endpoint of $\tilde{\gamma}$ with real part $p' > 0$ and define the line segments $\alpha = [0, p']$ and $\beta = [p', p]$ in the complex plane. As seen in Figure 4.2, the contour $\tilde{\gamma}$ is homotopic to $\alpha + \beta$, hence $\int_{\tilde{\gamma}} h(z)\, dz = \int_\alpha h(z)\, dz + \int_\beta h(z)\, dz$ for any analytic function h. On compact subsets of K, the real part of y^k is bounded from below by a positive constant, so there are positive constants C and ρ such that

$$\left|\tilde{A}(y) \exp(-\lambda y^k)\right| \leq C e^{-\rho\lambda}$$

on β (the reason we chose the principal value for our root was to ensure that β lies inside K). We conclude that

$$\mathcal{I}_+ = \int_\alpha \tilde{A}(y) \exp(-\lambda y^k)\, dy + R_\lambda$$

for a remainder R_λ that decays exponentially, and applying Theorem 4.6 (with complex amplitude) to the integral over α gives the asymptotic series

$$\mathcal{I}_+ \approx \sum_{j=\ell}^{\infty} a_j C(k,j)(c_k\lambda)^{-(1+j)/k}$$

satisfying conclusion (i) of Theorem 4.1.

Step 2: evaluation of the two-sided integral $\mathcal{I}$

In order to reduce the two-sided integral to the one-sided case, define the contour $\gamma_- : [0,\varepsilon] \to \mathbb{C}$ by $\gamma_-(t) = \gamma(-t)$. The curve γ_- is oriented from 0 to $-\varepsilon$ so it appears with sign reversed. In other words, $I = \mathcal{I}_+ - \mathcal{I}_-$, where

$$\mathcal{I}_- := \int_{\gamma_-} A(z)\exp(-\lambda\phi(z))\,dz\,.$$

The integral for $\mathcal{I}_-$ has nearly the same data as the integral for $\mathcal{I}_+$: the functions A and ϕ are identical so the only difference between the two integrals is the contour. The contour affects the integral only via the choice of η in (4.12). Denoting the two choices by η_+ and η_-, we know *a fortiori* that $\eta_- = \eta_+/\zeta$ for some ζ with $\zeta^k = 1$. Denoting the respective inverse functions of $f(x)$ in (4.12) by g_+ and g_- we see that $g_-(y) = g_+(\zeta y)$. The two changes of variables produce amplitudes $\tilde{A}_+$ and $\tilde{A}_-$ in (4.13) satisfying

$$\tilde{A}_+(y) = A(g_+(y)) \cdot g'_+(y)$$

and

$$\begin{aligned}\tilde{A}_-(y) &= A(g_-(y)) \cdot g'_-(y)\\ &= A(g_+(\zeta y)) \cdot \zeta g'_+(\zeta y)\\ &= \zeta\tilde{A}_+(\zeta y)\,.\end{aligned}$$

The coefficients of the power series for $\tilde{A}_+$ and $\tilde{A}_-$ are therefore related by $[y^j]\tilde{A}_-(y) = \zeta^{j+1}[y^j]\tilde{A}_+(y)$. The asymptotic expansions of $\tilde{A}_\pm$ are integrated term by term in (4.13), which implies that the coefficients α_j for the two-sided integral $\mathcal{I}$ are related to the coefficients a_j for the one-sided integral $\mathcal{I}_+$ via $\alpha_j = (1 - \zeta^{j+1})a_j$. Thus, part (iii) of Theorem 4.1 is reduced to the correct identification of ζ. The evaluation of ζ breaks into two cases, depending on the parity of k.

Suppose first that k is even. Since $\phi(z) \sim c_k z^k$, the image of the smooth curve γ under ϕ moves back in the same direction at the origin, with the tangents to the images $\phi(\gamma_-(t))$ and $\phi(\gamma_+(t))$ coinciding at $t = 0$ (see Figure 4.3). Because

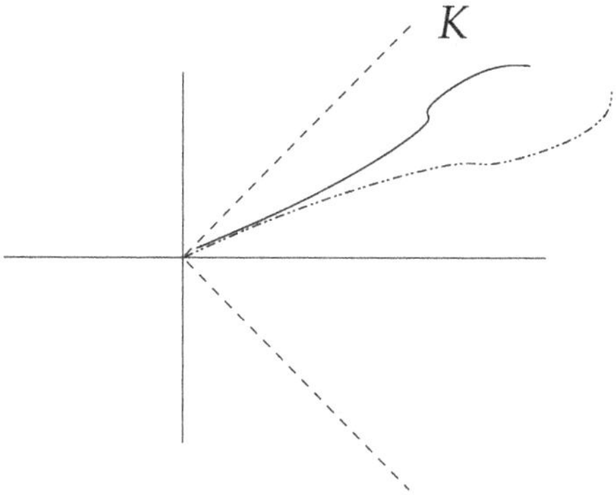

Figure 4.3 Illustration for even k: $\phi(\gamma_+)$ is a solid curve and $\phi(\gamma_-)$ is dotted.

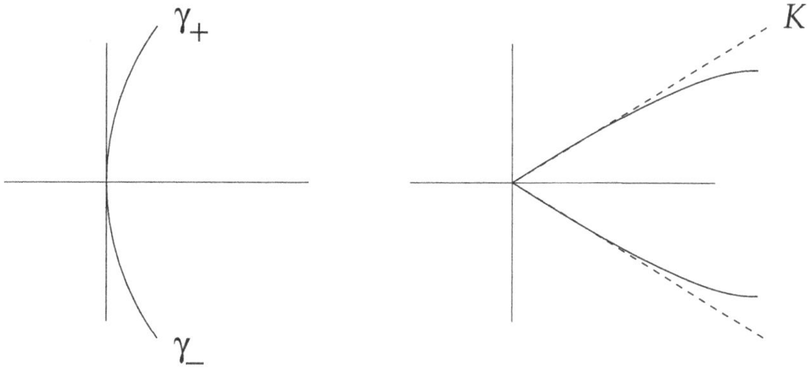

Figure 4.4 Illustration for odd k: $\phi(\gamma_+)$ and $\phi(\gamma_-)$ and their principal $1/k$ powers.

γ_- reverses the orientation of the parametrization, we see that $\boldsymbol{v}_- = \gamma_-'(0)$ and $\boldsymbol{v}_+ = \gamma_+'(0)$ satisfy $\boldsymbol{v}_- = -\boldsymbol{v}_+$. The powers $\boldsymbol{v}_-^k$ and $\boldsymbol{v}_+^k$ coincide, meaning

$$\eta_- = \boldsymbol{v}_-^{-1}\, \mathfrak{p}(c_k \boldsymbol{v}_-^k)^{1/k} = -\boldsymbol{v}_+^{-1}\, \mathfrak{p}(c_k \boldsymbol{v}_+^k)^{1/k} = -\eta_+ \,.$$

Thus, when k is even ζ takes the value -1. This leads to $\alpha_j = 2a_j$ for even j and $\alpha_j = 0$ for odd j, completing the proof of the theorem for even k.

When k is odd, the images of γ_+ and γ_- under ϕ point in opposite directions (see Figure 4.4). Since both are in the closed right half-plane, this implies that one points in the positive imaginary direction and the other points in the negative imaginary direction. Thus the argument of the tangent to $\phi(\gamma_+)$ at the origin is $\sigma\pi/2$, where the sign of σ is given by

$$\sigma = \operatorname{sgn} \operatorname{Im} \{\phi(\gamma'(0))\} \,.$$

The argument of the tangent to $\phi(\gamma_-)$ at the origin is $-\sigma\pi/2$ and thus differs from the argument of $\phi(\gamma_+)$ by $-\sigma\pi$. Mapping by the principal kth root shrinks the difference in arguments by a factor of k, thus

$$\mathfrak{p}(c_k \boldsymbol{v}_-^k)^{1/k} = e^{-i\pi\sigma/k}\, \mathfrak{p}(c_k \boldsymbol{v}_+^k)^{1/k} .$$

Again the reversal of parametrization implies $\boldsymbol{v}_- = -\boldsymbol{v}_+$, whence

$$\eta_- = (-1) \cdot e^{-i\pi\sigma/k} \eta_+ = \frac{\eta_+}{\zeta}$$

with $\zeta = -e^{i\pi\sigma/k}$ as in the statement of the theorem. □

Applications

The following classical result is a direct corollary of our machinery.

Proposition 4.7 (Watson's Lemma). *Let $A : \mathbb{R}_{>0} \to \mathbb{C}$ have asymptotic development*

$$A(t) \approx \sum_{m=0}^{\infty} b_m t^{\beta_m}$$

*for t near the origin, where $-1 < \mathrm{Re}\{\beta_0\} < \mathrm{Re}\{\beta_1\} < \cdots$ and $\mathrm{Re}\{\beta_m\} \to \infty$ as $m \to \infty$. Then the **Laplace transform***

$$L(\lambda) = \int_0^\infty A(t) e^{-\lambda t}\, dt$$

of A has an asymptotic series

$$L(\lambda) \approx \sum_{m=0}^{\infty} b_m \Gamma(\beta_m + 1) \lambda^{-(1+\beta_m)}$$

as $\lambda \to \infty$.

Proof As in our previous arguments, we may replace the integral in the Laplace transform by an integral on $[0, \varepsilon]$ while introducing only an exponentially small error. Writing

$$A(t) = \sum_{m=0}^{N} b_m t^{\beta_m} + R_N(t)$$

for $R_N = O\left(t^{\mathrm{Re}\,\{\beta_{m+1}\}}\right)$ at the origin, we may integrate term by term to get the first N terms of the expansion, up to an exponentially small correction from truncating the integral, then use Lemma 4.3 to see that the remainder satisfies

$$\left| \int_0^\varepsilon R_N(t) e^{-\lambda t}\, dt \right| = O\left(\lambda^{-\mathrm{Re}\{\beta_m\}-1}\right) ,$$

proving the proposition. □

Exercise 4.7. Let y be the positive real root of $x - 3xy - y^3 = 0$ satisfying $y \sim x^{1/3}$ as $x \downarrow 0$. Find the first two terms of the Puiseux series for y and use Watson's Lemma to compute the corresponding first two asymptotic terms of the Laplace transform $L(\lambda) = \int_0^\infty y(x)e^{-\lambda x}\,dx$.

As discussed in Section 3.5 of Chapter 3, Theorem 4.1 is a crucial component of the ***saddle point method***, which is also called the ***method of steepest descent***. If we consider an integral

$$\mathcal{I}(\lambda) = \int_\gamma A(z)\exp(-\lambda\phi(z))\,dz$$

where ϕ' does not vanish on γ then the idea is to deform γ to pass through a point x where ϕ' vanishes while introducing a negligible error. The phrase *steepest descent* comes from the fact that the real part of ϕ must have a local maximum on the contour at x, rather than a minimum or inflection point. Having deformed the contour to pass through x, Theorem 4.1 is then applied to determine asymptotics.

Example 4.8. Consider the univariate power series $f(z) = (1-z)^{-1/2}$, whose power series coefficients at the origin form the sequence $a_n = (-1)^n\binom{-1/2}{n}$. Instead of using a transfer theorem from Chapter 3 to obtain the asymptotic behavior $a_n \sim 1/\sqrt{\pi n}$, let us apply a saddle point approach to the Cauchy integral representation

$$a_n = \frac{1}{2\pi i}\int_C z^{-n-1}(1-z)^{-1/2}\,dz,$$

where C is any sufficiently small positively oriented circle around the origin. Since we understand meromorphic integrands the best, we make the change of variables $z = 1 - y^2$ to obtain

$$a_n = \frac{1}{2\pi i}\int_E (1-y^2)^{-n-1}y^{-1}\,(-2y)\,dy = \frac{i}{\pi}\int_E (1-y^2)^{-n-1}\,dy$$

where, as shown in Figure 4.5, E is a positively-oriented small circle in the y-plane around either the point $+1$ or the point -1 (since both of these contours map to a small contour around 0 in the z-plane). For concreteness, let us take E to be a small circle around $+1$. In the y-plane, there is a critical point for $\phi(y) = -\log(1-y^2)$ at the origin. Without crossing any singularities of our integrand, we can deform the contour E to a contour $\tilde{E}$ passing through the origin in the downward direction. Changing variables with $y = -it$ gives

$$a_n = \frac{i}{\pi}\int_{\mathbb{R}} (1+t^2)^{-n-1}\,(-i)\,dt = \frac{1}{\pi}\int_{\mathbb{R}} \frac{1}{1+t^2}\exp\left(-n\log(1+t^2)\right)\,dt,$$

and an application of Theorem 4.1 with the expansions $A(t) = 1/(1+t^2) = 1 + \cdots$ and $\phi(t) = \log(1+t^2) = t^2 + \cdots$ implies $a_n \sim \sqrt{2/(\pi n \phi''(0))} = 1/\sqrt{\pi n}$. ◁

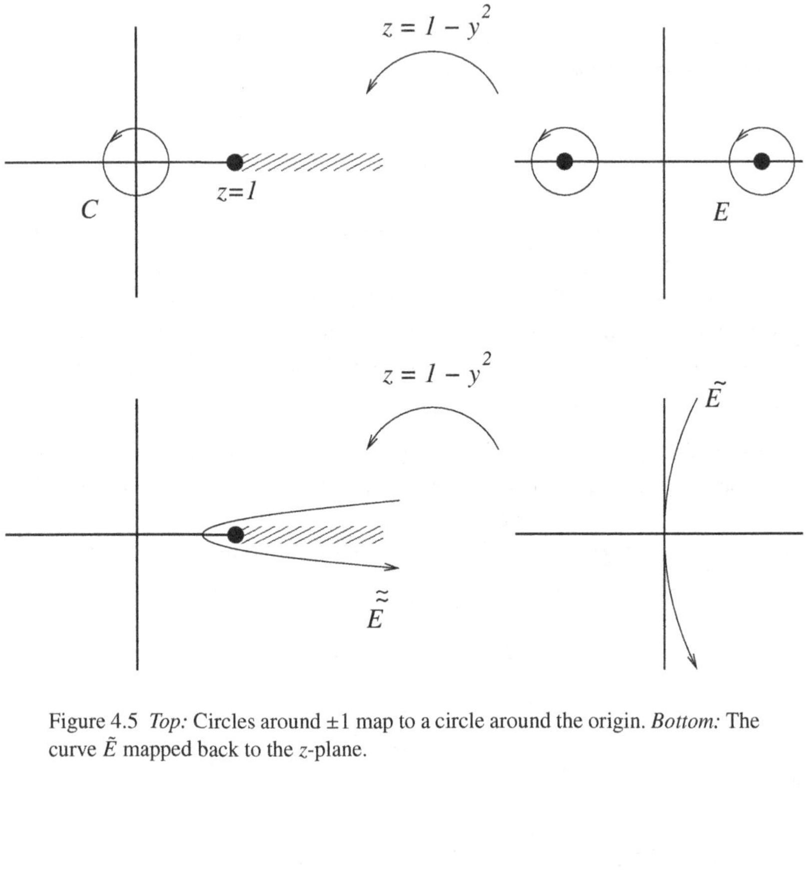

Figure 4.5 *Top:* Circles around ±1 map to a circle around the origin. *Bottom:* The curve $\tilde{E}$ mapped back to the z-plane.

Remark. The function $(1-z)^{-1/2}$ is analytic on the slit plane $\mathbb{C}\backslash\{x \in \mathbb{R} : x \geq 1\}$, which we may view as half of the Riemann surface $\mathfrak{R}$ obtained by gluing two copies of the slit plane, with the upper half of one attaching along the slit to the lower half of the other. The change of variables $z = 1 - y^2$ maps from $\mathfrak{R}$ to $\mathbb{C}$. The saddle point contour $\tilde{E}$, when mapped to the z-plane, comes in to $+1$ along one copy of the slit, does a U-turn, and goes back along the other copy of the slit. Perturbing slightly gives a hairpin-shaped contour $\tilde{\tilde{E}}$ that may be drawn in the slit plane. This shape near 1 reflects the design of the Δ-domain in Figure 3.1.

4.3 Analytic versus smooth functions

Although most of our results above are stated for analytic amplitudes and phases, their conclusions hold assuming only that these functions are C^∞-smooth. This indicates that there should be arguments using smooth techniques, such as partitions of unity and integration by parts, rather than contour deformation. Such an approach to evaluating saddle point integrals has been developed and used extensively by harmonic analysts, who are chiefly interested in the case where ϕ is purely imaginary. This is not covered by Theorem 4.1, which requires that $\text{Re}\{\phi\}$ be strictly positive away from zero for the contour decomposition used in our proof. Results in this section will be used only once or twice for the analysis of generating functions in this book; our chief reason for including this material is that any treatment of Fourier–Laplace integrals bypassing smooth methods is pedagogically and historically incomplete.

When the exponent ϕ is imaginary, the modulus of the Fourier–Laplace integrand is equal to $|A(z)|$, so it is no longer true that one may cut off the integral outside of an interval $[-\varepsilon, \varepsilon]$ and expect to introduce negligible remainders. Instead, one assumes that A has compact support, then uses smooth partitions of unity to reduce to integrals over small intervals. Note that neither partitions of unity nor compactly supported functions exist in the analytic category; however, when the contour of integration γ is a closed curve, any amplitude function has compact support on γ so both the analytic and the smooth methods apply and may be compared.

In this section we give asymptotics for the integral

$$I(\lambda) = \int_a^b A(z) \exp(i\lambda\phi(z))\, dz,$$

where A and ϕ are smooth, ϕ is real, and A is supported on a compact subinterval of (a, b). Matching existing literature we use the term *phase* to denote ϕ rather than $i\phi$. The overall argument in the smooth case is the same as the proof of Theorem 4.6, except for the insertion of a localization step at the beginning and the introduction of a damping term in the step where both amplitude and phase are monomial. Our process towards general asymptotics thus becomes: localization, big-O estimate, monomials (with damping), monomial phase, and then the full theorem.

Localization Lemma in C^∞

Following [Ste93], we begin with a localization principle.

Lemma 4.9 (Localization Lemma). *Suppose ϕ is real and $\phi'(x) \neq 0$ for all $x \in (a, b)$. Then $\mathcal{I}(\lambda)$ is rapidly decreasing, meaning that for any fixed $N \geq 0$ we have $\mathcal{I}(\lambda) = O(\lambda^{-N})$ as $\lambda \to \infty$.*

Proof Our assumption that A is supported on a subset of the domain of integration allows us to integrate by parts without introducing boundary terms. Integrating by parts with $U = A/(i\lambda\phi')$ and $dV = i\lambda\phi' e^{i\lambda\phi}\, dx$ gives

$$\mathcal{I}(\lambda) = -\int_a^b e^{i\lambda\phi(x)} \frac{d}{dx}\left(\frac{A}{i\lambda\phi'}\right)(x)\, dx\,,$$

and by repeating this $N \geq 1$ times we obtain

$$\mathcal{I}(\lambda) = \int_a^b e^{i\lambda\phi(x)}(-\lambda^{-N})\mathcal{D}^N(A)(x)\, dx\,, \tag{4.14}$$

where $\mathcal{D}$ is the differential operator $f \mapsto (d/dx)(f/i\phi')$. If

$$K_N = (b - a) \sup_{a \leq x \leq b} |\mathcal{D}^N A(x)| \tag{4.15}$$

then $|\mathcal{I}(\lambda)| \leq \lambda^{-N} K_N$, which proves that $\mathcal{I}$ is a rapidly decreasing function of λ. □

Remarks. (i) In the analytic case, if ϕ' is nowhere vanishing then the contour can be *pushed down* along a gradient flow so that the maximum of $\mathrm{Re}\{i\phi\}$ is strictly negative, resulting in an exponentially (rather than rapidly) decreasing integral.

(ii) Since ϕ' does not vanish, we may change variables to $y = \phi(x)$ and the conclusion of Lemma 4.9 is equivalent to the perhaps more familiar statement that the Fourier transform of the smooth function $\tilde{A}(y)$ obtained is rapidly decreasing.

(iii) While Lemma 4.9 is stated only for purely imaginary phase functions, the same argument shows that $\mathcal{I}(\lambda)$ is rapidly decreasing whenever the real part of ϕ is nonnegative and ϕ' is nonvanishing.

Exercise 4.8. Suppose $\phi(x) = x$ and $A(x)$ is periodic with fundamental domain $(-\pi, \pi]$.

(a) Show that there is no boundary term when integrating $\mathcal{I}$ over $[-\pi, \pi]$.
(b) Applying the Localization Lemma with $A(x) = 1 + \cos(x)$, what value do you get for K_2 in the bound $|\mathcal{I}(\lambda)| \leq K_2\lambda^{-2}$?

We call Lemma 4.9 the *Localization Lemma* for the following reason. Suppose we allow ϕ' to vanish on some finite set of points $x_1, \ldots, x_d \in [a, b]$. Then the contribution to $\mathcal{I}(\lambda)$ from any closed region not containing some x_i

is rapidly decreasing, so the asymptotics for $\mathcal{I}(\lambda)$ may be read off as the sum of contributions local to each x_i. Indeed, for each i let $[a_i, b_i]$ be a tiny interval containing x_i, with all intervals disjoint, and let $\xi_1, \ldots, \xi_d$ be a partition of unity subordinate to $\{[a_i, b_i] : 1 \le i \le d\}$. Once we see how to obtain asymptotics in a neighborhood of x_i containing no other critical points, we can write $A(x) = A_0(x) + \sum_{i=1}^d A(x)\xi_i(x)$, so that the support of A_0 contains no x_i. The integral $\int A_0(x)e^{i\lambda\phi(x)}\,dx$ is rapidly decreasing by the Localization Lemma, so as long as the integrals

$$\mathcal{I}_i(\lambda) = \int_a^b A(x)\xi_i(x)e^{i\lambda\phi(x)}\,dx = \int_{a_i}^{b_i} A(x)\xi_i(x)e^{i\lambda\phi(x)}\,dx$$

sum to something not rapidly decreasing, the asymptotic development of $\mathcal{I}(\lambda)$ is obtained by summing the developments of the $\mathcal{I}_i(\lambda)$.

We thus need only consider integrals over domains where the derivative of the phase vanishes at a single point. Our main univariate result for integrals with purely imaginary phase is the asymptotic development described in the following result.

Theorem 4.10. *Let ϕ and A be smooth real functions defined on $\mathbb{R}$ with A having compact support in an interval (a, b) whose closure contains zero. Let $k \ge 2$ and $\ell \ge 0$ be integers and suppose the power series coefficients for A and ϕ at the origin are $\{b_j\}$ and $\{c_j\}$ as in Theorem 4.6, with $c_k > 0$. Suppose that ϕ' vanishes in $[a, b]$ at the origin but nowhere else, and let $\tilde{A} = (A \circ g) \cdot g'$ where g is the inverse function to $x \mapsto (\phi/c_k)^{1/k}$. Then as $\lambda \to \infty$ there is an asymptotic development*

$$\mathcal{I}(\lambda) = \int_a^b A(x)\exp(i\lambda\phi(x))\,dx \approx \sum_{j=\ell}^{\infty} \alpha_j C(k, j)(i\,c_k\,\lambda)^{-(1+j)/k}.$$

The coefficients α_j are obtained from the power series coefficients $a_0, \ldots, a_j$ for $\tilde{A}$ exactly as in part (iii) of Theorem 4.1. If the asymptotic expansion is computed up to an $O\left(\lambda^{-(N+1)/k}\right)$ error term then this error term can be explicitly bounded by a continuous function of the suprema of the first $N+1$ derivatives of ϕ and A on the support of A. The kth root of $i\,c_k\lambda$ in this expansion is the principal root.

The rest of this chapter is devoted to proving Theorem 4.10, using the steps outlined above.

The C^∞ Big-O Lemma

The smooth counterpart to Lemma 4.3 is established by showing that the main contribution to $\mathcal{I}(\lambda)$ comes from an interval of size $\lambda^{-1/k}$. The increase in length for the proof of this result compared to the very short proof of Lemma 4.3 comes from the need to keep track of a partition of unity function and its derivatives.

Lemma 4.11. *If η is smooth and compactly supported and $\ell \geq 1$ and $k \geq 2$ are integers, then*

$$\left| \int_{-\infty}^{\infty} e^{i\lambda x^k} x^\ell \eta(x)\, dx \right| \leq C\lambda^{-(\ell+1)/k} \tag{4.16}$$

for a constant C depending only on k, ℓ, and the first ℓ derivatives of η.

Proof Let α be a smooth bump function taking values in $[0, 1]$ that is equal to 1 on $|x| \leq 1$ and vanishes on $|x| \geq 2$. Choose $\varepsilon > 0$ and rewrite (4.16) as

$$\int_{-\infty}^{\infty} e^{i\lambda x^k} x^\ell \eta(x)\alpha(x/\varepsilon)\, dx + \int_{-\infty}^{\infty} e^{i\lambda x^k} x^\ell \eta(x)[1 - \alpha(x/\varepsilon)]\, dx. \tag{4.17}$$

For any $S \subset \mathbb{R}$ let $\mathbf{1}_S(x)$ denote the indicator function that equals one when $x \in S$ and zero otherwise. The absolute value of the first integrand in (4.17) is everywhere bounded by $|x|^\ell \cdot \sup_{|x|\leq 2} |\eta(x)| \cdot \mathbf{1}_{|x|\leq 2\varepsilon}$, so its integral is bounded by $C_1\varepsilon^{1+\ell}$ where $C_1 = 2^{\ell+2} \sup_{|x|\leq 2} |\eta(x)|$.

The second integral will be done by parts, and to prepare for this we examine the iteration of the operator $D : f \mapsto (d/dx)(f/x^{k-1})$ applied to the function $f(x) = x^\ell \eta(x)(1 - \alpha(x/\varepsilon))$. The result will be a sum of monomials, each monomial being a product of a power of x, a derivative of η, a derivative of α, and a power of ε: if $[a, b, c, d]$ is shorthand for the term $x^a\, \eta^{(b)}(x)\, \alpha^{(c)}(x/\varepsilon)\, \varepsilon^d$ then

$$D\cdot[a, b, c, d] = (a-k+1)[a-k, b, c, d]+[a-k+1, b+1, c, d]+[a-k+1, b, c+1, d-1]$$

whenever $a \geq 0$. By induction, we see that $D^N \cdot [a, b, c, d]$ is the sum of terms of the form $C[r, s, t, u]$ where

$$r + u \geq a + d - kN, \qquad s \leq b + N, \qquad t \leq c + N,$$

and C is bounded above by the factorial of $\max\{kN, a\}$. In particular, since $\varepsilon \leq x$ we may replace positive powers of ε by the same power of x to arrive at the upper bound

$$\left| D^N \left[x^\ell \eta(x)(1 - \alpha(x/\varepsilon)) \right] \right| \leq \mathbf{1}_{|x|\geq\varepsilon} C_2 |x|^{\ell-kN}, \tag{4.18}$$

where C_2 is the product of $2 \cdot 3^N \cdot \sup_{j\leq N,|x|\in(1,2)} \eta^{(j)}(x)$ and the maximum of 1 and $\sup_{j\leq N,|x|\in(1,2)} \alpha^{(j)}(x)$.

Now we fix an N large enough so that $\ell - kN + 1 < 0$, and perform integration by parts on the second integral of (4.17) N times, each time integrating $-ik\lambda x^{k-1}e^{i\lambda x^k}$ and differentiating the rest, ending with

$$\int_{-\infty}^{\infty} e^{i\lambda x^k}(-ik\lambda)^{-N}D^N\left[x^\ell \eta(x)(1-\alpha(x/\varepsilon))\right]dx.$$

By (4.18) the modulus of this integrand is at most $C_2\mathbf{1}_{|x|\geq\varepsilon}|x|^{\ell-kN}(k\lambda)^{-N}$ so the integral is bounded by $C_3\lambda^{-N}\varepsilon^{\ell-kN+1}$, where $C_3 = 2C_2/(1-\ell+kN)$. Setting $\varepsilon = \lambda^{-1/k}$ and adding the bounds on the two integrals yields an upper bound of $(C_1+C_3)\lambda^{-(1+\ell)/k}$ on the modulus of the integral in (4.16). We have also shown that C_1 and C_3 depend only on k, ℓ, the first ℓ derivatives of η, and the first ℓ derivatives of α. Fixing any valid choice of the function α thus completes our proof. □

Exercise 4.9. For a smooth function α taking value zero on negative arguments and value one on arguments ε or greater, what is the least possible value of $\sup_x |\alpha''(x)|$? This question is motivated by the reliance on smooth partitions of unity and the derivatives of α appearing in (4.18).

Lemma 4.11 immediately implies the following result.

Corollary 4.12. *If a smooth function g vanishes in a neighborhood of the origin and decreases rapidly at infinity (or is compactly supported) then $\mathcal{I}(\lambda) = \int_{-\infty}^{\infty} g(x)e^{i\lambda x^k}\,dx$ is rapidly decreasing.* □

A and ϕ are monomials and A is damped

In this section we prove the C^∞ version of Lemma 4.4. Since a monomial amplitude function does not have compact support, we introduce a damping function which will later need to be removed. For parameters $\lambda, k, \ell, \delta \in \mathbb{R}$ with $k \geq 2$ define

$$\mathcal{I}(\lambda, k, \ell, \delta) = \int_{-\infty}^{\infty} e^{i\lambda x^k} e^{-\delta|x|^k} x^\ell\,dx. \tag{4.19}$$

Lemma 4.13. *An asymptotic expansion*

$$\mathcal{I}(\lambda, k, \ell, \delta) = \lambda^{-(1+\ell)/k}\sum_{j=0}^{\infty} C(j,k,\ell,\delta)\lambda^{-j}$$

holds as $\lambda \to \infty$, where the constants $C(j,k,\ell,\delta)$ with $j > 0$ go to zero as $\delta \to 0$.

Proof Let $\mathcal{I}_+(\lambda, k, \ell, \delta)$ denote the integral obtained by restricting the domain of integration in (4.19) to $[0, \infty)$ and define $\mathcal{I}_-(\lambda, k, \ell, \delta)$ by restricting the domain of integration to $(-\infty, 0]$. The change of variables $z = (\delta - i\lambda)^{1/k} x$ implies $\mathcal{I}_+(\lambda, k, \ell, \delta) = \lim_{M\to\infty} \mathcal{I}_+^{[M]}(\lambda, k, \ell, \delta)$, where

$$\mathcal{I}_+^{[M]}(\lambda, k, \ell, \delta) = (\delta - i\lambda)^{-(1+\ell)/k} \int_0^{M(\delta - i\lambda)^{1/k}} e^{-z^k} z^\ell dz$$

has as its domain of integration the line segment from the origin to $M(\delta - i\lambda)^{1/k}$ in $\mathbb{C}$. The integral over this line segment in complex space is the same as the integral over the real line segment from the origin to $M|\delta - i\lambda|^{1/k}$ followed by the imaginary line segment from $M|\delta - i\lambda|^{1/k}$ to $M(\delta - i\lambda)^{1/k}$. Thus, for fixed λ we can write

$$\mathcal{I}_+^{[M]}(\lambda, k, \ell, \delta) = (\delta - i\lambda)^{-\frac{1+\ell}{k}} \int_0^{M|\delta - i\lambda|^{1/k}} e^{-z^k} z^\ell dz + (\delta - i\lambda)^{-\frac{1+\ell}{k}} \int_{M|\delta - i\lambda|^{1/k}}^{M(\delta - i\lambda)^{1/k}} e^{-z^k} z^\ell dz.$$

Since $k \geq 2$ the modulus of the integrand $e^{-z^k} z^\ell$ on the imaginary line segment decays exponentially with M, and the length of this segment grows only linearly with M. Taking $M \to \infty$ thus implies

$$\mathcal{I}_+(\lambda, k, \ell, \delta) = (\delta - i\lambda)^{-(1+\ell)/k} \int_0^\infty e^{-z^k} z^\ell \, dz,$$

rotating our domain of integration back to the real axis. The definite integral in this expression has value $C(k, \ell) = k^{-1}\Gamma((1 + \ell)/k)$, and expanding $(\delta - i\lambda)^{-(1+\ell)/k} = (-i\lambda)^{-(1+\ell)/k}(1 + \delta i/\lambda)^{-(1+\ell)/k}$ using the binomial theorem gives

$$\mathcal{I}_+(\lambda, k, \ell, \delta) = C(k, \ell) e^{i\pi(1+\ell)/(2k)} \lambda^{-(1+\ell)/k} \sum_{j=0}^\infty (i\delta)^j \binom{-(1+\ell)/k}{j} \lambda^{-j}.$$

Writing

$$C_+(j, k, \ell, \delta) = k^{-1}\Gamma\left(\frac{\ell + 1}{k}\right) e^{i\pi(1+\ell+jk)/(2k)} \binom{-(1+\ell)/k}{j} \delta^j \tag{4.20}$$

and defining constants $C_-(j, k, \ell, \delta)$ by performing the analogous computation on $\mathcal{I}_-(j, k, \ell, \delta)$ proves the lemma by taking $C(j, k, \ell, \delta) = C_+(j, k, \ell, \delta) + C_-(j, k, \ell, \delta)$. □

ϕ is a monomial and A is anything

Theorem 4.14. *Let $\phi(x) = x^k$ and let A be a smooth function with compact support containing the origin. If the Taylor series of A at the origin* *has coefficients* $\{b_j\}$ *and* ℓ *denotes the smallest index such that* $b_\ell \neq 0$ *then* $\mathcal{I}_+ = \int_0^\infty A(x)\exp(i\lambda\phi(x))\,dx$ *has an asymptotic development*

$$\mathcal{I}_+ \approx \sum_{j=\ell}^{\infty} b_j\, C(k,j)\,(i\lambda)^{-(1+j)/k}.$$

The error obtained by taking the $(N-1)$*st partial sum is* $O(\lambda^{-N/k})$ *with the implied constant bounded in terms of the suprema of the first* N *derivatives of* A *near the origin. A similar result holds for the two-sided integral* $\mathcal{I}$ *with coefficients* α_j *obtained from conclusion* (iii) *in Theorem 4.1 with* $a_j = b_j$.

Proof Let U be a smooth function that is 1 on the support of A and vanishes outside of a compact set. Fix $N \geq 1$ and $\delta > 0$, and let $P(x) = P_{N,\delta}(x)$ be the Nth Taylor polynomial for $e^{\delta x^k} A(x)$ (obtained by truncating its Taylor series at the x^N term). If $b_{j,\delta}$ denotes the coefficient of x^j in $P(x)$ and the normalized remainder term $R(x) = R_{N,\delta}(x)$ is defined by $e^{\delta x^k} A(x) = P(x) + x^{N+1}R(x)$ then $\mathcal{I}_+ = B_1 + B_2 + B_3$, where

$$B_1 = \int_0^\infty e^{i\lambda x^k} e^{-\delta x^k} P(x)\,dx\,,$$

$$B_2 = \int_0^\infty e^{i\lambda x^k} x^{N+1} e^{-\delta x^k} R(x)U(x)\,dx\,,$$

$$B_3 = \int_0^\infty e^{i\lambda x^k} e^{-\delta x^k} P(x)(U(x)-1)\,dx\,.$$

By Lemma 4.11 with $\eta(x) = e^{-\delta x^k} R(x)U(x)$ and $\ell = N+1$, we know that $|B_2| \leq K\lambda^{-(\ell+2)/k}$ for some constant $K > 0$ that can be bounded in terms of k, ℓ, and the first ℓ derivatives of A, the bound being uniform over δ in a neighborhood of the origin.

Similarly, by Corollary 4.12 we see that B_3 is rapidly decreasing as $\lambda \to \infty$. It follows that the asymptotic series for $\mathcal{I}_+$ up to the $\lambda^{-(\ell+1)/k}$ term may be obtained by taking $\delta \to 0$ in B_1. Since P is a finite sum of monomials, we may use Lemma 4.13 to compute B_1 and prove the theorem. As before, we may sum results for $\mathcal{I}_+$ and the analogous integral $\mathcal{I}_-$ over the negative real half-line to prove the result for $\mathcal{I}$. □

General A and ϕ

Since $i\phi$ always lies along the imaginary axis, we may use a diffeomorphic change of variables to transform ϕ into $ic_k x^k$, under which the contour remains along the imaginary axis (thus there is no need for arguments about moving the contour).

Proof of Theorem 4.10. By assumption, we can write $\phi(x) = c_k x^k(1 + \theta(x))$, where $\theta(x) = O(|x|)$. If $y = x(1 + \theta(x))^{1/k}$ then $y(x)$ is a diffeomorphism in a neighborhood of the origin, and we write $x = g(y)$ for its inverse. Since $c_k y^k = \phi(x)$ we may change variables to see that

$$\int e^{i\lambda\phi(x)} A(x)\, dx = \int e^{i\lambda c_k y^k} \tilde{A}(y)\, dy\,,$$

and the result follows from Lemma 4.4. □

Notes

Our chief sources for Sections 1 and 2 were [BH86] and [Won01], along with [Hen91]. Although our main theorem follows from the extensive analyses in [BH86, Chapter 7], for example, asymptotics of Fourier and Laplace transforms are seldom treated together, and we have never seen the univariate Fourier–Laplace Theorem stated in exactly this form. We have also not seen a derivation by purely complex analytic methods.

Watson's Lemma may be found in many places. The version here agrees with the statements in [BH86, Section 4.1] and [Hen91, Section 11.5]. The saddle point method is described very nicely in [dBru81] and [FS09, Chapter VIII]. Our treatment is more akin to [Hen91, Section 11.8]; see also [BH86, Chapter 7], especially for some of our exercises.

Section 4.3 borrows heavily from [Ste93], although we have attempted to fill in some details. For instance, our proof of Lemma 4.11 is summarized as "A simple computation shows..." in [Ste93, page 335], which also omits details as to how the argument for $k = 2$ extends to greater values of k. Despite its omission of detail in elementary arguments, Stein's book is a beautifully written modern classic, and is a recommended addition to anyone's bookshelf.

Additional exercises

Exercise 4.10. Let $\phi : \mathbb{R} \to \mathbb{R}$ be defined by

$$\phi(x) = \begin{cases} \exp\left(-\frac{1}{x^2}\right) & \text{if } x \neq 0 \\ 0 & \text{if } x = 0. \end{cases}$$

Show that ϕ is of class C^∞ and that all derivatives of ϕ vanish at 0. Explain the relevance of this fact to the asymptotic formulae derived in the theorems in this chapter.

Exercise 4.11. If $k = 2$ and $\ell = 0$ then Theorem 4.1 gives

$$\mathcal{I} = b_0 \sqrt{\frac{\pi}{c_2}} \lambda^{-1/2} + \alpha_2 \lambda^{-3/2} + O(\lambda^{-5/2}).$$

Compute the coefficient α_2 in terms of b_0, b_1, b_2, c_2, c_3, and c_4.

Exercise 4.12. The ***Bessel function*** is defined by

$$J_m(r) = \frac{1}{2\pi} \int_0^{2\pi} \exp(ir \sin\theta - im\theta)\, d\theta,$$

where m is a fixed parameter (you may assume it is a positive integer). Use Theorem 4.10 to find the two leading terms of an asymptotic series for $J_m(r)$ in decreasing powers of r.

Exercise 4.13. The ***Airy function*** is defined by

$$\mathrm{Ai}(x) = \frac{1}{2\pi} \int_{-\infty}^{\infty} e^{i(xt+t^3/3)}\, dt.$$

Find an asymptotic expression for $\mathrm{Ai}(x)$ as $x \to \infty$ in $\mathbb{R}$ by (1) performing the change of variables $t = ix^{1/2}u$, (2) finding the critical points and deforming the contour of integration to pass through one or more of them, and (3) computing the expansion on a compactly supported interval and arguing that this converges as the limits of integration go to infinity.

Exercise 4.14. Continue the error estimation in Exercise 4.9.

(a) In terms of ε and j, what is the least value of $\sup_{0 \le x \le \varepsilon} |\alpha^{(j)}(x)|$ for a bump function α that is flat on the complement of $[0, \varepsilon]$ and goes from value zero to value one?
(b) What does this imply about a lower bound on the implied constant C_M in the $O(\lambda^M)$ error term?
(c) Assuming this lower bound can be achieved, for fixed λ what value of M optimizes the error bound $C_M \lambda^M$?

5

Multivariate Fourier–Laplace integrals

5.1 Overview

In this chapter we generalize the univariate saddle point techniques of Chapter 4 to ***multivariate Fourier–Laplace integrals*** of the form

$$\mathcal{I}(\lambda) = \int_C A(z)\exp(-\lambda\phi(z))\,dz\,, \tag{5.1}$$

where the ***amplitude*** A and ***phase*** ϕ are now analytic functions of a vector argument z in d variables and C is a d-chain in $\mathbb{C}^d$ (see Section A.3 of Appendix A for definitions involving integration of chains on manifolds). In one variable, the comprehensive Theorem 4.1 covers all degrees of vanishing of the phase and amplitude functions. The range of possibilities for the phase function ϕ in higher dimensions is much greater, however, and we restrict ourselves here to the case of ***nondegenerate phase*** where the $d \times d$ ***Hessian*** matrix $\mathcal{H} = \left(\dfrac{\partial^2\phi}{\partial z_j\,\partial z_k}\right)$ of ϕ is nonsingular at the points in the domain of integration determining asymptotics. The Taylor series for ϕ at a point $p \in \mathbb{C}^d$ is

$$\phi(z) = \phi(p) + (z-p)^T(\nabla\phi)(p) + \frac{1}{2}(z-p)^T\,\mathcal{H}(p)(z-p) + O\left(|z-p|^3\right),$$

hence the Hessian matrix $\mathcal{H}(p)$ represents (twice) the quadratic term in the phase, and nondegeneracy is a generalization of nonvanishing of the quadratic term for a univariate phase function.

Exercise 5.1. Determine whether the phase function $\phi(x,y,z) = z^2+(x+y)z+xy$ is degenerate at the origin.

We begin, analogously to the univariate case, by considering integrals whose phase is restricted to the ***standard quadratic*** $S(z) := z_1^2 + \cdots + z_d^2$. Asymptotic behavior when A is monomial and ϕ is the standard quadratic (Corollary 5.7

below) is coupled with a big-O bound (Lemma 5.8 below), allowing us to integrate term by term and obtain the following result. Recall that if $\boldsymbol{r} \in \mathbb{N}^d$ we write $|\boldsymbol{r}| = r_1 + \cdots + r_d$.

Theorem 5.1 (standard phase). *Let $A(\boldsymbol{x}) = \sum_{\boldsymbol{r}\in\mathbb{N}^d} a_{\boldsymbol{r}}\boldsymbol{x}^{\boldsymbol{r}}$ be a real analytic function defined on a neighborhood $\mathcal{N}$ of the origin in $\mathbb{R}^d$. If*

$$\mathcal{I}(\lambda) = \int_{\mathcal{N}} A(\boldsymbol{x})e^{-\lambda S(\boldsymbol{x})}\,d\boldsymbol{x} \tag{5.2}$$

then there is an asymptotic series expansion

$$\mathcal{I}(\lambda) \approx \sum_{n\geq 0}\sum_{|\boldsymbol{r}|=n} a_{\boldsymbol{r}}\beta_{\boldsymbol{r}}\lambda^{-(|\boldsymbol{r}|+d)/2},$$

where

$$\beta_{\boldsymbol{r}} = \begin{cases} 0 & \text{if any } r_j \text{ is odd} \\ \pi^{d/2}\displaystyle\prod_{j=1}^{d}\frac{(2m_j)!}{m_j!4^{m_j}} & \text{if } \boldsymbol{r} = 2\boldsymbol{m} \end{cases}.$$

After establishing Theorem 5.1 in Section 5.2, we use a change of variables and contour deformation to study the case of nondegenerate phase whose real part has a strict minimum at the origin. Our next result is proven in Section 5.3; note that we change our variables from $\boldsymbol{x}$ to $\boldsymbol{z}$ to reflect the fact that our proof works over the complex numbers.

Theorem 5.2 (Re$\{\phi\}$ has a strict minimum). *Let A and ϕ be complex-valued analytic functions on a compact neighborhood $\mathcal{N}$ of the origin in $\mathbb{R}^d$. Suppose that the real part of ϕ is nonnegative on $\mathcal{N}$ and vanishes only at the origin, and that the Hessian matrix $\mathcal{H}$ of ϕ at the origin is nonsingular. Then $\mathcal{I}(\lambda) = \int_{\mathcal{N}} A(\boldsymbol{z})e^{-\lambda\phi(\boldsymbol{z})}$ has an asymptotic expansion*

$$\mathcal{I}(\lambda) \approx \sum_{\ell\geq 0} c_\ell \lambda^{-d/2-\ell} \tag{5.3}$$

with leading coefficient

$$c_0 = A(\mathbf{0})\frac{(2\pi)^{d/2}}{\sqrt{\det\mathcal{H}}}, \tag{5.4}$$

where $\sqrt{\det\mathcal{H}}$ is the product of the principal square roots of the eigenvalues of $\mathcal{H}$.

Exercise 5.2. Show that the expansion in (5.3) can be written as

$$\mathcal{I}(\lambda) \approx \frac{A(\mathbf{0})}{\sqrt{\det 2\pi\lambda\mathcal{H}}}\sum_{\ell\geq 0} c'_\ell\lambda^{-\ell}, \tag{5.5}$$

where $c'_0 = 1$.

Exercise 5.3. Let $c > 0$, where c is independent of λ. What happens to the right side of (5.5) if we change ϕ to ϕ/c and λ to $c\lambda$?

When $\text{Re}\{\phi\}$ is strictly positive except at a finite number of points in the domain of integration of $\mathcal{I}(\lambda)$ then, up to an asymptotically negligible error, we can express $\mathcal{I}(\lambda)$ as a sum of integrals localized to neighborhoods of these vanishing points. A chain of integration in a manifold which can be localized to arbitrarily small neighborhoods can be pulled back to correspondingly localized integrals over $\mathbb{R}^d$, which is why the above results are stated for integrals over neighborhoods of the origin in $\mathbb{R}^d$. In dimension greater than one, however, it is possible for an analytic function to have a real part that vanishes along a set of positive dimension without vanishing everywhere, meaning it may not be possible to localize.

Exercise 5.4. Over which of the following chains in $\mathbb{C}^2$ does $\phi(x, y) = x^2 + y^2$ have a nonnegative real part and, among those where this holds, which have the real part of ϕ vanishing only at the origin?

(a) a small neighborhood of $\mathbf{0}$ in the real $\times$ real subspace of $\mathbb{C}^2$
(b) a small neighborhood of $\mathbf{0}$ in the imaginary $\times$ imaginary subspace of $\mathbb{C}^2$
(c) a small neighborhood of $\mathbf{0}$ in the diagonal subspace $\{(x, y) \in \mathbb{C}^2 : x = y\}$ of $\mathbb{C}^2$
(d) a small neighborhood of $\mathbf{0}$ in the linear subspace of $\mathbb{C}^2$ spanned over $\mathbb{R}$ by $(1 + i, 0)$ and $(0, 1 + i)$

Such difficulties lead us to state our most general results in the language of stratified spaces and vector flows. A ***vector flow*** on a space X is the solution $\Psi : X \times [0, T] \to X$ to a differential equation $(d/dt)\Psi(x, t) = \boldsymbol{v}(\Psi(\boldsymbol{x}, t))$, where $\boldsymbol{v}$ is a vector field on X (see, e.g., Lemma 5.14); an ***upward gradient flow*** is defined by the vector field $\boldsymbol{v} = \nabla\phi$, while a ***downward gradient flow*** is defined by $\boldsymbol{v} = -\nabla\phi$. These constructions and results, summarized in Appendix D, have been around for over 50 years, though they are not very well known outside of differential topology and singularity theory. To ease exposition we now state our main result using some terminology to be defined in Section 5.4, where the result is proved. If X is an oriented stratified space then $\boldsymbol{q} \in X$ is a ***critical point (in the stratified sense)*** of an analytic map $\phi : X \to \mathbb{C}$ if $\boldsymbol{q}$ lies in a stratum $\mathcal{S}$ and the differential $d\phi|_{\mathcal{S}}$ at $\boldsymbol{q}$ is zero. Appendix D contains a more complete explanation of the vector flows we use.

Theorem 5.3 (minimum of Re$\{\phi\}$ is not strict but there are finitely many critical points). *Let $\mathcal{V}$ be a smooth complex $(d-k)$-dimensional algebraic variety and suppose that*

(i) *$X = \Delta^p \times M^{d-k}$ is a stratified space of dimension $p+d-k$ in $\mathbb{C}^{p+d-k}$, where $\Delta^p \subseteq \mathbb{R}^p$ is the standard p-simplex and $M^{d-k} \subseteq \mathcal{V} \subseteq \mathbb{C}^d$ is a smooth $(d-k)$-dimensional analytic submanifold of $\mathcal{V}$,*
(ii) *the closure of X is represented by an analytic $(p+d-k)$-chain C,*
(iii) *$\phi : X \to \mathbb{C}$ is an analytic map with $\min_{x \in X} \operatorname{Re} \phi(x) = 0$, and*
(iv) *$\eta = A(\mathbf{z})\, d\mathbf{z}$ is a holomorphic $(p+d-k)$-form on X.*

Assume that the set G of critical points of ϕ on C is finite and that the subset $G' \subseteq G$ where $\operatorname{Re} \phi$ vanishes are all in strata of dimension $p+d-k$. Suppose also that

(v) *$\det \mathcal{H}(\mathbf{q}) \neq 0$ for all $\mathbf{q} \in G'$, where $\mathcal{H}$ is the Hessian matrix for ϕ in some local coordinates near $\mathbf{q}$,*
(vi) *the imaginary part (and thus all) of $\phi(\mathbf{q})$ is zero for all $\mathbf{q} \in G'$,*
(vii) *the boundary ∂C is supported on strata of dimension at most $p+d-k-1$, with its simplices σ_j analytic orientation preserving maps having disjoint interiors, and*
(viii) *the elements of G' lie in the interiors of distinct simplices σ_j.*

Then the integral

$$\mathcal{I}(\lambda) = \int_C e^{-\lambda \phi(\mathbf{z})}\, \eta$$

has an asymptotic expansion

$$\mathcal{I}(\lambda) \approx \sum_{\ell=0}^{\infty} c_\ell \lambda^{-(d-k)/2-\ell} \tag{5.6}$$

as $\lambda \to \infty$, with leading term

$$c_0 = (2\pi)^{d/2} \sum_{q \in G'} \frac{A(\mathbf{q})}{\sqrt{\det \mathcal{H}(\mathbf{q})}} . \tag{5.7}$$

The sign for the square root of the determinant is computed by choosing a parametrization Υ for X near $\mathbf{q}$ by a neighborhood of the origin in $\mathbb{R}^{p+d-k}$ and then multiplying the principal square roots of the eigenvalues of ϕ in these coordinates with the Jacobian determinant $\det d\Upsilon(\mathbf{q})$.

Without the assumption that the imaginary part of $\phi(\mathbf{q})$ vanishes for $\mathbf{q} \in G'$ the formula (5.7) holds when each summand is multiplied by the term $e^{-\lambda\phi(\mathbf{q})}$ with modulus 1, making $c_0 = c_0(\lambda)$ dependent on λ but having bounded modulus.

Remark 5.4. We apply Theorem 5.3 at two points in this book: first with $p = 0$ and $k = 1$ to derive Theorem 9.12, and second in its more general form to prove asymptotics for multiple points in Chapter 10. In the statement of the theorem, the first four conditions are geometric conditions ensuring the existence of the necessary deformations and analytic extensions, while the last four ensure we know how to do computations.

Example 5.5. Let $X = I \times S^1$, where I is the interval $[-1, 1]$ and S^1 is the unit circle parametrized by $\theta \in [-\pi, \pi]$ with the endpoints identified, so that X can be nicely embedded in $\mathbb{C}^2$. We apply Theorem 5.3 in the case $p = k = 1$ and $d = 2$, with $A(x, y) = 1$ and $\phi : X \to \mathbb{C}$ defined by

$$\phi(t, \theta) = K\theta^2 + iL\theta t \tag{5.8}$$

for real numbers $K > 0$ and L. The phase ϕ is analytic on X, and the 2-chain C representing X can be any cell complex with a subcomplex $I \times \mathcal{N}$ for a compact neighborhood $\mathcal{N}$ of $\theta = 0$ in S^1. There is a single critical point $\boldsymbol{p} = (0, 0)$, at which $\operatorname{Re}\phi$ vanishes, so $G' = G = \{\boldsymbol{p}\}$.

Note that the strip $I \times \{0\}$ on which the phase function vanishes extends out to the bounding circles of the cylinder X, so we are not in a case where the magnitude of the integrand is small away from $\boldsymbol{p}$, and Theorem 5.2 does not apply. The Hessian matrix of ϕ at $\boldsymbol{p}$ is $\left(\begin{smallmatrix} 2K & iL \\ iL & 0 \end{smallmatrix}\right)$, so Theorem 5.3 implies

$$\mathcal{I}(\lambda) = \int_{\mathcal{N}\times I} e^{-\lambda\phi(\boldsymbol{x})}\, d\boldsymbol{x} \sim \frac{2\pi}{\lambda\,|L|}.$$

The choice of sign on the term $\sqrt{L^2} = |L|$ is arbitrary and depends on properly orienting $\mathcal{N} \times I$ for the application at hand. ◁

Exercise 5.5. Let X be the real sphere $\{(x, y, z) \in \mathbb{R}^3 : x^2 + y^2 + z^2 = 1\} \subset \mathbb{C}^3$ and let $\phi(x, y, z) = z^2 + ix^2$.

(a) Identify the sets G and G'.
(b) Determine $\mathcal{H}(\boldsymbol{q})$ for $\boldsymbol{q} \in G'$.
(c) Determine $c_0(\lambda)$ when $A(x, y, z) = 1 + x + y$.

5.2 Standard phase

As in the one-dimensional case, we begin with the simplest phase function and a monomial amplitude. We first state a formula for the one-dimensional integral with amplitude $A(x) = x^{2n}$ and standard phase in terms of the explicit constants

$$\beta_{2n} = \sqrt{\pi}\,\frac{(2n)!}{n!\,4^n}.$$

Proposition 5.6. *For all $n \in \mathbb{N}$,*

$$\int_{-\infty}^{\infty} x^{2n} e^{-x^2}\,dx = \beta_{2n}.$$

Proof For $n = 0$ this is just the standard Gaussian integral

$$\int_{-\infty}^{\infty} e^{-x^2}\,dx = \sqrt{\pi},$$

and the general result follows by induction. Indeed, rewriting

$$\int_{-\infty}^{\infty} x^{2n} e^{-x^2}\,dx = \int_{-\infty}^{\infty} \frac{-x^{2n-1}}{2}\left(-2x\,e^{-x^2}\,dx\right)$$

and applying integration by parts gives

$$\begin{aligned}\int_{-\infty}^{\infty} x^{2n} e^{-x^2}\,dx &= \frac{2n-1}{2}\int_{-\infty}^{\infty} x^{2n-2}\,e^{-x^2}\,dx\\ &= \frac{2n-1}{2}\cdot\sqrt{\pi}\cdot\frac{(2n-2)!}{(n-1)!\,4^{n-1}}\\ &= \sqrt{\pi}\,\frac{(2n)!}{n!\,4^n}\end{aligned}$$

by induction, as claimed. □

Corollary 5.7 (monomial integral). *Let $S(\boldsymbol{z}) = \sum_{j=1}^{d} z_j^2$ and $\boldsymbol{r} \in \mathbb{N}^d$. Then*

$$\int_{\mathbb{R}^d} \boldsymbol{z}^{\boldsymbol{r}} e^{-\lambda S(\boldsymbol{z})}\,d\boldsymbol{z} = \beta_{\boldsymbol{r}}\lambda^{-(d+|\boldsymbol{r}|)/2}$$

for any $\lambda > 0$, where $\beta_{\boldsymbol{r}} = \prod_{j=1}^{d}\beta_{r_j}$ if all the components r_j are even and $\beta_{\boldsymbol{r}} = 0$ otherwise.

Proof If $n \in \mathbb{N}$ then making the change of variables $x = y\lambda^{-1/2}$ and applying Proposition 5.6 proves

$$\int_{-\infty}^{\infty} x^{2n}\,e^{-\lambda x^2}\,dx = \lambda^{-1/2-n}\int_{-\infty}^{\infty} y^{2n} e^{-y^2}\,dy = \lambda^{-1/2-n}\beta_{2n},$$

while $\int_{-\infty}^{\infty} x^{2n+1}\,e^{-\lambda x^2} = 0$ as its integrand is odd. The integral under consideration factors as

$$\int_{\mathbb{R}^d} \boldsymbol{z}^{\boldsymbol{r}} e^{-\lambda S(\boldsymbol{z})}\,d\boldsymbol{z} = \prod_{j=1}^{d}\left[\int_{-\infty}^{\infty} z_j^{r_j}\,e^{-\lambda z_j^2}\,dz_j\right],$$

and the result follows from simplifying each factor. □

Before establishing Theorem 5.1 we also need to bound the error terms that appear.

Lemma 5.8 (Big-O Lemma). *Let A be a measurable function satisfying $A(\boldsymbol{z}) = O(|\boldsymbol{z}|^r)$ at the origin. Then the integral of $A(\boldsymbol{z})e^{-\lambda S(\boldsymbol{z})}$ over any compact set K may be bounded from above by*

$$\int_K A(\boldsymbol{z})e^{-\lambda S(\boldsymbol{z})}\,d\boldsymbol{z} = O\left(\lambda^{-(d+r)/2}\right).$$

The implied constant on the right goes to zero as the implied constant in the hypothesis $A(\boldsymbol{z}) = O(|\boldsymbol{z}|^r)$ goes to zero.

Proof Because K is compact and $A(\boldsymbol{z}) = O(|\boldsymbol{z}|^r)$ at the origin, there exists a constant $C > 0$ such that $|A(\boldsymbol{z})| \leq C|\boldsymbol{z}|^r$ on all of K. Let

$$K_0 = \left\{\boldsymbol{z} \in K : |\boldsymbol{z}| \leq \lambda^{-1/2}\right\}$$

denote the intersection of K with the ball of radius $\lambda^{-1/2}$, and for $n \geq 1$ let

$$K_n = \left\{\boldsymbol{z} \in K : 2^{n-1}\lambda^{-1/2} \leq |\boldsymbol{z}| \leq 2^n\lambda^{-1/2}\right\}$$

denote the intersection of K with a shell. On K_0 we have $|A(\boldsymbol{z})| \leq C\lambda^{-r/2}$ and $|e^{-\lambda S(\boldsymbol{z})}| \leq 1$, so

$$\int_{K_0} A(\boldsymbol{z})e^{-\lambda S(\boldsymbol{z})}\,d\boldsymbol{z} \leq \text{Vol}(K_0)\,C\lambda^{-r/2} = \frac{(\pi/\lambda)^{d/2}}{\Gamma\left(\frac{d}{2}+1\right)}\,C\,\lambda^{-r/2}.$$

For $n \geq 1$, when $\boldsymbol{z} \in K_n$ we have the upper bounds

$$|A(\boldsymbol{z})| \leq 2^{rn}C\lambda^{-r/2} \qquad \text{by upper bound on } |\boldsymbol{z}|$$

$$e^{-\lambda S(\boldsymbol{z})} \leq e^{-2^{2n-2}} \qquad \text{by lower bound on } |\boldsymbol{z}|$$

$$\text{Vol}(K_n) \leq \frac{2^{dn}\pi^{d/2}}{\Gamma\left(\frac{d}{2}+1\right)}\lambda^{-d/2} \qquad \text{by upper bound on } |\boldsymbol{z}|.$$

Thus, if $C' = 1 + \sum_{n\geq 0} 2^{(d+r)n}e^{-2^{2n-2}} < \infty$ then

$$\left|\int_K A(\boldsymbol{z})e^{-\lambda S(\boldsymbol{z})}\,d\boldsymbol{z}\right| \leq \sum_{n=0}^{\infty}\left|\int_{K_n} A(\boldsymbol{z})e^{-\lambda S(\boldsymbol{z})}\,d\boldsymbol{z}\right| \leq \frac{\pi^{m/2}}{\Gamma\left(\frac{d}{2}+1\right)}\,C\,C'\,\lambda^{-(d+r)/2}$$

with the right-hand side going to zero with the implied constant C, as claimed. □

We are now ready to prove Theorem 5.1.

Proof of Theorem 5.1. Writing $A(\boldsymbol{z})$ as a power series up to degree N plus a remainder term,

$$A(\boldsymbol{z}) = \left(\sum_{n=0}^{N} \sum_{|\boldsymbol{r}|=n} a_{\boldsymbol{r}} \boldsymbol{z}^{\boldsymbol{r}} \right) + R(\boldsymbol{z}),$$

where $R(\boldsymbol{z}) = O\left(|\boldsymbol{z}|^{N+1}\right)$. Using Corollary 5.7 to integrate all the monomial terms and Proposition 5.8 to bound the integral of $R(\boldsymbol{z})e^{-\lambda S(\boldsymbol{z})}$ shows that

$$\mathcal{I}(\lambda) = \sum_{n=0}^{N} \sum_{|\boldsymbol{r}|=n} a_{\boldsymbol{r}} \beta_{\boldsymbol{r}} \lambda^{-(d+n)/2} + O\left(\lambda^{-(d+N+1)/2}\right),$$

proving the stated asymptotic expansion. □

Exercise 5.6. Let $M(t) = \sup\{|A(\boldsymbol{z})| : |\boldsymbol{z}| = t\}$ and let K_R be a ball of fixed radius R. Prove a version of Theorem 5.1 using the fact that if $M(t) = O(t^{\alpha})$ then

$$\int_{K_R} A(\boldsymbol{z})e^{-\lambda S(\boldsymbol{z})}\, d\boldsymbol{z} \leq \int_0^R e^{-\lambda t} M(t)\, dV(K_t),$$

where $dV(K_t) = c_t t^{d-1} dt$ is the volume of a spherical shell of thickness dt and radius t in d dimensions.

5.3 Real part of phase has a strict minimum

We extend our results beyond integrals with standard quadratic phases using complex analytic techniques. If $\mathcal{N}$ is a neighborhood of the origin in $\mathbb{R}^d$ and $\phi : \mathcal{N} \to \mathbb{C}$ is analytic on $\mathcal{N}$ then ϕ can be viewed as a complex analytic function on a neighborhood $\mathcal{N}_{\mathbb{C}}$ of the origin in $\mathbb{C}^d$ using its power series expansion. Suppose that $\phi(\mathbf{0}) = 0$ and the real part of ϕ is nonnegative on $\mathcal{N}$, so that the gradient of ϕ must vanish at the origin. Our first key lemma is that, under an assumption of nondegeneracy, we can change variables so that ϕ becomes the standard quadratic form.

Lemma 5.9 (Complex Morse Lemma). *If $\phi(\boldsymbol{x})$ has vanishing gradient and nonsingular Hessian $\mathcal{H}$ at the origin then there is a bi-holomorphic change of variables $\boldsymbol{x} = \psi(\boldsymbol{y})$ around $\boldsymbol{x} = \boldsymbol{y} = \mathbf{0}$ such that $\phi(\psi(\boldsymbol{y})) = S(\boldsymbol{y}) = \sum_{j=1}^{d} y_j^2$. The Jacobian matrix $J_\psi = d\psi(\mathbf{0})$ satisfies* $(\det J_\psi)^2 = \frac{2^d}{\det \mathcal{H}}$.

Our proof of Lemma 5.9 is adapted from the proof of the real version given in [Ste93, VIII:2.3.2].

Proof To prove the claim about the Jacobian determinant, we apply the chain rule to the equation $\phi(\psi(\boldsymbol{y})) = S(\boldsymbol{y})$ and conclude that the Hessian matrix of S at the origin equals $J_\psi^T \mathcal{H} J_\psi$. The Hessian of S is twice the identity matrix, so $2I = J_\psi^T \mathcal{H} J_\psi$ and taking determinants gives the stated result.

To prove the change of variables, we begin by writing

$$\phi(\boldsymbol{x}) = \sum_{j,k=1}^{d} x_j x_k \phi_{j,k}(\boldsymbol{x})$$

for analytic functions $\phi_{j,k} = \phi_{k,j}$ with constant terms $\phi_{j,k}(\mathbf{0}) = \mathcal{H}_{j,k}/2$. There is plenty of freedom, but a convenient choice is to take

$$x_j x_k \phi_{j,k}(\boldsymbol{x}) = \sum_{|\boldsymbol{r}|\geq 2} \frac{r_j(r_k - \delta_{j,k})}{|\boldsymbol{r}|(|\boldsymbol{r}|-1)} a_{\boldsymbol{r}} \boldsymbol{x}^{\boldsymbol{r}}, \tag{5.9}$$

where $a_{\boldsymbol{r}}$ are the Taylor coefficients of ϕ at $\boldsymbol{x}$ and $\delta_{j,k} = 1$ if $j = k$ and 0 otherwise. For fixed $\boldsymbol{r}$ it is easy to check that

$$\sum_{1\leq j,k\leq d} \frac{r_j(r_k - \delta_{j,k})}{|\boldsymbol{r}|(|\boldsymbol{r}|-1)} = 1,$$

so that $\phi(\boldsymbol{x}) = \sum_{j,k} x_j x_k \phi_{j,k}(\boldsymbol{x})$, and matching coefficients on the terms of order precisely two verifies $\phi_{j,k}(\mathbf{0}) = \mathcal{H}_{j,k}/2$. We may assume without loss of generality that $\phi_{j,j}(\mathbf{0}) = \mathcal{H}_{j,j} \neq 0$ for all j, because there is always a unitary map U such that the Hessian of $\phi \circ U$ has nonvanishing diagonal entries, and if $(\phi \circ U) \circ \psi_0 = S$ for some ψ_0 then $\phi \circ \psi = S$, where $\psi = U \circ \psi_0$.

We conclude with an induction. Since we are assuming that $\phi_{1,1}(\mathbf{0}) \neq 0$, the reciprocal $1/\phi_{1,1}(\boldsymbol{x})$ and a branch of $\sqrt{\phi_{1,1}(\boldsymbol{x})}$ are both analytic in a neighborhood of the origin. If

$$y_1 = \sqrt{\phi_{1,1}(\boldsymbol{x})}\left[x_1 + \sum_{k>1} \frac{x_k \phi_{1,k}(\boldsymbol{x})}{\phi_{1,1}(\boldsymbol{x})}\right]$$

then the terms of y_1^2 of total degree at most one in $x_2, \ldots, x_d$ match those of ϕ, since

$$\phi(\boldsymbol{x}) - y_1^2 = \sum_{j,k=2}^{d} x_j x_k \left(\phi_{j,k}(\boldsymbol{x}) - \frac{\phi_{1,j}(\boldsymbol{x})\phi_{1,k}(\boldsymbol{x})}{\phi_{1,1}(\boldsymbol{x})}\right). \tag{5.10}$$

In the new coordinates $y_1, x_2, x_3, \ldots, x_d$, the Hessian matrix of ϕ is a $(1, d-1)$ block matrix, where the submatrix $\mathcal{H}^{(1)}$ that corresponds to the variables $x_2, \ldots, x_d$ has determinant $\det \mathcal{H}^{(1)} = \det \mathcal{H}/\phi_{1,1} \neq 0$. In fact, if $\mathcal{H}$ is real positive definite then so is $\mathcal{H}^{(1)}$, provided the correct branch of the square root is chosen. Equation (5.10) thus writes ϕ in the form

$$\phi(\boldsymbol{x}) = y_1^2 + \sum_{j,k\geq 2} x_j x_k \phi_{j,k}^{(1)}(\boldsymbol{x}) \tag{5.11}$$

for some analytic functions $\phi_{j,k}^{(1)}$ satisfying $\phi_{j,k}^{(1)}(\mathbf{0}) = \mathcal{H}_{j,k}^{(1)}/2$. By induction, if we assume that

$$\phi(\boldsymbol{x}) = \sum_{j=1}^{r-1} y_j^2 + \sum_{j,k\geq r} x_j x_k \phi_{j,k}^{(r-1)}(\boldsymbol{x})$$

for some $1 \leq r \leq d$ then setting

$$y_r = \sqrt{\phi_{r,r}(\boldsymbol{x})}\left[x_r + \sum_{k>r} \frac{x_k \phi_{r,k}^{(r-1)}(\boldsymbol{x})}{\phi_{r,r}^{(r-1)}(\boldsymbol{x})}\right]$$

gives

$$\phi(\boldsymbol{x}) = \sum_{j=1}^{r} y_j^2 + \sum_{j,k\geq r+1} x_j x_k \phi_{j,k}^{(r)}(\boldsymbol{x})$$

for some analytic functions $\phi_{j,k}^{(r)}$ satisfying $\phi_{j,k}^{(r)}(\mathbf{0}) = \mathcal{H}_{j,k}^{(r)}/2$ with $\mathcal{H}^{(r)}$ nonsingular, leading in the end to a sequence of bi-holomorphic changes of variables writing $\phi(\boldsymbol{x}) = \sum_{j=1}^{d} y_j^2$ as claimed. □

Exercise 5.7. Use the Complex Morse Lemma to find a bi-holomorphic change of variables turning $\phi(x, y, z) = xy + yz + zx + xyz$ into the standard quadratic form $S(u, v, w)$.

We are now ready to prove Theorem 5.2.

Proof of Theorem 5.2. The convergent power series expansion of ϕ allows us to extend it to a neighborhood of the origin in $\mathbb{C}^d$. Under the change of variables ψ from Lemma 5.9,

$$\begin{aligned}\mathcal{I}(\lambda) &= \int_{\psi^{-1}\mathcal{N}} A(\psi(\boldsymbol{y}))e^{-\lambda S(\boldsymbol{y})}(\det d\psi(\boldsymbol{y}))\, d\boldsymbol{y} \\ &= \int_{\psi^{-1}\mathcal{N}} \tilde{A}(\boldsymbol{y})e^{-\lambda S(\boldsymbol{y})}\, d\boldsymbol{y}\,.\end{aligned}$$

We need to check that we can move the chain of integration $C = \psi^{-1}\mathcal{N}$ back to the real plane. If we can, then applying the expansion from Theorem 5.1 and noting that the terms with odd values of $|\boldsymbol{r}|$ all vanish yields the desired expansion in powers $\lambda^{-d/2-\ell}$.

Let $h(\boldsymbol{z}) = \text{Re}\{S(\boldsymbol{z})\}$. Our assumption that the real part of ϕ is nonnegative on $\mathcal{N}$ and vanishes only at the origin implies that the chain C lies in the region

$\{z \in \mathbb{C}^d : h(z) > 0\}$ except when $z = 0$, meaning there exists $\varepsilon > 0$ such that $h(z) \geq \varepsilon > 0$ for all $z \in \partial C$. Let

$$H(z,t) = \text{Re}\{z\} + (1-t)\, i \,\text{Im}\{z\}$$

be a homotopy from the identity map to the projection map $\pi(z) = \text{Re}\{z\}$. For any chain σ, the homotopy H induces a chain homotopy $H(\sigma)$ satisfying

$$\partial H(\sigma) = \sigma - \pi(\sigma) + H(\partial\sigma).$$

Taking $\sigma = C$, and using the fact that $h(H(z,t)) \geq h(z)$, we see there is a d-chain C' supported on $\{z \in \mathbb{C}^d : h(z) > \varepsilon\}$ and a $(d+1)$-chain $\mathcal{D}$ such that

$$\partial\mathcal{D} = C - \pi(C) + C'.$$

Stokes's Theorem (Theorem A.24 in Appendix A) implies that

$$\int_{\partial\mathcal{D}} \omega = \int_{\mathcal{D}} d\omega = 0$$

for any holomorphic d-form ω, which means

$$\int_C \omega = \int_{\pi(C)} \omega - \int_{C'} \omega .$$

When $\omega = \tilde{A}(\boldsymbol{y})e^{-\lambda S(\boldsymbol{y})}\, d\boldsymbol{y}$ the integral over C' is $O\left(e^{-\lambda\varepsilon}\right)$, giving

$$\mathcal{I}(\lambda) = \int_{\pi(C)} \tilde{A}(\boldsymbol{y})e^{-\lambda S(\boldsymbol{y})}\, d\boldsymbol{y} + O\left(e^{-\varepsilon\lambda}\right).$$

The projection π maps any real d-manifold in $\mathbb{C}^d$ locally diffeomorphically into $\mathbb{R}^d$ wherever its tangent space is not parallel to the imaginary subspace of $\mathbb{C}^d$. Because $h(z) \geq 0$ on C, the tangent space to the support of C at the origin is not parallel to the imaginary subspace. The tangent space varies continuously, so in a neighborhood of the origin π is a diffeomorphism. In particular, the chain $\pi(C)$ is a disk Δ in $\mathbb{R}^d$ plus a collection of points whose image under h is bounded above zero (which will contribute an exponentially negligible term to dominant asymptotic behavior). Observing that

$$\tilde{A}(\mathbf{0}) = A(\mathbf{0}) \det(d\psi(\mathbf{0})) = \frac{2^d A(\mathbf{0})}{\sqrt{\det \mathcal{H}}}$$

finishes the proof, up to the choice of sign of the square root corresponding to the orientation of Δ.

The stated sign choice of $\tilde{A}(\mathbf{0})$ in this theorem can be verified by proving that the linear map $d\pi \circ d\psi^{-1}$ at the origin sends the standard basis of $\mathbb{R}^d$ to another positively oriented basis if and only if $\det(d\psi(\mathbf{0}))$ is the product of the principal square roots of the eigenvalues of $\mathcal{H}$. We state and prove the necessary technical result in Lemma 5.10 below, which completes our proof of this theorem. □

Lemma 5.10. *Let* $W = \{z \in \mathbb{C}^d : \mathrm{Re}\{S(z)\} > 0\}$ *and suppose that* $\alpha \in GL_d(\mathbb{C})$ *maps* $\mathbb{R}^d$ *into* $\overline{W}$. *If* $M = \alpha^T \alpha$ *is the matrix representing the quadratic form* $S \circ \alpha$ *and* π *is the projection map from* $\mathbb{C}^d$ *onto* $\mathbb{R}^d$ *then* $\pi \circ \alpha$ *is orientation preserving on* $\mathbb{R}^d$ *if and only if* $\det \alpha$ *is the product of the principal square roots of the eigenvalues of* M *(rather than the negative of this).*

Proof First suppose $\alpha \in GL_d(\mathbb{R})$. Then M has positive eigenvalues, and the product of their principal square roots is positive. The map π is the identity on $\mathbb{R}^d$, so the claimed statement boils down to saying that α preserves orientation if and only if it has positive determinant, which is true by definition. In the general case, let $\alpha_t = \pi_t \circ \alpha$, where $\pi_t(z) = \mathrm{Re}\{z\} + (1 - t)\,\mathrm{Im}\{z\}$. Since $\pi_t(\mathbb{R}^d) \subseteq \overline{W}$ for all $0 \leq t \leq 1$, the matrix $M_t = \alpha_t^T \alpha_t$ always has eigenvalues with nonnegative real parts. The product of the principal square roots of the eigenvalues is a continuous function on the set of nonsingular matrices with no negative real eigenvalues. The determinant of α_t is a continuous function of t, and we have seen it agrees with the product of principal square roots of eigenvalues of M_t when $t = 1$ (the real case), so by continuity this is the correct sign choice for all $0 \leq t \leq 1$. Taking $t = 0$ proves the lemma. □

Exercise 5.8. Suppose ϕ is the logarithm of an analytic function, defined only up to the addition of $(2\pi i)n$ for $n \in \mathbb{Z}$. How does this affect the conclusion of Theorem 5.2?

5.4 General nondegenerate phase with finite critical set

In this section we prove Theorem 5.3 by moving the chain of integration of $\mathcal{I}$ so that $\mathrm{Re}\,\phi$ is minimized only at the finite set of critical points, then applying Theorem 5.2. We first remark on some differences from our previous arguments. While the chain of integration may be defined on a subspace $X \subseteq \mathbb{R}^d$, the form in the integrand will be extended to a neighborhood of X in $\mathbb{C}^d$ and the deformation in general will not be confined to X. To accomplish this, we define a complexification of X with analytic structure. Our deformation is defined by a smooth vector field $\boldsymbol{v}$ which, although not analytic, lies in the complex tangent bundle to the complexification of each stratum. The proof is broken into the following steps.

(1) Define the complexification $X \otimes \mathbb{C}$.
(2) Extend ϕ and η to a neighborhood U of X in $\mathbb{C}^d$.

(3) Construct a vector flow $\boldsymbol{v}$ on U, tangent to the strata of $X \otimes \mathbb{C}$ and vanishing precisely on G, such that $\langle \mathrm{Re}\, d\phi, \boldsymbol{v} \rangle > 0$ on $U \setminus G$.
(4) Show that $\int_C e^{-\lambda\phi}\eta = \int_{C'} e^{-\lambda\phi}\eta$, where C' is obtained from C by flowing along $\boldsymbol{v}$ for a short time.
(5) Use Theorem 5.2 on the deformed chain C'.

Step (2) follows in a straightforward manner from Step (1). Step (3) is where the special assumptions on X are used. This allows us to follow the methodology of locally constant vector fields and partitions of unity in [ABG70], rather than the more difficult methodology of controlled vector fields in [Mat70]. Step (4) is an application of Stokes's Theorem together with the crucial observation that part of the boundary of the chain representing the deformation lies in a complex manifold of dimension less than d. Step (5), once we reach it, is immediate.

Step 1: Complexification

Lemma 5.11. *Under the hypotheses of Theorem 5.3, there is a complex stratified space $X \otimes \mathbb{C}$ with strata $\mathcal{S} \otimes \mathbb{C}$ as $\mathcal{S}$ ranges over the strata of X, such that $X \otimes \mathbb{C}$ is a neighborhood of X in $\mathbb{C}^d$ and such that the chart maps $\psi \otimes \mathbb{C} : \mathbb{C}^k \to \mathcal{S} \otimes \mathbb{C}$ restricted to $\mathbb{R}^k$ are chart maps for $\mathcal{S}$.*

Proof We complexify Δ^p in $\mathbb{C}^p$ and M^{d-k} in $\mathbb{C}^d$ separately and take the product. First, we note that the set Δ^p is defined by $p + 1$ linear inequalities. Relaxing these by $\varepsilon > 0$ produces a neighborhood $\mathcal{D}$ of Δ^p in $\mathbb{R}^d$. Taking the product with a sufficiently small imaginary interval $[-\varepsilon i, \varepsilon i]$ in each of the p coordinates produces a complex stratified space that is a neighborhood of Δ^p in $\mathbb{C}^d$. Because of our assumptions, we can complexify M^{d-k} to $\mathcal{V}$. Taking the product of $F \otimes \mathbb{C}$ and $M^{d-k} \otimes \mathbb{C}$ as stratified spaces, where F is a face of Δ^p, produces $(F \times M) \otimes \mathbb{C}$; these are the strata $\mathcal{S} \otimes \mathbb{C}$ and fit together to form the stratified space $X \otimes \mathbb{C}$ satisfying the conclusion of the theorem. □

Remark 5.12. In Chapter 9, we always have $k = 1$ and M^{d-k} is a smooth open $(d-1)$-patch in $\mathcal{V}_Q$. In Chapter 10, we have $k = 0$ and M^d is a patch of a middle-dimensional torus in $\mathbb{C}^d \setminus \mathcal{V}$. In general, under the condition that the real tangent space spans a complex space of dimension $d - k$, one may define $M^{d-k} \otimes \mathbb{C}$ to be the *intrinsic complexification* of M, namely the smallest complex manifold containing M. This is known to exist [BER99], following a construction of [BW59], and will be a complex $(d-k)$-manifold.

Exercise 5.9. Suppose X is the circle in $\mathbb{R}^3$ defined by the real solutions to $x^2 + y^2 - 1 = z = 0$. Find the complexification $X \otimes \mathbb{C}$.

Step 2: Extending analytic maps

Proposition 5.13. *Let $\phi : X \to \mathbb{C}$ be an analytic map on an analytic stratified space X, and let ψ_1 and ψ_2 be chart maps whose ranges are overlapping domains $\mathcal{N}_1$ and $\mathcal{N}_2$ in X. As $\boldsymbol{y}$ takes values in the intersection of the ranges of $\psi_1 \otimes \mathbb{C}$ and $\psi_2 \otimes \mathbb{C}$, the function $\tilde{\phi} = \tilde{\phi}_j$ defined by*

$$\tilde{\phi}_j(\boldsymbol{y}) = \tau_j \circ (\psi_j \otimes \mathbb{C})^{-1}(\boldsymbol{y}) \tag{5.12}$$

is independent of j, where τ_j represents analytic continuation of the map $\phi \circ \psi_j$ from the real parameter space to a complex neighborhood of it. This common value defines an analytic extension of ϕ on a neighborhood of X in $\mathbb{C}^d$.

Proof The maps $\tilde{\phi}_1$ and $\tilde{\phi}_2$ agree when $\boldsymbol{y} \in \mathcal{N}_1 \cap \mathcal{N}_2 \cap X$. Being analytic, they must agree in a neighborhood of X in $X \otimes \mathbb{C}$. □

Step 3: Constructing the vector flow

Lemma 5.14. *Under the assumptions of Theorem 5.3, there is a vector field $\boldsymbol{v}$ on a neighborhood U of X in $\mathbb{C}^d$, tangent to each complexified stratum $\mathcal{S} \otimes \mathbb{C}$ of $X \otimes \mathbb{C}$ and vanishing only on G, with the property that $\langle \mathrm{Re}\, d\phi, \boldsymbol{v} \rangle > 0$ at every point of $U \setminus G$. For sufficiently small s, there is a well-defined differential flow $\Psi : [0, 1] \times |C| \to \mathbb{C}^d$ satisfying $(d/dt)\Psi(t, x) = s\,\boldsymbol{v}(\Psi(t, x))$, and the map $\boldsymbol{x} \mapsto \Psi(\varepsilon, \boldsymbol{x})$ is a local diffeomorphism for sufficiently small $\varepsilon > 0$.*

Proof As described in Proposition D.14 of Appendix D, there is a diffeomorphic local product structure under the assumptions of Theorem 5.3. The argument discussed in Step 2 of the proof of Proposition D.13 in Appendix D implies that there is a Lipschitz vector flow $\boldsymbol{v}$ on a neighborhood U of X in $\mathbb{C}^d$ defined by (D.2.1). This vector field $\boldsymbol{v}$ is tangent to each complexified stratum, vanishes precisely on G, and satisfies $\langle \mathrm{Re}\, d\phi, \boldsymbol{v} \rangle > 0$ on $U \setminus G$.

The map $\boldsymbol{x} \mapsto \Psi(\varepsilon, \boldsymbol{x})$ is a local diffeomorphism for sufficiently small $\varepsilon > 0$ because $\boldsymbol{v}$ is smooth and bounded (see, for example, [Lee03, Proposition 9.12]). □

Step 4: Deforming the contour

Composing the flow from Lemma 5.14 with the simplex σ_j gives a homotopy from each σ_j to a new analytic simplex σ'_j, and summing the σ'_j defines a new chain C'. Because $\boldsymbol{v}$ is tangent to $\mathcal{S} \otimes \mathbb{C}$ for each stratum $\mathcal{S}$, the flow preserves each complexified stratum $\mathcal{S} \otimes \mathbb{C}$, and because ∂C is contained in the union of strata of dimensions at most $p + d - k - 1$, it follows that $\Psi(\partial C)$ is contained in the union of complexified strata of dimensions at most $p + d - k - 1$.

Lemma 5.15. *Using the above notation,*

$$\int_C e^{-\lambda\phi}\eta = \int_{C'} e^{-\lambda\phi}\eta\,. \tag{5.13}$$

Proof The homotopy Ψ defined by Lemma 5.14 is a chain with boundary $C - C' - [0,1]\times\partial C$. Since $e^{-\lambda\phi}\eta$ is a holomorphic $p+d-k$-form, it is annihilated by the differential, and Stokes's Theorem implies

$$\begin{aligned} 0 &= \int_\Psi d(e^{-\lambda\phi}\eta) \\ &= \int_{\partial\Psi} e^{-\lambda\phi}\eta \\ &= \int_{C'} e^{-\lambda\phi}\eta - \int_C e^{-\lambda\phi}\eta - \int_{\mathcal{D}} e^{-\lambda\phi}\eta\,, \end{aligned}$$

where $\mathcal{D}$ is a chain representing $[0,1]\times\partial C$. Because $e^{-\lambda\phi}\eta$ is a holomorphic $(p+d-k)$-form, its integral vanishes over any $p+d-k$-chain supported on a $(p+d-k-1)$-dimensional complex manifold (see Exercise A.15 of Appendix A), finishing the proof. □

Exercise 5.10. A simpler version of the deformation argument can be illustrated in two real variables. Suppose that $V(x,y)$ is a smooth function on $\mathbb{R}^2$ and $\eta = V_x(x,y)dx + V_y(x,y)dy$. Assume that a homotopy $H : [0,T]\times[0,1]$ carries the path $\alpha : [0,T] \to \mathbb{R}^2$ to the path $\beta : [0,T] \to \mathbb{R}^2$, meaning $H(t,0) = \alpha(t)$ and $H(t,1) = \beta(t)$ for all $t \in [0,T]$. If $f(u) = H(0,u)$ and $g(u) = H(T,u)$ then what conditions on df and dg guarantee that $\int_\alpha \eta = \int_\beta \eta$ by implying that the integrals over the paths traced out by the endpoints of the path as it moves from α to β are everywhere zero?

Step 5: Evaluating the integral on C'

From the construction of $\boldsymbol{v}$ we see that $C' = \sum_{j=1}^m \sigma'_j$, where each simplex σ'_j contains the same critical point $\boldsymbol{q}$ in its interior as σ_j. By Lemma 5.14 we know that $\operatorname{Re}\phi(\sigma'_j(\boldsymbol{x})) \geq \operatorname{Re}\phi(\sigma_j(\boldsymbol{x})) \geq 0$ for all $j \leq m$ and $\boldsymbol{x} \in \Delta^p$, where the first inequality is strict unless $\sigma_j(\boldsymbol{x})$ is a critical point. Thus, the image $|C'|$ is a stratified space represented analytically by $(p+d-k)$-simplices σ'_j for $1 \leq j \leq m$ and the function $\operatorname{Re}\phi$ is nonnegative and vanishes precisely on G'.

If σ'_j contains no point of G' then the modulus of $\int_{\sigma'_j} e^{-\lambda\phi(\boldsymbol{z})}\eta$ is bounded above by $Me^{-\lambda K}$, where $M = \max\{|A(\boldsymbol{x})| : \boldsymbol{x} \in \Delta^d\}$ and $K = \min\{\operatorname{Re}\phi(\boldsymbol{x}) : \boldsymbol{x} \in \Delta^d\}$. If σ'_j contains $\boldsymbol{q} \in G'$ then translating the preimage of $\boldsymbol{q}$ to the origin turns Δ^{p+d-k} into a neighborhood $\mathcal{N}$ of the origin on which A and $\tilde{\phi} = \phi\circ\sigma'_j$ are analytic and $\operatorname{Re}\tilde{\phi}(\boldsymbol{z}) \geq 0$, with equality only at the origin. Theorem 5.2

then gives an asymptotic expansion for the integral over each simplex. Invariance of $\det J(\Upsilon)/\sqrt{\mathcal{H}}$ under coordinate transformations equates the summand in (5.7) with the result (5.4) in Theorem 5.2, provided the correct sign in (5.7) is chosen to match (5.4). Summing these expansions then finishes the proof of Theorem 5.3. □

Exercise 5.11. The simplest case of Theorem 5.3 occurs when $p = 0, d = 1$, and $k = 0$. Take $\phi(z) = -z^2$, so $e^{-\lambda\phi(z)} = e^{\lambda z^2}$, and let M be the 1-chain consisting of the imaginary axis. Which sign in the equality

$$\int_M e^{\lambda z^2}\, dz = \pm i \sqrt{\frac{2\pi}{\lambda}}$$

is correct when parametrizing M by $z = it$ for $t \in \mathbb{R}$, and which sign is correct when parametrizing M by $z = -it$?

5.5 Higher order terms in the expansions

All of our explicit asymptotic computations ultimately reduce to computing terms in the asymptotic expansions of Fourier–Laplace integrals. It is therefore useful to have a closed formula for the higher-order terms that can appear. The following result is derived in [Hör83, Thm. 7.7.5] using smooth methods, and we refer the reader to that source for a proof.

Lemma 5.16. *Let $X \subseteq \mathbb{R}^d$ be an open neighborhood of the origin and let ϕ and A be smooth functions on X such that* $\text{Re}\{\phi\} \geq 0$ *on X. Further suppose that ϕ has a unique critical point on the support of A at the origin, that $\phi(\mathbf{0}) = 0$, and that the Hessian $\mathcal{H}$ of ϕ at $\mathbf{0}$ is nonsingular. Then for any positive integer M and $\lambda > 0$ there exist constants $L_k(A, \phi)$ such that*

$$\left| \int_X A(\boldsymbol{x}) e^{-\lambda\phi(\boldsymbol{x})} d\boldsymbol{x} - \lambda^{-d/2} \frac{(2\pi)^{d/2}}{\sqrt{\det \mathcal{H}}} \sum_{0 \leq k < M} \lambda^{-k} L_k(A, \phi) \right|$$
$$\leq C(\phi) \lambda^{-d/2-M} \sum_{|\beta| \leq 2M} \sup |\mathcal{D}^\beta A|,$$

where the constant $C(\phi)$ has a uniform bound when ϕ stays in a bounded set of $(3N+1)$-differentiable functions on X for which $\|\boldsymbol{x}\|/\|\nabla\phi(\boldsymbol{x})\|$ has a uniform bound. Setting

$$\underline{\phi}(\boldsymbol{x}) = \phi(\boldsymbol{x}) - (1/2)\boldsymbol{x} \cdot \mathcal{H} \cdot \boldsymbol{x}^T,$$

which vanishes to order three at $\mathbf{0}$*, we have*

$$L_k(A,\phi) = (-1)^k \sum_{0\le \ell\le 2k} \left. \frac{\mathcal{D}^{\ell+k}\left(A(\boldsymbol{x})\cdot \underline{\phi}(\boldsymbol{x})^{\ell}\right)}{2^{\ell+k}\ell!(\ell+k)!}\right|_{\boldsymbol{x}=\mathbf{0}}, \tag{5.14}$$

where $\mathcal{D}$ *is the differential operator*

$$\mathcal{D} = -\sum_{1\le i,j\le d} \left(\mathcal{H}^{-1}\right)_{ij} \partial_i\partial_j.$$

The total number of derivatives of A *in the term* $L_k(A,\phi)$ *is at most* $2k$ *and the total number of derivatives of* ϕ *is at most* $2k+2$. □

Interpreting this in our context, we obtain the following.

Corollary 5.17 (full expansion of Fourier–Laplace integral)**.** *Assume the hypotheses of Theorem 5.2 and suppose further that* ϕ *has a single critical point on* $\mathcal{N}$ *which lies at the origin. Then the constants in* (5.3) *satisfy*

$$c_k = \frac{(2\pi)^{d/2}}{\sqrt{\det \mathcal{H}}} L_k(A,\phi)$$

for each $k \ge 0$*, where* L_k *is as defined by* (5.14).

The fact that Lemma 5.16 requires only smoothness also makes it easy to localize.

Lemma 5.18. *If* ϕ *has a finite number of critical points on* $\mathcal{N}$ *where the hypotheses of Corollary 5.17 hold, then an asymptotic expansion for* $I(\lambda)$ *is obtained by summing the expansions corresponding to each of these critical points.*

Proof The proof of Theorem 5.2 shows that the contribution from the boundary of the domain of integration may be ignored – we may localize to a neighborhood $\mathcal{N}'$ of the critical point that is diffeomorphic to an open ball in $\mathbb{R}^d$. Replacing A by the product $A\alpha$ of A with a compactly supported smooth function α that is equal to one on $\mathcal{N}'$, the result follows from Lemma 5.16. □

Exercise 5.12. Verify that when $A(\mathbf{0}) \neq 0$ the leading constant of (5.3) matches the expression in Corollary 5.17.

Notes

A number of the results in this section originally appeared in [PW10]. In the case of purely real or imaginary phase, the results of this chapter are fairly standard; see [BH86; Won01] for real phase or [Ste93] for imaginary phase. We have not seen the complex phase result Theorem 5.2 stated before, though such a result was certainly understood to be true. The remaining statements and proofs via complex deformation methods are new, though not surprising.

Theorem 5.4.8 from the first edition has been replaced by Theorem 5.3. The proof has been split into two parts. The first part, involving stratified Morse deformations, is a standard stratified construction and is summarized in Appendix D. The second part, deforming a chain by complexifying and manipulating the chain through complex space, is new as far as we know and spelled out in Section 5.4.

According to B. Lamel *(personal communication)*, intrinsic complexifications of strata of any stratified space (see Remark 5.12) should fit together to form a complex stratified space, provided the (real) tangent spaces E at every point satisfy $E \cap iE = \{0\}$. This would imply that Lemma 5.11 and hence Theorem 5.3 holds for any stratified space of dimension m analytically embedded in $\mathbb{C}^m$. However, this appears to require some condition about extending the (real) chart maps of each stratum beyond its boundary, which can be done with Δ^p and M^{d-1} but not for arbitrary stratified spaces (where the boundary might be a singularity of the stratum). We could not find such an argument in the existing literature, which is why Theorem 5.3 is restricted to products of stratified spaces for which we have explicit complexifications.

The general approach, namely to stay in the analytic category and use deformations suggested by stratified Morse theory for complex spaces, is an extension of our treatment of univariate Fourier–Laplace integrals in Chapter 4. The main motivation for doing things the way we have is that the analysis of multiple points in Chapter 10 requires us to integrate over the product of a chain in $\mathcal{V}$ with an abstract simplex; when integrating terms with imaginary phase over manifolds with boundary, one needs a way to eliminate boundary terms. A result similar to Theorem 5.3 was proved in [PV19, Theorem 4.2], via an approach which avoids Morse theoretic contour deformation arguments, replacing these by iterated integrals and single parameter steepest descent curves.

The first edition of the book assumed a strong torality hypothesis when deriving asymptotics (see, e.g., Corollaries 9.2.4 and 9.2.9 there), which allowed use of known techniques in the case where the phase is purely imaginary and the contour of integration has no boundary. In the present edition this overly strong hypothesis has been replaced by a weaker notion for which Theorem 5.3 has been specifically designed (see Theorem 9.12).

Additional exercises

Exercise 5.13. Consider an integrand of the form $A(x, y, z)e^{-\lambda S(x,y,z)}\,dz$, where $A(x, y, z) = (x^2 + y^2 + z^2)^\alpha$ is not smooth at the origin when $\alpha \notin \mathbb{Z}_{>0}$. What kind of asymptotic estimate or expansion can be obtained in this case?

Exercise 5.14. Prove that for critical points on the boundary of the chain of integration, where the chain is locally diffeomorphic to a halfspace and ϕ has vanishing one-sided normal derivative, the conclusion of Theorem 5.2 holds with the leading coefficient multiplied by $1/2$. (See Example 10.66 in Chapter 10 for an application of this result.)

Exercise 5.15. (non-isolated critical points) Consider the integral

$$\int_{-\varepsilon}^{\varepsilon} \int_0^1 e^{-\lambda\phi(\theta,t)}\,dt\,d\theta,$$

where $\phi(\theta, t) = (1-t)g_1(\theta) + tg_1(\theta)$ and each g_i is analytic and vanishes to order 2 at $\theta = 0$, with positive second derivative. Calculate the first-order asymptotic as $\lambda \to \infty$ in terms of derivatives of g_1 and g_2 at 0. (This foreshadows the computations in Section 10.5, in particular Proposition 10.62.)

Exercise 5.16. Use the vanishing to order 3 of $\underline{\phi}$ at the origin to write a simpler expression for $L_1(A, \phi)$. Further simplify it in the cases where A vanishes to orders one, two, and three, respectively, at the origin.

Exercise 5.17. (alternative method to compute higher order terms) This exercise outlines an alternative way of computing higher order terms in the asymptotic expansion of a Fourier–Laplace integral. Assume for simplicity that the phase ϕ has a single critical point, occurring at $\mathbf{0}$, and that the amplitude A vanishes outside the closure of a neighborhood of $\mathbf{0}$. Let S be the standard quadratic $S(z) = \sum_{i=1}^d z_i^2$.

(i) Prove that when $\phi = S$, the differential operator

$$\sum_{|r|=k} \frac{\partial_1^{2r_1} \cdots \partial_d^{2r_d}}{4^k r_1! \cdots r_d!},$$

when applied to A and evaluated at $\mathbf{0}$, gives the coefficient c_k from Theorem 5.3.

(ii) The Complex Morse Lemma gives an analytic change of variables $z = \psi(y)$ such that $S(y) = (\phi \circ \psi)(y)$. Apply the result of (i) and solve a triangular system to compute the derivatives of ψ at $\mathbf{0}$. Note that this changes the amplitude function A to $(A \circ \psi) \det \psi'$.

(iii) Use Corollary 5.7 and steps (i) and (ii) to derive an explicit formula for asymptotics in the case $d = k = 1$. Check your result against the formula given in Corollary 5.17.

6

Laurent series, amoebas, and convex geometry

The theory of analytic combinatorics works by encoding a sequence of interest as a convergent series and applying tools from complex analysis. Although our results in previous chapters have focused on power series expansions that converge near the origin, a single rational (or meromorphic) function can be represented by multiple convergent series expansions over different domains in $\mathbb{C}^d$.

Example 6.1. The function $F(z) = 1/(1-z)$ has the convergent power series representation

$$F(z) = \sum_{n \geq 0} z^n \quad \text{when} \quad |z| < 1$$

and also the convergent *Laurent series* representation

$$F(z) = \frac{-1/z}{1 - \frac{1}{z}} = -\sum_{n \geq 0} z^{-n-1} \quad \text{when} \quad |z| > 1.$$

◁

In this chapter we introduce several constructions that, in addition to being used in later arguments, expand our viewpoint and help motivate how to make certain choices while computing on actual examples. The first section of this chapter develops the theory of convergent Laurent series, their domains of convergence, and a generalization of the Cauchy Integral Formula. Not only do Laurent series allow us to deal more naturally with generating functions that are ratios of Laurent polynomials, which appear frequently in combinatorial applications, they are also the natural level of generality with which to discuss *polynomial amoebas* – algebro-geometric objects that allow us to visualize the relationship between a rational function and its convergent series expansions. Furthermore, the algebraic computations that we undertake in order to determine asymptotics of a series expansion of a rational (or meromorphic) function naturally capture properties related to all convergent series expansions of the function. Thus, even if one wishes to stay within the realm of power series, a knowledge of Laurent expansions provides insight into the analysis that can help in real applications.

The second section of this chapter surveys basic properties of amoebas, and describes their use in analytic combinatorics, while the third section introduces some basics from convex geometry and describes their connection to polynomial amoebas. The fourth section studies the boundary points of amoebas and *minimal singularities*, which are very important to our ACSV analysis. The final section collects some related topics that are needed in later chapters.

Notational remarks

To simplify notation we extend the logarithm and exponential functions coordinatewise to vectors,

$$\log(\mathbf{z}) := (\log z_1, \dots, \log z_d)$$
$$\exp(\mathbf{x}) := (e^{x_1}, \dots, e^{x_d}),$$

and define the coordinatewise log-modulus ***Relog map***

$$\operatorname{Relog}(\mathbf{z}) := (\log |z_1|, \dots, \log |z_d|).$$

The inverse image of $\mathbf{x} \in \mathbb{R}^d$ under the Relog map is the ***exponential torus***

$$\mathbf{T}_e(\mathbf{x}) := \operatorname{Relog}^{-1}(\mathbf{x}) = \{\mathbf{z} \in \mathbb{C}^d : |z_j| = e^{x_j} \text{ for all } j\}.$$

Note that the torus $\mathbf{T}(\mathbf{z})$ passing through $\mathbf{z}$ is equal to $\mathbf{T}_e(\operatorname{Relog}(\mathbf{z}))$. As we will soon see, the image of the singular set of a multivariate meromorphic function under the Relog map provides important information about its analytic behavior.

6.1 Laurent series

A ***Laurent polynomial*** is a finite complex linear combination of monomials $\mathbf{z}^{\mathbf{r}}$ with integer exponents $\mathbf{r}$. The ***ring of Laurent polynomials*** $\mathbb{C}\left[\mathbf{z}, \mathbf{z}^{-1}\right] = \mathbb{C}\left[z_1, z_1^{-1}, \dots, z_d, z_d^{-1}\right]$ consists of the set of all Laurent polynomials, with addition and multiplication defined analogously to the polynomial ring $\mathbb{C}[\mathbf{z}]$. Since a Laurent polynomial has only a finite number of terms, the algebraic structure of $\mathbb{C}[\mathbf{z}, \mathbf{z}^{-1}]$ is similar to the algebraic structure of $\mathbb{C}[\mathbf{z}]$. On the other hand, the ***space of formal Laurent expressions*** in $\mathbf{z}$ is the complex vector space $\mathbb{L}(\mathbf{z})$ of formal complex linear combinations of monomials $\mathbf{z}^{\mathbf{r}}$ as $\mathbf{r}$ ranges over all of

$\mathbb{Z}^d$. Although the space $\mathbb{L}(\boldsymbol{z})$ is a module over the ring of Laurent polynomials, it is not itself a ring. Indeed, because a formal Laurent expression can have indices with negative sign and arbitrarily large magnitude, there is no natural product structure on $\mathbb{L}(\boldsymbol{z})$.

Example 6.2. If $G(z) = \sum_{n\in\mathbb{Z}} z^n$ then attempting to use the usual definition of series multiplication gives, for instance, that the constant term of $G(z)^2$ is $\sum_{n\in\mathbb{Z}} 1$, which is not well defined. We also note that $(1-z)G(z) = 0$ so, as a $\mathbb{C}[\boldsymbol{z}, \boldsymbol{z}^{-1}]$ module, elements of $\mathbb{L}(\boldsymbol{z})$ can have non-trivial annihilators. ◁

One approach to giving $\mathbb{L}(\boldsymbol{z})$ a ring structure involves restricting the indices that can appear, for instance requiring that all coordinates of all indices have a fixed lower bound (this is sufficient to develop a rich theory in one variable, but the multivariate case is more delicate). Here we work around these formal difficulties by restricting ourselves to the study of ***convergent Laurent series***. Let $\mathcal{D}$ be an open simply connected subset of $\mathbb{C}_*^d$ and let $\mathbb{L}_{\mathcal{D}}(\boldsymbol{z})$ denote the subspace of $\mathbb{L}(\boldsymbol{z})$ consisting of series that are absolutely convergent, uniformly on compact subsets of $\mathcal{D}$.

When discussing convergence of Laurent series, we always mean uniform convergence on compact sets of $\mathbb{C}_*^d$ (we disallow zero coordinates so that we do not need to worry about dividing by zero in terms with negative indices).

Definition 6.3. The ***domain of convergence of a Laurent series*** $\sum_{\boldsymbol{r}\in\mathbb{Z}^d} a_{\boldsymbol{r}}\boldsymbol{z}^{\boldsymbol{r}}$ is its open domain of absolute convergence, equal to the interior of the set of $\boldsymbol{z} \in \mathbb{C}_*^d$ for which $\sum_{\boldsymbol{r}\in\mathbb{Z}^d} |a_{\boldsymbol{r}}\boldsymbol{z}^{\boldsymbol{r}}|$ is finite.

As usual, we define a ***convex set*** to be a subset S of $\mathbb{R}^d$ with the property that for each $x, y \in S$, and each $\lambda \in [0,1]$, necessarily $(1-\lambda)x + \lambda y \in S$. It follows easily by induction that each convex set is closed under taking a ***convex combination*** of elements and the Chain Rule. That is, if $x_1, \ldots, x_n \in S$ and $\lambda_1, \lambda_n \in [0,1]$ with $\sum_i \lambda_i = 1$, then $\sum_i \lambda_i x_i \in S$.

Theorem 6.4 (domains of convergence of Laurent series).

(1) Let $F(\boldsymbol{z}) = \sum_{\boldsymbol{r}\in\mathbb{Z}^d} a_{\boldsymbol{r}}\boldsymbol{z}^{\boldsymbol{r}}$ be a Laurent series. Then the domain of convergence of F has the form $\mathcal{D} = \mathrm{Relog}^{-1}(B)$ for some convex open set $B \subset \mathbb{R}^d$.

(2) The function f defined by $f(\boldsymbol{z}) = F(\boldsymbol{z})$ for $\boldsymbol{z} \in \mathcal{D}$ is analytic in $\mathcal{D}$.

(3) Conversely, if B is an open convex subset of $\mathbb{R}^d$ and $f(\boldsymbol{z})$ is an analytic function on $\mathcal{D} = \mathrm{Relog}^{-1}(B)$ then there is a unique Laurent series $F(\boldsymbol{z}) \in \mathbb{L}_{\mathcal{D}}(\boldsymbol{z})$ converging to $f(\boldsymbol{z})$. For each $\boldsymbol{r} \in \mathbb{Z}^d$, the coefficient $a_{\boldsymbol{r}} = [\boldsymbol{z}^{\boldsymbol{r}}]F(\boldsymbol{z})$ is given by the Cauchy integral

$$a_{\boldsymbol{r}} = \left(\frac{1}{2\pi i}\right)^d \int_{\mathrm{Relog}^{-1}(\boldsymbol{x})} f(\boldsymbol{z})\boldsymbol{z}^{-\boldsymbol{r}-\boldsymbol{1}}\, d\boldsymbol{z} \tag{6.1}$$

for any $\boldsymbol{x} \in B$.

Proof To be given after Exercise 6.1 below. □

Theorem 6.4 allows us to *define* multiplication in $\mathbb{L}_{\mathcal{D}}(\boldsymbol{z})$ using multiplication of analytic functions: if ι is the identification map from $\mathbb{L}_{\mathcal{D}}(\boldsymbol{z})$ to the space of analytic functions on $\mathcal{D}$ then ι is invertible, and for $G, H \in \mathbb{L}_{\mathcal{D}}(\boldsymbol{z})$ we set $G \cdot H := \iota^{-1}(\iota(G) \cdot \iota(H))$. Similarly, if Q is everywhere nonvanishing on $\mathcal{D}$ then there is a unique Laurent series identified with the analytic function $1/Q(\boldsymbol{z})$. We may therefore identify a quotient of Laurent polynomials $P(\boldsymbol{z})/Q(\boldsymbol{z})$ with a unique convergent Laurent series by specifying a point $\boldsymbol{w} \in \mathbb{C}_*^d$ such that Q does not vanish on $\mathbf{T}(\boldsymbol{w})$ or, equivalently, a point $\boldsymbol{x} \in \mathbb{R}^d$ such that Q does not vanish on $\mathbf{T}_e(\boldsymbol{x}) = \text{Relog}^{-1}(\boldsymbol{x})$.

Exercise 6.1. Let $f(x, y, z) = 1/(1-x-y-z)$ and let B be a small neighborhood of the point $(0, 0, 2)$. Find a Laurent series for f converging on $\text{Relog}^{-1}(B)$. *Hint:* You can make $1/z$, x/z, and y/z simultaneously less than $\frac{1}{3}$ on $\text{Relog}^{-1}(B)$.

Theorem 6.4 is classical, but a complete proof is difficult to find in the literature so we provide one here. The technicalities that arise are not essential to one's understanding of analytic combinatorics, and the proof may be skipped on a first reading without much worry.

Proof of Theorem 6.4

Our argument requires the development of a few well-known facts about series of analytic functions.

Proposition 6.5. *If* (f_k) *is a sequence of analytic functions on a domain* $U \subset \mathbb{C}^d$ *and* f_k *converges uniformly to a function* f *then* f *is analytic.*

Proof Let $\boldsymbol{p} \in U$ and let $D, T \subset \mathbb{C}^d$ be a polydisk and torus around $\boldsymbol{p}$ with radii $\boldsymbol{r} \in \mathbb{R}^d_{>0}$ such that $\overline{D} \subset U$. Without loss of generality, we may assume that $\boldsymbol{p} = \boldsymbol{0}$. If $\boldsymbol{w} \in D$ then the multivariate Cauchy Integral Formula for power series (Proposition A.28 in Appendix A) implies

$$
\begin{aligned}
f(\boldsymbol{w}) = \lim_{k\to\infty} f_k(\boldsymbol{w}) &= \lim_{k\to\infty} \left(\frac{1}{2\pi i}\right)^d \int_T \frac{f_k(\boldsymbol{z})}{(z_1 - w_1)\cdots(z_d - w_d)} d\boldsymbol{z} \\
&= \left(\frac{1}{2\pi i}\right)^d \int_T \lim_{k\to\infty} \frac{f_k(\boldsymbol{w})}{(z_1 - w_1)\cdots(z_d - w_d)} d\boldsymbol{z} \\
&= \left(\frac{1}{2\pi i}\right)^d \int_T \frac{f(\boldsymbol{z})}{(z_1 - w_1)\cdots(z_d - w_d)} d\boldsymbol{z},
\end{aligned}
$$

where the interchange of limit and integral is permissible because the fact that $f_k(\boldsymbol{z}) \to f(\boldsymbol{z})$ uniformly for $\boldsymbol{z} \in T$ implies $\frac{f_k(\boldsymbol{z})}{(z_1-w_1)\cdots(z_d-w_d)} \to \frac{f(\boldsymbol{z})}{(z_1-w_1)\cdots(z_d-w_d)}$ uniformly for $\boldsymbol{z} \in T$. Expanding the final integrand as a series in $\boldsymbol{z}$ gives

$$\frac{f(\boldsymbol{z})}{(z_1-w_1)\cdots(z_d-w_d)} = \sum_{\boldsymbol{i}\in\mathbb{N}^d} \frac{f(\boldsymbol{z})}{z_1\cdots z_d}\left(\frac{w_1}{z_1}\right)^{i_1}\cdots\left(\frac{w_d}{z_d}\right)^{i_d},$$

which converges absolutely and uniformly for $\boldsymbol{z} \in T$ since $|w_k| < r_k = |z_k|$ for each k. Thus, for all $\boldsymbol{w} \in D$ we obtain the convergent series expansion

$$\begin{aligned} f(\boldsymbol{w}) &= \left(\frac{1}{2\pi i}\right)^d \int_T \left(\sum_{\boldsymbol{i}\in\mathbb{N}^d} \frac{f(\boldsymbol{z})}{z_1\cdots z_d}\left(\frac{w_1}{z_1}\right)^{i_1}\cdots\left(\frac{w_d}{z_d}\right)^{i_d}\right) d\boldsymbol{z} \\ &= \sum_{\boldsymbol{i}\in\mathbb{N}^d}\left(\left(\frac{1}{2\pi i}\right)^d \int_T f(\boldsymbol{z})\boldsymbol{z}^{-\boldsymbol{i}-\boldsymbol{1}} d\boldsymbol{z}\right)\boldsymbol{w}^{\boldsymbol{i}}. \end{aligned}$$

□

Remark 6.6. If $f(\boldsymbol{z})$ is analytic at $\boldsymbol{w}$ and T is a sufficiently small torus around $\boldsymbol{w}$ then induction from the univariate setting implies

$$\begin{aligned} &\left(\frac{1}{2\pi i}\right)^d \int_T \frac{f(\boldsymbol{z})}{(z_1-w_1)^{i_1+1}\cdots(z_d-w_d)^{i_d+1}} d\boldsymbol{z} \\ &= \frac{1}{i_1!}\left(\frac{1}{2\pi i}\right)^{d-1} \int_{T'} \frac{(\partial_{z_1}^{i_1} f)(w_1, z_2, \ldots, z_d)}{(z_2-w_2)^{i_2+1}\cdots(z_d-w_d)^{i_d+1}} d\boldsymbol{z} \\ &\ \ \vdots \\ &= \frac{(\partial_{z_1}^{i_1}\cdots\partial_{z_d}^{i_d} f)(\boldsymbol{w})}{i_1!\cdots i_d!}, \end{aligned}$$

where T' is the projection of T onto its last $d-1$ coordinates and (as usual) ∂_{z_k} denotes the partial derivative $\partial/\partial z_k$.

Proposition 6.7 (identity theorem). *If f and g are analytic functions on a connected domain $D \subset \mathbb{C}^d$ which agree on an open subset of D then they agree on all of D.*

Proof Let $K \subset D$ be the interior of the set where f and g agree. It is sufficient to show $J = \overline{K} \cap D$ equals K, since then K is open and closed in the connected set D, meaning $K = D$. To that end, pick $\boldsymbol{z}_0 \in J$ and $\varepsilon > 0$ sufficiently small so that the polydisk of radius $(\varepsilon, \ldots, \varepsilon)$ is contained in D. Since $\boldsymbol{z}_0 \in \overline{K}$ there exists a point $\boldsymbol{z} \in K$ contained in the polydisk of radius $(\varepsilon/2, \ldots, \varepsilon/2)$ centered at $\boldsymbol{z}_0$. Then f and g are analytic and agree in a neighborhood of $\boldsymbol{z}$, so partial derivatives of all orders exist for each function at $\boldsymbol{z}_0$ and each function is equal to its power series expansion at $\boldsymbol{z}$. Because the domains of convergence for these power series contain $\boldsymbol{z}_0$, the functions f and g agree at $\boldsymbol{z}_0$ so $\boldsymbol{z}_0 \in K$. Since $\boldsymbol{z}_0 \in J$ was arbitrary, J equals K as desired. □

Proposition 6.8 (logarithmic convexity of domains of convergence). *Let* $F(\boldsymbol{z}) = \sum_{\boldsymbol{r}\in\mathbb{Z}^d} a_{\boldsymbol{r}} \boldsymbol{z}^{\boldsymbol{r}}$ *be a Laurent series with domain of convergence* $\mathcal{D}$. *Then* $\mathcal{D} = \text{Relog}^{-1}(B)$ *for some open convex set* $B \subset \mathbb{R}^d$.

Proof Convergence depends on $\boldsymbol{z}$ only through the moduli of the components, hence $\mathcal{D}$ is invariant under any rotation map $z_j \mapsto e^{i\theta} z_j$. The domain of convergence is thus the union of tori $\mathbf{T}(\boldsymbol{w}) = \mathbf{T}_{\mathrm{e}}(\text{Relog}\, \boldsymbol{w})$ and therefore equal to $\text{Relog}^{-1}(B)$ for some $B \subset \mathbb{R}^d$. We claim that the open set B is the interior of

$$S = \left\{ \boldsymbol{x} \in \mathbb{R}^d : \sup_{\boldsymbol{r}\in\mathbb{Z}^d} |a_{\boldsymbol{r}}| e^{\boldsymbol{r}\cdot\boldsymbol{x}} \text{ is finite} \right\}.$$

First, if $\boldsymbol{z} \in \text{Relog}^{-1}(\boldsymbol{x})$ and $\sum_{\boldsymbol{r}\in\mathbb{Z}^d} |a_{\boldsymbol{r}} \boldsymbol{z}^{\boldsymbol{r}}| = \sum_{\boldsymbol{r}\in\mathbb{Z}^d} |a_{\boldsymbol{r}}| e^{\boldsymbol{r}\cdot\boldsymbol{x}}$ converges then $\sup_{\boldsymbol{r}\in\mathbb{Z}^d} |a_{\boldsymbol{r}} e^{\boldsymbol{r}\cdot\boldsymbol{x}}|$ is finite, so $\boldsymbol{x} \in B$ implies $\boldsymbol{x} \in S$ and (since it is open) B is contained in the interior of S.

Now, suppose that $\boldsymbol{x}$ lies in the interior of S. Because we are considering absolute convergence, we may split our series into 2^d subseries depending on the signs of its indices. In particular, using the notation $\boldsymbol{v} \odot \boldsymbol{w} = (v_1 w_1, \ldots, v_d w_d)$ for coordinatewise vector product, we have

$$\sum_{\boldsymbol{r}\in\mathbb{Z}^d} |a_{\boldsymbol{r}}| e^{\boldsymbol{r}\cdot\boldsymbol{x}} \leq \sum_{\boldsymbol{\sigma}\in\{\pm 1\}^d} \sum_{\boldsymbol{s}\in\mathbb{N}^d} |a_{\boldsymbol{\sigma}\odot\boldsymbol{s}}|\, e^{(\boldsymbol{\sigma}\odot\boldsymbol{s})\cdot\boldsymbol{x}},$$

where the inequality reflects the fact that we have duplicated terms with indices equal to zero (this simplifies notation and does not affect our argument as we only need an upper bound). Because $\boldsymbol{x}$ lies in the interior of S, for fixed $\boldsymbol{\sigma} \in \{\pm 1\}$ and $\boldsymbol{\varepsilon} = (\varepsilon, \ldots, \varepsilon)$ for sufficiently small $\varepsilon > 0$ there exists some $C > 0$ such that

$$\begin{aligned}
\sum_{\boldsymbol{s}\in\mathbb{N}^d} |a_{\boldsymbol{\sigma}\odot\boldsymbol{s}}|\, e^{(\boldsymbol{\sigma}\odot\boldsymbol{s})\cdot\boldsymbol{x}} &= \sum_{\boldsymbol{s}\in\mathbb{N}^d} |a_{\boldsymbol{\sigma}\odot\boldsymbol{s}}| e^{(\boldsymbol{\sigma}\odot\boldsymbol{s})\cdot(\boldsymbol{x}+\boldsymbol{\sigma}\odot\boldsymbol{\varepsilon})}\, e^{-\boldsymbol{s}\cdot\boldsymbol{\varepsilon}} \\
&\leq \sup_{\boldsymbol{r}\in\mathbb{Z}^d} |a_{\boldsymbol{r}}| e^{\boldsymbol{r}\cdot(\boldsymbol{x}+\boldsymbol{\sigma}\odot\boldsymbol{\varepsilon})} \cdot \sum_{\boldsymbol{s}\in\mathbb{N}^d} e^{-\boldsymbol{s}\cdot\boldsymbol{\varepsilon}} \\
&< \frac{C}{(1-e^{-\varepsilon})^d}.
\end{aligned}$$

In particular, if $\boldsymbol{x}$ lies in the interior of S then $\sum_{\boldsymbol{r}\in\mathbb{Z}^d} |a_{\boldsymbol{r}}| e^{\boldsymbol{r}\cdot\boldsymbol{x}}$ is finite, so $\boldsymbol{x} \in B$ and we have proven that B equals the interior of S. Furthermore, if $\boldsymbol{x}, \boldsymbol{y} \in S$

then for any $\lambda \in [0,1]$

$$\sup_{\boldsymbol{r}\in\mathbb{Z}^d} |a_{\boldsymbol{r}}| e^{\boldsymbol{r}\cdot(\lambda\boldsymbol{x}+(1-\lambda)\boldsymbol{y})} \leq \sup_{\boldsymbol{r}\in\mathbb{Z}^d} |a_{\boldsymbol{r}}|^{\lambda} e^{\lambda(\boldsymbol{r}\cdot\boldsymbol{x})} \cdot \sup_{\boldsymbol{r}\in\mathbb{Z}^d} |a_{\boldsymbol{r}}|^{1-\lambda} e^{(1-\lambda)(\boldsymbol{r}\cdot\boldsymbol{y})}$$

is finite, so $\lambda\boldsymbol{x} + (1-\lambda)\boldsymbol{y} \in S$ and S is convex. Since B is the interior of S, it is also convex. □

Proposition 6.9 (uniqueness of expansion). *Let $\sum_{\boldsymbol{r}\in\mathbb{Z}^d} a_{\boldsymbol{r}}\boldsymbol{z}^{\boldsymbol{r}}$ be a Laurent series converging uniformly to zero on the torus* $\mathrm{Relog}^{-1}(\boldsymbol{x})$*. Then $a_{\boldsymbol{r}} = 0$ for all $\boldsymbol{r} \in \mathbb{Z}^d$.*

Proof Assume without loss of generality that $\boldsymbol{x} = \boldsymbol{0}$, so that $\sum_{\boldsymbol{r}\in\mathbb{Z}^d} a_{\boldsymbol{r}} e^{i\boldsymbol{r}\cdot\boldsymbol{y}} \to 0$ uniformly for $\boldsymbol{y} \in [-\pi,\pi)^d$. For any fixed $\boldsymbol{r} \in \mathbb{Z}^d$, uniform convergence implies

$$\begin{aligned} a_{\boldsymbol{r}} &= \sum_{\boldsymbol{s}\in\mathbb{Z}^d} \frac{1}{(2\pi i)^d} \int_{\mathbf{T}_e(\boldsymbol{x})} a_{\boldsymbol{s}} e^{i(\boldsymbol{s}-\boldsymbol{r})\cdot\boldsymbol{y}}\, d\boldsymbol{y} \\ &= \frac{1}{(2\pi i)^d} \int_{\mathbf{T}_e(\boldsymbol{x})} e^{-i\boldsymbol{r}\cdot\boldsymbol{y}} \cdot \sum_{\boldsymbol{s}\in\mathbb{Z}^d} a_{\boldsymbol{s}} e^{i\boldsymbol{s}\cdot\boldsymbol{y}}\, d\boldsymbol{y} \\ &\to \frac{1}{(2\pi i)^d} \int_{\mathbf{T}_e(\boldsymbol{x})} 0\, d\boldsymbol{y} \\ &= 0, \end{aligned}$$

as desired. □

We have now proven most of Theorem 6.4: part (1) is Proposition 6.8, while part (2) follows from Proposition 6.5 because $\iota(F)$ is the uniform limit of the series of partial sums. Uniqueness in part (3) is given by Proposition 6.9, so it remains only to prove that (6.1) defines a Laurent series $F(\boldsymbol{z}) = \sum_{\boldsymbol{r}\in\mathbb{Z}^d} a_{\boldsymbol{r}}\boldsymbol{z}^{\boldsymbol{r}}$ converging uniformly to f.

By analyticity, the integral (6.1) defining $a_{\boldsymbol{r}}$ is independent of the value of $\boldsymbol{x} \in B$ defining the chain of integration $\mathbf{T}_e(\boldsymbol{x})$. Following an argument analogous to the proof of Proposition 6.8 above, for $\varepsilon > 0$ sufficiently small there exists a neighborhood $\mathcal{N}(\boldsymbol{x})$ of $\boldsymbol{x}$ in $\mathbb{R}^d$ and constant $C > 0$ such that $|a_{\boldsymbol{r}}\boldsymbol{z}^{\boldsymbol{r}}| \leq C\exp(-\varepsilon|\boldsymbol{r}|)$ for any $\boldsymbol{z} \in \mathbf{T}_e(\boldsymbol{x}')$ with $\boldsymbol{x}' \in \mathcal{N}(\boldsymbol{x})$. If $K \subset B$ is any compact set, covering it with finitely many neighborhoods $\mathcal{N}(\boldsymbol{x})$ shows that such a bound holds for all $\boldsymbol{z} \in \mathrm{Relog}^{-1}(K)$. In particular, the series F converges uniformly on compact subsets of $\mathcal{D}$, so it is sufficient to prove that $\iota(F) = f$ on a collection of sets with nonempty interior covering $\mathcal{D}$.

Let $B' = \prod_{j=1}^d [\log a_j, \log b_j]$ be a product of closed rectangles with nonempty interior strictly contained in B. Proposition A.29 in Appendix A gives a representation

$$f(\boldsymbol{w}) = \left(\frac{1}{2\pi i}\right)^d \sum_{\eta\in\{a,b\}^d} \operatorname{sgn}(\eta) \int_{T_\eta} \frac{f(\boldsymbol{z})}{(z_1 - w_1)\cdots(z_d - w_d)} d\boldsymbol{z} \qquad (6.2)$$

for all $\boldsymbol{w} \in \operatorname{Relog}^{-1}(B')$, where $T_\eta = \operatorname{Relog}^{-1}(\boldsymbol{r})$ with $r_k = a_k$ if $\eta_k = a$ and $r_k = b_k$ if $\eta_k = b$. Since $a_k < |w_k| < b_k$ for all k, we have the series expansions

$$\frac{1}{z_k - w_k} = \begin{cases} -\sum_{n<0} z_k^n w_k^{-n-1} & \text{if } |z_k| = |a_k| \\ \sum_{n\geq 0} z_k^{-n-1} w_k^n & \text{if } |z_k| = |b_k| \end{cases}.$$

In particular, each summand in (6.2) is a distinct Laurent series in $\boldsymbol{w}$ uniformly converging on compact subsets and, taking the $\operatorname{sgn}(\eta)$ coefficient into account, we obtain the desired representation

$$f(\boldsymbol{w}) = \sum_{\boldsymbol{r}\in\mathbb{Z}^d} \left(\frac{1}{(2\pi i)^d} \int_{\mathbf{T}_e(\boldsymbol{x})} f(\boldsymbol{z}) \boldsymbol{z}^{-\boldsymbol{r}-\mathbf{1}}\right) \boldsymbol{w}^{\boldsymbol{r}}$$

for $\boldsymbol{w} \in B'$. By independence of the Cauchy integral on the point $\boldsymbol{x}$ in B, this holds for all $\boldsymbol{x} \in B$. □

Example 6.10. For $d = 1$, we obtain the classical representation

$$f(w) = \underbrace{\frac{1}{2\pi i} \int_{\gamma_2} \frac{f(z)}{z - w} dz}_{f_2(w)} - \underbrace{\frac{1}{2\pi i} \int_{\gamma_1} \frac{f(z)}{z - w} dz}_{f_1(w)},$$

where γ_1 and γ_2 are the inner and outer boundaries of an annulus $A = \{z : e^a \leq |z| \leq e^b\}$, the point $w \in A$, and the analytic function f is represented by the convergent series $\sum_{n\in\mathbb{Z}} a_n z^n$ on A. The function $f_2(w)$ is analytic on the disk of radius e^b while the function $f_1(1/w)$ is analytic on the disk of radius e^{-a}, so $f_2(w) = \sum_{n\geq 0} a_n w^n$ while $f_1(w) = -\sum_{n\leq -1} a_n w^n$. ◁

6.2 Polynomial amoebas

If $f(\boldsymbol{z})$ is a Laurent polynomial then the ***algebraic hypersurface*** or ***algebraic variety defined by*** f is the zero set $\mathcal{V}(f) = \{\boldsymbol{z} \in \mathbb{C}_*^d : f(\boldsymbol{z}) = 0\}$. The ***amoeba*** of f is the set

$$\texttt{amoeba}(f) := \left\{\operatorname{Relog} \boldsymbol{z} : \boldsymbol{z} \in \mathbb{C}_*^d \text{ and } f(\boldsymbol{z}) = 0\right\} \subset \mathbb{R}^d$$

consisting of the image under Relog of the non-zero points in the algebraic hypersurface defined by f.

Amoebas provide a crucial tool for the study of convergent Laurent expansions because they live in real space (which is much easier to picture than complex space) and, as we will see in Theorem 6.18 below, the connected components of $\texttt{amoeba}(f)^c = \mathbb{R}^d \setminus \texttt{amoeba}(f)$ are convex sets corresponding to the convergent Laurent expansions of $1/f(\boldsymbol{z})$.

Example 6.11. The amoebas of $f(x, y) = 2-x-y$ and $f(x, y) = (3-2x-y)(3-x-2y)$ are shown in Figure 6.1; note that the amoeba of a product is the union of amoebas. Algorithms for drawing amoebas are discussed in Chapter 8. ◁

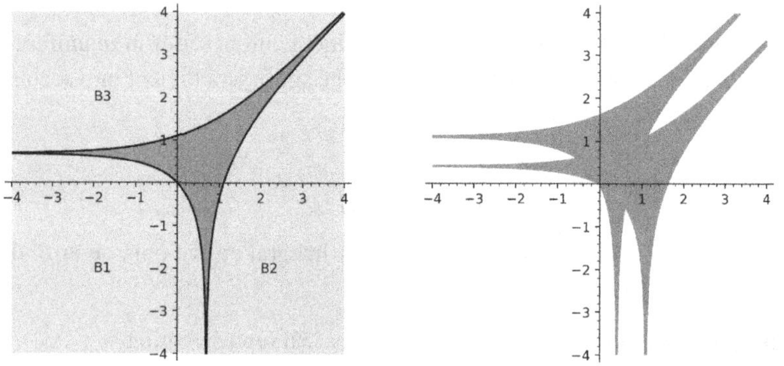

Figure 6.1 *Left:* $\texttt{amoeba}(2-x-y)$ with the components of its complement. *Right:* $\texttt{amoeba}((3-2x-y)(3-x-2y))$.

Exercise 6.2. Identify the convergent Laurent expansions corresponding to the connected components B_1, B_2, and B_3 of $\texttt{amoeba}(2-x-y)$ shown on the left side of Figure 6.1.

Exercise 6.3. Use the definition of an amoeba to transform the sketch of $\texttt{amoeba}(2-x-y)$ in Figure 6.1 into a sketch of $\texttt{amoeba}(2-x-y^2)$.

In order to study the connected components of $\texttt{amoeba}(f)^c$, we introduce some terminology.

Definition 6.12 (Newton polytopes)**.** The ***convex hull*** $\texttt{hull}(A)$ of a subset $A \subset \mathbb{R}^d$ is the smallest convex set in $\mathbb{R}^d$ containing A. The convex hull of A equals the convex hull of its ***extreme points***: those points $x \in A$ which do not lie in the interior of any line segment contained in A. A ***convex polytope*** P is the convex hull of any finite set $E \subset \mathbb{R}^d$ and the ***vertices of a polytope*** P are the extreme points of $\texttt{hull}(E)$. If $f(\boldsymbol{z})$ is a Laurent polynomial then the ***support of the Laurent polynomial*** f is the (finite) set of exponents of the monomials appearing in f, and the ***Newton polytope*** $\mathcal{N}_f$ of f is the convex hull of its support,

$$\mathcal{N}_f := \texttt{hull}\left\{\boldsymbol{r} \in \mathbb{Z}^d : a_{\boldsymbol{r}}\boldsymbol{z}^{\boldsymbol{r}} \text{ is a non-zero monomial of } f\right\}.$$

We now fix a Laurent polynomial $f(\boldsymbol{z}) = \sum_{\boldsymbol{i}} a_{\boldsymbol{i}}\boldsymbol{z}^{\boldsymbol{i}}$. The ***order map*** ν : $\texttt{amoeba}(f)^c \to \mathbb{R}^d$ sends a point $\boldsymbol{x} \in \texttt{amoeba}(f)^c$ to the vector $\nu(\boldsymbol{x})$ with kth coordinate

$$\nu(\boldsymbol{x})_k = \frac{1}{2\pi i}\int_{|z_k|=e^{x_k}} \frac{f_{z_k}(e^{x_1},\dots,z_k,\dots,e^{x_d})}{f(e^{x_1},\dots,z_k,\dots,e^{x_d})}dz_k,$$

where $(e^{x_1},\dots,z_k,\dots,e^{x_d})$ denotes the vector $\boldsymbol{z}$ with $z_j = e^{x_j}$ for $j \neq k$.

Remark 6.13. Recall that if g is any univariate meromorphic function inside and on a contour γ, with no zeros or poles on γ, then the classical *argument principle* from complex analysis states that (counting with multiplicity) the number of zeros of g inside γ minus the number of poles of g inside γ equals

$$\frac{1}{2\pi i}\int_\gamma \frac{g'(z)}{g(z)}dz.$$

In particular, $\frac{1}{2\pi i}\int_\gamma \frac{g'(z)}{g(z)}dz$ is an integer so the order map sends $\texttt{amoeba}(f)^c$ to $\mathbb{Z}^d$.

Exercise 6.4. Use Cauchy's residue theorem to prove the argument principle.

Exercise 6.5. Prove that for any $\boldsymbol{x} \in \texttt{amoeba}(f)^c$,

$$\nu(\boldsymbol{x})_k = \frac{1}{(2\pi i)^d}\int_{\text{Relog}^{-1}(\boldsymbol{x})} \frac{z_k f_{z_k}(\boldsymbol{z})}{f(\boldsymbol{z})}\frac{d\boldsymbol{z}}{z_1\cdots z_d}.$$

Note that the order map is continuous, and hence constant on the connected components of $\texttt{amoeba}(f)^c$, so the order of a component of $\texttt{amoeba}(f)^c$ is well defined. The order map was introduced by Forsberg et al. [FPT00] as a multivariate generalization of the argument principle. Our treatment of the order map closely follows the arguments given in [FPT00].

Lemma 6.14. *If $\boldsymbol{x} \in \texttt{amoeba}(f)^c$ and $\boldsymbol{s} \in \mathbb{Z}^d \setminus \{\boldsymbol{0}\}$ then $\boldsymbol{s} \cdot \nu(\boldsymbol{x})$ equals the number of zeros of $a_{\boldsymbol{s}}(w) = f(p_1 w^{s_1},\dots,p_d w^{s_d})$ inside the unit circle $|w| = 1$ minus the order of the pole of $a_{\boldsymbol{s}}(w)$ at the origin, where $\boldsymbol{p} \in \mathbb{C}_*^d$ is any point with $\boldsymbol{x} = \text{Relog}(\boldsymbol{p})$.*

Proof The argument principle implies that the number of zeros of $a_{\boldsymbol{s}}(w)$ inside the unit circle minus the order of the pole of $a_{\boldsymbol{s}}(w)$ at the origin equals

$$\frac{1}{2\pi i}\int_{|w|=1}\frac{\frac{d}{dw}a_{\boldsymbol{s}}(w)}{a_{\boldsymbol{s}}(w)}dw.$$

By Exercise 6.16 below, the image of $|w| = 1$ under the change of variables $\boldsymbol{z} = (p_1 w^{s_1}, \dots, p_d w^{s_d})$ is homotopic to the chain of integration $s_1\gamma_1 + \cdots + s_d\gamma_d$, where $\gamma_k = \{(p_1, \dots, w, \dots, p_d) : |w| = |p_k|\}$. Thus, the difference of zeros and poles is

$$\begin{aligned}
\sum_{k=1}^{d}\frac{s_k}{2\pi i}\int_{\gamma_k}\frac{\frac{d}{dw}a_{\boldsymbol{s}}(w)}{a_{\boldsymbol{s}}(w)}dw &= \sum_{k=1}^{d}\frac{s_k}{2\pi i}\int_{|z_k|=|p_k|}\frac{f_{z_k}(p_1,\dots,z_k,\dots,p_d)}{f(p_1,\dots,z_k,\dots,p_d)}dz_k \\
&= \sum_{k=1}^{d}\frac{s_k}{2\pi i}\int_{|z_k|=e^{x_k}}\frac{f_{z_k}(e^{x_1},\dots,z_k,\dots,e^{x_d})}{f(e^{x_1},\dots,z_k,\dots,e^{x_d})}dz_k \\
&= \sum_{k=1}^{d} s_k \nu(\boldsymbol{x})_k
\end{aligned}$$

as claimed, where the first equality follows from the chain rule, and the second follows from the expression for the order map in Exercise 6.5. □

Lemma 6.14 is powerful as it allows us to characterize the order map using the roots of a univariate polynomial; it forms the basis for all of the results about the order map that we require.

Corollary 6.15. *The image of* amoeba$(f)^c$ *under the order map lies in the Newton polytope of* f.

Exercise 6.6. Prove Corollary 6.15 by showing that $\boldsymbol{s} \cdot \nu(\boldsymbol{x}) \le \max_{\boldsymbol{n}\in N_f}(\boldsymbol{s} \cdot \boldsymbol{n})$ for all $\boldsymbol{s} \in \mathbb{Z}^d \setminus \{\boldsymbol{0}\}$.

Corollary 6.16. *If* $\boldsymbol{x}$ *and* $\boldsymbol{x}'$ *lie in distinct components of* amoeba$(f)^c$ *then* $\nu(\boldsymbol{x}) \neq \nu(\boldsymbol{x}')$.

Proof Writing $\boldsymbol{x}' = \boldsymbol{x} + t\boldsymbol{s}$ for some $\boldsymbol{s} \in \mathbb{Z}^d \setminus \{\boldsymbol{0}\}$ and $t > 0$, we have

$$\begin{aligned}
&\boldsymbol{s}\cdot\nu(\boldsymbol{x}') \\
&= \#\text{ zeros of } f(e^{x_1+ts_1}w^{s_1},\dots,e^{x_d+ts_d}w^{s_d}) \text{ inside } |w|=1 \text{ minus pole order at } \boldsymbol{0} \\
&= \#\text{ zeros of } f(e^{x_1}w^{s_1},\dots,e^{x_d}w^{s_d}) \text{ inside } |w|=e^{t} \text{ minus pole order at } \boldsymbol{0} \\
&> \#\text{ zeros of } f(e^{x_1}w^{s_1},\dots,e^{x_d}w^{s_d}) \text{ inside } |w|=1 \text{ minus pole order at } \boldsymbol{0} \\
&= \boldsymbol{s}\cdot\nu(\boldsymbol{x}'),
\end{aligned}$$

where the inequality follows from the fact that there is at least one root of

$f(e^{x_1}w^{s_1},\dots,e^{x_d}w^{s_d})$ with $1<|w|<e^t$ since $\boldsymbol{x}$ and $\boldsymbol{x}'$ lie in different components of $\mathtt{amoeba}(f)^c$. Thus $\nu(\boldsymbol{x}')\neq\nu(\boldsymbol{x})$. □

Corollaries 6.15 and 6.16 imply that we can uniquely identify the Laurent expansions of $1/f(\boldsymbol{z})$ with integer points in the Newton polytope of f.

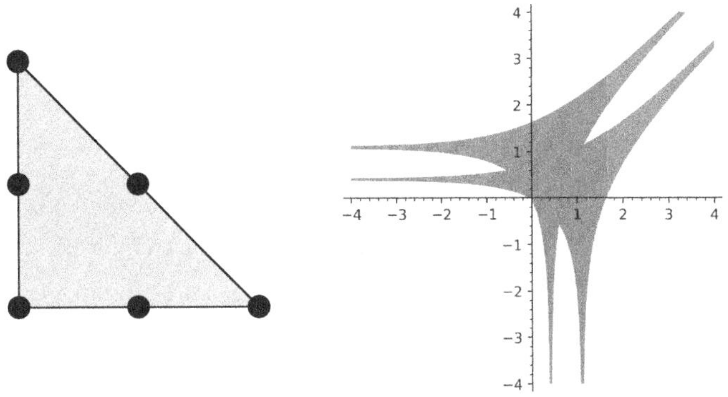

Figure 6.2 *Left:* The Newton polytope of $(3-2x-y)(3-x-2y)=9-9x-9y+2x^2+5xy+2y^2$. *Right:* Drawing of $\mathtt{amoeba}((3-2x-y)(3-x-2y))$, whose complement contains a connected component for each integer point of the Newton polytope.

It is of great interest to know which elements of $\mathcal{N}_f\cap\mathbb{Z}^d$ correspond to Laurent expansions of $1/f(\boldsymbol{z})$. In general this is a very difficult question, however in one important case we can be definitive.

Corollary 6.17. *If $\boldsymbol{v}\in\mathcal{N}_f\cap\mathbb{Z}^d$ and there exists $\boldsymbol{x}\in\mathtt{amoeba}(f)^c$ such that*

$$|a_{\boldsymbol{v}}\boldsymbol{z}^{\boldsymbol{v}}|>\left|f(\boldsymbol{z})-a_{\boldsymbol{v}}\boldsymbol{z}^{\boldsymbol{v}}\right|$$

whenever $\boldsymbol{z}\in\mathbf{T}_e(\boldsymbol{x})$ then the Laurent series

$$\frac{1}{f(\boldsymbol{z})}=\frac{1}{a_{\boldsymbol{v}}\boldsymbol{z}^{\boldsymbol{v}}}\sum_{k\geq 0}\left(\frac{a_{\boldsymbol{v}}\boldsymbol{z}^{\boldsymbol{v}}-f(\boldsymbol{z})}{a_{\boldsymbol{v}}\boldsymbol{z}^{\boldsymbol{v}}}\right)^k \tag{6.3}$$

has domain of convergence $\mathrm{Relog}^{-1}(B)$ *for the component $B\subset\mathtt{amoeba}(f)^c$ with order $\boldsymbol{v}$. In particular, the vertices of $\mathcal{N}_f$ are contained in the image of $\mathtt{amoeba}(f)^c$ under the order map.*

Proof The Laurent series (6.3) converges in a neighborhood of $\mathbf{T}_e(\boldsymbol{x})$ as a geometric series with ratio less than one, meaning it has domain of convergence $\mathrm{Relog}^{-1}(B)$ for the component $B\subset\mathtt{amoeba}(f)^c$ containing $\boldsymbol{x}$. If $\boldsymbol{s}$ is any vector

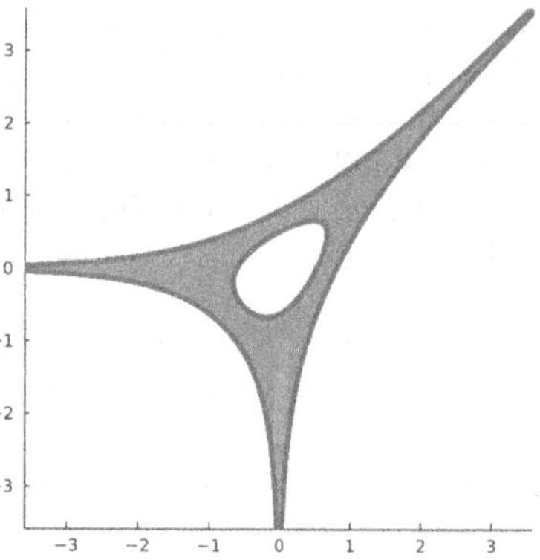

Figure 6.3 Sketch of amoeba$(1 - 4xy + x^3 + y^3)$.

in $\mathbb{Z}^d \setminus \{\mathbf{0}\}$, then Rouché's Theorem implies

$$\begin{aligned} \boldsymbol{s} \cdot \nu(\boldsymbol{x}) &= \#\text{ zeros} - \#\text{ poles of } f(p_1 w^{s_1}, \ldots, p_d w^{s_d}) \text{ inside } |w| = 1 \\ &= \#\text{ zeros} - \#\text{ poles of } a_{\boldsymbol{v}} \boldsymbol{p}^{\boldsymbol{v}} w^{\boldsymbol{s}\cdot\boldsymbol{v}} \text{ inside } |w| = 1 \\ &= \boldsymbol{s} \cdot \boldsymbol{v}. \end{aligned}$$

Since this holds for all $\boldsymbol{s} \in \mathbb{Z}^d \setminus \{\mathbf{0}\}$, we see that $\nu(\boldsymbol{x}) = \boldsymbol{v}$.

If $\boldsymbol{v}$ is a vertex of $\mathcal{N}_f$ then there exists $\boldsymbol{r} \in \mathbb{R}^d$ such that $\boldsymbol{v} \cdot \boldsymbol{r} > \boldsymbol{w} \cdot \boldsymbol{r}$ for all $\boldsymbol{w} \neq \boldsymbol{v}$ in the support of f. When $\boldsymbol{p} = e^{t\boldsymbol{r}}$ for $t > 0$ the term $|a_{\boldsymbol{v}} \boldsymbol{p}^{\boldsymbol{v}}| = |a_{\boldsymbol{v}}| e^{t(\boldsymbol{v}\cdot\boldsymbol{r})}$ grows exponentially faster, as $t \to \infty$, than $|a_{\boldsymbol{w}} \boldsymbol{p}^{\boldsymbol{w}}| = |a_{\boldsymbol{w}}| e^{t(\boldsymbol{w}\cdot\boldsymbol{r})}$ for all $\boldsymbol{w} \neq \boldsymbol{v}$ in the support of f. In particular, when t is sufficiently large then $\boldsymbol{p} = e^{t\boldsymbol{r}}$ satisfies $|a_{\boldsymbol{v}} \boldsymbol{z}^{\boldsymbol{v}}| > \left| f(\boldsymbol{z}) - a_{\boldsymbol{v}} \boldsymbol{z}^{\boldsymbol{v}} \right|$ for all $\boldsymbol{z} \in \mathbf{T}(\boldsymbol{p})$. □

Summarizing our results, we have shown the following.

Theorem 6.18. *The connected components of* amoeba$(f)^c$ *are convex subsets of* $\mathbb{R}^d$ *in one-to-one correspondence with the convergent Laurent expansions of* $1/f(\boldsymbol{z})$*. The order map sends each component of* amoeba$(f)^c$ *to a different element of* $\mathcal{N}_f \cap \mathbb{Z}^d$ *and every vertex of* $\mathcal{N}_f$ *lies in the image of* amoeba$(f)^c$.

Exercise 6.7. Figure 6.3 displays amoeba$(1-4xy+x^3+y^3)$. Use Corollary 6.17 to prove algebraically that the amoeba complement contains a bounded component and find its order. Prove that for $N \geq 0$ the coefficient of $x^N y^N$ in the Laurent expansion of $F(x, y) = 1/(1 - 4xy + x^3 + y^3)$ corresponding to the bounded component has the (convergent) infinite series representation

$$[x^N y^N]F(x,y) = -\sum_{k\geq 0} 4^{-2N+3k+3} \frac{(2N+3k+2)!}{k!\,(N+k+1)!^2}.$$

Give as simple a formula as you can for the coefficient of $x^N y^N$ in this Laurent expansion when $N < 0$.

6.3 Convex cones and exponential bounds

Let $f(\boldsymbol{z}) = \sum_{\boldsymbol{r}\in\mathbb{Z}^d} a_{\boldsymbol{r}} \boldsymbol{z}^{\boldsymbol{r}}$ be any convergent Laurent series with domain of convergence $\mathcal{D} = \text{Relog}^{-1}(B)$. The coarsest asymptotic information about the coefficients $\{a_{\boldsymbol{r}}\}$ is their exponential rate of growth (or decay) when $\boldsymbol{r} \to \infty$ in different directions. Recall the notation $|\boldsymbol{r}| = |r_1| + \cdots + |r_d|$.

Definition 6.19. A sequence $\{\boldsymbol{r}_n\} = \boldsymbol{r}_1, \boldsymbol{r}_2, \ldots$ in $\mathbb{Z}^d$ ***goes to infinity in the direction*** $\hat{\boldsymbol{r}}$ if $|\boldsymbol{r}_n| \to \infty$ and $\boldsymbol{r}_n/|\boldsymbol{r}_n| \to \hat{\boldsymbol{r}}$. We often drop the subscript n, writing $\boldsymbol{r} \to \infty$ to denote a sequence going to infinity and using shorthand like $\lim_{\boldsymbol{r}\to\infty} p(\boldsymbol{r})$ for an expression $\lim_{n\to\infty} p(\boldsymbol{r}_n)$ involving a function $p : \mathbb{R}^d \to \mathbb{R}$.

In the univariate case one can only move forwards or backwards on the real line, however in multivariate cases there are many directions to study. Typically the asymptotic behavior of $a_{\boldsymbol{r}}$ changes in a predictable manner with the direction of a limit $\boldsymbol{r} \to \infty$, however this is not always the case. It can therefore be useful to consider both the exponential growth in a fixed direction, and a smoothed version that can be easier to compute.

Definition 6.20. The (limsup logarithmic) ***growth rate*** of a sequence $a_{\mathbf{i}}$ in the direction $\hat{\boldsymbol{r}}$ is the quantity

$$\beta(\hat{\boldsymbol{r}}) := \limsup_{\substack{\boldsymbol{s}\to\infty \\ \boldsymbol{s}/|\boldsymbol{s}|=\hat{\boldsymbol{r}}}} \frac{1}{|\boldsymbol{s}|} \log |a_{\boldsymbol{s}}|,$$

while the (limsup logarithmic) ***neighborhood growth rate*** of $a_{\mathbf{i}}$ in the direction $\hat{\boldsymbol{r}}$ is

$$\overline{\beta}(\hat{\boldsymbol{r}}) := \limsup_{\substack{\boldsymbol{s}\to\infty \\ \boldsymbol{s}/|\boldsymbol{s}|\to\hat{\boldsymbol{r}}}} \frac{1}{|\boldsymbol{s}|} \log |a_{\boldsymbol{s}}|.$$

The (neighborhood) growth rate of a generating function is the (neighborhood) growth rate of its coefficient sequence.

Example 6.21. The growth rate of the power series expansion of

$$f(x,y) = \frac{x-y}{1-x-y} = \sum_{r,s\geq 0} \left[\binom{r+s-1}{s} - \binom{r+s-1}{r}\right] x^r y^s$$

in the direction $\hat{r} = (1,1)$ is $-\infty$, because $a_{n,n} = 0$ for all $n \geq 0$, however the neighborhood growth rate in the direction $\hat{r}$ is $\log 2$. For instance, $|a_{n,n+k}|^{1/n} \to 2$ as $n \to \infty$ whenever $k \neq 0$ is a fixed integer. ◁

Computing the growth rate of a multidimensional sequence can be difficult, however it is possible to bound the growth rate using the behavior of f on its domain of convergence $\mathcal{D}$. For any $\boldsymbol{w} \in \mathcal{D}$, bounding the Cauchy integral expression

$$|a_{\boldsymbol{r}}| = \left|\left(\frac{1}{2\pi i}\right)^d \int_{\mathrm{T}(\boldsymbol{w})} f(\boldsymbol{z})\boldsymbol{z}^{-\boldsymbol{r}-\boldsymbol{1}}\, d\boldsymbol{z}\right|$$

by the maximum modulus of its integrand times the surface area of its domain of integration gives

$$|a_{\boldsymbol{r}}| \leq |\boldsymbol{w}^{-\boldsymbol{r}}| \cdot \max_{\boldsymbol{z}\in\mathrm{T}(\boldsymbol{w})} |f(\boldsymbol{z})|. \tag{6.4}$$

The ***dual rate*** of f over $\mathcal{D}$ (or B) is the quantity

$$\beta^*(\hat{\boldsymbol{r}}) := \inf_{\boldsymbol{z}\in\mathcal{D}} -\hat{\boldsymbol{r}} \cdot \operatorname{Relog}(\boldsymbol{z}) = \inf_{\boldsymbol{x}\in B} -\hat{\boldsymbol{r}} \cdot \boldsymbol{x}\,. \tag{6.5}$$

Optimizing (6.4) over $\boldsymbol{w} \in \mathcal{D}$ implies

$$\limsup_{\boldsymbol{r}\to\infty} |a_{\boldsymbol{r}}|^{1/|\boldsymbol{r}|} \leq \inf_{\boldsymbol{w}\in\mathcal{D}} |\boldsymbol{w}|^{-\hat{\boldsymbol{r}}}\,, \tag{6.6}$$

so that

$$\beta(\hat{\boldsymbol{r}}) \leq \overline{\beta}(\hat{\boldsymbol{r}}) \leq \beta^*(\hat{\boldsymbol{r}}) \tag{6.7}$$

for any direction $\hat{\boldsymbol{r}}$.

Remark 6.22. Equation (6.7) is the multivariate analogue of relationship (3.5) between the radius of convergence of a univariate power series and exponential growth, however in the multivariate case we get an inequality instead of an equality. The dual rate is an example of a Legendre transform (see the notes at the end of this chapter).

Remark 6.23. To compute $\beta^*(\hat{\boldsymbol{r}})$ it is sufficient to minimize the function $|\boldsymbol{z}|^{\hat{\boldsymbol{r}}}$ on the closure $\overline{\mathcal{D}}$. Writing $\boldsymbol{z} = \boldsymbol{x} + i\boldsymbol{y}$ for real variables $\boldsymbol{x}$ and $\boldsymbol{y}$ implies $|\boldsymbol{z}|^{2\boldsymbol{v}} = (x_1^2+y_1^2)^{v_1} \cdots (x_1^2+y_d^2)^{v_d}$ for any $\boldsymbol{v} \in \mathbb{R}^d$. Thus, if $\hat{\boldsymbol{r}}$ has positive rational coordinates and f represents a rational function then (potentially after taking powers of some variables to remove fractional powers) we can represent the problem of minimizing $|\boldsymbol{z}|^{\hat{\boldsymbol{r}}}$ on $\overline{\mathcal{D}}$ as a polynomial optimization problem over a semialgebraic domain. Computing such a minimizer can be done algorithmically, although in general it is extremely expensive. See [Las15] for a detailed account of such optimization problems, and computationally feasible relaxations.

Because B is an open convex set in $\mathbb{R}^d$, either $\beta^*(\hat{\boldsymbol{r}}) = -\infty$ or the infimum in (6.5) is achieved by taking $\boldsymbol{x}$ to a point on the boundary ∂B. In the rest of this section we examine what happens when $\beta^*(\hat{\boldsymbol{r}}) = -\infty$, while the next section studies the case when $\beta^*(\hat{\boldsymbol{r}})$ is finite.

Proposition 6.24. *Let $f(\boldsymbol{z}) = P(\boldsymbol{z})/Q(\boldsymbol{z})$ be a ratio of polynomials P and Q. If $\boldsymbol{r}_1, \boldsymbol{r}_2, \ldots$ is a sequence of vectors in $\mathbb{Z}^d$ going to infinity in the direction $\hat{\boldsymbol{r}} \in \mathbb{R}^d$ and the set $\{-\hat{\boldsymbol{r}} \cdot \boldsymbol{x} : \boldsymbol{x} \in B\}$ is unbounded from below then $a_{\boldsymbol{r}_n} = 0$ for all but finitely many n.*

Proof If the linear function $-\hat{\boldsymbol{r}} \cdot \boldsymbol{x}$ is unbounded from below on B then, since P and Q are polynomials, we can find a sequence of points $\{\boldsymbol{x}_k\}$ in B and a polynomial $p(x)$ such that $-\hat{\boldsymbol{r}} \cdot \boldsymbol{x} \leq -k$ and

$$|f(\boldsymbol{z})| \leq p(e^k) \text{ on } \mathrm{Relog}^{-1}(\boldsymbol{x}_k).$$

Because $\boldsymbol{r}_n/|\boldsymbol{r}_n| = \hat{\boldsymbol{r}} + o(1)$ we have $|\boldsymbol{z}^{-\boldsymbol{r}_n}| = |e^{-\boldsymbol{x}\cdot\boldsymbol{r}_n}| \leq e^{-k(|\boldsymbol{r}_n|+o(|\boldsymbol{r}_n|))}$ when $\boldsymbol{z} \in \mathrm{Relog}^{-1}(\boldsymbol{x})$, so the maximum modulus bound (6.4) implies

$$|a_{\boldsymbol{r}_n}| \leq p(e^k) \exp\Big[-k(|\boldsymbol{r}_n| + o(|\boldsymbol{r}_n|))\Big].$$

If n is sufficiently large then $|\boldsymbol{r}_n| + o(|\boldsymbol{r}_n|)$ is positive and the upper bound on the right-hand side goes to zero as $k \to \infty$. Since this bound holds for all $k \in \mathbb{N}$, it follows that $a_{\boldsymbol{r}_n}$ is zero for all sufficiently large n. □

Remark 6.25. If $f(\boldsymbol{z})$ is a ratio of *analytic* functions with a Laurent series converging on $\mathcal{D} = \mathrm{Relog}^{-1}(B)$ and $-\hat{\boldsymbol{r}} \cdot \boldsymbol{x}$ is unbounded from below on B then the function p giving an upper bound on $|f(\boldsymbol{z})|$ may no longer be polynomial, and thus the coefficients $a_{\boldsymbol{r}_n}$ may not eventually be zero. However, our argument still proves that the coefficients decay faster than any exponential function in n. The methods of ACSV we develop are not concerned with this super-exponential regime, and if the reader encounters such a situation we recommend trying direct application of the multivariate saddle point methods described in Chapter 5.

Proposition 6.24 describes the extreme behavior that occurs when $\{-\hat{\boldsymbol{r}} \cdot \boldsymbol{x} : \boldsymbol{x} \in B\}$ is unbounded from below, and it is natural to wonder when such behavior can occur. Serendipitously, the connection between the Newton polytope

Figure 6.4 A convex set (light gray) with three selected boundary points. Parts of the corresponding tangent cones (facing into the set) and normal cones (facing away from the set) are displayed. Note that the rightmost normal cone is a one-dimensional line. The recession cone of this set is $\{\mathbf{0}\}$ since the set is bounded.

of Q and the components of $\texttt{amoeba}(Q)^c$ described in Theorem 6.18 above can be strengthened to help provide such information. In order to state this relationship we need some additional terminology.

Definition 6.26. Let B be a convex set. A ***convex cone*** is a subset of $\mathbb{R}^d$ that is closed under addition and closed under multiplication by positive scalars. The ***recession cone*** of B is the set of vectors $\{\boldsymbol{v} \in \mathbb{R}^d : \boldsymbol{x} + \boldsymbol{v} \in B \text{ for all } \boldsymbol{x} \in B\}$ and the (open) ***tangent cone*** to B at $\boldsymbol{x} \in B$ is

$$\texttt{tan}_{\boldsymbol{x}}(B) = \{\boldsymbol{s} \in \mathbb{R}^d : \boldsymbol{x} + \varepsilon\boldsymbol{s} \in B \text{ for all sufficiently small } \varepsilon > 0\}.$$

The tangent cone is a generalization of the tangent space of a manifold to spaces with singularities. The (closed) ***normal cone*** to B at $\boldsymbol{x} \in B$ is

$$\texttt{normal}_{\boldsymbol{x}}(B) = \{\boldsymbol{s} \in \mathbb{R}^d : \boldsymbol{s} \cdot \boldsymbol{x} \geq \boldsymbol{s} \cdot \boldsymbol{b} \text{ for all } \boldsymbol{b} \in B\}.$$

See Figure 6.4 for an illustration of these concepts.

Exercise 6.8. If K is an open convex cone in $\mathbb{R}^d$ then its ***dual cone*** $K^* \subset \mathbb{R}^d$ is the set of vectors $\{\boldsymbol{s} \in \mathbb{R}^d : \boldsymbol{s} \cdot \boldsymbol{x} \geq 0 \text{ for all } \boldsymbol{x} \in K\}$. Prove the following statements.

a) The recession cone, tangent cone, and normal cone are convex cones.
b) Let K and L be open convex cones. Then $(K \cap L)^* = \texttt{hull}(K^* \cup L^*)$ and if $K \subset L$ then $K^* \supset L^*$.

c) Every open convex cone is the intersection of all open half-spaces that contain it.
d) After a translation by $\boldsymbol{x}$, the tangent cone to B at $\boldsymbol{x}$ is the interior of the intersection of all halfspaces that contain B and have $\boldsymbol{x}$ on their boundary.
e) If $\boldsymbol{x}$ is an interior point of B then $\mathtt{normal}_{\boldsymbol{x}}(B) = \{\mathbf{0}\}$.
f) The normal cone is the dual cone to the negative of the tangent cone,

$$\mathtt{normal}_{\boldsymbol{x}}(B) = (-\mathtt{tan}_{\boldsymbol{x}}(B))^* .$$

The result we need comes from the following correspondence.

Proposition 6.27. *Let f be a Laurent polynomial and let $\boldsymbol{x}$ be a point in a component B of $\mathtt{amoeba}(f)^c$. Then the recession cone of B equals the normal cone of $\mathcal{N}_f$ at $\nu(\boldsymbol{x})$.*

Proof Let $\boldsymbol{s} \in \mathbb{Z}^d \setminus \{\mathbf{0}\}$. By definition, the ray $\{\boldsymbol{x} + t\boldsymbol{s} : t > 0\}$ stays in B if and only if all zeros of $f(p_1 w^{s_1}, \ldots, p_d w^{s_d})$ lie inside $|w| = 1$ for all $\boldsymbol{p} \in \mathrm{Relog}^{-1}(\boldsymbol{x})$, and Lemma 6.14 implies this occurs if and only if $\boldsymbol{s} \cdot \nu(\boldsymbol{x})$ equals the total number of zeros of $a_{\boldsymbol{s}}(w) = f(e^{x_1} w^{s_1}, \ldots, e^{x_d} w^{s_d})$ in the complex plane minus the order of the pole of $a_{\boldsymbol{s}}(w)$ at the origin. Since $a_{\boldsymbol{s}}(w)$ has degree $\max_{\boldsymbol{n} \in \mathcal{N}_f}(\boldsymbol{s} \cdot \boldsymbol{n})$, it follows that

$$\begin{aligned}
&\boldsymbol{s} \text{ is in the recession cone of } B \\
&\text{if and only if } \{\boldsymbol{x} + t\boldsymbol{s} : t > 0\} \subset B \text{ for all } \boldsymbol{x} \in B \\
&\text{if and only if } \boldsymbol{s} \cdot \nu(\boldsymbol{x}) = \max_{\boldsymbol{n} \in \mathcal{N}_f}(\boldsymbol{s} \cdot \boldsymbol{n}) \text{ for all } \boldsymbol{x} \in B \\
&\text{if and only if } \boldsymbol{s} \in \mathtt{normal}_{\nu(\boldsymbol{p})}\left(\mathcal{N}_f\right),
\end{aligned}$$

proving the desired equivalence. □

Exercise 6.9. Prove that the order of any bounded component of $\mathtt{amoeba}(f)^c$ is an interior point of $\mathcal{N}_f$, then prove that the recession cone of a component of $\mathtt{amoeba}(f)^c$ has nonempty interior if and only if its order is a vertex of $\mathcal{N}_f$.

Proposition 6.27 allows us to read off properties of the components of the complement $\mathtt{amoeba}(f)^c$ directly from the Newton polytope of f. It also has strong implications for the asymptotic behavior of sequences with multivariate rational generating functions.

Corollary 6.28. *Assume the hypotheses of Proposition 6.24 and let $\boldsymbol{x} \in B$. If the ray $\{\nu(\boldsymbol{x}) + t\hat{\boldsymbol{r}} : t > 0\}$ does not intersect $\mathcal{N}_Q$ then $a_{\boldsymbol{r}_n} = 0$ for all but finitely many n.*

Proof See Exercise 6.17. □

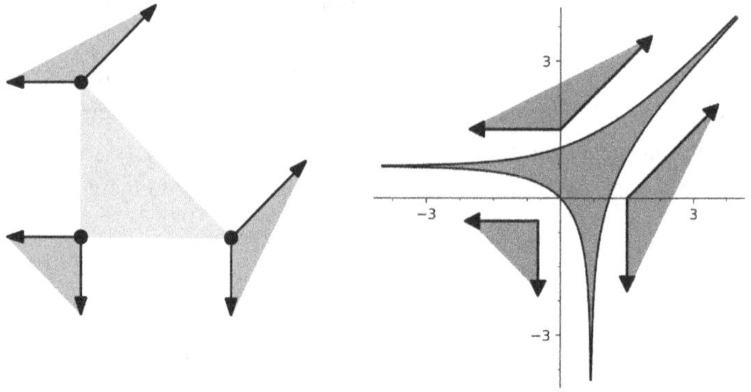

Figure 6.5 *Left:* The Newton polytope of $1 - x - y$ with the normal cones at each vertex. *Right:* A drawing of $\texttt{amoeba}(1 - x - y)$ with the recession cones of each component of the complement added.

Corollary 6.29. *Let $f(\boldsymbol{z})$ be a rational function and let $\{\boldsymbol{r}_n\}$ be a sequence of vectors in $\mathbb{Z}^d$ going to infinity in the non-zero direction $\hat{\boldsymbol{r}} \in \mathbb{R}^d$. Then there is a component B of $\texttt{amoeba}(f)^c$ such that the convergent Laurent expansion $f(\boldsymbol{z}) = \sum_{\boldsymbol{r}\in\mathbb{Z}^d} a_{\boldsymbol{r}} \boldsymbol{z}^{\boldsymbol{r}}$ with domain of convergence $\mathcal{D} = \mathrm{Relog}^{-1}(B)$ has $a_{\boldsymbol{r}_n} = 0$ for all but finitely many n.*

Proof For any non-zero vector $\hat{\boldsymbol{r}} \in \mathbb{R}^d$ there exists a vertex $\boldsymbol{n} \in \mathcal{N}_f$ such that $\{\boldsymbol{n} + t\hat{\boldsymbol{r}} \,:\, t > 0\}$ does not intersect $\mathcal{N}_f$ (any of the vertices of $\mathcal{N}_f$ on the hyperplane touching the boundary of $\mathcal{N}_f$ with outward normal $\hat{\boldsymbol{r}}$ have this property). The desired result then follows from Corollary 6.28 where B is the component of $\texttt{amoeba}(f)^c$ with order $\boldsymbol{n}$, whose existence is guaranteed by Theorem 6.18. □

Corollary 6.29 turns out to be crucial to our framework for ACSV. Roughly, we represent a sequence of interest as a Cauchy integral then move the point $\boldsymbol{x}$ in the domain of integration $\mathrm{Relog}^{-1}(\boldsymbol{x})$ from its starting component of $\texttt{amoeba}(f)^c$ to another component, picking up a *residue integral* over an *intersection cycle* along the way. By moving $\boldsymbol{x}$ into a component where series coefficients are eventually zero, we only need to approximate the residue integral, which can be done using saddle point techniques. Fully describing this argument takes up the majority of Chapter 7.

6.4 Singularities, amoeba boundaries, and minimal points

Let $f(z) = \sum_{r \in \mathbb{Z}^d} a_r z^r$ be any convergent Laurent series representing an analytic function f in the domain of convergence $\mathcal{D} = \text{Relog}^{-1}(B)$. In this section we use the analytic behavior of $f(z)$ to study what happens when the dual rate $\beta^*(\hat{r})$ in the upper bound (6.7) is finite.

Definition 6.30. A point $w \in \mathbb{C}^d$ is a ***singularity*** of f if f can be analytically continued to an open set $O \subset \mathbb{C}^d$ with w on its boundary but cannot be analytically continued to w. The set of all singularities of f forms its ***singular variety*** $\mathcal{V}$.

Although Definition 6.30 is stated for general Laurent series, in this text we almost always consider Laurent expansions of rational functions, in which case it is easy to characterize $\mathcal{V}$.

Lemma 6.31. *If $f(z) = P(z)/Q(z)$ is a ratio of coprime polynomials P and Q then the singular variety of f is $\mathcal{V} = \{z \in \mathbb{C}^d : Q(z) = 0\}$.*

Proof Let $w \in \mathbb{C}^d$ be any point such that $Q(w) = 0$. Our stated result holds if every neighborhood of w in $\mathbb{C}^d$ contains a point p with $P(p) \neq 0$ and $Q(p) = 0$, since then $|f(z)|$ is unbounded in any neighborhood of w. The existence of p follows from the fact that the intersection of algebraic varieties $\mathcal{V}_P \cap \mathcal{V}_Q$ has codimension at least two since P and Q are coprime (if Q is irreducible and P vanishes on a neighborhood of $w \in \mathcal{V}_Q$ then P vanishes on all of $\mathcal{V}_Q$, and thus is in the ideal generated by the radical of Q and not coprime to Q – if Q is reducible then apply this argument to its irreducible factors). □

See also Melczer [Mel21, Proposition 3.2] for an elementary proof of Lemma 6.31 using the division algorithm.

Remark 6.32. If $f(z)$ is a meromorphic function (i.e., can be locally represented by the ratio of analytic functions on its domain) then $w \in \mathbb{C}^d$ is a singularity of f when there exist analytic functions g and h defined in a neighborhood $\mathcal{N}$ of w such that $f(z) = g(z)/h(z)$ on the points of $\mathcal{N}$ where f is defined, h vanishes at w, and g does not identically vanish in any sufficiently small neighborhood of w. In particular, the singular variety of a meromorphic function forms an ***analytic variety***. Note that, in contrast to the univariate case, it is possible for g to vanish at w.

The singular variety of f is defined by its analytic behavior, independent of its various Laurent expansions. In order to discuss the singularities that are relevant to a specific expansion, we thus need further restrictions. First, we show that every point on the boundary of a Laurent series domain of convergence has the same coordinatewise modulus as a singularity of f.

Proposition 6.33. *If* $\boldsymbol{w} \in \partial\mathcal{D}$ *then* $\mathbf{T}(\boldsymbol{w}) \cap \mathcal{V} \neq \emptyset$. *Conversely, if* $\boldsymbol{p} \in \mathcal{D}$ *and* $\boldsymbol{w} \in \mathcal{V}$, *and* $\mathbf{T}(\boldsymbol{z}) \cap \mathcal{V} = \emptyset$ *for all* $\boldsymbol{z}$ *in the open line segment between* $(|p_1|, \ldots, |p_d|)$ *and* $(|w_1|, \ldots, |w_d|)$, *then* $\boldsymbol{w} \in \partial\mathcal{D}$.

Proof *(First statement)* For simplicity we first prove the result for power series expansions, then note how the argument changes for the general Laurent series case. Towards a contradiction, suppose that $\boldsymbol{w} \in \partial\mathcal{D}$ and $\mathbf{T}(\boldsymbol{w}) \cap \mathcal{V} = \emptyset$. The Cauchy integral expression

$$a_{\boldsymbol{r}} = \left(\frac{1}{2\pi i}\right)^d \int_{T(\boldsymbol{y})} f(\boldsymbol{z})\boldsymbol{z}^{-\boldsymbol{r}-\mathbf{1}}\, d\boldsymbol{z} \tag{6.8}$$

holds for all $\boldsymbol{y} \in \mathcal{D}$ and is unchanged if we move $\boldsymbol{y}$ through points where $\mathbf{T}(\boldsymbol{y}) \cap \mathcal{V} = \emptyset$. In particular, if $\mathbf{T}(\boldsymbol{w}) \cap \mathcal{V} = \emptyset$ then for some $\varepsilon > 0$ we have $\mathbf{T}(\boldsymbol{y}) \cap \mathcal{V} = \emptyset$ whenever $|y_k - w_k| < 2\varepsilon$ for all $k = 1, \ldots, d$ and (6.8) holds with $\boldsymbol{y} = \boldsymbol{w} + \varepsilon\mathbf{1}$. A maximum modulus bound then gives $a_{\boldsymbol{r}} \leq C|w_1 + \varepsilon|^{-r_1} \cdots |w_d + \varepsilon|^{-r_d}$ for some $C > 0$ and all $\boldsymbol{r} \in \mathbb{N}^d$. Our proof of Proposition 6.8 above showed that $\mathcal{D}$ is the interior of

$$\left\{ \boldsymbol{z} \in \mathbb{C}^d : \sup_{\boldsymbol{r} \in \mathbb{N}^d} |a_{\boldsymbol{r}}| \cdot |\boldsymbol{z}|^{\boldsymbol{r}} \text{ is finite} \right\}$$

and $a_{\boldsymbol{r}} \leq C|w_1 + \varepsilon|^{-r_1} \cdots |w_d + \varepsilon|^{-r_d}$ implies the supremum is finite for all $\boldsymbol{z}$ in a neighborhood of $\boldsymbol{w}$, contradicting that $\boldsymbol{w} \in \partial\mathcal{D}$. In the general Laurent series case, we need to take the sign of the coordinates of $\boldsymbol{r}$ into account. In particular, we want to prove that

$$\sup_{\boldsymbol{r} \in \mathbb{Z}^d} |a_{\boldsymbol{r}}| \cdot |\boldsymbol{z}|^{\boldsymbol{r}} = \sup_{\boldsymbol{\sigma} \in \{\pm 1\}^d} \sup_{\boldsymbol{r} \in \mathbb{N}^d} |a_{\boldsymbol{\sigma} \odot \boldsymbol{r}}| \cdot |\boldsymbol{z}|^{\boldsymbol{\sigma} \odot \boldsymbol{r}}$$

is finite for all $\boldsymbol{z}$ in some neighborhood of $\boldsymbol{w}$, where as above we use the notation $\boldsymbol{\sigma} \odot \boldsymbol{r} = (\sigma_1 r_1, \ldots, \sigma_d r_d)$. For fixed $\boldsymbol{\sigma} \in \{\pm 1\}^d$ we apply a maximum modulus bound to (6.8) with $\boldsymbol{y} = \boldsymbol{w} + \varepsilon\boldsymbol{\sigma}$ to get

$$a_{\boldsymbol{\sigma} \odot \boldsymbol{r}} \leq C|w_1 + \sigma_1\varepsilon|^{-\sigma_1 r_1} \cdots |w_d + \sigma_d\varepsilon|^{-\sigma_d r_d}$$

for some $C > 0$ and all $\boldsymbol{r} \in \mathbb{N}^d$, and thus

$$a_{\boldsymbol{\sigma} \odot \boldsymbol{r}} |\boldsymbol{z}|^{\boldsymbol{\sigma} \odot \boldsymbol{r}} \leq C \left(\frac{|z_1|}{|w_1 + \sigma_1\varepsilon|}\right)^{-\sigma_1 r_1} \cdots \left(\frac{|z_d|}{|w_d + \sigma_d\varepsilon|}\right)^{-\sigma_d r_d}. \tag{6.9}$$

For $\boldsymbol{z}$ in a sufficiently small neighborhood of $\boldsymbol{w}$ in $\mathbb{C}^d$, each term on the right-hand side of (6.9) is less than 1 if it has a positive power, and greater than 1 if it has a negative power, so the supremum over all $\boldsymbol{r} \in \mathbb{N}^d$ is finite. This holds for all $\boldsymbol{\sigma} \in \{\pm 1\}^d$, and once again we have shown $\boldsymbol{z} \in \mathcal{D}$ for $\boldsymbol{z}$ in a neighborhood of $\boldsymbol{w}$, contradicting $\boldsymbol{w} \in \partial\mathcal{D}$.

(Second statement) As in the proof of the first statement, if $\mathbf{T}(\boldsymbol{z}) \cap \mathcal{V} = \emptyset$ for all $\boldsymbol{z}$ in the open line segment between $(|p_1|, \dots, |p_d|)$ and $(|w_1|, \dots, |w_d|)$ then all points on this line segment lie in $\mathcal{D}$, so $\boldsymbol{w} \in \overline{\mathcal{D}}$. Since $\boldsymbol{w} \in \mathcal{V}$ we cannot have $\boldsymbol{w} \in \mathcal{D}$, and thus $\boldsymbol{w} \in \partial\mathcal{D}$. □

The singularities on the boundary of the domain of convergence of a Laurent expansion are very important to the asymptotic analysis of the corresponding series coefficients.

Definition 6.34 (minimal points). A point $\boldsymbol{w} \in \mathcal{V}$ is called a ***minimal point*** with respect to the Laurent expansion $f(\boldsymbol{z}) = \sum_{\boldsymbol{r} \in \mathbb{Z}^d} a_{\boldsymbol{r}} \boldsymbol{z}^{\boldsymbol{r}}$ (or the domain of convergence $\mathcal{D}$, or the convex set B) if $\boldsymbol{w} \in \mathcal{V} \cap \partial\mathcal{D}$. Equivalently, $\boldsymbol{w} \in \mathcal{V}$ is minimal when $\operatorname{Relog}(\boldsymbol{w}) \in \partial B$. We further say $\boldsymbol{w}$ is a ***finitely minimal point*** if $\mathbf{T}(\boldsymbol{w}) \cap \mathcal{V}$ is finite, and a ***strictly minimal point*** if $\mathbf{T}(\boldsymbol{w}) \cap \mathcal{V}$ is the singleton $\{\boldsymbol{w}\}$.

Example 6.35. Let $0 < p < 1$ and define

$$Q_1(x, y) = 1 - x - y \qquad \sigma_1 = (p, 1 - p)$$

$$Q_2(x, y) = 1 - x^2 - y \qquad \sigma_2 = \left(\sqrt{1 - p}, p\right)$$

$$Q_3(x, y) = 1 - xy \qquad \sigma_3 = (p, 1/p).$$

Consider $\boldsymbol{\sigma}_k$ with respect to the power series expansion of $1/Q_k(x, y)$. Then $\boldsymbol{\sigma}_1$ is a strictly minimal point since $|x + y| < |x| + |y|$ unless x is a nonnegative real multiple of y, and $\boldsymbol{\sigma}$ is the only point on $\mathbf{T}(\boldsymbol{\sigma}) \cap \mathcal{V}_{Q_1}$ satisfying that condition. Similarly, $\boldsymbol{\sigma}_2$ is a finitely minimal point, but it is not strictly minimal because $\left(-\sqrt{1 - p}, p\right) \in \mathcal{V}_{Q_2}$. Finally, $\boldsymbol{\sigma}_3$ is minimal but it is not finitely minimal as $(p\omega, 1/(p\omega))$ lies in $\mathbf{T}(\boldsymbol{\sigma}_3) \cap \mathcal{V}_{Q_3}$ whenever $|\omega| = 1$. ◁

Proposition 6.33 allows us to test minimality.

Corollary 6.36 (minimality test). *If $\boldsymbol{y} \in \mathcal{D}$ then $\boldsymbol{w} \in \mathcal{V}$ is minimal if and only if the open line segment*

$$\{t \operatorname{Relog}(\boldsymbol{y}) + (1 - t) \operatorname{Relog}(\boldsymbol{w}) : t \in (0, 1)\}$$

from $\operatorname{Relog}(\boldsymbol{y})$ *to* $\operatorname{Relog}(\boldsymbol{w})$ *stays in B. If we consider the power series expansion of f, then $\boldsymbol{w}$ is minimal if and only if $\mathbf{T}(t\boldsymbol{w}) \cap \mathcal{V} = \emptyset$ for all $t \in (0, 1)$.*

Proof The first statement follows from the convexity of B and the fact that $\boldsymbol{w}$ is minimal if and only if $\operatorname{Relog}(\boldsymbol{w}) \in \partial B$. The second condition follows from Proposition 6.33: since $\mathbf{0}$ lies in the power series domain of convergence, if

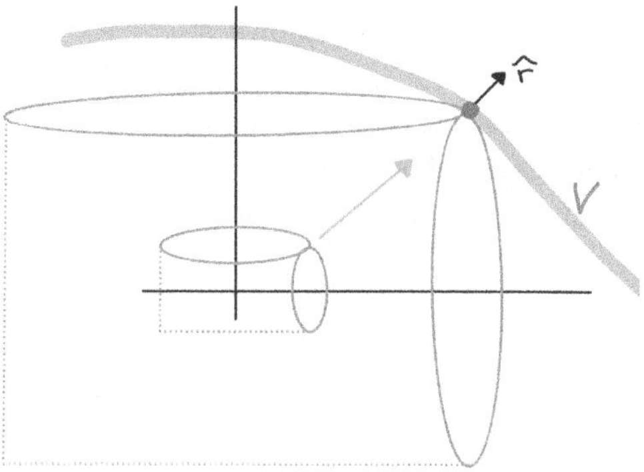

Figure 6.6 $\mathcal{V}$ does not intersect the open polydisk with the large torus on its boundary.

$\mathbf{T}(t\boldsymbol{w}) \cap \mathcal{V} = \emptyset$ for all $t \in (0, 1)$ then $\boldsymbol{w}$ is minimal, and if $\boldsymbol{w}$ is minimal then no point with coordinatewise smaller modulus can lie in $\mathcal{V}$. □

Although the hypotheses of Corollary 6.36 can be checked computationally, this requires one to decide the satisfiability of inequalities involving the moduli of coordinates, which is computationally expensive (see Chapter 8 for further discussion on algorithms to certify minimality). Luckily, there is one special case that occurs often in applications and greatly simplifies the analysis.

Definition 6.37. A ***combinatorial series*** is a Laurent series $\sum_{\boldsymbol{r}\in\mathbb{Z}^d} a_{\boldsymbol{r}} \boldsymbol{z}^{\boldsymbol{r}}$ such that $a_{\boldsymbol{r}} \geq 0$ for all but a finite number of $\boldsymbol{r} \in \mathbb{Z}^d$.

Proposition 6.38 (multivariate Vivanti–Pringsheim). *Let $f(\boldsymbol{z}) = \sum_{\boldsymbol{r}\in\mathbb{Z}^d} a_{\boldsymbol{r}} \boldsymbol{z}^{\boldsymbol{r}}$ be a convergent combinatorial power series expansion with domain of convergence $\mathcal{D}$. Then $\boldsymbol{w} \in \mathcal{V}_*$ is a minimal point if and only if $(|w_1|, \ldots, |w_d|)$ is also a minimal point (and, in particular, lies in $\mathcal{V}_*$).*

Proposition 6.38 is a multivariate generalization of the classical univariate case [FS09, Theorem IV.6], and is proved analogously.

Proof Towards a contradiction, suppose that $\boldsymbol{w} \in \partial\mathcal{D} \cap \mathbb{R}^d_{>0}$ and f is analytic at $\boldsymbol{w}$. Then there exists some $\varepsilon > 0$ such that f is analytic in a ball of radius $4d\varepsilon$ centered at $\boldsymbol{w}$. The Taylor expansion of f at $\boldsymbol{v} = \boldsymbol{w} - \varepsilon\mathbf{1}$ is

$$f(\boldsymbol{z}) = \sum_{\boldsymbol{r}\in\mathbb{N}^d} \frac{f^{(\boldsymbol{r})}(\boldsymbol{v})}{\boldsymbol{r}!}(\boldsymbol{z} - \boldsymbol{v})^{\boldsymbol{r}},$$

where $\boldsymbol{r}! = r_1! \cdots r_d!$. Differentiating the power series expansion of f at the origin, which is valid when $\boldsymbol{z} = \boldsymbol{v} \in \mathcal{D}$, gives

$$\frac{f^{(\boldsymbol{r})}(\boldsymbol{v})}{\boldsymbol{r}!} = \sum_{\boldsymbol{s}\in\mathbb{N}^d} \binom{\boldsymbol{s}}{\boldsymbol{r}} a_{\boldsymbol{s}} \boldsymbol{v}^{\boldsymbol{s}-\boldsymbol{r}}$$

where $\binom{\boldsymbol{s}}{\boldsymbol{r}} = \frac{\boldsymbol{s}!}{\boldsymbol{r}!(\boldsymbol{s}-\boldsymbol{r})!}$ if $s_i \geq r_i$ for all $1 \leq i \leq d$ and 0 otherwise, so

$$f(\boldsymbol{z}) = \sum_{\boldsymbol{r}\in\mathbb{N}^d} \sum_{\boldsymbol{s}\in\mathbb{N}^d} \binom{\boldsymbol{s}}{\boldsymbol{r}} a_{\boldsymbol{s}} \boldsymbol{v}^{\boldsymbol{s}-\boldsymbol{r}} (\boldsymbol{z}-\boldsymbol{v})^{\boldsymbol{r}}. \tag{6.10}$$

By construction the expansion (6.10) is valid at $\boldsymbol{z} = \boldsymbol{w}+\varepsilon\mathbf{1} = \boldsymbol{v}+2\varepsilon\mathbf{1}$, meaning

$$\begin{aligned} f(\boldsymbol{w}+\varepsilon\mathbf{1}) &= \sum_{\boldsymbol{r}\in\mathbb{N}^d} \sum_{\boldsymbol{s}\in\mathbb{N}^d} \binom{\boldsymbol{s}}{\boldsymbol{r}} a_{\boldsymbol{s}} \boldsymbol{v}^{\boldsymbol{s}-\boldsymbol{r}} (2\varepsilon\mathbf{1})^{\boldsymbol{r}} \\ &= \sum_{\boldsymbol{s}\in\mathbb{N}^d} a_{\boldsymbol{s}} \sum_{\boldsymbol{r}\in\mathbb{N}^d} \binom{\boldsymbol{s}}{\boldsymbol{r}} \boldsymbol{v}^{\boldsymbol{s}-\boldsymbol{r}} (2\varepsilon\mathbf{1})^{\boldsymbol{r}} \\ &= \sum_{\boldsymbol{s}\in\mathbb{N}^d} a_{\boldsymbol{s}} (\boldsymbol{v}+2\varepsilon\mathbf{1})^{\boldsymbol{r}}, \end{aligned}$$

where we can exchange the infinite series in $\boldsymbol{r}$ and $\boldsymbol{s}$ because all coefficients $a_{\boldsymbol{s}}$ are nonnegative. This series converges, so $\mathbf{T}(\boldsymbol{v}+2\varepsilon\mathbf{1}) = \mathbf{T}(\boldsymbol{w}+\epsilon\mathbf{1}) \subset \mathcal{D}$, contradicting the fact that $\boldsymbol{w} \in \partial\mathcal{D}$. □

Proposition 6.38 is valid for ordinary power series, without regard to the type of singularity. The proof for polar singularities is even easier, and generalizes to Laurent expansions.

Exercise 6.10. Let $f(\boldsymbol{z}) = \sum_{\boldsymbol{r}\in\mathbb{Z}^d} a_{\boldsymbol{r}} \boldsymbol{z}^{\boldsymbol{r}}$ be a convergent combinatorial Laurent expansion that represents a meromorphic function on its domain of convergence $\mathcal{D}$. Prove that $\boldsymbol{w} \in \mathcal{V}_*$ is a minimal point if and only if $(|w_1|, \ldots, |w_d|)$ is also a minimal point. *Hint:* First prove that $|f(\boldsymbol{z})| \leq f(|z_1|, \ldots, |z_d|)$ for all $\boldsymbol{z} \in \mathcal{D}$.

Combining this result with Proposition 6.33 immediately gives the following simplified test for minimality in the rational combinatorial case.

Corollary 6.39 (combinatorial minimality test). *Let B be a component of* $\texttt{amoeba}(Q)^c$ *and $f(\boldsymbol{z}) = P(\boldsymbol{z})/Q(\boldsymbol{z})$ be the ratio of coprime polynomials P and Q with combinatorial series expansion on $\mathcal{D} = \mathrm{Relog}^{-1}(B)$. If $\boldsymbol{x} \in \mathcal{D}$ then $\boldsymbol{w} \in \mathcal{V}$ is minimal if and only if*

$$Q\Big(|x_1|^t|w_1|^{1-t},\ \ldots,\ |x_d|^t|w_d|^{1-t}\Big) \neq 0 \textit{ for all } t \in (0,1).$$

When considering the power series expansion of $f(\boldsymbol{z})$, the point $\boldsymbol{w} \in \mathcal{V}$ is *minimal if and only if*

$$Q\left(t|w_1|, \ldots, t|w_d|\right) \neq 0 \textit{ for all } t \in (0, 1).$$

Remark 6.40. When f is a ratio of fixed analytic functions P and Q on $\mathcal{D}$, Corollary 6.39 can be adapted to give an analogous test for minimality, once a notion of *coprime germs of analytic functions* is defined. Of course, computation is generally harder when dealing with analytic functions because techniques for solving polynomial systems cannot be applied.

Proof Proposition 6.38 and Corollary 6.36 imply that for a combinatorial expansion $\boldsymbol{w}$ is minimal if and only if

$$\left(e^{t\log|x_1|+(1-t)\log|w_1|}, \ldots, e^{t\log|x_d|+(1-t)\log|w_d|}\right) \notin \mathcal{V}$$

for $t \in (0, 1)$ and, in the power series case, if and only if $\left(t|w_1|, \ldots, t|w_d|\right) \notin \mathcal{V}$ for $t \in (0, 1)$. The stated conclusion then follows from Lemma 6.31. □

Perhaps surprisingly, even in the univariate case it is unknown whether determining combinatoriality of a rational function is decidable. In practice, most multivariate generating functions we encounter are combinatorial, although there are many examples with combinatorially interesting nonnegative coefficients on a diagonal and negative coefficients in off-diagonal terms. In some circumstances, the structure of the function under consideration can also help characterize minimality.

Lemma 6.41. *Let* $f(\boldsymbol{z}) = P(\boldsymbol{z})/Q(\boldsymbol{z})$ *be the ratio of coprime polynomials* P *and* Q*, where* $Q(\boldsymbol{z}) = 1 - q(\boldsymbol{z})$ *for a combinatorial polynomial* q *such that* $q(\mathbf{0}) = 0$*. Then the following statements hold for the power series expansion of* f.

a) Every root of Q *with positive coordinates is a minimal point of* $\mathcal{V}$.

b) If $\boldsymbol{z}$ *is a minimal point then so is* $(|z_1|, \ldots, |z_d|)$.

c) If q *is* ***aperiodic*** *(meaning every element of* $\mathbb{Z}^d$ *can be written as an integer sum of the exponents appearing in* q*) then the only minimal points of* f *are the roots of* Q *with positive coordinates (and thus each is strictly minimal).*

Proof Part (b) follows from Proposition 6.38 applied to $1/Q(\boldsymbol{z})$. For the other parts, note that by assumption we can write $q(\boldsymbol{z}) = \sum_{\boldsymbol{r}\in\mathbb{N}^d} q_{\boldsymbol{r}}\boldsymbol{z}^{\boldsymbol{r}}$, where each $q_{\boldsymbol{r}} \geq 0$. Suppose that $\boldsymbol{w}$ is a root of Q and with positive coordinates and let $\boldsymbol{z}$

be any point with the same coordinate-wise modulus. Then

$$|q(t|z_1|, \dots t|z_d|)| = |q(t\boldsymbol{w})| = \sum_{\boldsymbol{r} \in \mathbb{N}^d} a_{\boldsymbol{r}} t^{|\boldsymbol{r}|} \boldsymbol{w}^{\boldsymbol{r}} < \sum_{\boldsymbol{r} \in \mathbb{N}^d} a_{\boldsymbol{r}} \boldsymbol{w}^{\boldsymbol{r}} = q(\boldsymbol{w}) = 1$$

for $t \in (0, 1)$, so $\boldsymbol{w}$ is minimal by Corollary 6.39. Furthermore, if q is aperiodic and $\boldsymbol{z}$ has the same coordinate-wise modulus as a minimal point $\boldsymbol{w}$ but does not have positive coordinates then the arguments of monomials $\boldsymbol{z}^{\boldsymbol{r}}$ appearing in $q(\boldsymbol{z})$ are not all equal, and the triangle inequality implies

$$|q(\boldsymbol{z})| = \left| \sum_{\boldsymbol{r} \in \mathbb{N}^d} q_{\boldsymbol{r}} \boldsymbol{z}^{\boldsymbol{r}} \right| < \sum_{\boldsymbol{r} \in \mathbb{N}^d} q_{\boldsymbol{r}} \boldsymbol{w}^{\boldsymbol{r}} = q(\boldsymbol{w}) = 1,$$

so $\boldsymbol{z}$ is not minimal. □

Remark 6.42. When f is a ratio of coprime analytic functions P and Q on $\mathcal{D}$, where $Q(\boldsymbol{z}) = 1 - q(\boldsymbol{z})$ for a combinatorial power series q, then the results of Lemma 6.41 still hold. Without aperiodicity, $\boldsymbol{w}$ may not even be finitely minimal, as the periodic example $q(x, y) = 1 - xy$ illustrates.

Lemma 6.43 (Grace-Walsh-Szegő theorem). *Let $Q(\boldsymbol{z})$ be a polynomial in $\mathbb{C}[\boldsymbol{z}]$ such that Q is invariant under permuting its variables and is linear in each variable. If $Q(\boldsymbol{w}) = 0$ for some $\boldsymbol{w} \in \mathbb{C}^d$ with $|w_1|, \dots, |w_d| < R$ then $Q(x, x, \dots, x) = 0$ for some $|x| < R$.*

Proof This classical lemma is proved, among other places, in [BB09, Theorem 1.1]. It also follows from the Borcea–Brändén symmetrization lemma, which we return to in Theorem 13.8 of Chapter 13. □

Exercise 6.11. Let $Q(z_1, z_2, z_3, z_4) = 3 - z_1 - z_2 - z_3 - z_4 + z_1 z_2 z_3 z_4$. Determine whether $(1, 1, 1, 1)$ is a minimal point

for Q by using the Grace-Walsh-Szegő theorem. As a challenge you can also try to prove minimality from first principles, but this is quite tricky.

Returning to our original motivation, we care about minimal points because they provide the best upper bound in the exponential growth bound (6.6). A ***supporting hyperplane*** to a convex set C at $\boldsymbol{y} \in C$ is a hyperplane through $\boldsymbol{y}$ such that all elements of C lie on one side of the hyperplane; a normal $\boldsymbol{v}$ to such a supporting hyperplane is an ***inward-facing normal*** if $(\boldsymbol{x} - \boldsymbol{y}) \cdot \boldsymbol{v} \geq 0$ for all $\boldsymbol{x} \in C$ and an ***outward-facing normal*** if $(\boldsymbol{x} - \boldsymbol{y}) \cdot \boldsymbol{v} \leq 0$ for all $\boldsymbol{x} \in C$ (one of these conditions must be true). The ***logarithmic gradient*** of a differentiable function Q at $\boldsymbol{w} \in \mathbb{C}^d$ is the vector $(\nabla_{\log} Q)(\boldsymbol{w}) = (w_1 Q_{z_1}(\boldsymbol{w}), \dots, w_d Q_{z_d}(\boldsymbol{w}))$.

Theorem 6.44. *Let $f(z) = P(z)/Q(z)$ be the ratio of coprime polynomials $P, Q \in \mathbb{R}[z]$ with series expansion $f(z) = \sum_{r \in \mathbb{Z}^d} a_r z^r$ on the domain of convergence $\mathcal{D} = \text{Relog}^{-1}(B)$.*

a) Let $\hat{r} \in \mathbb{R}^d \setminus \{0\}$. If the minimum of $-\hat{r} \cdot x$ over $x \in \overline{B}$ is finite then it is achieved by $x = \text{Relog}(w)$ for a minimal point $w \in \mathbb{C}_*^d$ and there exists a supporting hyperplane to $\overline{B}$ at $\text{Relog}(w)$ with outward normal $\hat{r}$.

b) If $w \in \mathbb{C}_*^d$ is a minimal point then $(\nabla_{\log} Q)(w) = \lambda v$ for some $\lambda \in \mathbb{C}_*$ and $v \in \mathbb{R}^d$. If $v \neq 0$ then the hyperplane through $\text{Relog}(w)$ with normal v is a supporting hyperplane to $\overline{B}$. If v is an outward-facing normal for this supporting hyperplane then $x = \text{Relog}(w)$ achieves the minimum of $-v \cdot x$ on $\overline{B}$ (if v is an inward-facing normal then the maximum of $-v \cdot x$ on $\overline{B}$ is achieved at this point).

The ***contour of an amoeba*** amoeba(Q), denoted $C(Q)$, is the collection of points in $\mathcal{V}_Q \cap \mathbb{C}_*^d$ where $\nabla_{\log} Q$ is a multiple of a real vector. Theorem 6.44 implies all boundary points of amoeba(Q) lie in the contour which, as we will see in Chapter 8, can be useful for sketching amoebas.

Proof Part (a) follows from the fact that on the closed convex set $\overline{B}$ the linear function $-\hat{r} \cdot x$ is either unbounded from below or achieves its minimum on at least one point $y \in \partial B$, which by Proposition 6.33 corresponds to at least one minimal point. If $\hat{r}$ is not an outward-facing normal to a supporting hyperplane of $\overline{B}$ through y then there exists $b \in B$ with $(b - y) \cdot \hat{r} > 0$, contradicting that y minimizes $-\hat{r} \cdot x$ on $\overline{B}$.

To prove part (b), let $\log : \mathbb{C}_*^d \to \mathbb{R}^d$ be a branch of the logarithm $\log(z) = (\log(z_1), \ldots, \log(z_d))$ and suppose $w \in \mathcal{V} \cap \mathbb{C}_*^d$ with $(\nabla Q)(w) \neq 0$. If w is a minimal point then the tangent spaces to $\mathcal{V}$ and $\mathbf{T}(w)$ at w can't jointly span all of $\mathbb{C}^d$ as a $2d$-dimensional real vector space. Thus, since the tangent space to $\mathcal{V}$ has real dimension $2d - 2$ and the tangent space to $\mathbf{T}(w)$ has real dimension d, if w is minimal then the tangent spaces of $\mathcal{V}$ and $\mathbf{T}(w)$ at w intersect in a subspace of dimension at least $d - 1$. This final condition is equivalent to the tangent spaces of $\log(\mathcal{V}_Q)$ and $\log(\mathbf{T}(w))$ at $\log(w)$ intersecting in a subspace of dimension at least $d - 1$. Because

$$\log(\mathbf{T}(w)) = \{(\log|w_1| + i\theta_1, \ldots, \log|w_d| + i\theta_d) \text{ for all } \theta_j \in (-\pi, \pi)\},$$

the tangent space to $\log(\mathbf{T}(w))$ at $\log(w)$ is $i\mathbb{R}^d \subset \mathbb{C}^d$, and w is a minimal point only if the tangent space to $\log(\mathcal{V}_Q)$ contains $d - 1$ linearly independent vectors in $i\mathbb{R}^d$. The tangent space to $\log(\mathcal{V}_Q)$ at $\log(w)$ is the hypersurface with normal $(\nabla_{\log} Q)(w) = a + ib$ for some $a, b \in \mathbb{R}^d$. When $(\nabla_{\log} Q)(w)$ is normal to $d - 1$ purely imaginary vectors then the real matrix with rows a and b has

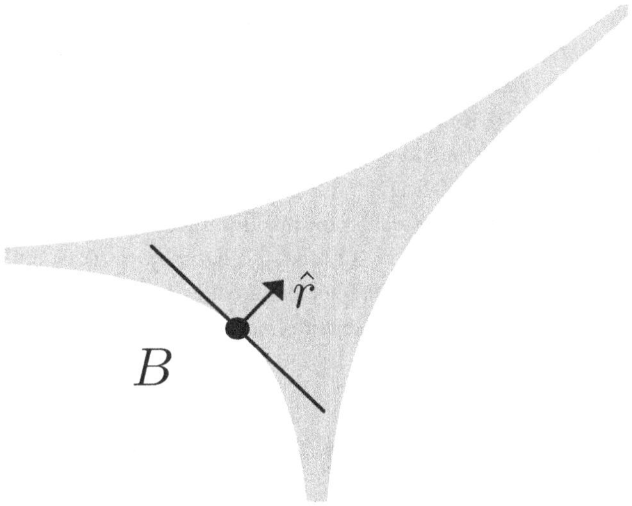

Figure 6.7 The dual rate of an expansion is determined by a point on the boundary of the amoeba where $\hat{r}$ is the normal to a supporting hyperplane.

rank one, so $\boldsymbol{b} = \tau\boldsymbol{a}$ for some $\tau \in \mathbb{R}$ and $(\nabla_{\log} Q)(\boldsymbol{w}) = (1+i\tau)\boldsymbol{a}$ is a multiple of a real vector. Thus $(\nabla_{\log} Q)(\boldsymbol{w})$ is a multiple of a real vector $\boldsymbol{v}$ whenever $\boldsymbol{w}$ is a minimal point with $(\nabla Q)(\boldsymbol{w}) \neq \boldsymbol{0}$. If $\boldsymbol{w}$ is a minimal point with $(\nabla Q)(\boldsymbol{w}) = \boldsymbol{0}$ then the same result holds with $\boldsymbol{v} = \boldsymbol{0}$. If the real vector $\boldsymbol{v}$ is non-zero then it is the normal vector to the tangent plane of a neighborhood of $\boldsymbol{w}$ in $\mathcal{V}$ under the Relog map, meaning it is the normal to a supporting hyperplane of the convex component B of $\texttt{amoeba}(Q)^c$ with $\boldsymbol{w}$ on its boundary. When $\boldsymbol{v}$ is an outward normal then the point $\text{Relog}(\boldsymbol{w})$ is a minimizer of $-\boldsymbol{v} \cdot \boldsymbol{x}$ on $\overline{B}$ for the same reason as our argument in part (a). □

Theorem 6.44 will be used in later chapters to help guide deformations of domains of integration for Cauchy integrals representing series coefficients of interest. Unfortunately, even if the minimum of $-\hat{\boldsymbol{r}} \cdot \boldsymbol{x}$ over $\boldsymbol{x} \in \overline{B}$ is finite and achieved, the bound it gives on exponential growth may not be tight. There are two main reasons for this: first, an amoeba is the "real shadow" of a variety living in complex space, and the real picture may not capture what happens in complex space; second, in non-generic cases the numerator polynomial P may cause the coefficients to grow slower than expected in fixed directions, as we saw already in Example 6.21.

Example 6.45. Let $F = 1/Q(x,y)$, where Q is the product of $Q_1(x,y) = 3 - x - 2y$ and $Q_2(x,y) = 3 + 2x + y$. The amoeba of Q is the same as the amoeba of

$(3-2x-y)(3-x-2y)$ and is shown on the right-hand side of Figure 6.2. Consider the main diagonal direction $\hat{\boldsymbol{r}} = (1/2, 1/2)$. On the component corresponding to the power series expansion of F, the minimum of $-\hat{\boldsymbol{r}} \cdot \boldsymbol{x}$ over $\boldsymbol{x} \in \overline{B}$ is 0, occurring at $\boldsymbol{x} = (0, 0)$, however we will see in Chapter 10 that the growth rate in this direction is $\log(8/9)$, not 0. Replacing Q_2 by the polynomial $3 - 2x - y$ does yield a growth rate of 0 (in fact, this growth rate holds for an entire cone of directions containing $\hat{\boldsymbol{r}}$). The reason for this difference is that in the first case the points on the lines defined by Q_1 and Q_2 that map to $(0, 0)$ under Relog have different signs, so the amoeba intersection does not reflect an intersection of varieties, while in the second case both lines contain the point $(1, 1)$. ◁

6.5 Additional constructions

We end this chapter by defining some additional constructions that will be required in Chapters 10 and 11.

Definition 6.46. If $f : \mathbb{C}^d \to \mathbb{C}$ is analytic at $\boldsymbol{w} \in \mathbb{C}^d$ then the ***order of vanishing*** of f at $\boldsymbol{w}$ is

$$\deg(f, \boldsymbol{w}) := \sup\left\{n \in \mathbb{N} : f(\boldsymbol{w} + \boldsymbol{z}) = O(|\boldsymbol{z}|^n) \text{ as } |\boldsymbol{z}| \to 0\right\},$$

and the ***homogeneous part*** $\texttt{hom}(f, \boldsymbol{w})$ of f at $\boldsymbol{w}$ is the sum of all terms of degree $\deg(f, \boldsymbol{w})$ in the power series for $f(\boldsymbol{w} + \boldsymbol{z})$ at $\boldsymbol{z} = \boldsymbol{0}$, so

$$f(\boldsymbol{w} + \boldsymbol{z}) = \texttt{hom}(f, \boldsymbol{w})(\boldsymbol{z}) + O\left(|\boldsymbol{z}|^{\deg(f, \boldsymbol{z})+1}\right).$$

The ***algebraic tangent cone*** of f at $\boldsymbol{w}$ is the set

$$\texttt{algtan}_{\boldsymbol{w}}(f) = \mathcal{V}_{\texttt{hom}(f, \boldsymbol{w})}.$$

When $\boldsymbol{w} = \boldsymbol{0}$ we omit $\boldsymbol{w}$ from the notation and write $\texttt{hom}(f) = \texttt{hom}(f, \boldsymbol{0})$.

Remark. The algebraic tangent cone is not always a cone. For instance, if $f(x, y) = xy$ then the algebraic tangent cone $\mathcal{V}_{xy}$ is the union of the x and y axes.

A more geometric definition of the algebraic tangent cone is that it consists of lines through $\boldsymbol{x}$ that are the limits of secant lines through $\boldsymbol{x}$. Thus, for a unit vector $\boldsymbol{u}$ the line $\boldsymbol{x} + t\boldsymbol{u}$ lies in the algebraic tangent cone if there are points $\boldsymbol{x}_n \in \mathcal{V}_f$ distinct from but converging to $\boldsymbol{x}$ for which $(\boldsymbol{x}_n - \boldsymbol{x})/\|\boldsymbol{x}_n - \boldsymbol{x}\| \to \pm\boldsymbol{u}$. Equivalence of the two definitions follows from our next result, which we state after introducing some new terminology.

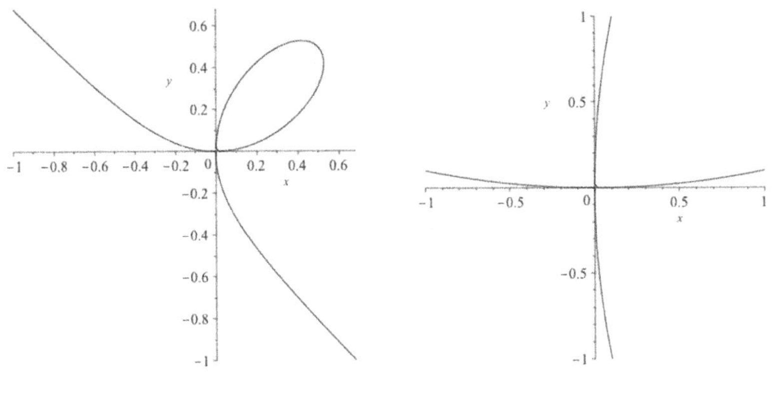

Figure 6.8 The real parts of $\mathcal{V}$ and $\mathcal{V}_\varepsilon$ with $\varepsilon = 1/10$.

Definition 6.47. If (X, d) is a metric space then the ***Hausdorff metric*** ρ on compact subsets of X is defined stepwise by

$$d(x, Y) = \inf_{y \in Y} d(x, y) \qquad (x \in X \text{ and } Y \subset X)$$

$$d(Z, Y) = \sup_{z \in Z} d(z, Y) \qquad (Z, Y \subset X)$$

$$\rho(Z, Y) = \max\{d(Z, Y), d(Y, Z)\} \qquad (Z, Y \subset X).$$

Lemma 6.48 (algebraic tangent cone is the limiting secant cone)**.** *Let $Q(\mathbf{z})$ be a polynomial vanishing to degree $m \geq 1$ at the origin. Define the m-homogeneous part $A(\mathbf{z}) = \mathtt{hom}(Q)$, the remainder*

$$R(\mathbf{z}) = Q(\mathbf{z}) - A(\mathbf{z}),$$

and, for $\varepsilon > 0$, the scaled polynomial $Q_\varepsilon(\mathbf{z}) = \varepsilon^{-m} Q(\varepsilon \mathbf{z})$. If $\mathcal{V}_\varepsilon$ denotes the intersection of the zero set of Q_ε with the unit sphere S_1 and $\mathcal{V}_0$ denotes the intersection of the zero set of A with S_1 then $\mathcal{V}_\varepsilon \to \mathcal{V}_0$ in the Hausdorff metric as $\varepsilon \to 0$.

Example 6.49. Let $Q(x, y) = x^3 - xy + y^3$, which vanishes to order $m = 2$. The leading homogeneous part is $A(x, y) = -xy$, the remainder is $R(x, y) = x^3 + y^3$, and $Q_\varepsilon(x, y) = \varepsilon(x^3 + y^3) - xy$. Figure 6.8 shows $\mathcal{V}$ and $\mathcal{V}_\varepsilon$ for $\varepsilon = 1/10$, illustrating that $\mathcal{V}_\varepsilon$ is close to the union of the x and y axes $\mathcal{V}_0 = \mathcal{V}_A$. ◁

Proof On any compact set, including S_1, the rescaled remainder $\varepsilon^{-m} R(\varepsilon \mathbf{z})$ goes to zero uniformly. If $\mathbf{z}_n \to \mathbf{z}$ and $\mathbf{z}_n \in \mathcal{V}_{1/n}$ then, for each n,

$$|A(\mathbf{z}_n)| = \left|Q_{1/n}(\mathbf{z}_n) - R_{1/n}(\mathbf{z}_n)\right| = \left|R_{1/n}(\mathbf{z}_n)\right| \to 0,$$

so $A(z) = 0$ by the continuity of A. Thus, any limit point of $\mathcal{V}_\varepsilon$ as $\varepsilon \to 0$ lies in $\mathcal{V}_0$.

Conversely, fix a unit vector $z \in \mathcal{V}_0$. The homogeneous polynomial A is not identically zero, therefore there is a projective line along which A has a zero of finite order k at z. Let $\gamma : \mathbb{C} \to S_1$ denote any analytic curve through z along which A has a zero of finite order at z. The univariate analytic function $\gamma \circ Q_\varepsilon$ converges uniformly to $\gamma \circ A$, and Hurwitz's Theorem [Con78b, p. 152] implies that for ε sufficiently small it has k zeros of Q_ε converging uniformly to z as $\varepsilon \to 0$. In particular, z is a limit point of $\mathcal{V}_\varepsilon$ as $\varepsilon \to 0$. □

Exercise 6.12. Is it true that the algebraic tangent cone $\mathcal{V}_{\text{hom}(f,z)}$ is always homeomorphic to $\mathcal{V}_f$ within a sufficiently small ball around z?

Finally, let A be a homogeneous polynomial with *square-free part* $\tilde{A}$ (equal to the product of the distinct irreducible factors of A), so $\mathcal{V}_A = \mathcal{V}_{\tilde{A}}$ but $\mathcal{V}_A$ is locally a manifold near any point where the gradient of $\tilde{A}$ doesn't vanish. Any hyperplane $H_n = \{z \in \mathbb{C}^d : z \cdot n = 0\}$ through the origin can be identified with its normal n, viewed as an element in complex projective space $\mathbb{CP}^{d-1}$. We say the hyperplane H_n is a ***tangent hyperplane*** to $\mathcal{V}_A$ if $n = (\nabla\tilde{A})(w)$ for some zero w of A in $\mathbb{CP}^{d-1}$. The closure of all normals n to tangent hyperplanes of $\mathcal{V}_A$ in $\mathbb{CP}^{d-1}$ is an algebraic variety $\mathcal{V}_A^\vee$, called the ***projective dual variety*** to A. In generic situations $\mathcal{V}_A^\vee$ is defined by the vanishing of a homogeneous polynomial A^*, called the ***algebraic dual polynomial*** to A.

Example 6.50. If $A(x,y,z) = x^2 + y^2 + z^2$ then $(X,Y,Z) \in \mathcal{V}_A^\vee$ if and only if $(X,Y,Z) = \lambda\nabla A(x,y,z) = \lambda(2x,2y,2z)$ and $x^2 + y^2 + z^2 = 0$ for some $(x,y,z) \in \mathcal{V}_A$ and $\lambda \neq 0$. Eliminating x, y, and z from these equations gives $X^2/4\lambda^2 + Y^2/4\lambda^2 + Z^2/4\lambda^2 = 0$, so the algebraic dual of $A(x,y,z)$ is $A^*(X,Y,Z) = X^2 + Y^2 + Z^2$. ◁

Exercise 6.13. Prove that if A is a nonsingular quadratic form $A(z) = z^T M z$ then A^* is the quadratic form $A^*(\mathbf{Z}) = \mathbf{Z}^T M^{-1}\mathbf{Z}$.

Exercise 6.14. Find the algebraic dual to $A(x,y,z) = xy$.

Notes

Perhaps the earliest use of amoebas is the algebraic work of Bergman [Ber71], although their application to the problems considered here, along with the coinage of the term *amoeba*, is generally credited to Gelfand, Kapranov, and Zelevinsky [GKZ08]. That seminal text on discriminants devotes much of its

Chapter 6 to amoebas and Newton polytopes. The development of basic results on amoebas [GKZ08, Section 6.1] begins by quoting without proof some basic facts about Laurent series akin to Theorem 6.4. The reference they give, namely [Kra01], proves these only for ordinary power series, and the resulting wild-goose chase led us to write down a more complete development (the first author learned much of the material used in our proofs while sitting in on a graduate course at the University of Pennsylvania given by L. Matusevich in Fall 2004). Additional background material on amoebas can be found in [FPT00; Mik00; The02; Mik04; dWol13; dWol17; Tim18; For+19]. As noted above, our treatment of the order map and connections between the Newton polytope and amoeba complement components is heavily based on Forsberg et al. [FPT00]. Exercise 6.16 is taken from Rudin [Rud69, Theorem 4.6.2]. Our proof of Theorem 6.44(b) is inspired by Mikhalkin [Mik00; Mik04].

A good part of the theory of amoebas of algebraic hypersurfaces goes through for analytic hypersurfaces. This and its applications to statistical physics, for example, make up the content of [PPT13]. Because the theory of amoebas of analytic hypersurfaces is still being formed, we mostly avoid its use.

The ***Legendre transform*** (or ***convex dual***) of a convex function $f : \mathbb{R}^d \to \mathbb{R}$ is the function $f^* : (\mathbb{R}^d)^* \to \mathbb{R}$ defined by

$$f^*(\boldsymbol{v}) = \sup_{\boldsymbol{x} \in \mathbb{R}^d} \langle \boldsymbol{v}, \boldsymbol{x} \rangle - f(\boldsymbol{x}),$$

which satisfies the duality relation $f^{**} = f$ – see, e.g., [Roc66], where f^* is also called the *convex conjugate* of f. Legendre transforms are intimately connected with exponential rates of growth and decay. For example, in probability theory the large deviation *rate function* $f(\lambda)$, defined as the Legendre transform of the logarithm of the moment generating function, gives the rate of exponential decay of the probability of the mean of n IID variables to exceed λ. In our setting, the dual rate $\beta^*(\boldsymbol{r}) = \inf_{\boldsymbol{x} \in B} -\boldsymbol{r} \cdot \boldsymbol{x}$ from (6.5) is the negative of the Legendre transform of the convex function that is 1 on B and ∞ on B^c.

Additional exercises

Exercise 6.15. Give a simple necessary condition on f for $\texttt{amoeba}(f)$ to fail to be strictly convex (in other words, for $\texttt{amoeba}(f)$ to be *flat* by containing a line segment in its boundary).

Exercise 6.16. If $\alpha, \beta : [0, 1] \to \mathbb{C}_*^2$ are curves with $\alpha(0) = \alpha(1) = \beta(0) = \beta(1)$ then the *product loop* $\alpha\#\beta : [0, 1] \to \mathbb{C}_*^2$ obtained by following α and then β is defined by

$$(\alpha\#\beta)(t) = \begin{cases} \alpha(2t) & : 0 \le t \le 1/2 \\ \beta(2t-1) & : 1/2 \le t \le 1 \end{cases}.$$

If $\alpha(t) = (a_1(t), a_2(t))$ and $\beta(t) = (b_1(t), b_2(t))$ then the *Hadamard loop* $\alpha \odot \beta : [0,1] \to \mathbb{C}_*^2$ obtained by coordinatewise product is defined by $(\alpha \odot \beta)(t) = \big(a_1(t)b_1(t),\ a_2(t)b_2(t)\big)$.

a) Prove that $\alpha \odot \beta$ is homotopic to $[\alpha\#u] \odot [u\#\beta]$, where u is the constant curve $u(t) = (1,1)$.
b) Prove that $[\alpha\#u] \odot [u\#\beta] = [\alpha \odot u]\#[u \odot \beta]$.
c) Fix $a, b \in \mathbb{Z}$ and let $p_1, p_2 : [0,1] \to \mathbb{C}_*^2$ be the curves $p_1(t) = (e^{a2\pi it}, 1)$ and $p_2(t) = (1, e^{b2\pi it})$. Prove that the image of $w = e^{2\pi it}$ under the map (w^a, w^b) is homotopic to $p_1\#p_2$.

Exercise 6.17. Use Propositions 6.27 and 6.24 to prove Corollary 6.28.

Exercise 6.18. Let P be a d-dimensional polytope with integer vertices and let $f(z) = \sum_{m \in P} c_m z^m$ where $c_m = \exp(-\lambda|m|^2)$.

(a) Show that if $\lambda > 0$ is sufficiently large then there is some $x = x(m) \in \mathbb{R}^d$ such that

$$c_m \exp(m \cdot x) \ge \sum_{\substack{m' \in P \\ m' \neq m}} c_{m'} \exp(m' \cdot x).$$

(b) Conclude that $x(m)$ is not in $\texttt{amoeba}(f)$.
(c) Show that each $x(m)$ is in a separate component of $\texttt{amoeba}(f)^c$ as m varies over P, establishing that for every polytope with integer coordinates there is a polynomial whose amoeba complement has a component for every integer point of the polytope.

Exercise 6.19. Prove that if λ is sufficiently large in Exercise 6.18 then the boundary $\partial\, \texttt{amoeba}(f)$ equals the contour $C(f)$. (Be warned this is difficult.)

PART III

MULTIVARIATE ENUMERATION

7

Overview of analytic methods for multivariate GFs

We now return to the problem at the heart of this book: asymptotically approximating the coefficients of a convergent Laurent series expansion $F(\boldsymbol{z}) = \sum_{\boldsymbol{r}\in\mathbb{Z}^d} a_{\boldsymbol{r}}\boldsymbol{z}^{\boldsymbol{r}}$ through the Cauchy integral representation

$$a_{\boldsymbol{r}} = \left(\frac{1}{2\pi i}\right)^d \int_T \boldsymbol{z}^{-\boldsymbol{r}-\boldsymbol{1}} F(\boldsymbol{z})\, d\boldsymbol{z}\,, \tag{7.1}$$

for a suitable domain of integration T. We accomplish this by deforming T and using residue computations to reduce the Cauchy integral into a finite sum of local integrals that can be asymptotically approximated using the results of Chapter 5. When this approach succeeds, which it does in *generic* situations, it provides asymptotic formulae of the form

$$a_{\boldsymbol{r}} \approx \sum_{\boldsymbol{w}\in\mathtt{critical}(\hat{\boldsymbol{r}})} n_{\boldsymbol{w}} \Phi_{\boldsymbol{w}}(\boldsymbol{r})\,, \tag{7.2}$$

where the sum is over a finite set of certain *critical points* $\boldsymbol{w}$, each $\Phi_{\boldsymbol{w}}$ is an asymptotic series that can be computed to any desired accuracy algorithmically, and the coefficients $n_{\boldsymbol{w}}$ are integers that may or may not be easy to compute.

This textbook is designed so that combinatorialists can find easy-to-apply results with hypotheses and conclusions that are comprehensible with a minimum of cross-referencing to lengthy definitions, while readers with topological background can see the larger framework behind the results using advanced methods, such as those described in Chapters 4–6 and the appendices. In order to achieve this goal, the current chapter gives an overview of our approach and its relationship to the higher-level theories we draw on. Chapter 8 takes a computational view of the same material, giving explicit descriptions of how to compute the quantities appearing in the analysis using a computer algebra system. This material out of the way, Chapters 9–11 give our asymptotic results for families of generating functions with increasingly complicated singular behavior, together covering most known examples of rational generating functions in the combinatorial literature. Chapter 12 then gives a large variety of examples and applications before Chapter 13 describes further extensions.

In order to guide intuition and introduce the high-level constructions to be used in later chapters of the book, the current chapter begins by sketching the analysis on some examples, showing how the computations in the simplest case are a straightforward generalization of the univariate methods from Part I, describing the limits of these methods, and illustrating why we require more advanced techniques for our strongest results. After this we introduce the algebraic and topological constructions necessary for our work, and prove the theoretical results underpinning later chapters.

The computation of asymptotics is considerably simpler, and easier to explain, when the set of singularities $\mathcal{V}$ of F is *smooth* (meaning it is a manifold, at least near points dictating asymptotics). Before going into technical details, we illustrate the smooth case through extended examples in Section 7.1. Readers who want to understand the method but not the details can quit after the examples and skip to Chapters 8 and 9. In Section 7.2 we describe the theory when $\mathcal{V}$ is smooth, allowing readers to understand the smooth point formulae of Chapter 9 without the greater overhead of stratified Morse theory.

Section 7.3 gives a parallel treatment of everything in the previous sections, without the assumption that $\mathcal{V}$ is smooth. This involves the introduction of stratified Morse theory to explain the corresponding notions of critical points and *quasi-local cycles* for non-smooth varieties. The quasi-local cycles are defined in terms of the *tangential cycles* γ_j and homology generators β_j for the *normal link* at z_j. Section 7.4 discusses the types of singular geometry that arise frequently in combinatorial applications.

The results of (stratified) Morse theory describe the topology of a surface using a *height function* mapping the surface to the real numbers. In classical Morse theory this height function is almost always assumed to be proper, meaning the set of points with heights in a closed interval forms a compact set. Unfortunately, we work in situations where the height function is usually *non-proper*. To get around this difficulty, Section 7.5 introduces the concept of *critical points at infinity* (CPAI) and *critical values at infinity* (CVAI), which help characterize when the results of Morse theory we need apply without an assumption of a proper height function. A fundamental lemma is stated concerning the existence of certain deformations, provided there are no critical values at infinity, and its proof is cited from the literature. This lemma is then used to prove the theorems previously stated in the chapter.

Notational conventions

For the rest of this book we use the following notational conventions. Bold quantities are reserved for vectors, such as $\boldsymbol{z} = (z_1, \ldots, z_d)$, and we define $\boldsymbol{z}^\circ := (z_1, \ldots, z_{d-1})$. The d-variate function $F(\boldsymbol{z})$ is a quotient of coprime polynomials $P(\boldsymbol{z})/Q(\boldsymbol{z})$, with the denominator Q vanishing on the ***singular variety*** $\mathcal{V} = \mathcal{V}_Q = \{\boldsymbol{z} \in \mathbb{C}^d : Q(\boldsymbol{z}) = 0\}$. We fix a component B of the complement of $\texttt{amoeba}(Q)$ and consider the Laurent series expansion $F(\boldsymbol{z}) = \sum_{\boldsymbol{r}\in\mathbb{Z}^d} a_{\boldsymbol{r}} \boldsymbol{z}^{\boldsymbol{r}}$ that converges on $\mathcal{D} = \text{Relog}^{-1}(B)$. As in previous chapters, for $\boldsymbol{w} \in \mathbb{C}_*^d$ we use the notation $\mathbf{T}(\boldsymbol{w})$ for the torus $\mathbf{T}(\boldsymbol{w}) = \{\boldsymbol{z} \in \mathbb{C}^d : |z_j| = |w_j| \text{ for all } j\}$. The simplest and most common case, of a convergent power series expansion, occurs when B is the component containing points of the form $(-N, \ldots, -N)$ for N sufficiently large, so that $\mathcal{D}$ is a neighborhood of the origin.

Remark 7.1. Although we mainly study rational generating functions, most of our results also hold for meromorphic functions. We point out as we go which major results still hold for meromorphic functions, and the small ways in which they differ from the rational case.

Given $\boldsymbol{r} \in \mathbb{Z}^d$ the d-form $\omega = \boldsymbol{z}^{-\boldsymbol{r}-\boldsymbol{1}} F(\boldsymbol{z}) d\boldsymbol{z}$ is the integrand of the Cauchy integral (7.1), with domain of analyticity $\mathcal{M} = \mathbb{C}_*^d \setminus \mathcal{V}$. Unless otherwise stated, we write $|\boldsymbol{r}|$ for the ℓ^1-norm $|\boldsymbol{r}| = \sum_{j=1}^d |r_j|$ and as above define the normalized vector $\hat{\boldsymbol{r}} = \boldsymbol{r}/|\boldsymbol{r}|$. We seek to compute asymptotics for the series coefficients $a_{\boldsymbol{r}}$ as $\boldsymbol{r} \to \infty$ with $\hat{\boldsymbol{r}}$ varying over a compact set, typically around some fixed direction.

7.1 Some illustrative examples

Example 7.2 (Binomial Coefficients). We start with perhaps the simplest non-trivial bivariate rational function for our purposes: $F(x, y) = 1/Q(x, y)$ with $Q(x, y) = 1 - x - y$. The amoeba of Q is pictured in Figure 7.1 (see Chapter 8 for methods to compute amoebas). Because there are three components in the amoeba complement, there are three convergent Laurent series expansions of $F(x, y)$. Consider the power series expansion $F(x, y) = \sum_{i,j\geq 0} \binom{i+j}{i} x^i y^j$, corresponding to the component of the amoeba complement that lies in the third quadrant. Since

$$\sum_{i,j\geq 0} \left| \binom{i+j}{i} x^i y^j \right| = \sum_{i,j\geq 0} \binom{i+j}{i} |x|^i |y|^j = \frac{1}{1 - |x| - |y|}, \tag{7.3}$$

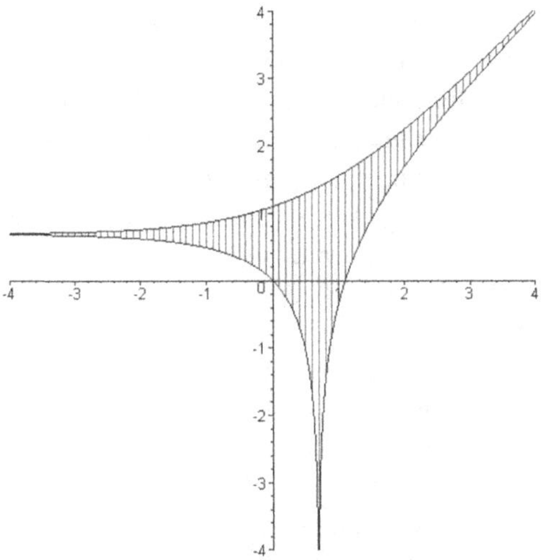

Figure 7.1 Amoeba of the function $1 - x - y$.

this series expansion has domain of convergence $\mathcal{D} = \{(x, y) \in \mathbb{C}^2 : |x| + |y| < 1\}$. For any $a, b \in (0, 1)$ with $a + b < 1$ we can write

$$\begin{aligned}\binom{i+j}{i} &= \frac{1}{(2\pi i)^2} \int_{\mathbf{T}(a,b)} \frac{1}{1-x-y} \frac{dxdy}{x^{i+1}y^{j+1}} \\ &= \frac{1}{(2\pi i)^2} \int_{\mathbf{T}(a,b)} \frac{1}{1-x-y} e^{-\phi(x,y)} \frac{dxdy}{xy},\end{aligned} \tag{7.4}$$

where $\phi(x, y) = i \log x + j \log y$. We aim to use residue computations to reduce the two-dimensional integral (7.4) to a one-dimensional integral over some path in the singular set $\mathcal{V} = \{(x, y) \in \mathbb{C}^2 : x + y = 1\}$, and then compute a saddle point integral. Thus, we set $y = 1 - x$ in $\phi(x, y)$ and solve for a saddle point, where the first derivative of the function vanishes. The equation

$$0 = \frac{d}{dx}\phi(x, 1 - x) = \frac{i}{x} - \frac{j}{1 - x}$$

implies $x = i/(i + j)$. Hence, we aim to determine asymptotic behavior by studying the Cauchy integral near $(x_*, y_*) = (i/(i + j), j/(i + j)) \in \mathcal{V}$. For this discussion we fix positive integers $r, s > 0$ and derive asymptotics of the coefficient sequence $(i, j) = n(r, s)$ as $n \to \infty$. To that end, define

$$I = \frac{1}{(2\pi i)^2} \int_{|x|=x_*} \left(\int_{|y|=y_*-\varepsilon} \frac{1}{1-x-y} \frac{dy}{y^{ns+1}} \right) \frac{dx}{x^{nr+1}}$$

$$I_{\text{loc}} = \frac{1}{(2\pi i)^2} \int_{\mathcal{N}} \left(\int_{|y|=y_*-\varepsilon} \frac{1}{1-x-y} \frac{dy}{y^{ns+1}} \right) \frac{dx}{x^{nr+1}}$$

$$I_{\text{out}} = \frac{1}{(2\pi i)^2} \int_{\mathcal{N}} \left(\int_{|y|=y_*+\varepsilon} \frac{1}{1-x-y} \frac{dy}{y^{ns+1}} \right) \frac{dx}{x^{nr+1}},$$

where $\mathcal{N} = \{x \in \mathbb{C} : |x| = x_* \text{ and } \arg(x) \in (-\delta, \delta)\}$ is any arbitrarily small neighborhood of x_* in the circle $\{|x| = x_*\}$. As we will see in Chapter 9, both $I - I_{\text{loc}}$ and I_{out} grow exponentially slower than I, so (7.4) implies

$$\binom{nr+ns}{rn} = I = I_{\text{loc}} - I_{\text{out}} + \text{exponentially negligible term.}$$

Thus, we can use the (univariate) residue theorem to approximate $\binom{nr+ns}{rn}$ by

$$\begin{aligned}
I_{\text{loc}} - I_{\text{out}} &= \frac{1}{(2\pi i)^2} \int_{\mathcal{N}} \left(\int_{|y|=y_*-\varepsilon} \frac{1}{1-x-y} \frac{dy}{y^{ns+1}} - \int_{|y|=y_*+\varepsilon} \frac{1}{1-x-y} \frac{dy}{y^{ns+1}} \right) \frac{dx}{x^{nr+1}} \\
&= \frac{-1}{(2\pi i)} \int_{\mathcal{N}} \underset{y=1-x}{\text{Res}} \frac{y^{-ns-1}}{1-x-y} \frac{dx}{x^{nr+1}} \\
&= \frac{1}{(2\pi i)} \int_{\mathcal{N}} \frac{dx}{x^{nr+1}(1-x)^{ns+1}}.
\end{aligned}$$

Making the change of variables $x = x_* e^{i\theta}$ results in the saddle point integral

$$I_{\text{loc}} - I_{\text{out}} = \frac{x_*^{-rn} y_*^{-sn}}{2\pi} \int_{-\delta}^{\delta} A(\theta) e^{-n\phi(\theta)},$$

where

$$A(\theta) = \frac{1}{1 - x_* e^{i\theta}} = \frac{r+s}{s} + O(\theta)$$

and

$$\phi(\theta) = r \log(x_* e^{i\theta}) + s \log(1 - x_* e^{i\theta}) - r \log(x_*) - s \log(y_*) = \frac{r(r+s)}{2s}\theta^2 + O\left(\theta^3\right).$$

Theorem 4.1 from Chapter 4 then gives an asymptotic expansion

$$\binom{nr+ns}{nr} = \left(\frac{r+s}{r}\right)^{rn} \left(\frac{r+s}{s}\right)^{sn} n^{-1/2} \left(\frac{\sqrt{r+s}}{2rs\pi n} + \cdots \right).$$

◁

The approach taken in Example 7.2 is known as the ***surgery method*** for multivariate asymptotics. It works by performing an explicit deformation to move the torus of integration in the Cauchy integral near a critical point, then changing the radius in one coordinate to enclose singularities. The ordinary (univariate) residue theorem, a localization argument, and the saddle point results from Chapters 4 and 5 then yield asymptotics.

Although this approach can be generalized successfully, as will be done in Section 9.1 of Chapter 9, such explicit deformations require additional assumptions on the singularities where local behavior of $F(\mathbf{z})$ determines asymptotics. In particular, such singularities need to be *minimal* in the sense of Section 6.4, meaning they lie on the boundary of the domain of convergence of the Laurent expansion being considered. In fact, we require *finite minimality*, meaning such singularities are minimal and only a finite number of other singularities have the same coordinatewise modulus. Although this is usually not an unreasonable assumption, in practice it can be very expensive to verify formally (see Chapter 8 for more details).

Exercise 7.1. Suppose we perturb Example 7.2 by taking $Q_\varepsilon(x,y) = 1 - x - y - \varepsilon y^2$ for some $\varepsilon > -1$. Let $D_\varepsilon = \{(x,y) \in \mathbb{C}^2 : |x|, |y| < \rho_\varepsilon\}$ where $\rho_\varepsilon = \left(\sqrt{1+\varepsilon} - 1\right)/\varepsilon$ the positive root of $Q_\varepsilon(x,x)$. When $\varepsilon = 0$, the function $1/Q_\varepsilon(x,y)$ is the function in (7.3), whose power series domain of convergence contains $D_0 = \{(x,y) \in \mathbb{C}^2 : |x|, |y| < 1/2\}$.

(a) As $\varepsilon \to 0$, determine the first two terms of the asymptotic behavior of ρ_ε.
(b) When $\varepsilon > 0$, is there an easy way to see that $1/Q(x,y)$ is analytic on D_ε?
(c) When $-1 < \varepsilon < 0$, can you show that $1/Q(x,y)$ is analytic on D_ε?
(d) What can you say when $\varepsilon \le -1$?

We now study an example where the surgery method does not directly apply, and sketch a more general ***topological method*** for multivariate asymptotics. Although the topological method applies in a wider variety of situations, as its name suggests it will require more advanced constructions from topology and differential geometry. Our next example also illustrates how the topological approach generalizes hands-on surgery in the smooth case to a topologically characterized contour integration.

Example 7.3 (Non-Minimal Contributing Points)**.** Consider the $(1,1)$-diagonal sequence $a_{n,n}$ of the power series expansion

$$F(x,y) = \frac{1}{Q(x,y)} = \sum_{i,j\ge 0} a_{i,j} x^i y^j,$$

where $Q(x, y) = (1 - x - y)(1 + 3x)$, so

$$a_{i,j} = \sum_{k=0}^{i} \binom{k+j}{k} (-3)^{i-k}.$$

The singular set $\mathcal{V} = \mathcal{V}_Q$ is the union of the hyperplane $\mathcal{V}_{1-x-y}$ from Example 7.2 with the hyperplane $\mathcal{V}_{1+3x}$. It contains the point $(x_{**}, y_{**}) = (-1/3, 4/3)$ on the intersection of the hyperplanes where $\mathcal{V}$ is not a manifold.

Since $\mathcal{V}$ still contains the hyperplane $\mathcal{V}_{1-x-y}$, the point $(x_*, y_*) = (1/2, 1/2)$ identified in Example 7.2 is still of interest for the asymptotic analysis. Furthermore, the topology of $\mathcal{V}$ changes at the non-smooth point (x_{**}, y_{**}), so this point is also of interest. The function $\phi(x, y) = \log x + \log y$ has nonvanishing derivative when restricted to $\mathcal{V}_{1+3x}$, hence there are no other points where we could restrict the Cauchy integrand to the singular variety and get a saddle point integral.

As we will see later, asymptotics of the coefficient sequence $a_{n,n}$ are still determined by reducing to an integral near (x_*, y_*). However, unlike Example 7.2 we cannot simply move the contour of integration in the Cauchy integral

$$a_{n,n} = \frac{1}{(2\pi i)^2} \int_{|x|=\varepsilon_1} \int_{|y|=\varepsilon_2} \frac{1}{(1-x-y)(1+3x)} \frac{dxdy}{x^{n+1}y^{n+1}}$$

to a torus $\{(x, y) : |x| = x_*, |y| = y_* - \varepsilon\}$ as we would cross the singular set $\mathcal{V}$ at points where $x = -1/3$. To work around this, we expand y *through* the singular variety, resulting in an integral over a tube around $\mathcal{V}_{1-x-y}$, reduce to an integral on $\mathcal{V}_{1-x-y}$ through a residue computation, and *then* move the contour of integration to the saddle point.

For concreteness, we now take $\varepsilon_1 = \varepsilon_2 = 1/10$, although any positive values satisfying $0 < \varepsilon_1 + \varepsilon_2 < 1$ and $\varepsilon_1 < 1/3$ would work. Let $T_0 = \{|x| = |y| = 1/10\}$ and, for any $M > 0$, define the map

$$\begin{aligned} K_M : T_0 \times [0, 1] &\to \mathbb{C}^2 \\ (x, y, t) &\mapsto (x,\ y(1 + Mt)). \end{aligned}$$

Then K_M is a homotopy from T_0 to the torus $T_1 = \{|x| = 1/10, |y| = (M+1)/10\}$. As long as $M > 10$ then $F(x, y)$ is analytic on T_0 and T_1, the image of K_M does not intersect the coordinate axes of $\mathbb{C}^2$, and this image intersects $\mathcal{V}$ in the set $C = \{(x, 1 - x) : |x| = 1/10\}$. Furthermore, the image of K_M intersects $\mathcal{V}$ *transversely*, meaning the tangent planes of these sets jointly span $\mathbb{C}^2$ at their common points. See Figure 7.2 for a visualization of the path of this homotopy after taking the Relog map.

Because $F(x, y)$ is analytic on $\mathbb{C}^2 \setminus \mathcal{V}$, Stokes's Theorem (Theorem A.24 in

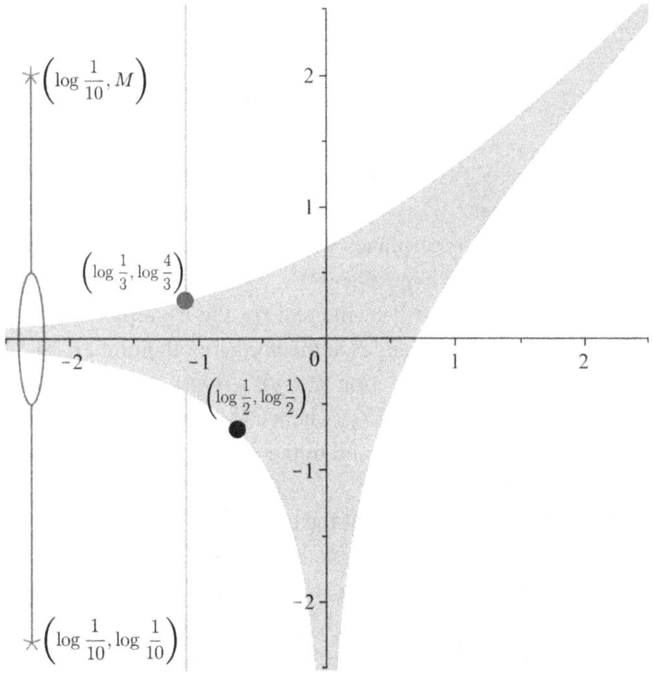

Figure 7.2 The amoeba of $(1 - x - y)(1 + 3x)$. We start by integrating over the torus defined by $|x| = 1/10$ and $|y| = 1/10$ and expand $|y|$ to $(M + 1)/10 > 11/10$, resulting in an integral over a *tubular neighborhood* of $\mathcal{V}$. Taking a residue reduces to an integral over a curve lying on the hyperplane $1 - x - y = 0$ then, avoiding the set of points where $1 + 3x = 0$, we slide this contour to a curve near the point $(1/2, 1/2)$ on $\mathcal{V}$ together with points that do not affect dominant asymptotics.

Appendix A) implies that the Cauchy integral over the boundary of any 3-cycle in $\mathbb{C}_*^2 \setminus \mathcal{V}$ is zero. In particular,

$$\int_{T_0} F(x, y)\frac{dxdy}{x^{n+1}y^{n+1}} = \int_{v} F(x, y)\frac{dxdy}{x^{n+1}y^{n+1}} + \int_{T_1} F(x, y)\frac{dxdy}{x^{n+1}y^{n+1}}, \qquad (7.5)$$

where v is a ***tubular neighborhood*** of C: the union of circles normal to the tangent plane of $\mathcal{V}$ with centers at the points of C (see Figure 7.3). Furthermore, because (7.5) holds for any $M > 10$, and

$$\int_{T_1} F(x, y)\frac{dxdy}{x^{n+1}y^{n+1}} = O\left(10^n(M + 1)^{-n}\right),$$

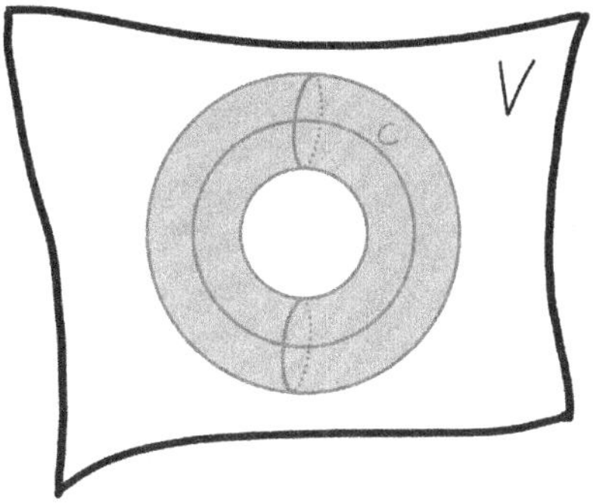

Figure 7.3 A visualization of the tubular neighborhood ν.

taking $M \to \infty$ shows that the integral over T_1 is zero for $n > 0$, and thus

$$a_{n,n} = \frac{1}{(2\pi i)^2} \int_\nu F(x,y) \frac{dxdy}{x^{n+1}y^{n+1}} = \frac{1}{(2\pi i)^2} \int_\nu \frac{1}{(1-x-y)(1+3x)} \frac{dxdy}{x^{n+1}y^{n+1}}.$$

The tubular neighborhood ν is the union of circles with centers on C, and each point of C corresponds to a simple pole of $F(x, y)$, where $1 - x - y = 0$, so a generalization of the classical univariate residue theorem implies

$$\begin{aligned} a_{n,n} &= \frac{1}{2\pi i} \int_{|x|=1/10} \operatorname*{Res}_{y=1-x} \frac{1}{(1-x-y)(1+3x)} \frac{dx}{x^{n+1}y^{n+1}} \\ &= \frac{1}{2\pi i} \int_{|x|=1/10} \frac{1}{1+3x} \frac{dx}{x^{n+1}(1-x)^{n+1}}. \end{aligned} \tag{7.6}$$

As in the last example, the integrand of (7.6) becomes a saddle point integral near $x = 1/2$. The difference is that while we previously used a residue to localize near the saddle point, this time we took a more "convenient" residue and obtained a univariate integral away from the saddle point. Because we are dealing with an integrand having a linear denominator, we can move our domain of integration to pass through the critical point without much difficulty. We now describe three methods for doing this, listed in decreasing order of explicitness but increasing order of generality.

Method One: Because the only singularity of the integrand in (7.6) between the circles $|x| = 1/10$ and $|x| = 1/2$ occurs at $x = -1/3$, the domain of integration in (7.6) can be replaced by the union of the circle $|x| = 1/2$ and a sufficiently small clockwise circle around $x = -1/3$ (see Figure 7.4 left). The

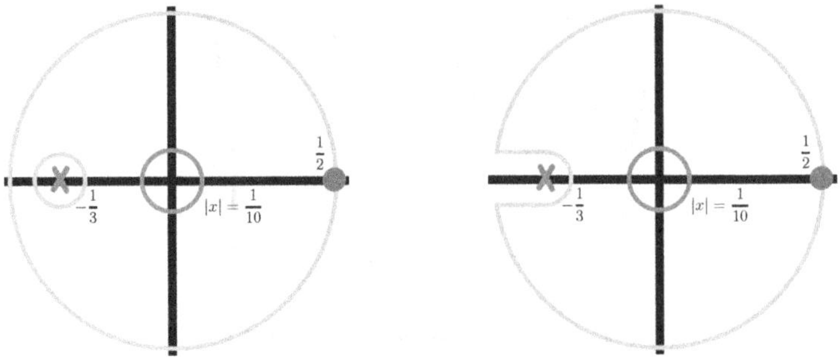

Figure 7.4 *Left:* The circle $|x| = 1/10$ can be expanded to $|x| = 1/2$ by introducing a circle around $x = -1/3$. This results in an extra residue integral which is exponentially negligible. *Right:* Alternatively, we can expand from $|x| = 1/10$ to hit $x = 1/2$ while stopping the increase in an arbitrarily small circle around $x = -1/3$.

residue theorem then implies

$$\begin{aligned} a_{n,n} &= \frac{1}{2\pi i}\int_{|x|=1/2} \frac{1}{1+3x}\frac{dx}{x^{n+1}(1-x)^{n+1}} + \frac{1}{2\pi i}\int_{|x+1/3|=\varepsilon} \frac{1}{1+3x}\frac{dx}{x^{n+1}(1-x)^{n+1}} \\ &= \frac{1}{2\pi i}\int_{|x|=1/2} \frac{1}{1+3x}\frac{dx}{x^{n+1}(1-x)^{n+1}} - \operatorname*{Res}_{x=-1/3}(x+1/3)^{-1}\frac{1/3}{x^{n+1}(1-x)^{n+1}} \\ &= \frac{1}{2\pi i}\int_{|x|=1/2} \frac{1}{1+3x}\frac{dx}{x^{n+1}(1-x)^{n+1}} - \frac{1}{3}\left(\frac{9}{4}\right)^{n+1}, \end{aligned}$$

and a change of variables yields the saddle point approximation

$$\begin{aligned} a_{n,n} &= \frac{4^n}{2\pi}\int_{-\pi}^{\pi} \frac{1}{(1+3e^{i\theta}/2)(1-e^{i\theta}/2)} e^{-ni\theta - n\log(2-e^{i\theta})}d\theta - \frac{1}{3}\left(\frac{9}{4}\right)^{n+1} \\ &= \frac{4^n}{\sqrt{\pi n}}\left(\frac{2}{5} + O\left(\frac{1}{n}\right)\right). \end{aligned}$$

Method Two: In general we cannot work around other singularities by taking residues in such an explicit manner. Although this means we cannot get an explicit representation for error terms coming from other singularities, all we really need to determine dominant asymptotics is to bound any potential asymptotic contributions from these singularities. The only factor of the integrand in (7.6) that depends on n is $x^{-n}(1-x)^{-n}$, so when n is large the modulus of the integrand is well approximated by $e^{nh(x)}$, where

$$h(x) = -\log|x| - \log|1-x|.$$

Points with smaller *height h* make the integrand of (7.6) exponentially smaller, so up to an exponentially negligible error we can ignore points with height bounded below $h(1/2) = \log 4$. Since $h(-1/3) = \log(9/4)$ we could proceed by expanding the circle $|x| = 1/10$ to the circle $|x| = 1/2$ while stopping in a tubular shape around $x = -1/3$ (see Figure 7.4 right). The integral over the resulting curve can be truncated to a neighborhood of $x = 1/2$ in $|x| = 1/2$ while introducing an exponentially negligible error. The integral over this neighborhood of $x = 1/2$ is again a saddle point integral.

Method Three: Although Method Two is more general than Method One, it still requires that we know how to explicitly deform around $\mathcal{V}$, which is not always possible. We thus move to an even more general argument, which will be fully described below. The key is to use the local geometry of $\mathcal{V}$ to describe how to move the domain of integration $|x| = 1/10$ to heights below $h(1/2) = \log 4$, except in a neighborhood of $x = 1/2$, while avoiding $\mathcal{V}$. This is accomplished using a *gradient flow*. Writing $x = a + ib$ for real variables a and b, so that $|x| = \sqrt{a^2 + b^2}$, we see that

$$h(a, b) = h(a + ib) = -\log\left(a^2 + b^2\right)/2 - \log\left((1 - a)^2 + b^2\right)/2.$$

We want to move an arbitrary point $a_\theta + ib_\theta = e^{i\theta}/10$ on our starting circle $|x| = 1/10$ down to points on $\mathcal{V}$ of lower height with respect to h. Since $(\nabla h)(a, b)$ gives the direction of greatest increase of h, we want to locally move a point (a_θ, b_θ) along the direction $-(\nabla h)(a_\theta, b_\theta)$. In other words, we want to solve the first-order differential system of equations

$$\begin{pmatrix} a'_\theta(t) \\ b'_\theta(t) \end{pmatrix} = -\nabla h(a_\theta(t), b_\theta(t)), \qquad a_\theta(0) = \cos(\theta)/10, \qquad b_\theta(0) = \sin(\theta)/10$$

for $a_\theta(t)$ and $b_\theta(t)$. Figure 7.5 shows the trajectories of points under this (negative) gradient flow. Here it can be verified in a computer algebra system that under the flow all points will go below height $h(1/2) = \log 4$, except in a neighborhood of $x = 1/2$. Near $x = 1/2$ the flow approaches a vertical line, ultimately resulting in a saddle point integral. The key reason this method can be generalized is that techniques from Morse theory allow us to know when such a flow exists, and characterize the resulting domains of integration, *without having to actually compute them.*

◁

In our last example the non-smooth point did not affect dominant asymptotics, but this will not always be the case.

Example 7.4 (Dealing with Multiple Points)**.** Consider now asymptotics in the

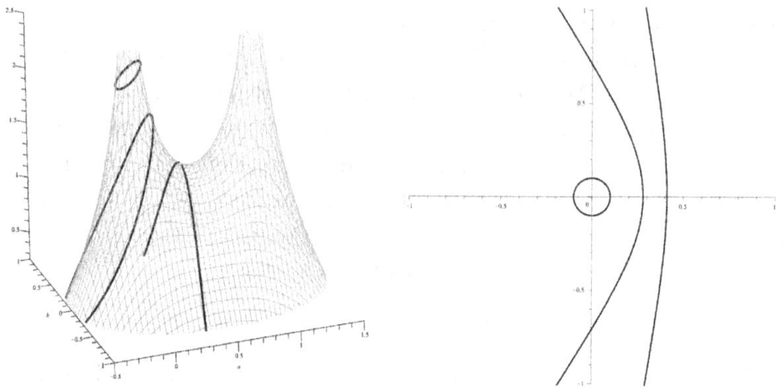

Figure 7.5 *Left:* The gradient flow of $|x| = 1/10$ at three points in time, plotted on $\mathbb{C}^2 \setminus \{0, 1\}$ when arranged by height $h(x) = h(a + ib)$. *Right:* The curves under the flow plotted in the complex plane.

main diagonal direction $\boldsymbol{r} = (1, 1)$ of the power series expansion of $F(x, y) = 1/Q(x, y)$ with $Q(x, y) = (1 - x - y)(1 - 3x)$. The factor $1 - x - y$ is the same as in the above examples, but having the second factor change from $1 + 3x$ to $1 - 3x$ moves the non-smooth point to $(1/3, 2/3)$. Because the height $h(x, y) = -\log x - \log y$ is now larger at $(1/3, 2/3)$ than $(1/2, 1/2)$, we can no longer easily rule out the non-smooth point. In fact, following Method One from the last example shows

$$
\begin{aligned}
a_{n,n} &= \frac{1}{2\pi i} \int_{|x-1/3|=\varepsilon} \frac{1}{1-3x} \frac{dx}{x^{n+1}(1-x)^{n+1}} + \frac{1}{2\pi i} \int_{|x|=1/2} \frac{1}{1-3x} \frac{dx}{x^{n+1}(1-x)^{n+1}} \\
&= \operatorname*{Res}_{x=1/3} (x - 1/3)^{-1} \frac{1/3}{x^{n+1}(1-x)^{n+1}} + \frac{1}{2\pi i} \int_{|x|=1/2} \frac{1}{1+3x} \frac{dx}{x^{n+1}(1-x)^{n+1}} \\
&= \frac{1}{3} \left(\frac{9}{2} \right)^{n+1} + O(4^n).
\end{aligned}
$$

More generally, if $\mathcal{V}$ is no longer a manifold then we compute a *Whitney stratification*, partitioning $\mathcal{V}$ into a finite collection of manifolds such that the local geometry of $\mathcal{V}$ is consistent near the points in any fixed element of the partition. We then perform an analysis similar to the smooth case on each of the manifolds, obtaining a set of equations for each manifold that characterizes the points of interest for our asymptotic calculations. The asymptotic contribution of such a point depends on the geometry near the singularity. In this text we study singularities where $\mathcal{V}$ is locally smooth (in Chapter 9), looks like the union of hyperplanes (in Chapter 10), or looks like a cone point (in

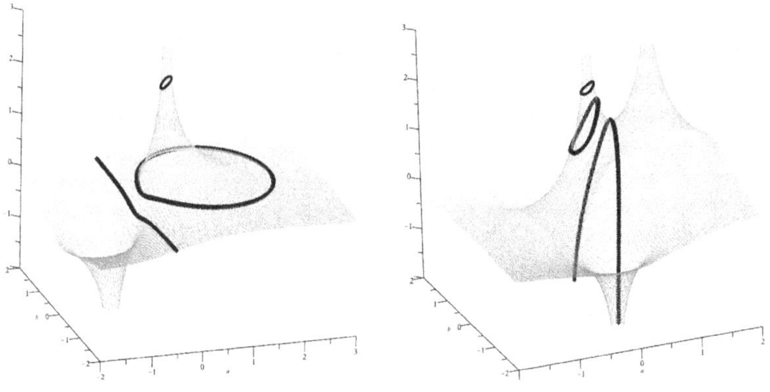

Figure 7.6 The gradient flow of $|x| = 1/10$ at three points in time, plotted on $\mathbb{C}^2 \setminus \{0, 1\}$ when arranged by height $h(x) = h(a + ib)$. *Left:* The flow on $\mathcal{V}_{1-y-xy}$. *Right:* The flow on $\mathcal{V}_{1-x-y-x^2y}$.

Chapter 11). Studying flows on general algebraic varieties requires us to adapt tools from *stratified* Morse theory. ◁

Exercise 7.2. Sketch the vector field $-(\nabla h)(a, b)$ on the right-hand side of Figure 7.5.

Gradient flows form an important component of our analytic toolbox. Indeed, rather than computing a flow for each example, standard results in Morse theory usually guarantee the existence of flows that push domains of integration down to points where saddle point approximations can be computed. Unfortunately, these results require the height map to be *proper* (meaning that the set of points with height in a closed interval is compact). Because this properness condition is usually not satisfied in our setting, it is possible for the desired flows not to exist.

Example 7.5 (Critical Points at Infinity). Consider the diagonal sequences $a_{n,n}$ of the power series expansions of $1/A(x, y)$ and $1/B(x, y)$, where $A(x, y) = 1 - y - xy$ and $B(x, y) = 1 - x - y - x^2y$. The negative gradient flows of the circle $|x| = 1/10$ on $\mathcal{V}_A$ and $\mathcal{V}_B$ are shown in Figure 7.6. Because $y = 1/(x + 1)$ on $\mathcal{V}_A$, the product $xy = x/(x + 1) \to 1$ as $x \to \infty$, and thus the height function $h(x, y) = -\log|x| - \log|y| \to 0$ as $x \to \infty$ on $\mathcal{V}_A$. Since $h(x, y)$ can stay bounded as (x, y) goes to infinity, the height function is not proper. As seen in the left of Figure 7.6, the circle $|x| = 1/10$ stays at bounded height but never reaches a saddle. In fact, $\mathcal{V}_A$ contains no saddles, and we say that it has a *critical point at infinity*.

Similarly, because $y = (1 - x)/(1 + x^2)$ on $\mathcal{V}_B$, the height function $h(x, y)$ approaches zero as $x \to \infty$ on $\mathcal{V}_B$. However, on $\mathcal{V}_B$ the circle $|x| = 1/10$ does flow to a saddle point of height greater than zero. Again we have a critical point at infinity, but this time it is of lower height than an actual saddle point on the variety. Thus, non-properness of the height function does not preclude an asymptotic analysis, as we can ignore points of height bounded below the saddle point if we care only about dominant asymptotic behavior. ◁

In Section 7.5 we discuss computable conditions, often satisfied in practice, that imply the conclusions of Morse theory we require apply even without a proper height function.

Exercise 7.3. Let $Q(x, y) = 1 - x - y - x^2y^2$ and $h_{\boldsymbol{r}}(x, y) = -r_1 \log |x| - r_2 \log |y|$ be the height function corresponding to the $\boldsymbol{r}$-diagonal sequence $(a_{r_1 n, r_2 n})$. Prove that when $\boldsymbol{r} = (2, 1)$ the height function $h_{\boldsymbol{r}}(x, y)$ approaches a finite limit as $y \to \infty$ and $x \to 0$, and evaluate the limit. Prove that when $\boldsymbol{r} = (1, 1)$ the height function $h_{\boldsymbol{r}}(x, y)$ has no finite limit as either x or y goes to infinity.

7.2 The smooth case

We now generalize the above argument to any rational function whose singular variety $\mathcal{V}$ is a complex manifold. The ***square-free part*** $\tilde{Q}$ of the polynomial Q is the product of its distinct irreducible factors over the complex numbers, and we say that Q is ***square-free*** if $\tilde{Q} = Q$. We call $\boldsymbol{z} \in \mathcal{V}$ a ***smooth point*** if $\nabla \tilde{Q}(\boldsymbol{z})$ is non-zero, and say that $\mathcal{V}$ is ***smooth*** if $\tilde{Q}$ and all its partial derivatives never simultaneously vanish. The implicit function theorem implies that a smooth singular variety can be viewed both as a $(d-1)$-dimensional complex manifold and as a $(2d-2)$-dimensional real manifold, and both of these viewpoints will be beneficial. We introduce the square-free part of Q so that the converse also holds.

Lemma 7.6. *The inequality $\nabla \tilde{Q}(\boldsymbol{z}) \neq \boldsymbol{0}$ holds for a point $\boldsymbol{z} \in \mathcal{V}_Q$ if and only if $\mathcal{V}$ is a smooth manifold in a neighborhood of $\boldsymbol{z}$.*

Proof Sketch The forward implication follows from the implicit function theorem. The converse, that $\nabla \tilde{Q}(\boldsymbol{z}) = \boldsymbol{0}$ implies a geometric singularity, is harder to prove. Let $m_{\boldsymbol{x}}$ denote the maximal ideal of functions vanishing at $\boldsymbol{x}$ in the ring of polynomial functions vanishing on $\mathcal{V}$, and let $n_{\boldsymbol{x}}$ denote the maximal ideal of functions vanishing at $\boldsymbol{x}$ in the ring of germs of analytic functions at $\boldsymbol{x}$ (as defined in Definition 10.42 below). Then $\nabla \tilde{Q}(\boldsymbol{x}) = \boldsymbol{0}$ implies $m_{\boldsymbol{x}}/m_{\boldsymbol{x}}^2$ has dimension d rather than $d-1$ (see [Sha13, Exercise 2.2 and Theorem 2.1]) so that $n_{\boldsymbol{x}}/n_{\boldsymbol{x}}^2$ has dimension d, and this property is invariant under bi-analytic mapping. At any smooth point of a complex hypersurface there is a coordinate neighborhood taking $\boldsymbol{x}$ to the origin and making the hypersurface into the coordinate plane where $z_1 = 0$. In this case $n_{\mathbf{0}}/n_{\mathbf{0}}^2$ has dimension $d-1$, which would be a contradiction, hence $\mathcal{V}$ is not a complex manifold in a neighborhood of $\boldsymbol{x}$. A little more work shows $\mathcal{V}$ is not locally a C^∞-manifold either. □

Our starting point, as always, is the multivariate Cauchy Integral Formula

$$a_{\boldsymbol{r}} = \left(\frac{1}{2\pi i}\right)^d \int_T \boldsymbol{z}^{-\boldsymbol{r}-\boldsymbol{1}} F(\boldsymbol{z})\, d\boldsymbol{z}\,, \tag{7.7}$$

which gives an exact representation for $a_{\boldsymbol{r}}$. We view this representation not as a standard integral from multivariate calculus, but as the integral of the differential form $\omega = \boldsymbol{z}^{-\boldsymbol{r}-\boldsymbol{1}} F(\boldsymbol{z})\, d\boldsymbol{z}$ over the d-chain T. The necessary background on differential geometry and the basics of integration of forms is discussed in Appendix A. Appendix B reviews concepts from algebraic topology, including homology and cohomology classes. In particular, since $\mathcal{M} = \mathbb{C}_*^d \setminus \mathcal{V}$ is the domain of holomorphicity for ω, the Cauchy integral depends only on the class of T in the singular homology group $H_d(\mathcal{M})$ and the class of ω in the singular cohomology group $H^d(\mathcal{M})$.

We break our argument into pieces, generally mirroring the final approach to Example 7.3 above. In this chapter we mainly stick to theoretical considerations; methods for computing the quantities that arise are discussed in Chapter 8.

Step 1: Characterize critical points

We begin by defining the ***height function***

$$h_{\boldsymbol{r}}(\boldsymbol{z}) := -\boldsymbol{r} \cdot \operatorname{Relog} \boldsymbol{z} = -\sum_{j=1}^d r_j \log |z_j|\,,$$

which captures the magnitude of the Cauchy integrand

$$\left|\boldsymbol{z}^{-\boldsymbol{r}-\boldsymbol{1}} F(\boldsymbol{z})\right| = e^{|\boldsymbol{r}|\, h_{\hat{\boldsymbol{r}}}(\boldsymbol{z})} \cdot \left|\boldsymbol{z}^{-\boldsymbol{1}} F(\boldsymbol{z})\right|$$

as $\left|\boldsymbol{z}^{-\boldsymbol{1}} F(\boldsymbol{z})\right|$ independent of $|\boldsymbol{r}|$. The ordering $h_{\boldsymbol{r}}$ gives to $\mathbb{C}_*^d$ does not change if $\boldsymbol{r}$ is multiplied by a positive scalar, so our arguments about the height function will hold whenever $\boldsymbol{r}$ is replaced by any positive multiple. This invariance property means that an analysis of $a_{\boldsymbol{r}}$ as $\boldsymbol{r} \to \infty$ with $\hat{\boldsymbol{r}} = \boldsymbol{r}/|\boldsymbol{r}|$ converging to some fixed $\hat{\boldsymbol{r}}_*$ can usually be accomplished with the fixed height function $h_{\hat{\boldsymbol{r}}_*}$. In particular, if $\hat{\boldsymbol{r}}_*$ is a fixed direction and $h_{\boldsymbol{r}_*}(\boldsymbol{x}) < h_{\boldsymbol{r}_*}(\boldsymbol{y})$ then, as $\boldsymbol{r} \to \infty$ with $\hat{\boldsymbol{r}} \to \boldsymbol{r}_*$, the Cauchy integrand is exponentially smaller at $\boldsymbol{z} = \boldsymbol{x}$ than at $\boldsymbol{z} = \boldsymbol{y}$. When $\boldsymbol{r}$ is understood we write simply h for $h_{\boldsymbol{r}}$.

Definition 7.7. A ***smooth critical point*** $\boldsymbol{z}$ of the rational function $F = P/Q$ in the direction $\hat{\boldsymbol{r}}$ is a smooth point of $\mathcal{V}_*$ that is a critical point of $h_{\hat{\boldsymbol{r}}} : \mathcal{V}_* \to \mathbb{R}$ as a smooth mapping of real manifolds. The set of critical points in the direction $\hat{\boldsymbol{r}}$ is denoted by $\texttt{critical}(\boldsymbol{r})$.

The height function $h_{\hat{\boldsymbol{r}}}$ is the real part of (a branch of) the analytic function $\phi(\boldsymbol{z}) = -\boldsymbol{r} \cdot \log \boldsymbol{z}$, and the Cauchy–Riemann equations imply that the critical points of F in the direction $\hat{\boldsymbol{r}}$ can also be computed as the critical points of $\phi : \mathcal{V}_* \to \mathbb{C}$ as a (locally) holomorphic mapping of complex manifolds. In particular, we have the following explicit definition of smooth critical points.

Lemma 7.8. *Assume that $\mathcal{V}$ is a smooth manifold and let $\tilde{Q}$ be the square-free part of the denominator Q. Then $\boldsymbol{w} \in \mathbb{C}_*^d$ is a critical point in the direction $\hat{\boldsymbol{r}}$ if and only if it satisfies the* ***smooth critical point equations***

$$\tilde{Q}(\boldsymbol{w}) = r_k w_1 \tilde{Q}_{z_1}(\boldsymbol{w}) - r_1 w_k \tilde{Q}_{z_k}(\boldsymbol{w}) = 0 \qquad (2 \le k \le d), \tag{7.8}$$

where $\tilde{Q}_{z_j}$ denotes the derivative of $\tilde{Q}$ with respect to the variable z_j.

Proof The point $\boldsymbol{w}$ is a critical point when $Q(\boldsymbol{w}) = 0$ and the differential of $\phi : \mathcal{V}_* \to \mathbb{C}$ is zero. Vanishing of this differential occurs exactly when the differential of ϕ as a map from $\mathbb{C}_*^d$ to $\mathbb{C}$ projects to zero on the tangent space of $\mathcal{V}_*$ at $\boldsymbol{w}$. Since the tangent space to $\mathcal{V}_*$ at $\boldsymbol{w}$ is the hyperplane with normal $(\nabla\tilde{Q})(\boldsymbol{w})$, the differential of ϕ projects to zero if and only if $(\nabla\phi)(\boldsymbol{w})$ is parallel to $(\nabla\tilde{Q})(\boldsymbol{w})$. These vectors are parallel if and only if all 2×2 minors of the matrix

$$\begin{pmatrix} \left(\nabla\tilde{Q}\right)(\boldsymbol{w}) \\ (\nabla\phi)(\boldsymbol{w}) \end{pmatrix} = \begin{pmatrix} \tilde{Q}_{z_1}(\boldsymbol{w}) & \cdots & \tilde{Q}_{z_d}(\boldsymbol{w}) \\ -r_1/w_1 & \cdots & -r_d/w_d \end{pmatrix}$$

vanish. Vanishing of the minors simplifies to give the smooth critical point equations. □

Remark 7.9. The smooth critical point equations (7.8) form a polynomial system with d equations in d variables. It is therefore unsurprising that *generically* Q has a finite number of critical points (i.e., this holds for all polynomials Q except for those whose coefficients come from a fixed algebraic set depending only on the degree of Q). This follows directly from an algebraic version of Sard's Theorem, which can be found in [BPR03, Theorem 5.56]; see also [Mel21, Section 5.3.4] for an explicit derivation.

Exercise 7.4. Continuing Exercise 7.3, let $\boldsymbol{r} = (2, 1)$ and find the critical points for $h_{\boldsymbol{r}}$ on $\mathcal{V}$. Compute the heights of these critical points and compare them to the limit height for the sequence approaching infinity in Exercise 7.3. Is the limit height larger than the heights of all critical points on $\mathcal{V}$?

Proposition 7.10. *Singularity of the Hessian matrix for $h_{\hat{r}}$ in local coordinates at a critical point for $h_{\hat{r}}$ on the smooth variety $\mathcal{V}_*$ is independent of the choice of coordinatization of $\mathcal{V}_*$ as a complex manifold.*

Proof At a point p where $\nabla h_{\hat{r}}$ vanishes, the chain rule under a coordinate change Ψ simplifies to $\mathcal{H}' = J_\Psi \mathcal{H}$, where $\mathcal{H}'$ is the new Hessian, $\mathcal{H}$ is the old Hessian, and J_Ψ is the Jacobian matrix of Ψ at p. The claim follows from nonsingularity of J_Ψ at p. □

Definition 7.11. A smooth critical point w of h is called a ***nondegenerate critical point*** if the Hessian matrix for h in local coordinates around w is nonsingular.

This definition is generalized to non-smooth points in Definition 7.34 below. Under our assumption that $\mathcal{V}$ is smooth, one of the partial derivatives of the square-free part of Q is nonvanishing at w. Without loss of generality, we assume that $\tilde{Q}_{z_d}(w) \neq 0$ is non-zero, so we can parametrize $\mathcal{V}$ near w as $z_d = g(z^\circ) = g(z_1, \ldots, z_{d-1})$ for some analytic function g defined in a neighborhood of w°. The critical point w is nondegenerate if and only if the Hessian matrix of $h(z^\circ, g(z^\circ))$ with respect to $z_1, \ldots, z_{d-1}$ has non-zero determinant at $z^\circ = w^\circ$. We say h is a ***Morse height function*** when all of its critical points are nondegenerate.

Remark. Most topological works, such as [Mil63; GM88], study spaces using Morse height functions. However, as discussed in Appendix C, as long as there are finitely many critical points the basic Morse decompositions hold whether or not h is Morse: the topology of the space is still generated by attachments at the critical points. However, the description of the attachments becomes more complicated for non-Morse height functions.

Step 2: Intersect the torus with the singular variety

The Cauchy integral representation (7.7) holds for any torus $T = \text{Relog}^{-1}(x)$ with x in the component B of $\texttt{amoeba}(Q)^c$ corresponding to the convergent Laurent expansion with coefficients a_r. We want to replace the domain of integration T with a domain of integration close to $\mathcal{V}$ that "wraps around" the singular variety, so we can use a residue computation in Step 3 below to reduce to an integral "on" $\mathcal{V}$.

If γ is any $(d-1)$-chain in $\mathcal{V}_*$ then the *Collar Lemma* (Lemma C.1 in Appendix C) shows how to construct the *tube* $\circ\gamma$ *around* γ, which is a d-chain in the domain $\mathcal{M}$ where the Cauchy integral ω is holomorphic. The tube $\circ\gamma$ can be viewed as a union of circles with centers at the points of γ, and the map $\gamma \mapsto \text{o}\gamma$ is well defined as a map from the homology group $H_{d-1}(\mathcal{V}_*)$ to $H_d(\mathcal{M})$.

Theorem C.2 of Appendix C implies that $\text{o} : H_{d-1}(\mathcal{V}_*) \to H_d(\mathcal{M})$ is injective, and if T' is any torus contained in $\mathcal{M}$ then pulling back $[T - T'] \in H_d(\mathcal{M})$ via o gives a well-defined class $\mathbf{INT}(T, T') \in H_{d-1}(\mathcal{V}_*)$ known as the *intersection class* of T and T'. By construction, $[T] - [T'] = \text{o}\,\mathbf{INT}(T, T')$ in $H_d(\mathcal{M})$, so that

$$a_{\mathbf{r}} = \left(\frac{1}{2\pi i}\right)^d \int_T \omega = \left(\frac{1}{2\pi i}\right)^d \int_{\text{o}\,\mathbf{INT}(T,T')} \omega + \left(\frac{1}{2\pi i}\right)^d \int_{T'} \omega .$$

One can picture $\text{o}\,\mathbf{INT}(T, T')$ by imagining a continuous deformation of T to T'. If this deformation is sufficiently generic it will intersect $\mathcal{V}_*$ transversely, with the intersection yielding $\mathbf{INT}(T, T')$. The tube around $\mathbf{INT}(T, T')$ is thus the chain that needs to be added to account for passing the deformation through $\mathcal{V}_*$. See Figure 7.7 for an illustration.

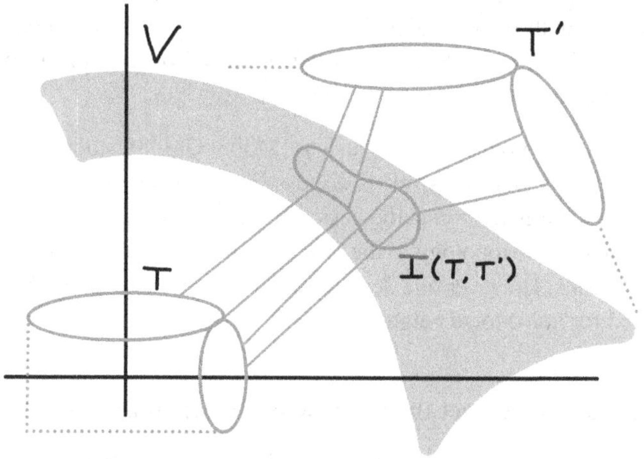

Figure 7.7 An intersection class of T and T' with respect to $\mathcal{V}$.

If we pick a torus T' so that $\int_{T'} \omega = 0$ then we have succeeded in expressing the Cauchy integral as an integral over a tube around a curve in $\mathcal{V}_*$. Corollary 6.29 implies the existence of such a torus, giving the following.

Proposition 7.12. *Assume F is the ratio of coprime polynomials $F(\mathbf{z}) = P(\mathbf{z})/Q(\mathbf{z})$. As $\mathbf{r} \to \infty$ in the direction $\hat{\mathbf{r}}$ there exists a torus T' such that*

$\int_{T'} \omega = 0$ *for all but finitely many* $\mathbf{r}$, *and*

$$a_{\mathbf{r}} = \left(\frac{1}{2\pi i}\right)^d \int_{\circ\,\mathbf{INT}(T,T')} \omega \tag{7.9}$$

whenever the integral over T' *is zero.* □

Exercise 7.5. Let $Q(x, y) = 1 - x - y - x^2y^2$, whose amoeba is shown in Figure 7.8. When $\mathbf{r} = (1, 1)$, which components of $\mathtt{amoeba}(Q)^c$ have $h_{\mathbf{r}}$ unbounded from below, and which vertices of the Newton polygon for Q do these regions correspond to under the relationship described in Theorem 6.18?

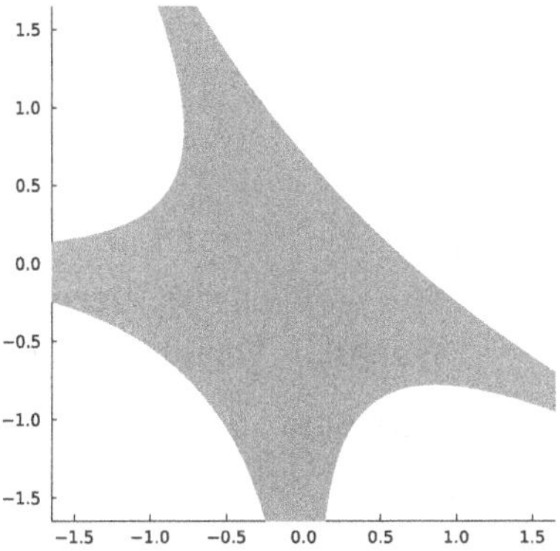

Figure 7.8 Amoeba of $Q(x, y) = 1 - x - y - x^2y^2$.

We can convert the d-dimensional integral in (7.12) to a $(d - 1)$-dimensional integral over the intersection cycle, which lies in $\mathcal{V}_*$. This is accomplished using the concept of *multivariate residues* (also called *Leray residues*). Appendix C.2 gives a summary of multivariate residues, but for this discussion it is sufficient to note that the residue form $\mathrm{Res}(\tau)$ of a meromorphic d-form τ with singularities contained in $\mathcal{V}$ is a $(d-1)$ form restricted to $\mathcal{V}$. Theorem C.9 implies that

$$\frac{1}{2\pi i} \int_{\circ\gamma} \tau = \int_{\gamma} \mathrm{Res}(\tau)$$

for any $(d - 1)$-chain γ in $\mathcal{V}_*$ and holomorphic d-form τ on $\mathcal{M}$.

In particular, combining the residue operator with Proposition 7.12 gives the following.

Proposition 7.13. *If T' is a torus described by Proposition 7.12 then*

$$a_{\mathbf{r}} = \left(\frac{1}{2\pi i}\right)^{d-1} \int_{\mathbf{INT}(T,T')} \operatorname{Res} \omega. \tag{7.10}$$

□

Exercise 7.6. Suppose $d = 1$ and $\mathcal{V} = \mathcal{V}_Q$ where $Q(z) = 2 - 3z + z^2$. Let T be a circle of some small positive radius ε and T' be a circle of some large radius M.

(a) What is the cycle $\mathbf{INT}(T, T')$?
(b) What is the form $\operatorname{Res}(\omega)$ when $\omega = Q(z)^{-1}z^{-n-1}dz$?
(c) What is $\int_{\mathbf{INT}(T,T')} \operatorname{Res}(\omega)$?
(d) What is $\circ\, \mathbf{INT}(T, T')$?
(e) Describe in words why $\circ\, \mathbf{INT}(T, T')$ is homologous to $T - T'$ in $H_1(\mathbb{C}_* \setminus \mathcal{V})$.

Step 3: Determine a Morse-Theoretic Decomposition of the Singular Variety

Having reduced the Cauchy integral to an integral over an intersection cycle $\gamma = \mathbf{INT}(T, T')$ lying *in* the singular variety $\mathcal{V}_*$, we now want to deform γ in $\mathcal{V}_*$ to represent the coefficient sequence of interest as a sum of saddle point integrals. Because we are currently assuming $\mathcal{V}$ is smooth, we could try to compute such a representation by taking a gradient flow of γ on $\mathcal{V}$ with respect to the height function $h_{\hat{\mathbf{r}}}$. If γ can be deformed so that it lies in the neighborhood of a nondegenerate critical point σ of $h_{\hat{\mathbf{r}}}$, except for points of height at most $\sigma - \varepsilon$ for some $\varepsilon > 0$, then we can apply the saddle point techniques of Chapter 5 to compute asymptotics (up to an exponentially negligible error, coming from ignoring points of lower height).

Actually computing such a gradient flow on real examples is usually not feasible. Fortunately, one of the most important consequences of Morse theory is that under reasonable conditions *there are only a finite number of possibilities for the long-term behavior of such a flow*. In particular, as detailed in Appendix C and summarized here, if the flow does not stay at bounded height while escaping to infinity on $\mathcal{V}_*$ then we can flow γ until it gets locally "stuck" on one of the critical points of $h_{\hat{\mathbf{r}}}$.

Our results are phrased in the language of *singular homology*, reviewed in Appendix B. Of particular use to us are the notions of *relative homology*, which allows us to discuss homology near a critical point while ignoring points

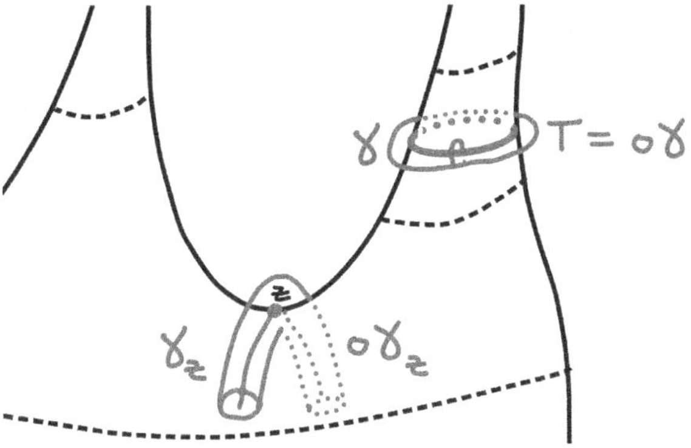

Figure 7.9 The curve γ is deformed to a curve $\gamma_{\mathbf{z}}$ locally draped over a saddle $\mathbf{z}$ centered at a critical point for the height function. The tubes around γ and $\gamma_{\mathbf{z}}$ are also pictured.

of lower height that do not affect dominant asymptotic behavior, and *attachments*, which describe how to decompose the singular variety by joining together topologically simpler spaces. Our discussion here summarizes the main points of the machinery developed in the appendices before applying them to our situation.

Morse theory represents the topology of a manifold X equipped with a smooth map $h : X \to \mathbb{R}$ in terms of successive attachments. The smooth function h is referred to as a *height function* on X. As discussed above, we say h is a *Morse* if its critical points are nondegenerate, and *proper* if the inverse image of any closed interval is compact. Let $X_{\leq c}$ denote the subspace of all points $\mathbf{z} \in X$ with $h(\mathbf{z}) \leq c$ and suppose that h is a proper Morse function. As described in Section C.3 of Appendix C, Morse theory describes the change in topology when the space $X_{\leq a}$ is increased to $X_{\leq b}$ using the language of attachments. Moving from $X_{\leq a}$ to $X_{\leq b}$ is a homotopy equivalence (no change in topology) unless h has critical values in $[a, b]$. When there is a single critical point $\mathbf{z}$ with height in this interval, the topology changes via a topological attachment: $X_{\leq b}$ is homotopy equivalent to $X_{\leq a}$ on which a λ-ball B is glued via an attaching map $\phi : \partial B \to X_{\leq a}$. The value of λ is the *Morse index* of the critical point $\mathbf{z}$, which can be thought of as the dimension of the downward facing part of the generalized saddle at $\mathbf{z}$ and computed in local coordinates using the Hessian of h at $\mathbf{z}$.

Figure C.3 in Appendix C shows how the decapitated unit sphere $S_{\leq 1-\varepsilon}$ becomes the full unit sphere by the attachment of a cap and the north pole (Morse index 2), while Figure C.5 in Appendix C shows how a contractible patch near the bottom of a torus becomes homotopy equivalent to a circle when a bridge (homotopy equivalent to an arc) is added at the first Morse index-1 critical point. These diagrams are reproduced here in Figure 7.10 for convenience.

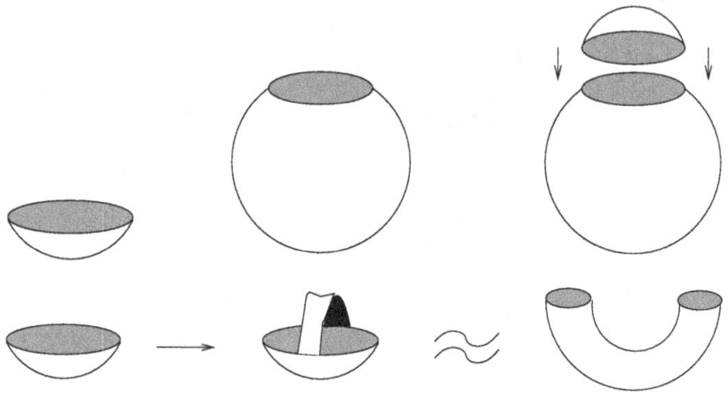

Figure 7.10 Two examples of attachments.

We now specialize to the case where $X = \mathcal{V}_*$ and $h = h_{\hat{r}}$ for some fixed unit vector $\hat{r}$. It is not true that h will always be proper, however we can work around this difficulty. If $\sigma \in \mathcal{V}_*$ is a critical point then the gradient $(\nabla h_{\hat{r}})(\sigma)$ projects to zero on the tangent plane $T_\sigma \mathcal{V}_* \subset \mathbb{C}^d$. Roughly speaking, a *critical point at infinity* is a sequence of points $z^{(k)} \in \mathcal{V}_*$ going off to infinity such that the projection of $(\nabla h_{\hat{r}})(z^{(k)})$ to $T_{z^{(k)}} \mathcal{V}_*$ approaches zero as $k \to \infty$; the associated *critical value at infinity* is the limit of $h_{\hat{r}}(z^{(k)})$ as $k \to \infty$. Critical points at infinity are defined formally in Definition 7.42 below. Provided there are no critical points at infinity, the classic results of Morse theory hold even when the height function is not proper.

Lemma 7.14. *Suppose $h_{\hat{r}}$ has no critical values at infinity in the interval $[a, b]$. If there are no critical values in $[a, b]$ then the inclusion $X_{\le a} \subseteq X_{\le b}$ is a homotopy equivalence. If there is a single critical point z with critical value $h_{\hat{r}}(z) = c \in (a, b)$, then the pair $(\mathcal{V}_{\le b}, \mathcal{V}_{\le a})$ is homotopy equivalent to a λ-cell relative to its boundary, where λ is the Morse index of the critical point z for $h_{\hat{r}}$.*

Exercise 7.7. What is λ in the attachment in the bottom row of Figure 7.10?

It is convenient to postpone the proof of Lemma 7.14 until the more general setting when we no longer require $\mathcal{V}$ to be smooth. After establishing additional results below, Lemma 7.14 follows directly from Lemma 7.25, which asserts the homotopy equivalence, and the identification of the attachment in

Theorem 7.35(b). In the present smooth case, a nice simplification occurs: because the height function is the real part of a complex (locally) analytic function, every critical point $z \in \mathcal{V}_*$ has Morse index $d - 1$.

Exercise 7.8. Prove that the real part of a complex analytic function defined on an open set in $\mathbb{C}^d$ has Morse index d, then prove that the real part of such a function restricted to a smooth hypersurface has Morse index $d - 1$. *Hint:* The Cauchy–Riemann equations yield a lot of information about the eigenvectors and eigenvalues of the Hessian.

This characterization of the index allows us to show that $H_{d-1}(\mathcal{V}_*)$ is homologically a *bouquet of* $(d-1)$*-spheres*, one quasi-local to each critical point. A version of the following theorem, with the stronger assumption that h is proper replacing the assumption of no CVAI, is stated and proved as Theorem C.39 in Appendix C (the appendices contain background material not specialized to ACSV). The restriction that the critical values are distinct is removed in Corollary 7.17.

Theorem 7.15. *Assume that $\mathcal{V}$ is smooth, $h_{\hat{r}}$ is a Morse height function, and there are no critical values at infinity (according to Definition 7.43 below). Assume further that the critical values $c_j = h_{\hat{r}}(z_j)$ are distinct and listed in descending order.*

(i) Each projection $H_{d-1}(\mathcal{V}_*) \to H_{d-1}(\mathcal{V}_{\leq c_j+\varepsilon}, \mathcal{V}_{\leq c_j-\varepsilon})$ is surjective. In other words, the relative homology generator at z_j can be chosen to be an absolute cycle.

(ii) Each inclusion $\mathcal{V}_{\leq c} \subseteq \mathcal{V}_*$ induces an injection on H_{d-1}. In other words, there are no relations: no homology generator ever gets killed.

It follows that $H_{d-1}(\mathcal{V}_*) \cong \mathbb{Z}^m$ and that a basis $\gamma_1, \ldots, \gamma_m$ for $H_{d-1}(\mathcal{V}_*)$ can be chosen so that each γ_j is a cycle on which $h_{\hat{r}}$ attains its maximum value at z_j.

Proof Part (*i*) of Theorem 7.44 below extends the fundamental Morse Lemma, namely homotopy equivalence of $\mathcal{M}_{\leq c}$ as c varies in an interval with no critical values (Lemma C.27) from the case where h is a proper Morse function to the case where h need not be proper but there are no CVAI in the interval. Part (*ii*) of Theorem 7.44 extends the smooth attachment theorem for a single critical value c (Theorem C.28) from the case where h is a proper Morse function to the case where h need not be proper but there are no CVAI in the interval. Accordingly, the conclusions of Theorems C.38 and C.39 hold for this case, via the same argument. Specifically, these follow from the identification of the attachment and from the homology long exact sequence for the filtration of pairs $(\mathcal{M}_{\leq b_j}, -\infty)$, where b_j are real numbers between each successive pair of critical values, b_0 is above the highest critical value, and $-\infty$ is $\mathcal{M}_b$ for any b less than the least critical value. See Section C.4 for details. □

Remark 7.16. The isomorphism $H_{d-1}(\mathcal{V}_*) \cong \bigoplus_{j=1}^m H_{d-1}(\mathcal{V}_{c_j+\varepsilon}, \mathcal{V}_{c_j-\varepsilon})$ is not natural. For each attachment at $\mathbf{z}_j$ there is an arbitrary choice of an absolute cycle γ_j that projects to the generator of the homology group for the attachment. The cycle $\gamma_j + \alpha$ would do equally well for any cycle α supported on $\mathcal{V}_{c_j-\varepsilon}$. One might say that the choice of $\gamma_1, \ldots, \gamma_m$, listed in decreasing order of height, can be altered by an arbitrary upper triangular map, replacing γ_j by $\gamma_j + \sum_{i>j} b_i\gamma_i$. This is the so-called ***Stokes phenomenon***, illustrated in Figure 7.11: the saddle point integral from $\mathbf{z}_j$ might pass on either side of $\mathbf{z}_i$ as it travels downward, with the integrals over the two choices of contour differing by the integral over γ_i. Thus, for a cycle C the decomposition $[C] = \sum_{k=1}^m n_k\gamma_k$ is not natural. It is important to note, however, that the leading coefficient n_{j_*} is well-defined independent of the chosen basis $\{\gamma_j\}$, where j_* is the least index such that $n_{j_*} \neq 0$.

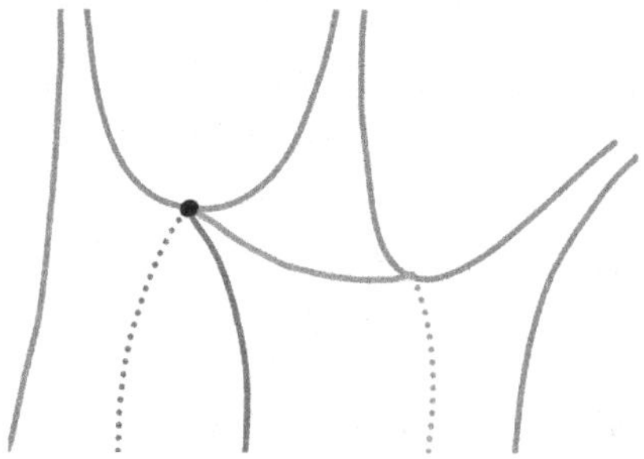

Figure 7.11 Stokes's phenomenon reflects the fact that a curve draped over the higher saddle can descend on either side of the lower saddle, as shown here by two possible branches. The difference between these two curves is a curve draped over the lower saddle.

The simplifying assumption of distinct critical values is not important. To get rid of this, we define the local homology pair $\mathcal{V}^{\mathbf{p},loc}$ at a critical point $\mathbf{p}$ at height c to be the pair (X, Y), where $Y = \mathcal{V}_{\leq c-\varepsilon/2}$ and X is the union of Y with the ball $B_\varepsilon(\mathbf{p})$ for $\varepsilon > 0$ sufficiently small (see Definition C.31 of Appendix C for full details). Any such pairs are homotopy equivalent as long as ε is small enough that the 2ε-balls about different critical points are disjoint.

Deformations defined in Appendix C show that if there is a unique critical point $\mathbf{p}$ with height $c \in [a, b]$ then, for small $\varepsilon > 0$, the local pair $\mathcal{V}^{\mathbf{p},loc}$ is homotopy equivalent to the *slab* $(\mathcal{V}_{\leq c+\varepsilon}, \mathcal{V}_{c-\varepsilon})$. The benefit to replacing the slab by the local pair occurs when there are multiple critical points sharing a critical value. If $h_{\hat{r}}(p) = c$ for all p in some finite set E then

$$(\mathcal{V}_{\le b}, \mathcal{V}_{\le a}) \simeq \bigoplus_{p \in E} \mathcal{V}^{p,\mathrm{loc}}, \tag{C.3.1}$$

giving the following.

Corollary 7.17. *Replacing $(\mathcal{V}_{c_j+\varepsilon}, \mathcal{V}_{c_j-\varepsilon})$ by $\mathcal{V}^{z,\mathrm{loc}}$ for each z, the conclusions of Theorem 7.15 hold without the assumption of distinct critical values.*
□

We end this subsection with some examples of this topological decomposition.

Example 7.18 (binomial coefficients). Recall that the binomial coefficients $a_{rs} = \binom{r+s}{r}$ have bivariate generating function $F(x, y) = 1/(1-x-y)$. If $\hat{r} = (r, s)$ with $r + s = 1$ and $r, s \in (0, 1)$ then as r varies from 0 to 1, the critical point $w(\hat{r})$ of F in the direction $\hat{r}$ slides from $(0, 1)$ to $(1, 0)$. The homology group $H_1(\mathcal{V}_*)$ has a single generator γ_{z_*}. The homology group $H_2(\mathcal{M})$ is cyclic as well, generated by $\mathrm{o}\gamma_{z_*}$. ◁

Example 7.19 (Delannoy numbers). The Delannoy number generating function from Example 2.7 in Chapter 2 is $1/(1-x-y-xy)$. The situation is similar to Example 7.18, except that as r varies from 0 to 1 the critical point $w(\hat{r})$ traverses the arc the other way from $(0, 1)$ to $(1, 0)$, and there is another critical point w' traversing a hyperbola in the third quadrant. ◁

Exercise 7.9. Consider the amoeba of the denominator $Q(x, y) = 1-x-y-xy$ of the Delannoy generating function, shown in Figure 7.12.

(a) Compute the critical points w and w' in the direction determined by $r = (2, 3)$, then draw dots where $p = \mathrm{Relog}(w)$ and $p' = \mathrm{Relog}(w')$ lie on the amoeba.
(b) Find a path β, from the power series component of the amoeba complement to a component where the Cauchy integral is zero, that enters the amoeba at p and exits it at p'.
(c) Describe $\gamma = \mathrm{Relog}^{-1}(\beta)$.
(d) State why $[\gamma] = \mathbf{INT}(T, T')$ in $H_1(\mathcal{V}_*)$ and why $\int_\gamma \mathrm{Res}\,\omega$ is easy to estimate, where

$$\omega = \frac{x^{-2n-1}y^{-3n-1}}{1-x-y-xy} dx \wedge dy\,.$$

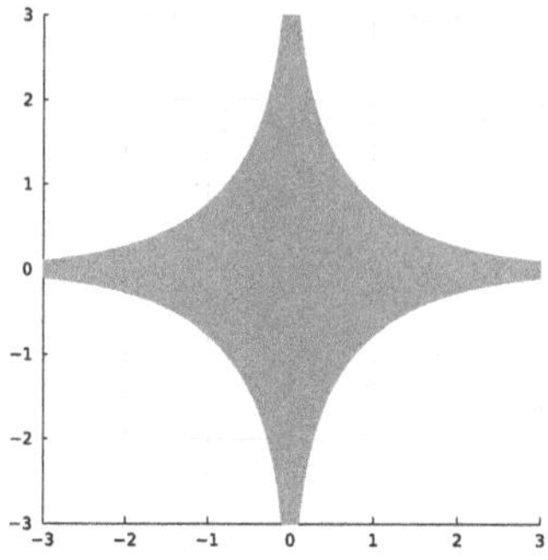

Figure 7.12 Amoeba for the Delannoy generating function $Q(x, y) = 1-x-y-xy$. Each point interior to the amoeba is the image of precisely two points of $\mathcal{V}$ under Relog.

Result: a saddle point integral decomposition in the smooth case

Theorem 7.15 and Corollary 7.17 give a basis of $H_{d-1}(\mathcal{V}_*)$ consisting of cycles that attain their maximum values at critical points. Vanishing of $dh_{\hat{\boldsymbol{r}}}|_{\mathcal{V}}$ at $\boldsymbol{z}$ is equivalent to $\boldsymbol{z}^{-\boldsymbol{r}}$ being in stationary phase at $\boldsymbol{z}$ for any $(d-1)$-chain γ_j supported on $\mathcal{V}_*$. Thus, combining Theorem 7.15 and Corollary 7.17 with the integral representation in Proposition 7.13 gives the following, our ultimate goal for generating functions with smooth singular varieties.

Theorem 7.20 (smooth saddle point integral decomposition). *Assume that $\mathcal{V}$ is smooth, $h_{\hat{\boldsymbol{r}}}$ is a Morse height function, and that there are no critical values at infinity (see Definition 7.43 below). Assume further that the critical values $c_j = h_{\hat{\boldsymbol{r}}}(\boldsymbol{z}_j)$ for $1 \le j \le m$ are listed in descending order. Then there exist integers $\kappa_j \in \mathbb{Z}$ and smooth chains of integration γ_j with heights uniquely maximized at $\boldsymbol{z}_j$, such that*

$$a_{\boldsymbol{r}} = \sum_{j=1}^{m} \frac{\kappa_j}{(2\pi i)^{d-1}} \int_{\gamma_j} \boldsymbol{z}^{-\boldsymbol{r}-\boldsymbol{1}} \operatorname{Res}(F(\boldsymbol{z})\, d\boldsymbol{z}). \tag{7.11}$$

The integral in the jth summand is in stationary phase at $\boldsymbol{z}_j$. The least j such that $\kappa_j \neq 0$, and the homology class $\sum_{j' \in E} \kappa_{j'} \gamma_{j'}$ for all j' such that $\boldsymbol{z}_{j'}$ has height c_j, are uniquely defined. □

There are two important tasks remaining: computing asymptotics of the saddle point integrals and determining the integers κ_j. While integral asymptotics (in this smooth case) follow in a straightforward manner from the results of Chapter 5, it can be very difficult to determine these unknown integers. We discuss both of these questions in Chapter 9, where we derive explicit asymptotic formulae for $a_{\mathbf{r}}$ in terms of the generating function $F(\mathbf{z})$. Readers who are interested only in smooth asymptotics (and do not need to see the technical discussion of critical points at infinity) may go directly to Chapters 8 and 9, after a brief discussion about removing our simplifying hypotheses.

The requirement of no critical value at infinity is essential: when there are critical points at infinity, asymptotics are in principle affected. Classifying these cases and computing the asymptotics remains an open problem discussed further in Chapter 13. Removing the smoothness assumption involves the apparatus of stratified Morse theory, which we make use of in the next section. The assumption that $h_{\hat{\mathbf{r}}}$ is nondegenerate is not essential, however in its absence there is no longer a unique cycle γ_j for each j. We handle this case, for now, by two examples.

Example 7.21 (cubic degeneracy)**.** The simplest degeneracy at a critical point, a so-called *monkey saddle*, leads to two independent homology generators as in Figure 7.13. ◁

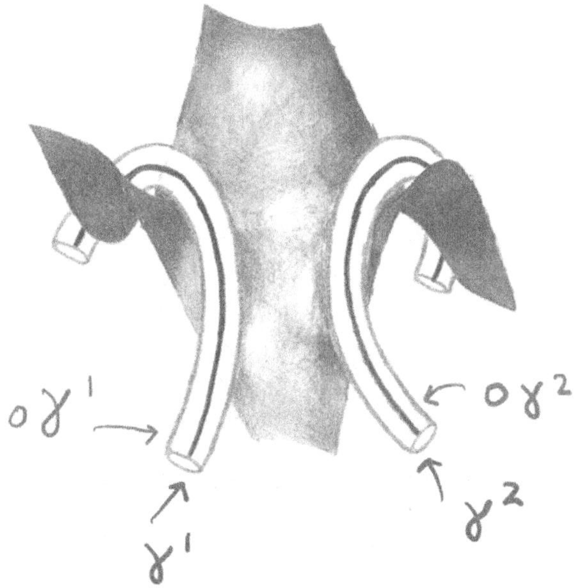

Figure 7.13 Two homology generators associated with a monkey saddle.

While critical points are generically nondegenerate, we now give one combinatorial example in which a critical point $\boldsymbol{w}$ is indeed degenerate and contributes more than one generator.

Example 7.22 (bi-colored supertrees)**.** Example 9.32 in Chapter 9 looks at a rational generating function counting bi-colored supertrees (certain planar binary trees that need not concern us here). The singular variety is the smooth surface defined by the vanishing of $Q(x, y) = x^5y^2 + 2x^2y - 2x^3y + 4y + x - 2$. When $\hat{r} = (1/2, 1/2)$ then there are two nondegenerate critical points, $\mathbf{u}$ and $\mathbf{v}$, together with a critical point $\boldsymbol{w}$ near which $h_{\hat{r}}$ is quartic (so doubly degenerate). Accordingly there is one cycle $\gamma_{\mathbf{u}}$, one cycle $\gamma_{\mathbf{v}}$, and three cycles $\gamma_{\boldsymbol{w}}^{(j)}$ which may be configured all to enter $\boldsymbol{w}$ along the solid arc and exit along one of the three dashed arcs shown in Figure 7.14. ◁

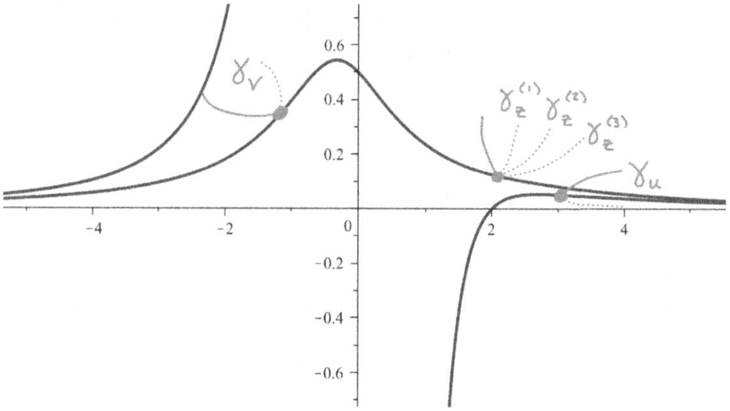

Figure 7.14 The supertree generating function yields two nondegenerate critical points and one doubly degenerate critical point.

7.3 The general case via stratified Morse theory

We now drop the assumption that the singular variety $\mathcal{V}$ is smooth. As detailed in Appendix D, the correct notion for us is the concept of a *Whitney stratified space*: every real or complex algebraic (or analytic) variety admits a *Whitney stratification*, and the analytic constructions we require for asymptotics can be built for stratified spaces. In this chapter we recount only the results from Appendix D that we directly require. Although we typically assume that F is a rational function to simplify our presentation, the results discussed here hold for general meromorphic functions with minor modifications.

Example 7.23. Figure 7.15 shows the zero set $\mathcal{V}$ of the polynomial $Q = (z^3 - x^2)(2 - x - y - z)$. We can split the variety $\mathcal{V}$ into a finite number of semi-algebraic strata (defined by polynomial equalities and inequalities): two strata of codimension 1,

$$S_1 = \left\{x + y + z = 2 \text{ and } z^3 - x^2 \neq 0\right\}$$
$$S_2 = \left\{z^3 - x^2 = 0 \text{ and } x + y + z \neq 2\right\} \setminus \{x = z = 0\},$$

two strata of codimension 2,

$$S_3 = \left\{z^3 - x^2 = 0 \text{ and } x + y + z = 2 \text{ and } (x, z) \neq (0, 0)\right\},$$
$$S_4 = \{(x, z) = 0 \text{ and } y \neq 2\},$$

and one stratum of codimension 3 at the point

$$S_5 = \{z^3 - x^2 = x + y + z - 2 = x = z = 0\} = \{(0, 2, 0)\}.$$

Note that we introduce additional strata both to account for multiple irreducible components of $\mathcal{V}$ and to account for singularities in individual components. ◁

As described in Appendix D, it is usually not sufficient to partition $\mathcal{V}$ into any general set of smooth manifolds – we must also make sure the elements in such a partition "fit together nicely." This concept is formalized by the notion of a Whitney stratification, given in Definition D.3 of Appendix D. For the rest of this chapter we fix a Whitney stratification of $\mathcal{V}$, which is a partition of $\mathcal{V}$ into manifolds $\{S_\alpha : \alpha \in I\}$ indexed by some partially ordered set I such that

(i) $S_\alpha \cap \overline{S_\beta} \neq \emptyset$ if and only if $S_\alpha \subset \overline{S_\beta}$ if and only if $\alpha \leq \beta$, and
(ii) if $\alpha < \beta$, if the sequences $\{x_i \in S_\beta\}$ and $\{y_i \in S_\alpha\}$ both converge to $y \in S_\alpha$, if the lines $\ell_i = \overline{x_i\, y_i}$ converge to a line ℓ, and if tangent planes $T_{x_i}(S_\beta)$ converge to a plane T, then both ℓ and $T_y(S_\alpha)$ are contained in T.

We always take *algebraic* stratifications defined by polynomial equalities and inequalities. In fact, we may assume that our Whitney stratification is defined by a finite sequence of nested algebraic sets $\mathcal{V} = F_0 \supset F_1 \supset \cdots \supset F_m = \emptyset$ such that the connected components of the sets $F_i \setminus F_{i+1}$ for all $1 \leq i \leq m-1$ form the strata. If S is a stratum defined as a connected component of $F_i \setminus F_{i+1}$ then the ***dimension of the stratum*** S (respectively the ***codimension of the stratum*** S) is the dimension (respectively codimension) of $F_i \subset \mathbb{C}^d$ as an algebraic set. Whitney stratifications exist for all algebraic (and analytic) varieties, and algorithms to compute them are discussed in Chapter 8 and Appendix D.

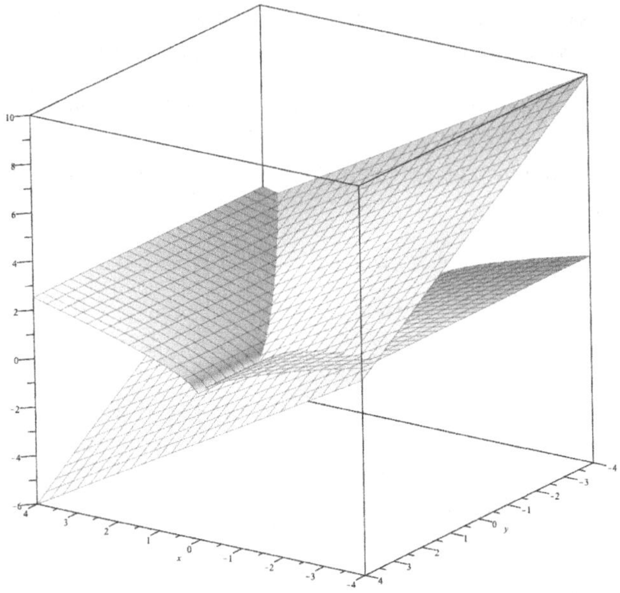

Figure 7.15 The zero set of $Q = (z^3 - x^2)(2 - x - y - z)$.

Stratified critical points

A point $\boldsymbol{p}$ in a stratum S is said to be a ***(stratified) critical point*** for the height function $h = h_{\hat{\boldsymbol{r}}}$ if the restriction $dh|_S$ vanishes at $\boldsymbol{p}$. Analogously to the smooth case above, because h is the real part of $\phi(\boldsymbol{z}) = -\boldsymbol{r} \cdot \log \boldsymbol{z}$ the Cauchy–Riemann equations imply that $\boldsymbol{p}$ is a critical point if the gradient of ϕ lies in the normal space to S at $\boldsymbol{p}$. If S has codimension k then there exists an open set $U \subset \mathbb{C}^d$ containing $\boldsymbol{p}$ and irreducible polynomials $g_1, \ldots, g_k$ such that $S \cap U = \mathcal{V}(g_1, \ldots, g_k) \cap U$ (i.e., S is locally defined by the polynomials g_i near $\boldsymbol{p}$). The point $\boldsymbol{p}$ is a critical point if and only if the gradient $(\nabla\phi)(\boldsymbol{p})$ lies in the complex span of the gradients $(\nabla g_1)(\boldsymbol{p}), \ldots, (\nabla g_k)(\boldsymbol{p})$. Although ϕ involves logarithms, its gradient is a rational function, so we may compute stratified critical point by solving polynomial systems. Computation of stratified critical points is discussed at greater length in Chapter 8.

Recall from Chapter 6 that the *logarithmic gradient* of a differentiable function f at $\boldsymbol{z} \in \mathbb{C}^d$ is the vector

$$\nabla_{\log} f(\boldsymbol{z}) = (z_1 f_{z_1}, \ldots, z_d f_{z_d}(\boldsymbol{z})) \,, \tag{7.12}$$

with the word *logarithmic* coming from the fact that the logarithmic gradient of $f(\boldsymbol{z})$ at $\boldsymbol{z} = \exp(\boldsymbol{x})$ is the gradient of $(f \circ \exp)(\boldsymbol{z})$ at $\boldsymbol{z} = \boldsymbol{x}$. If $\boldsymbol{p}$ is a smooth point of the algebraic hypersurface defined by the vanishing of Q, then the vanishing of $dh_{\hat{r}}|_{\mathcal{V}}$ at $\boldsymbol{p}$ is equivalent to the direction vector $\hat{\boldsymbol{r}}$ being parallel to $(\nabla_{\log} Q)(\boldsymbol{p})$. More generally, vanishing of $dh_{\hat{r}}|_S$ at $\boldsymbol{p}$ is equivalent to $\hat{\boldsymbol{r}}$ lying in the space spanned by the logarithmic gradients of the functions g_j locally defining the stratum S at $\boldsymbol{p}$.

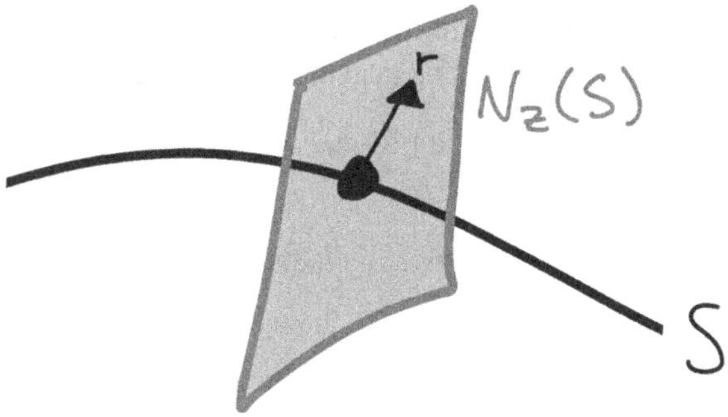

Figure 7.16 The point $\boldsymbol{z}$ on a stratum S defined by the vanishing of two transversely intersecting smooth sheets in three dimensions is a critical point in the direction $\hat{\boldsymbol{r}}$ if $\boldsymbol{r}$ lies in the log-normal plane to S at $\boldsymbol{z}$.

Example 7.24. If $\mathcal{V}$ is the union of two transversely intersecting smooth sheets defined by the vanishing of two polynomials g_1 and g_2 then $\boldsymbol{z}$ is a critical point on $\mathcal{V}_{g_1} \setminus \mathcal{V}_{g_2}$ in a direction $\hat{\boldsymbol{r}}$ if $\boldsymbol{r}$ is parallel to $(\nabla_{\log} g_1)(\boldsymbol{z})$, and the analogous criteria holds for critical points on $\mathcal{V}_{g_2} \setminus \mathcal{V}_{g_1}$. A critical point $\boldsymbol{z}$ on the intersection stratum $S = \mathcal{V}_{g_1} \cap \mathcal{V}_{g_2}$, pictured in Figure 7.16, has $\boldsymbol{r}$ lying somewhere in the log-normal plane spanned by $(\nabla_{\log} g_1)(\boldsymbol{z})$ and $(\nabla_{\log} g_2)(\boldsymbol{z})$. ◁

Exercise 7.10. Describe the set of directions $\boldsymbol{r} \in \mathbb{R}^2_*$ such that $dh_{\hat{r}}|_S = 0$ at a point (x, y, z) of the codimension 2 stratum S_3 in Example 7.23.

Obstructions are critical points

The fundamental lemma of Morse theory, described in Lemma 7.14 above, states that, in the absence of critical values at infinity, critical values are the only places the topology of the sublevel sets of a manifold can change. The fundamental lemma of *stratified* Morse theory says that (stratified) critical values are still the only places the topology of $\mathcal{M}_{\leq c}$ and $\mathcal{V}_{\leq c}$ can change, and thus are the only places obstructions to pushing down cycles of integration can occur. Lemma 7.14 also specifies the nature of the attachment at a critical point, but since this requires a more lengthy explanation, we state the stratified version of Lemma 7.14 without describing the attachment. As a reminder, we postpone the formal definitions of critical points at infinity and critical values at infinity until Definition 7.42 below.

Lemma 7.25. *If $h_{\hat{r}}$ has no critical values (including at infinity) in $[a,b]$, then the inclusion $\mathcal{M}_{\le a} \subseteq \mathcal{M}_{\le b}$ is a homotopy equivalence. The same is true of the inclusion $\mathcal{V}_{\le a} \subseteq \mathcal{V}_{\le b}$.*

Remark 7.26. The fact that stratified critical values isolate all the topological change in $\mathcal{V}_*$ may be less surprising than the fact that they do so in $\mathcal{M}$.

Let $\mathbf{p}$ be a stratified critical point for $h_{\hat{r}}$ in some stratum S. Mirroring our definition of $\mathcal{V}^{\mathbf{p},\mathrm{loc}}$ above, we let

$$\mathcal{M}^{\mathbf{p},\mathrm{loc}} := (\mathcal{M}_{\le c-\varepsilon} \cup B_{2\varepsilon}(\mathbf{p}), \mathcal{M}_{\le c-\varepsilon}),$$

for any sufficiently small $\varepsilon > 0$, which is defined up to homotopy equivalence.

The simplifying assumption of distinct critical values often fails in ACSV, for example if there is a pair of complex conjugate critical points, necessitating one further definition. Let c be a critical value, let $\mathbf{p}_1, \ldots, \mathbf{p}_m$ be the critical points at height c, and assume ε is sufficiently small so that the balls $B_{2\varepsilon}(\mathbf{p}_i)$ are disjoint.

Definition 7.27 (all attachments at height c)**.** Under the setup above, the ***total attachment pair*** at height c is

$$(\mathcal{M}_{c+}, \mathcal{M}_{c-}) := \left(\mathcal{M}_{\le c-\varepsilon} \cup \bigcup_{j=1}^{m} B_{2\varepsilon}(\mathbf{p}_j), \mathcal{M}_{\le c-\varepsilon} \right). \tag{7.13}$$

By disjointness of the balls $B_{2\varepsilon}(\mathbf{p}_j)$, this is a direct sum in the category of pairs of $\mathcal{M}^{\mathbf{p}_j,\mathrm{loc}}$, hence the homology $H_*(\mathcal{M}_{c+}, \mathcal{M}_{c-})$ is the direct sum $\bigoplus_{j=1}^{m} H_*(\mathcal{M}^{\mathbf{p}_j,\mathrm{loc}})$.

Lemma 7.28. *Suppose $h_{\hat{r}}$ has no critical values at infinity in $[a,b]$ and has a single critical value c in $[a,b]$, occurring in the interior (a,b). Then the pairs $(\mathcal{M}_{\le b'}, \mathcal{M}_{\le a'})$ are naturally homotopy equivalent for any $a \le a' < c < b' \le b$.*

Lemmas 7.25 and 7.28 are taken from [BMP22]; a sketch of the proof is reproduced in Section 7.5.

Building by attachment

We now fit together the attachments at critical points of all possible heights. This involves classical topological facts, and works without knowing the homotopy type of any individual attachment.

Let $c_1 > c_2 > \cdots > c_m$ denote the critical values in the interval $[c_m, \infty)$

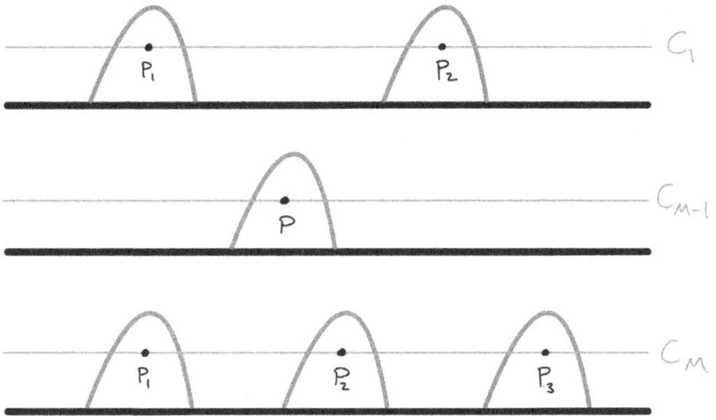

Figure 7.17 Building $\mathcal{M}$ via successive attachments at the critical values $c_1 > c_2 > \cdots > c_m$. In this case we attach three bumps to $\mathcal{M}_{c_m-}$ around critical points at height c_m, then attach a single bump, and so on until attaching two final bumps.

and assume there are no CVAI in $[c_m, \infty)$. For each j let c_{j-} denote $c_j - \varepsilon$ where $\varepsilon > 0$ is sufficiently small so that there are no critical values or CVAI in $[c_j-\varepsilon, c_j)$, and let c_{j+} denote $c_j+\varepsilon$ where $\varepsilon > 0$ is sufficiently small so that there are no critical values or CVAI in $(c, c_j + \varepsilon]$. Intuitively, we think of building up the space $\mathcal{M}$ from the space $\mathcal{M}_{c_m-}$ by successive attachment. First, we attach $(\mathcal{M}_{c_m+}, \mathcal{M}_{c_m-})$ to arrive at the space $\mathcal{M}_{c_m+}$, which, by Lemma 7.28, is homotopy equivalent to the space $\mathcal{M}_{c_{m-1}-}$. Next we attach the pair $(\mathcal{M}_{c_{m-1}+}, \mathcal{M}_{c_{m-1}-})$. Repeating this until the pair $(\mathcal{M}_{c_1+}, \mathcal{M}_{c_1-})$ has been attached, we have built the space $\mathcal{M}_{c_1+}$, which is homotopy equivalent to $\mathcal{M}_{\leq b}$ for all sufficiently large b, and hence to $\mathcal{M}$ itself. This process is illustrated in Figure 7.17. The "bump" $\mathcal{N}(\mathbf{p})$ near a point $\mathbf{p}$ is the intersection of $\mathcal{M}$ with a ball of sufficiently small radius δ. Shrinking ε if necessary, $(\mathcal{M}_{c_j-} \cup \mathcal{N}(\mathbf{p}), \mathcal{M}_{c_j})$ has the homotopy type of the local pair $\mathcal{M}^{\mathbf{p},\text{loc}}$ discussed above.

Each attachment has a long exact homology sequence. Because all of the spaces involved are cell complexes of real dimension at most d (see Section D.4 of Appendix D), the homology groups H_k of dimension $k \geq d+1$ vanish. Thus, the long exact sequence for any $j \leq m$ always begins

$$0 \to H_d(\mathcal{M}_{c_j-}) \to H_d(\mathcal{M}_{c_j+}) \to H_d(\mathcal{M}_{c_j+}, \mathcal{M}_{c_j-}) \to \cdots . \tag{7.14}$$

Definition 7.29. For each critical point $\mathbf{p}$ at height c_j, let $G(\mathbf{p})$ denote the image in $H_d(\mathcal{M}_{c_j+}, \mathcal{M}_{c_j-})$ of the map projecting $\mathcal{M}_{c_j-} \cup \mathcal{N}(\mathbf{p})$ to the pair $(\mathcal{M}_{c_j-} \cup \mathcal{N}(\mathbf{p}), \mathcal{M}_{c_j})$, as in Figure 7.17. In other words, $G(\mathbf{p})$ are those relative d-homology classes, once the bump $\mathcal{N}(\mathbf{p})$ near $\mathbf{p}$ is added, that are represented by absolute cycles. We further define $G = G(c_j) := \bigoplus_{h(\mathbf{p})=c_j} G(\mathbf{p})$.

The sequence (7.14) gives rise to the short exact sequence

$$0 \to H_d(\mathcal{M}_{c_j-}) \to H_d(\mathcal{M}_{c_j+}) \to G \to 0\,. \tag{7.15}$$

As we are working with coefficients in $\mathbb{C}$, there is no torsion, hence the short exact sequence implies a (not natural) direct sum

$$H_d(\mathcal{M}_{c_j+}) \cong H_d(\mathcal{M}_{c_j-}) \oplus H_d(\mathcal{M}_{c_j+}, \mathcal{M}_{c_j-})\,. \tag{7.16}$$

Assuming Lemmas 7.25 and 7.28, we have proved the following.

Theorem 7.30. *Suppose there are no critical values at infinity above height a and finitely many critical values $c_1 > \cdots > c_m$ in $[a, \infty)$. Then the homology of $\mathcal{M}$ is given by*

$$H_d(\mathcal{M}) \cong H_d\left(\mathcal{M}_{<a}\right) \oplus \bigoplus_{\mathbf{p}} G(\mathbf{p})\,,$$

where $G(\mathbf{p})$ is defined in Definition 7.29 and the sum is over critical points $\mathbf{p}$ such that $h_{\hat{\mathbf{r}}}(\mathbf{p}) \geq a$. If there are no critical values at infinity and finitely many critical values then

$$H_d(\mathcal{M}, \mathcal{M}_{-\infty}) \cong \bigoplus_{\mathbf{p}} G(\mathbf{p})$$

where $M_{-\infty}$ denotes $\mathcal{M}_{\leq a}$ for any a less than the least critical value, and the sum is over all critical points. □

Description of the attachments

Next we describe the attachment cycles for $\mathcal{M}$. We could also develop the attachment cycles for $\mathcal{V}_*$ in the stratified setting, however our asymptotic results don't need them so we skip this extra step.

The key to understanding attachments in Whitney stratified spaces is a local product structure described in Theorem D.9 of Appendix D, which follows from the famous (and somewhat difficult) Thom's Isotopy Lemma (Lemma D.16 in Appendix D). The Isotopy Lemma says that for a fixed stratum S of dimension j and any point $\mathbf{p} \in S$ there is a neighborhood of $\mathbf{p}$ where the space $\mathcal{V}$ looks like $\mathbb{R}^j \times N$ where N is the *normal slice* of the strata (see Definition 7.32 and Figure 7.18 below).

What does the local product structure imply for our attachments? Let $\mathbf{p} \in S$ be a critical point for the height function h, and consider S as a complex manifold of dimension i (where it has dimension $j = 2i$ as a real manifold). Because h is the real part of (a branch of) an analytic function, it is harmonic and all critical points have Morse index i. Thus, when S is arranged by height near $\mathbf{p}$ there is an i-dimensional part that 'bends downwards' and an i-dimensional part that 'bends upwards'. By the local product structure, the pair for the attachment of $\mathcal{M}$ at p is (homotopy equivalent to) the product of a pair $(B^i, \partial B^i)$ in the tangent space to S and a pair $(\mathcal{L}, \mathcal{L} \cap \mathcal{M}_{\leq c-\varepsilon})$, where c is the height of p, the constant $\varepsilon > 0$ is sufficiently small, and $\mathcal{L}$ denotes the *normal link* (the intersection of $\mathcal{M}$ with the normal space to S at p in a suitable small neighborhood of p, described in Definition 7.32 below).

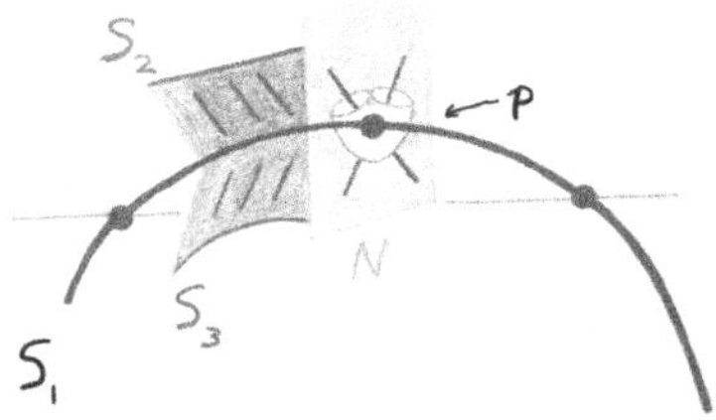

Figure 7.18 An example of an attachment given by the product of a torus with an arc (a relative 3-torus).

Example 7.31. Figure 7.18 shows an example of an attachment on a stratum S_1 with complex dimension 1 defined by the intersection of two transversely intersecting smooth sheets. In one direction it curves down, as shown; in the other direction it curves up (this is not shown). The level set defined by $h(z) = c - \varepsilon$ is the horizontal line and the pair $(B^1, \partial B^1)$ is the black arc modulo its endpoints. The normal link is the complement of two intersecting complex lines in complex 2-space, which is homotopy equivalent to a 2-torus. The 2-torus can be drawn arbitrarily close to p, so it can be chosen as an absolute cycle and the pair $(\mathcal{L}, \mathcal{L} \cap \mathcal{M}_{\leq c-\varepsilon})$ is simply $(\mathcal{L}, \emptyset) \simeq \mathcal{L}$. The attachment pair is obtained by sliding the 2-torus along the black arc from one endpoint to the other, with the second element in the pair being the starting and ending positions. Because an arc modulo its boundary is a circle, this means the attachment is a 3-torus, manifested as a 2-torus times an arc that localizes to a 1-torus. ◁

Formal statement of the attachments

The following definitions and results are special cases of material in Section D.3 of Appendix D. Attachments are defined in the category of (homotopy types of) topological pairs, as are both the tangential and normal Morse data. Products in this category are defined by

$$(A, B) \times (C, D) = (A \times C, A \times D \cup B \times C), \tag{7.17}$$

and the homology of a product obeys the usual Künneth formula for homology with complex coefficients,

$$H_k(U \times V) = \bigoplus_{j=0}^{k} H_j(U) \times H_{k-j}(V) . \tag{7.18}$$

For the space $\mathcal{M}$, we define the Morse data for the attachment at a critical point $\boldsymbol{p} \in \mathcal{V}_*$ by the following steps.

Definition 7.32 (Morse data)**.** Let S be a stratum of complex codimension k containing a critical point $\boldsymbol{p}$ at height c.

(i) The *tangential Morse data* $T(\boldsymbol{p})$ at $\boldsymbol{p}$ is the homotopy type of the pair $(B^{d-k}, \partial B^{d-k})$ consisting of a ball of codimension k modulo its boundary. A representative of this class is the *unstable manifold* for the negative gradient flow induced by $h_{\hat{\boldsymbol{r}}}$ on $\mathcal{V}$ (the set of points that flow into the critical point under the positive gradient flow, see [HPS77, Section 4] or [Con78a]).
(ii) The *normal plane* $N_{\boldsymbol{p}}(\mathcal{S})$ to $\mathcal{S}$ at $\boldsymbol{p}$ is the (complex) orthogonal complement of the tangent space $T_{\boldsymbol{p}}(\mathcal{S})$.
(iii) The *normal slice* $\mathtt{N}$ at $\boldsymbol{p}$ is the mutual intersection of $\mathcal{V}$, a sufficiently small ball about $\boldsymbol{p}$, and the normal plane $N_{\boldsymbol{p}}(\mathcal{S})$.
(iv) The *normal link* $\mathcal{L}(\boldsymbol{p})$ is the mutual intersection of $\mathcal{M}$, a sufficiently small ball about $\boldsymbol{p}$, and $\mathtt{N}$.
(v) The *normal Morse data* $L(\boldsymbol{p})$ is the pair $(\mathcal{L}(\boldsymbol{p})_{\geq c}, \mathcal{L}(\boldsymbol{p})_{=c})$, where $\mathcal{L}(\boldsymbol{p})_{=c}$ is the intersection of the normal link with the real codimension 1 surface where $h_{\hat{\boldsymbol{r}}}(\boldsymbol{z}) = c$.
(vi) The *Morse data* at $\boldsymbol{p}$ is the product of the tangential and normal Morse data.

The following result, which is the main result in the monograph [GM88], is stated as Theorem D.21 in Appendix D.

Theorem 7.33. *The homotopy type of the attachment pair $\mathcal{M}^{\boldsymbol{p},\mathrm{loc}}$ is the Morse data at $\boldsymbol{p}$.* □

Theorem 7.33 yields a general topological decomposition of $H_d(\mathcal{M}, -\infty)$, which is a stratified version of Theorem 7.15.

Definition 7.34. A critical point $\boldsymbol{p}$ in direction $\hat{\boldsymbol{r}}$ on a stratum $\mathcal{S}$ is called a ***nondegenerate critical point on*** $\mathcal{S}$ if $h_{\hat{\boldsymbol{r}}}|_{\mathcal{S}}$ is nondegenerate in the sense of Definition 7.11 (meaning the Hessian for $h_{\hat{\boldsymbol{r}}}|_{\mathcal{S}}$ in local coordinates around $\boldsymbol{p}$ is nonsingular).

Theorem 7.35. *Fix $\hat{\boldsymbol{r}}$ and assume there are no critical values at infinity. Let $\boldsymbol{z}_1, \ldots, \boldsymbol{z}_m$ enumerate the stratified critical points of $\mathcal{V}_*$ in (weakly) decreasing order of the height function $h_{\hat{\boldsymbol{r}}}$, where the stratum containing $\boldsymbol{z}_j$ has complex codimension k_j. If all critical points are quadratically nondegenerate then there are cycles $\gamma_1, \ldots, \gamma_m$ on $\mathcal{V}_*$, along with a basis $\beta_{j,1}, \ldots \beta_{j,s_j}$ for the k_j-homology of the normal Morse data, with the following properties.*

(a) $h_{\hat{\boldsymbol{r}}}$ achieves its maximum on γ_j at $\boldsymbol{z}_j$;
(b) $\gamma_j \simeq (B^{d-k_j}, \partial B^{d-k_j})$;
(c) A basis for the integer homology group $H_d(\mathcal{M}, -\infty)$ can be formed by cycles $\sigma_{j,i} = \gamma_j \times \beta_{j,i}$ which, for fixed j, form a basis for $G(\boldsymbol{z}_j)$.

Proof Theorem 7.33 and part (i) of Definition 7.32 imply (a) and (b). Comparing parts (v) and (vi) of the definition with Theorem 7.30 gives (c). The fact that $\{\sigma_{j,i}\}$ is an integer homology basis follows from the lack of torsion in $H_d(\mathcal{M})$, which follows from the fact that $\mathcal{V}_*$ and $\mathcal{M}$ have the homotopy type of a d-dimensional cell complex (see Theorem D.23), with no boundaries in dimension d. Because the Morse theoretic results identify the homotopy type of the attachments, not just the relative homology groups, the cycles $\sigma_{j,i}$ generate homology with both integer and rational coefficients. □

While this theorem may look somewhat abstract, its power lies in its generality, and typical applications can be simple. For instance, in Figure 7.19 we have a surface $\mathcal{V}$ with complement $\mathcal{M} := \mathbb{C}_*^2 \setminus \mathcal{V}$ where $H_2(\mathcal{M})$ has one generator local to a critical point in a stratum of complex dimension 0 and two quasi-local to critical points in strata of dimension 1; the former has a 2-torus for its normal link, while the latter have normal links of dimension 1 which may be taken to be topological circles.

A further generalization removes the assumption of quadratic nondegeneracy. We do not use this generalization in this text, as we directly compute integral manipulations for the few quadratically degenerate cases that arise.

Corollary 7.36. *Without the assumption of quadratic nondegeneracy of $h_{\hat{\boldsymbol{r}}}$ at each critical point $\boldsymbol{p}$, a modified version of Theorem 7.35 still holds. Instead of a pair $(B, \partial B)$ consisting of a ball and its boundary, the tangential Morse data is replaced by a more general collection of $(d-k_j)$-cycles $\{\gamma_{j,k} : 1 \le k \le r_j\}$ where r_j is the rank of $H_{d-k_j}(\mathcal{V}_{\le c}, \mathcal{V}_{\le c-\varepsilon})$. Consequently, the basis in part (d) of Theorem 7.35 is instead formed by cycles $\gamma_{j,k} \times \beta_{j,i}$ for $1 \le j \le m$ with $1 \le k \le r_j$ and $1 \le i \le s_j$.*

Exercise 7.11. Let $\mathcal{M}$ be a manifold of real dimension d in $\mathbb{R}^n$ for $d < n$. Many classical Morse-theoretic analyses use the height function $h(\boldsymbol{x}) = d(\boldsymbol{p}, \boldsymbol{x})$,

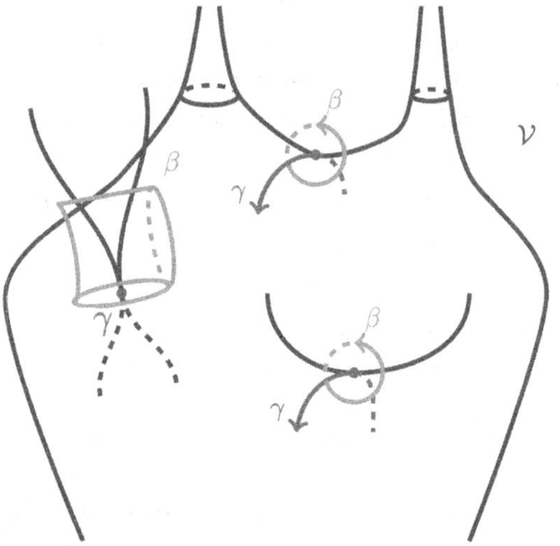

Figure 7.19 Critical points and their normal and tangential homology generators. On the right are two index-1 smooth critical points with tangential relative homology generators γ and normal homology generators $\beta \asymp S^1$. On the left is an isolated self-intersection point γ of $\mathcal{V}$, thus a zero-dimensional stratum, pictured with a two-dimensional normal link homotopy equivalent to (and in the picture homeomorphic to) a 2-torus.

where d is distance and $\mathbf{p}$ is a fixed point in $\mathbb{R}^n \setminus \mathcal{M}$. Explain why this is not a good Morse function to use if trying to establish the "bouquet of spheres" result for smooth varieties – described before Theorem 7.15 above – via Theorem 7.35(b).

7.4 Geometry

Theorems 7.15 and 7.35 allow us to express the Cauchy integral for coefficients as a finite sum of integrals localized near critical points (up to negligible error). Asymptotically approximating these integrals depends on the geometry of the singular set near the critical points, after which the coefficients $n_{\mathbf{z}}$ appearing in (7.2) must be determined. To make this process more concrete, and give an idea of its implications for coefficient asymptotics, we discuss some special cases arising often in combinatorial examples. These situations are covered in great detail in Chapters 9–11.

Smooth points

As seen above, the quasi-local cycle σ_w corresponding to a smooth critical point w is a tube $\circ\gamma_w$ around a $(d-1)$-chain γ_w in $\mathcal{V}_*$ such that $h_{\hat{r}}$ is maximized on γ_w at w.

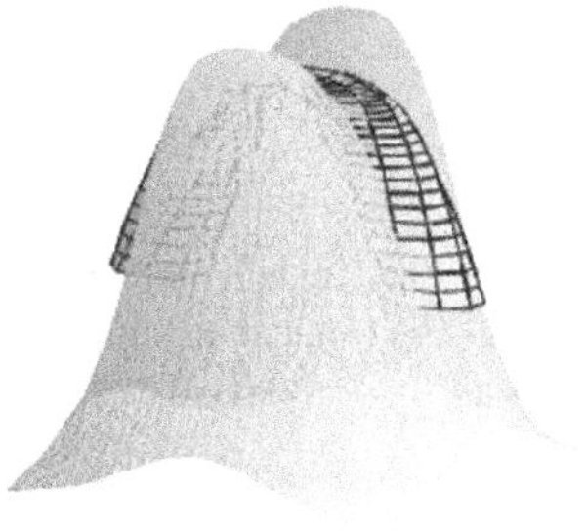

Figure 7.20 A quasi-local cycle near a smooth point.

The residue form is a complex $(d-1)$-dimensional saddle integral of the type discussed in Chapter 5. Assuming quadratic nondegeneracy, the asymptotic formula for the contribution to a_r from the integral over σ_w has the form

$$\Phi_w(r) = w^r \cdot |r|^{-(d-1)/2} \cdot \left(C(\hat{r}) + O\left(|r|^{-1}\right)\right), \tag{7.19}$$

where $C(\hat{r})$ is a constant arising from saddle point asymptotics and, as usual, $|r| = |r_1| + \cdots + |r_d|$. As $\hat{r}$ varies, the critical point w varies smoothly except for bifurcation values where $h_{\hat{r}}$ becomes quadratically degenerate. The amplitude C also varies smoothly with $\hat{r}$ away from bifurcation values where the topology may change, which are also points where the coefficient n_w in (7.2) may change. The values of $\hat{r}$ for which $h_{\hat{r}}$ is quadratically degenerate can be computed using the methods of Section 8.4 in Chapter 8. Removing "bad" directions partitions the set of directions into open cones over which the estimate (7.2) is uniform over compact subsets.

Transverse multiple points

When $Q(z) = \prod_{j=1}^{k} Q_j(z)$ is a product of (potentially non-polynomial) analytic functions in a neighborhood of some $w \in \mathbb{C}$ and the zero sets of Q_j are smooth and intersect transversely at w, then we call w a *transverse multiple point*; see Figure 7.21. Note that every smooth point is trivially a transverse multiple point.

The quasi-local cycle σ_w defined by such a point w is the product of a k-torus β_w and a $(d-k)$-chain γ_w. The torus β_w is a product of circles about w

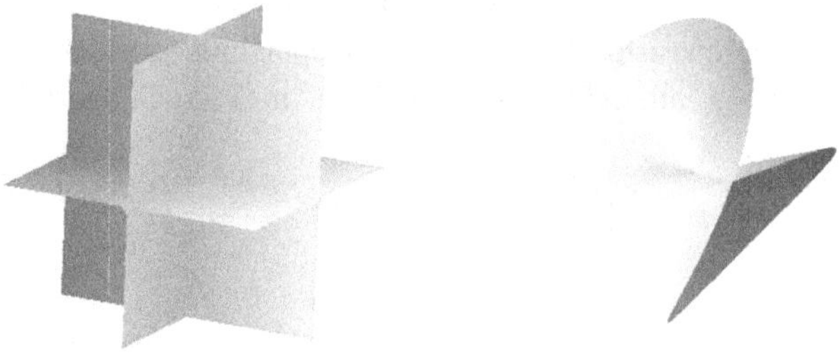

Figure 7.21 *Left:* A singular variety containing only transverse multiple points (including smooth points). *Right:* A singular variety with smooth points, a ray of (non-smooth) transverse multiple points, and a non-transverse multiple point (the origin of that ray).

in the complex normal space to each divisor Q_j. The chain $\gamma_{\boldsymbol{w}}$ is supported in the stratum defined as the common intersection of the varieties defined by the factors vanishing at $\boldsymbol{w}$, and achieves its maximum height at $\boldsymbol{w}$.

Example 7.37. If $\boldsymbol{p}$ is the common intersection of all three surfaces on the left of Figure 7.21 then $d = k = 3$ and the stratum of $\boldsymbol{p}$ is zero-dimensional. In this case $\sigma_{\boldsymbol{p}} = \beta_{\boldsymbol{p}}$ is a three torus defined by the product of circles about $\boldsymbol{p}$ in each of the three complex normal spaces to the surfaces. ◁

Example 7.38. Let $\mathcal{V}$ be the union of two complex hypersurfaces in dimension three. Any point $\boldsymbol{w}$ on the stratum S defined intersection of these two hypersurfaces is a (non-smooth) transverse multiple point. The stratum S has codimension $k = 2$ and the homology of the normal link is generated by a 2-torus $\beta_{\boldsymbol{w}}$. ◁

There is a theory of multiple residues for transverse multiple points, not too much more difficult than the residue forms already introduced, and asymptotics for an integral over a quasi-local cycle may be computed rather neatly using this approach. Such residues can be used even when the denominator is irreducible as a polynomial but locally factors into power series that converge in a neighborhood of the critical point $\boldsymbol{w}$ and each define smooth analytic varieties that intersect transversely at $\boldsymbol{w}$ (see, for instance, Example 10.4 in Chapter 10 for such a situation). The d-dimensional Cauchy integral over $\gamma_{\boldsymbol{w}} \times \beta_{\boldsymbol{w}}$ is reduced by residue computations to a $(d-k)$-dimensional integral over $\gamma_{\boldsymbol{w}}$. When $d = k$ the resulting residue integral is simply a function of $\boldsymbol{r}$, while if $k < d$ then the integral over $\gamma_{\boldsymbol{w}}$ is asymptotically approximated via the saddle point method. Ultimately, we typically obtain an asymptotic formula of the form

$$\Phi_{\boldsymbol{w}}(\boldsymbol{r}) = \boldsymbol{w}^{\boldsymbol{r}} \cdot |\boldsymbol{r}|^{-(d-k)/2} \cdot \left(C(\hat{\boldsymbol{r}}) + O\left(|\boldsymbol{r}|^{-1}\right)\right), \tag{7.20}$$

where again $\boldsymbol{w}$ varies smoothly with $\hat{\boldsymbol{r}}$ away from quadratic degeneracies and certain cone boundaries, and the value $\boldsymbol{w}(\hat{\boldsymbol{r}})$ is constant over a set of $\hat{\boldsymbol{r}}$ of dimension $k-1$. Since a smooth point is a special case of a transverse multiple point with $k = 1$, (7.19) is a special case of (7.20).

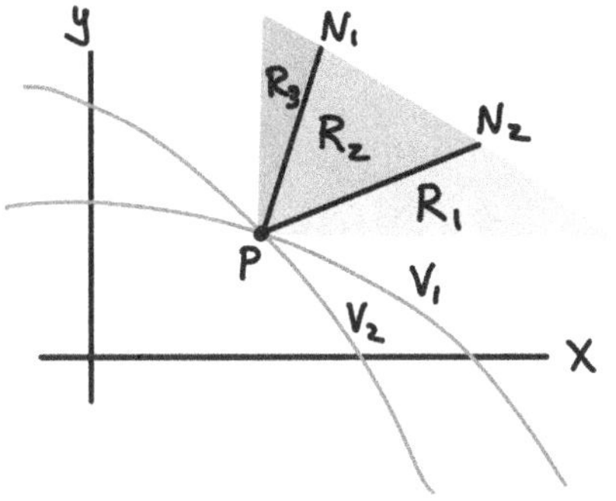

Figure 7.22 The logarithmic gradients of two transversely intersecting sheets at a critical point $\boldsymbol{p}$ decompose the first quadrant of $\mathbb{R}^2$ into three cones.

Example 7.39. Figure 7.22 illustrates two transversely intersecting smooth curves defined by the vanishing of two-dimensional functions $Q_1(x, y)$ and $Q_2(x, y)$ that meet at a single point $\boldsymbol{p}$. There are two one-dimensional strata $S_1 = \mathcal{V}(Q_1) \setminus \boldsymbol{p}$ and $S_2 = \mathcal{V}(Q_2) \setminus \boldsymbol{p}$, containing points on exactly one of the curves, together with a zero-dimensional stratum containing only $\boldsymbol{p}$. Given $\hat{\boldsymbol{r}}$ there is at most one critical point $\boldsymbol{z}_j$ on each stratum S_j, and $h_{\hat{\boldsymbol{r}}}$ is quadratically nondegenerate for any $\hat{\boldsymbol{r}}$ in the positive quadrant. As $\hat{\boldsymbol{r}}$ varies from $\boldsymbol{x} = (1, 0)$ to $\boldsymbol{y} = (0, 1)$ it crosses through two "bad" directions, given by the logarithmic gradients $N_1 = (\nabla_{\log} Q_1)(\boldsymbol{p})$ and $N_2 = (\nabla_{\log} Q_2)(\boldsymbol{p})$ shown emanating from $\boldsymbol{p}$. The zero-dimensional stratum $\boldsymbol{p}$ remains fixed, but the critical points $\boldsymbol{z}_j$ move smoothly with $\hat{\boldsymbol{r}}$ on their respective strata S_j. As $\hat{\boldsymbol{r}}$ crosses the log-normal direction N_2 the critical point $\boldsymbol{z}_2$ collides with $\boldsymbol{p}$, then when $\hat{\boldsymbol{r}}$ crosses N_1 the point $\boldsymbol{z}_1$ collides with $\boldsymbol{p}$. The positive quadrant in $\mathbb{R}^2$ can thus be broken into three regions: the cone R_1 defined by the positive real span of $\boldsymbol{x}$ and N_2, the cone R_2 defined by N_2 and N_1, and the cone R_3 defined by N_1 and $\boldsymbol{y}$. It turns out that $n_{\boldsymbol{z}_1} = n_{\boldsymbol{z}_2} = 1$ on all regions, but $n_{\boldsymbol{p}}$ is equal to one on R_2 and zero on $R_1 \cup R_3$. Accordingly, the asymptotic expansion expressed in (7.2) changes across the boundaries of these regions. ◁

Multiple and arrangement points

When varieties intersect tangentially instead of transversely, the resulting integrals are more challenging to asymptotically approximate. However, if the intersection lattice for smooth sheets of the variety coincides with the intersection lattice for the tangent planes of these sheets then non-transversality can be handled combinatorially. Such a point is called an *arrangement point*, after hyperplane arrangements such as the one in Figure 7.23.

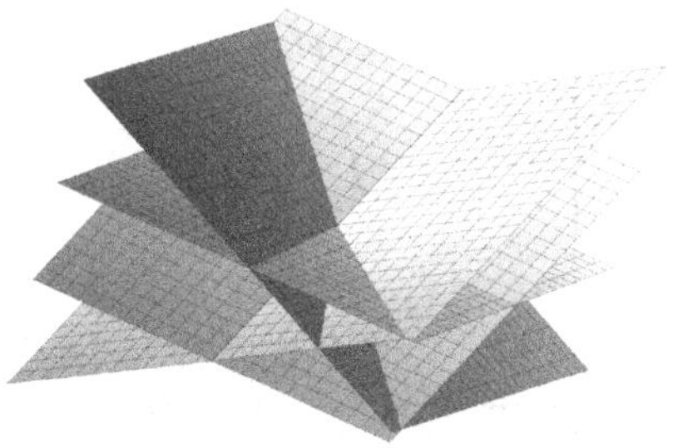

Figure 7.23 When $\mathcal{V}$ is a hyperplane arrangement, all points are arrangement points.

Exercise 7.12. For which of the polynomials $Q_1(x,y,z) = z(x-y)(x-y+z-x^2)$ and $Q_2(x,y,z) = z(x-y)(x-y+z-xyz)$ is the origin an arrangement point?

The generators $\beta_{\boldsymbol{p},j}$ for the normal link of an arrangement point are the same as for a transverse multiple point, only there are more of them.

Example 7.40. Figure 7.24 shows a case where $d = 2$ and $k = 3$. Here three one-dimensional sheets intersect pairwise transversely in a point $\boldsymbol{p}$. Instead of one two-torus $\beta_{\boldsymbol{p}}$ there are two tori $\beta_{\boldsymbol{p},1}$ and $\beta_{\boldsymbol{p},3}$, where $\beta_{\boldsymbol{p},1}$ is the product of circles about $\boldsymbol{p}$ in $\mathcal{V}_2$ and $\mathcal{V}_3$, and $\beta_{\boldsymbol{p},3}$ is the product of circles about $\boldsymbol{p}$ in $\mathcal{V}_1$ and $\mathcal{V}_2$. One might have expected a third torus $\beta_{\boldsymbol{p},2}$, a product of circles about $\boldsymbol{p}$ in $\mathcal{V}_1$ and $\mathcal{V}_3$, and indeed there is such a torus, however this final torus is not linearly independent of the first two because $\beta_{\boldsymbol{p},1} - \beta_{\boldsymbol{p},2} + \beta_{\boldsymbol{p},3} = 0$ in the relevant homology class. ◁

A similar multivariate residue computation as in (7.20) leads to a formula of the form

$$\Phi_{\boldsymbol{w}}(\boldsymbol{r}) = \boldsymbol{w}^{\boldsymbol{r}} \cdot |\boldsymbol{r}|^{-(d-k)/2} \cdot \left(P_{\boldsymbol{r}}(\boldsymbol{w}) + O\left(|\boldsymbol{r}|^{-1}\right)\right), \tag{7.21}$$

where $P_{\boldsymbol{r}}(\boldsymbol{w})$ is a polynomial of degree at most $m - k$ with m the number of sheets intersecting at $\boldsymbol{w}$ and k the codimension of the stratum containing

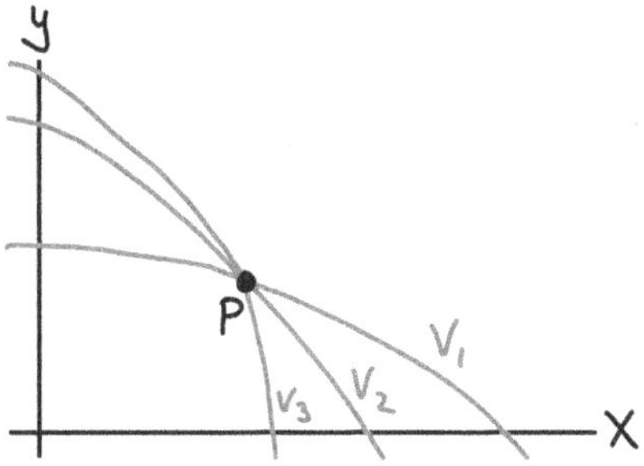

Figure 7.24 Quasi-local cycles at an arrangement point with $d = k = 2$.

$\boldsymbol{w}$. This approach also works when Q has repeated factors, provided that m is counted with the right multiplicity. Further details are given in Chapter 10.

Cone points

Beyond the above cases are critical points near which $\mathcal{V}$ does not look like a union of smooth sheets. We give a general analysis only in one case, namely when $\mathcal{V}$ is locally diffeomorphic to a cone $\sum_{j=1}^{d} z_j^2 = 0$. Such an isolated singularity is called a ***cone point singularity***, and is illustrated in Figure 7.25. Cone points arise, among other places, in statistical physics.

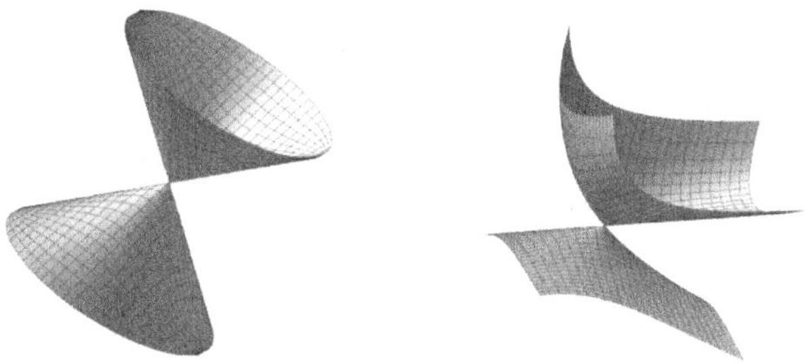

Figure 7.25 Two examples of cone point singularities.

Chapter 11 is devoted to the analysis of cone-point singularities. For any isolated singularity $\boldsymbol{w}$ we have $d = k$, so the stratum containing $\boldsymbol{w}$ is zero-dimensional, the cycle $\gamma_{\boldsymbol{w}}$ is just a point, and $\sigma_{\boldsymbol{w}} = \beta_{\boldsymbol{w}}$. General theory derived in [ABG70] indicates what to expect for the leading asymptotic term of $\int_{\beta_{\boldsymbol{w}}} \boldsymbol{z}^{-\boldsymbol{r}-\boldsymbol{1}} P(\boldsymbol{z})/Q(\boldsymbol{z})\, d\boldsymbol{z}$ at an isolated singularity $\boldsymbol{w}$: it is given by the inverse Fourier transform of the reciprocal of the leading homogeneous term of Q near

$\boldsymbol{w}$. For a cone point, the inverse Fourier transform yields an asymptotic contribution

$$\Phi_{\boldsymbol{w}}(\boldsymbol{r}) = C(\boldsymbol{w}) \cdot \tilde{q}(\boldsymbol{r})^{1-d/2} \cdot \left(1 + O\left(|\boldsymbol{r}|^{-1}\right)\right), \tag{7.22}$$

where $\tilde{q}$ is the dual quadratic form in $\boldsymbol{r}$-space to the quadratic leading term of Q at $\boldsymbol{w}$.

Example 7.41. The so-called ***cube grove*** creation generating function, analyzed in Example 11.43 of Chapter 11, is the rational function

$$F(x, y, z) = \frac{1}{1 + xyz - (1/3)(x + y + z + xy + xz + yz)}. \tag{7.23}$$

The variety $\mathcal{V}$ is smooth except at the single point $(1, 1, 1)$ where, after an orthogonal affine change of variables and a translation of the origin to $(1, 1, 1)$, the denominator of F looks asymptotically like the quadratic cone $2xy + 2xz + 2yz = 0$. The asymptotic formula given in Corollary 11.44 implies

$$a_{rst} \sim \frac{1}{\pi}\left[rs + rt + st - \frac{1}{2}(r^2 + s^2 + t^2)\right]^{-1/2}$$

when (r, s, t) lies inside the dual to the tangent cone to the denominator of F. ◁

For more general isolated singularities, analysis via inverse Fourier transforms lead to asymptotic contributions of the form

$$\Phi_{\boldsymbol{w}}(\boldsymbol{r}) = \boldsymbol{w}^{\boldsymbol{r}} \cdot |\boldsymbol{r}|^{-d-\kappa} \cdot \left(C(\hat{\boldsymbol{r}}) + O\left(|\boldsymbol{r}|^{-1}\right)\right), \tag{7.24}$$

which are valid as $\hat{\boldsymbol{r}}$ varies over the open dual cone to the tangent cone to $\mathcal{V}$ at $\boldsymbol{w}$, and uniform if $\hat{\boldsymbol{r}}$ is restricted to any compact subcone. The constant κ is the homogeneous degree of F at $\boldsymbol{w}$.

Exercise 7.13. Give a simple reason why cone points can never be multiple points.

Examples from the literature of these further variants include isolated singularities where Q is locally homogeneous of degree three [KP16] or four [BP21], or where $\mathcal{V}$ is the union of a quadratic cone with a smooth sheet passing through the cone point [BP11].

The first two of these examples are illustrated in Figure 7.26. Figure 7.27 shows the final example, where $\mathcal{V}$ is locally the union of a quadratic cone and a smooth sheet; an asymptotic formula is derived in [BP11]. In general asymptotics for this sort of geometry would be expressed in terms of an elliptic

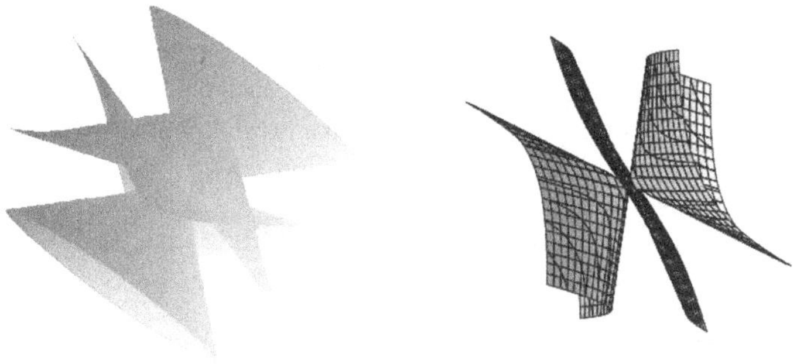

Figure 7.26 Isolated singularities of degree greater than 2.

integral, but in this case there is an explicit formula (see Theorem 11.49)

$$a_{rst} \sim \frac{1}{\pi} \arctan\left(\frac{\sqrt{1 - 2\hat{r}^2 - 2\hat{s}^2}}{1 - 2\hat{s}} \right).$$

A plot of this limiting behavior against $\hat{r}$ is shown on the right side of Figure 7.27.

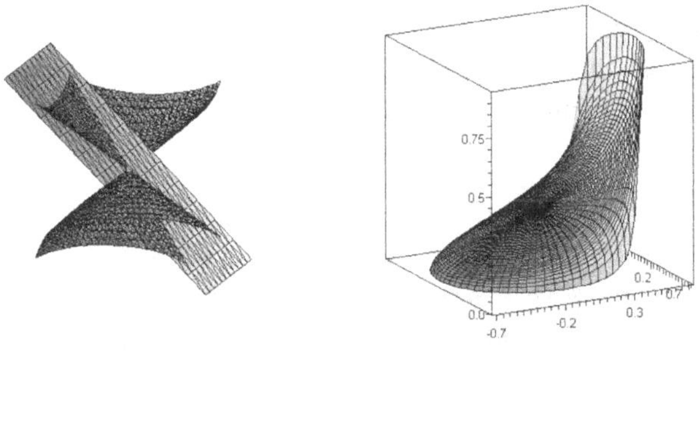

Figure 7.27 *Left:* A singular set consisting of a cone and a smooth sheet. *Right:* Asymptotic behavior of the corresponding coefficient sequence.

Exercise 7.14. Which of the two graphs in Figure 7.26 have arrangement points that are not smooth points?

7.5 Deformations

Finally, we end this chapter with a treatment of critical points and critical values at infinity. After giving a rigorous definition of such points we show how, in their absence, to construct the deformations that prove Lemmas 7.25 and 7.28. These two lemmas then imply Theorems 7.15 and 7.30.

7.5.1 Critical points at infinity

As described above and in Appendix C, the results of Morse theory typically require a *proper* height function so that certain gradient flows are guaranteed to reach points of low height, except when they get stuck near critical points. This properness condition is often satisfied in classical contexts by studying compact spaces, however our singular varieties are not compact and we often have non-proper height functions.

Our goal, therefore, is to formulate weaker but still sufficient conditions for there to be no topological obstructions to deforming our Cauchy integral to points of low height, proving results like Lemma 7.14 and Theorem 7.35 above. A considerable stream of topological research has gone into defining *bifurcation values*, at which the height function is not a locally trivial fibration and the topology of the space changes. While exact conditions for these topological obstructions remain murky, we care only about pushing down domains of integration to lower height, and may thus proceed by generalizing our definition of critical points (and critical values) to include "points at infinity."

We begin by defining the binary relation $\mathcal{R} \subseteq \mathbb{C}_*^d \times \mathbb{CP}^{d-1}$ that holds for a pair $(\boldsymbol{z}, \hat{\boldsymbol{r}})$ when the differential $dh_{\hat{\boldsymbol{r}}}|_S$ of the height function $h_{\hat{\boldsymbol{r}}}$ restricted to the stratum S containing $\boldsymbol{z}$ vanishes at $\boldsymbol{z}$. To facilitate computation we view $\hat{\boldsymbol{r}}$ as an element of $\mathbb{CP}^{d-1}$.

Definition 7.42 (CPAI). Let $\overline{\mathcal{R}}$ be the closure in $\mathbb{CP}^d \times \mathbb{CP}^{d-1}$ of the relation $\mathcal{R}$. A ***critical point at infinity*** (CPAI) in the direction $\hat{\boldsymbol{r}}_*$ is a limit point $(\boldsymbol{z}_*, \hat{\boldsymbol{r}}_*)$ in $\overline{\mathcal{R}}$ of points $(\boldsymbol{z}, \hat{\boldsymbol{r}}) \in \mathcal{R}$ such that $\boldsymbol{z}_* \notin \mathbb{C}_*^d$. When necessary we refer to our usual notion of critical points (not at infinity) as ***affine critical points*** to distinguish them from critical points at infinity.

In other words, a critical point at infinity in the direction $\hat{\boldsymbol{r}}_*$ is a limit, lying either at infinity or on a coordinate plane, of a sequence $\boldsymbol{z}^{(k)}$ of critical points contained in strata S_k such that the projection of $\hat{\boldsymbol{r}}_*$ to the tangent space of N_k at $\boldsymbol{z}^{(k)}$ converges to zero as $k \to \infty$ (i.e., $\hat{\boldsymbol{r}}_*$ lies in the "limit normal space" of the sequence $\boldsymbol{z}^{(k)}$) – see Figure 7.28.

To track the heights of CPAIs, given $\hat{\boldsymbol{r}} \in \mathbb{CP}^{d-1}$ we define the ternary relation

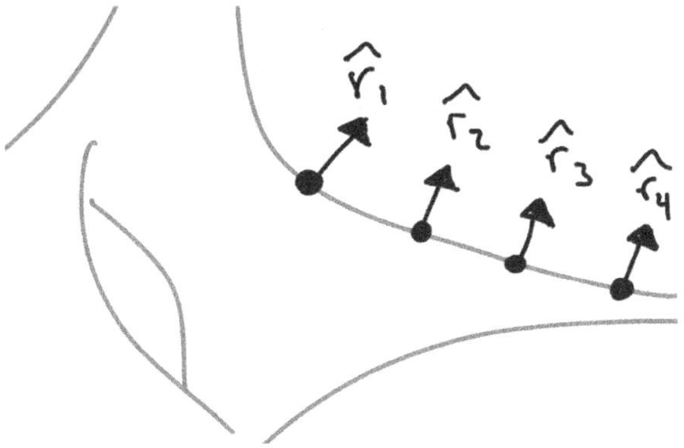

Figure 7.28 A sequence of points moving out to infinity, such that the logarithmic gradient of Q approaches the vector $\boldsymbol{v}$ pointing straight up. This sequence witnesses a critical point at infinity in the direction $\boldsymbol{v}$.

$\mathcal{T}(\hat{\boldsymbol{r}}) \subseteq \mathbb{C}_*^d \times \mathbb{CP}^{d-1} \times \mathbb{R}$ containing elements $(\boldsymbol{z}, \boldsymbol{y}, \eta)$ such that $(\boldsymbol{z}, \boldsymbol{y}) \in \mathcal{R}$ and $h_{\hat{\boldsymbol{r}}}(\boldsymbol{z}) = \eta$.

Definition 7.43 (CVAI). Let $\overline{\mathcal{T}} \subseteq \mathbb{CP}^d \times \mathbb{CP}^{d-1} \times \mathbb{R}$ be the closure of the ternary relation $\mathcal{T}$ in $\mathbb{CP}^d \times \mathbb{CP}^{d-1} \times \mathbb{R}$. We call η a ***critical value at infinity*** (CVAI) if some point $(\boldsymbol{z}_*, \hat{\boldsymbol{r}}_*, \eta)$ is in $\overline{\mathcal{T}}$ and $\boldsymbol{z}_* \notin \mathbb{C}_*^d$.

7.5.2 Vector fields and flows

The results we prove in this section are based on [BMP22, Theorem 1].

Theorem 7.44 (homotopy equivalences in the absence of CVAI). *Fix a direction $\hat{\boldsymbol{r}}$ and a Whitney stratification $\{S_\alpha : \alpha \in I\}$ of $(\mathbb{C}_*^d, \mathcal{M})$.*

(i) If there are neither affine critical values nor CVAI in the interval $[a, b]$ then the inclusion of $\mathcal{M}_{\le a}$ into $\mathcal{M}_{\le b}$ is a homotopy equivalence. The same is true replacing $\mathcal{M}$ by any stratum S of $\mathcal{V}_*$.

(ii) If there are no CVAI in $[a, b]$ but there is a single affine critical value $c \in [a, b]$ and it corresponds to the set of critical points $\boldsymbol{z}_1, \ldots, \boldsymbol{z}_m$ then there is a stratified flow deforming any chain C in $\mathcal{M}$ down to a chain in the union of $M_{<c}$ with sufficiently small balls about each $\boldsymbol{z}_i$. When $[c - \varepsilon, c + \varepsilon] \subseteq (a, b)$, this induces a homotopy equivalence between $(\mathcal{M}_{c+\varepsilon}, \mathcal{M}_{c-\varepsilon})$ and the direct sum of attachment spaces $\mathcal{M}^{\boldsymbol{z},loc}$ defined in Section 7.3 above (see also Definition C.31 and Figure C.6 in Appendix C).

The first part of Theorem 7.44 directly implies Lemma 7.25. The main work in proving Theorem 7.44 comes from establishing the following result, which requires bounding elements of a vector field. Although general stratified spaces do not come equipped with Riemannian metrics, we deal only with spaces embedded in $\mathbb{C}^d \cong \mathbb{R}^{2d}$ and the norm of any vector or covector refers to the norm inherited from this embedding.

Lemma 7.45 ([BMP22, Lemma 2]). *Suppose $a < b$ are real numbers such that $h_{\hat{r}}$ has no* CVAI *in the interval $[a, b]$ and one affine critical value in $c \in [a, b]$, not at either endpoint. Then there is a vector field $\boldsymbol{v}$ on $\mathbb{C}_*^d \cap h_{\hat{r}}^{-1}[a, b]$ with the following properties.*

(i) $\boldsymbol{v}$ is smooth on strata and continuous on $\mathbb{C}_*^d$;
(ii) $\boldsymbol{v}$ is tangent to strata;
(iii) $\boldsymbol{v}$ is a controlled vector field in the sense of [Mat70, Section 9];
(iv) $\boldsymbol{v}$ is bounded;
(v) $\boldsymbol{v}$ has unit downward speed, meaning $dh_{\hat{r}}(\boldsymbol{v}) \equiv -1$.

The first, relatively easy, step in establishing Lemma 7.45 is to show that $|dh_{\hat{r}}|$ is bounded away from zero except near critical points.

Lemma 7.46 ([BMP22, Lemma 1]). *Suppose h is a Morse function on a stratified space $\mathcal{V}_* \subseteq \mathbb{C}_*^d$, with no* CVAI *in $[a, b]$. Then, excluding an arbitrarily small neighborhood $\mathcal{N}$ of critical points, the differential dh has its magnitude bounded away from zero, meaning $|dh| \geq \delta(\mathcal{N}) > 0$.*

Proof It suffices to prove the result when dh is restricted to an arbitrary stratum S. Let $S_{[a,b]}$ denote the elements of S with height in $[a, b]$ and let $\mathcal{L} = \mathbb{R}^d \times (\mathbb{R}/2\pi\mathbb{R})^d$ denote the logarithmic parametrizing space for $\mathbb{C}_*^d$ via the exponential map $\exp : \mathcal{L} \to \mathbb{C}_*^d$. In this parametrization, $h_{\hat{r}}$ becomes $\tilde{h} = h_{\hat{r}} \circ \exp$. This parametrization is useful because $d\tilde{h}$ is the constant vector $\hat{r}$ (formally, $d\tilde{h} = \sum_{j=1}^d r_j dx_j + 0 dy_j$ is a constant with respect to the embedding in $\mathcal{L}$ obtained from the embedding in $\mathbb{C}_*^d$, pulled back via exp). We use tildes to denote inverse images under this parametrization, meaning $\widetilde{\texttt{critical}} = \exp^{-1}[\texttt{critical}]$ is the inverse image of the set of critical points of the Morse function h.

Assume towards a contradiction that the norm of the tangential differential is *not* bounded from below on $\widetilde{S}_{[a,b]} \backslash \mathcal{N}$, where $\mathcal{N}$ is a neighborhood of $\widetilde{\texttt{critical}}$ in $\mathcal{L}$, and let $\tilde{\boldsymbol{x}}_k$ be a sequence in $\widetilde{S}_{[a,b]} \setminus \mathcal{N}$ for which $|d\tilde{h}_{\widetilde{S}}(\tilde{\boldsymbol{x}}_k)|$ goes to zero. This sequence has no limit points whose height lies outside of $[a, b]$ and no limit points in $\widetilde{\texttt{critical}}_{[a,b]}$. There are also no affine limit points outside of $\widetilde{\texttt{critical}}$, because if $\boldsymbol{x} \to \boldsymbol{y}$ with $\boldsymbol{y}$ in a substratum $\widetilde{S}_y$ then $|d\tilde{h}_{\widetilde{S}_y}(\boldsymbol{y})| \leq \liminf_{\boldsymbol{x} \to \boldsymbol{y}} |d\tilde{h}_{\widetilde{S}}(\boldsymbol{x})|$ since the projection of the differential onto a substratum is at most the projection onto $\widetilde{S}$. By compactness, $\{\boldsymbol{x}_k\}$ must have a limit point $\boldsymbol{x} \in \mathbb{CP}^d$. It follows from ruling out noncritical points, and affine stationary points with heights inside or outside of $[a, b]$, that $\boldsymbol{x}$ lies at infinity. The sequence therefore defines a CPAI of height c, contradicting the hypothesis and proving the lemma. □

With Lemma 7.46 in hand, we outline the proof of Lemma 7.45 before returning to Theorem 7.44.

Sketch of proof of Lemma 7.45 This lemma for proper height functions is the usual Morse theoretic construction of stratified gradient-like vector fields, as found in standard references [GM88; ABG70]. It is proved by using a partition of unity to piece together the unit-speed downward gradient of $h_{\hat{\boldsymbol{r}}}$ restricted to each stratum and then extended to a neighborhood of the stratum in $\mathbb{C}_*^d$ via the local product structure. Property (*v*), unit downward speed, then follows from Lemma 7.46. In the nonproper case, the biggest headache is extending to the product; this was the motivation for the notion of *control data*, developed in [Mat70]. When $h_{\hat{\boldsymbol{r}}}$ is not proper, there is no *a priori* guarantee that the descent rate divided by the magnitude of the vector remains bounded away from zero, even locally. This is because the Whitney condition fits these C^∞ strata in a way that is in principle only C^0. In fact, Mather's argument contains the seeds of a proof for this fact, which is accomplished via some linear algebra and further explicit use of the Whitney conditions. See [BMP22, Lemma 2] for full details. □

Proof of Theorem 7.44 Let $\boldsymbol{v}$ be the vector field constructed in Lemma 7.45, altered so as to be zero on $\mathcal{M}_{\leq a}$. This vector field defines a flow $\Phi(x, t)$ such that

- $\frac{d}{dt}\Phi(x, t) = \boldsymbol{v}(x)$ when $h_{\hat{\boldsymbol{r}}}(x) \in (a, b]$;
- $\Phi(x, t)$ is defined for $0 \leq t \leq h_{\hat{\boldsymbol{r}}}(x) - a$ and, for t in this range, $h_{\hat{\boldsymbol{r}}}(\Phi(x, t)) = h_{\hat{\boldsymbol{r}}}(x) - t$;
- the map $\psi(x) = \Phi(x, b - a)$ is a continuous map on $\mathcal{M}_{\leq b}$ with range $\mathcal{M}_{\leq a}$ and fixing $\mathcal{M}_{\leq a}$.

It follows from these properties that the inclusion $\iota : \mathcal{M}_{\leq a} \to \mathcal{M}_{\leq b}$ is a homotopy equivalence: $\psi \circ \iota$ is the identity map on $\mathcal{M}_{\leq a}$ while $\iota \circ \psi$ is homotopic to the identity map on $\mathcal{M}_{\leq b}$ via the homotopy Φ on $\mathcal{M}_{\leq b} \times [0, b - a]$. This is sufficient to imply conclusion (*i*) of Theorem 7.44, and also proves the weaker version of (*ii*) found in [BMP22], namely that any cycle in $\mathcal{M}_{\leq b}$ may be deformed to lie in the union of $\mathcal{M}_{\leq c-\varepsilon}$ with arbitrarily small neighborhoods of each critical point. However, this argument does not imply that the deformation is induced by a homotopy equivalence.

To that end, let $\kappa(\boldsymbol{x}) = \min\{|\boldsymbol{x} - \boldsymbol{z}| : \boldsymbol{z} \in \texttt{critical}\}$ denote distance to the critical set. For $s > 0$, define a new vector field $\boldsymbol{v}_s$ by

$$\boldsymbol{v}_s(\boldsymbol{x}) = \begin{cases} \boldsymbol{v}(\boldsymbol{x}) & \kappa(\boldsymbol{x}) \geq s \\ \rho(\kappa(\boldsymbol{x}))\boldsymbol{v}(\boldsymbol{x}) & \kappa(\boldsymbol{x}) \in [s/2, s] \\ 0 & \kappa(\boldsymbol{x}) \leq s/2 \end{cases},$$

where ρ is a smooth nondecreasing function with $\rho(s/2) = 0$ and $\kappa(s) = 1$.

Fix $\varepsilon > 0$ with $[c - \varepsilon, c + \varepsilon] \subseteq (a, b)$ and let $\Phi_s(x, t)$ denote the flow defined similarly to Φ but with $\boldsymbol{v}_s$ in place of $\boldsymbol{v}$. Let $\tau(\boldsymbol{x}) = \Phi_s(\boldsymbol{x}, 2\varepsilon)$ denote the time 2ε map for the flow Φ_s. Because $h = h_{\hat{\boldsymbol{r}}}$ is nonincreasing along Φ_s, points in $\mathcal{M}_{\leq c-\varepsilon}$ remain inside $\mathcal{M}_{\leq c-\varepsilon}$, hence the flow defines a homotopy equivalence between the pairs $(\mathcal{M}_{c+\varepsilon}, \mathcal{M}_{c-\varepsilon})$ and $(X, \mathcal{M}_{c-\varepsilon})$, where $X = \tau[\mathcal{M}_{c+\varepsilon}]$.

Define a modified height function $g = h \circ \tau$. We claim that g has the same critical points as h in $h^{-1}([c-\varepsilon, c+\varepsilon])$. To see this, first observe that trajectories of Φ_s are either rest trajectories at points in the set $V_0 = \{\boldsymbol{x} : \boldsymbol{v}_s(\boldsymbol{x}) = 0\}$ or else never enter V_0. Indeed, this follows from the fact that $\boldsymbol{v}_s$ is tangent to all strata, smooth on every stratum, and that trajectories of a flow defined by a smooth vector field cannot merge. Inside V_0, the height functions h and g are equal, and hence have the same critical points. Outside V_0, the differential $dg|_S$ can never vanish because $dg|_S(\boldsymbol{v}) < 0$; this follows from the fact that for $\boldsymbol{v} \in S$, the map $dg(\boldsymbol{v})(\boldsymbol{x}) = d(h \circ \tau)(\boldsymbol{v})(\boldsymbol{x})$ sends $\boldsymbol{v}$ to $dh(D\tau(\boldsymbol{v}))$. On trajectories, the map $D\tau$ carries $\boldsymbol{v}_s \in T_{\boldsymbol{x}}(S)$ to $\rho(\tau(\boldsymbol{x})) \in T_{\tau(\boldsymbol{x})}(S)$, where $\rho(\tau(\boldsymbol{x})) > 0$. Thus

$$dg(\boldsymbol{v}_s)(\boldsymbol{x}) = \rho(\tau(\boldsymbol{x}))dh(\boldsymbol{v}_s)(\tau(\boldsymbol{x})) = -\rho(\tau(\boldsymbol{x})) < 0,$$

showing that $dg|_S$ is nonvanishing outside V_0 and proving the claim.

The map τ takes all points of $\mathcal{M}_{\leq c+\varepsilon}$ into $\mathcal{M}_{\leq c-\varepsilon}$, except possibly for those whose trajectories come within distance s of the critical set within time 2ε. Because $\boldsymbol{v}$ is bounded, trajectories coming within s of the (finite) critical set within time 2ε are all contained in some compact set K, independent of $s \in [0, 1]$. Therefore, the difference $X \setminus \mathcal{M}_{c-\varepsilon}$ is bounded and g is a proper height function on the pair $(X, \mathcal{M}_{c-\varepsilon})$, in the sense that the inverse image of a compact set in $X \setminus \mathcal{M}_{c-\varepsilon}$ is compact.

We may now apply the results of stratified Morse theory (see Theorem D.21 in Appendix D). The result is that the pair $(X, \mathcal{M}_{\leq c-\varepsilon})$ is homotopy equivalent to the direct sum of pairs $(\mathcal{N}(\boldsymbol{z}) \cup \mathcal{M}_{c-\delta}, \mathcal{M}_{c-\delta})$ as $\boldsymbol{z}$ varies over the critical points of height c and $\mathcal{N}(\boldsymbol{z})$ can be chosen to be arbitrarily small neighborhoods of these, after which δ is chosen sufficiently small. We have seen, in addition, that each cycle in X may be deformed into the union of $\mathcal{M}_{c-\delta}$ and the neighborhoods $\mathcal{N}(z)$ by running the flow v_s, hence the homotopy equivalence is induced by this flow. Finally, having shown that the flow induces a homotopy equivalence between $(\mathcal{M}_{c+\varepsilon}, \mathcal{M}_{c-\varepsilon})$ and $(X, \mathcal{M}_{c-\varepsilon})$, we can pick s sufficiently small and $\mathcal{N}(z)$ and δ so that $\delta < \varepsilon$, finishing the proof of part (*ii*). □

Notes

The rigorous foundation of the main theorems of this book in the framework of Morse theory is new to the second edition. Before the appearance of [BMP22], Morse-theoretic results were not available because $h_{\hat{r}}$ is not, in general, a proper function. Therefore, in the first edition, Morse theory was used only as a motivation and individual results were obtained via hands-on deformations and surgeries, informed by Morse theory but proved as special cases, tailored to the individual hypotheses.

Asymptotic formulae in the presence of smooth strictly minimal points first appeared in [PW02], followed by formulae for strictly minimal multiple points in [PW04]. Results proving the irrelevance of non-critical minimal points were derived in [Bar+10], and then in greater generality in [BP11], with an overview presented in [Pem10]. The proof sketch of Lemma 7.6 was suggested to us by Tony Pantev.

The second part of Theorem 7.44 is an improvement on the result originally published in [BMP22]. There, it was shown that cycles may be pushed down into the union of levels below the critical value and neighborhoods of the critical points, but not that this union is homotopy equivalent to the space at a level above the critical value. The sticking point is that the latter requires a deformation remaining at all times within the union, which requires geometric facts developed at length throughout [GM88]. The present proof avoids this by using the results of [BMP22] to eliminate escape to infinity, then finishing by using results of [GM88] as a black box.

Additional exercises

Exercise 7.15. When $d = 2$, the map Relog : $\mathcal{V}_* \to \mathbb{R}^d$ is locally one-to-one at most points. We say that amoeba(f) is a *doublet* if $\text{Relog}^{-1}(\boldsymbol{x})$ has cardinality precisely 2 for all $\boldsymbol{x}$ in the interior of amoeba(f). Give a proof by picture that if amoeba(f) is a doublet then there is a natural isomorphism κ between the reduced homology group $\tilde{H}_0(\text{amoeba}(f)^c)$ and $H_1(\mathcal{V}_*)$, defined by $\kappa([\boldsymbol{x}'] - [\boldsymbol{x}]) = \mathbf{INT}(\mathbf{T}(\boldsymbol{x}), \mathbf{T}(\boldsymbol{x}'))$.

Exercise 7.16. Let $Q(x, y) = 5 - x - x^{-1} - y - y^{-1}$ and $F(x, y) = 1/Q(x, y)$.

(a) Sketch $\texttt{amoeba}(Q)$ and mark a point in each component of $\texttt{amoeba}(Q)^c$.
(b) Recall that the *Fourier series* of F is the series $\hat{F}(x, y) = \sum_{a,b\in\mathbb{Z}} c_{a,b}e^{i(ax+by)}$, where

$$c_{a,b} = \frac{1}{(2\pi)^2}\int_{-\pi}^{\pi}\int_{-\pi}^{\pi} F(x, y)e^{-in(ax+by)}dxdy.$$

The Fourier series $\hat{F}$ is related to the Laurent expansion of F corresponding to one component of $\texttt{amoeba}(f)^c$. Identify this component, and describe the relation.
(c) Prove that $\texttt{amoeba}(Q)$ is a doublet.
(d) Let $\boldsymbol{x} = (0, 0)$ and $\boldsymbol{x}' = (0, 2)$. Show that $\boldsymbol{x}$ and $\boldsymbol{x}'$ are in different components of the complement of $\texttt{amoeba}(Q)$, and describe or sketch $\mathbf{INT}(\mathbf{T}(\boldsymbol{x}), \mathbf{T}(\boldsymbol{x}'))$.
(e) Find all critical points of $\mathcal{V}_*$ in the direction $\boldsymbol{r} = (1/3, 2/3)$ and mark them on your sketch from part (*a*).
(f) Deform the intersection cycle γ you found in part (*d*) until its highest and lowest points are critical points in direction $\boldsymbol{r} = (1/3, 2/3)$. At which of these points is the phase $h_{\hat{\boldsymbol{r}}}$ maximized on γ?
(g) What does your result tell you about the coefficients of the Fourier series for F?

Exercise 7.17. For $Q(x, y, z) = z(x - y)(x - y + z - xyz)$, as in Exercise 7.12, state the dimension of the stratum containing the origin, describe the normal link, and describe the local homology group $H_3(\mathcal{M}^{\boldsymbol{p},\mathrm{loc}})$ when $\boldsymbol{p}$ is the origin and $\boldsymbol{r}$ is a direction of your choice (as usual, $\mathcal{M} = \mathbb{C}_*^d \setminus \mathcal{V}_Q$).

8

Effective computations and ACSV

In this chapter we revisit the key ACSV concepts introduced in Chapters 6 and 7 — critical points, minimal points, Whitney stratifications, and polynomial amoebas — through the viewpoint of explicit computation. First, Section 8.1 describes techniques for manipulating the solutions of polynomial systems. Section 8.2 then illustrates how to use these techniques to characterize the critical points of a rational function by determining an algebraic Whitney stratification, computing a set of polynomial equalities and inequalities for the critical points on each stratum, and reducing these polynomial equalities into a convenient form. Our most explicit results hold for minimal critical points, which are the critical points on the boundary of the domain of convergence of the series under consideration. Section 8.3 describes how to verify minimality of critical points, which can be viewed as determining boundary points on amoeba complements. Finally, Section 8.4 describes additional computations that must be performed in order to find coefficient asymptotics.

Throughout this chapter we fix a ratio $F(\boldsymbol{z}) = P(\boldsymbol{z})/Q(\boldsymbol{z})$ of coprime polynomials P and Q and a convergent Laurent expansion $F(\boldsymbol{z}) = \sum_{\boldsymbol{r}\in\mathbb{Z}^d} a_{\boldsymbol{r}}\boldsymbol{z}^{\boldsymbol{r}}$ that holds in the domain of convergence $\mathcal{D} = \text{Relog}^{-1}(B)$ for a component B of $\texttt{amoeba}(H)^c$ (the most common case occurring when $\mathcal{D}$ is the domain of convergence of a power series expansion). We say that a property involving Q is a ***generic property*** if it holds for all choices of Q except those whose coefficients lie in a proper algebraic set depending only on the degree of Q (this is an algebraic analog of a property holding almost surely). Our goal is to determine asymptotics of a coefficient sequence $a_{\boldsymbol{r}} = [\boldsymbol{z}^{\boldsymbol{r}}]F(\boldsymbol{z})$ as $\boldsymbol{r} \to \infty$ with $\hat{\boldsymbol{r}}$ varying in some compact set of directions.

Because the vanishing set of a polynomial is the same as the vanishing set of any positive integer power of that polynomial, we recall from Chapter 7 that the *square-free* part $\tilde{Q}$ of a polynomial Q is the product of its distinct irreducible factors over the complex numbers.

Exercise 8.1. Show that $\tilde{Q}$ can be computed by dividing Q by the greatest common divisor of all of its first order partial derivatives.

8.1 Techniques for polynomial systems

Before turning to the computations used in our asymptotic analyses, we first introduce the algebraic background needed to describe our approach. The critical points of a multivariate generating function are described by a finite collection of polynomial equalities and inequalities, and we manipulate the solutions to these polynomial systems by encoding them in polynomial ideals and applying techniques from computer algebra. Once we have a convenient representation for the critical points of a rational function, related quantities can be computed via the same representation, as we show below.

We work here in a polynomial ring $K[\mathbf{z}]$ over a field K of characteristic zero: in applications K is typically the field $\mathbb{Q}$ or the field $\mathbb{Q}(\mathbf{r})$ of rational functions in the parameters $\mathbf{r}$ encoding a generic direction. A ***polynomial ideal*** in $K[\mathbf{z}]$ is a nonempty subset $I \subset K[\mathbf{z}]$ that is closed under addition of elements in I and closed under multiplication by any element of $K[\mathbf{z}]$. Hilbert's basis theorem [CLO07, Section 2.5] states that every ideal I has a finite ***generating set*** $\{g_1, \ldots, g_r\} \subset K[\mathbf{z}]$ such that the elements of I are precisely the $K[\mathbf{z}]$-linear combinations of $g_1, \ldots, g_r$. We write $I = \langle g_1, \ldots, g_r \rangle$ when I is generated by $g_1, \ldots, g_r$.

If $I \subset K[\mathbf{z}]$ is a polynomial ideal then we use the notation $\mathcal{V}(I)$ to denote the ***algebraic variety*** consisting of all roots of all elements of I over the algebraic closure of K. If $I = \langle g_1, \ldots, g_r \rangle$ then $\mathcal{V}(I)$ is the set of solutions, in an algebraic closure of K, to the polynomial system

$$g_1(\mathbf{z}) = \cdots = g_r(\mathbf{z}) = 0$$

defined by the generators. Typically, we start with an ideal I defined by a specific generating set and want to compute another generating set that allows us to better understand properties of $\mathcal{V}(I)$. This area of computer algebra is well-established, with algorithms for manipulating polynomial ideals already implemented in many computer algebra systems.

Term orders and polynomial division

One of the most basic questions we can ask about an ideal is whether or not it contains a specific element $f \in K[\mathbf{z}]$. In one variable every ideal I is generated by a single polynomial, $I = \langle g \rangle$, so $f \in I$ if and only if $f(z) = a(z)g(z)$ for some $a(z) \in K[z]$. The (univariate) polynomial division algorithm for $f(z)/g(z)$ produces a quotient $a(z)$ and a remainder $r(z)$ such that $f(z) = a(z)g(z) + r(z)$ and the degree of r is less than the degree of g. Thus, $f \in I$ if and only if the division algorithm returns a remainder of zero.

The division algorithm works because we can repeatedly divide the highest degree term of g into the highest degree term of f until the degree of the remainder is less than the degree of g. To duplicate this approach in several variables we therefore need to define an ordering on multivariate monomials. Our ordering must be compatible with multiplication, and there must be no infinite chains of ever-smaller elements, motivating the following definition.

Definition 8.1. A ***monomial order*** on $K[\boldsymbol{z}]$ is a relation $>$ on the set of monomials $\boldsymbol{z}^{\boldsymbol{r}}$ such that

(i) $>$ is a well-order (i.e., it is a total order and every nonempty subset of $K[\boldsymbol{z}]$ has a minimal element), and
(ii) if $\alpha, \beta, \gamma \in \mathbb{N}^d$ and $\boldsymbol{z}^\alpha > \boldsymbol{z}^\beta$ then $\boldsymbol{z}^{\alpha+\gamma} > \boldsymbol{z}^{\beta+\gamma}$.

One common term order is the ***lexicographic term order***, where $\boldsymbol{z}^\alpha > \boldsymbol{z}^\beta$ if and only if $\alpha_j > \beta_j$ for some index j and $\alpha_i = \beta_i$ for all $i < j$. Another common order is the ***total degree term order***, in which $\boldsymbol{z}^\alpha > \boldsymbol{z}^\beta$ if and only if either the degree of $\boldsymbol{z}^\alpha$ is greater than the degree of $\boldsymbol{z}^\beta$ or the degrees are equal and $\boldsymbol{z}^\alpha > \boldsymbol{z}^\beta$ in the lexicographic order.

Exercise 8.2. Rank the monomials

$$y^{10} \qquad xz^3 \qquad yz \qquad x^2z \qquad xyz$$

in increasing order, first under the lexicographic term order and then under the total degree term order.

Definition 8.2. For a polynomial $p(\boldsymbol{z}) = \sum_{\alpha \in A} c_\alpha \boldsymbol{z}^\alpha$ with non-zero coefficients c_α for $\alpha \in A$, the ***leading term of the polynomial*** p under a monomial ordering $>$ is the term $\mathrm{LT}(p) := c_\alpha \boldsymbol{z}^\alpha$ with maximum monomial $\boldsymbol{z}^\alpha$ according to $>$.

Monomial orders allow one to run a multivariate division algorithm to divide a polynomial f by a list of polynomials $L = [g_1, \ldots, g_m]$ to obtain quotients a_i and a remainder r such that $p(\boldsymbol{z}) = \sum_i a_i(\boldsymbol{z}) g_i(\boldsymbol{z}) + r(\boldsymbol{z})$ and $\mathrm{LT}(r)$ is not divisible by any $\mathrm{LT}(g_i)$. Roughly, we run through the elements of L in order, trying to divide the leading term of f by the leading term of the g_i under consideration. If $\mathrm{LT}(g_i)$ divides $\mathrm{LT}(f)$ then we add $c = \mathrm{LT}(f)/\mathrm{LT}(g_i)$ to the quotient a_i, replace f by $f - cg_i$, and repeat. If the leading term of f is not divisible by any $\mathrm{LT}(g_i)$ then we add the leading term of f to the remainder r, replace f by $f - \mathrm{LT}(f)$, and repeat from the previous step.

In order to make the multivariate division algorithm deterministic, it is important to check divisibility of $\mathrm{LT}(f)$ by $\mathrm{LT}(g_i)$ in the order the g_i appear in L, as the remainder returned by the division algorithm can be different for different orderings of the g_i. Related to this technicality is a large problem: even if f can be written as a polynomial combination of the g_i, meaning f lies in the ideal $I = \langle g_1, \ldots, g_m \rangle$, it is possible for the multivariate division algorithm to return a non-zero remainder. The trick to working around this difficulty is to compute a particularly nice generating set for the ideal $I = \langle g_1, \ldots, g_m \rangle$, which also allows us to resolve many other questions about polynomial ideals.

Gröbner bases

Fix a monomial order $>$ and let I be an ideal in $K[\mathbf{z}]$, where K is algebraically closed.

Definition 8.3. A ***Gröbner basis*** for I with respect to the monomial order $>$ is a generating set $G = \{g_1, \ldots, g_k\}$ for I with the property that for any non-zero $f \in I$ the leading term $\mathrm{LT}(f)$ is divisible by $\mathrm{LT}(g_i)$ for some i. The basis is a ***reduced Gröbner basis*** if for all $i \neq j$ no monomial of g_i is divisible by $\mathrm{LT}(g_j)$.

Every non-zero ideal has a Gröbner basis with respect to any ordering, reduced Gröbner bases are unique and algorithmically computable, and algorithms to find them have been implemented in most computer algebra systems. The size and computation time of finding a Gröbner basis can depend heavily on the term order used. If G is a Gröbner basis for I and $f \in K[\mathbf{z}]$ then

- the remainder when f is divided by the elements of G is unique (it doesn't depend on the order in which the elements of G are listed),
- $f \in I$ if and only if the remainder when f is divided by G is zero,
- $\mathcal{V}(I)$ is empty if and only if G contains the constant 1,
- $\mathcal{V}(I)$ is finite if and only if there are a finite number of monomials not divisible by a leading term of the elements of G, and
- if G is a Gröbner basis for the lexicographical order and $\mathcal{V}(I)$ is finite then G contains a non-zero univariate polynomial in $f \in K[z_d]$.

The first two items here follow directly from the properties of Gröbner bases, while the third item is Hilbert's Nullstellensatz [CLO07, Section 4.1]. The fourth item is proved in [CLO07, Section 3.5] and the final item, sometimes known as the elimination theorem, can be found in [CLO07, Section 3.1]. The elimination theorem, which shows how lexicographical Gröbner bases generalize the classical resultant beyond two bivariate polynomials, is key as it allows us to conveniently encode the solutions of a polynomial system.

Example 8.4. Consider the generating function

$$F(x,y) = \sum_{r,s\geq 0} a_{rs}x^r y^s = \frac{xy(1-x)^3}{(1-x)^4 - xy(1-x-x^2+x^3+x^2y)},$$

discussed further in Example 12.23 of Chapter 12. If P and Q denote the numerator and denominator of F, then the ideal I formed by Q and all its partial derivatives characterizes points where $\mathcal{V}$ could be non-smooth. Using the lexicographic term order with $x > y$ we obtain a Gröbner basis $B = \left[x^2 - 2x + y + 1, xy - y, y^2\right]$. Solving the last polynomial of B, which contains only y, implies that any point in $\mathcal{V}(I)$ must satisfy $y = 0$, and substituting this into the first element of B gives $\mathcal{V}(I) = \{(1,0)\}$. Thus, $\mathcal{V}$ is locally smooth near all of its points except $(1,0)$. A Taylor expansion of Q near $(1,0)$ shows that it locally looks like two curves touching at a point. ◁

Example 8.5. The fact that $\mathcal{V}(I)$ contains only points with explicit rational coordinates in Example 8.4 is not reflective of the general situation. Often, one has a point $\boldsymbol{w}$ of interest with (non-rational) algebraic coordinates defined only implicitly and wants to simplify another algebraic quantity depending on $\boldsymbol{w}$. When all the quantities involved are algebraic, the results can be simplified using the multivariate division algorithm and Gröbner bases.

For instance, suppose we want to compute an expansion of Q in Example 8.4 at one of the points $(a,b) \in \mathcal{V}$ where $a = \sqrt{2}$ (there are two such points, at each of which y is a degree-four algebraic number). Dividing the polynomial $Q(X+a, Y+b)$ by the ideal $[a^2 - 2, Q(a,b)]$ using, for instance, the Gröbner basis with respect to the lexicographic order where $a > b > X > Y$ gives the polynomial

$$-(6ab + 6b^2 - 20a - 5b + 28)X - (4ab - a + 2)Y + \cdots - X^4Y$$

describing the expansion of Q at (a,b) in terms of $X = x - a$ and $Y = y - b$. The coefficients have been reduced using the algebraic relations satisfied by a and b, in this case showing that Q vanishes to first order at (a,b). ◁

Gröbner basis computations also serve as primitives in algorithms to compute many other fundamental properties of ideals $I, J \subset K[\boldsymbol{z}]$, including

- the ***radical*** $\sqrt{I} = \left\{f \in K[\boldsymbol{z}] : f^k \in I \text{ for some } k \in \mathbb{N}\right\}$;
- the ***prime decomposition*** of I, which is the set of prime ideals $P_1, \ldots, P_r$ such that $\sqrt{I} = P_1 \cap \cdots \cap P_r$ (an ideal is prime if $fg \in I$ implies $f \in I$ or $g \in I$);

- the ***irreducible decomposition*** of $\mathcal{V}(I)$, which is the set of irreducible varieties $\mathcal{V}_1, \ldots, \mathcal{V}_s$ such that $\mathcal{V}(I) = \mathcal{V}_1 \cup \cdots \cup \mathcal{V}_s$ (an ***irreducible variety*** is a variety that cannot be written as the union of two proper subvarieties);
- the ***ideal quotient*** $I : J$, which is the ideal $\{f \in K[\boldsymbol{z}] : fg \in I \text{ for all } g \in J\}$;
- the ***ideal saturation*** $I : J^\infty$, which is defined as the largest ideal in the eventually stable increasing chain of ideals

$$I \subset I : J \subset I : J^2 \subset \cdots.$$

Geometrically, $\mathcal{V}(I : J^\infty)$ is the smallest algebraic set containing all points in $\mathcal{V}(I) \setminus \mathcal{V}(J)$;

- the ***dimension of an ideal*** I, equal to the *dimension* of the algebraic set $\mathcal{V}(I)$ and defined as the maximum $d \in \mathbb{N}$ such that there exist irreducible varieties $\mathcal{V}_1, \ldots, \mathcal{V}_d$ with $\mathcal{V}_0 \subsetneq \mathcal{V}_1 \subsetneq \cdots \subsetneq \mathcal{V}_d \subsetneq \mathcal{V}(I)$.

Describing algorithms for these operations is the domain of a computer algebra textbook, such as [BW93], and we simply use implementations in computer algebra software as black boxes.

Example 8.6. Consider the ideals $I = \langle 3 - x - 2y, 3 - 2x - y \rangle$ and $J = \langle x - y \rangle$. Then I is prime and $\sqrt{I} = I$, and the quotient and saturation of I by J both equal the ideal $\langle 1 \rangle = \mathbb{Q}[x, y]$. The ideal I has dimension 0, since $\mathcal{V}(I)$ is the single point $(1, 1)$. ◁

8.2 Computing critical points

The results of Chapter 7 illustrate that the main objects involved in our asymptotic analysis are the critical points of F in directions $\boldsymbol{r}$. In order to describe the critical points of F, we must first decompose the singular set $\mathcal{V}$ of F into a Whitney stratification. Recall from Section 7.3 (or Appendix D) that this is a partition of $\mathcal{V}$ into smooth manifolds of various dimensions satisfying specific conditions on how they fit together.

8.2.1 Effective Whitney stratification

The algebraic set $\mathcal{V}(Q) = \mathcal{V}(\tilde{Q})$ forms the singular variety $\mathcal{V}$. Lemma 7.6 from Chapter 7 implies that the algebraic set $\mathcal{W}_1$ defined by the polynomial system

$$\tilde{Q}(\boldsymbol{z}) = \tilde{Q}_{z_1}(\boldsymbol{z}) = \cdots = \tilde{Q}_{z_d}(\boldsymbol{z}) = 0 \tag{8.1}$$

encodes the points where $\mathcal{V}$ is not a smooth manifold. In particular, if a Gröbner basis of the ideal $I = \langle \tilde{Q}, \tilde{Q}_{z_1}, \ldots, \tilde{Q}_{z_d} \rangle$ contains the element 1 then $\mathcal{W}_1 = \emptyset$, so $\mathcal{V}$ is everywhere a smooth manifold and a Whitney stratification of $\mathcal{V}$ is just $\mathcal{V}$ itself.

Remark 8.7. The algebraic Sard Theorem [BPR03, Theorem 5.56] states that $\mathcal{V}$ is generically a manifold, which is unsurprising as (8.1) contains $d + 1$ polynomial equations in d variables; see also [Mel21, Section 5.3] for an elementary derivation in our context.

If $\mathcal{W}_1$ is nonempty, then it is an algebraic set whose dimension is smaller than the dimension of $\mathcal{V}$. The singularities of $\mathcal{W}_1$, in turn, form an algebraic set $\mathcal{W}_2$ of dimension smaller than the dimension of $\mathcal{W}_1$. Thus, we can obtain a nesting of structures

$$\mathcal{V} = \mathcal{W}_0 \supsetneq \mathcal{W}_1 \supsetneq \mathcal{W}_2 \supsetneq \cdots \supsetneq \mathcal{W}_r = \emptyset$$

such that each difference $\mathcal{W}_k \setminus \mathcal{W}_{k+1}$ is a smooth manifold. The polynomials defining the algebraic set $\mathcal{W}_{k+1}$ can be computed recursively from the polynomials defining $\mathcal{W}_k$ using algorithms for ideal dimension and prime decomposition, and the Jacobian Criterion [Eis95, Corollary 16.20] for singularity. Since $\mathcal{V}$ is the union of all consecutive differences $\mathcal{W}_k \setminus \mathcal{W}_{k+1}$, this gives a partition of $\mathcal{V}$ into smooth manifolds.

Although conceptually simple, this approach does not, in general, compute a Whitney stratification of $\mathcal{V}$. Thankfully, there is a Whitney stratification of the same form. In the early 1980s, Teissier [Tei82, Proposition VI.3.2] proved the existence of the *canonical Whitney stratification* of $\mathcal{V}$, defined by a nested sequence of algebraic sets

$$\mathcal{V} = \mathcal{F}_0 \supsetneq \mathcal{F}_1 \supsetneq \mathcal{F}_2 \supsetneq \cdots \supsetneq \mathcal{F}_m = \emptyset \tag{8.2}$$

such that the set of all connected components of $\mathcal{F}_k \setminus \mathcal{F}_{k+1}$ for all k form a Whitney stratification of $\mathcal{V}$ and every Whitney stratification of $\mathcal{V}$ is a refinement of this canonical stratification. Older algorithms to compute Whitney stratifications use quantifier elimination and cylindrical algebraic decomposition [Ran98; MR91], while recent work [DJ21; HN22] gives more practical algorithms[1] based around Gröbner basis computations.

Remark 8.8. All combinatorially interesting multivariate generating functions that the authors have encountered in the literature have Whitney stratifications that can be computed directly by hand. Often, this is because $\mathcal{V}$ is either a

[1] Helmer and Nanda [HN22] give Macaulay2 implementations at `http://martin-helmer.com/Software/WhitStrat/`.

smooth manifold or can be expressed as the union of smooth manifolds intersecting transversely. However, even in applications with more complicated singularities, such as the isolated quartic singularity of Example 11.47, a Whitney stratification is obvious from inspection.

Example 8.9. Recall the generating function

$$F(x,y) = \sum_{r,s\geq 0} a_{rs}x^r y^s = \frac{xy(1-x)^3}{(1-x)^4 - xy(1-x-x^2+x^3+x^2y)}$$

from Example 12.23. The singular variety $\mathcal{V}$ admits a Whitney stratification with strata $\{\boldsymbol{p}\}$ and $\mathcal{V} \setminus \{\boldsymbol{p}\}$ where $\boldsymbol{p} = (1,0)$. ◁

8.2.2 Critical point equations

From now on we assume that we have the canonical Whitney stratification (8.2) of $\mathcal{V}$, computed either by a computer algebra system or by hand. For each $1 \leq k \leq m$ let I_k denote the set

$$I_k = \{f \in K[\boldsymbol{z}] : f(\boldsymbol{z}) = 0 \text{ for } \boldsymbol{z} \in \mathcal{F}_k\}.$$

Exercise 8.3. Prove that, for each k, the set I_k is an ideal with $I_k = \sqrt{I_k}$.

Fixing a value of $k \in \{0, \ldots, m-1\}$, our goal is to compute polynomial equalities and inequalities characterizing the critical points on each connected component of $\mathcal{F}_k \setminus \mathcal{F}_{k+1} = \mathcal{V}(I_k) \setminus \mathcal{V}(I_{k+1})$. First, we use a Gröbner basis based algorithm to compute the prime decomposition $I_k = P_1 \cap \cdots \cap P_\ell$. Because all points in an intersection $\mathcal{V}(P_i) \cap \mathcal{V}(P_j)$ with $i \neq j$ are non-smooth points of $\mathcal{F}_k$, they are contained in $\mathcal{F}_{k+1}$. This means we can compute the critical points on the zero set $\mathcal{V}(P_j)$ of each prime ideal in the decomposition of I, remove the points of $\mathcal{F}_{k+1}$, then take the union of these sets.

The dimension δ of the prime ideal P_j can be computed using a Gröbner basis algorithm, from which we can compute the ***codimension*** $c = d - \delta$. If $P_j = \langle p_1, \ldots, p_s\rangle$ then, after translation to the origin, the tangent space to $\mathcal{V}(P_j)$ at each smooth point $\boldsymbol{p} \in \mathcal{V}(P_j)$ is the codimension c linear subspace of $\mathbb{C}^d$ normal to all of the gradients $(\nabla p_1)(\boldsymbol{p}), \ldots, (\nabla p_s)(\boldsymbol{p})$; see Mumford [Mum95, Section 1.A].

Recall the *logarithmic gradient* $\nabla_{\log} f = (z_1 f_{z_1}, \ldots, z_d f_{z_d})$. Because the height function $h_{\boldsymbol{r}}$ is the real part of $\tilde{h}_{\boldsymbol{r}}(\boldsymbol{z}) = -\boldsymbol{r} \cdot \log(\boldsymbol{z})$, the point $\boldsymbol{p}$ is a critical point lying in $\mathcal{V}(P_j) \setminus \mathcal{F}_{k+1}$ if and only if the gradient vector $-(r_1/p_1, \ldots, r_d/p_d)$ of $\tilde{h}_{\boldsymbol{r}}$ at $\boldsymbol{z} = \boldsymbol{p}$ is normal to the tangent space of $\mathcal{V}(P_j)$ at $\boldsymbol{p}$. Since the tangent space is a linear subspace of codimension c, and we consider only points with non-zero coordinates, we see that $\boldsymbol{p}$ is a critical point if and only if the matrix

$$M(\boldsymbol{z},\boldsymbol{r}) = \begin{pmatrix} & (\nabla_{\log} p_1)(\boldsymbol{z}) & \\ & \vdots & \\ & (\nabla_{\log} p_s)(\boldsymbol{z}) & \\ r_1 & \cdots & r_d \end{pmatrix} \tag{8.3}$$

has rank c when $\boldsymbol{z} = \boldsymbol{p}$, which occurs if and only if all $(c+1)\times(c+1)$ minors of $M(\boldsymbol{p},\boldsymbol{r})$ vanish. Thus, the critical points of F contained in $\mathcal{V}(P_j) \setminus \mathcal{F}_{k+1}$ are defined by the *vanishing* of the polynomials $p_1(\boldsymbol{z}),\ldots,p_s(\boldsymbol{z})$ together with the $(c+1)\times(c+1)$ minors of M and the *nonvanishing* of a generating set of I_{k+1} together with the polynomial $z_1\cdots z_d$ (since we only search for critical points with non-zero coordinates). Because our Whitney stratification is defined by a finite number of algebraic sets, we ultimately obtain the following.

Proposition 8.10. *The set* $\texttt{critical}(\boldsymbol{r})$ *of all (stratified) critical points of any rational function* $F \in \mathbb{Q}(\boldsymbol{z})$ *in a direction* $\hat{\boldsymbol{r}}$ *is defined by a finite collection of polynomial systems over* $\mathbb{Q}[\boldsymbol{r}][\boldsymbol{z}]$*, each involving polynomial equalities and inequalities. The polynomials in these (in)equalities can be computed explicitly.*

The complexity of computing critical points, discussed in Melczer and Salvy [MS21], is typically lower than the complexity of verifying which (if any) are minimal points.

Example 8.11. In the simplest case $\mathcal{V} = \mathcal{V}(Q)$ is a smooth manifold, so the canonical Whitney stratification of $\mathcal{V}$ is specified by $\mathcal{F}_0 = \mathcal{V}$ and $\mathcal{F}_1 = \emptyset$. Then $I_0 = (\tilde{Q})$ is generated by the square-free part of Q and $\texttt{critical}(\boldsymbol{r})$ is defined by $\tilde{Q}(\boldsymbol{z}) = 0$ and the vanishing of the 2×2 minors of

$$M(\boldsymbol{z},\boldsymbol{r}) = \begin{pmatrix} z_1\tilde{Q}_{z_1}(\boldsymbol{z}) & \cdots & z_d\tilde{Q}_{z_d}(\boldsymbol{z}) \\ r_1 & \cdots & r_d \end{pmatrix},$$

matching Lemma 7.8. Note that we can skip the step of decomposing $I_0 = \langle\tilde{Q}\rangle$ into its prime factors: $\mathcal{V}$ being smooth implies that the prime factors of I_0 are distinct and each corresponds to a variety of codimension 1. ◁

Remark 8.12. Given $\boldsymbol{z} \in \mathcal{V}$ let $\mathbf{L}(\boldsymbol{z})$ denote the logarithmic normal space to the stratum S containing $\boldsymbol{z}$, given by the span of the logarithmic gradients of the functions locally defining S at $\boldsymbol{z}$. Then $\boldsymbol{z} \in \texttt{critical}(\boldsymbol{r})$ if and only if $\boldsymbol{r} \in \mathbf{L}(\boldsymbol{z})$, and this defines a binary relation on $(\mathcal{V},\mathbb{R}^d)$.

We end this section with some examples in low dimensions.

Example 8.13 (binomial coefficients). Recall the binomial coefficient generating function

$$\frac{1}{1-x-y} = \sum_{r,s\geq 0} \binom{r+s}{r,\, s} x^r y^s.$$

Since $Q(x,y) = 1-x-y$ we find $\nabla Q = (-1,-1)$ never vanishes, so the variety $\mathcal{V}$ is a smooth manifold. The smooth critical point equations (7.8) form the polynomial system

$$\begin{aligned} 1 - x - y &= 0 \\ -sx &= -ry, \end{aligned}$$

which we solve to obtain the unique critical point $(x,y) = \left(\frac{r}{r+s}, \frac{s}{r+s}\right) = \hat{\boldsymbol{r}}$. ◁

Example 8.14 (Delannoy numbers). Recall from Example 2.7 that the denominator for the Delannoy generating function is given by $Q(x,y) = 1-x-y-xy$. To check whether $\mathcal{V}$ is a manifold, we form the ideal $I = \langle Q, Q_x, Q_y \rangle$ and compute the reduced Gröbner basis [1], meaning $\mathcal{V}(I) = \emptyset$ and $\mathcal{V}$ is a smooth manifold.

To characterize the critical points in a direction $\hat{\boldsymbol{r}}$, where $\boldsymbol{r} = (r,s)$, we form the ideal $J = \langle Q, sxQ_x - ryQ_y \rangle$ and compute the reduced Gröbner basis $G = \{p(x,y), g(y)\}$ of J, where $p(x,y) = rx - sy - r + s$ and $g(y) = sy^2 + 2ry - s$. Solving the quadratic $g(y)$ for y gives

$$y = \frac{-r \pm \sqrt{r^2+s^2}}{s}$$

and back substitution into $p(x,y)$ implies that $\texttt{critical}(r,s)$ consists of the two points

$$\left(\frac{\sqrt{r^2+s^2}-s}{r}, \frac{\sqrt{r^2+s^2}-r}{s}\right) \text{ and } \left(\frac{-\sqrt{r^2+s^2}-s}{r}, \frac{-\sqrt{r^2+s^2}-r}{s}\right). \tag{8.4}$$

We will see later that the first point is the one that determines the asymptotics of a_{rs} as $r, s \to \infty$. ◁

Example 8.15 (two intersecting planes). Let $Q(x,y,z) = A(x,y,z)B(x,y,z)$ with $A(x,y,z) = 4 - x - 2y - z$ and $B(x,y,z) = 4 - 2x - y - z$. The zero set $\mathcal{V} = \mathcal{V}(Q)$ is the union of two hyperplanes $\mathcal{V}_1 = \mathcal{V}(A)$ and $\mathcal{V}_2 = \mathcal{V}(B)$ that intersect transversely in the line $\ell = \mathcal{V}(A,B)$. The canonical Whitney stratification of $\mathcal{V}$ is given by the algebraic sets

$$\mathcal{V} = \mathcal{F}_0 \supset \mathcal{F}_1 \supset \mathcal{F}_2 = \emptyset$$

with $\mathcal{F}_1 = \ell$, and corresponding ideals $I_0 = \langle AB \rangle$ and $I_1 = \langle A, B \rangle$. The ideal I_0 has the prime decomposition $I_0 = \langle AB \rangle = P_1 \cap P_2$ with $P_1 = \langle A \rangle$ and $P_2 = \langle B \rangle$, and the strata of $\mathcal{V}$ are

$$S_1 = \mathcal{V}_1 \setminus \ell, \quad S_2 = \mathcal{V}_2 \setminus \ell, \quad \text{and } S_{1,2} = \ell,$$

see Figure 8.1.

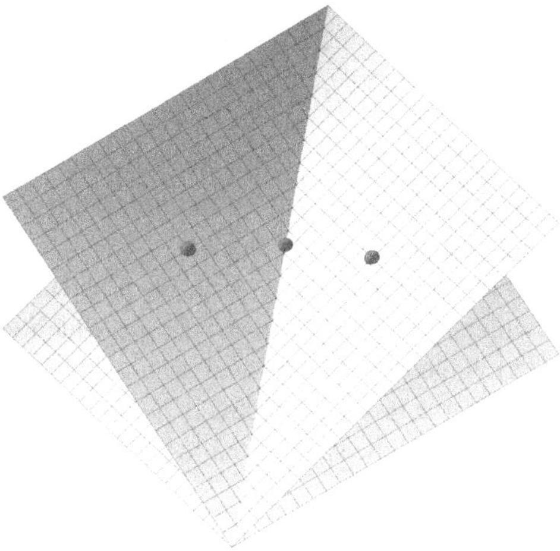

Figure 8.1 Critical points on different strata when $\mathcal{V}$ is two intersecting planes.

Critical points in the direction $\hat{\boldsymbol{r}}$, where $\boldsymbol{r} = (r, s, t)$, on the stratum S_1 are defined by the vanishing of $A(x, y, z)$ and the 2×2 minors of the matrix

$$\begin{pmatrix} xA_x & yA_y & zA_z \\ r & s & t \end{pmatrix}$$

together with the nonvanishing of $B(x, y, z)$. Solving

$$4 - x - 2y - z = tx - rz = 2ty - sz = 0$$

in (x, y, z) gives the point $\boldsymbol{\sigma} = \frac{1}{r+s+t}(4r, 2s, 4t) = (4\hat{r}, 2\hat{s}, 4\hat{t})$, which lies in S_1 unless $2r = s$, in which case the point is in ℓ. Finding the critical point on S_2 is analogous and gives the point $\boldsymbol{\tau} = (2\hat{r}, 4\hat{s}, 4\hat{t})$ unless $2s = r$. These critical points are the two points marked with dots that are not on the line of intersection in Figure 8.1.

Critical points on ℓ are defined by the vanishing of $A(x, y, z)$, $B(x, y, z)$, and the 3×3 minor (i.e., determinant) of the matrix

$$\begin{pmatrix} xA_x & yA_y & zA_z \\ xB_x & yB_y & zB_z \\ r & s & t \end{pmatrix}.$$

Solving the resulting polynomial system

$$\begin{aligned} 4 - x - 2y - z &= 0 \\ 4 - 2x - y - z &= 0 \\ ryz + sxz - 3txy &= 0 \end{aligned}$$

gives the point $\boldsymbol{\omega} = (4/3)(\hat{r} + \hat{s}, \hat{r} + \hat{s}, 3\hat{t})$, which is the unique point on ℓ at which (r, s, t) lies in the plane spanned by the logarithmic tangents $(x, 2y, z)$ and $(2x, y, z)$ to the planes defined by A and B.

Thus, `critical`(r, s, t) consists of the points $\boldsymbol{\sigma}, \boldsymbol{\tau}$, and $\boldsymbol{\omega}$, which are distinct unless $2r = s$ (in which case $\boldsymbol{\sigma} = \boldsymbol{\omega}$) or $r = 2s$ (in which case $\boldsymbol{\tau} = \boldsymbol{\omega}$). For $\boldsymbol{z} \in \ell$ the space $\mathbf{L}(\boldsymbol{z})$ is the two-dimensional span of the log-gradient vectors of A and B at $\boldsymbol{z}$, while for $\boldsymbol{z}$ in S_1 (respectively S_2) the space $\mathbf{L}(\boldsymbol{z})$ is the one-dimensional space defined as the span of the log-normal to A (respectively B) at $\boldsymbol{z}$. ◁

The last three examples have critical points that are either rational functions of $\boldsymbol{r}$ or require only square roots. In general we may obtain algebraic functions of arbitrary complexity, and it is often best to encode critical points via their minimal polynomials.

Example 8.16. Again consider the generating function

$$F(x, y) = \sum_{r,s \geq 0} a_{rs} x^r y^s = \frac{xy(1-x)^3}{(1-x)^4 - xy(1 - x - x^2 + x^3 + x^2 y)}.$$

The smooth critical points of F in the direction $\hat{\boldsymbol{r}}$, where $\boldsymbol{r} = (r, s)$, are encoded by the ideal $\langle Q, sxQ_x - ryQ_y \rangle$. Writing $\lambda = s/r$, the Gröbner basis of this ideal with respect to the lexicographic term order where $y > x$ (which gives a simpler result than $x > y$) contains a polynomial of the form $(1-x)^5 \beta_\lambda(x)$ where

$$\beta_\lambda(x) = (1+\lambda)x^4 + 4(1+\lambda)^2 x^3 + 10(\lambda^2 + \lambda - 1)x^2 + 4(2\lambda - 1)^2 x + (1-\lambda)(1-2\lambda)$$

is irreducible for generic λ. There is some simplification for particular values of λ, for instance the roots of the polynomial $\beta_{1/2}(x) = x^2(x^2 + 6x - 5/3)$ lie in a quadratic extension of $\mathbb{Q}$. ◁

8.2.3 Critical points at infinity

As seen in Chapter 7, our Morse-theoretic decompositions hold in the absence of critical points at infinity (CPAI). In this section we give an algebraic test, using Gröbner bases, to certify the absence of CPAI.

So far in this chapter we have dealt with affine critical points, and we now extend this to consider points in projective space. A polynomial $P(\boldsymbol{z})$ is a ***homogeneous polynomial*** if every monomial appearing in P has the same degree δ; equivalently, $P(\lambda \boldsymbol{z}) = \lambda^\delta P(\boldsymbol{z})$ when λ is a variable. The ***homogenization of a polynomial*** $P \in K[\boldsymbol{z}]$ is the homogeneous polynomial $P^h \in K[z_0, \boldsymbol{z}]$ defined by $P^h(z_0, \boldsymbol{z}) = z_0^{\deg P} P(z_1/z_0, \ldots, z_d/z_0)$, and the ***homogenization of an ideal*** $I \subset K[\boldsymbol{z}]$ is the ideal $I^h = \{P^h : P \in I\} \subset K[z_0, \boldsymbol{z}]$. If $\{g_1, \ldots, g_r\}$ is a Gröbner basis of I with respect to any monomial order where $\boldsymbol{z}^\alpha > \boldsymbol{z}^\beta$ whenever $|\alpha| > |\beta|$ then I^h is generated by the set of homogenizations $\{g_1^h, \ldots, g_r^h\}$ (see [CLO07, Section 8.4]). Because I^h is generated by homogeneous polynomials, $\boldsymbol{p} \in \mathcal{V}(I^h)$ implies $\lambda \boldsymbol{p} \in \mathcal{V}(I^h)$ for all $\lambda \in \mathbb{C}$ so we can view the variety $\mathcal{V}(I^h)$ as a subset of complex projective space $\mathbb{CP}^d$.

Remark 8.17. The elements of $\mathcal{V}(I)$ can be viewed as the elements of $\mathcal{V}(I^h)$ with $z_0 = 1$, while the limit of points in $\mathcal{V}(I)$ "at infinity" can be viewed as the elements of $\mathcal{V}(I^h)$ with $z_0 = 0$. If $I = \langle f_1, \ldots, f_r \rangle$ for an arbitrary generating set $\{f_1, \ldots, f_r\}$ then $J = \langle f_1^h, \ldots, f_r^h \rangle$ is an ideal of $K[z_0, \boldsymbol{z}]$ contained in I^h, however J may be strictly smaller than I^h. Nonetheless, this containment implies that $\mathcal{V}(I^h) = \emptyset$ whenever $\mathcal{V}(J) = \emptyset$.

Fix a direction $\boldsymbol{r}$ and let P be an element of the prime decomposition of an ideal $I(\mathcal{F}_k)$ coming from the canonical Whitney stratification $\mathcal{F}_0 \supset \mathcal{F}_1 \supset \cdots \supset \mathcal{F}_m$ of $\mathcal{V}$. As seen above, the critical points on a stratum of $\mathcal{V}(P) \setminus \mathcal{F}_{k+1} \subset \mathcal{F}_k \setminus \mathcal{F}_{k+1}$ in a direction $\boldsymbol{y}$ are encoded by the vanishing of a finite collection C of polynomials containing a generating set $p_1, \ldots, p_s$ of P together with appropriate minors of the matrix $M(\boldsymbol{z}, \boldsymbol{y})$ in (8.3). To detect CPAI in the direction $\boldsymbol{r}$, it is tempting to homogenize the polynomials in C to obtain a new set of polynomials C' and then search for any common solutions to the elements of C' where $z_0 = 0$ and $\boldsymbol{y} = \boldsymbol{r}$. Unfortunately, this approach is too weak to form a useful test: homogenizing the elements of C means that $\mathcal{V}(C')$ can have complicated behavior at infinity that is completely unrelated to the behavior of $\mathcal{V}(C)$.

We get around this difficulty by using ideal saturation. Recall from above that the saturation of an ideal I by another ideal J is an ideal $I : J^\infty$ whose zero set $\mathcal{V}(I : J^\infty)$ is the closure of $\mathcal{V}(I) \setminus \mathcal{V}(J)$. Saturating the ideal generated by the elements of C' with the ideal $\langle z_0 \rangle$ leaves only the points of $\mathcal{V}(C')$ that are limits of points in $\mathcal{V}(C)$. Note that it is crucial to leave the direction $\boldsymbol{y}$ as a vector of variables when performing the saturation: CPAI in the direction $\boldsymbol{r}$ are limits of critical points in directions *approaching* $\boldsymbol{r}$, and substituting $\boldsymbol{y} = \boldsymbol{r}$ before performing the saturation will leave only points that are limits of critical points in the precise direction $\boldsymbol{r}$. Performing this saturation, then substituting $\boldsymbol{y} = \boldsymbol{r}$ and searching for non-trivial solutions at infinity forms an effective test to certify the absence of CPAI. Note that CPAI also occur when one of the variables z_k is zero, and we test for this analogously. Ultimately we obtain Algorithm 1, taken from [BMP22], along with the simplified Algorithm 2 that works when the singular variety $\mathcal{V}$ is a smooth manifold.

Algorithm 1: Algorithm to compute CPAI.

Input: Polynomial $Q \in K[\boldsymbol{z}]$, direction $\boldsymbol{r}$ and polynomial generators of algebraic sets $\mathcal{F}_0 \supset \mathcal{F}_1 \supset \cdots \supset \mathcal{F}_m$ defining the canonical Whitney stratification of $\mathcal{V}(Q)$.

Output: Set $\mathcal{S}$ of pairs of ideals in $K[z_0, \boldsymbol{z}]$ such that if there is a CPAI in the direction $\boldsymbol{r}$ then there exists $(A, B) \in \mathcal{S}$ with $\mathcal{V}(A) \setminus \mathcal{V}(B) \neq \{\boldsymbol{0}\}$.

Set $\mathcal{S} = \emptyset$

For j from 1 to $m - 1$:

  Compute the prime decomposition $F_j = P_1 \cap \cdots \cap P_r$

  For each $I \in \{P_1, \ldots, P_r\}$:

   Let c be the codimension of I

   Let C be the ideal generated by I and the $(c + 1) \times (c + 1)$ minors of $M(\boldsymbol{z}, \boldsymbol{y})$

   Homogenize C in the $\boldsymbol{z}$ variables with the new variable z_0 to obtain the ideal C'

   Saturate C' by $z_0 z_1 \cdots z_d$ to obtain the ideal D

   Let A be the ideal generated by D and $z_0 z_1 \cdots z_d$ after substituting $\boldsymbol{y} = \boldsymbol{r}$

   Let B be the homogenization of $I(\mathcal{F}_{j+1})$

   Add (A, B) into $\mathcal{S}$

Return $\mathcal{S}$

Example 8.18. Let

$$F(z_1, z_2, z_3, z_4) = \frac{1}{1 - z_1 - z_2 - z_3 - z_4 + 27 z_1 z_2 z_3 z_4}$$

and consider the main diagonal direction. Applying Algorithm 2, we see that there are no CPAI. ◁

Algorithm 2: Simplified algorithm to compute CPAI assuming $\mathcal{V}$ is smooth.

Input: Polynomial $Q \in K[\boldsymbol{z}]$ with $\mathcal{V}(Q)$ smooth and direction $\boldsymbol{r}$.
Output: Ideal $I \subset K[z_0, \boldsymbol{z}]$ such that if there is a CPAI in the direction $\boldsymbol{r}$ then $\mathcal{V}(I) \neq \{\boldsymbol{0}\}$.
If Q is not square-free then replace it with its square-free part $\tilde{Q}$
Let C be the ideal generated by Q and the polynomials
$y_j z_1 Q_{z_1} - y_1 z_j Q_{z_j}$ for $2 \leq j \leq d$
Homogenize C in the $\boldsymbol{z}$ variables with the new variable z_0 to obtain the ideal C'
Saturate C' by $z_0 z_1 \cdots z_d$ to obtain the ideal D
Substitute $\boldsymbol{y} = \boldsymbol{r}$ into D and return the resulting ideal with the generator $z_0 z_1 \cdots z_d$ added

When $\boldsymbol{r} \in \mathbb{N}_*^d$ we can furthermore characterize the heights of any CPAI by adding the polynomial $p(\eta, z_0, \boldsymbol{z}) = \eta z_0^{|\boldsymbol{r}|} - \boldsymbol{z}^{\boldsymbol{r}}$ to the ideal C in Algorithm 1 (or Algorithm 2 in the smooth case), where η is a new variable. The heights of any CPAI are given by $-\log|\eta|$ as η ranges among its values at the CPAI encoded by the output of this modified algorithm. This technique can be extended to any $\boldsymbol{r} \in \mathbb{Q}_*^d$ by scaling variables – replacing each z_k by an integer power of z_k – and taking numerators of the (now rational function) generators of the ideal C.

Example 8.19. The following two examples are taken from [BMP22]. Let $Q = 1 - x - y - xy^2$ and consider the main diagonal direction. The adapted algorithm of the previous paragraph yields a CPAI $(0, 1)$ with height 0, and an affine critical point $(1/2, \sqrt{2} - 1)$ of greater height.

Now consider $Q = -x^2y - 10xy^2 - x^2 - 20xy - 9x + 10y + 20$, again with respect to the main diagonal. The algorithm in this case yields a critical point at infinity having height 0, and several affine points having lower height.

In the first case, the CPAI does not contribute to dominant asymptotics, which are determined by the affine critical point of higher height. In the second case, however, the CPAI does contribute to dominant asymptotics. ◁

8.3 Verifying minimal points

Our most explicit asymptotic results hold for minimal critical points. In a typical application we encode a finite number of critical points by polynomial equations computed through the techniques of Section 8.2, and want to verify which are minimal. Verifying minimality requires testing inequalities between the moduli of coordinate variables, a *real* algebraic problem that is more expensive than simply computing critical points, but effective tests for minimality follow from the results in Section 6.4 of Chapter 6.

Combinatorial expansions

The simplest situation holds for *combinatorial* power series expansions, which we recall are power series expansions with all but a finite number of coefficients being nonnegative. Corollary 6.39 implies that a point $\boldsymbol{w} \in \mathcal{V}$ is minimal with respect to such an expansion if and only if

$$Q(t|w_1|, \ldots, t|w_d|) \neq 0 \qquad (0 \leq t \leq 1),$$

which we can test algorithmically using Gröbner bases.

Example 8.20. We return again to the generating function

$$F(x,y) = \sum_{r,s \geq 0} a_{rs} x^r y^s = \frac{xy(1-x)^3}{(1-x)^4 - xy(1-x-x^2+x^3+x^2y)}$$

from Example 12.23 which, as a multivariate generating function, is combinatorial.

In the direction $\hat{\boldsymbol{r}} = (3/4, 1/4)$ there are four smooth critical points, corresponding to the 4 roots of the quartic for x derived in Example 8.16. The point $(1,0)$ also occurs as a solution to the critical point equations, but is not a smooth point, and we ignore it for now. Only one of the four smooth critical points $\boldsymbol{\sigma} = (0.45040\ldots, 0.39142\ldots)$ lies in the first quadrant, and there are no other critical points with the same coordinate-wise modulus. Thus $\boldsymbol{\sigma}$ is the only smooth critical point which *could* be minimal. To verify that it is minimal, we need only check that the variety defined by adding $Q(tx, ty) = 0$ to the critical point equations has no solutions with $(x,y) = \boldsymbol{\sigma}$ and $0 < t < 1$. Another Gröbner basis computation shows that this new variety has 21 solutions, including the five critical points which correspond to $t = 1$. The only element of the variety where $(x,y) = \boldsymbol{\sigma}$ has $t = 1$, proving that $\boldsymbol{\sigma}$ is minimal. ◁

The difficult part of this approach is working with the critical points implicitly through their defining algebraic equations. Gröbner basis computations can be slow, and theoretical complexity guarantees are weak in the worst case. Melczer and Salvy [MS21] give a complexity analysis for smooth varieties using a *Kronecker representation*, which parametrizes the generically finite set of critical points by the roots of a univariate polynomial in a new variable that is an integer linear combination of $z_1, \ldots, z_d, t$. Other alternatives include homotopy methods for polynomial system solving [LMS22], or proving minimality by hand using specialized results like the Grace–Walsh–Szegő theorem discussed in Section 6.4.

For a combinatorial Laurent expansion, Corollary 6.39 implies that $\boldsymbol{w} \in \mathcal{V}$ is minimal if and only if

$$Q\left(|x_1|^t|w_1|^{1-t}, \ \ldots, \ |x_d|^t|w_d|^{1-t}\right) \neq 0 \text{ for all } t \in (0, 1)$$

where $\boldsymbol{x}$ is a point inside the domain of convergence of the series. This non-equality can still be rigorously tested using truncation bounds for functions satisfying linear differential equations [Mez19] and zero certification of analytic functions based on Smale's α-theory [HL17]; however, this is more expensive than the algebraic methods in the power series case.

General expansions

If we consider a (not necessarily combinatorial) power series expansion then Corollary 6.36 implies that $\boldsymbol{w} \in \mathcal{V}$ is minimal if and only if

$$T(t|w_1|, \ldots, t|w_d|) \cap \mathcal{V} = \emptyset \qquad (0 \leq t < 1). \tag{8.5}$$

In order to describe the moduli of complex values using polynomial equations we write $\boldsymbol{z} = \boldsymbol{x} + i\boldsymbol{y}$ for real variables $\boldsymbol{x}, \boldsymbol{y} \in \mathbb{R}^d$ and decompose any polynomial $f(\boldsymbol{z}) \in \mathbb{R}[\boldsymbol{z}]$ into

$$f(\boldsymbol{z}) = f(\boldsymbol{x} + i\boldsymbol{y}) = f^R(\boldsymbol{x}, \boldsymbol{y}) + if^I(\boldsymbol{x}, \boldsymbol{y})$$

for $f^R, f^I \in \mathbb{R}[\boldsymbol{x}, \boldsymbol{y}]$. The condition (8.5) is then equivalent to the polynomial system

$$\begin{aligned} Q^R(\boldsymbol{x}, \boldsymbol{y}) &= 0 \\ Q^I(\boldsymbol{x}, \boldsymbol{y}) &= 0 \\ x_k^2 + y_k^2 &= t|w_k|^2 \ \ (1 \leq k \leq d) \end{aligned}$$

having no solution $(\boldsymbol{x}, \boldsymbol{y}, t) \in \mathbb{R}^{2d+1}$ with $0 < t < 1$. Melczer and Salvy [MS21] analyze algorithms testing this condition using techniques from real algebraic geometry, working implicitly with critical points through their defining algebraic equations.

Unfortunately, the complexity of these real algebraic methods and the increase in dimension corresponding to the change from $\mathbb{C}^d$ to $\mathbb{R}^{2d}$ mean that most current implementations fail to terminate for anything beyond low degree bivariate examples. An approach using efficient algorithms for homotopy continuation to solve polynomial systems can rule out non-minimal critical points for higher dimensional rational functions via a Julia implementation [LMS22], however this approach cannot usually prove minimality without additional, by hand, arguments. Minimality for general Laurent expansions can be tested using similar methods, but only becomes even more expensive to compute. As of the publication of this book, completely automatic detection of minimal critical points is practical only for combinatorial power series expansions.

8.4 Further computations for asymptotics

We end this section with additional computational results that will be necessary for our asymptotic arguments in later chapters.

8.4.1 Local Hessian and quadratic degeneracy

Let S be a stratum of codimension $s < d$, locally defined near some $\mathbf{w} \in S$ by the vanishing of functions $H_1, \ldots, H_s$, and assume that there exist distinct coordinates $\{\sigma_1, \ldots, \sigma_s\}$ such that the Jacobian matrix

$$\begin{pmatrix} H_1^{[\sigma_1]}(\mathbf{w}) & \cdots & H_1^{[\sigma_s]}(\mathbf{w}) \\ \vdots & \ddots & \vdots \\ H_s^{[\sigma_1]}(\mathbf{w}) & \cdots & H_s^{[\sigma_s]}(\mathbf{w}) \end{pmatrix}$$

is nonsingular, where $H_a^{[b]}$ denotes the partial derivative of H_a with respect to z_b. If $\pi = \{\pi_1, \ldots, \pi_{d-s}\} = \{1, \ldots, d\} \setminus \{\sigma_1, \ldots, \sigma_s\}$ denotes the remaining coordinates, then the implicit function theorem implies the existence of analytic functions $g_j(z_{\pi_1}, \ldots, z_{\pi_{d-s}})$ for $j \notin \pi$ and a sufficiently small neighborhood O of $\mathbf{w}$ in $\mathbb{C}^d$ such that for all $\mathbf{z} \in O$, $\mathbf{z}$ lies in S if and only if $z_j = g_j(z_{\pi_1}, \ldots, z_{\pi_{d-s}})$ for all $j \notin \pi$.

To derive asymptotics, we will ultimately apply the saddle point analyses of Chapters 4 and 5 to phase functions of the form

$$\phi(\boldsymbol{\theta}) = \sum_{j \notin \pi} r_j \log \left[\frac{g_j\left(w_{\pi_1} e^{i\theta_1}, \ldots, w_{\pi_{d-s}} e^{i\theta_{d-s}}\right)}{w_j} \right] + i \sum_{j=1}^{d-s} r_{\pi_j} \theta_j, \tag{8.6}$$

obtained by substituting $z_j = g_j(z_{\pi_1}, \ldots, z_{\pi_{d-s}})$ for $j \notin \pi$ into $\log(\mathbf{z}^{\mathbf{r}}/\mathbf{w}^{\mathbf{r}})$ and then switching to exponential coordinates near $\mathbf{w}$. In order to use saddle point techniques in later chapters, we verify the following.

Lemma 8.21. *If $\mathbf{w}$ is a critical point of S then $\phi(\mathbf{0}) = 0$ and $(\nabla\phi)(\mathbf{0}) = \mathbf{0}$.*

Proof By construction $g_j(w_{\pi_1}, \ldots, w_{\pi_{d-s}}) = w_j$ for all $j \notin \pi$, so $\phi(\mathbf{0}) = \sum_{j \notin \pi} r_j \log(w_j/w_j) = 0$.

By permuting variables if necessary we may assume, without loss of generality, that $(\sigma_1, \ldots, \sigma_s) = (1, \ldots, s)$ and, to simplify notation, we write $\hat{\mathbf{z}} = (z_{s+1}, \ldots, z_d)$. Since, by construction,

$$H_1(g_1(\hat{\mathbf{z}}), \ldots, g_s(\hat{\mathbf{z}}), \hat{\mathbf{z}}) = \cdots = H_s(g_1(\hat{\mathbf{z}}), \ldots, g_s(\hat{\mathbf{z}}), \hat{\mathbf{z}}) = 0$$

for all $\hat{\mathbf{z}}$ sufficiently close to $(w_{s+1}, \ldots, w_d)$, we can differentiate these equations with respect to z_{s+1} and use the implicit function theorem to get the linear system

$$\underbrace{\begin{pmatrix} H_1^{[1]}(\boldsymbol{w}) & \cdots & H_1^{[s]}(\boldsymbol{w}) \\ \vdots & \ddots & \vdots \\ H_s^{[1]}(\boldsymbol{w}) & \cdots & H_s^{[s]}(\boldsymbol{w}) \end{pmatrix}}_{M} \begin{pmatrix} g_1^{[s+1]}(\hat{\mathbf{w}}) \\ \vdots \\ g_s^{[s+1]}(\hat{\mathbf{w}}) \end{pmatrix} = -\underbrace{\begin{pmatrix} H_1^{[s+1]}(\boldsymbol{w}) \\ \vdots \\ H_s^{[s+1]}(\boldsymbol{w}) \end{pmatrix}}_{b}.$$

Because M is nonsingular, if we let M_k denote M after replacing its kth column with the vector b then Cramer's rule implies

$$g_k^{[s+1]}(\hat{\mathbf{w}}) = -\frac{\det(M_k)}{\det(M)}$$

for all $1 \leq k \leq s$. If $\boldsymbol{w}$ is a critical point on S then the matrix

$$\begin{pmatrix} & (\nabla_{\log} H_1)(\boldsymbol{w}) & \\ & \vdots & \\ & (\nabla_{\log} H_s)(\boldsymbol{w}) & \\ r_1 & \cdots & r_d \end{pmatrix}$$

has rank s, so the determinant of the $(s+1) \times (s+1)$ sub-matrix

$$\begin{pmatrix} H_1^{[1]}(\boldsymbol{w}) & \cdots & H_1^{[s]}(\boldsymbol{w}) & H_1^{[s+1]}(\boldsymbol{w}) \\ \vdots & \ddots & \vdots & \vdots \\ H_s^{[1]}(\boldsymbol{w}) & \cdots & H_s^{[s]}(\boldsymbol{w}) & H_s^{[s+1]}(\boldsymbol{w}) \\ \frac{r_1}{w_1} & \cdots & \frac{r_s}{w_s} & \frac{r_{s+1}}{w_{s+1}} \end{pmatrix}$$

vanishes. A cofactor expansion of the determinant along the bottom row gives, after carefully accounting for signs, that

$$0 = \frac{r_{s+1}}{w_{s+1}} \det(M) - \sum_{j=1}^{s} \frac{r_j}{w_j} \det(M_k),$$

so

$$\sum_{j=1}^{s} \frac{r_j}{w_j} g_j^{[s+1]}(\hat{\mathbf{w}}) = -\frac{r_{s+1}}{w_{s+1}}$$

and thus

$$\left(\frac{\partial \phi}{\partial \theta_{s+1}}\right)(\mathbf{0}) = i\sum_{j=1}^{s} \frac{r_j}{w_j} g_j^{[s+1]}(\hat{\mathbf{w}}) w_{s+1} + i r_{s+1} = 0.$$

Repeating this argument with indices $s+2, \ldots, d$ shows that the gradient of ϕ vanishes at $\boldsymbol{\theta} = \mathbf{0}$. □

Recall from Definitions 7.11 and 7.34 that a nondegenerate critical point $\boldsymbol{w}$ is a critical point where the Hessian of ϕ at the origin is nonsingular; integrals near nondegenerate critical points are those to which we can apply the techniques of Chapter 5. Although the Hessian matrix can always be computed in terms of derivatives of the g_j, which themselves can be determined through implicit differentiation or the determinant expressions derived in our proof of Lemma 8.21, in the smooth case things are simple enough to derive an explicit expression for the Hessian matrix itself.

Lemma 8.22. *Suppose $\boldsymbol{w}$ is a critical point where $H_{z_d}(\boldsymbol{w}) \neq 0$ and let*

$$\phi(\boldsymbol{\theta}) = \log\left(\frac{g\left(\boldsymbol{w}^{\circ} e^{i\boldsymbol{\theta}}\right)}{w_d}\right) + i\frac{(\boldsymbol{r}^{\circ} \cdot \boldsymbol{\theta})}{r_d}$$

in the variables $\boldsymbol{\theta} = (\theta_1, \ldots, \theta_{d-1})$, where $\left(\boldsymbol{w}^{\circ} e^{i\boldsymbol{\theta}}\right) = \left(w_1 e^{i\theta_1}, \ldots, w_{d-1} e^{i\theta_{d-1}}\right)$. Then the $(d-1) \times (d-1)$ Hessian matrix of ϕ at $\boldsymbol{\theta} = \mathbf{0}$ has (i,j)th entry

$$\mathcal{H}_{i,j} = \begin{cases} V_i V_j + U_{i,j} - V_j U_{i,d} - V_i U_{j,d} + V_i V_j U_{d,d} & : i \neq j \\ V_i + V_i^2 + U_{i,i} - 2V_i U_{i,d} + V_i^2 U_{d,d} & : i = j \end{cases}, \tag{8.7}$$

where

$$U_{i,j} = \frac{w_i w_j H_{z_i z_j}(\boldsymbol{w})}{w_d H_{z_d}(\boldsymbol{w})} \quad \text{and} \quad V_i = \frac{r_i}{r_d} \tag{8.8}$$

for all $1 \leq j \leq d$. If, more generally, $H_{z_k}(\boldsymbol{w}) \neq 0$ then the same formula holds when d is replaced by k.

Exercise 8.4. Use the chain rule to prove Lemma 8.22.

8.4.2 Numeric analytic continuation of D-finite functions

Recall from Chapter 2 that a *D-finite equation* is a linear differential equation with polynomial coefficients, and a D-finite function is one that satisfies a D-finite equation. Because the diagonal of any multivariate rational function is D-finite, we can use D-finite equations as data structures to manipulate and compute with diagonals. Given a multivariate rational function with a convergent power series near the origin, an annihilating D-finite equation for the diagonal in any fixed integer direction $\boldsymbol{r} \in \mathbb{Z}^d$ can be computed using the methods of *creative telescoping*. Briefly, one represents the diagonal $D(z)$ as a parameterized complex integral $D(z) = \int_\gamma I(z, \boldsymbol{x})d\boldsymbol{x}$ over a closed contour γ, with the integrand I constructed from the rational function F and direction $\boldsymbol{r}$. If $\mathcal{L}$ is a linear differential operator in z with polynomial coefficients such that $\mathcal{L}(I) = \partial_{x_1} A_1(z, \boldsymbol{x}) + \cdots + \partial_{x_d} A_d(z, \boldsymbol{x})$ for suitable rational functions $A_1, \ldots, A_d$ then, under mild conditions, the fact that γ is closed implies that $D(z)$ satisfies the D-finite equation $\mathcal{L}(D) = 0$. Methods to compute such differential operators have been the source of immense study in the computer algebra community, with currently optimal techniques relying on a reduction framework known as the *Griffiths–Dwork method* [BLS13; Lai16]. An implementation of this method in the MAGMA computer algebra system was created by Lairez [Lai16].

Consider a D-finite equation

$$c_r(z)f^{(r)}(z) + \cdots + c_1(z)f'(z) + c_0(z)f(z) = 0, \tag{8.9}$$

which admits a generating function F of interest as a solution $f(z) = F(z)$, and assume that F is analytic at the origin. The points $\rho \in \mathbb{C}$ where the leading coefficient $c_r(z)$ vanishes are called ***singular points of the ODE*** (8.9), and nonsingular points are called ***ordinary points of the ODE***. Because D-finite equations are linear, the solutions to (8.9) form an r-dimensional complex vector space, and the Cauchy existence theorem for ordinary differential equations [Poo60, Chapter 1.2] implies that in any sufficiently small neighborhood of an ordinary point there exists a basis of analytic solutions to (8.9).

A collection of deep results due to André, the Chudnovsky brothers, and Katz jointly imply that for a D-finite function with integer coefficients and at most exponential growth (i.e., the types of generating functions we consider), a minimal order defining equation admits only *regular* singular points; see [And89, Section VI]. Skipping over the definition of a regular singular point, this is important because if ρ is a regular singular point then in any sufficiently small slit disk centered at ρ (a disk with a ray from its center to its boundary removed to account for branch cuts) there is a basis of solutions of the form

$$(z-\rho)^\nu \sum_{k=0}^{K} \phi_k(z) \log^k\left(\frac{1}{1-z/\rho}\right),$$

where ν is algebraic and each ϕ_k is analytic at ρ; see [Poo60, Chapter 5] for details.

To determine asymptotic behavior of the power series coefficients of F we need to represent F in terms of the local basis expansions near the singular points of (8.9) (which can contain the singularities of F, the singularities of other solutions to the differential equation, and spurious points). In practice, this is accomplished using the series expansion of F at the origin to express it in terms of a basis near the origin, then approximating change of basis matrices from analytically continuing series centered at the origin to series centered at the singular points of (8.9). Using ideas going back to Cauchy's method of majorants, implemented by Mezzarobba [Mez16] in the Sage `ore_algebra` package, it is possible to rigorously compute numerical approximations of the entries of the change of basis matrices to any desired accuracy.

Ultimately, one can use these techniques to compute expansions of F near the singularities of (8.9) with rigorously approximated coefficients. This can often be used to identify the dominant singularities of F, and compute asymptotics of its power series coefficients up to constants that are rigorously approximated. Unfortunately, it is an open problem to identify when approximated coefficients that are zero to many decimal places are identically zero. This *connection problem* means that it is not always possible to identify the dominant asymptotics of a D-finite function directly from a D-finite equation it satisfies (although using current implementations one can get the relevant constants to thousands of decimal places, so even if a rigorous proof can't be found very strong heuristic evidence is always possible).

Example 8.23. Let a_n denote the number of lattice paths with the set of allowable steps $\{(-1,-1),(1,-1),(-1,1),(1,1)\}$ that start at the origin and stay in the nonnegative quadrant $\mathbb{N}^2$. The *kernel method* [Mel21, Chapter 4] implies that the generating function $A(t)$ of (a_n) satisfies the D-finite equation

$$\begin{aligned} 0 &= t^2(4t+1)(1-4t)^2A'''(t) + t(1-4t)(112t^2-5)A''(t) \\ &\quad + 4(8t-1)(20t^2-3t-1)A'(t) + (128t^2-48t-4)A(t). \end{aligned}$$

The dominant singularities of A occur at $t = \pm 1/4$, and the Sage package of Mezzarobba computes expansions

$$\begin{aligned} A(t) &= \left(\left[-0.636 \pm 3\cdot 10^{-3}\right] + \left[\pm 4\cdot 10^{-12}\right] i\right)\, \log(t-1/4) \\ &\quad + \left(\left[2.54 \pm 7\cdot 10^{-3}\right] + \left[\pm 1.56\cdot 10^{-11}\right] i\right)\, (t-1/4)\log(t-1/4) + \cdots \end{aligned}$$

near $t = 1/4$, where the notation $[a \pm b]$ denotes a constant in the interval $[a-b, a+b]$, and

$$A(t) = \left[1 \pm 7\cdot 10^{-18}\right] + \left[1.273 \pm 3\cdot 10^{-3}\right]\, (t-1/4)\log(t-1/4) + \cdots$$

near $t = -1/4$. Translating these singular expansions into asymptotic behavior implies

$$a_n = \frac{4^n}{n}\left(\left[0.636 \pm 3 \cdot 10^{-3}\right] + O\left(n^{-1}\right)\right).$$

The constants here are given to three decimal places for readability, but can be computed to hundreds of decimal places in under a second on modern computers. ◁

Example 8.24. Let b_n denote the number of lattice paths with the set of allowable steps $\{(0,1),(-1,-1),(0,-1),(1,-1)\}$ that start at the origin and stay in the nonnegative quadrant $\mathbb{N}^2$. The kernel method now implies that the generating function $B(t)$ of (b_n) satisfies a D-finite equation of order 3 whose leading coefficient has a root at $t = 1/4$, and no non-zero roots closer to the origin, suggesting that this might be a dominant singularity of $B(t)$. The Sage package of Mezzarobba can be used to compute an expansion

$$B(t) = \left[\pm 10^{-1000}\right](t - 1/4)^{-1/2} + \cdots$$

near $t = 1/4$, whose leading coefficient is zero to 1000 decimal places. In fact, although $t = 1/4$ is a singular point of this ODE satisfied by B, it is not a singularity of B. The dominant singularities of $B(t)$ occur at $t = \pm\sqrt{3}/6$ and $b_n \sim c_n(2\sqrt{3})^n n^{-2}$ where c_n takes one value when n is even, and another when n is odd, but this cannot be proven without using additional information beyond properties of this ODE. ◁

Asymptotics for many lattice path problems, including those in Examples 8.23 and 8.24, can be derived using ACSV [Mel21, Chapters 6 and 10].

Notes

Modern methods for solving polynomial systems can be traced back at least to work of Newton in the seventeenth century. After the work of Newton, and later work of Cramer and Bézout bounding the number of solutions to multivariate polynomial systems, the nineteenth century work of Sylvester illustrated the power of the resultant in effectively solving polynomial systems. Further work by Macaulay and his contemporaries developed a mature theory of multivariate resultants for use beyond bivariate systems; see [GKZ08] for a modern and far-reaching generalization of these techniques. Gröbner bases were introduced by Buchberger [Buc65] in his 1965 PhD thesis, and form the backbone of much modern work in computer algebra. The theory of effective methods for algebraic geometry is eloquently presented in the splendid volumes [CLO07;

CLO05] and the monograph [Stu02], and further historical details can be found in the survey [Cox20].

The theory of D-finite functions has a long history, going back at least to Abel. Major contributions were made by Fuchs, Frobenius, and G. D. Birkhoff, among others. Their use in combinatorial applications was pioneered by Zeilberger. The book [Mel21] gives further details on computational approaches to D-finite generating functions.

We have tried to present this book in such a way as to ensure a long shelf-life. In the present chapter it was useful for concreteness to mention specific implementations, even though these implementations may no longer be the state of the art in the future. Nevertheless, the specific packages we mention, in addition to the entire Sage computer algebra system, have open source code and practice modern version control and archiving. We thus expect they will prove useful for a long time to come.

Additional exercises

Exercise 8.5. Let R be an integral domain. The *resultant* of two polynomials $P(x) = p_0 + \cdots + p_r x^r$ and $Q(x) = q_0 + \cdots + q_s x^s$ in $R[x]$ with $p_r q_s \neq 0$ is a polynomial $\text{res}_x(p, q)$ in $p_0, \dots, p_r, q_0, \dots, q_r$ that vanishes if and only if P and Q share a root in the algebraic closure of R. Let α and β be algebraic numbers with $P(\alpha) = 0$ and $Q(\beta) = 0$ for $P, Q \in \mathbb{Z}[x]$. Find annihilating polynomials in $\mathbb{Z}[x]$ for the sum $\alpha + \beta$, the product $\alpha\beta$, and, when $\beta \neq 0$, the quotient α/β in terms of the resultants of explicit polynomials depending on P and Q.

Exercise 8.6. Suppose that a set of algebraic equations has a single solution $\boldsymbol{z} = (z_1, \dots, z_d)$. Prove that $\boldsymbol{z}$ is a rational point by showing that each coordinate z_j is the solution to a univariate algebraic equation having only one solution. Similarly, prove that if a zero-dimensional variety consists of two points $\{\boldsymbol{z}, \boldsymbol{w}\}$ then either it is reducible and the two points are rational, or it is irreducible and both are quadratic and algebraically conjugate.

9

Smooth point asymptotics

After discussing the overall framework of ACSV in Chapter 7, and the computational tools needed to carry out the analysis in Chapter 8, we are now ready to prove asymptotic theorems. As usual, we begin with a convergent Laurent expansion $F(\boldsymbol{z}) = \sum_{\boldsymbol{r}\in\mathbb{Z}^d} a_{\boldsymbol{r}} \boldsymbol{z}^{\boldsymbol{r}}$ in some domain $\mathcal{D} \subset \mathbb{C}^d$ and try to determine asymptotic behavior of $a_{\boldsymbol{r}}$ as $\boldsymbol{r} \to \infty$ with the normalized vector $\hat{\boldsymbol{r}} = \frac{\boldsymbol{r}}{|\boldsymbol{r}|} = \frac{\boldsymbol{r}}{|r_1|+\cdots+|r_d|}$ restricted to compact sets. In this chapter we give results when dominant asymptotic behavior is determined by the local behavior of F near a finite set of points where its set of singularities $\mathcal{V}$ forms a manifold. Typically we assume F is rational, although we also state results when F is meromorphic.

Remark 9.1. The smoothness assumption of this chapter is *generic*, meaning (for instance) that it holds for all rational functions except for those whose coefficients lie in a fixed proper algebraic set depending only on the degree of the denominator. Although this might suggest that every example encountered in practice is handled by the techniques of this chapter, non-generic behavior does occur in many combinatorial applications. Nonetheless, a large fraction of the multivariate generating functions encountered by the authors can be handled by the techniques presented here, without going into the more general theory of Chapters 10 and 11.

The Main Results of Smooth ACSV

We begin by stating the main theorems of this chapter. Let $F(\boldsymbol{z}) = P(\boldsymbol{z})/Q(\boldsymbol{z})$ be the ratio of coprime polynomials, where $Q \in \mathbb{C}[\boldsymbol{z}]$ has square-free part $\tilde{Q}$ (equal to the product of its distinct irreducible factors). Recall from past chapters that $\boldsymbol{w} \in \mathbb{C}_*^d$ is a *smooth critical point* for the direction $\hat{\boldsymbol{r}} \in \mathbb{R}^d$ if and only if $(\nabla\tilde{Q})(\boldsymbol{w}) \neq \mathbf{0}$ and

$$\tilde{Q}(\boldsymbol{w}) = \hat{r}_i w_d \tilde{Q}_{z_d}(\boldsymbol{w}) - \hat{r}_d w_i \tilde{Q}_{z_i}(\boldsymbol{w}) = 0 \qquad (1 \le i \le d-1). \tag{9.1}$$

The case when the direction vector $\hat{r}$ is the zero vector is trivial, so we always assume that $\hat{r}$ has a non-zero coordinate. When the series expansion of F under consideration is a power series we can further assume the stronger condition that $\hat{r}$ has no zero coordinates, because asking for terms where (say) $r_d = 0$ corresponds to extracting terms from the $(d-1)$-variate series obtained by setting $z_d = 0$. In this case, because our results hold only for critical points with non-zero coordinates, the smooth critical point equations imply that *none* of the partial derivatives of $\tilde{Q}$ vanish.

For Laurent expansions, on the other hand, there are combinatorially interesting cases where $\hat{r}$ has zero coordinates. Even so, if there are to be smooth critical points with non-zero coordinates then the critical point equations imply the existence of a coordinate k such that $r_k \neq 0$ and $\tilde{Q}_{z_k}(\boldsymbol{w}) \neq 0$. Without loss of generality, we may assume this coordinate k is the final coordinate d.

Consider a Laurent expansion of F with domain of convergence $\mathcal{D}$. Theorem 6.44 from Chapter 6 implies that if $\boldsymbol{w}$ is a smooth minimal critical point (see Definition 7.7) for the direction $\hat{r}$ then the hyperplane with normal $\hat{r}$ going through the point $\text{Relog}(\boldsymbol{w})$ is a support hyperplane to $B = \text{Relog}(\mathcal{D})$.

Definition 9.2 (contributing and nondegenerate points)**.** The smooth minimal critical point $\boldsymbol{w}$ described above is called a ***contributing point*** for the direction $\hat{r}$ if $\hat{r}$ points away from B at $\text{Relog}(\boldsymbol{w})$, meaning $\boldsymbol{x} \cdot \hat{r} < \text{Relog}(\boldsymbol{w}) \cdot \hat{r}$ for all $\boldsymbol{x} \in B$. Recall that the point $\boldsymbol{w}$ is nondegenerate if the Hessian matrix $\mathcal{H}$ defined by Lemma 8.22 in Chapter 8 is nonsingular with $H = \tilde{Q}$.

Remark 9.3. If we consider the power series expansion of $F(\boldsymbol{z})$, where $\hat{r}$ has positive coordinates, then every smooth minimal critical point is contributing. Recall that nondegeneracy is equivalent to previous definitions in terms of the Hessian of the height function $h(\boldsymbol{z}) = -\boldsymbol{r} \cdot \log \boldsymbol{z}$ restricted to $\mathcal{V}$.

Definition 9.2 is constructed so that contributing points are minimizers of the height function $h_{\hat{r}}$ on $\overline{\mathcal{D}}$, which turn out to be the points determining asymptotic behavior. Conversely, non-contributing smooth minimal critical points are *maximizers* of $h_{\hat{r}}$ on $\overline{\mathcal{D}}$; see Figure 9.1.

Exercise 9.1. Which of the components of the complement of the amoeba in Figure 9.1 have a contributing point in the direction $(1, -1)$?

We break our main result into three versions, depending on the assumptions required and the proof techniques used. Our first version is the most restrictive, however it still holds in a wide variety of applications and has the advantage that it can be derived purely through complex analysis and classical saddle point techniques, without the need for the homological framework of

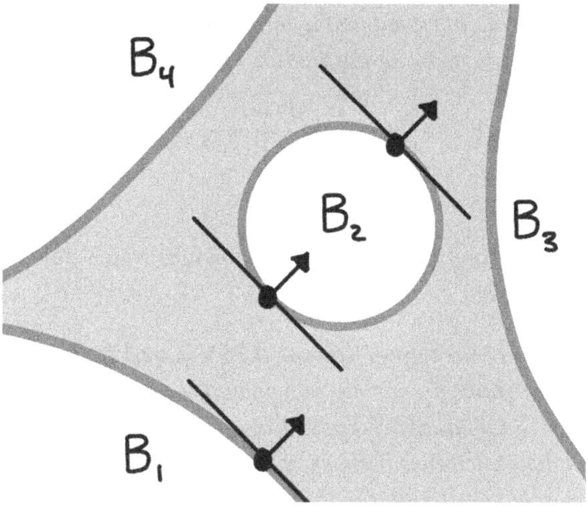

Figure 9.1 The amoeba complement component B_1, corresponding to a power series expansion, has one point on its boundary where its support hyperplane has normal $\boldsymbol{r} = (1, 1)$, which corresponds to contributing points. On the other hand, the component B_2 has two boundary points with support hyperplanes having normals $\boldsymbol{r} = (1, 1)$, only one of which (the upper-most one) corresponds to contributing singularities.

Chapter 7. In order to simplify our presentation, we begin by stating it in the common special case where Q is square-free and there is a single minimal contributing point.

Theorem 9.4 (Main Theorem of Smooth ACSV (Local Version, Square-Free Case)). *Let $F(\boldsymbol{z}) = P(\boldsymbol{z})/Q(\boldsymbol{z})$ be the ratio of coprime polynomials P and Q with convergent Laurent expansion $F(\boldsymbol{z}) = \sum_{\boldsymbol{r}\in\mathbb{Z}^d} a_{\boldsymbol{r}} \boldsymbol{z}^{\boldsymbol{r}}$. Suppose that there is a compact set $\mathcal{R} \subset \mathbb{R}^d$ of non-zero directions such that if $\hat{\boldsymbol{r}}$ lies in $\mathcal{R}$ then F has a smooth strictly minimal nondegenerate contributing point $\boldsymbol{w} = \boldsymbol{w}(\hat{\boldsymbol{r}}) \in \mathbb{C}_*^d$, and let $\mathcal{H} = \mathcal{H}(\hat{\boldsymbol{r}})$ be the Hessian matrix defined by (8.7) and (8.8) when $H = Q$. If $Q_{z_d}(\boldsymbol{w}) \neq 0$ then*

$$a_{\boldsymbol{r}} \approx \Phi_{\boldsymbol{w}}(\boldsymbol{r})$$

uniformly as $\boldsymbol{r} \to \infty$ with $\hat{\boldsymbol{r}} \in \mathcal{R}$, where $\Phi_{\boldsymbol{w}}(\boldsymbol{r})$ is an asymptotic series

$$\Phi_{\boldsymbol{w}}(\boldsymbol{r}) = \boldsymbol{w}^{-\boldsymbol{r}} |r_d|^{(1-d)/2} \frac{(2\pi)^{(1-d)/2} \operatorname{sgn}(r_d)}{\sqrt{\det(\operatorname{sgn}(r_d)\mathcal{H})}} \sum_{\ell=0}^{\infty} C_\ell(\hat{\boldsymbol{r}}) r_d^{-\ell}. \tag{9.2}$$

The square-root of the matrix determinant is the product of the principal branch square-roots of its eigenvalues (which will have nonnegative real parts). The *constants C_ℓ are explicitly computable in terms of the derivatives of $P(\boldsymbol{z})$ and $Q(\boldsymbol{z})$ evaluated at $\boldsymbol{z} = \boldsymbol{w}(\hat{\boldsymbol{r}})$. In particular,*

$$C_0 = \frac{P(\boldsymbol{w})}{-w_d Q_d(\boldsymbol{w})}. \tag{9.3}$$

□

Theorem 9.4 is a special case of the following, which holds for poles of general order.

Theorem 9.5 (Main Theorem of Smooth ACSV (Local Version)). *Let $F(\boldsymbol{z}) = P(\boldsymbol{z})/Q(\boldsymbol{z})$ be the ratio of coprime polynomials with convergent Laurent expansion $F(\boldsymbol{z}) = \sum_{\boldsymbol{r}\in\mathbb{Z}^d} a_{\boldsymbol{r}} \boldsymbol{z}^{\boldsymbol{r}}$. Suppose there exists a compact set $\mathcal{R} \subset \mathbb{R}^d$ of non-zero directions such that if $\hat{\boldsymbol{r}}$ lies in $\mathcal{R}$ then F has a smooth strictly minimal nondegenerate contributing point $\boldsymbol{w} = \boldsymbol{w}(\hat{\boldsymbol{r}}) \in \mathbb{C}_*^d$, and let $\mathcal{H} = \mathcal{H}(\hat{\boldsymbol{r}})$ be the Hessian matrix defined by* (8.7) *and* (8.8) *when $H = \tilde{Q}$. If $(\partial_d^p Q)(\boldsymbol{w}) \neq 0$ and $(\partial_d^q Q)(\boldsymbol{w}) = 0$ for all $0 \leq q < p$ then*

$$a_{\boldsymbol{r}} \approx \Phi_{\boldsymbol{w}}(\boldsymbol{r})$$

uniformly as $\boldsymbol{r} \to \infty$ with $\hat{\boldsymbol{r}} \in \mathcal{R}$, where $\Phi_{\boldsymbol{w}}(\boldsymbol{r})$ is an asymptotic series

$$\Phi_{\boldsymbol{w}}(\boldsymbol{r}) = \boldsymbol{w}^{-\boldsymbol{r}} |r_d|^{p-1+(1-d)/2} \frac{(2\pi)^{(1-d)/2}\,\mathrm{sgn}(r_d)^p}{\sqrt{\det(\mathrm{sgn}(r_d)\mathcal{H})}} \sum_{\ell=0}^{\infty} C_\ell(\hat{\boldsymbol{r}}) r_d^{-\ell}. \tag{9.4}$$

The square-root of the matrix determinant is the product of the principal branch square-roots of its eigenvalues (which will have nonnegative real parts). The constants C_ℓ are explicitly computable in terms of the derivatives of $P(\boldsymbol{z})$ and $Q(\boldsymbol{z})$ evaluated at $\boldsymbol{z} = \boldsymbol{w}(\hat{\boldsymbol{r}})$. In particular,

$$C_0 = \frac{(-1)^p P(\boldsymbol{w}) p}{w_d^p (\partial_d^p Q)(\boldsymbol{w})}.$$

If $\boldsymbol{w}$ is a finitely minimal point (instead of being strictly minimal) such that all points in the set $\mathbf{W}(\hat{\boldsymbol{r}}) = T(\boldsymbol{w}) \cap \mathcal{V}$ vary smoothly with $\hat{\boldsymbol{r}}$ in $\mathcal{R}$ and are contributing points satisfying the conditions above then

$$a_{\boldsymbol{r}} \approx \sum_{\boldsymbol{y}\in\mathbf{W}(\hat{\boldsymbol{r}})} \Phi_{\boldsymbol{y}}(\boldsymbol{r}),$$

where each $\Phi_{\boldsymbol{y}}$ is given by (9.4).

Exercise 9.2. What, in general, can go wrong pulling a factor of $\mathrm{sgn}(r_d)^{d-1}$ out of the square-root in the denominator of (9.4)?

Remark 9.6. An explicit (but unwieldy) formula for all coefficients in (9.4) is given in Section 9.4 below. If $Q = H^p$ for some square-free H with ∇H nonvanishing at $\boldsymbol{w}$ then

$$C_0 = \frac{(-1)^p P(\boldsymbol{w})}{(p-1)!\,(w_d \partial_d H(\boldsymbol{w}))^p}.$$

Remark 9.7. Our surgery approach below singles out the coordinate z_d for a residue computation, leading to an asymptotic expansion in powers of the non-zero coordinate r_d. With some extra work, the Fourier–Laplace integral used to deduce asymptotics can be modified to provide an asymptotic series in powers of $|\boldsymbol{r}|$, giving an expansion of the form

$$\Phi_{\boldsymbol{w}}(\boldsymbol{r}) = \boldsymbol{w}^{-\boldsymbol{r}} \frac{1}{\sqrt{\det(2\pi|\boldsymbol{r}|\mathcal{H}')}} \sum_{\ell=0}^{\infty} C'_\ell(\hat{\boldsymbol{r}})|\boldsymbol{r}|^{-\ell} \tag{9.5}$$

for a new Hessian matrix $\mathcal{H}'$. We leave details of such symmetric formulae to Chapter 10, where asymptotics are computed using multivariate residue forms that do not privilege individual coordinates.

Remark 9.8. If $\mathcal{R} = \{\boldsymbol{s}\}$ contains a single point with $s_d > 0$ then $\boldsymbol{r} = n\boldsymbol{s}$ and

$$\Phi_{\boldsymbol{w}}(n\boldsymbol{s}) \approx \boldsymbol{w}^{-n\boldsymbol{s}}\, n^{p-1+(1-d)/2} \frac{(2\pi)^{(1-d)/2}}{\sqrt{\det(\mathcal{H})}} \sum_{\ell=0}^{\infty} D_\ell n^{-\ell}$$

for constants D_ℓ with $D_0 = (s_d)^{p-d} C_0$.

Example 9.9. The hypotheses of Theorem 9.5 can be simplified for bivariate power series. In particular, suppose that $F(x,y) = \frac{P(x,y)}{Q(x,y)} = \sum_{i,j\geq 0} a_{ij} x^i y^j$ admits a strictly minimal critical point $\boldsymbol{w} = \boldsymbol{w}(\boldsymbol{r}) \in \mathbb{C}_*^2$ that varies smoothly as $\hat{\boldsymbol{r}}$ varies in a compact neighborhood $\mathcal{R}$ of directions. If both $P(x,y)$ and the expression

$$\underline{Q}(x,y) = -xy^2 Q_y^2 Q_x - x^2 y Q_y Q_x - x^2 y^2 (Q_y^2 Q_{xx} + Q_x^2 Q_{yy} - 2Q_x Q_y Q_{xy}) \tag{9.6}$$

are non-zero when $(x,y) = \boldsymbol{w}(\hat{\boldsymbol{r}})$ for each $\hat{\boldsymbol{r}} \in \mathcal{R}$ then

$$\begin{aligned} a_{r,s} &\sim \frac{P(x,y)}{-yQ_y} \frac{1}{\sqrt{2\pi}} x^{-r} y^{-s} \sqrt{\frac{(-yQ_y)^3}{s\underline{Q}}} \\ &= \frac{P(x,y)}{-xQ_x} \frac{1}{\sqrt{2\pi}} x^{-r} y^{-s} \sqrt{\frac{(-xQ_x)^3}{r\underline{Q}}} \end{aligned} \tag{9.7}$$

as $|\boldsymbol{r}| \to \infty$ after setting $(x,y) = \boldsymbol{w}(\hat{\boldsymbol{r}})$, uniformly over $\hat{\boldsymbol{r}} \in \mathcal{R}$. ◁

Exercise 9.3. Prove (9.7) by simplifying (9.2) in the bivariate case.

Example 9.10 (binomial coefficients continued)**.** If $F(x, y) = 1/(1-x-y)$ then the coefficient of $x^r y^s$ in the power series expansion of F is $\binom{r+s}{s}$. Solving the smooth critical point equations yields the unique critical point

$$\boldsymbol{w} = \left(\frac{r}{r+s}, \frac{s}{r+s}\right) = (\hat{r}, \hat{s}),$$

which is strictly minimal by Lemma 6.41. We obtain

$$\binom{r+s}{s} \sim \frac{(r+s)^{r+s}}{r^r s^s} \sqrt{\frac{r+s}{2\pi rs}}$$

as $r, s \to \infty$ with r/s bounded away from zero and infinity. For example, the central binomial coefficients given by $r = s = n$ satisfy $\binom{2n}{n} \sim 4^n / \sqrt{\pi n}$. ◁

Example 9.11 (Delannoy numbers continued)**.** If $F(x, y) = 1/(1 - x - y - xy)$ then we have the critical points

$$(x_*, y_*) = \left(\frac{\sqrt{r^2+s^2} - s}{r}, \frac{\sqrt{r^2+s^2} - r}{s}\right) \text{ or } \left(\frac{-\sqrt{r^2+s^2} - s}{r}, \frac{-\sqrt{r^2+s^2} - r}{s}\right),$$

the first of which is strictly minimal by Lemma 6.41. Writing $d = \sqrt{r^2 + s^2}$, we directly compute

$$a_{rs} \sim \left(\frac{r}{d-s}\right)^r \left(\frac{s}{d-r}\right)^s \sqrt{\frac{rs\,(d^2 + (r-s))}{2\pi\,(r+s-d)\,(d^2 + d(r-s))}}$$

as $r, s \to \infty$ with r/s bounded away from zero and infinity. ◁

We give a proof of Theorem 9.5 in Section 9.1 using the *surgery method* for ACSV, which works in the presence of smooth finitely minimal contributing points. Although the requirement of finite minimality makes proofs simpler, it is computationally difficult to check, and rules out cases that can be handled by our other results. In Section 9.2 we use more advanced techniques (including the theory of *hyperbolic polynomials* developed in Chapter 11) to prove an extension of Theorem 9.5 that ignores non-critical points and only requires that the torus $T(\boldsymbol{w})$ contains a *finite number of critical points.* This gives a large computational advantage, because generically there are a finite number of critical points described by a zero-dimensional algebraic set.

Theorem 9.12 (Main Theorem of Smooth ACSV (Minimal Point Version))**.** *Let $F(\boldsymbol{z}) = P(\boldsymbol{z})/Q(\boldsymbol{z})$ be the ratio of coprime polynomials with convergent Laurent expansion $F(\boldsymbol{z}) = \sum_{\boldsymbol{r}\in\mathbb{Z}^d} a_{\boldsymbol{r}} \boldsymbol{z}^{\boldsymbol{r}}$. Suppose there exists a compact set $\mathcal{R} \subset \mathbb{R}^d$ of non-zero directions such that F has a smooth minimal nondegenerate contributing point $\boldsymbol{w} = \boldsymbol{w}(\hat{\boldsymbol{r}}) \in \mathbb{C}_*^d$ whenever $\hat{\boldsymbol{r}} \in \mathcal{R}$. If the set $\mathbf{W}(\hat{\boldsymbol{r}})$ of solutions* *to* (9.1) *with the same coordinatewise modulus as* $\boldsymbol{w}(\hat{\boldsymbol{r}})$ *is finite and contains only smooth nondegenerate contributing points that vary smoothly with* $\hat{\boldsymbol{r}}$ *then*

$$a_{\boldsymbol{r}} \approx \sum_{\boldsymbol{y} \in \mathbf{W}(\hat{\boldsymbol{r}})} \Phi_{\boldsymbol{y}}(\boldsymbol{r}), \tag{9.8}$$

uniformly as $\boldsymbol{r} \to \infty$ *with* $\hat{\boldsymbol{r}} \in \mathcal{R}$, *where* $\Phi_{\boldsymbol{y}}$ *is defined in* (9.4).

Example 9.13 (negative binomial coefficients)**.** If $F(x, y) = -x/(1-x-y)$ then the coefficient of $x^{-r}y^s$ in the Laurent series expansion of F converging in the domain $1 + |y| < |x|$ is $(-1)^s\binom{r}{s}$. There is a unique critical point

$$\boldsymbol{w} = \left(\frac{-r}{-r+s}, \frac{s}{-r+s}\right),$$

where now, because $r > s$, the first coordinate of $\boldsymbol{w}$ is positive while the second is negative. This point is minimal, since it lives on the boundary $\{(x, y) \in \mathbb{C}^2 : 1 + |y| = |x|\}$ of the domain of convergence of this Laurent series, and contributing. Ultimately, we obtain

$$[x^{-r}y^s]F \sim (-1)^s \frac{r^r}{(r-s)^{r-s}s^s}\sqrt{\frac{r}{2\pi(r-s)s}}.$$

Note that if we replace x by $1/x$ in F, we obtain $G = 1/(1 - x + xy)$, whose (r, s)-coefficient is $(-1)^s\binom{r}{s}$. This is consistent with the usual identity

$$\binom{-r}{s} = (-1)^s\binom{r+s-1}{s}$$

for binomial coefficients when $r, s > 0$. Replacing y by $-y$, we are led back to the generating function $1/(1 - x - xy)$ for binomial coefficients examined above. ◁

Example 9.14 (Chebyshev polynomials)**.** Let $F(z, w) = 1/(1-2zw+w^2)$ be the generating function for ***Chebyshev polynomials*** of the second kind [Com74]. To use Theorem 9.12 for an arbitrary direction (r, s) with nonnegative indices and $r/s \in (0, 1)$, we first compute the critical points $\boldsymbol{w}_\pm = \left(i\left(\beta - \beta^{-1}\right)/2, i\beta\right)$, where $\beta = \pm\sqrt{\frac{s-r}{s+r}}$. These points are minimal by Corollary 6.36 because if we substitute $(z, w) = (tx, ty)$ in the denominator then $|2xy - y^2|$ is at most $t^2\left(1 - \beta^2 + \beta^2\right) < 1$, and hence $T(t\boldsymbol{w}_\pm) \cap \mathcal{V} = \varnothing$ for all $t \in (0, 1)$. These points are contributing because any smooth minimal critical points are contributing for power series expansions.

Summing the asymptotic contributions given by the two points implies

$$a_{rs} \sim \sqrt{\frac{2}{\pi}}(-1)^{(s-r)/2}\left(\frac{2r}{\sqrt{s^2-r^2}}\right)^{-r}\left(\sqrt{\frac{s-r}{s+r}}\right)^{-s}\sqrt{\frac{s+r}{r(s-r)}}$$

when $r + s$ is even, while $a_{rs} = 0$ when $r + s$ is odd. These asymptotics are uniform as r/s varies over any compact subset of $(0, 1)$. ◁

Exercise 9.4. Redo Examples 9.13 and 9.14 using Theorem 9.5 instead of Theorem 9.12. What extra conditions do you need to check?

In the presence of minimal critical points we do not need to rule out the critical points at infinity (CPAI) discussed in previous chapters. However, if we do rule out CPAI then Theorem 7.20 applies and we get the following.

Theorem 9.15 (Main Theorem of Smooth ACSV (No CPAI Version)). *Suppose that, as $\hat{\boldsymbol{r}}$ varies over a compact set $\mathcal{R} \subset \mathbb{R}^d$ of non-zero directions, the function F has no CPAI with height at least $M \in \mathbb{R}$, and that the set $\mathbf{W} = \mathbf{W}(\hat{\boldsymbol{r}})$ of critical points with height larger than M is finite and consists of smooth non-degenerate points. Then there exist $\kappa_{\boldsymbol{w}} \in \mathbb{Z}$ for $\boldsymbol{w} \in \mathbf{W}$ with*

$$a_{\boldsymbol{r}} \approx \sum_{\boldsymbol{w} \in \mathbf{W}} \kappa_{\boldsymbol{w}} \, \Phi_{\boldsymbol{w}}(\boldsymbol{r}) + O(e^{M|\boldsymbol{r}|}), \tag{9.9}$$

where each $\Phi_{\boldsymbol{w}}$ is the asymptotic series defined by (9.4).

To determine dominant asymptotic behavior, it is necessary to identify the highest critical points $\boldsymbol{w}$ with non-zero coefficients $\kappa_{\boldsymbol{w}}$. This seems to be a very difficult task in general, but we can say more in some circumstances. For instance, $\kappa_{\boldsymbol{w}} = 1$ for any smooth minimal contributing points, and if $a_{\boldsymbol{r}}$ is not eventually zero and $M = -\infty$ then at least one $\kappa_{\boldsymbol{w}}$ is non-zero. If the exponential growth of a sequence can be determined or bounded using other means, this can also be used to identify the highest coefficients which are non-zero, and thus pin down asymptotics up to these unknown integers.

Although Theorem 9.15 is the most abstract of our main theorems, it follows directly from the large amount of technical background in Chapter 7 and the appendices, and some computations from the proof of Theorem 9.12 below.

Proof of Theorem 9.15 Fix a direction $\hat{\boldsymbol{r}}$. In the absence of CPAI at height M or above, Theorem 7.20 in Chapter 7 shows that, for some $\varepsilon > 0$, the homology group $H_d(\mathcal{M}, \mathcal{M}_{\leq M-\varepsilon})$ has a basis indexed by the critical points $\boldsymbol{\sigma}_1, \ldots, \boldsymbol{\sigma}_m$ for Q whose elements are smooth cycles γ_j such that $h_{\hat{\boldsymbol{r}}}$ attains its maximum on γ_j at $\boldsymbol{\sigma}_j$ and

$$a_{\boldsymbol{r}} = \sum_{j=1}^{m} \frac{\kappa_j}{(2\pi i)^{d-1}} \int_{\gamma_j} \operatorname{Res}\left(F(\boldsymbol{z})\boldsymbol{z}^{-\boldsymbol{r}-\boldsymbol{1}} \, d\boldsymbol{z}\right) + O(e^{M|\boldsymbol{r}|}).$$

We will determine this residue integral and its uniform error term with $\hat{\boldsymbol{r}}$ in our proof of Theorem 9.12 below, giving the stated expansion. □

Sections 9.3.1 and 9.3.2 complement the decomposition (9.9) by presenting an algorithm to determine the integer coefficients $k_{\boldsymbol{w}}$ for bivariate series, the only case beyond minimal points and rational functions with linear denominators where we know a general strategy for their calculation. Section 9.3.3 also gives an asymptotic formula for degenerate critical points in the bivariate case. Finally, Section 9.4 ends this chapter with some related results, including explicit formulae for higher-order terms and a coordinate-free formula (9.23) in terms of geometric invariants such as the Gaussian curvature.

9.1 Finitely minimal points and the surgery method

To prove Theorem 9.5 we show that the Cauchy integral representation for series coefficients is negligible outside a small neighborhood of $\boldsymbol{w}$, reduce to a lower-dimensional integral using a univariate residue computation, parametrize the simplified integral to obtain a saddle point integral, and apply the theorems of Chapter 5 to the result.

Localization and residue

We start by assuming that $\mathbf{W}(\hat{\boldsymbol{r}})$ contains a strictly minimal contributing singularity $\boldsymbol{w} = \boldsymbol{w}(\hat{\boldsymbol{r}})$.

Definition 9.16. For simplicity, we write $\boldsymbol{v}^\circ = (v_1, \ldots, v_{d-1})$ for any vector $\boldsymbol{v} \in \mathbb{C}^d$.

Our hypotheses imply that $\tilde{Q}_{z_d}(\boldsymbol{w}) \neq 0$, so the implicit function theorem states that z_d is locally analytically parametrized by $\boldsymbol{z}^\circ$ near $\boldsymbol{w}$ on $\mathcal{V}$. More specifically, if $r_d > 0$ and we define $\rho = |w_d|$ then there exist a sufficiently small real number $\delta > 0$, a neighborhood $\mathcal{N}$ of $\boldsymbol{w}^\circ$ in $T(\boldsymbol{w}^\circ)$, and an analytic function $g : \mathcal{N} \to \mathbb{C}$ such that for $\boldsymbol{z}^\circ \in \mathcal{N}$,

(i) $Q(\boldsymbol{z}^\circ, g(\boldsymbol{z}^\circ)) = 0$,
(ii) $\rho \leq |g(\boldsymbol{z}^\circ)| < \rho + \delta$ with equality only if $\boldsymbol{z}^\circ = \boldsymbol{w}^\circ$, and
(iii) $Q(\boldsymbol{z}^\circ, t) \neq 0$ if $t \neq g(\boldsymbol{z}^\circ)$ and $|t - w_d| < \delta$.

If $r_d < 0$ then the same conditions hold except $\boldsymbol{w}$ being contributing means the inequality in (ii) is replaced by $\rho - \delta < |g(\boldsymbol{z}^\circ)| \leq \rho$.

Let C_1 denote the circle of radius $\rho - \delta$ centered at the origin of the complex plane and let C_2 denote the circle of radius $\rho + \delta$. The fact that $\boldsymbol{w}$ is contributing, combined with the Cauchy integral formula, implies that the series coefficients of interest can be represented by an iterated integral

$$a_{\boldsymbol{r}} = \begin{cases} \left(\frac{1}{2\pi i}\right)^d \int_{\mathbf{T}(\boldsymbol{w}^\circ)} \left[\int_{C_1} F(\boldsymbol{z}^\circ, t) t^{-r_d - 1}\, dt\right] (\boldsymbol{z}^\circ)^{-\boldsymbol{r}^\circ} \frac{d\boldsymbol{z}^\circ}{z_1 \cdots z_{d-1}} & \text{if } r_d > 0 \\ \left(\frac{1}{2\pi i}\right)^d \int_{\mathbf{T}(\boldsymbol{w}^\circ)} \left[\int_{C_2} F(\boldsymbol{z}^\circ, t) t^{-r_d - 1}\, dt\right] (\boldsymbol{z}^\circ)^{-\boldsymbol{r}^\circ} \frac{d\boldsymbol{z}^\circ}{z_1 \cdots z_{d-1}} & \text{if } r_d < 0 \end{cases}. \tag{9.10}$$

In either case, the key observation is that the inner integral is exponentially smaller than ρ^{-r_d} away from $\boldsymbol{w}^\circ$. Indeed, if $r_d > 0$ under our assumptions then for each fixed $\boldsymbol{z}^\circ \neq \boldsymbol{w}^\circ$ the function $f(t) = F(\boldsymbol{z}^\circ, t)$ has radius of convergence greater than ρ and the inner integral is $O((\rho+\varepsilon)^{-r_d})$ for some $\varepsilon > 0$; by continuity of the radius of convergence, a single $\varepsilon > 0$ may be chosen for all compact subsets of $\mathbf{T}(\boldsymbol{w}^\circ)$ not containing $\boldsymbol{w}^\circ$. Similarly, if $r_d < 0$ then the inner integral is $O((\rho + \varepsilon)^{-r_d})$ for some $\varepsilon \in (-\rho, 0)$. Thus,

$$|\boldsymbol{w}^{\boldsymbol{r}} (a_{\boldsymbol{r}} - I)| \to 0 \tag{9.11}$$

exponentially quickly, where I is any integral in (9.10) with $\mathbf{T}(\boldsymbol{w}^\circ)$ replaced by any neighborhood of $\boldsymbol{w}^\circ$ in $\mathbf{T}(\boldsymbol{w}^\circ)$. We now take the neighborhood defining I to be the set $\mathcal{N}$ on which the properties (i)–(iii) for the parametrization g hold, and compare the inner integral in (9.10) to one pushed 'beyond' the singular set. Note that in general we cannot do this without first 'cutting out' the small neighborhood $\mathcal{N}$.

Assume that $r_d > 0$ and compare

$$I = \left(\frac{1}{2\pi i}\right)^d \int_{\mathcal{N}} \left[\int_{C_1} F(\boldsymbol{z}^\circ, t) t^{-r_d - 1}\, dt\right] (\boldsymbol{z}^\circ)^{-\boldsymbol{r}^\circ} \frac{d\boldsymbol{z}^\circ}{z_1 \cdots z_{d-1}}$$

to the integral

$$I' = \left(\frac{1}{2\pi i}\right)^d \int_{\mathcal{N}} \left[\int_{C_2} F(\boldsymbol{z}^\circ, t) t^{-r_d - 1}\, dt\right] (\boldsymbol{z}^\circ)^{-\boldsymbol{r}^\circ} \frac{d\boldsymbol{z}^\circ}{z_1 \cdots z_{d-1}}$$

with the inner contour C_1 replaced by C_2. Because the points on C_2 have larger modulus than ρ,

$$\left|\boldsymbol{w}^{\boldsymbol{r}} I'\right| \to 0 \tag{9.12}$$

exponentially quickly. Furthermore, our assumption of strict minimality implies that the common inner integrand of I and I' has a unique pole in the annulus $\rho - \delta \le |t| \le \rho + \delta$, occurring at $t = g(\boldsymbol{z}^\circ)$. If

$$\Psi(\boldsymbol{z}^\circ) = \text{Res}\left(F(\boldsymbol{z}^\circ, t) t^{-r_d - 1}\, ; \, t = g(\boldsymbol{z}^\circ)\right) \tag{9.13}$$

then the difference of I and I' can be computed in terms of Ψ. If $r_d < 0$ the argument is the same, with the roles of C_1 and C_2 reversed, changing the sign in front of the residue integral. Ultimately, we obtain the following, which may be thought of as the computational analog of the fact that one can integrate in relative homology at the expense of an exponentially small error (see Proposition B.10 in Appendix B).

Theorem 9.17 (reduction to residue integral). *Let*

$$\chi = I - I' = \frac{-\operatorname{sgn}(r_d)}{(2\pi i)^{d-1}} \int_{\mathcal{N}} \Psi(z^\circ)(z^\circ)^{-r^\circ} \frac{dz^\circ}{z_1 \cdots z_{d-1}}, \tag{9.14}$$

with Ψ given by (9.13). *Assuming the hypotheses of Theorem 9.5 when $\mathbf{W}(r) = \{w(\hat{r})\}$,*

$$|w^r (a_r - \chi)| \to 0$$

exponentially in $|r|$, uniformly as $r \to \infty$ with $\hat{r}$ varying over $\mathcal{M}$.

The fact that we can obtain explicit asymptotic expansions is a consequence of the following result.

Lemma 9.18. *Under the hypotheses of Theorem 9.5, the residue Ψ has the form $\Psi(z^\circ) = -g(z^\circ)^{-r_d}\Psi_p(z^\circ)$ where*

$$\Psi_p(z^\circ) = \sum_{k=0}^{p-1} \frac{(r_d + 1)_{(p-k-1)}}{k!(p-k-1)!} R_k(z^\circ). \tag{9.15}$$

Here $(a)_{(b)} = a(a-1)\cdots(a-b+1)$ and

$$R_k(z^\circ) = (-g(z^\circ))^{-p+k} \lim_{z_d \to g(z^\circ)} \partial_d^k \left((z_d - g(z^\circ))^p F(z)\right).$$

In particular, Ψ_p is a polynomial of degree $p-1$ in r_d with leading coefficient

$$(-1)^p g(z^\circ)^{-p}\, p \frac{P(z^\circ, g(z^\circ))}{(\partial_d^p Q)(z^\circ, g(z^\circ))}.$$

Proof Our assumptions imply that $F(z^\circ, t)$ has a pole of order p at $t = g(z^\circ)$, and (9.15) comes from the classic residue formula

$$\operatorname{Res}\left(F(z^\circ, t)t^{-r_d-1}\,;\, t = g(z^\circ)\right) = \frac{1}{(p-1)!} \lim_{z_d \to g(z^\circ)} \partial_d^{p-1} \left((z_d - g(z^\circ))^p F(z) z_d^{-r_d-1}\right)$$

together with Leibniz's rule for derivatives. The leading term in r_d comes from the summand where $k = 0$. □

Remark 9.19. The results of this section only require that F be meromorphic in a neighborhood of the domain of convergence $\mathcal{D}$. If F is locally the ratio of analytic functions P and Q in a neighborhood of w then all formulae are still valid, provided $\tilde{Q}$ is interpreted to be a square-free factorization in the local ring of germs of analytic functions (see Definition 10.42 below).

Exercise 9.5. Let $F(x, y) = 1/(e^x + e^y - 1)$. What can you deduce from Theorem 9.5 about the power series coefficients of F?

Proof of Theorem 9.5

Making the change of variables $z_j = w_j e^{i\theta_j}$ for $1 \leq j \leq d-1$ turns χ into a saddle point integral

$$\chi = \frac{\operatorname{sgn}(r_d)}{(2\pi)^{d-1}} \boldsymbol{w}^{-\boldsymbol{r}} \int_{\mathcal{N}'} A(\boldsymbol{\theta}) e^{-|r_d| \phi(\boldsymbol{\theta})} d\boldsymbol{\theta} \tag{9.16}$$

with amplitude $A(\boldsymbol{\theta}) = \Psi_p\left(\boldsymbol{w}^\circ e^{i\boldsymbol{\theta}}\right)$ for Ψ_p defined in (9.15) and phase

$$\begin{aligned}\phi(\boldsymbol{\theta}) &= \frac{r_d}{|r_d|} \log\left(\frac{g\left(\boldsymbol{w}^\circ e^{i\boldsymbol{\theta}}\right)}{w_d}\right) + i\frac{(\boldsymbol{r}^\circ \cdot \boldsymbol{\theta})}{|r_d|} \\ &= \operatorname{sgn}(r_d)\left[\log\left(\frac{g\left(\boldsymbol{w}^\circ e^{i\boldsymbol{\theta}}\right)}{w_d}\right) + i\frac{(\boldsymbol{r}^\circ \cdot \boldsymbol{\theta})}{r_d}\right]\end{aligned}$$

in the variables $\boldsymbol{\theta} = (\theta_1, \ldots, \theta_{d-1})$, where $\left(\boldsymbol{w}^\circ e^{i\boldsymbol{\theta}}\right) = \left(w_1 e^{i\theta_1}, \ldots, w_{d-1} e^{i\theta_{d-1}}\right)$ and $\mathcal{N}'$ is a neighborhood of the origin in $\mathbb{R}^d$. Lemma 8.21 implies that this integral satisfies the conditions necessary to apply Theorem 5.2 in Chapter 5 (note that the real part of ϕ has a strict minimum at the origin by our conditions on g). Lemma 8.22 applied to $\operatorname{sgn}(r_d)\phi$ simplifies the Hessian and finishes the proof.

Modification for finitely minimal points

When $\boldsymbol{w}(\hat{\boldsymbol{r}})$ is finitely minimal then the Cauchy integral decays exponentially away from any element of $\mathbf{W}(\hat{\boldsymbol{r}})$. We can thus restrict the domain of integration to a disjoint union of neighborhoods $\mathcal{N}_k$ around the elements of $\mathbf{W}(\hat{\boldsymbol{r}})$. The residue computation in Theorem 9.17 results in a sum as k varies of integrals over neighborhoods $\mathcal{N}_{\boldsymbol{w}_k}$. The asymptotic contributions of each of the integrals in the sum can be computed in the same way as the strictly minimal case.

Modification under strong torality hypothesis

Because our residue computations are so explicit, they also hold under the following *strong torality hypothesis*. This hypothesis is important when studying generating functions whose singularities have many symmetries, for instance in the case of quantum random walks (see Exercise 9.12).

Definition 9.20 (strong torality). We say Q satisfies the ***strong torality hypothesis*** on the torus $\mathbf{T}(\boldsymbol{w})$ if $Q(\boldsymbol{z}) = 0$ and $|z_j| = |w_j|$ for $1 \le j \le d-1$ implies that $|z_d| = |w_d|$.

Exercise 9.6. Suppose that the function $Q(x, y) = a + bx + cy + dxy$ is bilinear. What conditions on the constants a, b, c, d are equivalent to strong torality of Q?

In the following proposition g is the multivalued function solving for z_d as a function of $\boldsymbol{z}^\circ$; the number of values, counted with multiplicities, is the degree m of z_d in Q, except on a lower dimensional set where two values coincide. The multivalued integrand should be interpreted as a sum over all m values.

Corollary 9.21 (reduction under strong torality). *Suppose $\boldsymbol{w}$ satisfies all of the hypotheses of Theorem 9.5 except that instead of $\boldsymbol{w}$ being finitely minimal, it is minimal and Q satisfies the strong torality hypothesis on $\mathbf{T}(\boldsymbol{w})$. If all poles of F on $\mathbf{T}(\boldsymbol{w})$ are simple (i.e., $p = 1$) then*

$$a_{\boldsymbol{r}} = \left(\frac{1}{2\pi i}\right)^{d-1} \int_{\mathbf{T}(\boldsymbol{w}^\circ)} (\boldsymbol{z}^\circ)^{-\boldsymbol{r}^\circ} g(\boldsymbol{z}^\circ)^{-r_d} \Psi(\boldsymbol{z}^\circ) \frac{d\boldsymbol{z}^\circ}{\boldsymbol{z}^\circ},$$

where Ψ is given by (9.13).

Proof This time we may take C_1 to be the circle of radius of $\rho - \delta$ and C_2 to be the circle of radius $\rho + \delta$ for any $\delta \in (0, \rho)$. The inner integral will be the sum of simple residues at points $g(\boldsymbol{z}^\circ)$ for any $\boldsymbol{z}^\circ$ and the proof is completed the same way as Theorem 9.17. □

In this case dimension is reduced by one without localizing. The localization occurs when we apply the multivariate saddle point results of Chapter 5, which implies that this $(d-1)$-dimensional integral is determined by the behavior of g and Ψ near the critical points on $T(\boldsymbol{w})$.

Corollary 9.22. *Suppose $\boldsymbol{w}$ satisfies all of the hypotheses of Theorem 9.5 except that instead of $\boldsymbol{w}$ being finitely minimal, Q satisfies the torality hypothesis on $\mathbf{T}(\boldsymbol{w})$. Then the conclusions of Theorem 9.5 still hold.*

9.2 The method of residue forms

In this section we use the homological framework of previous chapters, together with the appendices, to prove Theorem 9.12. For convenience, we begin by naming the minimality property assumed in Theorem 9.12.

Definition 9.23 (finite criticality). We say Q is ***finitely critical*** on the torus $\mathbf{T}(\boldsymbol{w})$ (in the direction $\hat{\boldsymbol{r}}$) if the intersection of $\mathcal{V}_Q$ with $\mathbf{T}(\boldsymbol{w})$ contains finitely many critical points of Q in the direction $\hat{\boldsymbol{r}}$.

Exercise 9.7. For which directions $\hat{\boldsymbol{r}}$ (if any) does the function $Q(x, y) = (1 + x)(1 + y)$ satisfy finite criticality on the unit torus $\mathbf{T}(1, 1)$?

Suppose that $\boldsymbol{p}$ is a minimal point of $\mathcal{V}$ that is critical in direction $\hat{\boldsymbol{r}}$ and lies in the exponential torus $\mathbf{T}_e(\boldsymbol{x}) = \text{Relog}^{-1}(\boldsymbol{x})$ defined by some $\boldsymbol{x} \in \partial B$, where B is a component of the complement of $\texttt{amoeba}(Q)$. Further assume that $\mathbf{T}_e(\boldsymbol{x}) \cap \mathcal{V}$ contains only finitely many critical points $\boldsymbol{p}_1, \ldots, \boldsymbol{p}_m$.

Proposition 9.24 (stratified flow). *For $\boldsymbol{x}' \in B$ arbitrarily close to $\boldsymbol{x}$, the torus $\mathbf{T}_e(\boldsymbol{x}')$ may be deformed in $\mathcal{M}$ so that it remains fixed in a neighborhood of each critical point $\boldsymbol{p}_j$ but moves to a height less than $-\hat{\boldsymbol{r}} \cdot \boldsymbol{x}$ outside of a larger neighborhood of each.*

Proof This is a consequence of Theorem 11.5, which uses *cones of hyperbolicity* to create a deformation based on Theorem 11.1. In the case that the points $\boldsymbol{p}_1, \ldots, \boldsymbol{p}_m$ are all smooth points, the cones and vectors can be constructed by the simpler and more explicit Theorem 11.9 and Corollary 11.10. □

We remark that because $\tilde{Q}$ is *hyperbolic* at all minimal points (see Proposition 11.26), the vector flow used in the proof of Proposition 9.24 can also be used to construct the general homotopy equivalence (C.3.1), giving the relative homology attachment groups up to points of height just below the minimal points. An important practical consequence is the following principle, stating that local integral formulae may be summed when finite criticality holds. It follows immediately from the deformation in Proposition 9.24.

Theorem 9.25 (finite criticality implies sum of local contributions). *Suppose that $\boldsymbol{w}$ is a minimal point satisfying finite criticality, with all critical points on $T(\boldsymbol{w})$ enumerated $\boldsymbol{p}_1, \ldots, \boldsymbol{p}_m$. If each of the $\boldsymbol{p}_j$ are nondegenerate contributing points and the Cauchy integral over a quasi-local cycle maximized near $\boldsymbol{p}_j$ has asymptotic expansion $\Phi_{\boldsymbol{p}_j}(\boldsymbol{r})$ then*

$$a_{\boldsymbol{r}} \approx \sum_{j=1}^{m} \Phi_{\boldsymbol{p}_j}(\boldsymbol{r}) + E(\boldsymbol{r}) \tag{9.17}$$

where $E(\boldsymbol{r})$ grows exponentially slower than the common value of the $|\boldsymbol{p}_j^{-\boldsymbol{r}}|$. □

Theorem 9.25 holds for general rational functions, not just those with smooth denominators, for more general definitions of *contributing points* that are discussed in later chapters.

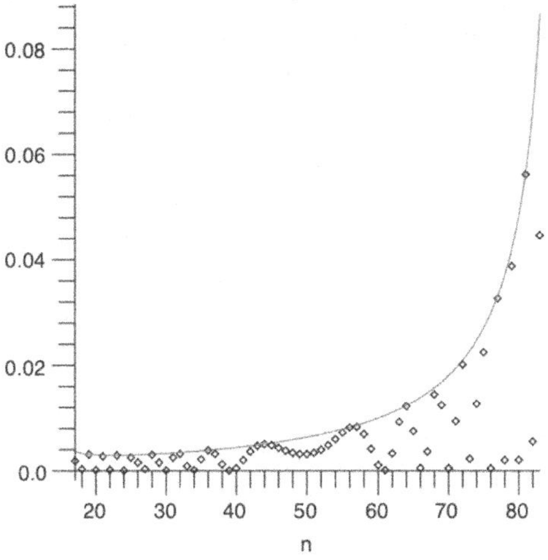

Figure 9.2 Spacetime generating function for a one-dimensional quantum walk.

Exercise 9.8. Let $Q(x,y) = 1 - cy(1+x) - xy^2$, where $c \in (0,1)$. The function $1/Q$ is the spacetime generating function for the simplest non-trivial one-dimensional quantum walk [BP07].

(a) Show that all singularities on the unit torus are minimal points.
(b) Show that the singularities on the unit torus are not finitely minimal.
(c) Show that for $|a - 1/2| < c/2$ and $\hat{r} = \left[\frac{a}{a+1}, \frac{1}{a+1}\right]$ there are two critical points in the direction $\hat{r}$ on the unit torus.
(d) Explain why (9.17) produces the picture in Figure 9.2 for the generating function $F(x,y) = 1/Q(x,y)$.

9.2.1 Theorem 9.12 via residue integrals

We are now ready to prove the *minimal point version* of the Main Theorem of Smooth ACSV.

Proof of Theorem 9.12 Assume the hypotheses of Theorem 9.12 and let B be the component of $\texttt{amoeba}(Q)^c$ corresponding to the convergent Laurent series under consideration. We use the homological constructions and terminology introduced in Appendix C. If $T = \mathbf{T}_e(\boldsymbol{x})$ for some $\boldsymbol{x} \in B$ and $T' = \mathbf{T}_e(\boldsymbol{x}')$ for some $\boldsymbol{x}' \in B'$, where B' is one of the components of $\texttt{amoeba}(Q)^c$ on which $h_{\hat{r}}$

is not bounded from below (whose existence is guaranteed by Theorem 6.29), then the intersection class $\mathbf{INT}(T, T')$ is represented by the intersection of $\mathcal{V}$ with any homotopy from T to T' intersecting $\mathcal{V}$ transversely. Choosing such a homotopy whose time-t cross-sections are tori that expand with t and go through $\boldsymbol{w}$, perhaps slightly perturbed to intersect $\mathcal{V}$ transversely, the class $\mathbf{INT}(T, T')$ can be represented by a smooth $(d-1)$-chain γ on $\mathcal{V}$ on which $h_{\hat{\boldsymbol{r}}}$ reaches its (not necessarily unique) maximum at $\boldsymbol{w}$. The Cauchy integral formula and the residue theorems from Chapter C imply

$$\begin{aligned} a_{\boldsymbol{r}} &= \frac{1}{(2\pi i)^d} \int_T F(\boldsymbol{z}) \boldsymbol{z}^{-\boldsymbol{r}-\boldsymbol{1}}\, d\boldsymbol{z} \\ &= \frac{1}{(2\pi i)^{d-1}} \int_\gamma \operatorname{Res}(F(\boldsymbol{z}) \boldsymbol{z}^{-\boldsymbol{r}-\boldsymbol{1}}\, d\boldsymbol{z}) + \frac{1}{(2\pi i)^{d-1}} \int_{T'} F(\boldsymbol{z}) \boldsymbol{z}^{-\boldsymbol{r}-\boldsymbol{1}}\, d\boldsymbol{z} \\ &= \frac{1}{(2\pi i)^{d-1}} \int_\gamma \operatorname{Res}(F(\boldsymbol{z}) \boldsymbol{z}^{-\boldsymbol{r}-\boldsymbol{1}}\, d\boldsymbol{z}). \end{aligned} \tag{9.18}$$

Assume first that Q is square-free and $r_d > 0$, so that (C.2.1) in Proposition C.8 implies

$$a_{\boldsymbol{r}} = \frac{e^{-h_{\boldsymbol{r}}(\boldsymbol{w})}}{(2\pi i)^{d-1}} \int_\gamma e^{-\lambda \phi(\boldsymbol{z})} \frac{P(\boldsymbol{z})}{Q_{z_d}(\boldsymbol{z}) \prod_{j=1}^d z_j}\, d\boldsymbol{z}^\circ$$

with $\lambda = r_d$. Applying Theorem 5.3 with a generic triangulation of $C = \gamma$ gives an asymptotic expansion of $a_{\boldsymbol{r}}$ which, after the change of variables $z_j = w_j e^{i\theta_j}$ and algebraic simplification, gives the expression for $\Phi_{\boldsymbol{w}}$ in (9.2). Note that the Hessian determinant of $h_{\hat{\boldsymbol{r}}}$ on $\mathcal{V}$ with respect to the θ_j variables equals the Hessian determinant with respect to the z_j variables multiplied by the Jacobian for the change of variables because the gradient of $h_{\hat{\boldsymbol{r}}}$ restricted to $\mathcal{V}$ vanishes at $\boldsymbol{w}$.

This completes the proof of Theorem 9.12 in the case that $p = 1$ and $r_d > 0$. The derivation for $p > 1$ is similar, with Lemma C.13 describing the residue and leading to (9.4). Likewise, accounting for the sign change in $\lambda = -r_d$ when $r_d < 0$ produces the sign factors in (9.4). □

9.2.2 Homological decompositions

Our results above help us prove that there is at most one torus containing smooth nondegenerate contributing points. A ***minimal torus*** with respect to a component B and direction $\hat{\boldsymbol{r}}$ is a torus $\mathbf{T}_e(\boldsymbol{x})$ for some $\boldsymbol{x} \in \partial B$ minimizing $\hat{\boldsymbol{r}} \cdot \boldsymbol{x}$ on $\overline{B}$, containing at least one point $\boldsymbol{w} = \exp(\boldsymbol{x} + i\boldsymbol{y})$ that is critical in direction $\hat{\boldsymbol{r}}$.

Proposition 9.26. *Let $F(z) = P(z)/Q(z)$ be the ratio of coprime polynomials P and Q. Fix a direction $\hat{r}$, a component B of the complement of* amoeba(Q) *on which $h_{\hat{r}}$ is bounded from below, and a component B' on which $h_{\hat{r}}$ is not bounded from below.*

(i) There is at most one minimal torus with respect to B and $\hat{r}$ satisfying finite criticality and on which each critical point is smooth, contributing, and nondegenerate.

(ii) Let $T = \mathbf{T}_e(\boldsymbol{x})$ for some $\boldsymbol{x} \in B$ and let $T' = \mathbf{T}_e(\boldsymbol{x}')$ for some $\boldsymbol{x}' \in B'$. Given the existence of the torus described in (i), the projection of $\mathbf{INT}(T, T')$ *to the relative homology group $H_{d-1}(\mathcal{V}_*, \mathcal{V}_{\leq c-\varepsilon})$, for sufficiently small $\varepsilon > 0$ and $c = -\hat{r} \cdot \boldsymbol{x}$, equals $\sum_{\boldsymbol{z} \in \mathrm{W}} \gamma_{\boldsymbol{z}}$, where the cycle $\gamma_{\boldsymbol{z}}$ is a generator for the cyclic local homology group $H_{d-1}(\mathcal{V}_*^{\boldsymbol{z},\mathrm{loc}})$.*

(iii) The projection of $[T]$ to $(\mathcal{M}, \mathcal{M}_{\leq c-\varepsilon})$ is equal to $\sum_{\boldsymbol{z} \in \mathrm{W}} o\gamma_{\boldsymbol{z}}$, where $\gamma_{\boldsymbol{z}}$ is a generator of the cyclic group $H_{d-1}\left(\mathcal{V}_*^{\boldsymbol{z},\mathrm{loc}}\right)$.

Proof To prove (*i*), suppose there are two such tori $\mathbf{T}_e(\boldsymbol{x})$ and $\mathbf{T}_e(\boldsymbol{x}')$. Applying Theorems 9.12 and 9.25 to the rational function $\tilde{F}(\boldsymbol{z}) = 1/\tilde{Q}(\boldsymbol{z})$ at the points in each torus gives two, necessarily equal, asymptotic series estimating the coefficients $\{\tilde{a}_{\boldsymbol{r}}\}$ uniformly as $|\boldsymbol{r}| \to \infty$ with $\boldsymbol{r}/|\boldsymbol{r}|$ remaining in some neighborhood $\mathcal{R}$ of $\hat{r}$ (we replace P by 1 and Q by its square-free part as this does not change the minimal tori or our nondegeneracy assumptions, but simplifies the asymptotic formulae). In particular, the leading term of each expansion $\Phi_{\boldsymbol{w}}(\boldsymbol{r})$ in (9.4) has the form $C(\boldsymbol{w}) \exp(-\boldsymbol{r} \cdot \boldsymbol{x}) \exp(-i\boldsymbol{r} \cdot \boldsymbol{y}) r_d^{(1-d)/2}$ with C nonvanishing. Summing the contributions of the finitely many points on $\mathbf{T}_e(\boldsymbol{x})$ (respectively $\mathbf{T}_e(\boldsymbol{x}')$) gives a function of $\boldsymbol{r}$ that is nonvanishing at least on some finite-index sublattice of $\mathbb{Z}^d$. Furthermore, the terms given by the elements of $\mathbf{T}_e(\boldsymbol{x})$ and the elements of $\mathbf{T}_e(\boldsymbol{x}')$ differ from each other in exponential growth, because $-\boldsymbol{x} \cdot \boldsymbol{r}$ and $-\boldsymbol{x}' \cdot \boldsymbol{r}$ disagree on $\mathcal{R}$ except possibly for a set of codimension 1. This contradicts the fact that both expansions represent asymptotics for the same sequences, so two such tori cannot exist.

To prove (*ii*), the deformation used to prove Theorem 5.3 shows that the intersection cycle may be deformed to a sum of elements of local homology groups. None of these can be zero because there is a term corresponding to each in (9.8). Similarly, each is a relative homology generator: this can be seen from the deformation, but an easier argument is that the corresponding term $\Phi_{\boldsymbol{w}}(\boldsymbol{r})$ is, up to sign, the integral obtained from a small $(d-1)$-patch and we know the local homology generator is a $(d-1)$-ball modulo its boundary (see, for example, Theorem C.38).

Conclusion (*iii*) can be argued similarly to (*ii*), using the stratified description of attachments from Theorem D.25 in place of Theorem C.38. Alternatively, the Thom isomorphism (Theorem C.2) says that o induces an injection from $H_{d-1}(\mathcal{V}_*)$ to $H_d(\mathbb{C}_*^d \setminus \mathcal{V})$. Being functorial, it commutes with π_* where $\pi : \mathcal{V}_* \to (\mathcal{V}_*, \mathcal{V}_{\leq c-\varepsilon})$ is projection. The Thom isomorphism carries $\mathbf{INT}(T, T')$ to $T - T'$, which is equal to T in $H_d(\mathcal{V}_*, \mathcal{V}_{\leq c-\varepsilon})$, proving *(iii)*. □

We remark that it is possible to have a minimal smooth contributing point $\boldsymbol{p}$ in the direction $\hat{\boldsymbol{r}}$, and another smooth critical (but not contributing) point $\boldsymbol{p}'$ in the direction $\hat{\boldsymbol{r}}$ *that is not minimal* but has the same height as $\boldsymbol{p}$.

9.3 Smooth bivariate functions

This section further explores bivariate rational functions, for which we can be more explicit and give stronger results.

9.3.1 Smooth bivariate power series

We first present a complete algorithm for bivariate power series that finds all smooth contributing critical points, without any assumption of minimality, following the techniques of [DeV11; DvdHP11].

Assumption 9.1. *In this section we always assume that $\mathcal{V}$ is smooth and Q is square-free, so that for every $(x, y) \in \mathcal{V}$, at least one of $Q_x(x, y)$ and $Q_y(x, y)$ is non-zero, and that the set of critical points is finite. If Q is not square-free then our arguments characterizing the singularities that determine asymptotics still hold when Q is replaced by its square-free part.*

In any number of variables, a potential program to determine asymptotics is the following.

1. Explicitly compute a cycle representing the intersection class.
2. Try to push the cycle below each critical point, starting at the highest.
3. When it is not possible to push past a point, describe the local cycle that is 'snagged' on the critical point.
4. Check whether this is a quasi-local cycle of the form we have already described and, if so, read off the estimate from saddle point asymptotics.

This program is not generally effective because the step of 'pushing the cycle down' is not algorithmic, which is why we use the framework of stratified Morse theory. However, an exception occurs when $d = 2$, since the cycle C has codimension 1 in $\mathcal{V}$ and thus, up to a time change, there is only one way for it to flow downward.

Fix a direction $\hat{\mathbf{r}} = (\hat{r}, \hat{s})$ with $\hat{r}$ and $\hat{s}$ positive (otherwise all series coefficients in this direction are zero, or given by a univariate rational function, since we consider the power series expansion of F). Because Q does not vanish at the origin, there exists some $\varepsilon > 0$ such that $\mathcal{V} = \mathcal{V}_Q$ does not intersect the set $\{(x, y) : |x| \leq \varepsilon, |y| \leq \varepsilon\}$. Now, for any $c \in \mathbb{R}$ if the height $h(x, y) = -\hat{r} \log |x| - \hat{s} \log |y|$ of a point (x, y) is at least c then either $|x| \leq e^{-c}$ or $|y| \leq e^{-c}$. Taking $c \geq \log(1/\varepsilon)$ thus shows that no connected component of $\mathcal{V}^{\geq c}$ contains both points with $|x| \leq \varepsilon$ and points with $|y| \leq \varepsilon$.

On the other hand, for sufficiently large c every connected component of $\mathcal{V}^{\geq c}$ contains points with arbitrarily large height, and hence points with either $|x| \leq \varepsilon$ or $|y| \leq \varepsilon$. Thus, we may decompose $\mathcal{V}^{\geq c}$ for sufficiently large c into a disjoint union $X^{\geq c} \cup Y^{\geq c}$, where $X^{\geq c}$ is the union of connected components containing points with arbitrarily small x-coordinates and $Y^{\geq c}$ is the union of connected components containing points with arbitrarily small y-coordinates. Puiseux's Theorem states that in a sufficiently small neighborhood of the origin in x, with a ray from the origin removed to account for branch cuts, every branch $y(x)$ of $Q(x, y) = 0$ has a representation

$$y(x) = \sum_{j \geq j_0} c_j x^{j/k}$$

for a fixed branch of the kth root, where $j_0 \in \mathbb{Z}$ and k is a positive integer (and analogous representations for the branches of x in terms of y also hold). By Rouché's Theorem, projection of such a connected component to its x-value is diffeomorphic as a covering to the projection of the graph of $y^k = Cx^j$ for some constant C, such a covering space being diffeomorphic to a punctured disk. Thus, for any sufficiently large c, the connected components of $X^{\geq c}$ and $Y^{\geq c}$ are diffeomorphic to disjoint open disks with their origins removed. The values of c such that this decomposition holds form an interval $[c_{xy}, \infty)$ for some $c_{xy} \in \mathbb{R}$.

Critical points at infinity

Puiseux's Theorem also helps characterize critical points at infinity. In particular, any branch $y(x)$ of $Q(x, y) = 0$ near the origin $x = 0$ satisfies

$$y(x) = Cx^{\alpha}(1 + o(1))$$

for some $C \in \mathbb{C}$ and $\alpha \in \mathbb{Q}$, and any branch $x(y)$ near the origin $y = 0$ satisfies

$$x(y) = C' y^{\beta}(1 + o(1))$$

for some $C' \in \mathbb{C}$ and $\beta \in \mathbb{Q}$. If $F(x, y)$ has a CPAI in the direction $\hat{\mathbf{r}}$ then either $\alpha = -\frac{\hat{r}}{\hat{s}}$ for some branch $y(x)$ or $\beta = -\frac{\hat{s}}{\hat{r}}$ for some branch $x(y)$. We thus make the following assumption to rule out the existence of CPAI.

Assumption 9.2 (No CPAI). *For any branch $y(x) = Cx^\alpha(1+o(1))$ of $Q(x,y) = 0$ as $x \to 0$ we have $\alpha \neq -\frac{\hat{r}}{\hat{s}}$ and for any branch $x(y) = C'y^\beta(1+o(1))$ of $Q(x,y) = 0$ as $y \to 0$ we have $\beta \neq -\frac{\hat{s}}{\hat{r}}$.*

When Assumption 9.2 holds we can be very explicit about the behavior of the height function near the coordinate axes.

Lemma 9.27. *Assume there are no CPAI. For any sufficiently small $\varepsilon > 0$, fixed $\theta \in [-\pi, \pi]$, and branch $y(x)$ of $Q(x,y) = 0$ near $x = 0$, the parametrized height function*

$$h_\theta(t) = h\left(te^{i\theta}, y\left(te^{i\theta}\right)\right)$$

is monotonic for $t \in [0, \varepsilon]$. Furthermore, if $y(x) \sim Cx^\alpha$ as $x \to 0$ then

$$\lim_{t \to 0^+} h_\theta(t) = \begin{cases} \infty & \text{if } \alpha > -\hat{r}/\hat{s} \\ -\infty & \text{if } \alpha < -\hat{r}/\hat{s} \end{cases}.$$

Proof Puiseux's theorem implies we can always find $\alpha \in \mathbb{Q}, C \in \mathbb{C}$, and a function ϕ with $\phi(x)$ and $x\phi'(x)$ in $o(1)$ such that $y(x) = Cx^\alpha(1 + \phi(x))$ as $x \to 0$. The height function is the real part of $H(x,y) = -\hat{r}\log x - \hat{s}\log y$, so

$$\frac{d}{dt}h_\theta(t) = \cos(\theta)\text{Re}\left[H_x(x, y(x))\right] - \sin(\theta)\text{Im}\left[H_x(x, y(x))\right],$$

where

$$H_x(x, y(x)) = \frac{-\hat{r} - \hat{s}\alpha}{x} - \frac{\phi'(x)}{1 + \phi(x)} \sim \frac{-\hat{r} - \hat{s}\alpha}{x}.$$

Thus

$$\frac{d}{dt}h_\theta(t) \sim \frac{-\hat{r} - \hat{s}\alpha}{|x|},$$

which is strictly positive or strictly negative under Assumption 9.2. Finally, we note

$$h_\theta(t) \sim \log\left(Ct^{-\hat{r}-\hat{s}\alpha}\right)$$

for t sufficiently small, giving the stated asymptotic behavior. □

Corollary 9.28. *Under Assumption 9.2 the connected components of $X^{\geq c}$ are diffeomorphic to disjoint open disks with their origins removed, corresponding to the branches $y(x) \sim Cx^\alpha$ of $Q(x,y) = 0$ as $x \to 0$ with $-\hat{r}/\hat{s} < \alpha \leq 0$.*

Intersection cycles and flows

Fixing $|x|$ small and expanding $|y|$ gives a homotopy that (up to minor perturbation) intersects $\mathcal{V}$ transversely. In particular, the intersection cycle C created from this operation contains a positively oriented circle around the removed origin from each of the punctured disks in $X^{\geq c}$ for c sufficiently large. As usual, we get a residue integral expression

$$a_{r,s} = \frac{1}{2\pi i} \int_C \text{Res}\left(\frac{P(x,y)}{Q(x,y)} x^{-r-1} y^{-s-1} dx \wedge dy \right)$$

when P and Q are polynomials. More generally, when P is an analytic function over appropriate regions of $\mathbb{C}^d$ we get

$$a_{r,s} = \frac{1}{2\pi i} \int_C \text{Res}\left(\frac{P(x,y)}{Q(x,y)} x^{-r-1} y^{-s-1} dx \wedge dy \right) + O(\delta^{r+s})$$

as $r, s \to \infty$, for any $\delta > 0$.

As we have already seen multiple times, in the absence of CPAI the topology of $\mathcal{V}^{\geq c}$ cannot change with c except at critical values. Because we work in two dimensions, we can be very explicit about the change in topology as c passes through a critical value.

Definition 9.29. Suppose $\sigma = (x_0, y_0)$ is a critical point where $Q_x(\sigma) \neq 0$, so that we can parametrize $y = y(x)$ in a neighborhood of σ in $\mathcal{V}$. The ***degree of degeneracy*** of Q at σ is the integer k such that there is a series expansion

$$h(x, y(x)) = h(\sigma) + \text{Re}\left[\sum_{j \geq k} c_j (x - x_0)^j \right]$$

in a neighborhood of x_0 with $c_k \neq 0$. Because σ is a critical point of the height function h, the degree of degeneracy is always at least 2, and σ is a nondegenerate critical point precisely when the degree of degeneracy is equal to 2. Because $\hat{r}$ has no zero coordinate and σ is a critical point, $Q_y(\sigma) \neq 0$ and the degree of degeneracy is the same parametrizing by y instead of x.

If $\sigma = (x_0, y_0)$ is a critical point with degree of degeneracy k then we can substitute $y = y(x)$ and expand $H(x,y) = -\hat{r} \log x - \hat{s} \log y$ near x_0 to obtain

$$H(x, y(x)) = C + (x - x_0)^k g(x)$$

for some $C \in \mathbb{C}$ and analytic function g with $g(x_0) \neq 0$. In particular, if $w = (x - x_0) g(x)^{1/k}$ then $(dw/dx)(x_0) \neq 0$ and we can parametrize the height function h in the local coordinate w near σ as

$$h(x(w), y(x(w))) = \text{Re}\left[H(x(w), y(x(w))) \right] = h(\sigma) + \text{Re}\left[w^k \right].$$

Thus, near $\boldsymbol{\sigma}$ the set $\mathcal{V}$ contains k disjoint *ascent regions*, where h increases while moving towards $\boldsymbol{\sigma}$, which alternate with k disjoint *descent regions*, where h decreases while moving towards $\boldsymbol{\sigma}$; this is illustrated in Figure 9.5 below.

Definition 9.30. Let $c_{xy} \in [-\infty, \infty)$ be the infimum of all values c such that $X^{>c} \cap Y^{>c} = \emptyset$ (which is also the smallest value c such that $X^{>c}$ and $Y^{>c}$ are well-defined). If $c_{xy} = -\infty$ then let $\mathbf{W} = \emptyset$, otherwise let $\mathbf{W}$ be the nonempty set of critical points $\boldsymbol{\sigma}$ such that $h(\boldsymbol{\sigma}) = c_{xy}$ and, for any sufficiently small neighborhood U of $\boldsymbol{\sigma}$ in $\mathcal{V}$, the sets $U \cap X^{>c_{xy}}$ and $U \cap Y^{>c_{xy}}$ are nonempty.

Our choice of the notation $\mathbf{W}$ comes from the following result.

Theorem 9.31. *Suppose Assumptions 9.1 and 9.2 hold. If* $\mathbf{W}$ *is empty then the intersection cycle* C *is in the same homology class as a cycle with maximum height* $-m$ *for all sufficiently large* $m \in \mathbb{R}$ *(it can be pushed down forever). If* $\mathbf{W}$ *is nonempty then* C *is in the same homology class as a cycle* κ *such that*

(i) The points of κ *with maximum height are precisely the points of* $\mathbf{W}$.
(ii) For $\boldsymbol{\sigma} \in \mathbf{W}$ *and a sufficiently small neighborhood* U *of* $\boldsymbol{\sigma}$ *in* $\mathcal{V}$*, if* $A_0, \ldots, A_{k-1}$ *and* $D_0, \ldots, D_{k-1}$ *denote the ascent and descent regions of* $\kappa \cap U$ *enumerated counterclockwise such that* D_j *lies between* A_j *and* $A_{j+1 \bmod k}$ *then*

$$\kappa \cap U = \sum_{j=0}^{k-1} \big[X(j+1) - X(j)\big] \gamma_j,$$

where each γ_j *is a curve traveling downward in* D_j *starting at* $\boldsymbol{\sigma}$ *and*

$$X(j) = \begin{cases} 1 & \text{if } A_{j \bmod k} \subset X^{>c_{xy}} \\ 0 & \text{if } A_{j \bmod k} \subset Y^{>c_{xy}} \end{cases}.$$

In particular, $\kappa \cap U$ *projects to a non-trivial cycle in the relative homology group* $H_1(U, U \cap \mathcal{V}^{\leq c_{xy}-\varepsilon})$ *for any* $\varepsilon > 0$ *sufficiently small (so the intersection cycle gets stuck at height* c_{xy}*).*

Proof Let $M \in \mathbb{R}$ be larger than all critical values of h. Then C is homologous to closed curves in each component of $X^{\geq M}$, and in fact it is homologous to the boundary $\partial X^{\geq M}$. First, we show that we can push down the intersection cycle until arriving at c_{xy}. The topology of $\mathcal{V}^{\geq c}$ only changes at critical values c, so let σ be a critical value in $(c_{xy}, M]$ and suppose that $\boldsymbol{\sigma}$ is the only critical point with $h(\boldsymbol{\sigma}) = \sigma$.

Figure 9.3 shows $\mathcal{V}^{\geq c}$ (shaded) for three values of c when the degree of degeneracy of $\boldsymbol{\sigma}$ is $k = 2$ and a circle enclosing a region where the parametrization $h = \sigma + \mathrm{Re}[w^k]$ holds (higher degrees of degeneracy are similar, just with

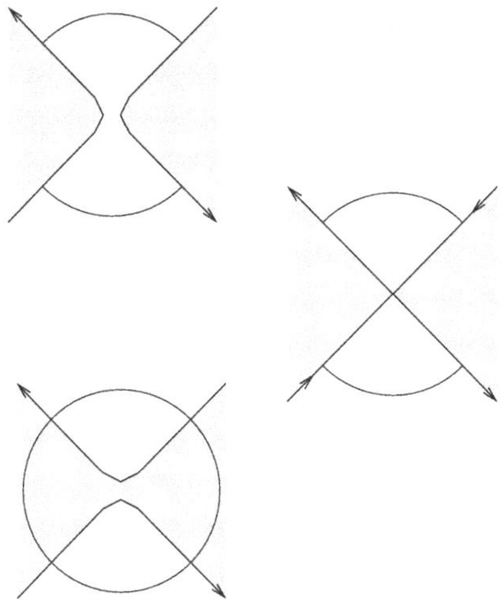

Figure 9.3 $\mathcal{V}^{\geq c}$ and its boundary for three values of c.

more components). In the top diagram $c > \sigma$, in the middle diagram $c = \sigma$, and in the bottom $c < \sigma$, with the arrows showing the orientation of $\partial\mathcal{V}^{\geq c}$ inherited from the complex structure of $\mathcal{V}$.

Consider the first picture where $c > \sigma$. Because $c > c_{xy}$ each of the k shaded regions is in $X^{\geq c}$ or $Y^{\geq c}$, but not both. In fact, since $\sigma > c_{xy}$ this persists in the limit as $c \downarrow \sigma$, so either all k regions are in $Y^{\geq c}$ or all k regions are in $X^{\geq c}$. In the first case $\partial X^{\geq c}$ does not contain any critical points of h on $\mathcal{V}$ with height in an interval $(\sigma - 2\varepsilon, \sigma + 2\varepsilon)$, so the first Morse Lemma implies $\partial X^{\geq \sigma+\varepsilon}$ is homotopic to $\partial X^{\geq \sigma-\varepsilon}$ as desired. In the latter case, the difference between $\partial\mathcal{V}^{\geq \sigma+\varepsilon}$ and $\partial\mathcal{V}^{\geq \sigma-\varepsilon}$ is a boundary ∂B (see Figure 9.4, or Figure 9.5 below) so these sets are still homologous. In fact, one can show they are still homotopic.

Thus, we can push the intersection cycle below any critical value above c_{xy} that has a single corresponding critical point, and the same argument holds generally by working locally around each critical point of fixed height larger than c_{xy}. In particular, if $\mathbf{W}$ is empty then we can push the intersection cycle down to arbitrarily low height.

It remains only to show that the intersection cycle can be represented by the stated cycle κ. Just as in Figure 9.3, the connected components of $\partial X^{\geq c_{xy}+\varepsilon}$ will contain curves moving through a descent region of $\mathcal{V}$ near $\boldsymbol{\sigma}$, crossing over an ascent region, then going back out through an adjacent descent region. However, unlike for critical points at higher height where all ascent regions are

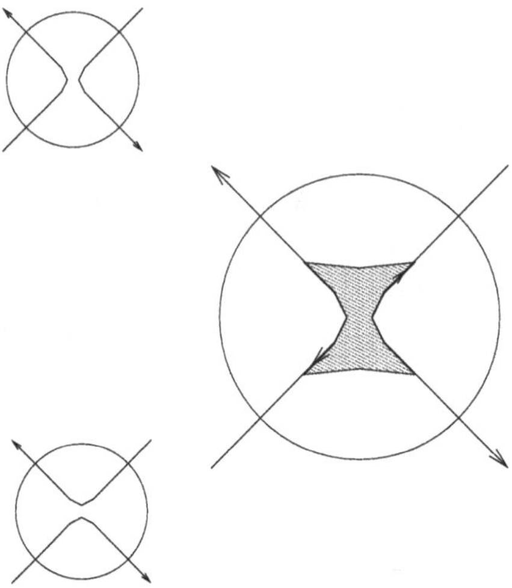

Figure 9.4 $\partial\mathcal{V}^{\geq\sigma+\varepsilon}$ and $\partial\mathcal{V}^{\geq\sigma-\varepsilon}$ differ locally by a boundary.

covered or none were, in this case the curves will only cross the ascent regions containing points of $X^{>c_{xy}}$; see Figure 9.5.

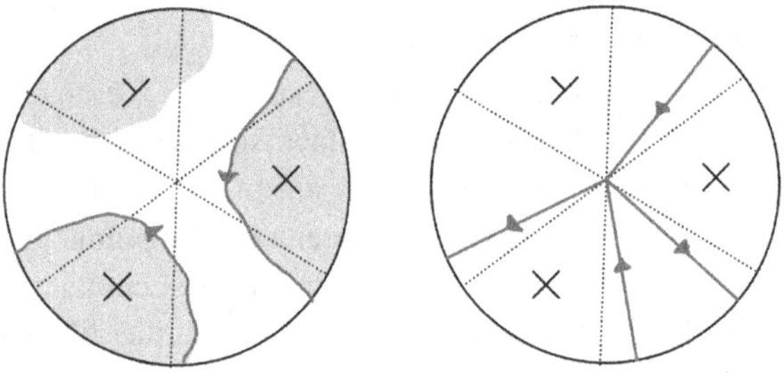

Figure 9.5 *Left:* Plot of $\mathcal{V}$ near a critical point with degree of degeneracy $k = 3$, where $\mathcal{V}^{>c_{xy}}$ contains two ascent regions with points in $X^{>c_{xy}}$ and one region with points in $Y^{>c_{xy}}$. The set $\mathcal{V}^{>c_{xy}+\varepsilon}$ is colored gray and the part of $\partial X^{\geq c_{xy}+\varepsilon}$ in view is drawn. *Right:* Straightening out the connected components of $\partial X^{\geq c_{xy}+\varepsilon}$ near this critical point gives the stated curves γ_j.

The connected components of $\partial X^{\geq c_{xy}+\varepsilon}$ can be straightened into rays γ_j (in terms of the local coordinate w) that stay in each descending region adjacent to an ascending region containing points of $X^{>c_{xy}}$. If the descending region is between ascending components containing points of $Y^{>c_{xy}}$ and $X^{>c_{xy}}$, working counterclockwise, then γ_j will start at the critical point and move down the descending region. Conversely, if the descending region is between ascending components containing points of $X^{>c_{xy}}$ and $Y^{>c_{xy}}$, working counterclockwise, then γ_j will move up the descending region to peak at the critical point. Descending regions between two ascending regions both containing points of $X^{>c_{xy}}$ have γ_j twice with opposite orientations, which can be joined and pushed to lower height. Descending regions between two ascending regions both containing points of $Y^{>c_{xy}}$ are not touched by the intersection cycle. Taking these sign considerations into account gives the stated formula for κ. □

Theorem 9.31 immediately gives an algorithm for the bivariate case.

Algorithm 3: Determination of **W** in the smooth, bivariate case.

Input: Bivariate rational function $F(x, y)$ and direction (r, s).
Output: Set of critical points **W** determining coefficient asymptotics of F in the (r, s) direction.

1 Verify that Assumptions 9.1 and 9.2 hold using Gröbner bases and Puiseux expansions
2 List the critical value in order of decreasing height
3 Set the provisional value of c_{xy} to the highest critical value
4 For each critical point at height c_{xy} do
   (a) Compute the order k of the critical point
   (b) Follow each of the k ascent paths until it is clear whether the x-coordinate or the y-coordinate goes to zero
   (c) Add the point to the set **W** if and only if at least one of the k paths has x-coordinate going to zero and at least one of the k paths has y-coordinate going to zero
5 If **W** is nonempty then terminate and output c_{xy} and **W**
6 Else, if c_{xy} is not the least critical value then replace c_{xy} by the next lower critical value and go to step 4
7 Else, if no critical values remain then $c_{xy} = -\infty$, **W** is empty, and the asymptotics decay super-exponentially

---

The doctoral dissertation [DeV11] discusses how to turn this breakdown into effective steps, and [MS22] give an implementation in Sage using interval arithmetic. The trickiest part is Step 4b. Ascent paths could conceivably get caught in a trap, approaching a critical point rather than continuing to height $+\infty$. However, this is a higher critical point, hence already known to be in an x- or y-component. Therefore, one only needs to know a radius ε for each higher critical point $\boldsymbol{p}$ such $|\boldsymbol{w} - \boldsymbol{p}| < \varepsilon$ implies $\boldsymbol{w}$ is in the same component as $\boldsymbol{p}$, which can be done with interval arithmetic. We conclude this section with an example of the evaluation of the intersection class for a particular smooth bivariate generating function whose analysis first appeared in [DeV10].

Example 9.32 (bi-colored supertrees)**.** A ***bi-colored supertree*** [FS09, Example VI.10] is a planar binary tree with each node replaced by a bi-colored rooted planar binary tree. The class of bi-colored supertrees is counted by the main diagonal of the bivariate function

$$F(x, y) = \frac{P(x, y)}{Q(x, y)} = \frac{2x^2y(2x^5y^2 - 3x^3y + x + 2x^2y - 1)}{x^5y^2 + 2x^2y - 2x^3y + 4y + x - 2},$$

and we give asymptotics following the algorithm above.

First, we note that there is one branch $y(x) \sim (-4)x^{-5}$ of y as $x \to 0$ and four branches $x(y)$ of x as $y \to 0$, two of which satisfy $x(y) \sim y^{-1/2}$ and two of which have $x(y) \sim -y^{-1/2}$. In particular, there are no critical points at infinity in the main diagonal direction $(r, s) = (1, 1)$. A quick Gröbner basis computation further verifies that the system

$$Q(x, y) = Q_x(x, y) = Q_y(x, y) = 0$$

has no solution, and the smooth critical point system

$$Q(x, y) = xQ_x(x, y) - yQ_y(x, y) = 0$$

has three solutions

$$\left(1 - \sqrt{5}, \frac{3 + \sqrt{5}}{16}\right), \quad \left(2, \frac{1}{8}\right), \quad \left(1 + \sqrt{5}, \frac{3 - \sqrt{5}}{16}\right),$$

listed here in order of decreasing height under $h_{1/2,1/2}$.

The highest critical point is nondegenerate, meaning $\mathcal{V}$ locally has two ascent paths. Following both ascent paths using, for instance, the Sage package of [MS22] shows that both contains points arbitrarily close to the x-axis, so the intersection cycle can be pushed lower. In this case we could also simply observe that the highest critical point cannot contribute to the asymptotics because the coordinates are real and of opposite sign. The factor $x^{-n}y^{-n}$ in the asymptotic formula for $a_{n,n}$ would then force the signs to alternate on the diagonal, whereas we know the diagonal terms to be positive.

Continuing to the next-highest point we consider the point $(2, 1/8)$. This point has degree of degeneracy four, of which three climb to the x-axis and one climbs to the y-axis. In particular, the point $(2, 1/8)$ determines dominant diagonal asymptotics and, using the notation of Theorem 9.31, the intersection cycle is homologous to $\gamma = \gamma_j - \gamma_{j-1}$ where j is the index of the region whose ascent region goes to the y-axis. Among the four descent regions, this path inhabits two consecutive ones, making a right-angle turn as it passes through the saddle.

Finally, we evaluate the univariate integral over this cycle. To compute the residue form in this example it is easiest to parametrize a neighborhood of $(2, 1/8)$ in $\mathcal{V}$ by the x-coordinate and use Proposition C.8 from Appendix C with $j = 2$ to see that

$$\omega = \text{Res}\left(F(x,y)x^{-n-1}y^{-n-1}dx \wedge dy\right) = \frac{-P(x,y)}{xy(x)Q_y(x,y(x))}x^{-n}y(x)^{-n}\,dx.$$

Moving the origin to $x = 2$, equals

$$\frac{1}{2\pi i}\int_\gamma \omega = 4^n \int_\gamma A(x)e^{-n\phi(x)}\,dx$$

where the series expansions for A and ϕ are given by

$$A(x) = -\frac{x^3}{8} - \frac{x^4}{16} + O(x^5)$$

$$\phi(x) = -\frac{x^4}{16} + O(x^6)\,.$$

Applying Theorem 4.1 to evaluate the integral on the segment $-\gamma_{j+1}$ using the parametrization $x = (i-1)t$ for $0 \le t \le \varepsilon$ gives a series for $\frac{1}{2\pi i}\int \omega$ that begins

$$4^n\left(\frac{-i}{4\pi}n^{-1} + \frac{(1+i)\sqrt{2}\Gamma(5/4)}{8\pi}n^{-5/4} + O(n^{-3/2})\right).$$

Similarly, on γ_j we parametrize by $x = (-i-1)t$ and obtain the complex conjugate of the previous expansion,

$$4^n\left(\frac{i}{4\pi}n^{-1} + \frac{(1-i)\sqrt{2}\Gamma(5/4)}{8\pi}n^{-5/4} + O(n^{-3/2})\right).$$

When the two contributions are summed the first terms cancel and we are left with

$$a_{n,n} \sim \frac{4^n\sqrt{2}\Gamma(5/4)}{4\pi}n^{-5/4}\,.$$

◁

9.3.2 Laurent series

In this section we discuss what can be done when the hypotheses of Algorithm 3 are satisfied, except that the series in question is a Laurent series rather than an ordinary power series. We first revisit Algorithm 3 from a different point of view, involving intersection numbers of middle-dimensional cycles.

Definition 9.33. Let X be a smooth, oriented real $2k$-manifold, and let γ_1 and γ_2 be two smooth, oriented k-cycles on X, intersecting transversely at finitely many points $\boldsymbol{x}_1, \ldots, \boldsymbol{x}_m$. The ***signed intersection number*** of γ_1 and γ_2 is the integer

$$\#(\gamma_1, \gamma_2) = \sum_{j=1}^{m} \operatorname{sgn}(\boldsymbol{x}_j),$$

where $\operatorname{sgn}(\boldsymbol{x}_j) = 1$ if the oriented bases B_1 and B_2 for the tangent spaces $T_p(\gamma_1)$ and $T_p(\gamma_2)$ form (in this order) a positively oriented basis for the tangent space $T_p(X)$, and $\operatorname{sgn}(\boldsymbol{x}_j) = -1$ otherwise.

The following construction can be found in [GP74] or [BJ82, pages 151–152].

Proposition 9.34. *Let X be an oriented real manifold of dimension $2k$ and let α and β be smooth oriented compact cycles of dimension k. Then generic perturbations of α and β will intersect transversely in a finite number of points [GP74, Section 2.3], and the resulting signed intersection number does not depend on the generic perturbation. In fact, the signed intersection number is an invariant [GP74, Section 3.3] of the homology classes $[\alpha]$ and $[\beta]$ in $H_k(X)$.* □

Let h be a (not necessarily proper) smooth Morse function on a complex k-manifold X with finitely many critical points $\boldsymbol{x}_1, \ldots, \boldsymbol{x}_m$, listed in order of decreasing height $h(\boldsymbol{x}_1) \geq \cdots \geq h(\boldsymbol{x}_m)$, such that all critical points have middle index k. For each $j \leq m$, let γ_j be a smooth cycle agreeing with the stable manifold of the (upward) gradient flow of h in a neighborhood of $\boldsymbol{x}_j$ having $\boldsymbol{x}_j$ as its highest point. Similarly, let γ^j be a smooth cycle agreeing with the unstable manifold of the gradient flow of h in a neighborhood of $\boldsymbol{x}_j$ with $\boldsymbol{x}_j$ as its lowest point. The γ_j are absolute cycles representing attachments in the Morse filtration at $\boldsymbol{x}_j$ (described in Appendix C). Similarly, the γ^j are absolute cycles representing attachments in the reverse Morse filtration at $\boldsymbol{x}_j$, obtained by replacing h by $-h$.

Proposition 9.35. *Let L be the subspace of $H_k(X)$ generated over the complex numbers by $\{[\gamma_j] : j \leq m\}$ and let L^* denote the dual space to L. Then $\{[\gamma^j] : j \leq m\}$ is a basis for L^* and the signed intersection number $\#(\gamma_i, \gamma^j)$ is a nonsingular pairing whose representing matrix M is upper triangular.*

Proof When $i = j$ the cycles γ_i and γ^j represent the stable and unstable manifolds of the gradient flow for a Morse function at the critical point $\boldsymbol{x}_j$. Morse functions are quadratically nondegenerate, therefore locally these intersect transversely at a single point, and they cannot intersect anywhere else due to the height restrictions. Hence the intersection number is ± 1. When $i > j$ the height restrictions prevent γ_i and γ^j from intersecting at all, whence $M_{ij} = 0$, so M is an upper triangular matrix with ± 1 diagonal entries, and thus nonsingular. □

Remark 9.36. If $h(\boldsymbol{x}_j) = h(\boldsymbol{x}_{j+r})$ for $r \geq 1$ then again the height restrictions prevent γ_i from intersecting γ^ℓ when i and ℓ are distinct elements of $\{j, \ldots, j+r\}$, hence the only non-zero entries in the submatrix $M_{[j,j+r],[j,j+r]}$ are those on the diagonal.

Algorithm 3 may be understood in terms of this pairing, as we now sketch.

Sketched alternative proof of correctness for Algorithm 3 Suppose that some component B' of the complement of the amoeba of $Q(x, y)$ contains a ray with small x coordinate that points up in the y direction, let T be the torus of integration for the bivariate Cauchy integral, with both x- and y-radii arbitrarily small, and let γ be the intersection cycle $\mathbf{INT}(T, T')$, where the basepoint of T' still has small x-coordinate but has sufficiently large y-coordinate to be in B'. Then γ consists of small cycles around the points $(0, s)$, as s ranges over the roots of $Q(0, y)$. Assuming these to be simple roots, the circles wind once about the origin.

The key is to interpret Steps 4(b-c) in Algorithm 3 using intersection numbers. Suppose $\boldsymbol{p}_j$ is a nondegenerate critical point reached by the algorithm, with corresponding ascent path γ^j. Steps 4(b-c) compute the intersection number of γ with γ^j. If, among the two branches of γ, one goes to the x-axis and one goes to the y-axis, then γ will intersect precisely one of the circles around a point $(0, s)$ and the intersection number will be ± 1. If both branches go to the x-axis then the intersection number is zero because they cannot intersect any of the small circles around the points $(0, s)$. Furthermore, the intersection number depends only on the homology class of the intersection cycle γ and, as shown in Corollary C.5 of Appendix C, the homology class of the intersection cycle obtained by keeping $|x|$ small and taking $|y|$ to infinity is the same as the intersection cycle obtained by keeping $|y|$ small and expanding $|x|$ to infinity. Interpreting the intersection cycle using this second construction shows that if both branches go to the y-axis then the intersection number is also zero.

The upshot is that in Step 4(c), the point $\boldsymbol{p}_j$ is added to $\mathbf{W}$ if and only if $\#(\gamma, \gamma^j) = \pm 1$ is non-zero. If any point at a given height is added, then all points at that height are added for which the intersection number of the ascent path with γ is ± 1 and no lower points are added. Inverting the dual basis shows that $\gamma - \sum_{i:\boldsymbol{p}_i \in \mathbf{W}} \pm \gamma_i$ is zero in $H_1(\mathcal{V}_*, \mathcal{V}_{<c_{xy}})$. □

We now return to the case of more general Laurent series. The difference between these and ordinary power series is that we can no longer count on the intersection cycle γ to be the union of small circles around points of intersection of $\mathcal{V}$ with one of the coordinate axes. The solution is to consider the components B where the series is defined and B' where height goes to $-\infty$ and trace an explicit intersection path γ between two points in these components. One can then try to infer the intersection numbers $\#(\gamma, \gamma^j)$ between γ and every ascent path γ^j from every critical point $\mathbf{p}_j$. If successful, this identifies $\mathbf{W}$ as the set of $\mathbf{p}_j$ such that $h(\mathbf{p}_j)$ is maximized among $\mathbf{p}_j$ such that $\#(\gamma, \gamma^j) = \pm 1$.

Example 9.37. The generating function $1/Q(x, y)$, where

$$Q(x, y) = 3 + x + x^{-1} + y + y^{-1} + \frac{1}{2}(x + x^{-1})(y + y^{-1}) + \frac{1}{5}(x - x^{-1} + y - y^{-1}),$$

appears in the analysis of certain matrix inversions arising from Green's function computations [Wan22]. Figure 9.6 shows a plot of the amoeba of Q. Its Newton polygon is the convex hull of the 3×3 grid of lattice points with $|x| \le 1$ and $|y| \le 1$.

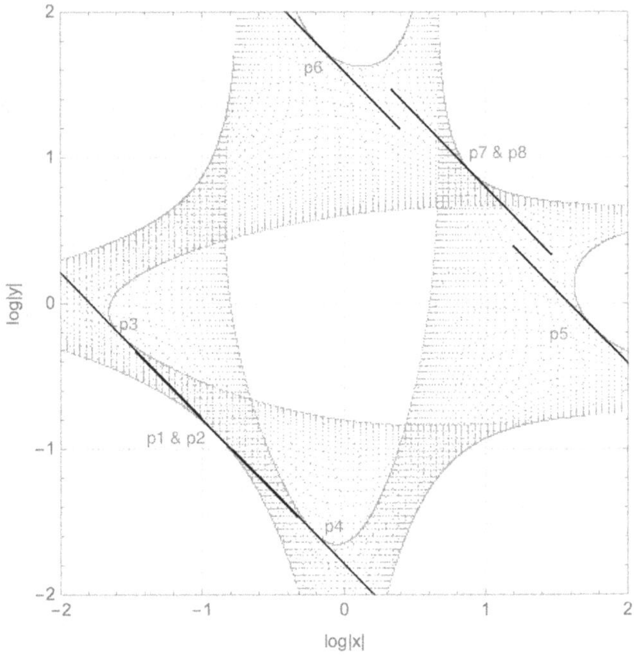

Figure 9.6 The amoeba of Q (reproduced with permission of Hong-Yi Wang).

The complement of the amoeba of Q has seven components, illustrated in

Figure 9.6. Six of the components are unbounded and correspond to vertices on the perimeter of the Newton polygon, and the seventh component is a bounded component corresponding to the origin, which is an interior lattice point of the Newton polygon.

Exercise 9.9. Laurent polynomials whose Newton polygons are as in Example 9.37 can have as many as nine components in the amoeba complement. However the specific Laurent polynomial $Q(x, y)$ under consideration only admits seven.

(a) Let $Q_{x+}(x, y)$ denote the sum of the three monomials in Q that have x-degree 1. How many distinct values do the magnitudes of the roots of $Q_{x+}(y)$ take?
(b) Let $Q_{x-}(x, y)$ denote the sum of the three monomials of Q that have x-degree -1. How many distinct values do the magnitudes of the roots of $Q_{x-}(y)$ take?
(c) Explain why there is only one amoeba 'tentacle' in the negative x direction whereas there are two in the positive x direction.

Continuing our current example, the component unbounded in the $(-1, -1)$ direction corresponds to a power series expansion. However, the series of combinatorial interest in this case is the one corresponding to the bounded component, which we call B. Specifically, asymptotics of this series in the $(1, 1)$ direction are desired. For the component B' we may choose any where $h_{(1,1)}$ is unbounded from below, and for specificity we choose the component in the upper right.

A quick computation shows the variety $\mathcal{V}_Q$ to be smooth and identifies eight critical points in the direction $\hat{\mathbf{r}}$ parallel to $(1, 1)$. Their projections under the Relog map are shown on the amoeba in Figure 9.6 and denoted by $\mathbf{p}_1, \ldots, \mathbf{p}_8$. All but two of the points, $\mathbf{p}_3$ and $\mathbf{p}_4$, are on the boundary of the amoeba, and the four points $\mathbf{p}_2 = \overline{\mathbf{p}_1}$ and $\mathbf{p}_8 = \overline{\mathbf{p}_7}$ come in conjugate pairs.

As described immediately prior to the example, we choose an explicit intersection cycle γ by moving the product of circles represented by the point $\boldsymbol{x} \in B$ to one represented by a point in B'. The size of a fiber $\texttt{amoeba}^{-1}(\boldsymbol{x})$ for $\boldsymbol{x} \in B$ only changes when crossing a point of the amoeba contour drawn in Figure 9.6, so by sampling points and performing algebraic computations it is possible to determine that the interior of the amoeba has four regions on which the log-modulus map from $\mathcal{V}$ to $\texttt{amoeba}(Q)$ is two-to-one, while the map is four-to-one on the remainder of the amoeba (the four-to-one regions are more heavily shaded in Figure 9.6).

To construct the intersection cycle γ, we first choose $\boldsymbol{x}$ to be the origin and move it in the $(1,0)$ direction halfway to p_5, then up and around the boundary of the component containing p_5 moving rightward to the edge of the picture and then upward into B'. Until the very end, this traces a single path in each of the two preimages of the log-modulus map. Therefore γ will be a single arc, centered on the preimage of the point x' where the amoeba was first entered (this boundary point having a single preimage) and extending downward until another piece of arc appears when passing through where the preimage size is four. This second arc can be made to occur below every critical point, therefore as a cycle relative to $-\infty$ the cycle γ is a simple arc in the preimage size 2 region with p_5 on its boundary.

We conclude immediately that p_1, p_2, p_3, p_4, p_5, and p_6 do not contribute. The first four are in fact higher than the origin, so the upward trajectories cannot possibly intersect the intersection cycle. For p_5, it suffices to check that the two upward trajectories can be drawn to be disjoint from our choice of γ. Indeed, the two ascent arcs, projected to the amoeba, move initially in the $(-1,-1)$ direction, and can do so until they are higher than the highest point on the intersection cycle. Where they go after that is unclear, because upon entering the preimage size 4 region it is no longer clear which is the increasing time direction, so the image of the arcs may no longer be able to move in the $(-1,-1)$ direction. However, they are already high enough that they cannot intersect γ. By symmetry, an identical argument (choosing a different γ) shows that p_6 cannot contribute.

By symmetry, p_7 contributes if and only if p_8 contributes. By process of elimination, because we know the asymptotics are non-zero, these both contribute. To argue this geometrically, one needs to understand where the two ascent arcs from p_7 go. The description in terms of the four preimages is a little complicated, but one finds in the end that the projections of the two ascent arcs to the amoeba pass around the hole (the region B) in opposite ways, one to the north and one to the south. This forces the intersection number with γ to be ± 1; see [Wan22] for details.

We conclude that the intersection cycle γ is homologous to the sum of a homology generator going downward from p_7 and a homology generator going downward from p_8, with properly chosen signs. The coordinates of the critical points are algebraic numbers satisfying

$$55\,x^8 + 664\,x^7 + 2840\,x^6 + 5780\,x^5 + 5610\,x^4 + 2520\,x^3 + 440\,x^2 - 44\,x - 45 = 0\,.$$

The points p_7 and p_8 are on the diagonal, conjugate to each other, with coordinates $-2.19\ldots \pm i1.10\ldots$. The two contributions have opposite phases, ultimately giving that

$$a_{n,n} \sim Cn^{-1/2}\alpha^{-n}\cos(n\theta),$$

where $\alpha = 6.03\ldots$ is the absolute value of the product of the coordinates in $\boldsymbol{p}_7$ (or equivalently the coordinates of $\boldsymbol{p}_8$) and C and θ are non-zero constants. ◁

Exercise 9.10. Find the constants C and θ for this example.

9.3.3 Smooth Bivariate Generating Functions with Degeneracies

Using the results of Chapter 4 we can give asymptotics for bivariate smooth point asymptotics in directions where the phase function ϕ vanishes to arbitrary order. For simplicity, we consider a power series expansion and assume that the dominant singularities are finitely minimal points where the numerator P is nonvanishing. It is also possible to derive (more complicated) results when these conditions fail: for instance, they fail in Example 9.32 above.

Let (x_*, y_*) be a smooth minimal critical point in the direction $\hat{\boldsymbol{r}}$ and assume that $Q_y(x_*, y_*) \neq 0$ so that we can parameterize $y = g(x)$ on $\mathcal{V}$ near (x_*, y_*). Theorem 9.17 and (9.16) define functions A and ϕ such that

$$x_*^r y_*^s (a_{rs} - \chi) = O\left(e^{-\varepsilon s}\right),$$

where

$$\chi(r,s) = x_*^{-r} y_*^{-s} \frac{1}{2\pi} \int_{-\varepsilon}^{\varepsilon} e^{-s\phi(\theta)} A(\theta)\, d\theta\,. \tag{9.19}$$

Let $c = c_\kappa$ denote the leading non-zero series coefficient in the expansion $\phi(x) \sim c_\kappa x^\kappa$ as $x \to 0$ and define the quantity

$$\Phi_{x_*,y_*}(\boldsymbol{r}) = -\frac{\Gamma(1/\kappa)}{2\kappa\pi}(1-\zeta)\frac{P(x_*, y_*)}{y_* Q_y(x_*, y_*)} c^{-1/\kappa} s^{-1/k} x_*^{-r} y_*^{-s}, \tag{9.20}$$

where, as in Theorem 4.1(*iii*), $\zeta = -1$ if κ is even and $\zeta = \exp(\sigma i\pi/\kappa)$ if κ is odd.

Theorem 9.38. *If (x_*, y_*) is a strictly minimal critical point in the direction $\hat{\boldsymbol{r}}$ and satisfies the conditions above then as $(r,s) \to \infty$ with the distance from (r,s) to the ray $\{t\hat{\boldsymbol{r}} : t \geq 0\}$ remaining bounded, there is an asymptotic series of the form*

$$a_{rs} \approx x_*^{-r} y_*^{-s} \sum_{j=0}^{\infty} v_j s^{(-1-j)/k}$$

with leading term $\Phi_{x_*,y_*}(\boldsymbol{r})$. If (x_*, y_*) is a finitely minimal point and all critical points with the same coordinate-wise modulus satisfy the same conditions as

(x_*, y_*) then an asymptotic series for a_{rs} is obtained by adding the contributions of each of the critical points.

Proof The asymptotic development follows from (9.19) and Theorem 4.1. It remains to check that the leading term is given by (9.20). Starting from (9.19) use Theorem 4.1 with $\ell = 0$ to get, in the notation of Theorem 4.1, the leading term

$$\begin{aligned}\chi &\sim \frac{x_*^{-r} y_*^{-s}}{2\pi} \int_{-\varepsilon}^{\varepsilon} A(x) e^{-s\phi(x)}\, dx \\ &= \frac{x_*^{-r} y_*^{-s}}{2\pi} I(s) \\ &= \frac{x_*^{-r} y_*^{-s}}{2\pi} (1-\zeta) C(\kappa, 0) A(0) (cs)^{-1/\kappa}.\end{aligned}$$

Parametrizing by y means choosing coordinate $k = 2$ giving sign $(-1)^{k-1} = -1$ in $A(0) = -\dfrac{P(x_*, y_*)}{y_* Q_y(x_*, y_*)}$, and the fact that $C(\kappa, 0) = \dfrac{\Gamma(1/\kappa)}{\kappa}$ gives (9.20). □

Example 9.39 (Cube root asymptotics). Let $F(x, y) = 1/(3 - 3x - y + x^2)$ so that the set $\mathcal{V}$ is parametrized by $y = g(x) = x^2 - 3x + 3$. If $\hat{r} = (a, 1-a)$ then asymptotic behavior depends on whether a is less than, equal to, or greater than $1/2$. When $a < 1/2$ there are two real critical points on the curve $y = g(x)$ – as a increases from 0 to $1/2$ one approaches $(1, 1)$ from the left, and the other approaches $(1, 1)$ from the right (see Figure 9.7). Only the critical point on the right of $(1, 1)$ is minimal, and it determines asymptotics. When $a = 1/2$, the two critical points meet and h becomes quadratically degenerate. Once $a > 1/2$, the critical points have complex conjugate coordinates and are both minimal.

Because $(1, 1)$ is a minimal point, the main diagonal has exponential rate zero, while all other directions have exponential decay at a rate that is uniform over compact subsets of directions not containing the diagonal. Implicit differentiation implies

$$g''(x) = -3\frac{x(x^2 - 4x + 3)}{(x^2 - 3x + 3)^2},$$

which vanishes when $x = 1$ as the critical point $(1, 1)$ in the main diagonal direction is degenerate. Computing further, we find that $g(x) - g(1)$ vanishes to order $\kappa = 3$ here, with $c = c_3 = g'''(1)/3! = i$. Checking the signs gives $\zeta = -e^{i\pi/3}$ and therefore

$$i^{-1/3}(1 - \zeta) = e^{i\pi/6} + e^{-i\pi/6} = 2\cos(\pi/6) = \sqrt{3}.$$

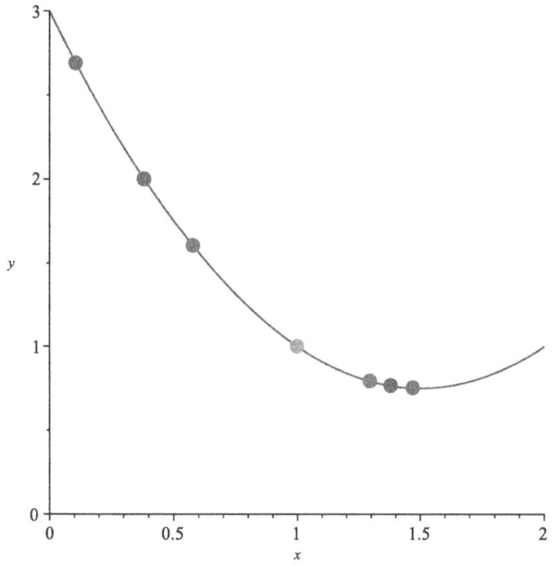

Figure 9.7 The two critical points for $Q(x, y) = 3 - 3x + x^2 - y$ in directions $(a, 1 - a)$ with $a < 1/2$, which approach the same degenerate 'double point' $(1, 1)$ when $a \to 1/2$. Only the critical points with $x > 1$ are minimal for such directions.

Evaluating $A(0) = -P(1, 1)/yQ_y(1, 1) = -1/(-1) = 1$, Theorem 9.38 gives

$$a_{r,r} \sim \frac{1}{2\pi} C(3, 0) i^{-1/3} (1 - \zeta) r^{-1/3} = \frac{\sqrt{3}\Gamma(1/3)}{6\pi} r^{-1/3} .$$

◁

Remark. We have given a formula holding only very near a fixed direction $\hat{\boldsymbol{r}}$. Because the results for nondegenerate smooth points hold in neighborhoods of directions, it remains to be seen whether asymptotics can be worked out that "bridge the gap" and hold when the distance $\delta = \|\boldsymbol{r} - |\boldsymbol{r}| \cdot \hat{\boldsymbol{r}}\|$ to the ray $\{\lambda \cdot \hat{\boldsymbol{r}}\}$ satisfies $|\boldsymbol{r}| \gg \delta \gg 1$. See Section 13.2 for further discussion.

9.4 Additional formulae for asymptotics

It is sometimes useful to have alternative or more detailed formulae for the coefficients of the asymptotic expansions derived above. We collect some such formulae in this section.

9.4.1 Higher order terms

We first examine explicit expressions for the higher order asymptotic coefficients.

Theorem 9.40. *Under the hypotheses of Theorem 9.5, the asymptotic series $\Phi_{\boldsymbol{w}}(\boldsymbol{r})$ can be expressed as*

$$\Phi_{\boldsymbol{w}}(\boldsymbol{r}) = \frac{\boldsymbol{w}^{-\boldsymbol{r}}}{\sqrt{\det(2\pi r_d \mathcal{H})}} \sum_{k\geq 0} \sum_{j=0}^{p-1} \frac{(r_d+1)_{(p-1-j)}}{(p-1-j)!j!} r_d^{-k} L_k(A_j, \underline{\phi}),$$

where

$$L_k(A,\phi) = (-1)^k \sum_{0\leq \ell \leq 2k} \left. \frac{\mathcal{D}^{\ell+k}\left(A(\boldsymbol{x})\cdot \underline{\phi}(\boldsymbol{x})^{\ell}\right)}{2^{\ell+k}\ell!(\ell+k)!} \right|_{\boldsymbol{x}=\boldsymbol{0}}$$

for the functions

$$\underline{\phi}(\boldsymbol{x}) = \phi(\boldsymbol{x}) - (1/2)\boldsymbol{x}\cdot \mathcal{H} \cdot \boldsymbol{x}^T$$

$$A_j(\boldsymbol{\theta}) = R_j(\boldsymbol{w}^{\circ} e^{i\boldsymbol{\theta}})$$

$$R_j(\boldsymbol{z}) = (-g(\boldsymbol{z}^{\circ}))^{-p+j} \lim_{z_d\to g(\boldsymbol{z}^{\circ})} \partial_d^j \left((z_d - g(\boldsymbol{z}^{\circ}))^p F(\boldsymbol{z})\right),$$

and $\mathcal{D}$ is the differential operator

$$\mathcal{D} = -\sum_{1\leq i,j\leq d} \left(\mathcal{H}^{-1}\right)_{ij} \partial_i \partial_j .$$

Proof Theorem 9.17 and Lemma 9.18 above imply that $\Phi_{\boldsymbol{w}}(\boldsymbol{r})$ is obtained from an asymptotic expansion of the saddle point integral (9.16), where the residue Ψ is a weighted sum of terms R_k occurring in the explicit formula (9.15). Distributing the integral over the sum of residue terms, Lemma 5.16 from Chapter 5 gives an asymptotic expansion of each. The stated result follows from simplifying the sum of these expansions, taking into account a subtle interplay between the lower order terms in (9.15) and the higher order terms in Lemma 5.16. □

Remark 9.41. It is possible to expand the falling factorials in terms of the Stirling numbers of the second kind and collect powers of r_d, to give an explicit formula for the coefficients C_k in (9.4) at the cost of an even more unwieldy formula.

Exercise 9.11. In the cases $p = 1$ and $p = 2$, explicitly compute the second term in the asymptotic expansion of $\Phi_{\boldsymbol{w}}(\boldsymbol{r})$ in descending powers of r_d.

9.4.2 A geometric formula for the leading term

Complementing the explicit expressions for asymptotics given above, it is possible to write down a coordinate-free representation for the leading term using the curvature of $\mathcal{V}$ near the contributing singularities determining asymptotics. In addition to an alternative, sometimes more compact, expression, coordinate-free representations can also help with conceptual understanding, such as in Example 9.47 below. We begin by reviewing the definition of the Gaussian curvature of a smooth hypersurface, before extending it to certain points of complex algebraic hypersurfaces.

Gaussian curvature of real hypersurfaces

For a smooth orientable hypersurface $\mathcal{V} \subset \mathbb{R}^{d+1}$, the ***Gauss map*** $\mathcal{G}$ sends each point $\mathbf{p} \in \mathcal{V}$ to a normal vector $\mathcal{G}(\mathbf{p})$, which we identify with an element of the d-dimensional unit sphere S^d. For a given patch $P \subset \mathcal{V}$ containing $\mathbf{p}$, let $\mathcal{G}[P] = \cup_{q \in P} \mathcal{G}(\mathbf{q})$. The ***Gaussian curvature*** (also called *Gauss–Kronecker curvature*) of $\mathcal{V}$ at $\mathbf{p}$ is defined as the limit

$$\mathcal{K} = \lim_{P \to p} \frac{A(\mathcal{G}[P])}{A[P]} \tag{9.21}$$

as P shrinks to the single point p, where $A(\mathcal{G}[P])$ is the area of $\mathcal{G}[P]$ in S^d and $A[P]$ is the area of P in $\mathcal{V}$. When d is odd, the antipodal map on S^d has determinant -1, whence the particular choice of unit normal will influence the sign $\mathcal{K}$, which is therefore only well defined up to sign. When d is even, we take the numerator to be negative if the map $\mathcal{G}$ is orientation reversing and we have a well defined signed quantity. The curvature $\mathcal{K}$ is equal to the Jacobian determinant of the Gauss map at the point $\mathbf{p}$.

For computational purposes, it is convenient to use standard formulae for the curvature of the graph of a function from $\mathbb{R}^d$ to $\mathbb{R}$. If η is a homogeneous quadratic form, we let $||\eta||$ denote the determinant of the Hessian matrix of η computed with respect to any orthonormal basis.

Proposition 9.42 ([Bar+10, Corollary 2.4]). *Let $\mathcal{P}$ be the tangent plane to $\mathcal{V}$ at $\mathbf{p}$ and let $\mathbf{v}$ be a unit normal vector. Suppose that $\mathcal{V}$ is the graph of a smooth function h over $\mathcal{P}$, meaning*

$$\mathcal{V} = \{\mathbf{p} + \mathbf{u} + h(\mathbf{u})\mathbf{v} : \mathbf{u} \in U \subseteq \mathcal{P}\}.$$

If η is the quadratic part of h, so that $h(\mathbf{u}) = \eta(\mathbf{u}) + O(|\mathbf{u}|^3)$, then the curvature of $\mathcal{V}$ at $\mathbf{p}$ is $\mathcal{K} = ||\eta||$. □

Corollary 9.43 (curvature of the zero set of a polynomial). *Suppose that $\mathcal{V} = \{\mathbf{x} \in \mathbb{R}^d : Q(\mathbf{x}) = 0\}$ and that $\nabla Q(\mathbf{p}) \neq \mathbf{0}$. If η is the quadratic part of Q at $\mathbf{p}$*

and $\eta_\perp$ is the restriction of η to the hyperplane orthogonal to $\nabla Q(\mathbf{p})$ then the curvature of $\mathcal{V}$ at $\mathbf{p}$ is given by

$$\mathcal{K} = \frac{\|\eta_\perp\|}{\|(\nabla Q)(\mathbf{p})\|_2^{d-1}}, \tag{9.22}$$

where $\|(\nabla Q)(\mathbf{p})\|_2$ denotes the Euclidean norm of the gradient of Q at $\mathbf{p}$.

Proof Replacing Q by $\|(\nabla Q)(\mathbf{p})\|_2^{-1} Q$ leaves $\mathcal{V}$ unchanged and reduces to the case $\|(\nabla Q)(\mathbf{p})\|_2 = 1$, so we assume without loss of generality that $\|(\nabla Q)(\mathbf{p})\|_2 = 1$. Given an arbitrary vector $\boldsymbol{u}$ we write $\boldsymbol{u} = \boldsymbol{u}_\perp + \lambda(\boldsymbol{u})(\nabla Q)(\mathbf{p})$ to denote the decomposition of $\boldsymbol{u}$ into its components orthogonal to, and contained in, the span of $(\nabla Q)(\mathbf{p})$. The Taylor expansion of Q near $\mathbf{p}$ is

$$Q(\mathbf{p} + \boldsymbol{u}) = (\nabla Q)(\mathbf{p}) \cdot \boldsymbol{u} + \eta_\perp(\boldsymbol{u}) + R,$$

where $R = O(|\boldsymbol{u}_\perp|^3 + |\lambda(\boldsymbol{u})||\boldsymbol{u}_\perp|)$. Near the origin, we can solve for λ to obtain

$$\lambda(\boldsymbol{u}) = \eta_\perp(\boldsymbol{u}) + O(|\boldsymbol{u}|^3),$$

and the result follows from Proposition 9.42. □

Gaussian curvature at minimal points of complex hypersurfaces

Suppose now that Q is a real polynomial in d variables and that $\mathbf{p}$ is a minimal smooth point of the corresponding complex algebraic hypersurface. We are interested in the curvature at $\log \mathbf{p}$ of the logarithmic image $\log \mathcal{V} = \{\boldsymbol{z} \in \mathbb{C}^d : (Q \circ \exp)(\boldsymbol{z}) = 0\}$ of $\mathcal{V}$ (this image is similar to the amoeba of Q except we do not take moduli). When $\mathbf{p}$ is a point with positive real coordinates then the curvature at $\log \mathbf{p}$ can be defined (up to a factor of ± 1) directly using (9.22) from Corollary 9.43. In fact, we use this formula to define curvature in the general complex case as it is invariant under scalar multiplications of Q and Theorem 6.44 from Chapter 6 implies that the normal $(\nabla_{\log} Q)(\mathbf{p})$ to $Q \circ \exp$ at a minimal point $\mathbf{p}$ is a scalar multiple of a real vector.

It is useful to observe that the curvature $\mathcal{K}$ is a reparametrization of the Hessian determinant in our asymptotic theorems, in the sense that they vanish together.

Proposition 9.44. *The quantity $\mathcal{K}$ defined by* (9.22) *vanishes if and only if the determinant of the Hessian matrix $\mathcal{H}$ in Theorem 9.5 vanishes.*

Proof Going back to its original definition in Lemma 8.22, the matrix $\mathcal{H}$ in Theorem 9.5 is the Hessian matrix for the function g expressing $\log \mathcal{V}$ as a graph over the first $(d-1)$ coordinates. At such a point, the tangent plane to $\log \mathcal{V}$ is not perpendicular to the dth coordinate plane, and reparametrizing the graph to be over the tangent plane does not change whether the Hessian is singular. The Hessian matrix obtained from such a reparametrization represents the quadratic form η in Proposition 9.42, so singularity of the Hessian matrix from Theorem 9.5 is equivalent to singularity of η in Proposition 9.42. □

Theorem 9.45 (Main Theorem of Smooth ACSV (Curvature Version)). *Let $F(\boldsymbol{z}) = P(\boldsymbol{z})/Q(\boldsymbol{z})$ be the ratio of coprime polynomials with convergent Laurent series expansion $F(\boldsymbol{z}) = \sum_{\boldsymbol{r}\in\mathbb{Z}^d} a_{\boldsymbol{r}}\boldsymbol{z}^{\boldsymbol{r}}$. Suppose there exists a compact set $\mathcal{R} \subset \mathbb{R}^d$ of directions such that F has a smooth strictly minimal nondegenerate contributing point $\boldsymbol{w} = \boldsymbol{w}(\hat{\boldsymbol{r}}) \in \mathbb{C}_*^d$, where $Q_{z_d}(\boldsymbol{w}) \neq 0$ whenever $\hat{\boldsymbol{r}} \in \mathcal{R}$. Let $\mathcal{K}(\hat{\boldsymbol{r}})$ denote the Gaussian curvature of* $\log \mathcal{V}$ *at* $\log \boldsymbol{w}(\hat{\boldsymbol{r}})$. *Then*

$$a_{\boldsymbol{r}} = \left(\frac{1}{2\pi\|\boldsymbol{r}\|_2}\right)^{(d-1)/2} \boldsymbol{w}^{-\boldsymbol{r}}\,\mathcal{K}(\hat{\boldsymbol{r}})^{-1/2}\left(\frac{P(\boldsymbol{w})}{\|\nabla_{\log} Q(\boldsymbol{w})\|_2^2} + O\left(\|\boldsymbol{r}\|_2^{-1}\right)\right) \tag{9.23}$$

uniformly as $\|\boldsymbol{r}\|_2 \to \infty$ with $\hat{\boldsymbol{r}} \in \mathcal{R}$. The square-root of the matrix determinant is the product of the principal branch square-roots of the Jacobian of the Gauss map when the Gauss map is oriented towards $-\hat{\boldsymbol{r}}$.

Proof As in the proofs above, we let $\omega = \boldsymbol{z}^{-\boldsymbol{r}-1}F(\boldsymbol{z})d\boldsymbol{z}$ so that

$$a_{\boldsymbol{r}} = \left(\frac{1}{2\pi i}\right)^{d-1} \int_\sigma \operatorname{Res}(\omega)$$

where σ is an intersection class on $\mathcal{V}$. To work in log space we let $\boldsymbol{z} = \exp(\boldsymbol{\zeta})$, so $d\boldsymbol{z} = \boldsymbol{z}d\boldsymbol{\zeta}$ and

$$a_{\boldsymbol{r}} = \left(\frac{1}{2\pi i}\right)^{d-1} \int_{\tilde{\sigma}} \operatorname{Res}\left(\exp(-\boldsymbol{r}\cdot\boldsymbol{\zeta})\tilde{F}(\boldsymbol{\zeta})\,d\boldsymbol{\zeta}\right),$$

where $\tilde{F} = F \circ \exp$ and $\tilde{\sigma} = \log \sigma$. In fact, our assumptions imply that we have a simple pole, so we can pull out the factor of $\boldsymbol{z}^{-\boldsymbol{r}} = \exp(-\boldsymbol{r}\cdot\boldsymbol{\zeta})$ to obtain

$$a_{\boldsymbol{r}} = \left(\frac{1}{2\pi i}\right)^{d-1} \int_{\tilde{\sigma}} \exp(-\boldsymbol{r}\cdot\boldsymbol{\zeta})\operatorname{Res}(\tilde{F}(\boldsymbol{\zeta})\,d\boldsymbol{\zeta}). \tag{9.24}$$

Let $\mathcal{P}$ be the tangent space to $\log \mathcal{V}$ at the point $\boldsymbol{\zeta}_* = \log \boldsymbol{w}$. This tangent space consists of the vectors orthogonal to $\hat{\boldsymbol{r}}$, so we may locally parameterize $\log \mathcal{V}$ near $\boldsymbol{\zeta}_*$ by $\mathcal{P}$ using a representation

$$\log \mathcal{V} = \{\boldsymbol{\zeta}_* + \boldsymbol{\zeta}_{\|} + h(\boldsymbol{\zeta}_{\|})\hat{\boldsymbol{r}} : \boldsymbol{\zeta}_{\|} \in \mathcal{P}\}.$$

Pick an orthonormal basis $\boldsymbol{v}^{(2)}, \ldots, \boldsymbol{v}^{(d)}$ for $\mathcal{P}$ so that a general point $\boldsymbol{\zeta} \in \mathbb{C}^d$ in a neighborhood of $\boldsymbol{\zeta}_*$ has a representation

$$\boldsymbol{\zeta} = \boldsymbol{\zeta}_* + u_1\hat{\boldsymbol{r}} + \sum_{j=2}^{d} u_j \boldsymbol{v}^{(j)}.$$

Proposition C.8 in Appendix C implies

$$\operatorname{Res}(\tilde{F}(\zeta)\,d\zeta) = \frac{P \circ \exp}{\partial(Q \circ \exp)/\partial u_1}\, du_2 \wedge \cdots \wedge du_{d+1}\,,$$

and the partial derivative in the direction of the gradient is the square of the magnitude of the gradient. Thus,

$$\operatorname{Res}(\tilde{F}(\zeta)\,d\zeta)(\zeta_*) = \frac{P(\boldsymbol{w})}{\|\nabla_{\log} Q(\boldsymbol{w})\|_2^2}\, dA, \tag{9.25}$$

where $dA = d\boldsymbol{u}_{\|} = du_2 \wedge \cdots \wedge du_d$ is equal to the oriented holomorphic $(d-1)$-area form for $\log \mathcal{V}$ as it is immersed in $\mathbb{C}^d$.

Let $\lambda = |\boldsymbol{r}|$ and $\phi(\zeta) = \hat{\boldsymbol{r}} \cdot \zeta$ so that (9.24) becomes

$$a_{\boldsymbol{r}} = \left(\frac{1}{2\pi i}\right)^d \int_{\tilde{\sigma}} \exp(-\lambda\phi(\zeta)) \operatorname{Res}(\tilde{F}(\zeta)\,d\zeta)\,, \tag{9.26}$$

and let η denote the quadratic part of h. By Proposition 9.42 (or Corollary 9.43) and the subsequent discussion, we see that the curvature $\mathcal{K}$ of $\log \mathcal{V}$ at the point ζ_* with respect to the unit normal $\hat{\boldsymbol{r}}$ is given by $\|\eta\|$.

To proceed, we describe a logspace intersection cycle $\tilde{\sigma}$. One way to construct $\tilde{\sigma}$ is to pick a point $\boldsymbol{x}'$ in the component of $\texttt{amoeba}(Q)^c$ giving the series expansion under consideration, and a point $\boldsymbol{x}''$ in a component of $\texttt{amoeba}(Q)^c$ on which the height function h is unbounded, and take the intersection cycle of $\log \mathcal{V}$ with a homotopy $\mathbf{H}$ obtained by taking a straight line from $\boldsymbol{x}'$ to $\boldsymbol{x}''$ and mapping by $\operatorname{Relog}^{-1}$. A convenient choice is to make the segment $\overline{\boldsymbol{x}'\boldsymbol{x}''}$ parallel to $\hat{\boldsymbol{r}}$. The real tangent space to $\mathbf{H}$ is then the sum of the imaginary d-space and the real 1-space in direction $\hat{\boldsymbol{r}}$. The tangent space to $\log \mathcal{V}$ is the sum of the real $(d-1)$-space orthogonal to $\hat{\boldsymbol{r}}$ and the imaginary $(d-1)$-space orthogonal to $\hat{\boldsymbol{r}}$. The tangent space to $\tilde{\sigma}$ is the intersection of these, which is the imaginary $(d-1)$-space orthogonal to $\hat{\boldsymbol{r}}$ – in other words, just $\operatorname{Im} \mathcal{P}$.

Because $\tilde{\sigma}$ is contained in the linear space $\operatorname{Im} \mathcal{P} + \mathbb{C} \cdot \hat{\boldsymbol{r}}$, we see that locally there is a unique analytic function $\alpha : \operatorname{Im} \mathcal{P} \to \mathbb{C} \cdot \hat{\boldsymbol{r}}$ such that $\zeta + \alpha(\zeta) \in \tilde{\sigma}$. Comparing to our parametrization above, we see that $\alpha = h$, so the quadratic part of α is therefore equal to η. Because our multivariate integral formulae are in terms of real parametrizations, we reparametrize $\operatorname{Im} \mathcal{P}$ by $\zeta = i\boldsymbol{y}$ and $d\zeta = i^d\, d\boldsymbol{y}$. In these coordinates, locally

$$\tilde{\sigma} = \{i\boldsymbol{y} + h(i\boldsymbol{y}) : \boldsymbol{y} \in \operatorname{Re} \mathcal{P}\}\,. \tag{9.27}$$

Using $\hat{\boldsymbol{r}} \cdot \boldsymbol{y}_{\parallel} = 0$ and $\hat{\boldsymbol{r}} \cdot \hat{\boldsymbol{r}} = 1$, we obtain

$$\begin{aligned}\phi(i\boldsymbol{y} + h(i\boldsymbol{y})) &= \phi(\boldsymbol{\zeta}_*) + h(i\boldsymbol{y}) \\ &= \phi(\boldsymbol{\zeta}_*) + \eta(i\boldsymbol{y}) + O(|\boldsymbol{y}|^3) \\ &= \phi(\boldsymbol{\zeta}_*) - \eta(\boldsymbol{y}) + O(|\boldsymbol{y}|^3)\,.\end{aligned}$$

We know, by our assumptions, that ϕ is a smooth phase function whose real part has a minimum on $\tilde{\sigma}$ at $\boldsymbol{\zeta}_*$, which is $\boldsymbol{y} = 0$ in the parametrization (9.27). Applying Theorem 5.3 to (9.26) using the evaluation (9.25) then gives (9.23), where the square-root of the curvature is taken to be the reciprocal of the product of the principal square-roots of the eigenvalues of $-\eta$ in the positive $\hat{\boldsymbol{r}}$-direction, all of which have nonnegative real parts. The eigenvalues of $-\eta$ in direction $\hat{\boldsymbol{r}}$ are the same as the eigenvalues of η in direction $-\hat{\boldsymbol{r}}$, which finishes the proof of the theorem. □

Again, minor modifications to proof extend to include the case where there are finitely many critical points on a minimizing torus.

Corollary 9.46. *Let $F(\boldsymbol{z}) = P(\boldsymbol{z})/Q(\boldsymbol{z})$ be the ratio of coprime polynomials with convergent Laurent series expansion $F(\boldsymbol{z}) = \sum_{\boldsymbol{r} \in \mathbb{Z}^d} a_{\boldsymbol{r}} \boldsymbol{z}^{\boldsymbol{r}}$ corresponding to the amoeba complement component $B \subset$ amoeba$(Q)^c$. Suppose there exists a compact set $\mathcal{R} \subset \mathbb{R}^d$ of directions such that for each $\hat{\boldsymbol{r}} \in \mathcal{R}$ the function $\hat{\boldsymbol{r}} \cdot \boldsymbol{x}$ is uniquely maximized at $\boldsymbol{x}_{\min} \in \overline{B}$, and that the set $\mathbf{W}$ of critical points in $T_e(\boldsymbol{x}_{\min})$ is finite, nonempty, and consists of smooth nondegenerate points where some partial derivative of Q does not vanish. For each $\boldsymbol{z} \in \mathbf{W}(\hat{\boldsymbol{r}})$ write $\boldsymbol{z} = \exp(\boldsymbol{x}_{\min} + i\boldsymbol{y})$. Then*

$$a_{\boldsymbol{r}} = \left(\frac{1}{2\pi|\boldsymbol{r}|}\right)^{(d-1)/2} e^{-\boldsymbol{r}\cdot\boldsymbol{x}} \left[\sum_{\boldsymbol{z} \in \mathbf{W}(\hat{\boldsymbol{r}})} e^{-i\boldsymbol{r}\cdot\boldsymbol{y}} \frac{P(\boldsymbol{z})}{\|\nabla_{\log} Q(\boldsymbol{z})\|_2^2} \mathcal{K}(\boldsymbol{z})^{-1/2} + O(|\boldsymbol{r}|^{-1})\right]$$

uniformly as $|\boldsymbol{r}| \to \infty$ with $\hat{\boldsymbol{r}} \in \mathcal{R}$. □

Example 9.47 (Quantum walk)**.** A ***quantum walk*** or ***quantum random walk*** (QRW) is a model for a particle moving in $\mathbb{Z}^d$ under a quantum evolution in which the randomness is provided by a unitary evolution operator on a hidden variable taking k states. States and position are simultaneously measurable, but one must not measure either until the final time n or the quantum interference is destroyed. A one-dimensional quantum walk was briefly presented in Exercise 9.8 as an example of torality. Here we illuminate the general form of asymptotics for a QRW, using Theorem 9.45 and Corollary 9.46 to qualitatively describe the probability profile of the particle at large time n. Further examples of QRWs are given in Chapter 12.

A QRW is defined by a $k \times k$ unitary matrix U along with k vectors $\boldsymbol{v}^{(1)}, \ldots, \boldsymbol{v}^{(k)}$

in $\mathbb{Z}^d$ representing possible steps of the walk. At each time step, the particle chooses a new state $j \in [k]$ and then moves by a jump of $\boldsymbol{v}^{(j)}$. The amplitude of a transition from state i to j is the entry $U_{i,j}$, while the amplitude of a path of n steps, starting in state i_0 and ending in states i_n is $\prod_{t=0}^{n-1} U_{i_t,i_{t+1}}$. Suppose the particle starts, at time zero, at the origin in state i. The amplitude of moving from $\mathbf{0}$ to a point $\boldsymbol{p}$ in n time steps and ending in state j is obtained by summing the amplitudes of all paths of n steps having total displacement $\boldsymbol{p}$ and ending in state j. This description gives us everything we need to compute asymptotics of QRWs – for more on the interpretation of quantum walks, see [Amb+01; Bar+10].

The multiplicative nature of the amplitudes makes QRW a perfect candidate for the transfer matrix method, the univariate version of which was discussed in Section 2.2 and whose multivariate version will be discussed at length in Section 12.4. Let M denote the $k \times k$ diagonal matrix whose (j, j)-entry is the monomial $\boldsymbol{z}^{\boldsymbol{v}^{(j)}}$ and let $P(\boldsymbol{p}, n)$ be the matrix whose (i, j)-entry is the amplitude to go from the origin in state i at time zero to $\boldsymbol{p}$ in state j at time n. Define the ***spacetime generating function***

$$F(\boldsymbol{z}) = \sum_{\substack{\boldsymbol{p}\in\mathbb{Z}^d \\ n\geq 0}} P(\boldsymbol{p}, n)(\boldsymbol{z}^\circ)^{\boldsymbol{p}} z_{d+1}^n \tag{9.28}$$

where $\boldsymbol{z}^\circ = (z_1, \ldots, z_d)$ are d variables tracking walk position and z_{d+1} is a variable tracking walk length. The transfer matrix method easily gives

$$F(\boldsymbol{z}) = (I - z_{d+1} M U)^{-1},$$

and the entries F_{ij} are rational functions with common denominator

$$Q = \det(I - z_{d+1} M U)\,. \tag{9.29}$$

Exercise 9.12. Prove that Q in (9.29) satisfies the strong torality hypothesis from Definition 9.20.

The component B of the amoeba complement that yields a series in z_{d+1} whose coefficients are Laurent polynomials in $z_1, \ldots, z_d$ is contained in the negative z_{d+1} halfspace and has the origin on its boundary. Its boundary is smooth everywhere except the origin, where its dual cone K has nonempty interior; see Figure 9.8. Recall the dual rate function β^* on directions from (6.5). Whenever $\mathbf{0} \in \partial B$ we may deduce that $\beta^*(\hat{\boldsymbol{r}}) \leq 0$ with equality only possible if $\hat{\boldsymbol{r}} \in \texttt{normal}_0(B)$. The ***feasible velocity region of the QRW*** is the set $R \subseteq \mathbb{R}^d$ consisting of all $(r_1, \ldots, r_d)$ such that exponential growth rate $\overline{\beta}(r_1, \ldots, r_d, 1)$ from Definition 6.20 vanishes (in other words, it is the set of directions in which the chance of finding the particle roughly at that rescaled point after a long time decays slower than exponentially). Then $R \subseteq \Xi$, where Ξ is the $r_d = 1$ slice of $\texttt{normal}_0(B)$ and is computed as an algebraic dual (see Example 6.50).

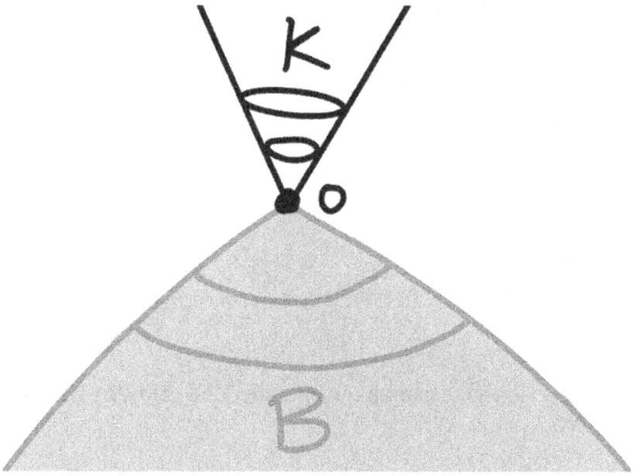

Figure 9.8 The component B and its dual cone K at the origin.

Smooth boundary points correspond to directions $\hat{r} \notin K$ satisfying $|r_1| + \cdots + |r_d| < |r_{d+1}|$. For each such $\hat{r}$ there is one or more minimal smooth critical points of $\mathcal{V}$. To compute R, we start by computing $\mathcal{V}_0 = \mathcal{V} \cap \mathbf{T}(\mathbf{0})$. For many QRW's one finds this to be a smooth manifold diffeomorphic to one or more d-tori. At any smooth point $\boldsymbol{z} \in \mathcal{V}_0$, the space $\mathbf{L}(\boldsymbol{z})$ is the line in the direction of $\nabla_{\log} Q(\boldsymbol{z})$. Thus, $\boldsymbol{r} \in R$ if and only if $\boldsymbol{r}$ is in the closure of the image when the logarithmic Gauss map $\nabla_{\log}$ is applied to $\mathcal{V}_0$, so that

$$R = \overline{\nabla_{\log}[\mathcal{V}_0]} .$$

This allows us to plot the feasible region by parametrizing $\mathcal{V}_0$ by an embedded grid and applying $\nabla_{\log}$ to each point of the embedded grid, an example of which is shown on the left of Figure 9.9.

The right of Figure 9.9 shows an intensity plot of the magnitude of the probability amplitude for the particle at time 200 for a QRW known as $S(1/8)$. The agreement of the shape of the empirically plotted feasible region (right) with the theoretical prediction based on the Gauss map (left) is apparent. What is also apparent is that not only do the regions agree but their fine structure of darker bands and light areas agree as well.

In particular, the image of $\mathcal{V}_0$ under $\nabla_{\log}$ will be more intense in places where the Jacobian determinant of $\nabla_{\log}$ is small because the density of the image of an embedded grid is proportional to the inverse of the Jacobian determinant. The Jacobian determinant of the logarithmic Gauss map is precisely the curvature, as discussed following (9.21). In Theorem 9.45, while

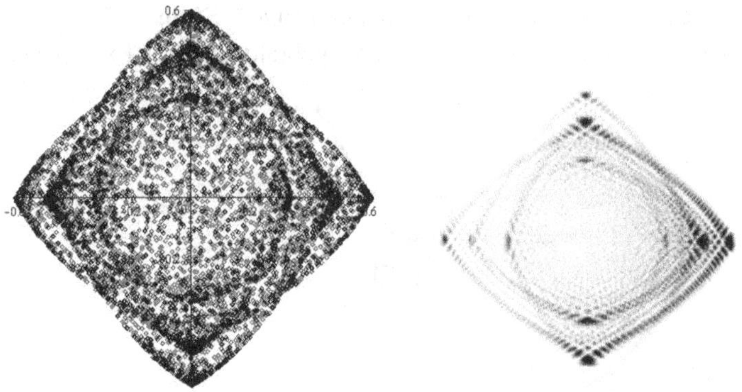

Figure 9.9 *Left:* The log-Gauss map on an embedded grid. *Right:* Probability amplitudes of a QRW.

the $P/\|\nabla_{\log} Q\|_2^2$ term varies a little, the dominant factor is the curvature term $\mathcal{K}^{-1/2}$. This explains why the density of the Gauss-mapped grid is a good surrogate for the probability amplitudes. ◁

Changing the matrix U or the vectors $\boldsymbol{v}^{(j)}$ changes the walk, hence there are many quantum walks, most of which don't have the symmetries of the $S(1/8)$ walk. Figure 9.10 shows the feasibility region for a more-or-less generic quantum walk. Again, one sees an image of the logarithmic Gauss map. It is notable that, as for many quantum random walks, the feasible region is nonconvex, indicating that parts of the cone $\texttt{normal}_0(B)$ do not correspond to any minimal points, but are instead in the region of exponential decay (the infeasible region).

Notes

Precursors to the derivations of the saddle point residue integrals in this chapter were the multivariate asymptotic results [BR83]. Breaking the symmetry among the coordinates, they wrote

$$F(\boldsymbol{z}) = \sum_{n=0}^{\infty} f_n(\boldsymbol{z}^\circ) z_d^n$$

for $(d-1)$-dimensional series $f_n(\boldsymbol{z}^\circ)$ and then used the fact that f_n is sometimes asymptotic to an nth power $f_n \sim C \cdot g \cdot h^n$ to obtain Gaussian asymptotics when certain minimality conditions are satisfied near a smooth critical point. Their

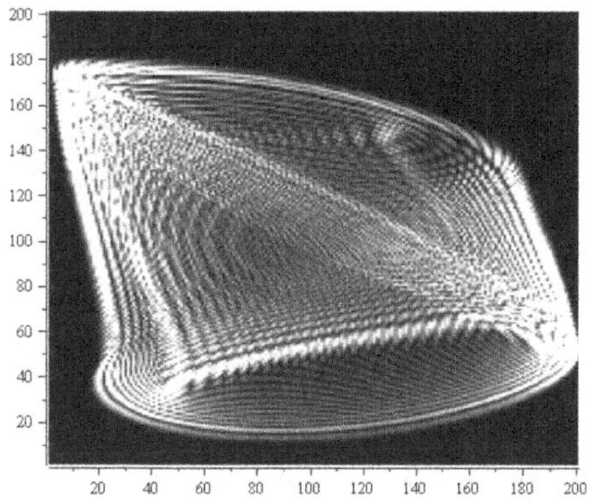

Figure 9.10 Intensity plot for a quantum walk without symmetries.

language is inherently one-dimensional, so geometric concepts such as smooth point did not arise explicitly.

The results presented in this chapter were first obtained via a direct surgery approach in [PW02], and are valid only for finitely minimal critical points. In addition, the minimality hypothesis in [PW02, Theorem 3.5] (and in many other results there) assumes an ordinary power series. The residue version of these computations appeared in print first in [Bar+10]. Extending the validity of the coordinate version beyond the case of finite intersection of $\mathcal{V}$ with $T(\boldsymbol{x}_{\min})$ was accomplished in [BP11].

Between the first and second editions of this book, a rigorous Morse theoretic foundation developed in [BMP22] streamlined some of the presentation of this chapter. The finite criticality hypothesis in this second edition replaces the strong torality hypothesis from the first edition; there, the latter is simply called *torality*. The explicit formula for higher order terms in Theorem 9.40 was first given by Raichev and Wilson [RW08]. Attempts to extend Algorithm 3 are an ongoing topic of research by an AMS Mathematics Research Community started in 2021.

The pictures in Figure 9.9 were first produced by a Penn graduate student, Wil Brady, in an attempt to produce rigorous computations verifying the limit shapes of feasible regions that were suspected from simulations. At that time, Theorem 9.45 was not known. The fact that the fine structure of the two plots agreed was a big surprise, and led to reformulated estimates such as (9.23) in terms of curvature.

Another rewriting of the leading term of the basic nondegenerate smooth point asymptotic formula is given in [Ben+12, Appendix B].

Additional exercises

Exercise 9.13. Let Res be the residue map on meromorphic forms with simple poles on a smooth variety V, as defined in Proposition C.6. Prove that Res is functorial, meaning it commutes with bi-holomorphic changes of coordinate.

Exercise 9.14. Let $f(x, y) = x^2 - 3x + 3 - y$. In Example 9.39, asymptotics in the diagonal direction reveal a quadratic degeneracy. To see what a quadratic degeneracy means topologically, begin by computing the critical points in the direction $\boldsymbol{r} = (r, 1 - r)$ as a function of r on the unit interval. There should usually be two critical points. At what value r_* of r is there a single critical point of multiplicity 2? Check whether this is the same r for which the quadratic term of $h_{\hat{\boldsymbol{r}}}$ near the critical point $\boldsymbol{z}(\hat{\boldsymbol{r}})$ vanishes.

Exercise 9.15. Let $F(x, y) = 1/(1 - x - y)^\ell$. Compute the asymptotics for the power series coefficients $a_{rs}^{(\ell)}$ and find the relation between these and the asymptotics of the binomial coefficients $a_{rs}^{(1)} = \binom{r+s}{r,s}$. Verify this combinatorially by finding the exact value of $a_{rs}^{(\ell)}$. *Hint:* When $\ell = 2$, the bivariate convolution of the binomial array with itself can be represented as divisions of $r + s$ ordered balls into r balls of one color and s of another, with a marker inserted somewhere dividing the balls into the two parts.

Exercise 9.16. (higher-order cube root asymptotics) In Example 9.39, dividing the error when approximating the sequence by its leading asymptotic term by the leading asymptotic term gives $0.00111\ldots$ when $r = 100$, hinting at the fact that the next nonvanishing asymptotic term is $r^{-m/3}$ for some m greater than 2. Compute enough derivatives of A and ϕ at zero to determine the next nonvanishing asymptotic term for a_{rr}.

10

Multiple point asymptotics

The asymptotic behavior of the coefficients $a_{\boldsymbol{r}}$ of a convergent Laurent expansion $F(\boldsymbol{z}) = \sum_{\boldsymbol{r} \in \mathbb{Z}^d} a_{\boldsymbol{r}} \boldsymbol{z}^{\boldsymbol{r}}$ in a domain $\mathcal{D} \subset \mathbb{C}^d$ is determined by certain *contributing points* of the singular set $\mathcal{V}$ of F. In Chapter 9 we derived asymptotics in the presence of *smooth* contributing points, where $\mathcal{V}$ is locally a manifold. In this chapter we derive asymptotics for the simplest non-smooth geometry, where $\mathcal{V}$ locally decomposes as a union of smooth sets (see Figure 10.1).

Definition 10.1 (multiple points)**.** A point $\boldsymbol{p} \in \mathcal{V}$ is said to be a ***multiple point*** of $\mathcal{V}$ (or F) if there exist complex manifolds $\mathcal{V}_1, \ldots, \mathcal{V}_n$ such that

$$U \cap \mathcal{V} = (U \cap \mathcal{V}_1) \cup \cdots \cup (U \cap \mathcal{V}_n)$$

for every sufficiently small neighborhood U of $\boldsymbol{p}$ in $\mathbb{C}^d$.

A taxonomy of multiple points, with examples, is described in Section 10.1 below. Combinatorial applications of generating functions whose asymptotics are controlled by multiple points include multi-server queues (Example 10.28), lattice points in polytopes (Example 12.28), binomial sums (Examples 10.11

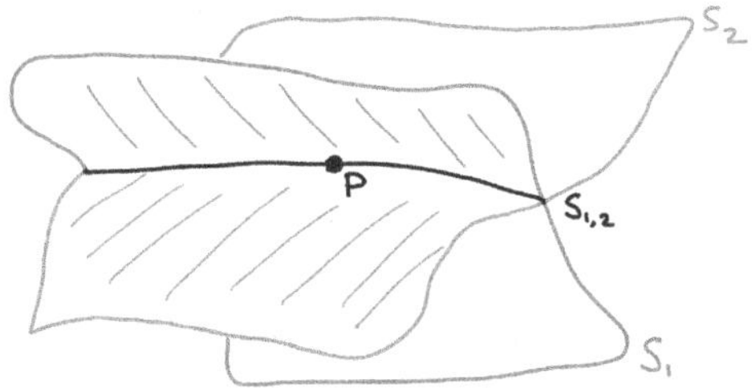

Figure 10.1 A singular variety containing multiple points.

and 10.69), discrete probability (Example 12.25), and lattice paths (Example 12.26).

Exercise 10.1. Prove directly from the definition that the origin is not a multiple point of $\mathcal{V}(z^2 - x^2 - y^2)$.

Recall from Chapter 7 that every complex algebraic variety (respectively, its complement) has a Whitney stratification and is locally the product of a stratum with its normal space (respectively, its normal link). Throughout this chapter we use n for the number of manifolds, also called ***divisors***, intersecting at a multiple point $\boldsymbol{p}$, and k for the codimension of the stratum containing $\boldsymbol{p}$. Every smooth point is a multiple point with $n = 1$. Mirroring Definition C.14 in Appendix C, when $n \leq d$ we call $\boldsymbol{p}$ a ***transverse multiple point*** if the normals to the divisors vanishing at $\boldsymbol{p}$ are linearly independent, and say the divisors are *transverse* or *intersect transversely* if they are transverse at every point where they intersect. Thus, n divisors intersecting transversely necessarily have codimension $k = n$. No point with $n > d$ is considered a transverse multiple point. A ***complete intersection*** in dimension d is a stratum of codimension d, which necessarily consists of isolated points.

One important simplification throughout most of the chapter is the assumption that the direction $\hat{\boldsymbol{r}}$ does not lie on a boundary between two regions of coefficient behavior.

Definition 10.2 (generic directions). Let $\{S_\alpha : \alpha \in A\}$ be a Whitney stratification of the variety $\mathcal{V}$. We call $\hat{\boldsymbol{r}}$ a ***generic direction*** if each critical point for $h_{\hat{\boldsymbol{r}}}$ in the closure $\overline{S_\alpha}$ of a stratum S_α lies in S_α rather than in a proper substratum $S_\beta \subset \overline{S_\alpha}$.

Non-generic directions occur when a critical point in a stratum "disappears" because it enters a substratum, leading to non-uniform behavior around non-generic directions. If a direction is generic, then Condition (2) of Definition D.12 of a stratified Morse function will be satisfied because the negation of Condition (2) implies that $\boldsymbol{p}$ is a critical point in direction $h_{\hat{\boldsymbol{r}}}$ for the stratum $\mathcal{S}_\beta$ but it lies in the proper substratum $\mathcal{S}_\alpha$. As our naming suggests, most directions are generic.

Exercise 10.2. Let $\mathcal{V}$ be the variety defined by the vanishing of $Q(x, y) = \left(1 - \frac{2}{3}x - \frac{1}{3}y\right)\left(1 - \frac{1}{3}x - \frac{2}{3}y\right)$. Give a stratification of $\mathcal{V}$ and find all non-generic directions $\hat{\boldsymbol{r}}$ for this stratification.

The results of Section 7.3 imply that (under mild conditions, such as having no critical values at infinity) we can write

$$a_{\mathbf{r}} = \sum_{\mathbf{w}\in\texttt{critical}(\hat{\mathbf{r}})} \frac{\kappa_{\mathbf{w}}}{(2\pi i)^d} \int_{C_{\mathbf{w}}} F(\mathbf{z})\mathbf{z}^{-\mathbf{r}-\mathbf{1}} d\mathbf{z} \tag{10.1}$$

where $\texttt{critical}(\hat{\mathbf{r}})$ is the set of critical points of the generating function F in the direction $\hat{\mathbf{r}}$, each $\kappa_{\mathbf{w}}$ is an integer, and each $C_{\mathbf{w}}$ is a chain of integration defined by *tangential* and *normal* Morse data at critical points of F. Roughly speaking, the normal Morse data describes an integral that can be simplified using residues, while the tangential data results in a saddle point integral that can be asymptotically approximated.

Throughout this chapter we make various assumptions that simplify both the determination of contributing critical points and the computation of integrals over tangential and normal Morse data. This was not necessary for smooth points because all smooth points look the same topologically, but multiple points can vary substantially in their local topology and geometry. We thus begin in Section 10.1 by introducing the types of multiple points we consider, before Section 10.2 derives asymptotics for the integrals that arise and Section 10.3 gives our main results on coefficient asymptotics. Section 10.4 discusses the algebraic techniques used to identify multiple points in more detail, and we conclude in Section 10.5 by examining asymptotics in non-generic directions and the difficult issue of tangential intersections.

10.1 A taxonomy of multiple points

Figure 10.2 displays the taxonomy we describe in this section. Effective classification within this taxonomy is discussed in Section 10.4.

The simplest multiple point case occurs when $\mathcal{V}$ is a ***hyperplane arrangement***, meaning a finite union of hyperplanes (a simple example is shown in Figure 10.3). If $\mathcal{V}$ is the union of transversely intersecting hyperplanes then the normal data of a critical point on a stratum of codimension k is a k-torus, both the normal and tangential Morse data can be described explicitly, and the critical points describing asymptotics can be determined algorithmically (see Theorem 10.23 below).

Example 10.3. Figure 10.4 shows the pole set of the generating function

$$F(x, y, z) = \frac{16}{(4 - 2x - y - z)(4 - x - 2y - z)},$$

whose divisors are two planes meeting at the complex line $S = \{(1, 1, 1) + \lambda(-1, -1, 3) \colon \lambda \in \mathbb{C}\}$.

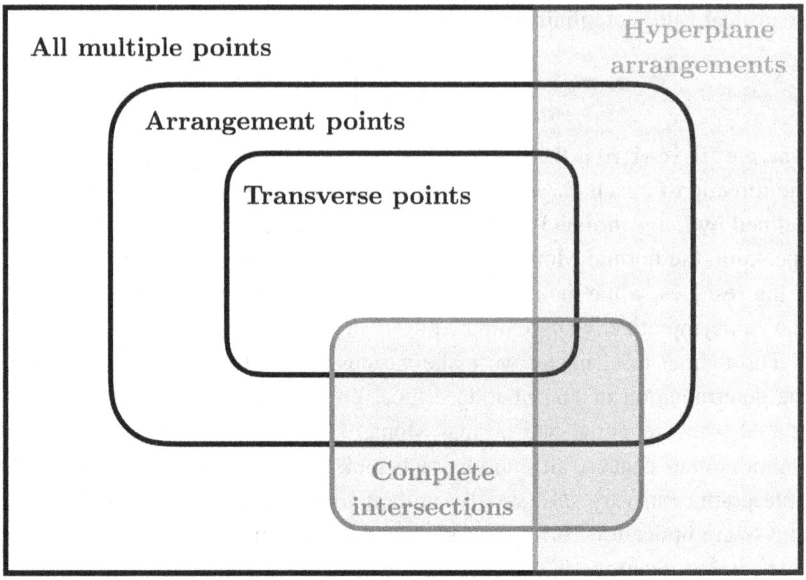

Figure 10.2 A taxonomy of multiple points.

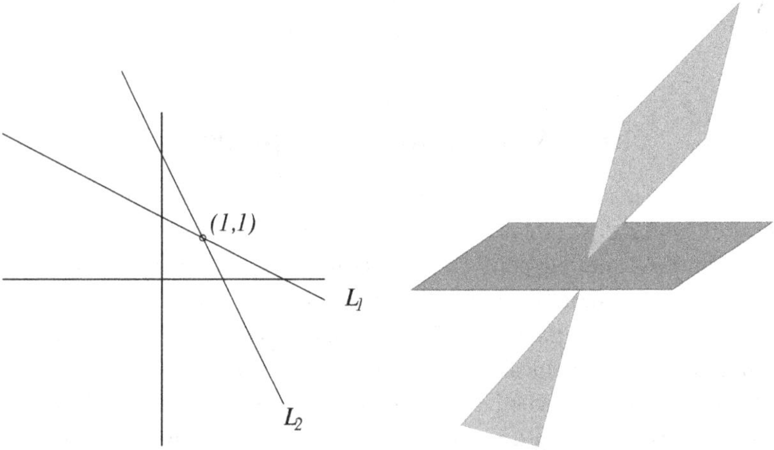

Figure 10.3 A transverse multiple point on a hyperplane arrangement, shown in two depictions.

For each direction $\hat{r}$ in the positive orthant, there are critical points $z_1(\hat{r})$ and $z_2(\hat{r})$ on the respective planes, and a critical point $p(\hat{r})$ on the line S. The normal data at p is a 2-torus and the asymptotic behavior of the Cauchy integral

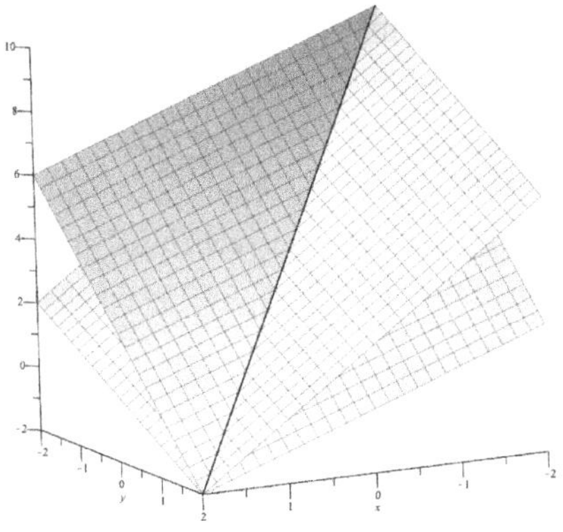

Figure 10.4 A singular variety defined by two hyperplanes in three dimensions.

in (10.1) with domain $C_{\mathbf{p}}$ is determined by analyzing a one-dimensional saddle point integral over a curve maximized at $\mathbf{p}$, while the normal data at $\mathbf{z}_1$ (or $\mathbf{z}_2$) is a 1-torus and the asymptotic behavior of the Cauchy integral with domain $C_{\mathbf{z}_1}$ (respectively $C_{\mathbf{z}_2}$) is a two-dimensional saddle point integral over an open disk maximized at $\mathbf{z}_1$ (respectively $\mathbf{z}_2$). ◁

As we will see below, partial fraction decomposition can always be used to reduce analysis of an arbitrary set of linear divisors to an analysis of transversely intersecting linear divisors. Figure 10.5 illustrates an example of three linear divisors in two dimensions that are pairwise transverse but not jointly transverse, resulting in a stratum of codimension $k = 2$.

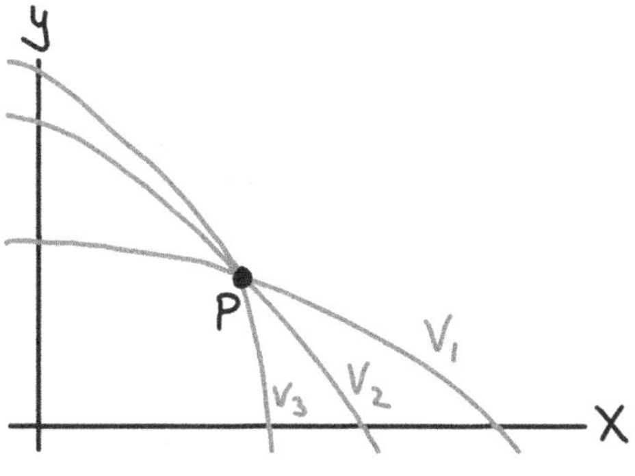

Figure 10.5 Three linear divisors intersecting in two dimensions.

Moving on from linear divisors, if $\mathcal{V}$ consists of transverse multiple points then it is locally diffeomorphic to the transverse linear case, and once again the topological normal data at a critical point is a k-torus.

Example 10.4 (lemniscate)**.** Let $F = 1/Q(x,y)$, where $Q(x,y) = 19 - 20x - 20y + 5x^2 + 14xy + 5y^2 - 2x^2y - 2xy^2 + x^2y^2$ is the polynomial whose real zeros are shown in Figure 10.6. The curve $\mathcal{V}$ intersects itself at the point $(x,y) = (1,1)$, near which it is a union of two distinct segments of smooth curves intersecting only at $(1,1)$. ◁

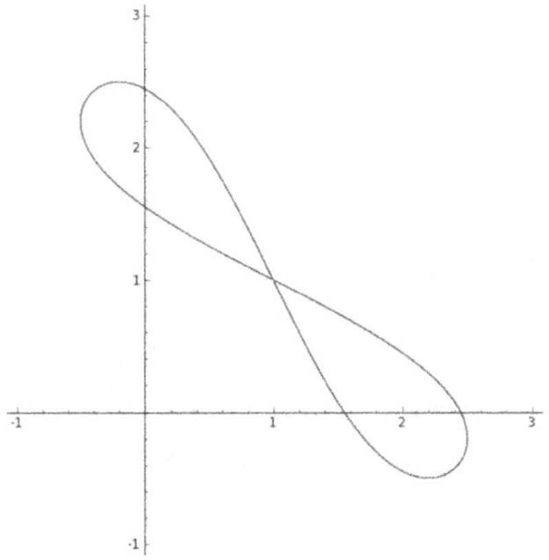

Figure 10.6 A local self-intersection.

Exercise 10.3. Make the change of coordinates $Q(x,y) = Q(1+u, 1+v)$ in Example 10.4 and write formulae $v = f(u)$ for the two curves of $\mathcal{V}$ that intersect at the origin in uv-coordinates.

Remark 10.5. If the transverse multiple point $\boldsymbol{p}$ lies in a complete intersection then there is no tangential integral and, provided $F(\boldsymbol{z}) = P(\boldsymbol{z})/Q(\boldsymbol{z})$ with $P(\boldsymbol{p}) \neq 0$, we will be able to compute the integral over the normal data explicitly as a polynomial times an exponential term (see Theorem 10.12 below).

Moving beyond transverse multiple points with non-linear divisors, the most general class of multiple points for which we have a general method of determining asymptotics is that of *arrangement points*, where the lattice of intersections of the divisors defining a stratum is the same as the lattice of intersections of the tangent planes to the divisors (this being a hyperplane arrangement). A ***central hyperplane arrangement*** is a hyperplane arrangement in which each hyperplane passes through the origin. Each central hyperplane arrangement $\mathcal{A} = \mathcal{W}_1 \cup \cdots \cup \mathcal{W}_n$ possesses a natural structure as a *matroid* $M(\mathcal{A})$, an axiomatic system discussed further below, whose elements index sets of hyperplanes $\mathcal{W}_{k_1}, \ldots, \mathcal{W}_{k_s}$ with linearly independent normal vectors. The set of all intersections $L_T = \bigcap_{j \in T} \mathcal{W}_j$ for $T \subset [n]$ is called the ***lattice of flats*** of $\mathcal{A}$ as it possesses a natural lattice structure [OT92, Lemma 2.3], and we let

$$\overline{T} = \{j \in [n] : L_T \subset \mathcal{W}_j\} \tag{10.2}$$

encode the maximal set of hyperplanes having intersection L_T. If $\overline{T} = T$ whenever L_T is nonempty then we call the arrangement ***transverse***.

Definition 10.6 (arrangement points)**.** The point $\mathbf{p} \in \mathcal{V}$ is an ***arrangement point*** of order n if there exist complex manifolds $\mathcal{V}_1, \ldots, \mathcal{V}_n$ such that

$$U \cap \mathcal{V} = (U \cap \mathcal{V}_1) \cup \cdots \cup (U \cap \mathcal{V}_n)$$

and the intersection lattice of the surfaces $\{\mathcal{V}_j\}$ within U coincides with the intersection lattice of their tangent planes, for every sufficiently small neighborhood U of $\mathbf{p}$ in $\mathbb{C}^d$.

Example 10.7. If Q is a product of linear factors then every point $\mathbf{p} \in \mathcal{V}$ is an arrangement point because the collection of surfaces $\{\mathcal{V}_j\}$ is a translation of the collection of tangent planes. ◁

Example 10.8. Let $F(x, y) = 1/(1-x)(1-y)(1-xy)$ be the generating function with coefficients $a_{rs} = 1 + \min\{r, s\}$. The three divisors are pairwise transverse, intersecting at the single point $\mathbf{p} = (1, 1)$ which is not transverse (as we are in two dimensions). The point $\mathbf{p}$ is an arrangement point because the intersection lattice (each pair of divisors intersects at $\mathbf{p}$, hence all three do) is the same as one would get by replacing the curve $\mathcal{V}(1 - xy)$ by its tangent line $\mathcal{V}(2 - x - y)$ at $\mathbf{p}$. This is illustrated in Figure 10.7. ◁

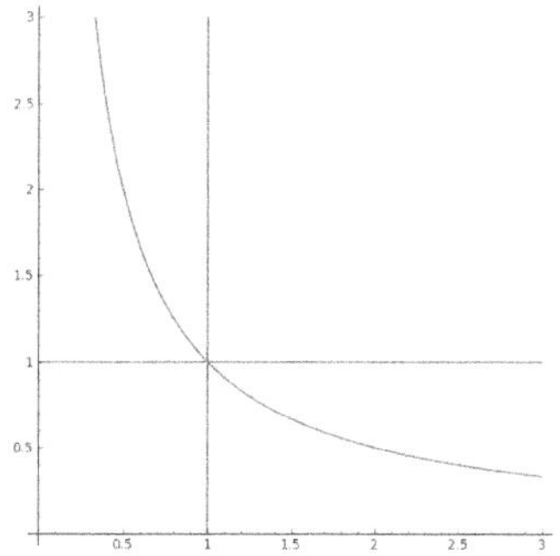

Figure 10.7 An arrangement point in dimension two where three curves intersect.

Example 10.9. The zero set of $Q(x, y, z) = (1-z)(1-z+(1-y)^2-(1+x)^3)$ is the union of the plane $\mathcal{V}_1$ defined by $1-z$ and a smooth surface $\mathcal{V}_2$ defined by the vanishing of the other factor $p(x, y, z)$. These two surfaces intersect in the curve $\mathcal{S}$ defined by translating the cusp of $\mathcal{V}(z, x^3-y^2)$ to the point $(-1, 1, 1)$. Because the expansion of p at $(-1, 1, 1)$ begins $p(x, y, z) = 1-z+(y-1)^2+\cdots$ the tangent planes of $\mathcal{V}_1$ and $\mathcal{V}_2$ coincide at $(-1, 1, 1)$, which is not an arrangement point. Note that the intersection $\mathcal{S}$ is not smooth, so the sets $\mathcal{V}_1 \setminus \mathcal{V}_2$, $\mathcal{V}_2 \setminus \mathcal{V}_1$, and $\mathcal{S}$ do not form a Whitney stratification. To refine these sets into a Whitney stratification requires decomposing $\mathcal{S}$ further into the singleton $\{(-1, 1, 1)\}$ and the remainder $\mathcal{S} \setminus \{(-1, 1, 1)\}$. ◁

Every transverse multiple point is an arrangement point. Although there are arrangement points that are not transverse points, arrangement points have a transversality in successive intersections.

Proposition 10.10. *Suppose* $\mathbf{p}$ *is an arrangement point of order* n, *with divisors* $\mathcal{V}_1, \ldots, \mathcal{V}_n$ *as above, and let* $i \notin T \subset [n]$. *Then either* $\mathcal{V}_T \subset \mathcal{V}_i$ *in a neighborhood of* $\mathbf{p}$ *or* $\mathcal{V}_T$ *intersects* $\mathcal{V}_i$ *transversely at* $\mathbf{p}$. *Consequently, if* $\mathcal{P}_i$ *denotes the tangent space of* $\mathcal{V}_i$ *at* $\mathbf{p}$ *then any intersection* $\mathcal{V}_T$ *is a manifold in a neighborhood of* $\mathbf{p}$ *with tangent space* $L_T = \cap_{j \in T} \mathcal{P}_j$ *at* $\mathbf{p}$.

Proof We induct on the codimension of $\mathcal{V}_T$. If $\mathcal{V}_T$ has codimension 1 then $\mathcal{V}_T = \mathcal{V}_i$ for all $i \in \overline{T}$, with tangent space L_T at $\mathbf{p}$. If $i \notin \overline{T}$ then $\mathcal{P}_i$ is distinct from the tangent plane to $\mathcal{V}_T$ at $\mathbf{p}$, hence the intersection of $\mathcal{V}_i$ and $\mathcal{V}_T$ is transverse.

Now suppose $\mathcal{V}_T$ has codimension $k \geq 1$ and $i \notin T$. By induction, $\mathcal{V}_T$ is smooth with tangent space L_T at $\mathbf{p}$. If $i \in \overline{T}$ then there is nothing to prove, so assume $i \notin \overline{T}$ and let $T' = T \cup \{i\}$. The normal vector to $\mathcal{V}_i$ at $\mathbf{p}$ is not in $(L_T)^\perp$, and hence the surface $\mathcal{V}_i$ intersects $\mathcal{V}_T$ transversely. The transverse intersection of smooth varieties is smooth with tangent space given by the intersection of the tangent spaces, hence $\mathcal{V}_{T'}$ is smooth with tangent space $L_{T'}$, completing the induction. □

We end this section with an example illustrating the most complicated behavior that can occur for multiple points: tangential intersections.

Example 10.11 (two curves intersecting tangentially). Consider the generating function

$$F(x, y) = \frac{1}{(2-x-y)(1-xy)} = \sum_{i,j \geq 0} a_{ij} x^i y^j,$$

with the divisors $\mathcal{V}(2-x-y)$ and $\mathcal{V}(1-xy)$ intersecting tangentially at the point $\mathbf{p} = (1, 1)$ as shown in Figure 10.8.

The multiple point at $\boldsymbol{p}$ is not an arrangement point because the lattice of intersections of the two curves has rank 2 while the tangent lines to the curves at $\boldsymbol{p}$ coincide, and thus form a rank 1 lattice. Although $\boldsymbol{p}$ is not an arrangement point, this rational function is simple enough to allow us to compute asymptotics of its coefficients, which we do in Example 10.69 at the end of this chapter. ◁

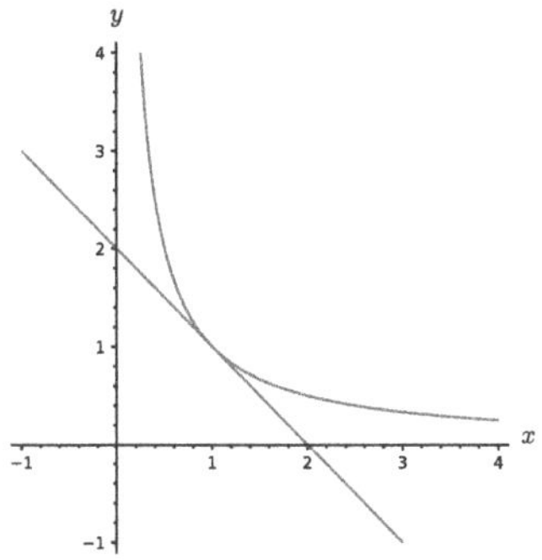

Figure 10.8 Two curves intersecting tangentially.

10.2 Main results on integrals

Having surveyed the different types of multiple points we will analyze, we now present our main asymptotic results. Effective methods to compute the set `critical`($\hat{\boldsymbol{r}}$) were described in Chapter 8, generically giving a finite set defined by explicit algorithmically computable polynomial equalities and inequalities. Thus, it is enough to compute asymptotic approximations of the integrals that appear in (10.1), which is done in this section, and to determine (at least some of) the coefficients $\kappa_{\boldsymbol{w}}$ in (10.1), which is done in Section 10.3. In the important case of linear divisors there is a simple test to determine the coefficients $\kappa_{\boldsymbol{w}}$ (which are all 0 or ± 1), while in general we give results only in the presence of minimal critical points.

Our first goal is to compute integrals over relative homology generators near multiple points. Suppose that $\boldsymbol{p}$ is a *transverse* multiple point on a stratum S of codimension k, so we can write

$$F(\boldsymbol{z}) = \frac{P(\boldsymbol{z})}{Q_1(\boldsymbol{z})^{m_1} \cdots Q_k(\boldsymbol{z})^{m_k}} \tag{10.3}$$

for a vector $\boldsymbol{m} = (m_1, \ldots, m_k)$ of positive integer exponents and *analytic* functions $P, Q_1, \ldots, Q_k$ at $\boldsymbol{z} = \boldsymbol{p}$ with the gradients of the Q_j linearly independent at $\boldsymbol{z} = \boldsymbol{p}$. As in the remark following Theorem C.24, we assume the factors of the denominator have been normalized so that each $\nabla_{\log} Q_i$ is real and P/Q has been normalized so that $Q(0) = 1$. Following Section C.2.3 of Appendix C, we parametrize $\mathcal{S}$ near $\boldsymbol{p}$ using $d-k$ coordinates $\boldsymbol{\pi} = \{\pi_1, \ldots, \pi_{d-k}\}$ so that, writing $\boldsymbol{z_\pi} = (z_{\pi_1}, \ldots, z_{\pi_{d-k}})$, there exist analytic functions $\zeta_i(\boldsymbol{z_\pi})$ for $i \notin \boldsymbol{\pi}$ with $\boldsymbol{z} \in \mathcal{S}$ near $\boldsymbol{p}$ if and only if $z_i = \zeta_i(\boldsymbol{z_\pi})$ for all $i \notin \boldsymbol{\pi}$. The map

$$\Psi(\boldsymbol{z}) = (Q_1(\boldsymbol{z}), \ldots, Q_k(\boldsymbol{z}), z_{\pi_1} - p_{\pi_1}, \ldots, z_{\pi_{d-k}} - p_{\pi_{d-k}}) \tag{10.4}$$

is a bi-analytic change of coordinates taking a neighborhood of $\boldsymbol{p}$ in $\mathcal{S}$ to a neighborhood of the origin in $\{\boldsymbol{0}\}\times\mathbb{C}^{d-k}$. If $T_\varepsilon \subset \mathbb{C}^k\times\{\boldsymbol{0}\}$ denotes the product of circles of radius ε in each of the first k coordinates then, when ε is sufficiently small, the cycle $\mathcal{T} = \Psi^{-1}(T_\varepsilon)$ is a generator for the normal Morse data. Our asymptotic formulae involve the *augmented lognormal matrix* $\boldsymbol{\Gamma}_\Psi(\boldsymbol{p})$ whose first k rows are the log-gradients of $Q_1, \ldots, Q_k$ at $\boldsymbol{p}$ and whose last $d-k$ rows consist of the vectors $z_{\pi_1}\boldsymbol{e}_{\pi_1}, \ldots, z_{\pi_{d-k}}\boldsymbol{e}_{\pi_{d-k}}$ (see Definition C.16 in Appendix C).

Exercise 10.4. Let $Q(x, y) = 19-20x-20y+5x^2+14xy+5y^2-2x^2y-2xy^2+x^2y^2$ be the lemniscate function from Example 10.4. What are Ψ, $\boldsymbol{\Gamma}_\Psi$, and $\det \boldsymbol{\Gamma}_\Psi$ at a point $(a, b) \in \mathcal{V}$? How do these change as (a, b) moves in $\mathcal{V}$?

The simplest case occurs for a complete intersection, because there is only a normal integral and no tangential integral.

Theorem 10.12 (complete intersection asymptotics). *Let F have the form* (10.3), *with $k = d$ and $Q_j(\boldsymbol{p}) = 0$ for all $j \le k$; in this case $\boldsymbol{\pi}$ is empty and the map Ψ defined by* (10.4) *coordinatizes $\mathbb{C}^d$ near $\boldsymbol{p}$ by $(Q_1, \ldots, Q_d)$. Define the cycle $\mathcal{T} = \Psi^{-1}(T_\varepsilon)$ and the augmented lognormal matrix $\boldsymbol{\Gamma}_\Psi(\boldsymbol{p})$ as above. If $P(\boldsymbol{p}) \neq 0$, then*

$$\frac{1}{(2\pi i)^d}\int_{\mathcal{T}} F(\boldsymbol{z})\boldsymbol{z}^{-\boldsymbol{r}-\boldsymbol{1}}\,d\boldsymbol{z} = \boldsymbol{p}^{-\boldsymbol{r}}\mathcal{P}(\boldsymbol{r}, \boldsymbol{p}), \tag{10.5}$$

where $\mathcal{P}(\boldsymbol{r}, \boldsymbol{p})$ is a polynomial in $\boldsymbol{r}$ of degree $|\boldsymbol{m}| - d$ with leading term

$$\mathcal{P}(\boldsymbol{r}, \boldsymbol{p}) \sim \frac{(-1)^{|\boldsymbol{m}-\boldsymbol{1}|}}{(\boldsymbol{m}-\boldsymbol{1})!}\;\frac{P(\boldsymbol{p})}{\det \boldsymbol{\Gamma}_\Psi(\boldsymbol{p})}\;(\boldsymbol{r}\boldsymbol{\Gamma}_\Psi^{-1})^{\boldsymbol{m}-\boldsymbol{1}}$$

for $(\boldsymbol{r}\boldsymbol{\Gamma}_\Psi^{-1})^{\boldsymbol{m}-\boldsymbol{1}} = \prod_{i=1}^d (\boldsymbol{r}\boldsymbol{\Gamma}_\Psi^{-1})_i^{m_i-1}$ and $(\boldsymbol{m}-\boldsymbol{1})! = \prod_{i=1}^d (m_i - 1)!$. When all poles are simple ($\boldsymbol{m} = \boldsymbol{1}$), the matrix $\boldsymbol{\Gamma}_\Psi$ is the $d \times d$ matrix of log gradients and

$$\mathcal{P}(\boldsymbol{r}, \boldsymbol{p}) = \frac{P(\boldsymbol{p})}{\det \boldsymbol{\Gamma}_\Psi(\boldsymbol{p})}.$$

Remark 10.13. The signs of the determinant of Γ_Ψ and of the orientation of $\mathcal{T}$ depend on the order in which the factors Q_j are listed. This is addressed further in the proof of Theorem 10.25 below.

Proof The claimed result follows from the definition of iterated residues, Theorem C.17 in Appendix C for the simple pole case, the reduction of the residue in the non-simple pole case to the simple pole case, and the formulae computed in Theorem C.24. □

Exercise 10.5. Apply Theorem 10.12 to $F(\boldsymbol{z}) = 1/Q(\boldsymbol{z})$ at the point $(1, 1)$, where Q is the polynomial in Exercise 10.4. How does the result differ from $a_{\boldsymbol{r}}$?

In the *simple pole* case where $\boldsymbol{m} = \mathbf{1}$ the polynomial $\mathcal{P}(\boldsymbol{r}, \boldsymbol{p})$ is just a constant. In two dimensions there is an explicit formula in terms of the numerator and denominator of F evaluated at $\boldsymbol{z} = \boldsymbol{p}$.

Corollary 10.14. *If the hypotheses of Theorem 10.12 hold in dimension $d = 2$ with $\boldsymbol{m} = \mathbf{1}$, so that $F(x,y) = P(x,y)/Q(x,y$ for $Q(x,y) = Q_1(x,y)Q_2(x,y)$, then*

$$\frac{1}{(2\pi i)^d} \int_{\mathcal{T}} F(\boldsymbol{z})\boldsymbol{z}^{-\boldsymbol{r}-\mathbf{1}}\, d\boldsymbol{z} = \boldsymbol{p}^{-\boldsymbol{r}} \frac{P(\boldsymbol{p})}{p_1 p_2 \sqrt{Q_{xy}^2(\boldsymbol{p}) - Q_{xx}(\boldsymbol{p})Q_{yy}(\boldsymbol{p})}} \tag{10.6}$$

for a suitable branch of the square-root.

Exercise 10.6. Prove Corollary 10.14 by computing $\mathcal{P}(\boldsymbol{r}, \boldsymbol{p})$ in terms of Q_{xx}, Q_{yy}, and Q_{xy} at $\boldsymbol{p}$.

When $k < d$ the stratum has positive dimension and, after computing the integral over the normal Morse data, there is an integral over a $(d-k)$-chain in the stratum S containing $\boldsymbol{p}$. The leading term of this integral on the stratum is straightforward to write down except for a multiplicative constant that depends on the quadratic part of $-\boldsymbol{r} \cdot \log \boldsymbol{z}$ when restricted to the stratum S. Computing this quadratic contribution requires logarithmic coordinates, and we define

$$g(\boldsymbol{\theta}) = \sum_{j \notin \pi} \hat{\boldsymbol{r}}_j \log\left[\zeta_j\left(p_{\pi_1}e^{i\theta_1}, \ldots, p_{\pi_{d-k}}e^{i\theta_{d-k}}\right)\right],$$

where, as above, $z_j = \zeta_j(\boldsymbol{z}_\pi)$ for $j \in \pi$ locally parametrizes $\boldsymbol{z}$ near $\boldsymbol{p}$ on $\mathcal{S}$. Let $\mathcal{H} = \mathcal{H}(\boldsymbol{p})$ denote the Hessian matrix of g at the origin.

Theorem 10.15 (partial intersection asymptotics). *Assume the same notation and hypotheses as Theorem 10.12, except that $k < d$ so π is nonempty and Ψ parametrizes $\mathbb{C}^d$ near $\boldsymbol{p}$ as the product of a normal slice containing the k-torus*

T_ε and a coordinatization of $\mathcal{S}$ as a complex $(d-k)$-space. Let $\mathcal{T} = \Psi^{-1}(T_\varepsilon)$, and let γ be a $(d-k)$-chain in $\mathcal{S}$ on which $h(\boldsymbol{z}) = -\hat{\boldsymbol{r}} \cdot \mathrm{Relog}(\boldsymbol{z})$ is maximized at $\boldsymbol{p}$. If $\det \mathcal{H}(\boldsymbol{p}) \neq 0$ then the integral of the Cauchy integrand over $\mathcal{T}$ has the asymptotic expansion

$$\frac{1}{(2\pi i)^d}\int_{\mathcal{T}\times\gamma} \boldsymbol{z}^{-\boldsymbol{r}-\boldsymbol{1}}F(\boldsymbol{z})\,d\boldsymbol{z} \approx \boldsymbol{p}^{-\boldsymbol{r}}|\boldsymbol{r}|^{-(d-k)/2+|\boldsymbol{m}-\boldsymbol{1}|}\frac{(2\pi)^{(k-d)/2}}{\sqrt{\det \mathcal{H}(\boldsymbol{p})}}\sum_{\ell=0}^{\infty} C_\ell(\hat{\boldsymbol{r}})|\boldsymbol{r}|^{-\ell} \tag{10.7}$$

with

$$C_0 = (-1)^{|\boldsymbol{m}-\boldsymbol{1}|}\,\frac{(\hat{\boldsymbol{r}}\boldsymbol{\Gamma}_\Psi^{-1})^{\boldsymbol{m}-\boldsymbol{1}}}{(\boldsymbol{m}-\boldsymbol{1})!}\;\frac{P(\boldsymbol{p})\prod_{j\in\pi} p_j}{\det \boldsymbol{\Gamma}_\Psi}\,.$$

When all poles are simple ($\boldsymbol{m} = \boldsymbol{1}$), the matrix $\boldsymbol{\Gamma}_\Psi$ is the $d \times d$ matrix of log gradients and the leading term of the asymptotic expansion (10.7) *becomes*

$$\frac{1}{(2\pi i)^d}\int_{\mathcal{T}\times\gamma} \boldsymbol{z}^{-\boldsymbol{r}-\boldsymbol{1}}F(\boldsymbol{z})\,d\boldsymbol{z} \sim \frac{P(\boldsymbol{p})\prod_{j\in\pi} p_j}{(2\pi)^{(d-k)/2}\det \boldsymbol{\Gamma}_\Psi \sqrt{\mathcal{H}(\boldsymbol{p})}}|\boldsymbol{r}|^{-(d-k)/2}\boldsymbol{p}^{-\boldsymbol{r}}\,.$$

Remark 10.16. In the case of linear divisors, $\mathcal{S}$ is the intersection of hyperplanes and some linear algebra proves that $\mathcal{H}(\boldsymbol{p})$ is always nonsingular; see [BMP23, Proposition 4.13].

Proof As in the proof of Theorem 10.12, combining Theorems C.17 and C.24 from Appendix C computes the integral over $\mathcal{T}$, reducing now to an integral of an iterated residue over γ. Using the formula for the residue in Theorem C.24 and changing coordinates via $z_j = p_j \exp(i\theta_j)$ for $j \in \pi$ implies

$$\frac{1}{(2\pi i)^d}\int_C \boldsymbol{z}^{-\boldsymbol{r}-\boldsymbol{1}}F(\boldsymbol{z})\,d\boldsymbol{z} = \frac{1}{(2\pi i)^{d-k}}\int_\gamma \boldsymbol{z}^{-\boldsymbol{r}}\frac{\mathcal{P}(\boldsymbol{r},\boldsymbol{z})}{\prod_{j\in\pi} z_j}\Bigg|_{z_i=\zeta_i(\boldsymbol{z}_\pi)\,:\,i\notin\pi} d\boldsymbol{z}_\pi$$

$$= \frac{1}{(2\pi)^{d-k}}\int_{\mathcal{N}} \exp(-|\boldsymbol{r}|\phi(\boldsymbol{\theta}))\mathcal{P}(\boldsymbol{r},\boldsymbol{z}(\boldsymbol{\theta}))\,d\boldsymbol{\theta}\,,$$

where $\mathcal{P}(\boldsymbol{r},\boldsymbol{z})$ is the polynomial in $\boldsymbol{r}$ from Theorem C.24, $\mathcal{N}$ is a neighborhood of the origin in $\mathbb{R}^{d-k}$, and $\phi(\boldsymbol{\theta}) = \hat{\boldsymbol{r}} \cdot \log \boldsymbol{z}(\boldsymbol{\theta})$. Lemma 8.21 shows that this is a standard multivariate saddle point integral that can be evaluated using Theorem 5.2. The coefficient C_0 comes from the leading term of $\mathcal{P}(\boldsymbol{r},\boldsymbol{z}(\boldsymbol{\theta}))$, which is described in Equation (C.2.15) of Appendix C. □

10.3 Main results on coefficient asymptotics

Having approximated the asymptotics of integrals near critical multiple points, to determine asymptotics of $a_{\boldsymbol{r}}$ it remains only to determine the coefficients

κ_w in (10.1). We first give results when $\mathcal{V}$ is a hyperplane arrangement: Section 10.3.1 assumes transverse intersections and gives a method to compute all coefficients, while Section 10.3.2 reduces a general hyperplane arrangement to transverse arrangements using algebraic reductions. Next, we show how to compute dominant asymptotics for non-linear divisors. As in the smooth case, we need to characterize which minimal critical points contribute to dominant asymptotic behavior. Section 10.3.3 deals with transverse intersections and Section 10.3.4 deals with non-transverse intersections that are arrangement points.

Notation and terminology

For each point $\boldsymbol{\sigma}$ on the boundary of the amoeba of a smooth algebraic hypersurface $\mathcal{V}_Q$, some complex scalar multiple of $(\nabla_{\log} Q)(\boldsymbol{\sigma})$ will be real and point into $\texttt{amoeba}(Q)$. We use this to define contributing points on the boundary of the amoeba.

Definition 10.17 (lognormal cones and contributing multiple points)**.** Suppose that $\boldsymbol{\sigma}$ is a transverse multiple point in a stratum of codimension k, such that $F(\boldsymbol{z}) = P(\boldsymbol{z})/Q_1(\boldsymbol{z})^{m_1} \cdots Q_k(\boldsymbol{z})^{m_k}$ near $\boldsymbol{\sigma}$ with the gradients of the Q_j linearly independent at $\boldsymbol{z} = \boldsymbol{\sigma}$. If $\text{Relog}(\boldsymbol{\sigma}) \in \partial\,\texttt{amoeba}(Q)$ then the ***lognormal cone*** $\mathtt{N}(\boldsymbol{\sigma})$ to $\mathcal{V}$ at $\boldsymbol{\sigma}$ is the positive real span of the lognormals to the Q_j at $\boldsymbol{\sigma}$, scaled so that the lognormal of Q_j is a real vector pointing into $\texttt{amoeba}(Q_j)$ at $\text{Relog}(\boldsymbol{\sigma})$. The set $\mathbf{W}(\hat{\boldsymbol{r}})$ of ***contributing multiple points*** in the direction $\hat{\boldsymbol{r}}$ consists of the critical points $\boldsymbol{\tau}$ for which $\text{Relog}(\boldsymbol{\tau}) \in \partial\,\texttt{amoeba}(Q)$ and $\hat{\boldsymbol{r}} \in \mathtt{N}(\boldsymbol{\tau})$.

Remark 10.18. The height function $h_{\hat{\boldsymbol{r}}}(\boldsymbol{z}) = -\hat{\boldsymbol{r}} \cdot \text{Relog}(\boldsymbol{z})$ is minimized on the closure $\overline{B}$ of a component of $\texttt{amoeba}(Q)^c$ at $\boldsymbol{\sigma} \in \partial B$ if and only if $\boldsymbol{\sigma}$ is a contributing multiple point. Definition 10.17 fails when $\texttt{amoeba}(Q)$ has codimension one near $\boldsymbol{\sigma}$, for instance in the case of the line $\texttt{amoeba}(1 - xy)$. However, when the component B of $\texttt{amoeba}(Q)^c$ is fixed then contributing multiple points can still be defined in this pathological case for any $\boldsymbol{\sigma} \in \partial B$ by scaling the lognormals defining $\mathtt{N}(\boldsymbol{\sigma})$ so that they point away from B.

Exercise 10.7. Let $P = 1$ and $Q = (1-2x/3-y/3)(1-x/3-2y/3)$. First, determine the directions $\hat{\boldsymbol{r}}$ for which the point $(1, 1)$ is a contributing multiple point. Second, find all contributing multiple points in the main diagonal direction.

10.3.1 Linear divisors: transverse arrangements

Suppose that $F(\boldsymbol{z}) = P(\boldsymbol{z})/Q(\boldsymbol{z})$ such that for polynomials P and Q such that $Q(\boldsymbol{z}) = \ell_1(\boldsymbol{z})^{m_1} \cdots \ell_n(\boldsymbol{z})^{m_n}$ for positive integers m_k and real linear factors $\ell_1, \ldots, \ell_n$ vanishing on distinct hyperplanes $\mathcal{V}_1, \ldots, \mathcal{V}_n$ in $\mathbb{C}^d$, with the factors scaled so that $\ell_j(\boldsymbol{z}) = 1 - \boldsymbol{z} \cdot \boldsymbol{b}^{(j)}$ for a real vector of coefficients $\boldsymbol{b}^{(j)}$. Let $\mathcal{M} = \mathbb{C}_*^d \setminus \mathcal{V}$ denote the set where F is analytic, and define $\mathcal{M}_{\mathbb{R}} = \mathcal{M} \cap \mathbb{R}^d$.

A stratum $\mathcal{S}$ in $\mathcal{V}$ can be represented by its closure, which is an intersection $\overline{\mathcal{S}} = \mathcal{V}_{k_1} \cap \cdots \cap \mathcal{V}_{k_s}$ of some of the hyperplanes defining $\mathcal{V}$ with the hyperplanes not in $\overline{\mathcal{S}}$ removed. By the convexity of the height function, each stratum contains at most one critical point in each orthant of $\mathbb{R}^d$, and if $\xi \in \{\pm 1\}^d$ then we let $\boldsymbol{\sigma}_{\mathcal{S}} = \boldsymbol{\sigma}_{\mathcal{S},\xi}(\hat{\boldsymbol{r}})$ denote the unique critical point of $\mathcal{S}$ in the same orthant as ξ. We further define $\mathcal{M}_{\mathbb{R}}(\mathcal{S})$ to be $\mathbb{R}_*^d$ with the hyperplanes defining $\mathcal{S}$ removed, and let $B_{\mathcal{S}} = B_{\mathcal{S},\xi}$ denote the unique component of $\mathcal{M}_{\mathbb{R}}(\mathcal{S})$ in the same orthant as ξ whose closure contains the origin.

The homology and cohomology of (complements of) hyperplane arrangements has been well studied. Recall the relative homology group $H_d(\mathcal{M}, -\infty)$ that describes the homology of $\mathcal{M}$ relative to any sublevel set $\mathcal{M}_{<c} = \{\boldsymbol{z} \in \mathcal{M} : h_{\hat{\boldsymbol{r}}}(\boldsymbol{z}) < c\}$, where c is smaller than the heights of all critical points.

Definition 10.19 (imaginary fibers). For each component B of $\mathcal{M}_{\mathbb{R}}$, the ***imaginary fiber*** with respect to B is any chain $C_B = \boldsymbol{x} + i\mathbb{R}^d$ with $\boldsymbol{x} \in B$, oriented so that each copy of $\mathbb{R}$ goes from $-\infty$ to $+\infty$.

The imaginary fiber C_B is well defined in $H_d(\mathcal{M})$, as if $\boldsymbol{x}, \boldsymbol{y} \in B$ then $f(t) = (1-t)\boldsymbol{x} + t\boldsymbol{y} + i\mathbb{R}^d$ defines a homotopy between $f(0) = \boldsymbol{x} + i\mathbb{R}^d$ and $f(1) = \boldsymbol{y} + i\mathbb{R}^d$ which stays in $\mathcal{M}$.

Proposition 10.20. *If $\hat{\boldsymbol{r}}$ is a generic direction then the set of fibers C_B as B varies over components of $\mathcal{M}_{\mathbb{R}}$ on which $h_{\hat{\boldsymbol{r}}}$ is bounded from below is a basis for $H_d(\mathcal{M}, -\infty)$.*

Proof See Varchenko and Gelfand [VG87, page 268]. □

Proposition 10.20 implies that for linear divisors the domains of integration $C_{\boldsymbol{w}}$ in (10.1) can be expressed in terms of imaginary fibers. We now specialize to the case of a transverse arrangement, where all nonempty intersections of k hyperplanes have codimension k. In this case we can explicitly compute cycle representatives for another homology basis using the following construction.

Definition 10.21 (linking tori). For any $\boldsymbol{w} \in \mathcal{V}_{\mathbb{R}}$ let $\text{Adj}(\boldsymbol{w})$ be the set of components B in $\mathcal{M}_{\mathbb{R}}$ with $\boldsymbol{w} \in \overline{B}$. If $\boldsymbol{w}$ is contained in the stratum $\mathcal{S}$ with closure $\overline{\mathcal{S}} = \mathcal{V}_{k_1} \cap \cdots \cap \mathcal{V}_{k_s}$ then the ***linking torus*** $\tau_{\boldsymbol{w}}$ is the relative cycle

$$\tau_{\boldsymbol{w}} = \sum_{B \in \text{Adj}(\boldsymbol{w})} \text{sgn}_{\boldsymbol{w}}(B)\, C_B \tag{10.8}$$

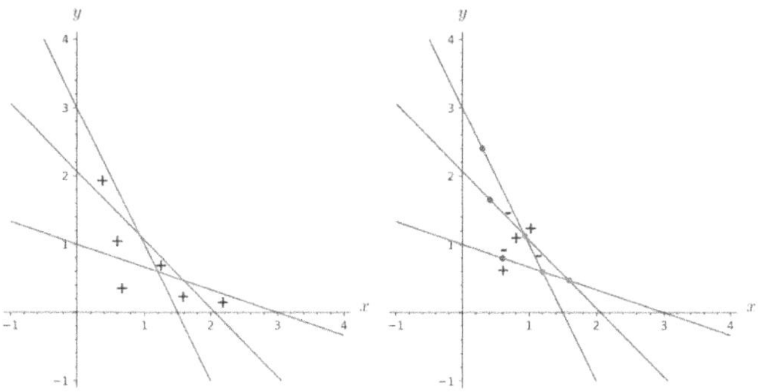

Figure 10.9 *Left:* The fiber basis for $H_2(\mathcal{M}, -\infty)$ for three lines not intersecting at a common point. *Right:* Linking tori are shown for the three critical points on these lines, and one of the codimension-2 critical points.

in $H_d(\mathcal{M}, -\infty)$, where $\operatorname{sgn}_{\boldsymbol{w}}(B)$ denotes the sign of $x_1 \cdots x_d \ell_{k_1}(\boldsymbol{x}) \cdots \ell_{k_s}(\boldsymbol{x})$ for any $\boldsymbol{x} \in B$.

Example 10.22. Figure 10.9 shows the zero set $\mathcal{V}_{\mathbb{R}}$ defined by the three linear functions $\ell_1(x, y) = 3-2x-y$, $\ell_2(x, y) = 33/16-x-y$, and $\ell_3(x, y) = 3-x-3y$. The real plane is split into six bounded components by $\mathcal{V}_{\mathbb{R}}$ and the coordinate axes, together with ten unbounded components. In any direction $\hat{\boldsymbol{r}}$ with positive coordinates the height function $h_{\hat{\boldsymbol{r}}}$ is unbounded from below on the unbounded regions, so a basis for $H_2(\mathcal{M}, -\infty)$ is given by imaginary fibers through points in each of the six bounded regions (illustrated on the left side of the figure).

The right side of Figure 10.9 shows the six critical points of $\mathcal{V}$ in the direction $\hat{\boldsymbol{r}} = (1/5, 4/5)$, three of which are smooth points (on codimension one strata) and three of which form their own strata of codimension two. Two linking tori are shown, one on a stratum of codimension two that consists of four fibers and another at a smooth point that consists of two fibers. ◁

Asymptotics of interest will be given by a sum of integrals over linking tori, but we still need to determine *which* linking tori appear in the sum, corresponding to the coefficients $\kappa_{\boldsymbol{w}}$ in (10.1) that are non-zero. In general this is a very difficult question, but in the linear divisor case we can answer it exactly. Note that in this linear case the lognormal cone $\mathtt{N}(\boldsymbol{\sigma})$ at a point $\boldsymbol{\sigma}$ in the stratum $\mathcal{S}$ defined by the intersection $\overline{\mathcal{S}} = \mathcal{V}_{k_1} \cap \cdots \cap \mathcal{V}_{k_s}$ is the positive span of the vectors $\tilde{\boldsymbol{b}}^{(k_1)}, \ldots, \tilde{\boldsymbol{b}}^{(k_s)}$ with coordinates $\tilde{b}_i^{(k_j)} = b_i^{(k_j)} \sigma_i$.

Theorem 10.23 (change of basis to linking tori)**.** *Fix a generic direction $\hat{\boldsymbol{r}}$ and*

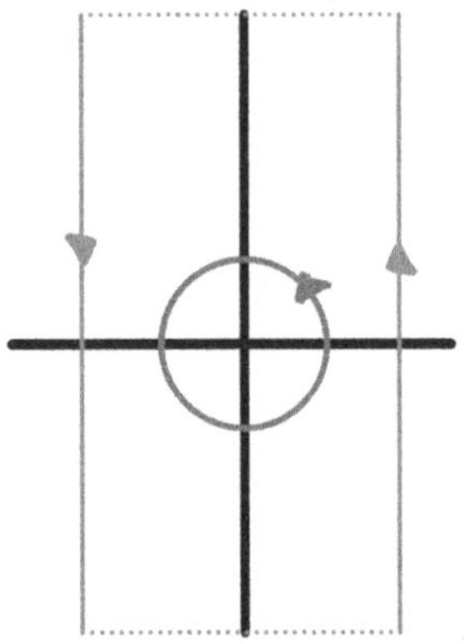

Figure 10.10 Because the integrand of the Cauchy integral decays sufficiently quickly away from the origin, the ends of the imaginary fibers defining a linking torus can be joined and the resulting curve deformed to give an actual torus. Similarly, a torus can be broken apart into a linking torus.

suppose that the ℓ_j *define a transverse arrangement. Then*

$$T(\varepsilon, \ldots, \varepsilon) = \sum_{w \in \mathbf{W}(\hat{r})} \tau_w \tag{10.9}$$

in $H_d(\mathcal{M}, -\infty)$ *for all sufficiently small* $\varepsilon > 0$*. In particular, the power series coefficients* a_r *of* F *satisfy*

$$a_r = \frac{1}{(2\pi i)^d} \sum_{\sigma \in \mathbf{W}(\hat{r})} \int_{\tau_\sigma} F(z) z^{-r-1} dz .$$

The decomposition (10.9) varies with $\hat{r}$ but is unique for each generic $\hat{r}$ because the linking tori form a basis. For simplicity we restrict Theorem 10.23 to power series expansions, however decompositions analogous to (10.9) can be computed for any torus $T(w)$ with $\text{Relog}(w) \notin \texttt{amoeba}(Q)$.

Proof Sketch Because any torus in $\mathcal{M}$ can be expressed as a sum of imaginary fibers, it is sufficient to prove that any imaginary fiber can be expressed as a sum of linking tori. For each bounded component B of $\mathcal{M}_{\mathbb{R}}$ there exists a critical point σ_B that minimizes the height function $h_{\hat{r}}$ on $\overline{B}$, and this critical point is a contributing point. If x lies in B then we can replace the fiber C_x by the τ_{σ_B} plus a sum of fibers $C_{x'}$,where $h_{\hat{r}}(x') < h_{\hat{r}}(\sigma_B)$. Repeatedly applying this rewriting process results in a sum of linking tori plus imaginary fibers over points in components of $\mathcal{M}_{\mathbb{R}}$ on which $h_{\hat{r}}$ is unbounded from below. These final imaginary fibers are null homologous in $(\mathcal{M}, -\infty)$, and thus may be discarded. The result then follows by writing $T(\varepsilon, \ldots, \varepsilon)$ as a "linking

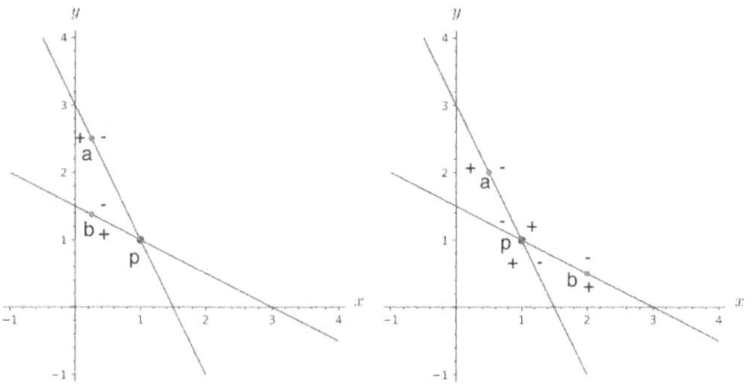

Figure 10.11 When a saddle crosses $\boldsymbol{p}$ the linking torus changes by $\tau_{\boldsymbol{p}}$.

torus" around the origin (see Figure 10.10), then proving inductively that the coefficient of every contributing point is 1. See [BMP23] for a full proof. □

Example 10.24. Figure 10.11 shows three critical points $\mathbf{a}, \mathbf{b}$, and $\boldsymbol{p}$ on two crossing lines. As the direction $\hat{\boldsymbol{r}}$ varies from a slope of 3 on the left of the figure to a slope of $2/5$ on the right of the figure, the critical point $\mathbf{b}$ crosses ρ. As illustrated in the figure, on the left-hand side $T(\varepsilon, \varepsilon) = \tau_{\mathbf{a}} + \tau_{\mathbf{b}}$ while on the right side $T(\varepsilon, \varepsilon) = \tau_{\mathbf{a}} + \tau_{\mathbf{b}} + \tau_{\boldsymbol{p}}$. Varying $\hat{\boldsymbol{r}}$ farther to a slope of $1/5$, the critical point $\mathbf{a}$ crosses $\boldsymbol{p}$, whereupon the coefficient of $\tau_{\boldsymbol{p}}$ becomes zero again and $T(\varepsilon, \varepsilon) = \tau_{\mathbf{a}} + \tau_{\mathbf{b}}$, this time with $\mathbf{a}$ and $\mathbf{b}$ both to the right of $\boldsymbol{p}$. ◁

Combining Theorems 10.15 and 10.23 gives coefficient asymptotics for functions whose denominators consist of transversely intersecting linear divisors.

Theorem 10.25 (transverse hyperplane arrangement asymptotics). *Let $Q(\boldsymbol{z}) = \prod_{j=1}^{n} \ell_j(\boldsymbol{z})^{m_j}$ for positive integers m_j and real linear functions ℓ_j defining a transverse hyperplane arrangement, normalized so that $Q(\mathbf{0}) = 1$ (as is the case when each $\ell_j(\boldsymbol{z}) = 1 - \mathbf{b}^{(j)} \cdot \boldsymbol{z}$). Let $\sum_{\boldsymbol{r}} a_{\boldsymbol{r}} \boldsymbol{z}^{\boldsymbol{r}}$ be a convergent power series expansion for $F(\boldsymbol{z}) = P(\boldsymbol{z})/Q(\boldsymbol{z})$ and $\hat{\boldsymbol{r}}$ be a generic direction. Then*

$$a_{\boldsymbol{r}} \approx \sum_{\sigma \in \mathbf{W}(\hat{\boldsymbol{r}})} \Phi_\sigma(\boldsymbol{r}), \tag{10.10}$$

where Φ_σ is an infinite asymptotic expansion when $\boldsymbol{\sigma}$ lies in a stratum of dimension at least one, and a finite expansion (an exponential times a polynomial in $\boldsymbol{r}$) when $\boldsymbol{\sigma}$ lies in a complete intersection. These expansions are computed (up to sign) in Theorems 10.12 and 10.15, and have the following leading terms.

(i) For complete intersection points σ,

$$\Phi_\sigma(r) \sim p^{-r} \frac{(-1)^{|m-1|}}{(m-1)!} \frac{P(p)}{|\det \Gamma_\Psi(p)|} (r\Gamma_\Psi^{-1})^{m-1} . \tag{10.11}$$

(ii) For critical points σ in strata of positive dimension,

$$\Phi_\sigma(r) \sim p^{-r} \frac{(-1)^{|m-1|}}{(m-1)!} \frac{P(p) \prod_{j\in\pi} |p_j|}{|\det \Gamma_\Psi(p)|} \frac{(\hat{r}\Gamma_\Psi^{-1})^{m-1}}{(2\pi)^{(d-k)/2} \sqrt{\det \mathcal{H}(p)}} |r|^{-(d-k)/2+|m-1|} . \tag{10.12}$$

In both (i) and (ii), asymptotics are uniform as $\hat{r}$ varies over compact sets of generic directions where the leading terms of the summands do not cancel.

Proof At this point, most of our work consists of getting the correct signs in (10.11) and (10.12). The linking torus τ_σ and the torus $\mathcal{T}(\sigma)$ are generators for the rank-d local homology of $\mathcal{M}$ at a complete intersection point σ. For such a point, define $\text{sgn}(\tau_\sigma, \mathcal{T}(\sigma)) = \pm 1$ according to whether these generators are equally or oppositely oriented. More generally, if γ is a generator of the relative homology group at σ within the stratum, then both τ_σ and $\mathcal{T}(\sigma) \times \gamma$ generate the rank-d local homology of $\mathcal{M}$ and we again let $\text{sgn}(\tau_\sigma, \mathcal{T}(\sigma), \gamma) = \pm 1$ according to whether these generators are equally or oppositely oriented.

Casting the conclusion of Theorem 10.23 in terms of the tori $\mathcal{T}$ gives

$$a_r \approx \frac{1}{(2\pi i)^d} \sum_{\sigma \in \mathbf{W}(\hat{r})} \text{sgn}(\tau_\sigma, \mathcal{T}(\sigma), \gamma) \int_{\mathcal{T}(\sigma)\times\gamma} F(z) z^{-r-1} dz ,$$

and plugging in the results of Theorems 10.12 and 10.15, respectively, produces (10.11) and (10.12), up to sign. We proceed to check the signs. Recall that the definition of $\mathcal{T}$ required us to normalize the factors Q_i so that the entries of Γ_Ψ are real.

Case 1 (complete intersection): For (10.11) we need to check that

$$\frac{\text{sgn}(\tau_\sigma, \mathcal{T})}{\det \Gamma_\Psi} = \frac{1}{|\det \Gamma_\Psi|} . \tag{10.13}$$

Let D be the cell in the complement of $\mathbb{R}_*^d \setminus \mathcal{A}(\sigma)$ whose closure contains both the origin and σ, where $\mathcal{A}(\sigma)$ denotes the sub-arrangement of all hyperplanes passing through σ. The tangent planes to both τ_σ and $\mathcal{T}$ can be made to pass through any point $p' = e^{-\varepsilon} p$ for ε sufficiently close to zero. At such a point, the two tangent planes, call them K and L respectively, are both subspaces of the purely imaginary space $p' + i\mathbb{R}^d$, with K oriented by the standard orientation of $\mathbb{R}^d$ and L oriented by $y \mapsto \mathbf{Q}^{-1}(p' + iy)$, with

$\mathbf{Q} : (z_1, \ldots, z_d) \mapsto (Q_1(\boldsymbol{z}), \ldots, Q_d(\boldsymbol{z}))$. The Jacobian of this change of coordinates is $d\mathbf{Q}/d\boldsymbol{z}$, hence $\mathrm{sgn}(\tau_{\boldsymbol{\sigma}}, \mathcal{T}) = \mathrm{sgn}(\det \boldsymbol{\Gamma}_\Psi)$ and establishing (10.13) and (10.11).

Case 2 (positive dimension): To establish (10.12), we need to check that

$$\mathrm{sgn}(\tau_{\boldsymbol{\sigma}}, \mathcal{T}, \gamma) \frac{\prod_{j \in \pi} p_j}{\det \boldsymbol{\Gamma}_\Psi} = \frac{\prod_{j \in \pi} |p_j|}{|\det \boldsymbol{\Gamma}_\Psi|}. \tag{10.14}$$

The linking torus $\tau_{\boldsymbol{\sigma}}$ is homologically a product of a k-dimensional torus τ' in the normal slice with the $(d-k)$-dimensional imaginary contour γ' through $\boldsymbol{\sigma}$ within the stratum. The tangent plane to τ' coincides up to orientation with the tangent plane to $\mathcal{T}$. As we have seen above, the orientation $\mathrm{sgn}(\tau', \mathcal{T})$ is the sign of $\det \boldsymbol{\Gamma}_\Psi$. Therefore it remains to show that the orientations to γ' and γ of the imaginary space in the stratum are related by $\prod_{j \in \pi} \mathrm{sgn}(p_j)$. The plane γ' is oriented by the imaginary coordinates $y_{\pi(1)}, \ldots, y_{\pi(d-k)}$ listed in that order and all increasing. The plane γ is parametrized by these same coordinates in the same order, inheriting the orientation from $dz_{\pi(1)} \wedge \cdots \wedge dz_{\pi(d-k)}$. Thus, the imaginary part is increasing when the real part is positive, and decreasing when the real part is negative. This establishes (10.14), hence (10.12), finishing the proof of the theorem. □

Remark 10.26. Theorem 10.25 holds when the numerator $P(\boldsymbol{z})$ is any analytic function, provided that an error term that decreases faster than any exponential function is added to the right-hand side of (10.10).

Example 10.27. Suppose that a and b are positive integers and let $F(x,y) = 1/Q(x,y)$ where $Q(x,y) = \left(1 - \frac{2}{3}x - \frac{1}{3}y\right)^a \left(1 - \frac{1}{3}x - \frac{2}{3}y\right)^b$. For a general direction $(\hat{r}, \hat{s}) = (r,s)/(r+s)$ there are contributing points

$$\boldsymbol{p}_1 = \left(\frac{3r}{r+s}, \frac{3s}{2(r+s)}\right) \text{ and } \boldsymbol{p}_2 = \left(\frac{3r}{2(r+s)}, \frac{3s}{r+s}\right)$$

on each of the individual smooth divisors. Letting $\lambda = r/s$, the exponential rates contributed by these smooth critical points are

$$R_1 = \frac{2^\lambda (1+\lambda)^{1+\lambda}}{3(3\lambda)^\lambda)} = \frac{1}{3}\left(\frac{2(1+1/\lambda)}{3}\right)^\lambda (1+\lambda)$$

$$R_2 = \frac{2(1+\lambda)^{1+\lambda})}{3(3\lambda)^\lambda)} = \frac{2}{3}\left(\frac{1+1/\lambda}{3}\right)^\lambda (1+\lambda).$$

The detailed asymptotic contributions of $\boldsymbol{p}_1$ and $\boldsymbol{p}_2$ can be worked out either from the results of Chapter 9 or from (10.7). The point $\boldsymbol{p} = (1,1)$ is a minimal critical multiple point for every direction $\hat{\boldsymbol{r}}$, but is only contributing when $\lambda \in$

$[1/2, 2]$, since the lognormal cone at $\boldsymbol{p}$ is spanned by the vectors $(2, 1)$ and $(1, 2)$ with slopes $1/2$ and 2.

When (r, s) lies in the open cone defined by $\lambda \in (1/2, 2)$, the exponential contributions R_1 and R_2 can be bounded below 1 as they are increasing functions of λ. Thus, Theorem 10.25 with $\det \boldsymbol{\Gamma}_\Psi = 1/3$ implies that when $1/2 < \lambda < 2$,

$$a_{rs} \sim \frac{3\,(r-2s)^{b-1}\,(s-2r)^{a-1}}{(a-1)!(b-1)!}.$$

Note that the boundary directions $\lambda \in \{1/2, 2\}$ are not generic and so Theorem 10.25 does not apply. In these cases $\boldsymbol{p}_1$ and $\boldsymbol{p}_2$ are distinct, but one of them coincides with $\boldsymbol{p}$ and the other corresponds to an exponentially decaying contribution. We deal with this boundary case in Example 10.66 via the surgery method. It is also possible for λ to approach the boundary of the interval rather than being constant, and in this case the analysis of asymptotics is more delicate – see Section 13.2 for a brief discussion. ◁

Example 10.28 (queueing partition function)**.** Let a, b, c, and d be positive real constants, and consider the partition generating function

$$F(x, y) = \frac{\exp(x+y)}{(1-ax-by)(1-cx-dy)}$$

for a closed multi-class queueing network with one infinite server [BM93; Kog02]. The most interesting case in applications occurs, without loss of generality, when $a > c$ and $b < d$, whence $D = ad - bc > 0$. The two linear divisors intersect in the positive real quadrant at the point $\boldsymbol{p} = (x_0, y_0) = \left(\frac{d-b}{D}, \frac{a-c}{D}\right)$, which is a critical transverse multiple point. In the main diagonal direction, there are also critical points at $(1/2a, 1/2b)$ and $(1/2c, 1/2d)$, contributing exponential growth rates of $4ab$ and $4cd$; the first of these coincides with the double point $\boldsymbol{p}$ if and only if $ad + bc - 2ab = 0$, and the second if and only if $ad + bc - 2cd = 0$. Thus, if we consider the case where the two points listed above are distinct and each different from the double point, the calculation

$$\begin{aligned}(ad-bc)^2 - 4ab(a-c)(d-b) &= [a(d-b) + b(a-c)]^2 - 4ab(a-c)(d-b) \\ &= [a(d-b) - b(a-c)]^2 \\ &= [ad + bc - 2ab]^2 > 0,\end{aligned}$$

and an analogous one with $4cd$ in place of $4ab$, shows that the double point has larger exponential growth rate than either of the smooth points. Leading asymptotics are determined by (10.5) throughout the cone of directions (r, s)

for which

$$\frac{b(a-c)}{a(d-b)} < \frac{s}{r} < \frac{d(a-c)}{c(d-b)}.$$

In particular,

$$a_{nn} \sim \frac{D^{2n-1} \exp\left(\frac{a+d-(b+c)}{D}\right)}{(a-c)^{n+1}(d-b)^{n+1}}.$$

◁

Example 10.29. Consider again the rational function $1/(Q_1 Q_2)$ from Example 10.3. The intersection of the zero sets of Q_1 and Q_2 contains the single critical point

$$(x_0, y_0, z_0) = \left(\frac{4(r+s)}{3(r+s+t)}, \frac{4(r+s)}{3(r+s+t)}, \frac{4t}{r+s+t}\right)$$

in the direction (r, s, t). The lognormal cone at this point is two-dimensional, so the same point will be critical for many directions: for instance, the point $(x_0, y_0, z_0) = (1, 1, 1)$ is critical in directions where $r + s = 3t$. When (r, s, t) lies in the lognormal cone at (x_0, y_0, z_0) then this point determines dominant asymptotics, and we recover expansions such as $a_{3t,3t,2t} \sim (48\pi t)^{-1/2}$ as $t \to \infty$. ◁

Exercise 10.8. Let $Q = (3-x-2y)(3-2x-y)(5-3x-3y)$. Sketch the real part of $\mathcal{V}$, show the positions of all the critical points in the main diagonal direction, state which one(s) contribute to the leading term of $a_{n,n}$, and determine whether the leading term is computed by Theorem 10.12 or by Theorem 10.15.

Exercise 10.9. Theorem 10.15 handles smooth point asymptotics in the special case where $k = 1$. Reconcile the formulae given here with the different-looking formulae from Chapter 9. Start with a concrete example such as Delannoy numbers $F(x, y) = (1 - x - y - xy)^{-1}$.

10.3.2 Linear divisors: non-transverse arrangements

A rational function whose singular variety forms a non-transverse hyperplane arrangement can be analyzed using a form of partial fraction decomposition.

For a hyperplane arrangement $\mathcal{A}$ defined by real linear functions $\ell_1, \ldots, \ell_n$, the ***matroid*** they define is the set of all subsets $T \subset [n]$ of indices corresponding to linearly independent collections $\{\ell_j : j \in T\}$. Subsets of $[n]$ not in the matroid are ***dependent sets***, and minimal dependent sets are called ***circuits***. Circuits need not all have the same cardinality. Following Brylawski [Bry77], we define a ***broken circuit*** to be a circuit with its greatest element deleted, and call a set containing no broken circuit ***χ-independent*** (note that χ-independence implies independence, but not vice versa). The ***support of a rational function*** is the set of divisors – without multiplicities – appearing in its denominator when in lowest terms.

Example 10.30. Let ℓ_1, ℓ_2, and ℓ_3 vanish on concurrent lines, as in Figure 10.12. Each pair of lines is linearly independent, but the set of all three is linearly dependent, so there is a unique circuit $[3] = \{1, 2, 3\}$ and a unique broken circuit $\{1, 2\}$. A subset of $[3]$ is independent if its cardinality is at most 2, and is χ-independent if it does not contain both 1 and 2. ◁

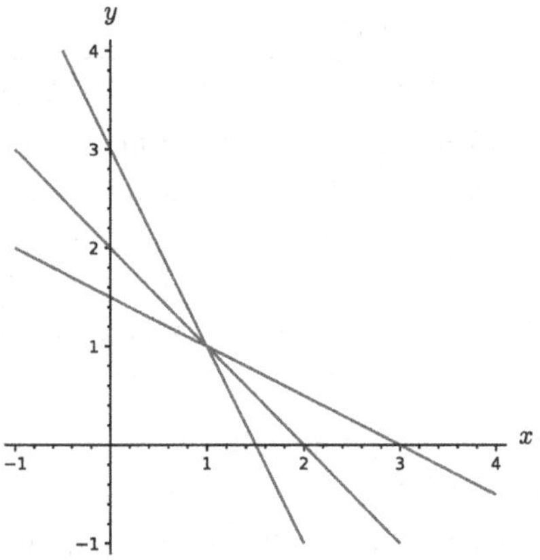

Figure 10.12 Three concurrent lines.

Let $S_{\mathcal{A}}$ denote the set of meromorphic functions whose pole variety is contained in the union of hyperplanes of $\mathcal{A}$. Broken circuits allow us to characterize $S_{\mathcal{A}}$ as a module over holomorphic functions.

Proposition 10.31. *Let $\mathcal{A} = \{\ell_1, \ldots, \ell_n\}$ be a hyperplane arrangement in $\mathbb{C}^d$.*

(i) The set of rational functions $(\ell_{i_1}(\boldsymbol{z}) \cdots \ell_{i_s}(\boldsymbol{z}))^{-1} \in S_{\mathcal{A}}$ with χ-independent support is linearly independent over $\mathbb{C}$.
(ii) The span over $\mathbb{C}$ of the rational functions $(\ell_{i_1}(\boldsymbol{z})^{m_1} \cdots \ell_{i_s}(\boldsymbol{z})^{m_s})^{-1} \in S_{\mathcal{A}}$ with χ-independent support and total degree $-\sum_{j=1}^{s} m_j$ contains the inverses of all products of the ℓ_j over multisets of cardinality $\sum_{j=1}^{s} m_j$.

Proof The claimed results follow from properties of hyperplane arrangements derived in [OT92] (see [BMP23, Proposition 5.2]). □

If $\mathfrak{w}$ is a circuit and $\sum_{j\in\mathfrak{w}} a_j\ell_j(z) = 0$ then dividing by $\prod_{j\in\mathfrak{w}} \ell_j(z)^{m_j}$ yields a linear relation among elements of $S_{\mathcal{A}}$ of total degree $1 - \sum_{j=1}^{s} m_j$,

$$\frac{a_j}{\prod_{i\in\mathfrak{w}} \ell_i(z)^{m_i-\delta_{ij}}} = -\sum_{\substack{j'\in\mathfrak{w}\\ j'\neq j}} \frac{a_{j'}}{\prod_{i\in\mathfrak{w}} \ell_i(z)^{m_i-\delta_{ij'}}}. \tag{10.15}$$

Such relations allow for a reduction to the transverse case above.

Algorithm 4: Decomposition into transverse hyperplane arrangements.

Input: Rational or meromorphic function $F(z) = P(z)/Q(z)$ whose singular variety is the hyperplane arrangement defined by the vanishing of $Q(z) = \prod_{j=1}^{n} \ell_j(z)^{m_j}$.

Output: Rational or meromorphic functions $F_1, \ldots, F_s$ such that $F(z) = F_1(z) + \cdots + F_s(z)$ and the singular variety of each F_j is a transverse hyperplane arrangement contained in the singular variety of F.

Set $S = F(z)$.

While there is a summand $s(z)$ of S with a broken circuit $\{\ell_{j_1}, \ldots, \ell_{j_{k-1}}\}$:

Apply (10.15) to $s(z)$ with $j_k > j_{k-1}$ such that $\{\ell_{j_1}, \ldots, \ell_{j_k}\}$ is dependent, and replace

$s(z)$ in S by the result

Return S

Example 10.32. Let $Q = \ell_1(x,y)\ell_2(x,y)\ell_3(x,y) = (3-2x-y)(3-x-2y)(2-x-y)$. The variety $\mathcal{V}$ consists of three concurrent lines, as shown in Example 10.30 and Figure 10.12. By examination we find that $2\ell_1(x,y)+2\ell_2(x,y)-6\ell_3(x,y) = 0$. Dividing through by $6Q$ gives

$$\frac{1}{\ell_1(x,y)\ell_2(x,y)} = \frac{1/3}{\ell_1(x,y)\ell_3(x,y)} + \frac{1/3}{\ell_2(x,y)\ell_3(x,y)}.$$

Starting with any fraction $F(x,y) = g(x,y)/\ell_1(x,y)^a\ell_2(x,y)^b\ell_3(x,y)^c$, this relation may be used in Algorithm 4 to reduce factors $\ell_1\ell_2\ell_3$ in the denominator until there are none remaining, resulting in a sum

$$F(x,y) = \sum_{a+b\geq j\geq 0} \frac{e_j(x,y)}{\ell_1(x,y)^j\ell_3(x,y)^{a+b+c-j}} + \sum_{a+b\geq j\geq 0} \frac{f_j(x,y)}{\ell_2(x,y)^j\ell_3(x,y)^{a+b+c-j}}.$$

Whereas $F(x,y)$ had a non-transverse multiple point at $(1,1)$, each summand now has a transverse multiple point at $(1,1)$. ◃

Theorem 10.33. *Let $Q(z) = \prod_{j=1}^{n} \ell_j(z)_j^m$ be a product of real linear functions* *defining a hyperplane arrangement. Given a meromorphic function $F(\mathbf{z}) = P(\mathbf{z})/Q(\mathbf{z})$ with poles on the zero set of Q, Algorithm 4 terminates and produces the claimed representation of F.*

Proof Let $\mathfrak{w}$ be a circuit and consider the weight function

$$w\left(\prod_{j=1}^{n} \ell_j(\mathbf{z})^{a_j}\right) = \sum_{j=1}^{n} j a_j .$$

The reduction described in (10.15) replaces one term by a sum of terms whose denominators have greater weight but the same total degree (the sum of the exponents that appear on the linear factors). Repeatedly applying the decomposition gives terms with the same total degree but successively higher weights. Because the weight function is bounded among terms whose total degree is fixed, the loop in the algorithm must terminate, and when it terminates every summand has a denominator whose linear factors form a χ-independent set. Thus, the pole set of each summand returned forms a transverse hyperplane arrangement contained in $\mathcal{V}(Q)$. □

Corollary 10.34. *If $F(\mathbf{z}) = \sum_{\mathbf{r}} a_{\mathbf{r}} \mathbf{z}^{\mathbf{r}}$ is a convergent power series expansion of a meromorphic function having poles on a hyperplane arrangement then asymptotics of $a_{\mathbf{r}}$ can be computed by decomposing F into a sum of functions having poles on transverse hyperplane arrangements using Theorem 10.33 and applying Theorem 10.25 to each summand.*

We note that the sum may have a large number of terms, the computations can be difficult to carry out by hand, and a computer algebra system is often useful.

Example 10.35. Let $F = P(x,y)/Q(x,y)$, where $P(x,y) = 1$ and $Q(x,y) = (3-2x-y)(3-x-2y)(2-x-y)^2$, a simple variation of Example 10.32. The decomposition algorithm yields

$$F = \frac{1/3}{(3-x-2y)(2-x-y)^3} + \frac{1/3}{(3-2x-y)(2-x-y)^3}$$

so that $a_{\mathbf{r}} \sim (r-2s)^2/2$ for directions $\hat{\mathbf{r}} = (r,s)$ with $s/r \in (1/2, 1)$, while $a_{\mathbf{r}} \sim (s-2r)^2/2$ when $s/r \in (1,2)$, and $a_{\mathbf{r}}$ is exponentially small when s/r is bounded strictly away from these two intervals. ◁

Up to this point, our roadmap for general hyperplane arrangements has been to use matroid theory and algebra to reduce to the transverse case. We finish our treatment of linear divisors by describing the general topological story. We will not be using these results, but include some discussion because it can help understand the non-linear case. The topology local to a critical point is the same for an arrangement point as it is for the hyperplane arrangement defined by the tangent planes of the surfaces intersecting near the point. However, when Q factors locally but not globally near several different multiple points, it is not clear how multiple local partial fraction decompositions of the type produced in Theorem 10.33 can be used together to give the correct integral representations for coefficients. Therefore, though we don't have an application that makes essential use of the homological decomposition in the non-transverse case, we will go ahead and describe it.

Without a transversality assumption, there is no longer a single generator per critical point in the Morse theoretic decomposition of $H^d(\mathcal{M})$. Instead of a single linking torus τ_σ whose coefficient will be zero or non-zero depending on whether $\boldsymbol{r}$ is in the normal cone, there will be more than one homology generator. Different regions of $\hat{\boldsymbol{r}}$ may yield different non-zero homology elements local to $\boldsymbol{\sigma}$ which combine with linking tori elsewhere to produce the Cauchy domain of integration. We sum this up as follows.

Let $\mathcal{A}$ denote the hyperplane arrangement defined by the linear divisors ℓ_j and suppose that $\boldsymbol{\sigma}$ is a critical point in the direction $\hat{\boldsymbol{r}}$ on some flat E of codimension k in a generic direction $\hat{\boldsymbol{r}}$. Let $\mathcal{N}(\boldsymbol{\sigma})$ denote a sufficiently small ball around $\boldsymbol{\sigma}$ in $\mathcal{M}$ and consider the rank-d homology group $G(\boldsymbol{\sigma}) = H_d(\mathcal{N}(\boldsymbol{\sigma}), \mathcal{N}(\boldsymbol{\sigma})_{\leq h(\boldsymbol{\sigma})-\varepsilon})$, where $\varepsilon > 0$ is small enough that this pair is the attachment pair at $\boldsymbol{\sigma}$.

Proposition 10.36. *Let $\mathcal{A}(E) \subseteq \mathcal{A}$ be the subarrangement of $\mathcal{A}$ consisting of all hyperplanes containing E and let $\mathcal{A}' \subseteq \mathcal{A}(E)$ be any collection of k hyperplanes of $\mathcal{A}(E)$ whose (necessarily transverse) common intersection is E. Let ι denote the inclusion $\mathbb{C}_*^d \setminus \mathcal{A}(E) \hookrightarrow \mathbb{C}_*^d \setminus \mathcal{A}'$ and let $\tau_{\boldsymbol{\sigma},\mathcal{A}'}$ denote the linking torus at $\boldsymbol{\sigma}$ in $\mathbb{C}_*^d \setminus \mathcal{A}'$. There is a unique homology element $\beta(\boldsymbol{\sigma}, \hat{\boldsymbol{r}}) \in G(\boldsymbol{\sigma})$ such that for every such collection $\mathcal{A}'$ we have $\iota_*\beta = \tau_{\boldsymbol{\sigma},\mathcal{A}'}$ if $\hat{\boldsymbol{r}} \in \mathrm{N}_{\mathcal{A}'}(\boldsymbol{\sigma})$ and $\iota_*\beta = 0$ otherwise. These homology elements satisfy*

$$[T(\varepsilon, \ldots, \varepsilon)] = \sum_{\boldsymbol{\sigma} \in \texttt{critical}(\hat{\boldsymbol{r}})} \beta(\boldsymbol{\sigma}, \hat{\boldsymbol{r}}) \tag{10.16}$$

for any $\varepsilon > 0$ sufficiently small. □

Example 10.37 (Example 10.32, continued). If $\hat{\boldsymbol{r}}$ has slope $3/2$, as shown in Figure 10.13, then $\beta(\boldsymbol{\sigma}, \hat{\boldsymbol{r}})$ is the sum of imaginary fibers with signs shown in the figure. To see this, observe that if the flattest line, called ℓ_2 in Example 10.32, is removed then the + and – fibers cancel in pairs, corresponding to $\iota_*\beta$ vanishing because $\hat{\boldsymbol{r}}$ is not in the normal cone spanned by the other two normals (1, 1) and (2, 1). Conversely, if either of the other two lines is removed then $\iota_*\beta$ equals τ_σ, corresponding to the fact that $\hat{\boldsymbol{r}}$ is in the normal

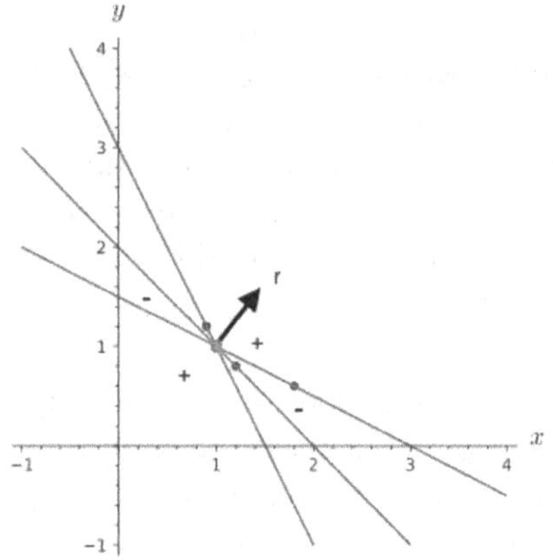

Figure 10.13 The linking torus at σ for a non-transverse intersection, with $\hat{r}$ having slope 3/2. The codimension 2 critical point σ is at the base of the arrow, while the codimension 1 critical points are depicted by smaller dots.

cone spanned by $(1, 2)$ and either $(1, 1)$ or $(2, 1)$. If we change $\hat{r}$ to have slope a little less than 1, then the critical point on the middle line moves to the right of σ and the negative fibers move from the regions between the two lines of least slope to the regions between the two lines of greatest slope. When $\hat{r}$ does not have slope between 1/2 and 2 then $\hat{r} \notin \mathtt{N}(\sigma)$ and $\beta(\sigma, \hat{r}) = 0$, again consistent with the fact that $\hat{r} \notin \mathtt{N}_{\mathcal{A}(E')}(\sigma)$ for any E'. ◁

10.3.3 Non-linear divisors: transverse multiple points

In the case of non-linear divisors, we determine the critical point(s) contributing to the asymptotics only when there are minimal critical points. As above, let p be a transverse multiple point on a stratum $\mathcal{S}$ of codimension k, so that we can write

$$F(z) = \frac{P(z)}{Q_1(z)^{m_1} \cdots Q_k(z)^{m_k}}$$

for positive integers m_j and analytic functions $P, Q_1, \ldots, Q_k$ at $z = p$ with the gradients of the Q_j linearly independent at $z = p$.

Near p the singular variety $\mathcal{V}$ is locally homeomorphic to the hyperplane arrangement $\mathcal{P}$ defined by the first-order expansions of the Q_j at p. In fact, these sets are locally isotopic in the sense that for some neighborhood $\mathcal{N}$ of p there is a homotopy of the pair $(\mathcal{N}, \mathcal{N} \cap \mathcal{V})$ fixing $\partial\mathcal{N}$ such that inside a smaller neighborhood $\mathcal{N}'$ the homotopy takes $\mathcal{V} \cap \mathcal{N}'$ to $\mathcal{P} \cap \mathcal{N}'$. It follows from this isotopy that the local homology at p has rank 1, and is generated by the linking torus $[\tau_{\mathbf{p}}]$. A representative of this homology class is $T \times \gamma$ where T is an actual torus in the normal link and γ is homeomorphic to a $(d-k)$-disk, lies in the stratum containing $\mathbf{p}$, and has a boundary lying below height $h_{\hat{\mathbf{r}}}(\mathbf{p})$. In the case of non-linear divisors, we use the term "linking torus" to refer to $T \times \gamma$; when $k = d$ this is an absolute cycle, namely a d-torus, while for $k < d$ it is a relative cycle. Combining this observation with the expansion computed in Theorem 10.15 gives the following.

Theorem 10.38 (strictly minimal transverse multiple point asymptotics). *Let $\mathbf{p}$ be a transverse multiple point of F with local decomposition*

$$F(\mathbf{z}) = \frac{P(\mathbf{z})}{Q_1(\mathbf{z})^{m_1} \cdots Q_k(\mathbf{z})^{m_k}}$$

as described above. Suppose that $\mathbf{p} = \mathbf{p}(\hat{\mathbf{r}})$ is a minimal point for the Laurent expansion $F(\mathbf{z}) = \sum_{\mathbf{r}} a_{\mathbf{r}} \mathbf{z}^{\mathbf{r}}$ with domain of convergence $\mathcal{D} = \mathrm{Relog}^{-1}(B)$, is a critical point in the generic direction $\hat{\mathbf{r}}$, and is the unique critical point with its height. Let $\mathtt{N}(\mathbf{p})$ be the lognormal cone described in Definition 10.17 above.

(i) If $\hat{\mathbf{r}} \in \mathtt{N}(\mathbf{p})$ and $\det \mathcal{H}(\mathbf{p}) \neq 0$ then

$$a_{\mathbf{r}} \approx \Phi_{\mathbf{p}}(\mathbf{r}), \tag{10.17}$$

where $\Phi_{\mathbf{p}}(\mathbf{r})$ is the asymptotic expansion defined in Theorem 10.25 by (10.11) *(for a complete intersection) or* (10.12) *(otherwise). This asymptotic expansion holds uniformly as $|\mathbf{r}| \to \infty$ with $\hat{\mathbf{r}}$ varying such that it and $\mathbf{p}(\hat{\mathbf{r}})$ satisfy the assumptions above.*

(ii) If $\hat{\mathbf{r}} \notin \mathtt{N}(\mathbf{p})$ then $a_{\mathbf{r}}$ grows exponentially smaller than $\mathbf{p}^{-\mathbf{r}}$.

Proof We have already seen that the local homology group of $\mathcal{M}$ at $\mathbf{p}$ has a single generator C, and that $\frac{1}{(2\pi i)^d} \int_C \mathbf{z}^{-\mathbf{r}-\mathbf{1}} F(\mathbf{z})\, d\mathbf{z} \sim \Phi_{\mathbf{z}}(\mathbf{r})$. To prove conclusion (i) it remains to show that the domain of integration T in the Cauchy integral for $a_{\mathbf{r}}$ satisfies $[T] = \pm\tau_{\mathbf{p}}$ in $H_d(\mathcal{M}, \mathcal{M}_{\leq c-\varepsilon})$, where $c = h_{\hat{\mathbf{r}}}(\mathbf{p})$, and that the modulus / absolute value signs in the middle term on the right of (10.12) capture the orientation.

In Section 10.5 we will see that this is indeed true by following the original derivation in [PW04]. That derivation involves the residue sum identity (10.29), but the topological fact may also be seen directly by applying the Thom isomorphism one divisor at a time. For the orientation, note that

$$\frac{\prod_{j \in \pi} |p_j|}{|\det \boldsymbol{\Gamma}_\Psi|} = \left| \frac{\prod_{j \in \pi} p_j}{\det \boldsymbol{\Gamma}_\Psi} \right| = \left| \frac{1}{\det \boldsymbol{\Gamma}'_\Psi} \right|,$$

where the rows of $\boldsymbol{\Gamma}'_\Psi$ are augmented by elementary basis vectors $e_{\pi(j)}$ instead of $z_j e_{\pi(j)}$. This matrix is real, so again the absolute value multiplies by

$\operatorname{sgn}(\det \mathbf{\Gamma}')$. This is equal to $\operatorname{sgn}(\tau_{\boldsymbol{p}}, \mathcal{T}, \gamma)$ because the picture is the same, except rotated in the j-coordinate by the argument of p_j.

Conclusion (*ii*) is a consequence of the Paley–Wiener theorem for ACSV, described in Theorem 11.42. Briefly, the lognormal cone $\mathtt{N}(\boldsymbol{p})$ is the dual of the geometric tangent cone to B at $\boldsymbol{p}$ and, for any minimal point, the class $[T]$ is represented as a fiber $\exp(\boldsymbol{x} + i\boldsymbol{y})$ over some point $\boldsymbol{x} \in B$. Because $\hat{\boldsymbol{r}}$ is not in the dual to the geometric tangent cone to B at $\boldsymbol{p}$, there is some point $\boldsymbol{p}' \in B$ with $-\hat{\boldsymbol{r}} \cdot \boldsymbol{p}' < -\hat{\boldsymbol{r}} \cdot \boldsymbol{p}$, implying (*ii*). □

Example 10.39 (lemniscate, continued). Let $F(x, y) = 1/Q(x, y)$ be the generating function from Example 10.4. To apply Theorem 10.38 we first prove that the critical point $(1, 1)$ is strictly minimal. The parametrization (10.18) below shows that $\mathcal{V}$ does not enter the unit polydisk near the point $(1, 1)$. If this curve enters the unit polydisk anywhere else, then there is a point (x_0, y_0) with one coordinate having modulus 1 and the other having modulus less than 1. By symmetry, it suffices to check the points on $\mathcal{V}$ with $|x| = 1$ to see if $|y| \le 1$ can hold. Using a rational parametrization of the unit circle, such as $x(t) = (1 - t^2)/(1 + t^2) + i(2t)/(1 + t^2)$, we can solve for the square norm of $y(t)$ as an algebraic function of t. Studying the zeroes of the derivative $|y(t)|^2$ allows one to prove that $|y(t)|^2 \ge 1$ with equality only at $t = 0$, meaning $(1, 1)$ is strictly minimal.

Applying Theorem 10.38 is now straightforward. The cone $\mathtt{N}(1, 1)$ consists of the directions (r, s) between the logarithmic normals to the two branches of $\mathcal{V}$; the lognormal directions were computed in Example 10.43 to be the rays with slopes 2 and 1/2. The point $(1, 1)$ is a square-free complete intersection so we may apply Corollary 10.14, with the values of x_0, y_0 and $P(x_0, y_0)$ all equal to 1. We obtain

$$a_{rs} = \frac{1}{6} + R_{rs},$$

where R_{rs} is a remainder converging to zero exponentially as $(r, s) \to \infty$ with r/s remaining in a compact subset of $(1/2, 2)$. The exponential decay of the error term is detectable numerically: for instance, the asymptotic approximation of $a_{30,30} \approx 0.1652$ has a relative error rate around 0.8%, while the relative error approximating $a_{60,60}$ is around 0.04%. ◁

Example 10.40. The number of nearest-neighbor walks of length n on the integer lattice that start at the origin, take steps in $\{NE, NW, S\}$, and remain confined to the nonnegative quadrant is shown in [MW19] to be the main diagonal of

$$\frac{(1 + x)(1 - 2xy^2z)}{(1 - y)(1 - z(1 + x^2 + xy^2))(1 - zxy^2)}.$$

A short analysis shows that dominant asymptotics are determined by the contributing point $(1, 1, 1/3)$ where the first two of the divisors above vanish (but the third does not) and Theorem 10.38 implies

$$a_{nnn} \sim \frac{3^n \sqrt{3}}{2\sqrt{\pi n}}.$$

◁

Because the homology group $H_d(\mathcal{M}, \mathcal{M}_{\leq c-\varepsilon})$ is the direct sum of local homology groups near critical points, similar results hold when the minimal critical multiple point determining asymptotics has a finite number of critical points with the same coordinate-wise modulus (this mirrors the *finite criticality* hypothesis discussed in Chapter 9).

Corollary 10.41 (minimal transverse multiple point asymptotics)**.** *Suppose that $\boldsymbol{w}$ is a minimal point such that the torus $T(\boldsymbol{w})$ contains a finite number of critical points $\mathbf{p}_1, \ldots, \mathbf{p}_m$, and suppose that each $\mathbf{p}_j$ satisfies the conditions of Theorem 10.38 (aside from strict minimality).*

(i) If $\hat{\mathbf{r}} \in \mathtt{N}(\boldsymbol{p}_j)$ for some j then

$$a_{\boldsymbol{r}} \approx \sum_{\boldsymbol{p} \in \mathbf{W}(\hat{\boldsymbol{r}})} \Phi_{\boldsymbol{p}}(\boldsymbol{r}),$$

where $\Phi_{\boldsymbol{p}}(\boldsymbol{r})$ is the expansion described in (10.17).

(ii) If $\hat{\mathbf{r}} \notin \mathtt{N}(\boldsymbol{p}_j)$ for all $1 \leq j \leq m$ then $a_{\boldsymbol{r}}$ grows exponentially smaller than $\boldsymbol{w}^{-\boldsymbol{r}}$. □

Exercise 10.10. Let $\boldsymbol{m}^{(1)}, \ldots, \boldsymbol{m}^{(d)}$ be positive integer vectors forming a lattice basis of $\mathbb{Z}^d$ and let $F(\boldsymbol{z}) = 1/Q(\boldsymbol{z})$ for $Q(\boldsymbol{z}) = \prod_{j=1}^{d}(1 - \boldsymbol{z}^{\boldsymbol{m}^{(j)}})$. Compare the behavior of $a_{\boldsymbol{r}}$ as $\boldsymbol{r} \to \infty$ when $\boldsymbol{r}$ is in the open positive hull of $\{\boldsymbol{m}^{(j)} : 1 \leq j \leq d\}$ and when it is not, using Corollary 10.41. Why does this make sense combinatorially?

10.3.4 Non-linear divisors: arrangement points

Similar to the linear case, near a general arrangement point $\boldsymbol{p}$ we can decompose $F(\boldsymbol{z})$ into a finite sum of functions whose singular varieties have transverse multiple points at $\boldsymbol{p}$. Unlike the linear case, however, we need to carefully consider the ring in which we perform this decomposition.

Definition 10.42. The ***local ring of analytic germs $O_{\boldsymbol{p}}$*** at $\boldsymbol{p}$ consists of all equivalence classes of analytic functions on neighborhoods of $\boldsymbol{p}$ under the relation of agreement on some neighborhood of $\boldsymbol{p}$. Because analytic functions are determined by their values in a neighborhood, all functions in such an equivalence class are analytic continuations of each other, making the situation somewhat simpler than for germs of smooth functions: $\mathcal{O}_{\mathbf{p}}$ is isomorphic to the ring $\mathbb{C}_{\mathbf{p}}\{\mathbf{z}\}$ of power series centered at $\mathbf{p}$ that converge in some neighborhood of $\mathbf{p}$. The ring $\mathcal{O}_{\mathbf{p}}$ is a local ring whose unique maximal ideal is the germs of functions vanishing at $\mathbf{p}$, and we consider the local ring as lying between the polynomial ring and the formal power series ring: $\mathbb{C}[\mathbf{z}] \subset \mathcal{O}_{\mathbf{p}} \subset \mathbb{C}_{\mathbf{p}}[[\mathbf{z}]]$.

Example 10.43 (lemniscate, continued). Let Q be the polynomial from Example 10.4, whose zero set in $\mathbb{R}^2$ has the shape of a figure eight. The polynomial Q is irreducible in $\mathbb{C}[x, y]$ but according to its geometry it must factor in $\mathcal{O}_{(1,1)}$. A computer algebra system computes parametrizations $(x, y_1(x))$ and $(x, y_2(x))$ for $\mathcal{V}$ near $(1, 1)$ with

$$y_1(x) - 1 = \frac{x^2 - (x-1)\sqrt{-4x^2 + 8x + 5} - 7x + 10}{x^2 - 2x + 5} \tag{10.18}$$

$$= -2(x-1) + \frac{2}{3}(x-1)^3 + \cdots; \tag{10.19}$$

$$y_2(x) - 1 = \frac{x^2 + (x-1)\sqrt{-4x^2 + 8x + 5} - 7x + 10}{x^2 - 2x + 5} \tag{10.20}$$

$$= -\frac{1}{2}(x-1) - \frac{1}{24}(x-1)^3 + \cdots. \tag{10.21}$$

The two branches have slopes -2 and $-1/2$ at the point $(1, 1)$. ◁

The decomposition of a function near an arrangement point is facilitated by the following result.

Lemma 10.44. *Let $\mathbf{p}$ be an arrangement point of $F(\mathbf{z})$ where $\mathcal{V}$ is locally the union of smooth sets $\mathcal{V}_1, \ldots, \mathcal{V}_n$ defined by the vanishing of analytic functions $Q_1, \ldots, Q_n$. If $\mathfrak{w}$ is a circuit in the matroid defined by the tangent hyperplanes of the varieties $\mathcal{V}_1, \ldots, \mathcal{V}_n$ at $\mathbf{z} = \mathbf{p}$ then there is a collection $\{g_i(\mathbf{z}) : i \in \mathfrak{w}\}$ of invertible elements of $\mathcal{O}_{\mathbf{p}}$ such that $\sum_{i \in \mathfrak{w}} g_i(\mathbf{z}) Q_i(\mathbf{z}) = 0$ in $\mathcal{O}_{\mathbf{p}}$.*

Our proof of Lemma 10.44 relies on the following result, whose proof is outlined in Exercise 10.16. Recall from (10.2) that for a set T indexing some divisors Q_i the symbol $\overline{T}$ denotes the set of all j such that $\mathcal{V}_j$ contains $\bigcap_{i \in T} \mathcal{V}_i$.

Lemma 10.45. *Suppose $\mathbf{p}$ is a multiple point with local irreducible factors $Q_1, \ldots, Q_n$ and for an index set $T \subset [n]$ let J_T denote the ideal in $\mathcal{O}_{\mathbf{p}}$ generated by $\{Q_i : i \in T\}$. If $\mathbf{p}$ is an arrangement point then for all $T \subset [n]$ the ideal J_T is radical. It follows that $J_T = J_{\overline{T}}$ for all T, and that the codimension of this ideal is equal to the codimension of L_T.*

Proof of Lemma 10.44 Fix any $i \in \mathfrak{w}$ and let $\mathcal{S} = \mathcal{V}_1 \cap \cdots \cap \mathcal{V}_n$. Because $\mathbf{p}$ is an arrangement point, near $\mathbf{p}$ we have the containment $\bigcap_{j \in \mathfrak{w} \setminus \{i\}} \mathcal{V}_j \subset \mathcal{V}_i$, which is equivalent to Q_i being in the radical of the ideal of $\mathcal{O}_{\mathbf{p}}$ generated by $\{Q_j : j \neq i\}$. By Lemma 10.45, this ideal is already radical, hence

$$Q_i(\mathbf{z}) = \sum_{j \in \mathfrak{w} \setminus \{i\}} g_j(\mathbf{z}) Q_j(\mathbf{z}) \tag{10.22}$$

for some functions $g_j \in \mathcal{O}_{\mathbf{p}}$. Taking gradients at $\mathbf{p}$ we have

$$(\nabla Q_i)(\mathbf{p}) = \sum_{j \in \mathfrak{w} \setminus \{i\}} g_j(\mathbf{p}) (\nabla Q_j)(\mathbf{p}) .$$

Because $\mathfrak{w} \setminus \{i\}$ is a circuit, the gradients on the right-hand side are linearly independent, hence this equation uniquely determines the values $\{g_j(\mathbf{p}) : j \in \mathfrak{w} \setminus \{i\}\}$. Also, the fact that $\mathfrak{w}$ is a circuit implies that none of these values $g_j(\mathbf{p})$ is zero, for that would imply a linear dependence among $\mathfrak{w} \setminus \{j\}$. By continuity, there is a neighborhood of $\mathbf{p}$ where none of the functions g_j vanishes, hence these functions are all units, and the representation (10.22) along with $g_i \equiv -1$ proves the lemma. □

Corollary 10.46. *Let $F(\mathbf{z}) = \sum_{\mathbf{r}} a_{\mathbf{r}} \mathbf{z}^{\mathbf{r}}$ be a convergent Laurent expansion. If $\mathbf{p}$ is a strictly minimal arrangement point then modifying Algorithm 4 to work over the local ring using Lemma 10.44 decomposes $F(\mathbf{z})$ near $\mathbf{p}$ as a finite sum of meromorphic functions with transverse multiple points at $\mathbf{p}$. Asymptotics of $a_{\mathbf{r}}$ can be computed using Theorem 10.38 when the summands satisfy the hypotheses of that theorem at $\mathbf{p}$. If $\mathbf{p}$ is not strictly minimal but the torus $T(\mathbf{p})$ through $\mathbf{p}$ has only a finite number of critical points then asymptotics of $a_{\mathbf{r}}$ can be determined by performing this decomposition and analysis at all critical points in $T(\mathbf{p})$, provided the hypotheses of Theorem 10.38 are always satisfied.*

We conclude this section with an example involving a *binomial variety*; for a similar application concerning lattice point enumeration, see Example 12.28.

Example 10.47. Continuing the analysis of the GF in Example 10.8, we consider the power series expansion of $F(x, y) = 1/Q(x, y)$, where

$$Q(x, y) = Q_1(x, y) Q_2(x, y) Q_3(x, y) = (1 - x)(1 - y)(1 - xy).$$

Localizing at the common intersection point $(1, 1)$, we know that each factor should be in the ideal generated by the others in $\mathcal{O}_{(1,1)}$, and in fact

$$Q_3(x, y) = Q_1(x, y) + Q_2(x, y) - Q_1(x, y) Q_2(x, y) . \tag{10.23}$$

We have a broken circuit $\{1,2\}$, so we eliminate the factor of Q_3 in the denominator of $F(x,y)$ by dividing (10.23) by Q to obtain

$$\frac{1}{Q_1(x,y)Q_2(x,y)} = \frac{1}{Q_2(x,y)Q_3(x,y)} + \frac{1}{Q_1(x,y)Q_3(x,y)} - \frac{1}{Q_3(x,y)}.$$

We may write the last term as $-\frac{Q_2(x,y)}{Q_2(x,y)Q_3(x,y)}$ so that

$$\frac{1}{(1-x)(1-y)} = \frac{y}{(1-y)(1-xy)} + \frac{1}{(1-x)(1-xy)}$$

and thus

$$F(x,y) = F_1(x,y) + F_2(x,y) = \frac{y}{(1-y)(1-xy)^2} + \frac{1}{(1-x)(1-xy)^2}.$$

The generating functions F_1 and F_2 have a transverse multiple point at $(1,1)$, which is the only critical point when $\hat{\boldsymbol{r}} \notin \{(0,1),(1,0),(1,1)\}$. Asymptotics can be obtained by applying Theorem 10.38 to these two rational functions at the point $\boldsymbol{p} = (1,1)$. The lognormal directions for the divisors $1-x$, $1-y$, and $1-xy$ are, respectively, horizontal, vertical, and on the main diagonal; $\hat{\boldsymbol{r}}$ thus lies in the dual normal cone to F_1 at $\boldsymbol{p}$ but not the dual normal cone to F_2 when $\hat{\boldsymbol{r}}$ has positive slope greater than 1, and vice versa when $\hat{\boldsymbol{r}}$ has positive slope less than 1. We deduce that

$$a_{rs} = \begin{cases} r + O(1) & \text{if } \varepsilon^{-1} > \dfrac{s}{r} > 1+\varepsilon \\ s + O(1) & \text{if } 1-\varepsilon > \dfrac{s}{r} > \varepsilon \end{cases},$$

uniformly for any $\varepsilon > 0$. A glance at the generating function shows that in fact $a_{rs} = 1 + \min\{r,s\}$. ◁

10.4 Classifying multiple points

Having discussed the coefficient asymptotics of generating functions with multiple points, we turn to the problem of recognizing when they arise.

Proposition 10.48. *The point $\boldsymbol{p} \in \mathcal{V}$ is a multiple point if and only if there is a factorization*

$$Q(\boldsymbol{z}) = \prod_{j=1}^{n} Q_j(\boldsymbol{z})^{m_j} \tag{10.24}$$

in $\mathcal{O}_{\boldsymbol{p}}$ with $\nabla Q_j(\boldsymbol{p}) \neq 0$ and $Q_j(\boldsymbol{p}) = 0$. The point $\boldsymbol{p}$ is a transverse multiple point with n divisors if and only if in addition the gradient vectors ∇Q_j for $1 \le j \le n$ are linearly independent.

When $m_j = 1$ for all j in the decomposition (10.24), we say that Q is ***square-free at*** p.

Proof The ring $O_{\boldsymbol{p}}$ is a unique factorization domain [Hör90], so Q has some factorization into powers of distinct irreducibles. The variety $\mathcal{V}$ is locally the union of the vanishing sets of the irreducibles Q_j, each of which defines a smooth hypersurface if and only if its gradient at $\boldsymbol{p}$ is non-zero. The transversality assertion follows immediately from the definition. □

To give an algebraic criterion for the multiple point $\boldsymbol{p}$ to be an arrangement point, let $Q_j(\boldsymbol{z})$ be as in (10.24), and let $\ell_j(\boldsymbol{z})$ denote the linear polynomial

$$\ell_j(\boldsymbol{z}) = \nabla Q_j(\boldsymbol{p}) \cdot (\boldsymbol{z} - \boldsymbol{p})$$

forming the leading homogeneous part of Q_j at $\boldsymbol{p}$. The leading homogeneous part of Q at $\boldsymbol{p}$ is the homogeneous polynomial of degree n

$$\mathtt{hom}(Q, \boldsymbol{p}) = \prod_{j=1}^{n} \ell_j(\boldsymbol{z})^{m_j}$$

and therefore the zero set of $\mathtt{hom}(Q, \boldsymbol{p})$, which is the *algebraic tangent cone* $\mathtt{algtan}_{\boldsymbol{p}}(Q)$ discussed previously in Chapter 6, equals the hyperplane arrangement $\mathcal{A}$ defined by the tangent planes to the Q_j at $\boldsymbol{p}$. We remark that the intersection lattice of this arrangement remains the same as $\boldsymbol{p}$ varies over the stratum S of $\mathcal{V}$ containing $\boldsymbol{p}$. By definition, for $\boldsymbol{p}$ to be an arrangement point such a product decomposition must hold and the lattice of flats of $\mathcal{A}$ must be isomorphic to the intersection lattice of the local surfaces $\mathcal{V}_j = \mathcal{V}_{Q_j}$. Repeated factors are allowed, so we may assume without loss of generality that each m_j is equal to 1, arriving at the following algebraic criterion.

Proposition 10.49. *The point $\boldsymbol{p} \in \mathcal{V}$ is an arrangement point if and only if both $\mathtt{hom}(Q, \boldsymbol{p})$ and Q factor into smooth factors (the former will be linear polynomials in $\mathbb{C}[\boldsymbol{z}]$ and the latter will be in $O_{\boldsymbol{p}}$) and the two intersection lattices agree.* □

Example 10.50 (Example 10.4, continued)**.** Let $Q(x, y)$ be the polynomial from Example 10.4 and let $\boldsymbol{p} = (1, 1)$. Taking the monomials of least degree of $Q(1 + x, 1 + y)$ gives

$$\mathtt{hom}(Q, \boldsymbol{p}) = 4x^2 + 10xy + 4y^2 .$$

Every homogeneous quadratic in two variables is the product of linear factors, which are distinct unless the discriminant of the quadratic vanishes. In this example the discriminant is 36, hence $\boldsymbol{p}$ is an arrangement point. In fact it is a transverse multiple point, and the local intersection lattice for both the divisors and their tangent planes is a Boolean lattice of rank 2. ◁

It is important to work over the correct field when analyzing multiple points.

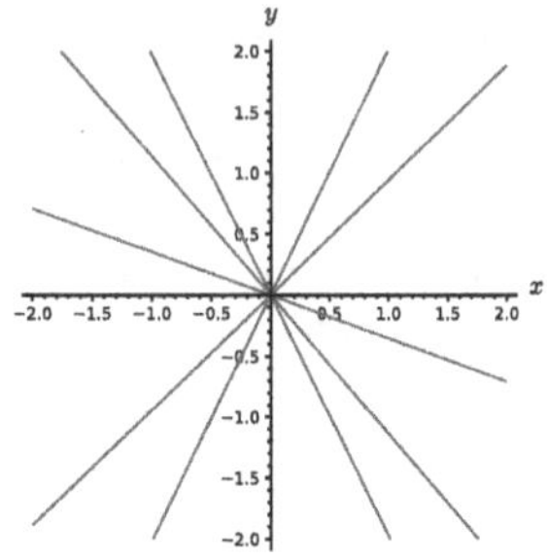

Figure 10.14 A homogeneous polynomial whose zero set is five lines.

Example 10.51. Let $f(x, y) = 2y^5 + y^4x - 10y^3x^2 - 5y^2x^3 + 8yx^4 + 3x^5$. This homogeneous polynomial has a zero set consisting of five distinct lines through the origin whose slopes are the five roots (all real) of the quintic $2t^5 + t^4 - 10t^3 - 5t^2 + 8t + 3$. This quintic does not factor over the rationals, so one must work over the splitting field of f to obtain the necessary factorization. This can be handled in many computer algebra systems, but is more complicated than factoring directly over the rationals. See also Proposition 10.53 below for useful criteria to determine multiple and arrangement points. ◁

Effective computation of multiple points

The foregoing discussion being of a theoretical nature, we pause to consider how the classification of multiple points might be computed. When given a polynomial Q, one imagines being able to "look at its zero set" to see by inspection whether it is locally the union of smooth sheets, but this is difficult when the dimension is high, the polynomial has many terms, or the procedure is automated. What, then, is an effective way to determine whether any of the singular points of Q is in fact a multiple point?

In many combinatorial applications, Q factors into polynomials with nonzero gradients at the origin, and Proposition 10.48 can be directly applied. The difficulty comes when an irreducible polynomial factor Q_j of Q has vanishing gradient: it is possible that the gradient of Q_j vanishes at $\boldsymbol{p}$ because $\boldsymbol{p}$ is not a multiple point, but it could also happen because Q_j is irreducible in $\mathbb{C}[\boldsymbol{z}]$

but reducible in O_p. Computing in O_p is more difficult than working with polynomials, but we can often determine what we need.

Proposition 10.52. *For any positive integers n and d, determining whether a homogeneous polynomial $f \in \mathbb{Q}[z]$ of degree n in d variables has a factorization into n (possibly repeated) linear terms is decidable via computer algebra. If f is a product of linear factors, then whether these factors are distinct is also decidable by computer algebra.*

Proof The set of coefficients of products of n homogeneous linear polynomials in d dimensions forms an algebraic variety. Various sets of defining polynomials are known [Bri10], and containment of a point in an algebraic variety is decidable. Such linear factors are distinct if and only if f and all its partial derivatives have no common linear factors. □

Exercise 10.11. In the special case $d = 2$, what is a quick method of deciding whether f factors into linear factors with no repeated factor?

Replacing the polynomial Q by its square-free part if necessary, we may assume without loss of generality that Q is square-free.

Proposition 10.53. *Let Q be a square-free polynomial and let $\tilde{Q}(z) = \mathtt{hom}(Q)$ be the lowest degree homogeneous part of Q.*

(i) If $\mathbf{0}$ is a multiple point for Q then $\tilde{Q}$ is the product of linear factors.
(ii) If $\mathbf{0}$ is an arrangement point for Q then $\tilde{Q}$ is the product of distinct linear factors.

Proof (*i*) Suppose $\mathbf{0}$ is a multiple point for Q. By Proposition 10.48, there is a factorization $Q(z) = \prod_{j=1}^n Q_j(z)$ in $O_\mathbf{0}$ with each $\nabla Q_j(\mathbf{0})$ nonvanishing. If $\ell_j(z) = \nabla Q_j(\mathbf{0}) \cdot z$ denotes the linear homogeneous term of Q_j then, because $\tilde{Q}$ is homogeneous of degree n, we see that $\tilde{Q} = \prod_{j=1}^n \ell_j$ factors into linear terms.

(*ii*) Suppose that $\mathbf{0}$ is an arrangement point. For each distinct i and j, examination of the sublattice $\{\mathcal{V}, \mathcal{V}_i, \mathcal{V}_j, \mathcal{V}_i \cap \mathcal{V}_j\}$ shows that the gradients at $\mathbf{0}$, which define the linear factors, are distinct. □

Exercise 10.12. If p is a non-smooth multiple point for Q in dimension 2 then must it be an isolated singularity of $\mathcal{V}_Q$? What about in dimension $d > 2$?

Example 10.54. Let

$$Q(x, y, z) = 1 + xyz - \frac{1}{3}(x + y + z + xy + xz + yz)$$

be the denominator of the cube grove generating function (7.23). The ideal

$$\left\langle Q, \frac{\partial Q}{\partial x}, \frac{\partial Q}{\partial y}, \frac{\partial Q}{\partial z} \right\rangle$$

has a Gröbner basis consisting of the polynomials $[x-1, y-1, z-1]$, so the point $(1,1,1)$ is a non-smooth point of $\mathcal{V}$. Translating this point to the origin replaces Q by $h(u,v,w) = Q(1+u, 1+v, 1+w) = 2uv + 2uw + 2vw + 3uvw$, and we let $\tilde{h} = 2(uv+uw+vw)$ be the leading homogeneous part of h. By writing down an attempted factorization we see that $\tilde{h}$ does not factor into linear factors, hence h does not and $(1,1,1)$ is not a multiple point. Asymptotics for the coefficients of the cube grove generating function are discussed in Example 11.43. ◁

Unfortunately the converse to part (*i*) of Proposition 10.53 does not hold. Linear factorization of the homogeneous part of Q does not imply multiple point geometry, even if the factors are distinct and linearly independent (see Example 10.55 below), so Proposition 10.53 helps certify that $\mathbf{p}$ is not a multiple or arrangement point, but doesn't help to certify that it is one.

Another approach to certifying multiple points is a characterization using *local monodromy*. Certification in the positive direction relies on conjectured properties of the relevant algorithms, but for those interested in the problem, this will at least give a start.

Up to taking an invertible linear change of coordinates, we may assume that there exists a coordinate j such that $\pi^{-1}(\mathbf{p})$ is always finite, where $\pi : \mathcal{V} \to \mathbb{C}^{d-1}$ is the projection map onto the coordinate plane $z_j = 0$. Let $\mathcal{B}$ denote the *branch locus* of π, consisting of the points $\mathbf{w} \in \mathbb{C}^{d-1}$ for which the cardinality of the fiber $\pi^{-1}(\mathbf{w})$ does not take its maximum value k. A path $\gamma : [0,1] \to \mathbb{C}^{d-1}$ that avoids $\mathcal{B}$ defines a map from the fiber $\pi^{-1}(\gamma(0))$ to the fiber $\pi^{-1}(\gamma(1))$ by so-called *homotopy continuation*, making continuous choices of $\pi^{-1}(\gamma(t))$ for $0 \le t \le 1$. Homotopic paths in $U = \mathbb{C}^{d-1} \setminus \mathcal{B}$ define the same map, and fixing a base point $\mathbf{z}_0 \notin \mathcal{B}$ and letting γ vary over loops based at $\mathbf{z}_0$ defines a map from the fundamental group of U to the permutations of the fiber at $\mathbf{z}_0$, which we identify with the symmetric group $\mathcal{S}_k$. The image in $\mathcal{S}_k$ is what is known as the ***monodromy group*** of π, and is intrinsic to $\mathcal{V}$ in the sense that it does not depend on the projection π. The ***local monodromy group*** at $\mathbf{p}$ is the image in $\mathcal{S}_k$ of the subgroup of the fundamental group consisting of loops that can be drawn so as never to leave an arbitrarily small neighborhood of $\mathbf{p}$. A point $\mathbf{p}$ is a multiple point if and only if the local monodromy group at $\mathbf{p}$ is trivial (that is, contains only the identity permutation)[1].

[1] If $\mathcal{V}$ is locally the union of smooth surfaces then the local monodromy group is trivial because, in a small path avoiding collisions of roots, each solution remains in the same smooth local piece. We don't prove the converse as it is not used in this book.

Algorithms to compute global monodromy – typically using *numerical algebraic geometry* to approximate continuations along paths with rigorous error bounds – are known [HRS18, Section 3]. Unfortunately, modified methods working to compute local monodromy are currently only conjectured, the difficulty being a way to certify that one has sampled enough curves to have a set of generators for the local monodromy group (so that if all generators are the identity then $\boldsymbol{p}$ is a multiple point).

Example 10.55. Let $Q_1(x, y) = xy - z^3$ and $Q_2(x, y) = xy$. Both varieties contain the origin, where they have leading homogeneous term xy. However, the origin is a multiple point of $\mathcal{V}_{Q_2}$ but not a multiple point of $\mathcal{V}_{Q_1}$. This can be established, for instance, by varying a line parallel to the z-axis through $x = y = \alpha$ as α traces out a small circle around the origin. The three roots are the three values of $\alpha^{2/3}$, and permute cyclically when α makes one loop around the origin. Hence the local monodromy group is not the trivial group. This example illustrates the value of the monodromy characterization, as showing local monodromy to be non-trivial may be easier than showing irreducibility in the local ring. ◁

10.5 Surgery, non-generic directions, and non-arrangement points

Instead of the iterated residue approach described above, multiple points were originally studied using a *surgery method* involving only univariate residues. While iterated residues provide a more direct path to coefficient asymptotics for multiple points, and work under weaker assumptions, the surgery method does have some advantages. For instance, in addition to more elementary computations, the explicitness of intermediate steps allows one to analyze certain non-generic and degenerate cases.

In this final section we follow the surgery approach, computing an explicit univariate residue and employing an analytic trick to asymptotically evaluate the resulting integrals. To simplify our discussion, we consider only power series expansions and assume that each smooth divisor meeting at a multiple point is locally parametrized by the same coordinates $z_1, \dots, z_{d-1}$. Recall from previous chapters that for any d-vector $\boldsymbol{v}$ we write $\boldsymbol{v}^\circ = (v_1, \dots, v_{d-1})$.

Consider a multivariate generating function $F(\boldsymbol{z}) = \sum_{\boldsymbol{r}\in\mathbb{N}^d} a_{\boldsymbol{r}} \boldsymbol{z}^{\boldsymbol{r}}$ with domain of convergence $\mathcal{D}$. For the remainder of this section, we make the following assumptions.

(1) F is meromorphic on some open set $\mathcal{D}'$ containing $\overline{\mathcal{D}}$.

(2) $p \in \mathbb{C}_*^d$ is a strictly minimal contributing point of F in a direction $\hat{r}_*$.
(3) In a neighborhood of p, we can write

$$F(z) = \frac{P(z)}{Q_1(z) \cdots Q_n(z)}$$

with each Q_j having nonvanishing gradient at p.

The implicit function theorem implies the *Weierstrass factorization*

$$F(z) = \frac{\psi(z)}{\prod_{j=1}^{n}(1 - z_d v_j(z^\circ))} \tag{10.25}$$

at p, where $\psi(p) \neq 0$ and the v_j are analytic at p° with $v_j(p^\circ) = 1/p_d$. We call the v_j the *inverse roots* of the Q_j at p, and their reciprocals $u_j(z^\circ) = 1/v_j(z^\circ)$ the *roots* of the Q_j at p.

Remark 10.56. More generally, we could have

$$F(z) = \frac{P(z)}{Q_1(z)^{m_1} \cdots Q_n(z)^{m_n}}$$

for any positive integers m_j. We stick to the *simple pole* case where $\mathbf{m} = \mathbf{1}$ for two reasons. First, the general case is already covered above. Second, it is always possible to reduce to the simple pole case using cohomological reductions (see [AY83, Theorem 17.6] or the definition of iterated residues for non-smooth higher order poles in Appendix C). Laurent series can also be treated in an analogous manner.

Exercise 10.13. Show that $Q(x, y) = 2 - x - 2y + y^2$ has a multiple point at $(1, 1)$, and find a Weierstrass factorization of $1/Q$ there.

A residue sum

Let $w \in \mathcal{D}$ be sufficiently close to w and let $C = T(w_d)$, so that breaking apart the Cauchy integral for a_r gives

$$a_r = \frac{1}{(2\pi i)^d} \int_{\mathbf{T}(w^\circ)} \left(\int_C F(z) \frac{dz_d}{z_d^{r_d+1}} \right) \frac{dz^\circ}{(z^\circ)^{r^\circ - 1}} .$$

Analogously to the smooth case described in Chapter 9, for a sufficiently small neighborhood $\mathcal{N}$ of p° in $\mathbf{T}(p^\circ)$ we can find a point w' sufficiently close to p such that $h_{\hat{r}}(p) > h_{\hat{r}}(w')$ and $\mathcal{N} \times C'$ is disjoint from $\mathcal{V}$ for $C' = T(w_d')$. Defining

$$\chi = \frac{1}{(2\pi i)^d} \int_{\mathcal{N}} \left(\int_{C - C'} F(z) \frac{dz_d}{z_d^{r_d+1}} \right) \frac{dz^\circ}{(z^\circ)^{r^\circ - 1}} \tag{10.26}$$

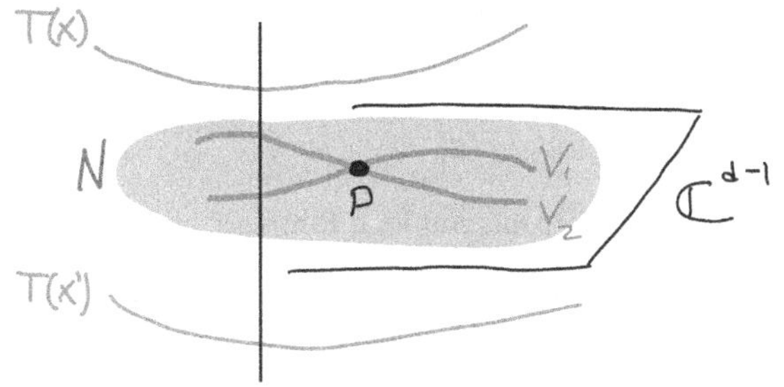

Figure 10.15 Two tori around a multiple point.

then gives the following result, mirroring the smooth case.

Lemma 10.57. *Under the assumptions above,* $|\boldsymbol{p}^{\boldsymbol{r}}\,(a_{\boldsymbol{r}} - \chi)| \to 0$ *exponentially fast in* $|\boldsymbol{r}|$ *as* $\boldsymbol{r} \to \infty$ *with* $\hat{\boldsymbol{r}} \to \hat{\boldsymbol{r}}_*$.

Proof If

$$R_1 = \frac{1}{(2\pi i)^d} \int_{\mathbf{T}(\boldsymbol{p}^\circ)\setminus\mathcal{N}} \left(\int_C F(\boldsymbol{z}) \frac{dz_d}{z_d^{r_d+1}} \right) \frac{d\boldsymbol{z}^\circ}{(\boldsymbol{z}^\circ)^{\boldsymbol{r}^\circ - \boldsymbol{1}}}$$

and

$$R_2 = \frac{1}{(2\pi i)^d} \int_{\mathcal{N}} \left(\int_{C'} F(\boldsymbol{z}) \frac{dz_d}{z_d^{r_d+1}} \right) \frac{d\boldsymbol{z}^\circ}{(\boldsymbol{z}^\circ)^{\boldsymbol{r}^\circ - \boldsymbol{1}}}$$

then

$$\begin{aligned} a_{\boldsymbol{r}} &= \frac{1}{(2\pi i)^d} \int_{T(\boldsymbol{p}^\circ)} \left(\int_C F(\boldsymbol{z}) \frac{dz_d}{z_d^{r_d+1}} \right) \frac{d\boldsymbol{z}^\circ}{(\boldsymbol{z}^\circ)^{\boldsymbol{r}^\circ - \boldsymbol{1}}} \\ &= R_1 + \frac{1}{(2\pi i)^d} \int_{\mathcal{N}} \left(\int_C F(\boldsymbol{z}) \frac{dz_d}{z_d^{r_d+1}} \right) \frac{d\boldsymbol{z}^\circ}{(\boldsymbol{z}^\circ)^{\boldsymbol{r}^\circ - \boldsymbol{1}}} \\ &= R_1 + R_2 + \chi. \end{aligned}$$

The quantity $|\boldsymbol{p}^{\boldsymbol{r}} R_1|$ is exponentially small because $F(\boldsymbol{z}^\circ, t)$ is analytic for t in an annulus extending beyond $|p_d|$ by an amount bounded away from zero when $\boldsymbol{z}^\circ \notin \mathcal{N}$. The quantity $|\boldsymbol{p}^{\boldsymbol{r}} R_2|$ is exponentially small because the integrand of R_2 is exponentially smaller than $\boldsymbol{p}^{-\boldsymbol{r}}$ on its domain of integration. □

Let Λ denote the set of values $\boldsymbol{z}^\circ \in \mathcal{N}$ for which the roots $u_1(\boldsymbol{z}), \ldots, u_n(\boldsymbol{z})$

are not all distinct, and for $z^\circ \notin \Lambda$ define

$$R(z^\circ) = -\sum_{j=1}^{n} \operatorname*{Res}_{t=u_j(z^\circ)} \left(\frac{F(z^\circ, t)}{t^{r_d+1}} \right). \tag{10.27}$$

Because (10.26) is absolutely integrable, it follows from the univariate residue theorem that

$$\chi = \frac{1}{(2\pi i)^{d-1}} \int_{\mathcal{N}} R(z^\circ) \frac{dz^\circ}{(z^\circ)^{r^\circ+1}} \tag{10.28}$$

when Λ has measure zero.

If R had the form $A(z^\circ)e^{-\lambda\phi(z^\circ)}$, which it does in the smooth case, then we could asymptotically approximate (10.28) as a Fourier–Laplace integral using the results of Chapter 5. Unfortunately, the n summands defining R have this form, but each on its own is not integrable when there is more than one sheet meeting at p. To work around this difficulty, we introduce a new integral expression representing R as a symmetric rational sum.

Let Δ_{n-1} denote the standard $(n-1)$-simplex

$$\Delta_{n-1} = \left\{ \mathbf{t} \in \mathbb{R}^n_{\geq 0} : \sum_{j=1}^{n} t_j = 1 \right\},$$

which is parametrized by its projection

$$\pi\Delta_{n-1} = \left\{ \mathbf{t} \in \mathbb{R}^{n-1}_{\geq 0} : \sum_{j=1}^{n-1} t_j \leq 1 \right\}$$

onto its first $n-1$ coordinates. We write ι for the map $\iota(\mathbf{t}) = (t_1, \ldots, t_{n-1}, 1 - \sum_{j=1}^{n-1} t_j)$ that inverts the projection, so that $\iota(\mathbf{t}) \cdot v$ is a convex combination:

$$\iota(\mathbf{t}) \cdot v = t_1 v_1 + \cdots + t_{n-1} v_{n-1} + \left(1 - \sum_{j=1}^{n-1} t_j \right) v_n .$$

Lemma 10.58. *Let f be an analytic function of one complex variable and let $v_1, \ldots, v_n \in \mathbb{C}$ be distinct. Then*

$$\sum_{j=1}^{n} \frac{f(v_j)}{\prod_{r \neq j}(v_j - v_r)} = \int_{\pi\Delta_{n-1}} f^{(n-1)}(\iota(\mathbf{t}) \cdot v)\, d\mathbf{t}. \tag{10.29}$$

Applying this with $f(v) = v^{r_d+n-1}\psi(z^\circ, 1/v)$ to the inverse roots $v_j(z^\circ)$ yields

$$R(z^\circ) = -\int_{\pi\Delta_{n-1}} f^{(n-1)}(\iota(\mathbf{t}) \cdot v(z^\circ))\, d\mathbf{t}. \tag{10.30}$$

Proof The identity (10.29) is proved in [DL93, p. 121]. The second conclusion follows from (10.27), since each $z_d = u_j(\boldsymbol{z}^\circ)$ is a simple pole and

$$\operatorname*{Res}_{t=u_j(\boldsymbol{z}^\circ)} \left(\frac{F(\boldsymbol{z}^\circ, t)}{t^{r_d+1}} \right) = \frac{v_j(\boldsymbol{z}^\circ)^{r_d+n-1} \psi(\boldsymbol{z}^\circ, 1/v_j(\boldsymbol{z}^\circ))}{\prod_{r \neq j} [v_r(\boldsymbol{z}^\circ) - v_j(\boldsymbol{z}^\circ)]}$$
$$= \frac{f(v_j(\boldsymbol{z}^\circ))}{\prod_{r \neq j} [v_r(\boldsymbol{z}^\circ) - v_j(\boldsymbol{z}^\circ)]}.$$

□

Exercise 10.14. Use the Fundamental Theorem of Calculus to verify (10.29) in the case $n = 2$.

Theorem 10.59. *Under the assumptions above*

$$a_{\boldsymbol{r}} \sim \left(\frac{1}{2\pi i} \right)^{d-1} \int_{\mathcal{N}} \left(\int_{\pi \Delta_{n-1}} h\left[\iota(\mathbf{t}) \cdot \boldsymbol{v}(\boldsymbol{z}^\circ)\right] \, d\mathbf{t} \right) \frac{d\boldsymbol{z}^\circ}{(\boldsymbol{z}^\circ)^{\boldsymbol{r}^\circ + \mathbf{1}}}, \tag{10.31}$$

where

$$h(y) = \frac{d^{n-1}}{dy^{n-1}} \left(y^{r_d+n-1} \psi\left(\boldsymbol{z}^\circ, \frac{1}{y} \right) \right).$$

This estimate holds uniformly as $\boldsymbol{r} \to \infty$ with $\boldsymbol{r} = \lambda \hat{\boldsymbol{r}}_* + O(1)$.

Proof If $\boldsymbol{r} = \lambda \hat{\boldsymbol{r}}_*$ then the result follows from Lemmas 10.57 and 10.58. If $\boldsymbol{r} = \lambda \hat{\boldsymbol{r}} + \boldsymbol{\alpha}$ then there is an extra factor of $\boldsymbol{z}^{-\boldsymbol{\alpha}}$ in the integrand. If $\boldsymbol{\alpha}$ is bounded then this extra factor and all its derivatives are uniformly bounded, meaning the remainder in an asymptotic expansion changes in a uniform manner. □

Asymptotic formulae

Having derived (10.31), we are now ready to compute asymptotics. First, we make a change of variables $\boldsymbol{z}^\circ = \boldsymbol{p}^\circ e^{i\boldsymbol{\theta}} = (p_1 e^{i\theta_1}, \ldots, p_{d-1} e^{i\theta_{d-1}})$ to convert the neighborhood $\mathcal{N}$ of $\boldsymbol{p}^\circ$ in $\mathbf{T}(\boldsymbol{p}^\circ)$ to a neighborhood $\mathcal{N}'$ of the origin in $\mathbb{R}^{d-1}$. Let $\Delta = \pi \Delta_{n-1}$, let $d\mathbf{t}$ denote the push-forward of Lebesgue measure by π, and let $\mathcal{E} = \mathcal{N}' \times \Delta$, which we equip with the measure $d\boldsymbol{\theta} \times d\mathbf{t}$. Define the functions

$$\phi(\boldsymbol{\theta}, \mathbf{t}) = \frac{i\boldsymbol{r}^\circ \cdot \boldsymbol{\theta}}{r_d} - \log\left(p_d \iota(\mathbf{t}) \cdot \boldsymbol{v}(\boldsymbol{p}^\circ e^{i\boldsymbol{\theta}}) \right)$$

$$\beta_j(s) = \frac{(n-1)! \Gamma(s+n)}{j!(n-1-j)! \Gamma(s+j+1)}$$

$$A_j(\boldsymbol{\theta}) = \left(\frac{d}{dy} \right)^j \psi(\boldsymbol{p}^\circ e^{i\boldsymbol{\theta}}, 1/y) \Big|_{y = v_j(\boldsymbol{p}^\circ e^{i\boldsymbol{\theta}})}$$

for $0 \leq j \leq n-1$, and observe that $\beta_j(s)$ is a constant multiple of a falling factorial with $n-j-1$ terms, and therefore has degree $n-j-1$ in s.

Lemma 10.60. *The right side of* (10.31) *in Theorem 10.59 can be rewritten as*

$$\chi = \left(\frac{1}{2\pi}\right)^{d-1} \mathbf{p}^{-\mathbf{r}} \sum_{j=0}^{n-1} \beta_j(r_d) \int_{\mathcal{E}} e^{-r_d \phi(\boldsymbol{\theta},t)} A_j(\boldsymbol{\theta})\, d(\boldsymbol{\theta}) \times d\mathbf{t}, \tag{10.32}$$

and the phase ϕ *satisfies* $\operatorname{Re}\phi \geq 0$ *on* $\mathcal{E}$.

Proof Applying the Leibniz rule for differentiating a product to f in (10.30) yields

$$\begin{aligned} f^{(n-1)}(y) &= \sum_{j=0}^{n-1} \binom{n-1}{j} \left(\frac{d}{dy}\right)^{n-1-j} y^{r_d+n-1} \left(\frac{d}{dy}\right)^{j} \psi(\mathbf{p}^\circ e^{i\boldsymbol{\theta}}, 1/y) \\ &= \sum_{j=0}^{n-1} \frac{(n-1)!(r_d+n-1)!}{j!(n-1-j)!(r_d+j)!} y^{r_d+j} \left(\frac{d}{dy}\right)^{j} \psi(\mathbf{p}^\circ e^{i\boldsymbol{\theta}}, 1/y) \\ &= y^{r_d} \sum_{j=0}^{n-1} \beta_j(r_d) y^j \left(\frac{d}{dy}\right)^{j} \psi(\mathbf{p}^\circ e^{i\boldsymbol{\theta}}, 1/y)\,. \end{aligned}$$

Plugging this into (10.30) and using the definitions of A_j, ϕ, and p_j yields the stated formula. By strict minimality of $\mathbf{p}$, for each j the modulus of $v_j(\mathbf{z}^\circ)$ achieves its maximum only when $\mathbf{z} = \mathbf{p}^\circ$. Thus each convex combination of $v_j(\mathbf{z}^\circ)$ with $\mathbf{z} \neq \mathbf{p}$ has modulus less than $|v_j(\mathbf{p}^\circ)|$, meaning that $\operatorname{Re}\phi \geq 0$ on $\mathcal{E}$ with equality only on $\{\mathbf{0}\} \times \Delta$. □

Analysis of the integral in (10.32) requires the advanced Theorem 5.3 from Chapter 5, as the simpler Theorem 5.2 assumes that the real part of the negative phase function has a strict minimum in the interior of the domain of integration. Our present assumptions imply that $\operatorname{Re}\phi$ will be strictly positive away from the origin in the $\boldsymbol{\theta}$-coordinates, but allow the real part of ϕ to vanish up to the boundary in the $\mathbf{t}$-coordinates.

To apply Theorem 5.3 we need to determine the set G' of stratified critical points for which $\operatorname{Re}\phi$ is maximized, i.e., the critical points in $\{\mathbf{0}\} \times \Delta$. Let $\boldsymbol{C}$ be the $n \times d$ matrix with entries

$$C_{ij} = \begin{cases} p_j p_d \frac{\partial v_i}{\partial z_j}(\mathbf{p}) & \text{if } j < d \\ 1 & \text{if } j = d. \end{cases}$$

Exercise 10.15. What is the exact relation between $\boldsymbol{C}$ used in this section and $\boldsymbol{\Gamma}_\Psi$ used in the residue approach?

Proposition 10.61. *Fix a vector $\boldsymbol{r} \in \mathtt{N}(\boldsymbol{p})$. The set G' of critical points of ϕ in the direction $\hat{\boldsymbol{r}}$ on $\mathcal{E}$ is the set of points $(\mathbf{0}, \mathbf{t})$ with $\mathbf{t} \in \mathcal{S}(\boldsymbol{r})$, where*

$$\mathcal{S}(\boldsymbol{r}) = \left\{ \mathbf{t} \in \pi\Delta_{n-1} \mid \iota(\mathbf{t})\boldsymbol{C}(\boldsymbol{p}) = \frac{\boldsymbol{r}}{r_d} \right\}.$$

Proof Since $\operatorname{Re}\phi(\boldsymbol{\theta}) > 0$ when $\boldsymbol{\theta} \neq \mathbf{0}$, all critical points in G' are of the form $(\mathbf{0}, \mathbf{t})$ for $\mathbf{t} \in \Delta_{n-1}$. In fact, ϕ is somewhat degenerate: $\phi(\mathbf{0}, \mathbf{t}) = 0$ for all $\mathbf{t} \in \Delta_{n-1}$, so not only does the real part of ϕ vanish when $\boldsymbol{\theta} = \mathbf{0}$, but also the $\mathbf{t}$-gradient of ϕ vanishes there. This is something of a blessing: checking for non-zero derivatives in directions for which the point $(\mathbf{0}, \mathbf{t})$ is interior to $\mathcal{E}$ reduces to checking whether or not the $\boldsymbol{\theta}$-gradient vanishes at $(\mathbf{0}, \mathbf{t})$. For $j \neq d$,

$$\frac{\partial \phi}{\partial \theta_j} = i \left(\frac{r_j}{r_d} - \frac{z_j}{\iota(\mathbf{t}) \cdot \boldsymbol{v}(\boldsymbol{p}^\circ)} \iota(\mathbf{t}) \cdot \left(\frac{\partial}{\partial z_j} \right) \boldsymbol{v}(\boldsymbol{z}^\circ) \right) \Bigg|_{\boldsymbol{z}=\boldsymbol{p}}$$

and the result follows because $v_j(\boldsymbol{p}^\circ) = 1/p_d$ for all j. □

Specialization to known cases

Proposition 10.61 may look ugly, but it simplifies in several situations. For instance, if $\boldsymbol{p}$ is a transverse multiple point then the rows of $\boldsymbol{C}$ are linearly independent and there is at most one point in $\mathcal{S}$. The normal cone $\mathtt{N}(\boldsymbol{p})$ is the set of convex combinations of the lognormal vectors, so there is at least one point in $\mathcal{S}$ and $\mathcal{S}$ is a singleton $\{\mathbf{t}\}$. If $\boldsymbol{r}$ is interior to $\mathtt{N}(\boldsymbol{p})$ then $\mathbf{t}$ is in the interior of Δ.

Conversely, if two sheets of singularities $\mathcal{V}_1 = \mathcal{V}(Q_1)$ and $\mathcal{V}_2 = \mathcal{V}(Q_2)$ are tangent then the first two rows of $\boldsymbol{C}$ are equal, so any solution $\mathbf{t}$ leads to a line of solutions $(t_1 - s, t_2 + s, t_3, \ldots, t_n)$. For example, if $d = n = 2$, then $\mathtt{N}(\boldsymbol{p})$ is a singleton and $\mathcal{S}$ is the whole unit interval.

Proposition 10.62 (Hessian of ϕ in the singleton case). *Suppose that $\mathcal{S}$ consists of the single point $(\mathbf{0}, \mathbf{t}_*)$. Then the Hessian matrix of $r_d\phi(\boldsymbol{\theta}, \mathbf{t})$ at $(\mathbf{0}, \mathbf{t}_*)$ has the block form*

$$\mathcal{H}(\mathbf{0}, \mathbf{t}_*) = \begin{pmatrix} H(\mathbf{0}, \mathbf{t}_*) & -i\overline{\boldsymbol{C}}^T \\ -i\overline{\boldsymbol{C}} & 0 \end{pmatrix}, \tag{10.33}$$

where

- *the $(d-1) \times (d-1)$ block $H(\mathbf{0}, \mathbf{t}_*)$ is the Hessian of the restriction of $r_d\phi$ to the $\boldsymbol{\theta}$-variables,*
- *the zero block has dimensions $(n-1) \times (n-1)$, and*
- *the $(n-1) \times (d-1)$ matrix $\overline{\boldsymbol{C}}$ is formed by subtracting the last row of $\boldsymbol{C}$ from each other column, and stripping the last row and column.*

Remark 10.63. The last column of $\mathcal{H}$ in (10.33), corresponding to the dth coordinate, is stripped off because this is a function of the others in the $\boldsymbol{\theta}$ parametrization; the last row is stripped off because we parametrize the simplex by its first $n-1$ coordinates.

Proof Let $\boldsymbol{v}$ denote the vector function $(v_1, \ldots, v_n)$. Constancy of ϕ in the $\mathbf{t}$-directions at $\boldsymbol{\theta} = \mathbf{0}$ shows that the second partial derivatives in those directions vanish, giving the block of zeros. Computing $(\partial/\partial\theta_j)\phi$ gives, up to a constant,

$$-\frac{i}{\iota(\mathbf{t})\boldsymbol{v}(\boldsymbol{z}^\circ)} \iota(\mathbf{t}) z_j \frac{\partial}{\partial z_j} \boldsymbol{v}(\boldsymbol{z}^\circ)\bigg|_{\boldsymbol{z}=\boldsymbol{p}} .$$

Because $\iota(\mathbf{t})\boldsymbol{v}(\boldsymbol{\theta}, \mathbf{t})$ is constant in $\mathbf{t}$ when $\boldsymbol{\theta} = \mathbf{0}$, differentiating in the $\mathbf{t}$ directions recovers the blocks $-i\overline{C}$ and $-i(\overline{C})^T$. The second partial derivatives in the $\boldsymbol{\theta}$ directions are unchanged. □

Specializing further yields a proof of Theorem 10.38 via surgery.

Surgery proof of Theorem 10.38 Under our assumptions $\mathcal{S}$ is a singleton, so we may apply Theorem 5.3 to the integral over $\mathcal{E}$ in each summand of (10.32) giving an asymptotic series in decreasing powers of r_d. The polynomials β_j decrease in degree from $n-1$ to zero and integration may be carried out term by term because the remainders are uniformly one power of r_d lower as long as $\hat{\boldsymbol{r}}$ is restricted to a compact subset of $\mathtt{N}(\boldsymbol{p})$. The leading term comes from a careful examination of the terms appearing in the saddle point integral. □

Remark 10.64. Theorem 5.3 allows us to localize as long as we don't put a boundary somewhere that the height is maximized while simultaneously creating a critical point somewhere in a stratum of this boundary. This allows us to go beyond the standing hypothesis of strict minimality in Theorem 10.59 and Lemma 10.60, however the analysis of maximal height critical points in Proposition 10.61 is more complicated.

The surgery method handles complicated geometries of the stationary set better, in principle, than the method of iterated residues. One example of this advantage appears when $\hat{\boldsymbol{r}}$ is on a *facet* of $\mathtt{N}(\boldsymbol{p})$ (a face of the cone whose dimension is one less than the dimension of the whole cone).

Theorem 10.65. *Fix a direction $\hat{\boldsymbol{r}} = \hat{\boldsymbol{r}}_*$ and assume that the hypotheses of Theorem 10.38 hold with $\mathbf{m} = \mathbf{1}$, except that $\hat{\boldsymbol{r}}_*$ lies on a facet of* $\mathtt{N}(\boldsymbol{p})$ *instead of being in the interior. If $\mathbf{r} \to \infty$ such that $\mathbf{r} - |\mathbf{r}|\hat{\boldsymbol{r}}_*$ stays bounded, an asymptotic expansion of the form* (10.17) *still holds. The leading term in this expansion is half the value Theorem 10.38 would predict if $\hat{\boldsymbol{r}}_*$ were in the interior of* $\mathtt{N}(\boldsymbol{p})$.

Proof The condition that $\hat{\boldsymbol{r}}_*$ is on a facet of $\mathtt{N}(\boldsymbol{p})$ is equivalent to the condition that $\iota(\mathbf{t}(\boldsymbol{r}_*))$ is on a facet of Δ_{n-1}. The main contribution to the inner integral in (10.32) comes from an ε-ball about $\mathbf{t}(\boldsymbol{r})$ for any sufficiently small $\varepsilon > 0$, and when $\mathbf{t}$ is on a facet this integral is over a halfspace neighborhood rather than a ball neighborhood. The leading term in such an integral is precisely half of the leading term for a ball neighborhood. After introducing a factor of $1/2$ in the inner integral of (10.32), the new hypotheses leave everything else unchanged. This proves the result when $\boldsymbol{r}$ is a precise multiple of $\hat{\boldsymbol{r}}_*$. When $\boldsymbol{r} - |\boldsymbol{r}|\hat{\boldsymbol{r}}_*$ is bounded, the uniformity statement in Theorem 10.59 shows that (10.31) and hence (10.32) remain valid, introducing the factor of $1/2$ and completing the proof. □

Example 10.66 (two lines, boundary directions)**.** Suppose $F = 1/Q(x, y)$, where

$$Q(x, y) = \left(1 - \frac{1}{3}x - \frac{2}{3}y\right)\left(1 - \frac{2}{3}x - \frac{1}{3}y\right).$$

As we have seen in Example 10.27, $a_{rs} \sim 3$ when $r, s \to \infty$ with r/s in a compact subset of $(1/2, 2)$. By Theorem 10.65, the coefficients along the $(1, 1)$-diagonal are asymptotically half of the coefficients on the interior of the cone $1/2 < r/s < 2$, thus

$$a_{2s,s} \sim 3/2$$

as $s \to \infty$. In fact, this approximation holds for any a_{rs} with $r - 2s = O(1)$. Note that as we approach the boundary of the cone from the interior, the convergence to the limiting value becomes slower and slower; on the boundary the relative error term is no longer exponentially small, but of order $1/s$. ◁

Remark 10.67. The requirement that $\boldsymbol{r} - |\boldsymbol{r}|\hat{\boldsymbol{r}}_*$ remain bounded is not best possible. In [BMP23, Section 6.2] it is shown that, for linear divisors, uniform estimates hold when $|\boldsymbol{r} - |\boldsymbol{r}|\,\hat{\boldsymbol{r}}_*| = o(|\boldsymbol{r}|^{1/2})$, yielding a Gaussian transition regime whose width scales as $|\boldsymbol{r}|^{1/2}$.

Non-arrangement points

As a final application of the surgery method, we consider the case where $n = d = 2$ and the two divisors $\mathcal{V}_1$ and $\mathcal{V}_2$ meet tangentially. In this case, the meeting point $\boldsymbol{p}$ is not an arrangement point because the intersection lattice of the divisors has order 2, whereas the intersection lattice of the tangent planes has order 1 (both tangent lines coincide). Furthermore, the normal cone at $\boldsymbol{p}$ degenerates to a single ray, so that this direction is a boundary direction and hence is not generic. For these reasons we turn to surgery. The Fourier–Laplace integral in Theorem 10.59 is more degenerate than can be handled by Theorem 5.3 of Chapter 5. However, we can still complete the analysis for this specific situation.

Proposition 10.68 (two curves intersecting tangentially in two dimensions). *Suppose that $F(x,y) = P(x,y)/(Q_1(x,y)Q_2(x,y))$, where Q_1 and Q_2 define smooth divisors $\mathcal{V}_j$ intersecting at a strictly minimal multiple point* $(1,1)$. *Suppose further that $\mathcal{V}_1$ and $\mathcal{V}_2$ intersect tangentially at* $(1,1)$, *where $\nabla_{\log} Q_1$ and $\nabla_{\log} Q_2$ are both multiples of some direction $(\lambda, 1)$. Let u_1 and u_2 denote analytic functions such that $\mathcal{V}_j = \{(x, u_j(x)) : x \in \mathbb{C}\}$ in a neighborhood of* $(1,1)$, *let $v_j(x) = 1/u_j(x)$ denote the inverse roots, and let $g_j(\theta) = \log v_j(e^{i\theta})$ be the parametrization of the logarithmic inverse roots in polar coordinates. If*

$$g_j(\theta) = ir_*\theta - \frac{\kappa_j}{2}\theta^2 + O(\theta)^3$$

for each j with $\kappa_1 + \kappa_2 > 0$, the κ_j nonnegative, and $P(1,1) \neq 0$ then

$$a_{rs} \sim \frac{P(1,1)}{\sqrt{2\pi}} \frac{2}{\sqrt{\kappa_1} + \sqrt{\kappa_2}} s^{1/2}$$

as $(r,s) \to \infty$ with $r = \lambda s + O(1)$.

Proof Lemma 10.60 implies that

$$a_{rs} \sim \frac{s}{(2\pi i)^2} \int_{\mathcal{E}} e^{-s\phi(\theta,t)} A_0(\theta)\, d\theta\, dt$$

for $(r,s) = \lambda(\hat{r}, \hat{s}) + O(1)$, where $\mathcal{E} = \mathcal{N} \times \pi\Delta_1$ with $\mathcal{N}$ a neighborhood $(-\delta, \delta)$ of zero in $\mathbb{R}$, the amplitude $A_0(x,y) = P(x,y)/(xy)$, and $\pi\Delta_1 = [0,1]$. The phase function ϕ is the convex combination

$$-\phi(\theta,t) = tg_1(\theta) + (1-t)g_2(\theta),$$

which has nonnegative real part on $(-\delta,\delta) \times [0,1]$, which vanishes on the line segment $\mathcal{S} = \{0\} \times [0,1]$. As opposed to previous cases we have considered, the entire line segment is now critical for ϕ. Integrating $e^{-s\phi(\theta,t)}\, d\theta$ over $(-\delta,\delta)$ for fixed t gives

$$(2\pi s\kappa_t)^{-1/2} + O\left(s^{-3/2}\right),$$

where $\kappa_t = t\kappa_1 + (1-t)\kappa_2$ is the quadratic term of $\phi(\theta,t)$. The $O(s^{-3/2})$ error term is uniformly $o(s^{-1/2})$, therefore the expression for the leading term may be integrated over [0, 1]. Using the change of variables $y = \kappa_t$ leads to

$$\begin{aligned} a_{rs} &\sim \frac{P(1,1)\sqrt{s}}{\sqrt{2\pi}} \int_0^1 \kappa_t^{-1/2}\, dt \\ &= \frac{P(1,1)\sqrt{s}}{\sqrt{2\pi}} \int_{\kappa_2}^{\kappa_1} y^{-1/2} \frac{dy}{\kappa_1 - \kappa_2} \\ &= \frac{P(1,1)\sqrt{s}}{\sqrt{2\pi}} \frac{2}{\sqrt{\kappa_1} + \sqrt{\kappa_2}}. \end{aligned}$$

The substitution $y = \kappa_t, dy = (\kappa_1 - \kappa_2)\,dt$ is valid only when $\kappa_1 \neq \kappa_2$ but the resulting expression for $\int_0^1 \kappa_t^{-1/2}$ is valid for all $\kappa_1, \kappa_2 \geq 0$ with $\kappa_1 + \kappa_2 > 0$. □

Example 10.69 (partial sums of normalized binomials, continued)**.** Recall the generating function $F = \frac{2}{(2-x-y)(1-xy)}$ from Example 10.11, whose (s, s)-coefficient is the normalized binomial sum $\sum_{i=0}^{s} 4^{-i}\binom{2i}{i}$ computing the expected number of returns of a simple random walk to zero by time $2s$. Expanding functions $g_1(\theta) = -\log(2 - e^{i\theta})$ and $g_2(\theta) = i\theta$ around $\theta = 0$ gives quadratic coefficients $\kappa_1 = 2$ and $\kappa_2 = 0$, so the identity $P(1, 1) = 2$ implies

$$a_{s,s} \sim 2\sqrt{\frac{s}{\pi}}.$$

This result can be verified by univariate methods, as the main diagonal has the algebraic generating function $f(z) = (1 - z)^{-3/2}$. Note that even though $(1, 1)$ is a complete intersection, there is no longer an exponentially decreasing error term. Instead the relative error is $O(1/s)$. ◁

Notes

Computations for multiple points based on the theory of iterated residues first appeared in the work of [BM93] on queueing models. More information on iterated residues can be found in [AY83; Pha11]. As noted above, the coefficient asymptotics in this chapter were first computed via the surgery method in [PW02; PW04]. The surgery method is only valid in its original form for finitely minimal points, which is why more advanced methods using hyperbolic polynomials and Morse theoretic arguments were introduced. The special case where all divisors are linear, and the pole variety is a hyperplane arrangement, is worked out at length in [BMP23].

The computation of the residue from an explicit factorization via Algorithm 4 is effective, assuming the factors Q_j are in a nice class of functions such as polynomials or algebraic functions. The efficiency of such algorithms is a topic of ongoing study. Some specific problems that can be studied with the results of this chapter can likely be improved with a more focused study. For instance, De Loera and Sturmfels [DS03] discuss alternate approaches to the enumeration of lattice points in a polytope, one involving computation of the Todd class of an associated toric variety and another pioneered by Barvinok et al. [Bar94].

The decomposition of the positive orthant into regions (chambers) in which the counts vary polynomially are objects of classical study, as are the counts themselves; one example is the enumeration of lattice points in the dilated *Birkhoff polytope*, where even the leading term asymptotic was found only in 2009 [CM09].

The fact that the iterated residue of a rational function is a polynomial for complete intersections, and its consequences for generating function asymptotics, are discussed in [Pem00].

The proof of Lemma 10.45 suggested by Exercise 10.16 was supplied to us by Frank Sottile.

Additional exercises

Exercise 10.16. Assume the hypotheses of Lemma 10.45. Prove the following statements, then prove Lemma 10.45.

(a) The gradients $\nabla Q_i(\mathbf{p})$ are non-zero for all $i \leq n$.
(b) For any $U \subseteq \overline{T}$, let $\mathcal{V}_U = \bigcap_{i \in U} \mathcal{V}_i$. If the set of gradients $\{\nabla Q_i(\mathbf{p}) : i \in U\}$ are linearly independent then $\mathcal{V}_U$ is smooth near $\mathbf{p}$ and equal to $\mathcal{V}_{\overline{T}}$.
(c) For such sets U, the ideal $\langle f_i : i \in U\rangle$ is *reduced* in $\mathcal{O}_{\mathbf{p}}$.
(d) For such sets U, the ideal $\langle f_i : i \in U\rangle$ is prime in $\mathcal{O}_{\mathbf{p}}$.

Exercise 10.17. Suppose that $d = 2$ and let $\mathbf{p}$ be a homogeneous point of $\mathcal{V}$ of degree k. Prove that $\mathbf{p}$ is a multiple point or is a cusp whose tangents are all equal.

Exercise 10.18. Let $F(x, y, z) = \frac{z/2}{(1-yz)(1-(x+x^{-1}+y+y^{-1})z/2+z^2)}$ be the Aztec diamond placement generating function, to be studied in Example 11.45 and Theorem 11.49. For what directions $\hat{\mathbf{r}}$ are there critical points on the stratum which is the intersection of the two divisors in the denominator of F? Determine, up to a constant factor, the asymptotic contribution at such a point. (These points never contribute to the leading asymptotics of a_{rst}, their contribution being too small to contribute in the cases when there is no exponential decay as in Theorem 11.49 and too large to contribute when there is, as in Example 11.45.)

11

Cone point asymptotics

Once again, let $F(z) = P(z)/Q(z) = \sum_r a_r z^r$ be a convergent Laurent expansion with domain of convergence $\mathcal{D} = \text{Relog}^{-1}(B)$ for a component B of $\texttt{amoeba}(Q)^c$. In Chapter 6 we defined the logarithmic growth rate

$$\beta(\hat{r}) = \limsup_{\substack{s\to\infty \\ s/|s|=\hat{r}}} \frac{1}{|s|} \log |a_s|,$$

the neighborhood growth rate

$$\overline{\beta}(\hat{r}) = \limsup_{\substack{s\to\infty \\ s/|s|\to\hat{r}}} \frac{1}{|s|} \log |a_s|,$$

and the dual rate

$$\beta^*(\hat{r}) = \inf_{x\in B} -\hat{r}\cdot x\,,$$

and showed that

$$\beta(\hat{r}) \le \overline{\beta}(\hat{r}) \le \beta^*(\hat{r})$$

for every direction $\hat{r}$. In this chapter we study conditions under which the equality $\overline{\beta}(\hat{r}) = \beta^*(\hat{r})$ holds – a useful criterion because β^* is much easier to compute than β or $\overline{\beta}$ (which usually require determining dominant asymptotic behavior). Our framework also allows us to give asymptotics in the presence of *cone points*, going beyond the smooth and multiple point cases of the last two chapters.

Deformations and vector fields: a third approach

Deformation of cycles underlies every result in ACSV. The surgery method, sketched in Chapter 9 for smooth points and Chapter 10 for multiple points, is very explicit but applies only to minimal points. In contrast, the powerful Morse-theoretic decompositions in Chapter 7 use Thom's Isotopy Lemma

(Lemma D.16), which requires the advanced apparatus of controlled vector fields. In this chapter we chart a course between these extremes, using the theory of hyperbolic polynomials to build a smooth vector field to deform a torus of integration below a given height except at minimal critical points. This is reasonably general, deforming any torus defined by an element of B up to and just beyond the amoeba boundary except arbitrarily close to critical points.

Proving all necessary results from scratch would necessitate adding (at least) another appendix to this book, so this chapter is not entirely self-contained. Instead, some of the technical background is cited from [BP11], which itself condenses a fair amount of background from [ABG70]. We give a complete, albeit telegraphic, rendition of the geometric results, extending the foundation built in Chapter 6 to the context of hyperbolic polynomials, but largely quote without proof the generalized function theory from [BP11, Section 6] necessary for rigorous justification of certain Fourier transforms.

Organization of the chapter

The remainder of this chapter proceeds as follows. First, we state Theorem 11.1, which asserts the existence of certain semi-continuously varying cones. In this context, a function from points $\boldsymbol{x} \in \mathbb{C}^d$ to open cones $K(\boldsymbol{x}) \subseteq \mathbb{R}^d$ is ***lower semi-continuous*** if for all sequences $\boldsymbol{x}_n \to \boldsymbol{x}$ we have $K(\boldsymbol{x}) \subseteq \liminf K(\boldsymbol{x}_n)$, where the limit inferior of a sequence (S_n) of sets contains the point $\boldsymbol{y}$ if and only if $\boldsymbol{y}$ is in all but finitely many of the sets S_n.

Exercise 11.1. Let H_n be the open halfspace $\{(x, y) \in \mathbb{R}^2 : x + ny > 0\}$. What is $\liminf H_n$?

Several deformation results are derived from Theorem 11.1. In order to illustrate the concepts involved, a proof of Theorem 11.1 is first provided in the simplest case when the intersection of the minimal torus and the variety consists entirely of smooth points. The second section of this chapter contains a full proof of Theorem 11.1, divided into a number of subsections: using cones to construct deformations, cones of hyperbolicity in the homogeneous case, cones of hyperbolicity in the general case, strong and weak hyperbolicity, semi-continuity of cones, the final steps of the proof, and a coda in which a projective (log-linear) version of the deformation is constructed.

Having localized all integrals using these deformations, in the third section we compute asymptotics for the local integrals using results from [BP11] and [ABG70]. These computations are simplified using log-linear deformations. The final section of the chapter applies this computational apparatus to a number of examples including cube groves, Szegő functions, Aztec diamonds, the fortress generating function, and the GKZ symmetric rational function, whose coefficients exhibit a *lacuna phenomenon*.

11.1 Results on cones and deformations

Let Q be a Laurent polynomial in d variables defining a variety $\mathcal{V}$ and set $\mathcal{M} = \mathbb{C}_*^d \setminus \mathcal{V}$. In order to work with polynomial amoebas, it is convenient to define and manipulate cones in *logarithmic space*. To that end, define the ***flat torus*** $\mathbf{T}_{\text{flat}} = (\mathbb{R}/(2\pi\mathbb{R}))^d$, let $q(\boldsymbol{x}) = (Q \circ \exp)(\boldsymbol{x}) = Q(e^{x_1}, \ldots, e^{x_d})$, and recall the notation $\mathbf{T}_e(\boldsymbol{x}) = \{\exp(\boldsymbol{x} + i\boldsymbol{y}) : \boldsymbol{y} \in \mathbf{T}_{\text{flat}}\}$ for any $\boldsymbol{x} \in \mathbb{R}^d$.

Exercise 11.2. Is the map $\exp : \mathbb{R}^d + i\,\mathbf{T}_{\text{flat}} \to \mathbb{C}_*^d$ a diffeomorphism?

Let B be a component of the complement of $\texttt{amoeba}(Q)$, fix a direction $\hat{\boldsymbol{r}}_*$, and suppose that $-\hat{\boldsymbol{r}}_* \cdot \boldsymbol{x}$ is minimized on $\overline{B}$ at a unique point $\boldsymbol{x}_*$, necessarily on the boundary ∂B. Recall from Definition 6.26 and Exercise 6.8 the open geometric tangent cone $\texttt{tan}_{\boldsymbol{x}}(B)$ to B at a point $\boldsymbol{x}$, with closed dual cone $\texttt{normal}_{\boldsymbol{x}}(B) = (-\texttt{tan}_{\boldsymbol{x}}(B))^*$. These cones are *projective*: they are cones over the origin, closed under multiplication by positive real constants. The following result is a combination of [BP11, Corollary 2.15 (i), Corollary 2.16, Proposition 2.22, Theorem 5.4].

Theorem 11.1. *Under the setup above, there exists a bundle $\mathcal{Z}$ of convex projective cones $\{Z(\boldsymbol{y}) : \boldsymbol{y} \in \mathbf{T}_{\text{flat}}\}$ with the following properties.*

(i) For all $\boldsymbol{y} \in \mathbf{T}_{\text{flat}}$ the cone $Z(\boldsymbol{y})$ contains $\tan_{\boldsymbol{x}_*}(B)$.

(ii) Let $\vec{V}$ be a smooth nonvanishing section of $\mathcal{Z}$, so that $\vec{V} : \mathbf{T}_{\text{flat}} \to \mathbb{R}^d \setminus \{\mathbf{0}\}$ is smooth and $\vec{V}(\boldsymbol{y}) \in Z(\boldsymbol{y})$ for all $\boldsymbol{y}$. If $\boldsymbol{x} \in B$ then there exists an $\varepsilon > 0$ such that the set

$$\left\{\exp\left(\boldsymbol{x}_* + \delta[t\vec{V}(\boldsymbol{y}) + (1-t)(\boldsymbol{x} - \boldsymbol{x}_*)] + i\boldsymbol{y}\right) : \boldsymbol{y} \in \mathbf{T}_{\text{flat}}, 0 \le t \le 1, 0 < \delta < \varepsilon\right\} \tag{11.1}$$

is disjoint from $\mathcal{V}$.

(iii) For each $\boldsymbol{y} \in \mathbf{T}_{\text{flat}}$ let $N(\boldsymbol{y}) = (-Z(\boldsymbol{y}))^*$ be the outward dual cone to $Z(\boldsymbol{y})$, and let $\hat{\boldsymbol{r}}$ be any real unit vector. If $\hat{\boldsymbol{r}} \in N(\boldsymbol{y})$ then $\exp(\boldsymbol{x}_* + i\boldsymbol{y})$ *is a stratified critical point for $h_{\hat{\boldsymbol{r}}}$. If $\hat{\boldsymbol{r}} \notin N(\boldsymbol{y})$ then there is some vector $\boldsymbol{v} \in Z(\boldsymbol{y})$ such that $\hat{\boldsymbol{r}} \cdot \boldsymbol{v} = 1$.*

(iv) The cones $Z(\boldsymbol{y})$ vary lower-semicontinuously in $\boldsymbol{y}$. In particular, if $\boldsymbol{y}_n \to \boldsymbol{y}$ in $\mathbf{T}_{\text{flat}}$ then $Z(\boldsymbol{y}) \subseteq$ $\liminf Z(\boldsymbol{y}_n)$.

Remark 11.2. The cones $Z(\boldsymbol{y})$ will typically be open cones, such as intersections of open halfspaces. We will see below that we can choose $Z(\boldsymbol{y})$ to be *cones of hyperbolicity* of the homogenization of q at $\boldsymbol{x} + i\boldsymbol{y}$, picking the one that contains $\mathtt{tan}_{\boldsymbol{x}}(B)$. While the theorem does not rule out other choices, choosing smaller cones would be wasteful because then $N(\boldsymbol{y})$ would be larger and the last part of conclusion (iii) would apply less often, while choosing larger cones turns out to make conclusion (ii) impossible to satisfy. The hypothesis that $\boldsymbol{x}_* \in \partial B$ is important: for a counterexample when this fails, see Exercise 11.10.

The cones $N(\boldsymbol{y})$ allow us to define a set of points that locally look like minimal critical points, except that they "don't see" points of Q.

Definition 11.3. The set $\mathtt{local}(\hat{\boldsymbol{r}})$ of ***locally minimal*** points of Q with respect to $\boldsymbol{x}_*$ in the direction $\hat{\boldsymbol{r}}$ consists of the points $\boldsymbol{z} = \exp(\boldsymbol{x}_* + i\boldsymbol{y}) \in \mathbf{T}_{\mathrm{e}}(\boldsymbol{x}_*)$, where Q vanishes and $\hat{\boldsymbol{r}} \in N(\boldsymbol{y})$. We also define the set of ***logarithmic locally minimal arguments*** $\mathtt{localarg}(\hat{\boldsymbol{r}}) = \{\boldsymbol{y} \in \mathbf{T}_{\mathrm{flat}} : \exp(\boldsymbol{x} + i\boldsymbol{y}) \in \mathtt{local}(\hat{\boldsymbol{r}})\}$.

Locally minimal points are the only points that can cause an obstruction when trying to deform the Cauchy domain of integration to lower height.

Theorem 11.4. *Assume the hypotheses of Theorem 11.1 and let $T = \mathbf{T}_{\mathrm{e}}(\boldsymbol{x})$ for any $\boldsymbol{x} \in \partial B$. If $\mathtt{local}(\hat{\boldsymbol{r}})$ is empty then T is homotopic in $\mathcal{M}$ to a cycle C_* whose maximum height is less than $h_* = -\hat{\boldsymbol{r}} \cdot \boldsymbol{x}_*$.*

Proof of Theorem 11.4 from Theorem 11.1 Fix $\boldsymbol{x} \in \partial B$ and for each $\boldsymbol{y} \in \mathbf{T}_{\mathrm{flat}}$ let $\boldsymbol{v}(\boldsymbol{y})$ denote a vector $\boldsymbol{v}$ as in conclusion (iii) of Theorem 11.1. By semicontinuity, for each $\boldsymbol{y} \in \mathbf{T}_{\mathrm{flat}}$ there is a neighborhood $\mathcal{N}(\boldsymbol{y})$ of $\boldsymbol{y}$ in $\mathbf{T}_{\mathrm{flat}}$ such that $\boldsymbol{v}(\boldsymbol{y}) \in Z(\boldsymbol{y}')$ when $\boldsymbol{y}' \in \mathcal{N}(\boldsymbol{y})$. Cover $\mathbf{T}_{\mathrm{flat}}$ with finitely many of the neighborhoods $\{\mathcal{N}(\boldsymbol{y}_j) : 1 \le j \le m\}$ by compactness, let $\{\psi_j : 1 \le j \le m\}$ be a partition of unity subordinate to this finite cover, and for $\boldsymbol{y} \in \mathbf{T}_{\mathrm{flat}}$ define

$$\vec{V}(\boldsymbol{y}) = \sum_{j=1}^{m} \psi(\boldsymbol{y})\boldsymbol{v}(\boldsymbol{y}_j) . \tag{11.2}$$

For each j such that $\boldsymbol{y} \in \mathcal{N}(\boldsymbol{y}_j)$, the vector $\boldsymbol{v}(\boldsymbol{y}_j)$ lies in the cone $Z(\boldsymbol{y})$. The vector $\vec{V}(\boldsymbol{y})$ is a convex combination of these vectors and therefore, by convexity of the cone $Z(\boldsymbol{y})$, we see that $\vec{V}(\boldsymbol{y}) \in Z(\boldsymbol{y})$. By linearity of the dot product with $\hat{\boldsymbol{r}}_*$, we also have that $\vec{V}(\boldsymbol{y}) \cdot \hat{\boldsymbol{r}}_* > 0$. In particular, $\vec{V}$ is a smooth nonvanishing section of $\mathcal{Z}$. Let $\varepsilon > 0$ be as in conclusion (ii) of Theorem 11.1. Again using convexity of $Z(\boldsymbol{y})$ and the fact that $Z(\boldsymbol{y}) \supseteq \mathtt{tan}_{\boldsymbol{x}_*}(B) \supseteq B - \boldsymbol{x}$ (the set B with $\boldsymbol{x}_*$ translated to the origin), we see that the line segment between and $\vec{V}(\boldsymbol{y})$ and any $\boldsymbol{x} \in B$ is contained in $Z(\boldsymbol{y})$.

Define the map $\phi : \mathbf{T}_{\mathrm{flat}} \times [0,1]$ by

$$\phi_t(\boldsymbol{y}) = \exp\left\{i\boldsymbol{y} + \boldsymbol{x}_* + \delta\left[(1-t)(\boldsymbol{x} - \boldsymbol{x}_*) + t\vec{V}(\boldsymbol{y})\right]\right\} . \tag{11.3}$$

By construction ϕ is continuous with $\phi_0(\mathbf{T}_{\text{flat}})$ being the chain $\mathbf{T}_e(\boldsymbol{u})$ for $\boldsymbol{u} = \boldsymbol{x}_* + \delta(\boldsymbol{x} - \boldsymbol{x}_*) \in B$. Because each line segment from $\boldsymbol{u}$ to $\vec{V}(\boldsymbol{y})$ is in $Z(\boldsymbol{y})$, and $\exp(i\boldsymbol{y} + \boldsymbol{x}_* + \boldsymbol{v}) \notin \mathcal{V}$ for any $\boldsymbol{v}$ in any $Z(\boldsymbol{y})$, we see that the homotopy ϕ avoids $\mathcal{V}$. The homotopy ϕ deforms $\mathbf{T}_e(\boldsymbol{u})$ to the cycle $\phi_1(\mathbf{T}_{\text{flat}})$, with the height of a point $\phi_1(\boldsymbol{y})$ given by $-\hat{\boldsymbol{r}}_* \cdot \text{Re}\{\boldsymbol{x}_* + \vec{V}(\boldsymbol{y})\}$. Because $\vec{V}(\boldsymbol{y})$ was picked according to conclusion (iii) of Theorem 11.1, we see that $\hat{\boldsymbol{r}}_* \cdot \vec{V}(\boldsymbol{y})$ is strictly positive, and therefore that $h_{\hat{\boldsymbol{r}}_*}(\phi_1(\boldsymbol{y})) < -\hat{\boldsymbol{r}}_* \cdot \boldsymbol{x}_*$, finishing the proof. □

Theorem 11.5. *Assume the hypotheses of Theorem 11.1 and let* $T = \mathbf{T}_e(\boldsymbol{x})$ *for any* $\boldsymbol{x} \in \partial B$. *If* $\texttt{local}(\hat{\boldsymbol{r}})$ *is finite and nonempty then for any* $\delta > 0$ *the torus* T *is homotopic in* $\mathcal{M}$ *to a cycle whose maximum height is less than* $h_* = -\hat{\boldsymbol{r}} \cdot \boldsymbol{x}_*$ *outside of balls of radius* δ *around the points of* $\mathbf{W} = \texttt{local}(\hat{\boldsymbol{r}})$.

Proof of Theorem 11.5 from Theorem 11.1 Repeat the proof of Theorem 11.4, with the following modifications. For $\boldsymbol{y} \notin \mathbf{W}$, choose a smaller $\mathcal{N}(\boldsymbol{y})$ if necessary so that $\mathbf{W} \cap \mathcal{N}(\boldsymbol{y}) = \emptyset$. For $\boldsymbol{y} \in \mathbf{W}$, define $\mathcal{N}(\boldsymbol{y})$ to be the preimage under the exponential map of the ball of radius δ centered at $\exp(\boldsymbol{x}_* + i\boldsymbol{y})$ and set $\boldsymbol{v}(\boldsymbol{y}) = \mathbf{0}$. Choose again a finite open subcover $\mathcal{N}(\boldsymbol{y}_j : 1 \le j \le m\}$ of $\mathbf{T}_{\text{flat}}$ and construct $\vec{V}$ by (11.2) and ϕ_t by (11.3) as before. The cycle $C_\delta = \phi_1(\mathbf{T}_{\text{flat}})$ is homotopic in $\mathcal{M}$ to $\mathbf{T}(\boldsymbol{u})$ and satisfies the conclusion of the theorem. □

Corollary 11.6. *The conclusions of Theorems 11.4 and 11.5 can be strengthened to hold simultaneously for all* $\hat{\boldsymbol{r}}$ *in some neighborhood.*

Proof This follows from the finiteness of the open cover in each case and the fact that the strict inequality over each $\mathcal{N}(\boldsymbol{y})$ actually bounds $\hat{\boldsymbol{r}}_* \cdot \vec{V}(\boldsymbol{y})$ away from zero. □

A simplified argument for smooth points

It will take a lot of work to prove Theorem 11.1 below, so we first illuminate the argument by proving a special case that does not need the full theory of cones of hyperbolicity. Assume for the remainder of this section that

$$\mathbf{T}(\boldsymbol{x}_*) \cap \mathcal{V} \text{ consists only of smooth points of } \mathcal{V}. \qquad (*)$$

It turns out that cones of hyperbolicity at smooth points are always halfspaces: Theorem 6.44 in Chapter 6 implies that for a minimal point $\boldsymbol{z} \in \mathcal{V}$ the vector $\nabla_{\log} Q(\boldsymbol{z})$ is a complex scalar multiple of a real unit vector $\boldsymbol{\lambda}$, and this real vector, with proper orientation, defines the correct cone.

Definition 11.7. For $\boldsymbol{y} \in \mathbf{T}_{\text{flat}}$ with $q(\boldsymbol{x}_* + i\boldsymbol{y}) \neq 0$ define $Z(\boldsymbol{y}) = \mathbb{R}^d$. For $\boldsymbol{y} \in \mathbf{T}_{\text{flat}}$ with $q(\boldsymbol{x}_* + i\boldsymbol{y}) = 0$ and $\boldsymbol{z} = \exp(\boldsymbol{x}_* + i\boldsymbol{y})$ a smooth point of $\mathcal{V}$, let

$$Z(\boldsymbol{y}) = \{\boldsymbol{x} \in \mathbb{R}^d : \boldsymbol{x} \cdot \boldsymbol{\lambda}(\boldsymbol{z}) < 0\}. \qquad (11.4)$$

Lemma 11.8. *Assume the hypotheses of Theorem 11.1 as well as the smooth point assumption (*). Then the family* $\{Z(\mathbf{y}) : \mathbf{y} \in \mathbf{T}_{\text{flat}}\}$ *in Definition 11.7 satisfies conclusions* (i), (iii), *and* (iv) *of Theorem 11.1.*

Note that we do not require conclusion (ii) of Theorem 11.1 for our special arguments about deformations in the smooth case below.

Proof Conclusion (i) follows from part (*b*) of Theorem 6.44. For conclusion (iii), note that $N(\mathbf{y})$ is by definition empty if $q(\mathbf{x}_* + i\mathbf{y}) \neq 0$, in which case $Z(\mathbf{y})$ is all of $\mathbb{R}^d$ and the conclusion is trivial. Assume therefore that $q(\mathbf{x}_* + i\mathbf{y}) = 0$ and let $\boldsymbol{\lambda} = \boldsymbol{\lambda}(\mathbf{y})$ denote the vector $(\nabla_{\log} Q)(\exp(\mathbf{x}_* + i\mathbf{y}))$ after normalizing to a unit vector and orienting to point away from B, so that $N(\mathbf{y})$ is the ray parallel to $\boldsymbol{\lambda}$. The point $\exp(\mathbf{x}_* + i\mathbf{y})$ is critical in directions precisely $\pm\boldsymbol{\lambda}$, because smooth points are critical precisely for directions parallel to the logarithmic gradient; this verifies that $\hat{\mathbf{r}} \in N(\mathbf{y})$ implies $\hat{\mathbf{r}}$ is a critical direction. If $\hat{\mathbf{r}} \neq \pm\boldsymbol{\lambda}$ then the dot product with $\hat{\mathbf{r}}$ is unbounded both above and below on the halfspace $Z(\mathbf{y})$ with outward normal $\boldsymbol{\lambda}$, so we can choose $\mathbf{v} \in \mathbb{Z}(\mathbf{y})$ with $\hat{\mathbf{r}} \cdot \mathbf{v} = 1$. If $\hat{\mathbf{r}} = -\lambda$ then $\hat{\mathbf{r}} \in \mathtt{tan}_{\mathbf{x}}(B) = Z(\mathbf{y})$ and we can choose $\mathbf{v} = c\hat{\mathbf{r}}$, where $c = \|\hat{\mathbf{r}}\|_2^{-1} > 0$. Conclusion (iv) follows from the fact that $S = \mathbf{T}(\mathbf{x}_*) \cap \mathcal{V}$ is closed and $\nabla_{\log} Q$ is continuous and nonvanishing on S. □

Theorem 11.9 (Theorem 11.4 for smooth points, self-contained)**.** *Assume the hypotheses of Theorem 11.4 as well as the smooth point assumption (*). Then for* $\mathbf{x} \in B$, *the torus* $\mathbf{T}(\mathbf{x})$ *is homotopic in* $\mathcal{M}$ *to a cycle* C_* *with maximum height less than* $h_* = -\hat{\mathbf{r}}_* \cdot \mathbf{x}_*$.

Proof Fix $\mathbf{x} \in B$ and mimic the proof of Theorem 11.4 up through (11.2). That is, for each $\mathbf{y} \in \mathbf{T}_{\text{flat}}$ choose vectors $\mathbf{v}(\mathbf{y}) \in Z(\mathbf{y})$ and neighborhoods $\mathcal{N}(\mathbf{y})$ with $\mathbf{v}(\mathbf{y}) \in Z(\mathbf{y}')$ when $\mathbf{y}' \in \mathcal{N}(\mathbf{y})$; then define $\vec{V}(\mathbf{y})$ via (11.2). We claim, picking $\mathcal{N}'(\mathbf{y})$ smaller if necessary, that over each $\mathcal{N}'(\mathbf{y})$ there is a sufficiently small $\delta > 0$ such that for all $0 \leq t \leq 1$,

$$q\left(\mathbf{x}_* + i\mathbf{y}' + \delta[t\vec{V}(\mathbf{y}) + (1-t)(\mathbf{x} - \mathbf{x}_*)]\right) \neq 0 . \tag{11.5}$$

If $q(\mathbf{x}_* + i\mathbf{y}) \neq 0$ then we can pick $\mathcal{N}'$ and δ to keep q from vanishing by continuity, so we assume that $q(\mathbf{x}_* + i\mathbf{y}) = 0$. Nonvanishing of (11.5) for sufficiently small δ follows from the Taylor expansion of the left-hand side in δ. Specifically, because the gradient of q is a multiple of $\boldsymbol{\lambda}$, we have

$$\begin{aligned} &q\left(\mathbf{x}_* + i\mathbf{y} + \delta[t\vec{V}(\mathbf{y}) + (1-t)(\mathbf{x} - \mathbf{x}_*)]\right) \\ &= 0 + \delta \underbrace{\boldsymbol{\lambda}(\mathbf{y}) \cdot (t\vec{V}(\mathbf{y}) + (1-t)(\mathbf{x} - \mathbf{x}_*))}_{g(t)} + O(\delta^2). \end{aligned}$$

Lemma 11.8 implies that $g(t)$ has modulus bounded away from zero over $0 \le t \le 1$ for fixed $\boldsymbol{x}$ and choice of $\vec{V}(\boldsymbol{y})$, because both $\vec{V}(\boldsymbol{y})$ and $\boldsymbol{x} - \boldsymbol{x}_*$ are in the interior of the negative halfspace for $\boldsymbol{\lambda}$.

We have now shown that δ can be chosen so that (11.5) holds with $\boldsymbol{y}$ in place of $\boldsymbol{y}'$. Choosing $\mathcal{N}'$ small enough completes the claim for the actual (11.5). This implies that the homotopy defined by (11.3) remains in $\mathcal{M}$, completing the proof of Theorem 11.9. □

The same argument gives the following version of Theorem 11.5 for smooth points.

Corollary 11.10. *Assume the hypotheses of Theorem 11.5 as well as the smooth point assumption (*). Then for $\boldsymbol{x} \in B$ and any $\delta > 0$ the torus $\mathbf{T}(\boldsymbol{x})$ is homotopic in $\mathcal{M}$ to a cycle C_* with maximum height less than $h_* = -\hat{\boldsymbol{r}}_* \cdot \boldsymbol{x}_*$ except in a δ-neighborhood of* $\mathbf{W}$. □

Exercise 11.3. Let $Q(x, y) = 5 - x - x^{-1} - y - y^{-1}$, whose amoeba is displayed below with the boundary points $\boldsymbol{p} = (-\log 2, -\log 2)$ and $\boldsymbol{q} = (\log 2, \log 2)$ specified. For $\boldsymbol{x} = \boldsymbol{p}$ and $\boldsymbol{x} = \boldsymbol{q}$ find $\mathbf{T}(\boldsymbol{x}) \cap \mathcal{V}_Q$, find $Z(\boldsymbol{y})$ for each $\boldsymbol{x} + i\boldsymbol{y}$ with $\exp(\boldsymbol{x} + i\boldsymbol{y}) \in \mathcal{V}$, and determine whether or not `local(`$\hat{\boldsymbol{r}}$`)` is empty.

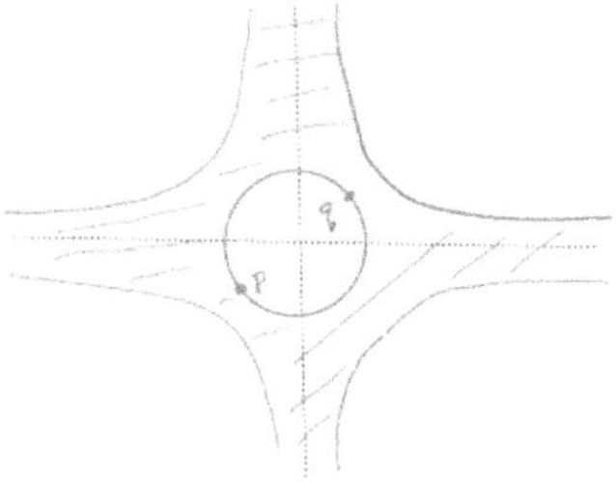

Example 11.11. Consider the bivariate rational function

$$F(x, y) = \frac{1}{(1 + 2x)(1 - x - y)}$$

in the main diagonal direction $\hat{\boldsymbol{r}} = (1/2, 1/2)$. The singular variety of F has one non-smooth point $(x, y) = (-1/2, 3/2)$, which is not minimal as it has strictly greater coordinate-wise modulus than the minimal critical point $\boldsymbol{\sigma} = (1/2, 1/2)$. The point $\boldsymbol{\sigma}$ is not finitely minimal because

$$\mathcal{V} \cap T(\boldsymbol{\sigma}) = \{(1/2, 1/2)\} \quad \cup \quad \{(-1/2, e^{i\theta_2}/2) : \theta_2 \in (-\pi, \pi)\},$$

but it does satisfy finite criticality. Let us construct the homotopy $\Phi_t(\boldsymbol{y})$ for this example. For $(\theta_1, \theta_2) \in \mathbf{T}_{\text{flat}}$ we have $Q(e^{-\ln(2)+i\theta_1}, e^{-\ln(2)+i\theta_2}) = 0$ only when $\theta_1 = \theta_2 = 0$ or $\theta_1 = \pi$. Since

$$(\nabla_{\log} Q)(1/2, 1/2) = -(1,1) \quad \text{and} \quad (\nabla_{\log} Q)(-1/2, y) = (y - 3/2) \times (1,0),$$

we have

$$Z(\theta_1, \theta_2) = \begin{cases} \{\boldsymbol{z} : \boldsymbol{z} \cdot (1,1) < 0\} & \text{if } \theta_1 = \theta_2 = 0 \\ \{\boldsymbol{z} : \boldsymbol{z} \cdot (1,0) < 0\} & \text{if } \theta_1 = \pi \\ \mathbb{R}^2 & \text{otherwise.} \end{cases}$$

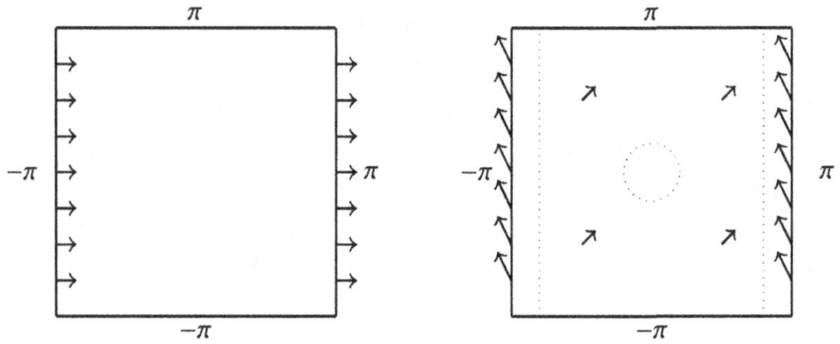

Figure 11.1 *Left:* The elements of $\mathbf{T}_{\text{flat}}$, drawn in $[-\pi, \pi]^2$, for which $Q(e^{-\ln(2)+i\theta_1}, e^{-\ln(2)+i\theta_2}) = 0$, together with the real vectors equal to $(\nabla_{\log} Q)(x, y)$ at these points, up to a non-zero scalar multiple. *Right:* The open sets $\mathcal{N}_1$ and $\mathcal{N}_2$ consist of the small circle of radius ε and the rectangular strips on either side (identified as one large strip since $-\pi = \pi$ in $\mathbf{T}_{\text{flat}}$). The open set $\mathcal{N}_3$ consists of the remaining points and some small overlap with $\mathcal{N}_1$ and $\mathcal{N}_2$ to make $\mathcal{N}_1 \cup \mathcal{N}_2 \cup \mathcal{N}_3$ an open cover. The vectors $\boldsymbol{v}_j$ are drawn in each region.

Fix $\varepsilon > 0$ sufficiently small, let

$$\mathcal{N}_1 = \{-\pi < \theta_1, \theta_2 \le \pi : \theta_1^2 + \theta_2^2 < \varepsilon\} \subset \mathbb{R}^2,$$

let

$$\mathcal{N}_2 = (\pi - \varepsilon, \pi + \varepsilon) \times (-\pi, \pi] \subset \mathbb{R}^2,$$

and let $\mathcal{N}_3$ be an open set containing $\mathbf{T}_{\text{flat}} \setminus (\mathcal{N}_1 \cup \mathcal{N}_2)$ having arbitrarily small intersection with $\mathcal{N}_1 \cup \mathcal{N}_2$. We identify $\mathcal{N}_1, \mathcal{N}_2$, and $\mathcal{N}_3$ with their cosets in $\mathbf{T}_{\text{flat}} = (\mathbb{R}/2\pi\mathbb{Z})^2$; Figure 11.1 gives a picture of the situation. If

$$\boldsymbol{v}_1 = (-1,-1), \quad \boldsymbol{v}_2 = (-1,2), \quad \text{and} \quad \boldsymbol{v}_3 = (1/2, 1/2)$$

then $\boldsymbol{v}_j \in K(\boldsymbol{y})$ for $\boldsymbol{y} \in \mathcal{N}_j$, and $\boldsymbol{v}_2 \cdot \boldsymbol{r} = \boldsymbol{v}_3 \cdot \boldsymbol{r} = 1$ as desired (in this case one can simply take $v_3 = v_2$, but we mirror the general construction described above). Hence,

$$\eta(\theta_1,\theta_2) = \psi_1(\theta_1,\theta_2)v_1 + \psi_2(\theta_1,\theta_2)v_2 + \psi_3(\theta_1,\theta_2)v_3$$
$$= \big(-\psi_1 - \psi_2 + \psi_3/2\,,\ -\psi_1 + 2\psi_2 + \psi_3/2\big),$$

where ψ_1, ψ_2, and ψ_3 form a partition of unity subordinate to $\mathcal{N}_1, \mathcal{N}_2$, and $\mathcal{N}_3$, and

$$\Phi_t(\theta_1,\theta_2) = \exp\left[i(\theta_1,\theta_2) - (\log 2, \log 2) - \delta(1-t, 1-t) + \delta t\eta(\theta_1,\theta_2)\right]$$

for some $\delta > 0$. When $t = 1$ and $(\theta_1,\theta_2) \in \mathcal{N}_1$ then $\psi_2(\theta_1,\theta_2) = 0$ since $\mathcal{N}_1$ and $\mathcal{N}_2$ are disjoint. Thus, the cycle C in Corollary 11.10 is the image of $\mathcal{N}_1$ under the map

$$\Phi_1(\theta_1,\theta_2) = \left(\frac{1}{2}e^{i\theta_1}\, e^{-\delta\psi_1+\delta\psi_3/2}, \frac{1}{2}e^{i\theta_2}\, e^{-\delta\psi_1+\delta\psi_3/2}\right).$$

When (θ_1,θ_2) is sufficiently close to the origin then it lies in $\mathcal{N}_1 \setminus \mathcal{N}_3$ and

$$\Phi_1(\theta_1,\theta_2) = \left(\frac{1}{2}e^{i\theta_1}\, e^{-\delta}, \frac{1}{2}e^{i\theta_2}\, e^{-\delta}\right)$$

is contained in $\mathcal{D}$. As (θ_1,θ_2) leaves the origin and approaches the boundary, however, the coordinate-wise moduli of $\Phi_1(\theta_1,\theta_2)$ increase, eventually leaving $\mathcal{D}$ without intersecting $\mathcal{V}$. Ultimately, we can apply the asymptotic arguments for smooth points to analyze the integral near $\boldsymbol{\sigma}$ and obtain an expansion

$$f_{n,n} = \frac{4^n}{\sqrt{\pi n}}\left(\frac{1}{2} - \frac{1}{8n} + \frac{1}{256n^2} + \frac{5}{256n^3} - \frac{819}{65536n^4} + O\left(\frac{1}{n^5}\right)\right).$$

◁

11.2 Proof of Theorem 11.1

In this section we prove Theorem 11.1. Our proof is split into the following steps.

- In Section 11.2.1 we introduce the hyperbolicity of a homogeneous polynomial and cones of hyperbolicity for a homogeneous polynomial. We show how a family of cones of hyperbolicity for a homogeneous polynomial can be defined so as to vary semi-continuously. The definition of hyperbolicity at this stage is valid only for homogeneous polynomials.

- In Section 11.2.2 we generalize to certain *log-Laurent polynomials* by defining the cone of a function f at one of its zeros $\boldsymbol{x} + i\boldsymbol{y}$ to be a choice of a cone of hyperbolicity for the leading homogeneous part of f at $\boldsymbol{x} + i\boldsymbol{y}$. Our constructions are well defined only for certain values of $\boldsymbol{x}$.
- Section 11.2.3 introduces notions of weak and strong hyperbolicity. The weaker notion is equivalent to hyperbolicity of a homogenization, and the main result of the section is that the stronger notion holds in similar circumstances.
- Section 11.2.4 establishes lower semi-continuity for a family of cones introduced in Section 11.2.3.
- Section 11.2.5 completes the proof of Theorem 11.1.
- Section 11.2.6 extends the construction of the deformations C_δ in Theorem 11.5 to a limiting log-linear deformation. This will be important later, when computing integrals.

11.2.1 Cones of hyperbolicity in the homogeneous case

The following definition of hyperbolicity, originally developed in order to classify and study wavelike second-order partial differential equations, goes back to [Går50].

Definition 11.12 (hyperbolicity). A homogeneous complex polynomial A of degree $m \geq 1$ is a ***hyperbolic polynomial*** in the direction $\boldsymbol{v} \in \mathbb{R}^d$ if $A(\boldsymbol{v}) \neq 0$ and for all $\boldsymbol{x} \in \mathbb{R}^d$ the polynomial $t \mapsto A(\boldsymbol{x} + t\boldsymbol{v})$ has only real roots. A seemingly weaker but actually equivalent condition is that $A(\boldsymbol{v} + i\boldsymbol{y}) \neq 0$ for all $\boldsymbol{y} \in \mathbb{R}^d$.

The set of $\boldsymbol{v}$ for which A is hyperbolic in direction $\boldsymbol{v}$ is an open set whose components are convex cones in $\mathbb{R}^d$, and each of these cones is called a ***cone of hyperbolicity*** for the homogeneous polynomial A. Denote by $\mathbf{K}^{\boldsymbol{v}}(A)$ the cone of hyperbolicity of A containing a given vector $\boldsymbol{v}$. Some multiple of A is positive on $\mathbf{K}^{\boldsymbol{v}}(A)$ and vanishing on $\partial\mathbf{K}^{\boldsymbol{v}}(A)$, and for $\boldsymbol{x} \in \mathbf{K}^{\boldsymbol{v}}(A)$ the roots of $A(\boldsymbol{x}+t\boldsymbol{v})$ will all be negative. These properties are proved, among other places, in [Gül97, Theorem 3.1].

Example 11.13 (hyperbolicity of a linear function). If $A(\boldsymbol{x}) = \boldsymbol{v} \cdot \boldsymbol{x}$ is a real linear function then its cones of hyperbolicity are the halfspaces $\{\boldsymbol{x} \in \mathbb{R}^d : \boldsymbol{v} \cdot \boldsymbol{x} > 0\}$ and $\{\boldsymbol{x} \in \mathbb{R}^d : \boldsymbol{v} \cdot \boldsymbol{x} < 0\}$. ◁

Example 11.14 (hyperbolicity of a quadratic function). If $A(\boldsymbol{x}) = x_1^2 - \sum_{j=2}^d x_j^2$ is the ***standard Lorentzian quadratic*** then its cones of hyperbolicity are the so-called positive and negative *time-like cones*

$$C_+ = \left\{ \boldsymbol{x} \in \mathbb{R}^d \ : x_1 > \sqrt{x_2^2 + \cdots + x_d^2} \right\}$$

and

$$C_- = \left\{ \boldsymbol{x} \in \mathbb{R}^d \ : x_1 < -\sqrt{x_2^2 + \cdots + x_d^2} \right\}.$$

◁

The following proposition and definition define the first version of the cones of hyperbolicity that will ultimately lead to the family of cones in Theorem 11.1.

Proposition 11.15. *Let A be a homogeneous polynomial, fix $\boldsymbol{x}$ with $A(\boldsymbol{x}) = 0$, and let $\tilde{A} = \mathtt{hom}(A, \boldsymbol{x})$ denote the leading homogeneous part of A at $\boldsymbol{x}$. If A is hyperbolic in the direction $\boldsymbol{u}$ then $\tilde{A}$ is also hyperbolic in the direction $\boldsymbol{u}$. Consequently, if C is any cone of hyperbolicity for A then there is some cone of hyperbolicity for $\tilde{A}$ that contains C.*

Proof If P is a polynomial of degree k then we can recover its leading homogeneous part $\mathtt{hom}(P)$ by

$$\mathtt{hom}(P)(\boldsymbol{y}) = \lim_{\lambda \to \infty} \lambda^k P(\lambda^{-1} \boldsymbol{y}).$$

This limit is uniform as $\boldsymbol{y}$ varies over compact sets, since monomials of degree k are invariant under the scaling on the right-hand side, while monomials of degree $k + j$ scale by λ^{-j}, uniformly over compact sets. Thus, in our situation

$$\tilde{A}(\boldsymbol{y} + t\boldsymbol{u}) = \lim_{\lambda \to \infty} \lambda^k A(\boldsymbol{x} + \lambda^{-1}(\boldsymbol{y} + t\boldsymbol{u}))$$

uniformly as t varies over compact sub-intervals of $\mathbb{R}$. If A is hyperbolic in direction $\boldsymbol{u}$ then, for any fixed λ, all the zeros of this polynomial in t are real. Hurwitz's Theorem on the continuity of zeros [Con78b, Corollary 2.6] says that a (uniform on bounded intervals) limit of polynomials having all real zeros will either have all real zeros or vanish identically. The limit $\tilde{A}(\boldsymbol{y} + t\boldsymbol{u})$ has degree $k \geq 1$; it does not vanish identically, so if A is hyperbolic in the direction $\boldsymbol{u}$ then $\tilde{A}$ has only real roots and is also hyperbolic. □

Definition 11.16 (family of cones in the homogeneous case)**.** Let A be a homogeneous polynomial and let C be a cone of hyperbolicity for A. If $A(\boldsymbol{x}) = 0$ then we define $\mathbf{K}^{A,C}(\boldsymbol{x})$ to be the cone of hyperbolicity of $\mathtt{hom}(A, \boldsymbol{x})$ containing C, whose existence we have just proved. If $A(\boldsymbol{x}) \neq 0$ then we define $\mathbf{K}^{A,C}(\boldsymbol{x})$ to be all of $\mathbb{R}^d$.

Remark 11.17. When $\boldsymbol{x} \neq 0$ the cone $\mathbf{K}^{A,C}(\boldsymbol{x})$ depends only on the direction $\widehat{\mathbf{x}} = \boldsymbol{x}/|\boldsymbol{x}|$.

Example 11.18. Let $S(\boldsymbol{x}) = x_1^2 - x_2^2 - \cdots - x_d^2$ be the standard Lorentzian quadratic and let $C = C_+$ be the cone of hyperbolicity at the origin described in Example 11.14. For $\boldsymbol{x} \in \partial B$, if $\boldsymbol{x} \neq 0$ then $\mathbf{K}^{S,C}(\boldsymbol{x})$ is the tangent halfspace $\{\boldsymbol{x}+\boldsymbol{y} : \boldsymbol{y} \cdot (\nabla S)(\boldsymbol{x}) \geq 0\}$. If $\boldsymbol{x} = \mathbf{0}$ then $\mathbf{K}^{S,C}(\boldsymbol{x}) = C$. A ***Lorentzian quadratic*** is a quadratic polynomial obtained from S by a real linear transformation. The boundary of the cone of hyperbolicity for any Lorentzian quadratic is an algebraic tangent cone and the cones of hyperbolicity at these points are halfspaces whose boundaries are support hyperplanes to time-like cones. ◁

Exercise 11.4. Describe the two cones of hyperbolicity of $A(x, y) = x + y$ at $(x, y) = (5, -5)$ and, letting C denote one of these cones, find $K^{A,C}(5, -5)$.

11.2.2 Cones of hyperbolicity in the general case

Although the definition of hyperbolicity in the homogeneous case is valid for complex polynomials and involves complex roots, the role played by the real subspace is essential. In this section we generalize by defining cones of hyperbolicity $\mathbf{K}^{q,B}(\boldsymbol{z})$ for functions $q = Q \circ \exp$ where Q is a Laurent polynomial. Our definitions are valid only at points $\boldsymbol{z} = \boldsymbol{x} + i\boldsymbol{y}$ for some $\boldsymbol{x}$ on the boundary of a component B of the complement of $\texttt{amoeba}(Q)$. We do not know whether $\mathbf{K}^{q,B}(\boldsymbol{z})$ may be defined in such a way that semi-continuity results still hold when $\boldsymbol{z}$ is not on the amoeba boundary, or when q is not a log-Laurent polynomial.

Definition 11.19 (hyperbolicity and normal cones)**.** Let Q be a Laurent polynomial, let B be a component of $\mathbb{R}^d \setminus \texttt{amoeba}(Q)$, and let $\exp(\boldsymbol{z}) = \exp(\boldsymbol{x}+i\boldsymbol{y}) \in \mathcal{V}_Q$ with $\boldsymbol{x} \in \partial B$. If $q = Q \circ \exp$ then we define

$$\mathbf{K}(\boldsymbol{z}) = \mathbf{K}^{q,B}(\boldsymbol{z}) = \mathbf{K}^{\boldsymbol{u}}(\texttt{hom}(q, \boldsymbol{x} + i\boldsymbol{y})) \text{ for any } \boldsymbol{u} \in B \qquad (11.6)$$

to be the (open) ***cone of hyperbolicity*** of $A = \texttt{hom}(q, \boldsymbol{x} + i\boldsymbol{y})$ that contains B. The existence of this cone is guaranteed by Proposition 11.26 below.

Remarks. (i) Defining $\mathbf{K}^{q,B}$ using $\texttt{hom}(q, \boldsymbol{x}+i\boldsymbol{y})$ may seem obvious, but the difficulty proving properties in this case illustrates that the extension is quite non-trivial.

(ii) We may extend the definition of $\mathbf{K}$ to all of $\mathbf{T}(\boldsymbol{x})$ by taking $\mathbf{K}(\boldsymbol{z}) = \mathbb{R}^d$ when $q(\boldsymbol{z}) \neq 0$.

(iii) When $\boldsymbol{z} = \boldsymbol{x} + i\boldsymbol{y}$ and $\boldsymbol{x}$ is understood, we sometimes write $\mathbf{K}^{q,B}(\boldsymbol{y})$.

Before going further we give a few examples.

Example 11.20 (cones of hyperbolicity at smooth points)**.** Suppose $\exp(\boldsymbol{z}) =$

$\exp(\boldsymbol{x} + i\boldsymbol{y})$ is a minimal smooth point of $\mathcal{V}_Q$. Then the leading homogeneous part $\tilde{q}$ of q at $\boldsymbol{z}$ is a linear function $\tilde{q}(\boldsymbol{x}) = \boldsymbol{v} \cdot \boldsymbol{x}$, and the assumption of minimality implies that the vector $\boldsymbol{v}$ is a complex scalar multiple of a real vector. As in Example 11.13 above, the cones of hyperbolicity of $\tilde{q}$ at $\boldsymbol{z}$ are halfspaces normal to $\boldsymbol{v}$. ◁

Example 11.21 (cones of hyperbolicity at multiple points). Suppose $\exp(\boldsymbol{z}) = \exp(\boldsymbol{x} + i\boldsymbol{y})$ is a multiple point of $\mathcal{V}_Q$. Then $\tilde{q}$ is a product of linear functions defining a central hyperplane arrangement. The normal vectors to the factors of $\tilde{q}$ are scalar multiples of real vectors and the corresponding real hyperplanes divide $\mathbb{R}^d$ into projective cones, each of which is a cone of hyperbolicity for $\tilde{q}$. ◁

Exercise 11.5. Let $Q = (1 - 2x - y)(1 - x - 2y)$. What are the cones of hyperbolicity $\mathbf{K}(\boldsymbol{z})$ when $\boldsymbol{z} = \exp(1/3, 1/3)$? What about when $\boldsymbol{z} = \exp(1/4, 1/2)$?

Example 11.22 (cones of hyperbolicity at quadratic cone points). If $Q(\boldsymbol{x}) = z_1^2 - \sum_{j=2}^{d} z_j^2$ is the standard Lorentzian quadratic then at the origin q has the two cones of hyperbolicity C_+ and C_- discussed above, while the cones of hyperbolicity of q at any other point are two halfspaces. ◁

11.2.3 Strong and weak hyperbolicity

To prove the existence result Proposition 11.26 and the semi-continuity result Lemma 11.27, we define intermediate notions of strong and weak hyperbolicity. These definitions are somewhat less natural than those above, and are not used for any other purpose, so readers not interested in the proofs of our semi-continuity results may safely skip to Section 11.2.5.

Definition 11.23 (strong and weak hyperbolicity). Let $q : \mathbb{C}^d \to \mathbb{C}$ vanish at $\boldsymbol{z}$ and be holomorphic in a neighborhood of $\boldsymbol{z}$. We say that q is ***strongly hyperbolic*** at $\boldsymbol{z}$ in the direction of the unit vector $\hat{\boldsymbol{v}}$ if there is an $\varepsilon > 0$ such that $q(\boldsymbol{z} + t\boldsymbol{v}' + i\boldsymbol{u}) \neq 0$ whenever $0 < t < \varepsilon$, the vector $\boldsymbol{v}'$ is at distance at most ε from $\hat{\boldsymbol{v}}$, and $\boldsymbol{u} \in \mathbb{R}^d$ has magnitude at most ε. The supremum of the ε for which this holds is called the ***radius of strong hyperbolicity*** at $\boldsymbol{z}$ in the direction $\hat{\boldsymbol{v}}$. We say that q is ***weakly hyperbolic*** in the direction $\hat{\boldsymbol{v}}$ if for every $M > 0$ there is an $\varepsilon > 0$ such that $q(\boldsymbol{z} + t\hat{\boldsymbol{v}} + i\boldsymbol{u}) \neq 0$ whenever $0 < t|\hat{\boldsymbol{v}}| < \varepsilon$ and $\boldsymbol{u} \in \mathbb{R}^d$ has magnitude at most ε with $\frac{|\boldsymbol{u}|}{t|\hat{\boldsymbol{v}}|} \leq M$.

Proposition 11.24. *The radius of strong hyperbolicity is Lipschitz continuous with constant 1. Thus, for fixed $\boldsymbol{x}$ strong hyperbolicity at $\boldsymbol{z} = \boldsymbol{x} + i\boldsymbol{y}$ in the direction $\hat{\boldsymbol{v}}$ is a neighborhood property in $\boldsymbol{y}$ and $\hat{\boldsymbol{v}}$.*

Proof Suppose that q is strongly hyperbolic at $\boldsymbol{x} + i\boldsymbol{y}$ in direction $\hat{\boldsymbol{v}}$ with radius ε and choose $\boldsymbol{y}'$ and $\hat{\boldsymbol{v}}'$ with $\max\{|\boldsymbol{y}' - \boldsymbol{y}|, |\hat{\boldsymbol{v}}' - \hat{\boldsymbol{v}}|\} = \delta < \varepsilon$. It follows from the definition that q is strongly hyperbolic at $\boldsymbol{y}'$ in direction $\hat{\boldsymbol{v}}'$ with radius at least $\varepsilon - \delta$. Hence the radius function $R(\boldsymbol{y}, \hat{\boldsymbol{v}})$ for q over $\boldsymbol{x}$ is Lipschitz in both arguments with constant at most 1. In particular, q is continuous in both arguments and the set of $(\boldsymbol{y}, \boldsymbol{v})$ for which q is strongly hyperbolic is open. □

As mentioned in the end of chapter notes, the next two propositions correct the statement of [BP11, Proposition 2.11] and the proof of [BP11, Proposition 2.12].

Proposition 11.25. *Let $q = Q \circ \exp$ with Q a Laurent polynomial, fix $\boldsymbol{z} = \boldsymbol{x} + i\boldsymbol{y}$, and let $A = \mathtt{hom}(q, \boldsymbol{z})$ be the leading homogeneous term of $q(\boldsymbol{z} + \boldsymbol{w})$ in $\boldsymbol{w}$. Then each of the following properties implies the next.*

(i) Strong hyperbolicity of q at $\boldsymbol{z}$ in the direction $\hat{\boldsymbol{v}}$.
(ii) Hyperbolicity of A in the direction $\hat{\boldsymbol{v}}$.
(iii) Weak hyperbolicity of q at $\boldsymbol{z}$ in the direction $\hat{\boldsymbol{v}}$.

Proof Assume without loss of generality that $\boldsymbol{z} = 0$.

For the contrapositive of the first implication, suppose that A is not hyperbolic in direction $\hat{\boldsymbol{v}}$. Then there is $\boldsymbol{u} \in \mathbb{R}^d$ such that $A(\hat{\boldsymbol{v}} + i\boldsymbol{u}) = 0$. Because the tangent cone is the limiting secant cone, by Lemma 6.48 there are two $\boldsymbol{v}_n, \boldsymbol{w}_n$ sequences of vectors going to zero in $\mathbb{R}^d$ such that $q(\boldsymbol{v}_n + i\boldsymbol{w}_n) = 0$, with $\boldsymbol{v}_n/|\boldsymbol{v}_n| \to \hat{\boldsymbol{v}}$ and $\boldsymbol{w}_n/|\boldsymbol{v}_n| \to \boldsymbol{u}$. This contradicts the definition of strong hyperbolicity.

For the second implication, suppose that q is not weakly hyperbolic at the origin in direction $\hat{\boldsymbol{v}}$. This means there is some $M > 0$ such that for any $\varepsilon > 0$ there are $t \in (0, \varepsilon)$ and $\boldsymbol{u} \in \mathbb{R}^d$ with $|\boldsymbol{u}| \leq M$ such that $q(t(\hat{\boldsymbol{v}} + i\boldsymbol{u})) = 0$. By compactness of the ball of radius M, this is equivalent to the existence of sequences of real numbers t_n and vectors $\boldsymbol{u}_n$ such that $t_n \to 0$, $\boldsymbol{u}_n \to \boldsymbol{u}$ and $q(t_n(\hat{\boldsymbol{v}} + i\boldsymbol{u}_n)) = 0$. Again, because the tangent cone is the limiting secant cone, this implies $A(\hat{\boldsymbol{v}} + i\boldsymbol{u}) = 0$, meaning A is not hyperbolic in direction $\hat{\boldsymbol{v}}$. □

Proposition 11.26. *Let Q be a Laurent polynomial, let B be a component of $\mathtt{amoeba}(Q)$, and let $\boldsymbol{x} \in \partial B$, so that $q = Q \circ \exp$ vanishes at some point $\boldsymbol{z} = \boldsymbol{x} + i\boldsymbol{y}$. Fix $\boldsymbol{u} \in \mathtt{tan}_{\boldsymbol{x}}(B)$ and let $\tilde{q} = \mathtt{hom}(q, \boldsymbol{x} + i\boldsymbol{y})$ denote the leading homogeneous part of $q(\boldsymbol{x} + i\boldsymbol{y} + \boldsymbol{w})$ with respect to $\boldsymbol{w}$. Then*

(i) q is strongly hyperbolic at $\boldsymbol{z}$ in direction $\boldsymbol{u}$;
(ii) $\tilde{q}$ is hyperbolic in direction $\boldsymbol{u}$;
(iii) $\mathbf{K}^{\boldsymbol{u}}(\tilde{q})$ contains $\mathtt{tan}_{\boldsymbol{x}}(B)$;
(iv) some complex scalar multiple of $\tilde{q}$ is real.

Proof Begin by observing a property of tangent cones in real space: if a unit vector $\hat{v}$ is in $\mathtt{tan}_{x}(B)$ and $v_n \to 0$ with $v_n/|v_n| \to \hat{v}$ then $x + v_n \in B$ for sufficiently large n. Strong hyperbolicity of q in any direction $u \in \mathtt{tan}_{x}(B)$ then follows from the definitions of strong hyperbolicity and B.

To derive the second conclusion, suppose for contradiction that $\tilde{q}$ is not hyperbolic in unit direction $u \in \mathbb{R}^d$, meaning that $\tilde{q}(u + i\theta) = 0$ for some $\theta \in \mathbb{R}^d$. Because the algebraic tangent cone is the limiting secant cone by Lemma 6.48, there must be sequences of real vectors $\{\alpha_n\}$ and $\{\beta_n\}$ such that $(\alpha_n + i\beta_n)/|\alpha_n + i\beta_n| \to (u + i\theta)/|u + i\theta|$ and $\tilde{q}(z + \alpha_n + i\beta_n) = 0$. Because $\alpha_n \to \mathbf{0}$ with $\alpha_n/|\alpha_n| \to u$, the above observation with $v_n = \alpha_n$ shows that $x + \alpha_n \in B$ for all sufficiently large n. But B is in the complement of the amoeba, contradicting $\tilde{q}(z + \alpha_n + i\beta_n) = 0$ and proving hyperbolicity of $\tilde{q}$ in direction u.

The third conclusion, $\mathbf{K}^{u}(\tilde{q}) \supseteq \mathtt{tan}_{x}(B)$, follows immediately. For the final conclusion, fix any non-zero real vector ξ. Then $\tilde{q}(\xi + tu)$ is a polynomial of degree m with leading coefficient $\tilde{q}(u)$ and constant term $\tilde{q}(\xi)$. The ratio $\tilde{q}(\xi)/\tilde{q}(u)$ is $(-1)^m$ times the product of all the roots. By definition of cones of hyperbolicity, all the roots are real, therefore $\tilde{q}(\xi)/\tilde{q}(u)$ is real. Because ξ was arbitrary, we conclude that $\tilde{q}(x)/\tilde{q}(u)$ maps $\mathbb{R}^d$ to $\mathbb{R}$. □

11.2.4 Semi-continuity

Near any point in any stratum of a complex algebraic variety there are one or more cones contained in the complement of the variety. Hyperbolicity may be thought of as a kind of orientability for families of such cones, ensuring a consistent choice of "inward tangent cone" (see Exercise 11.10 for a related perspective). This provides some intuition as to why it's easier to define hyperbolicity for minimal points, since the meaning of "inward" can be inferred. Hyperbolicity is also used to ensure the cones are convex.

Aiming at conclusion (iv) of Theorem 11.1, we quote and briefly outline proofs of the following semi-continuity results from [BP11, Theorem 2.14 and Corollary 2.15], which are also discussed in [ABG70, Lemma 3.22] and [Går50; Hör83].

Lemma 11.27 (semi-continuity). *(i) Let A be a homogeneous polynomial and C a cone of hyperbolicity for A. Then the cone $\mathbf{K}^{A,C}(y)$ is lower semi-continuous in y.*

(ii) Let $q = Q \circ \exp$ for some Laurent polynomial Q, let B be a component of the complement of $\mathtt{amoeba}(Q)$, and let $x \in \partial B$. Then $\mathbf{K}^{q,B}(z)$ is lower semi-continuous as z varies over $\mathbf{T}(x)$.

(iii) *With q as in (ii) and $\tilde{q} = \mathtt{hom}(q, z)$ for fixed z, suppose further that for u in some cone of hyperbolicity $\mathbf{K}$ of $\tilde{q}$ corresponding to the component B of $\mathtt{amoeba}(Q)^c$,*

$$q(z + iy + su) \neq 0 \text{ for real non-zero } s \text{ in some interval } [-c, c] \text{ and all } y\,. \tag{11.7}$$

Then

$$K^{A,B}(z + \widehat{\mathbf{y}}) \subseteq \liminf K^{q,B}(z + y_n)$$

as $y_n \to 0$ with $y_n/|y_n| \to \widehat{\mathbf{y}}$.

Remark 11.28. The first conclusion in Lemma 11.27 is nearly a specialization of the second conclusion to homogeneous functions, except that not every homogeneous function is the homogenization of a Laurent polynomial composed with the exponential function. Assumption (11.7) is satisfied whenever a generating function is symmetric under mapping all coordinates to their reciprocals. This seems to be a common feature of recursions arising from cluster algebras. In particular, this is true for the quadratic cone functions (Aztec diamond, cube grove, fortress) arising later in this chapter, as well as their products with binomial such as $(1 - yz)$. Though not symmetric under reciprocals, the spacetime generating function for quantum walks also satisfies Assumption (11.7) as a consequence of the defining matrix being unitary.

Proof Sketch Suppose an analytic function q is strongly hyperbolic in the direction v at the point $z = x + iy$ and let $A = \mathtt{hom}(\tilde{q}, z)$. Theorem 2.14 of [BP11] implies that if $u \in \mathbf{K}^v(q)$ then q is strongly hyperbolic in direction $tv + (1 - t)u$ for any $t \in [0, 1]$. The first two conclusions follow from this, and are stated as [BP11, Corollary 2.15].

To prove (*iii*), let m be the lowest homogeneous degree of $\tilde{q}$ and write $q = \tilde{q} + R$ with R analytic and vanishing to degree $m+1$ at the origin. By our choice of u we know $\tilde{q}(u) \neq 0$, and by homogeneity we know that $\tilde{q}(su) = s^m \tilde{q}(u)$. Using the Weierstrass Preparation Theorem, then expanding, gives

$$q(z + su) = h(z)\left(s^m + a_1(z)s^{m-1} + \cdots + a_m(z)\right) \tag{11.8}$$

$$= h(z)\prod_{k=1}^{m}(s + \mu_k(z, u))\,, \tag{11.9}$$

where $h, a_1, \ldots, a_m$ are holomorphic functions of z with all a_j vanishing at the origin, h does not vanish at the origin, and the roots $-\mu_k(z, u)$ tend to zero as $z \to 0$. Specializing to some line $z = tv$, we write (with a slight abuse of notation)

$$f(t\boldsymbol{v} + s\boldsymbol{u}) = h(t) \prod_{k=1}^{m} (s + \mu_k(t)) .$$

We claim that the $\mu_k(t)$ are real analytic functions of t in a neighborhood of 0. To see this, write μ_k as a Puiseux series $\sum_{r \in A_k} c_{r,k} t^r$, where A_k is an arithmetic progression of rational numbers, containing positive entries because $\lim_{t\to 0} \mu_k(t) = 0$. Suppose some exponent $r \in A_k$ is not an integer, and let b be the denominator of the least non-integral $r \in A_k$. Then either for t small and positive, or t small and negative, there will be b roots with arguments asymptotically distributed as $\{\pi/b + 2\pi j/b : 0 \le j \le b-1\}$, one of which is non-real, contradicting the real-rootedness of $q(i(\boldsymbol{y} + z\boldsymbol{u}))$. We have thus established our claim that the μ_k are real functions of t.

It follows that $\mu_k(t) \sim t\lambda_k(t)$, where λ_k are the slopes of the lines into which the bivariate function $\tilde{q}(t\boldsymbol{v} + s\boldsymbol{u})$ factors,

$$\tilde{q}(t\boldsymbol{v} + s\boldsymbol{u}) = \tilde{q}(\boldsymbol{u}) \prod_{k=1}^{m} (s + t\lambda_k(t)) .$$

When $\boldsymbol{v}$ is chosen in the same cone of hyperbolicity as $\boldsymbol{u}$, this further implies that all λ_k are positive and that μ_k are increasing functions of t in a neighborhood of 0. This is sufficient to derive, in the same manner as [ABG70, Lemma 5.1, 5.9], that the local branches μ_k are still increasing after perturbation, which implies semi-continuity of the cones. □

Example 11.29. Suppose that $\exp(\boldsymbol{z})$ is a multiple point of Q, so that the leading homogeneous part $\tilde{q}$ of q at $\exp(\boldsymbol{z})$ is the product of linear factors. The cones of hyperbolicity of q at $\boldsymbol{z}$ are projective cones that are the components of $\mathbb{R}^d$ when the hyperplanes on which the linear factors vanish are removed. If $q = Q \circ \exp$, where Q has only multiple point singularities and $\{\boldsymbol{z}_n\}$ is a sequence with $\exp(\boldsymbol{z}_n) \to \exp(\boldsymbol{z})$ in $\mathcal{V}_Q$ while remaining in a single stratum S, then $\exp(\boldsymbol{z})$ is in either S or ∂S. In the latter case, the partition near $\exp(\boldsymbol{z})$ of $\mathbb{R}^d$ into projective cones is finer at $\exp(\boldsymbol{z})$ than at the points $\exp(\boldsymbol{z}')$, and the semi-continuity described in Lemma 11.27 is strict. ◁

Exercise 11.6. Let X be a Whitney stratified space and let $m : X \to \mathbb{Z}_{>0}$ be the dimension function $m(\boldsymbol{x}) = \dim(\mathcal{S}(\boldsymbol{x}))$. Is m lower semi-continuous, upper semi-continuous, or neither?

11.2.5 Proof of Theorem 11.1

We now complete the proof of Theorem 11.1.

Proof For every $y \in \mathbf{T}_{\text{flat}}$, define

$$Z(y) = \mathbf{K}^{q,B}(x_* + iy) = \mathbf{K}^{u}(\texttt{hom}(q, x_* + iy)),$$

where u is any vector in $\texttt{tan}_{x_*}(B)$. By Proposition 11.26 this cone contains $\texttt{tan}_{x_*}(B)$, establishing conclusion (i) of Theorem 11.1. Conclusion (iv) is part of the conclusion of Lemma 11.27. For conclusion (iii), first suppose that $z = \exp(x_* + iy)$ is not a stratified critical point for $h_{\hat{r}}$ on $\mathcal{V}$. Pulling back by the exponential map, we see that $x_* + iy$ is not a critical point for the log-linear map $\hat{r}(w) = \hat{r} \cdot w$ on the stratum $\mathcal{S}$ of $\exp^{-1}[\mathcal{V}]$ containing $x_* + iy$. Thus $d(h_{\hat{r}} \circ \exp)|_S \neq 0$, meaning that $\hat{r}$ is not orthogonal to the tangent space $T_{x+iy}(S)$, so $\hat{r} \notin N(y)$. By the contrapositive, this establishes the first part of conclusion (iii). Conversely, if $\hat{r} \notin N(y)$ then since $N(y) = (-Z(y))^*$ there is some $v \in Z(y)$ with $\hat{r} \cdot v > 0$. The cone $Z(y)$ is projective, so v can be normalized so that $\hat{r} \cdot v = 1$.

To prove conclusion (ii), begin by fixing t, x, and y. By Proposition 11.26, q is strongly hyperbolic at $x + iy$ in any direction $u \in \tan_{x_*}(B)$ with some radius $\varepsilon(u, y) > 0$ that is Lipschitz-1 continuous in both arguments. This is (11.1) with t and y fixed, $u = u(t, y)$ taken to be $t\vec{V}(y) + (1 - t)(x - x_*)$, and ε set to $\varepsilon(u(t, y), y)$. Let (t, y) vary over the compact set $[0, 1] \times \mathbf{T}_{\text{flat}}$ and use continuity to see that $\varepsilon(u(t, y), y) \geq \varepsilon_0 > 0$ for some ε_0 and all (t, y). Equation (11.1) is then satisfied for any $\delta \leq \varepsilon_0$. □

11.2.6 Projective deformations

This section address a special but reasonably common case where the dominant critical point $z = \exp(x + iy)$ governs a set of coefficient asymptotics whose directions $\texttt{N}(z)$ form a set with nonempty interior. Because $\texttt{N}(z)$ is a subset of the lognormal space to the stratum containing z, this implies the stratum is the 0-dimensional singleton $\{z\}$. The prototypical examples are quadratic cone points, with or without additional linear divisors passing through the cone point (Aztec diamond, cube groves), as well as the other applications mentioned in the preliminary section of this chapter: Szegő functions, the GKZ symmetric rational functions, and the Kauers-Zeilberger 4-variable function. Looking ahead at Figures 11.2 and 11.3 may help to give a mental picture of the variety and the deformations used. A quick outline of this section and the next goes as follows.

(i) For a hyperbolic homogeneous polynomial $\tilde{q}$, construct a deformation from the imaginary fiber $u + i\mathbb{R}^d$ over a point u in a cone of hyperbolicity $\mathbf{K}$ to a *projective contour*: a set closed under multiplication by positive real numbers. We will apply this in logarithmic space to the function $\tilde{q} = \mathtt{hom}(q, \boldsymbol{x} + i\boldsymbol{y})$ where $q = Q \circ \exp$ and $\exp(\boldsymbol{x} + i\boldsymbol{y})$ is a minimal point of $\mathcal{V}$. The deformation will avoid the zero set $\mathcal{V}_{\tilde{q}}$ of $\tilde{q}$ except for touching it at a single point (the origin) at a single time ($t = 1$). This step is accomplished by Lemma 11.32.

(ii) Letting $q = Q \circ \exp$, we apply Step (i) to the homogenization $\tilde{q}$ of q near each point $\boldsymbol{x}+i\boldsymbol{y}$ with $\boldsymbol{y} \in \mathtt{localarg}$ (so that $\exp(\boldsymbol{x}+i\boldsymbol{y})$ is a critical point in direction $\hat{\boldsymbol{r}}_*$ for Q and $\hat{\boldsymbol{r}}_*$ lies in the normal cone there). We then use a partition of unity to piece these local log-projective contours together with the one from Theorem 11.4. This is accomplished in Theorem 11.33, resulting in a deformation of the contour that avoids $\mathcal{V}_q$ except at $t = 1$ with $\boldsymbol{y} \in \mathtt{localarg}$ and is projective in a neighborhood of $\mathtt{localarg}$.

(iii) We stop the homotopy a bit early near $\mathtt{localarg}$ in order to obtain a cycle that avoids $\mathcal{V}_q = \log \mathcal{V}$ entirely. This is discussed in Definition 11.34.

(iv) In the next section, we will see that on the domains of integration of interest any rational function $F = P/Q$ has local expansions in powers of $1/A$, where A is a homogenization of Q. This leads to a powerful theorem showing the asymptotics in this case to be given essentially by the Fourier transform of $\tilde{q}$.

As above, assume we have fixed Q along with a component B of $\mathtt{amoeba}(Q^c)$ and a point $\boldsymbol{x} \in \partial B$, let $q = Q \circ \exp$, and fix a point $\boldsymbol{x}+i\boldsymbol{y}_0$ where q vanishes. We assume without loss of generality that $\boldsymbol{y}_0 = \boldsymbol{0}$, with everything in this section working equally well for any non-zero $\boldsymbol{y}_0 \in \mathbf{T}_{\text{flat}}$. Because we want a log-projective contour, we work mainly in logarithmic coordinates. The log space $\mathcal{L} = \mathbb{R}^d \oplus i(\mathbb{R}/2\pi)^d$ is a discrete quotient of $\mathbb{R}^d \oplus i\mathbb{R}^d$, so for analyses taking place in a sufficiently small neighborhood in $\mathcal{L}$ it makes sense to consider the computations in $\mathbb{C}^d$. In particular, while q is not a polynomial on the log space $\mathcal{L}$, it is analytic at $\boldsymbol{x}$ and its leading homogeneous term $\lambda\tilde{q}$ is a complex scalar multiple of a real homogeneous polynomial $\tilde{q}$ of the same degree as the leading homogeneous part of $q(\boldsymbol{x} + i\boldsymbol{y} + \boldsymbol{w})$ in $\boldsymbol{w}$. Let $\mathcal{V}_{\tilde{q}}$ denote the zero set of $\tilde{q}$.

Because the homogeneous polynomial $\tilde{q}$ is hyperbolic with some cone of hyperbolicity $\mathbf{K}_0$ containing $\tan_{\boldsymbol{x}}(B)$, Definition 11.16 and part (i) of Lemma 11.27 imply the existence of a lower semi-continuous family of real cones $\mathbf{K}^{\tilde{q},B}(\boldsymbol{y})$ of hyperbolicity for the homogenizations of $\tilde{q}$ at $\boldsymbol{y}$ as $\boldsymbol{y}$ varies, with all cones containing $\tan_{\boldsymbol{x}}(B)$. For $\boldsymbol{y} \notin \mathcal{V}_{\tilde{q}}$, the cone $\mathbf{K}^{\tilde{q},B}(\boldsymbol{y})$ may be taken to be all of $\mathbb{R}^d$. By homogeneity, $\mathbf{K}^{\tilde{q},B}(\lambda\boldsymbol{y}) = \mathbf{K}^{\tilde{q},B}(\boldsymbol{y})$ for any real $\lambda > 0$.

Definition 11.30. Recall the normal cone $\mathtt{normal}_{\boldsymbol{x}}(B) = -\mathtt{tan}_{\boldsymbol{x}}(B)^*$. We say

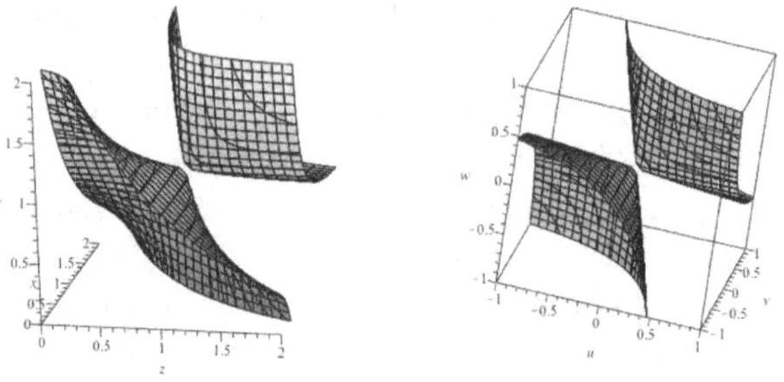

Figure 11.2 *Left:* A variety $\mathcal{V} = \mathcal{V}_Q$. *Right:* The corresponding variety $\log \mathcal{V} = \mathcal{V}_q$.

that a direction $\hat{\boldsymbol{r}}$ in the interior of $\mathtt{normal}_{\boldsymbol{x}}(B)$ is a ***non-obstructed direction*** if for any $\boldsymbol{y} \in \mathcal{V}_{\tilde{q}}$ the cone $\mathbf{K}^{\tilde{q},B}(\boldsymbol{y})$ contains a vector $\boldsymbol{v}$ with $\hat{\boldsymbol{r}} \cdot \boldsymbol{v} > 0$.

Note that obstruction is a property of the homogenization $\tilde{q}$.

Lemma 11.31. *Suppose that $\hat{\boldsymbol{r}}_*$ is a non-obstructed direction in the interior of $\mathtt{normal}_{\boldsymbol{x}}(B)$. Then there is a 1-homogeneous vector field η on $\mathbb{R}^d$, vanishing and continuous at the origin and smooth elsewhere, such that $\eta(\boldsymbol{y}) \in \mathbf{K}^{\tilde{q},B}(\boldsymbol{y})$ and $\hat{\boldsymbol{r}} \cdot \eta(\boldsymbol{y}) \geq |\hat{\boldsymbol{r}}||\boldsymbol{y}|$ for all $\boldsymbol{y} \in \mathbb{R}^d$ and all $\hat{\boldsymbol{r}}$ in a neighborhood of $\hat{\boldsymbol{r}}_*$.*

Proof It suffices to define η smoothly on the set of unit vectors, then extend by $\eta(\lambda \boldsymbol{y}) = \lambda \eta(\boldsymbol{y})$: this extension will be smooth away from the origin and continuous at the origin, and if $\hat{\boldsymbol{r}} \cdot \eta(\boldsymbol{y})$ is positive on the unit ball then the inequality in the conclusion of the lemma follows.

By the definition of non-obstruction, there is always a $\boldsymbol{v}$ depending on $\boldsymbol{y}$ with $\boldsymbol{v} \in \mathbf{K}^{\tilde{q},B}(\boldsymbol{y})$ and $\boldsymbol{v} \cdot \hat{\boldsymbol{r}} > 0$. Because of the strict inequality and semi-continuity of the cones, this same vector $\boldsymbol{v}$ works for nearby $\hat{\boldsymbol{r}}'$ and nearby $\boldsymbol{y}'$. One may therefore cover the unit ball with finitely many open sets on which $\boldsymbol{v}$ may be taken to be a constant. Construct η from these by a partition of unity, noting that convex combinations preserve both $\boldsymbol{v} \cdot \hat{\boldsymbol{r}} > 0$ and $\boldsymbol{v} \in \mathbf{K}^{\tilde{q},B}(\boldsymbol{y})$. □

We finish step (i) of our outline by finding a projective deformation with the required properties. To accomplish step (ii) we then piece these together for finitely many values of $\boldsymbol{y}_0$ and form a single deformation. We introduce a parameter ε that shrinks the entire deformation so that we can later keep a piece of it close to the origin.

Lemma 11.32. *Let A be a hyperbolic homogeneous polynomial with cone of hyperbolicity $\mathbf{K}$. Let $\hat{\boldsymbol{r}}_*$ be a fixed non-obstructed direction in the interior of the* *normal cone* $-\mathbf{K}^*$, *and let* η *be the projective vector field from Lemma 11.31. For* $\varepsilon > 0$ *and fixed* $\boldsymbol{u} \in \mathbf{K}$, *let* Φ_t *be the homotopy from* $\mathbb{R}^d \times [0,1]$ *into* $\mathbb{C}^d$ *defined by*

$$\Phi_t(\boldsymbol{y}) = \Phi_t^{\varepsilon,\boldsymbol{u},\eta} = i\boldsymbol{y} + \varepsilon\left[(1-t)\boldsymbol{u} + t\eta(\boldsymbol{y})\right] . \tag{11.10}$$

Then there exists $c > 0$ *such that* $\hat{\boldsymbol{r}} \cdot \Phi_1(\boldsymbol{y}) \geq c|\boldsymbol{y}|$ *for all* $\boldsymbol{y}$ *as* $\hat{\boldsymbol{r}}$ *varies over some neighborhood of* $\hat{\boldsymbol{r}}_*$. *Furthermore,* $A(\Phi_t(\boldsymbol{y})) \neq 0$ *for all* $0 \leq t \leq 1$ *except when* $t = 1$ *and* $\boldsymbol{y} = 0$. *Consequently, the chain* $\boldsymbol{u} + i\mathbb{R}^d$ *is homotopic in the complement of* $\mathcal{V}_A$ *to the projective chain* $\overline{C} = \Phi_1[\mathbb{R}^d]$ *on which* $\hat{\boldsymbol{r}} \cdot \boldsymbol{y}$ *grows linearly in* $|\boldsymbol{y}|$, *uniformly when* $\hat{\boldsymbol{r}}$ *varies over some neighborhood* $\mathcal{N}$ *of* $\hat{\boldsymbol{r}}_*$.

Remark. The parameter ε that shrinks the deformation toward $\boldsymbol{x} + i\boldsymbol{y}$ has no consequence in Lemma 11.32, but will be useful later when we want the homotopy to avoid a function f whose homogenization is A.

Proof By definition of cones of hyperbolicity for homogeneous functions, $\tilde{q}(\Phi_t(\boldsymbol{y}))$ cannot vanish except if $\boldsymbol{y} = 0$ and $t = 1$. The rest is Lemma 11.31 and the definition of Φ_t in (11.10). □

The next step is to glue together constructions in a neighborhood of log space $\mathcal{L}$ near each point of `localarg`($\hat{\boldsymbol{r}}_*$). We solve several problems at the same time. First, in order to remain in $\mathcal{M}$ we stop the homotopy slightly before $t = 1$ when $\boldsymbol{y}$ is near the set `localarg`($\hat{\boldsymbol{r}}_*$). Second, we do this construction simultaneously for all $\boldsymbol{w} \in$ `localarg`($\hat{\boldsymbol{r}}_*$), gluing together the constructions in these neighborhoods with a standard construction outside of these neighborhoods. Third, we do this in small neighborhoods of $\mathbb{R}^d$ that can be identified with neighborhoods in $\mathcal{L}$ so as to cover the space, thus ensuring that mapping forward by the exponential map produces a homotopy of the torus to a contour that is loglinear, with height $-\hat{\boldsymbol{r}}_* \cdot \boldsymbol{x}$ decreasing linearly on the contour with distance from the nearest point in `local`($\hat{\boldsymbol{r}}_*$) and has height bounded above by $-\boldsymbol{x} \cdot \hat{\boldsymbol{r}}_* - c$ outside of these neighborhoods for some positive constant c. Fourth, we ensure that the homotopy remains in $\mathcal{M}$. Finally, we take the opportunity to correct the double duty played by the parameter δ in [BP11, Section 5]: we will use the subscript ρ when restricting a chain to a neighborhood of radius ρ and keep a superscript δ for how early to stop the homotopy and how near to `localarg` to stop it. We always choose $\delta < \rho$, ensuring the contour looks like Figure 11.3.

Theorem 11.33 (local projective deformation). *Let* B *be a component of the complement of* `amoeba`(Q). *Suppose* Q *satisfies conclusion (iii) of Lemma 11.27 and that* $-\hat{\boldsymbol{r}}_* \cdot \boldsymbol{x}$ *is minimized on* $\overline{B}$ *uniquely at some point* $\boldsymbol{x} \in \partial B$. *Assume further that* `local`($\hat{\boldsymbol{r}}_*$) *is finite and that* $\hat{\boldsymbol{r}}_*$ *is a non-obstructed direction in the* *interior of* $\mathtt{N}(\boldsymbol{z})$ *for each* $\boldsymbol{z} \in \mathtt{local}(\hat{\boldsymbol{r}}_*)$. *Fixing* $\boldsymbol{x} + \boldsymbol{u} \in B$, *there is a vector field* $\overline{\eta} : \mathbf{T}_{\text{flat}} \to \mathbb{R}^d$ *and an* $\varepsilon > 0$ *such that the following are true.*

(i) $\overline{\eta}$ *is a smooth section of* $\mathbf{K}^{q,B}$.
(ii) For each $\boldsymbol{y} \in \mathtt{localarg}(\hat{\boldsymbol{r}}_*)$, *the vector field* $\overline{\eta}(\boldsymbol{y} + \cdot)$ *is 1-homogeneous in a neighborhood of* $\mathbf{0}$.
(iii) There is an $\varepsilon > 0$ *such that the homotopy* $\Phi_t = \Phi_t^{\varepsilon,\boldsymbol{u},\overline{\eta}}$ *avoids* $\log \mathcal{V}$ *except at* $t = 1$ *and* $\boldsymbol{y} \in \mathtt{local}(\hat{\boldsymbol{r}}_*)$.
(iv) There is a constant $c > 0$ *and a neighborhood* $\mathcal{R}$ *of* $\hat{\boldsymbol{r}}_*$ *such that* $-\hat{\boldsymbol{r}} \cdot \eta(\boldsymbol{w}) < -c|\boldsymbol{w} - \boldsymbol{y}|$ *for every* $\boldsymbol{w}$ *in some neighborhood of each* $\boldsymbol{y} \in \mathtt{local}(\hat{\boldsymbol{r}}_*)$ *and every* $\hat{\boldsymbol{r}} \in \mathcal{R}$.

Proof We glue together vector fields on $\mathbf{T}_{\text{flat}}$ in small balls around each point using a partition of unity. Fix $\rho > 0$ to be determined later. For a point $\boldsymbol{y} \in \mathtt{localarg}(\hat{\boldsymbol{r}})$ define $\mathcal{N}(\boldsymbol{y})$ to be the ball of radius ρ centered at $\boldsymbol{y}$, and define the vector field $\vec{V}_{\boldsymbol{y}}$ on $\mathcal{N}(\boldsymbol{y})$ to be the projective vector field η from Lemma 11.31 for the homogenization of $\tilde{q}$ at $\boldsymbol{y}$. Suppose now that $\boldsymbol{y} \notin \mathtt{localarg}(\hat{\boldsymbol{r}}_*)$. As in the proofs of Theorems 11.4 and 11.5, we pick a vector $\boldsymbol{v}$ in the cone $Z(\boldsymbol{y})$ from Theorem 11.1 such that $\boldsymbol{v} \cdot \hat{\boldsymbol{r}} \geq 1$ for every $\hat{\boldsymbol{r}} \in \mathcal{R}$. Define the vector field $\vec{V}_{\boldsymbol{y}}(\cdot)$ to be the constant vector $\boldsymbol{v}$ on a neighborhood $\mathcal{N}(\boldsymbol{y})$ of $\boldsymbol{y}$ sufficiently small to avoid $\mathtt{local}(\hat{\boldsymbol{r}})$ for every $\hat{\boldsymbol{r}} \in \mathcal{R}$, to avoid proper substrata of the stratum containing $\boldsymbol{y}$, and such that $\boldsymbol{v} \in Z(\boldsymbol{y}')$ for every $\boldsymbol{y}' \in \mathcal{N}(\boldsymbol{y})$.

We now have neighborhoods $\mathcal{N}(\boldsymbol{y})$ and vector fields $\vec{V}_{\boldsymbol{y}}$ on $\mathcal{N}(\boldsymbol{y})$ chosen for all $\boldsymbol{y} \in \mathbf{T}_{\text{flat}}$ whether or not $\boldsymbol{y} \in \mathtt{localarg}(\hat{\boldsymbol{r}}_*)$. Fix a subcover $\{\mathcal{N}(\boldsymbol{w}) : \boldsymbol{w} \in E\}$ indexed by some finite set E. Choose a partition of unity $\{\psi_{\boldsymbol{w}}\}$ subordinate to this cover and define the vector field $\overline{\eta} : \mathbf{T}_{\text{flat}} \to \mathbb{R}^d$ by

$$\overline{\eta}(\boldsymbol{y}) = \sum_{\boldsymbol{w} \in E} \psi_{\boldsymbol{w}}(\boldsymbol{y}) \vec{V}(\boldsymbol{w}),$$

mirroring the definition of $\vec{V}$ in (11.2).

Conclusion *(i)* is immediate from the partition of unity construction and the fact that $\overline{\eta}$ was built from vectors $\vec{V}_{\boldsymbol{w}}(\boldsymbol{y}) \in Z(\boldsymbol{y})$ for $\boldsymbol{w} \notin \mathtt{localarg}$ or $\vec{V}_{\boldsymbol{w}}(\boldsymbol{y}) \in \mathbf{K}^{\tilde{q},\boldsymbol{y}} \subseteq Z(\boldsymbol{y})$ for $\boldsymbol{w} \in \mathtt{localarg}$, provided that ρ was chosen sufficiently small; see part *(iii)* of Lemma 11.27. Conclusions *(ii)* and *(iv)* are immediate from the conclusion of Lemma 11.31 and the fact that each element of $\mathtt{local}(\hat{\boldsymbol{r}}_*)$ is forced to be in E. We prove Conclusion *(iii)* in three cases.

The easiest case occurs when $\boldsymbol{y}$ is not in $\mathcal{N}(\boldsymbol{w})$ for any $\boldsymbol{w} \in \mathtt{local}(\hat{\boldsymbol{r}}_*)$. Then by convexity $\overline{\eta}(\boldsymbol{y}) \in Z(\boldsymbol{y})$ and we already know Conclusion *(iii)* for such $\boldsymbol{y}$ from Conclusion *(ii)* of Theorem 11.1. The second case occurs when $\boldsymbol{y} \in \mathcal{N}'(\boldsymbol{w}) = \mathcal{N}(\boldsymbol{w}) \setminus \bigcup_{\boldsymbol{w}' \neq \boldsymbol{w}} \mathcal{N}(\boldsymbol{w}')$ for some $\boldsymbol{w} \in \mathtt{local}(\hat{\boldsymbol{r}}_*)$. Then $\overline{\eta}$ is defined as in Lemma 11.31 applied to the homogenization A of $\tilde{q}$ near $\boldsymbol{w}$. We quote the argument from [BP11, Theorem 5.8]. The range of the homotopy $\Phi_t^{\varepsilon,\boldsymbol{u},\bar{\eta}}$ is a projective set in a neighborhood of $\boldsymbol{w}$, meaning locally it coincides with a conical set of the form $\{\lambda \boldsymbol{b} : \boldsymbol{b} \in K, \lambda > 0\}$ for some closed subset K of the unit sphere. The variety $\mathcal{V}_{\tilde{q}}$ defined by $\tilde{q}$ is also a closed conical set intersecting the unit sphere in some set L. On the unit sphere these closed sets do not intersect and hence are separated by a positive distance $\rho_{\boldsymbol{w}}$. The set of normalized secants $(\boldsymbol{z} - \boldsymbol{w})/|\boldsymbol{z} - \boldsymbol{w}|$ for $\boldsymbol{z} \in \mathcal{V}_q$ and $|\boldsymbol{z} - \boldsymbol{w}| < s$ converges as $s \downarrow 0$ to L, hence, once the parameter ρ is chosen smaller than half each $\rho_{\boldsymbol{w}}$ the homotopy $\Phi_t(\boldsymbol{y})$ must avoid $\mathcal{V}_q$ for all $\boldsymbol{y}$ in this case.

The final case occurs when $\boldsymbol{y} \in \mathcal{N}(\boldsymbol{w}) \cap \mathcal{N}(\boldsymbol{w}')$ for some $\boldsymbol{w} \in \texttt{local}(\hat{\boldsymbol{r}}_*)$ and $\boldsymbol{w}' \notin \texttt{local}(\hat{\boldsymbol{r}}_*)$. Reasoning as in Conclusion (i), we observe that $\mathbf{K}^{\tilde{q},B}(\boldsymbol{y})$ is a subset of the cone $Z(\boldsymbol{y})$ once ρ is sufficiently small as a consequence of part (iii) of Lemma 11.27, and the reasoning of the first conclusion applies again. □

Let $\overline{c} = \Phi_1(\mathbf{T}_{\text{flat}})$ be the final cycle in the homotopy described in Theorem 11.33. In order to produce a homotopy and a final cycle that completely avoids $\mathcal{V}_q$, we stop the homotopy early near $\texttt{localarg}$. Define $\rho(\boldsymbol{y})$ by $\rho(\boldsymbol{y}) = \min_{\boldsymbol{w} \in \texttt{localarg}} |\boldsymbol{y} - \boldsymbol{w}|$, the distance from $\boldsymbol{y}$ to the closest point of $\texttt{localarg}$.

Definition 11.34 (chains and stopped homotopy). Consider the following definitions.

(i) Fix $\delta > 0$ and define the stopped homotopy $\Phi_t^\delta(\boldsymbol{y}) := \Phi_t^{\varepsilon,\boldsymbol{u},\eta,\delta}(\boldsymbol{y})$ by

$$\Phi_t^\delta(\boldsymbol{y}) = i\boldsymbol{y} + \boldsymbol{x} + \varepsilon\left[(1 - t\left[1 - (\delta - \rho(\boldsymbol{y}))^+\right])\, \boldsymbol{u} + t\left[1 - (\delta - \rho(\boldsymbol{y}))^+\right] \eta(\boldsymbol{y})\right], \tag{11.11}$$

where the notation x^+ for a real number x denotes $x^+ = \max\{0, x\}$.

(ii) Let $\overline{c}^\delta = \Phi_1^\delta(\mathbf{T}_{\text{flat}})$ denote the cycle resulting from the stopped homotopy; see Figure 11.3 for depictions of $\overline{c}$ and $\overline{c}^\delta$.

(iii) For $\boldsymbol{w} \in \texttt{localarg}$ and $0 < \delta < \rho$, where ρ is less than half the distance from $\boldsymbol{w}$ to the nearest other point in $\texttt{localarg}$, define $\overline{c}_\rho^\delta(\boldsymbol{w})$ to be the chain obtained by restricting Φ_1 to the ball of radius ρ centered at $\boldsymbol{w}$.

11.3 Evaluating asymptotics

Let $F(\boldsymbol{z}) = P(\boldsymbol{z})/Q(\boldsymbol{z}) = \sum_{\boldsymbol{r}} a_{\boldsymbol{r}} \boldsymbol{z}^{\boldsymbol{r}}$ be a rational Laurent series whose open logarithmic domain of convergence is the component B of the complement of $\texttt{amoeba}(Q)$. Let $\hat{\boldsymbol{r}}_*$ be a direction such that the height function $h(\boldsymbol{x}) = -\hat{\boldsymbol{r}}_* \cdot \boldsymbol{x}$ achieves a unique minimum m on $\overline{B}$ at some point $\boldsymbol{x}_* \in \partial B$, and suppose

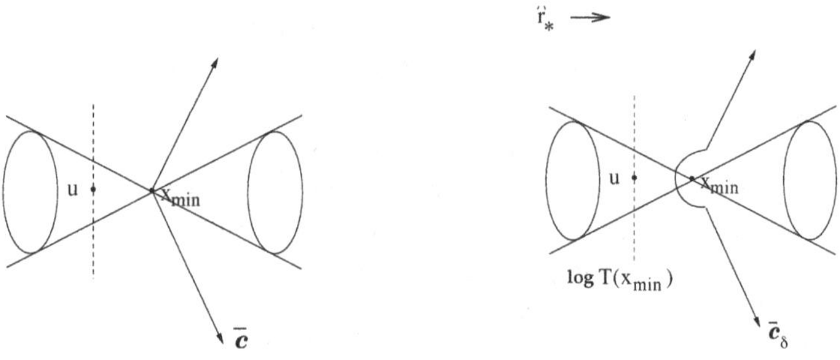

Figure 11.3 Deforming the chain: $\hat{\boldsymbol{r}}_*$ points to the right; the dotted plane is $\boldsymbol{x}-\boldsymbol{u}+i\mathbb{R}^d$. The figure on the left shows that this can be deformed to the projective contour $\overline{c}$. The figure on the right shows that stopping early produces a contour differing in an arbitrary small neighborhood of any point of `localarg` and completely avoiding $\mathcal{V}_q$.

that the set of local points $\texttt{local}(\hat{\boldsymbol{r}}_*)$ with respect to $\boldsymbol{x}_*$ is nonempty. Theorem 11.33 and Definition 11.34 show how to deform $T = \boldsymbol{x}_* - \boldsymbol{u} + i(\mathbb{R}/2\pi)^d$ in $\mathcal{L}$ with $\boldsymbol{x}_* - \boldsymbol{u} \in B$ into a cycle $\overline{c}^\delta$, which is the sum over $\boldsymbol{w} \in \texttt{localarg}(\hat{\boldsymbol{r}}_*)$ of locally projective chains $\overline{c}^\delta_\rho(\boldsymbol{w})$ near $\boldsymbol{x}_* + i\boldsymbol{w}$ together with a chain in $\mathcal{M}_{\le m-s}$ for some $s > 0$. Applying Cauchy's integral formula to F, changing coordinates via the exponential map, and breaking into pieces near each local point gives the following corollary.

Corollary 11.35. *Assume the hypotheses in the previous paragraph and let* $f = F \circ \exp$. *There is a constant* $c < m$, *a neighborhood* $\mathcal{R}$ *of* $\hat{\boldsymbol{r}}_*$, *and constants* $0 < \delta < \rho$ *such that*

$$a_{\boldsymbol{r}} = \sum_{\boldsymbol{w} \in \texttt{localarg}(\boldsymbol{r})} \frac{1}{(2\pi i)^d} \int_{\overline{c}^\delta_\rho(\boldsymbol{w})} e^{-\boldsymbol{r}\cdot\boldsymbol{z}} f(\boldsymbol{z})\, d\boldsymbol{z} + O(e^{c|\boldsymbol{r}|})$$

$$= \sum_{\boldsymbol{w} \in \texttt{localarg}(\boldsymbol{r})} \frac{\boldsymbol{z}^{-\boldsymbol{r}}}{(2\pi)^d} \int_{B(\boldsymbol{w},\rho)} e^{-i\boldsymbol{r}\cdot\boldsymbol{y}} f(i\boldsymbol{y} + \Phi^\delta_1(\boldsymbol{y})) J(\boldsymbol{y})\, d\boldsymbol{y} + O(e^{c|\boldsymbol{r}|})$$

uniformly as $\hat{\boldsymbol{r}} = \boldsymbol{r}/|\boldsymbol{r}|$ *ranges over* $\mathcal{R}$, *where* Φ^δ_1 *is specified in Definition 11.34 and* J *is the Jacobian determinant of the map* $\boldsymbol{y} \mapsto \Phi^\delta_1(\boldsymbol{y})$. □

We are left with the task of asymptotically evaluating the integrals in Corollary 11.35. While the use of the locally projective deformation in the last corollary is only to save a few lines in the estimate (Theorem 11.5 would have done instead), Lemmas 11.37 and 11.39 and Theorem 11.40 do require it.

When asymptotically evaluating integrals, the first step is typically to approximate the amplitude via a series, with each summand having a canonical form that is particularly easy to integrate; for instance, in Chapters 4 and 5 saddle point integrals with general amplitudes are reduced to those whose amplitudes are monomials. The key result for doing this is a *Big-O Lemma*, which tells us that functions differing by a small quantity will have integrals differing by a correspondingly small quantity. First, we show how to approximate the reciprocal of a function by its leading homogeneous part.

Lemma 11.36 (Straightening Lemma). *Suppose that* $q(\boldsymbol{x}) = \tilde{q}(\boldsymbol{x}) + R(\boldsymbol{x})$, *where* $\tilde{q}$ *is a homogeneous polynomial of degree* α *and* R *is analytic in a neighborhood of the origin with* $R(\boldsymbol{x}) = O(|\boldsymbol{x}|^{\alpha+1})$. *Let* K *be any closed cone on which* $\tilde{q}$ *does not vanish. Then on the intersection of* K *with some neighborhood of the origin the function* q *does not vanish and there is a convergent expansion*

$$q(\boldsymbol{x})^{-s} = \sum_{n=0}^{\infty} \tilde{q}(\boldsymbol{x})^{-s-n} \left[\sum_{|\boldsymbol{m}| \geq n(\alpha+1)} c(\boldsymbol{m}, n) \boldsymbol{x}^{\boldsymbol{m}} \right]. \tag{11.12}$$

Furthermore,

$$q(\boldsymbol{x})^{-s} - \sum_{|\boldsymbol{m}| - \alpha n < N} c(\boldsymbol{m}, n) \boldsymbol{x}^{\boldsymbol{m}} \tilde{q}(\boldsymbol{x})^{-s-n} = O\left(|\boldsymbol{x}|^{-\alpha s + N}\right). \tag{11.13}$$

Proof Let $R(\boldsymbol{x}) = \sum_{|\boldsymbol{m}| \geq \alpha+1} b(\boldsymbol{m}) \boldsymbol{x}^{\boldsymbol{m}}$ be a power series expansion for R, absolutely convergent in some ball B_ε centered at the origin, and let

$$M = \frac{\sup_{|\boldsymbol{x}| \in B_\varepsilon} \sum_{|\boldsymbol{m}| \geq \alpha+1} |b(\boldsymbol{m})| |\boldsymbol{x}|^{\boldsymbol{m}}}{\inf_{|\boldsymbol{x}| \in \partial B_\varepsilon \cap K} \tilde{q}(\boldsymbol{x})}.$$

By homogeneity

$$\sum_{|\boldsymbol{m}| \geq \alpha+1} \frac{|b(\boldsymbol{m}) \boldsymbol{x}^{\boldsymbol{m}}|}{|\tilde{q}(\boldsymbol{x})|} \leq 1/2$$

on the ball $B_{\varepsilon/(2M)}$ of radius $\varepsilon/(2M)$ centered at the origin, and the binomial expansion $(1+u)^{-s} = \sum_{n \geq 0} \binom{-s}{n} u^n$ converges for $|u| < 1/2$. Thus, the series expansion

$$\left(1 + \frac{R(\boldsymbol{x})}{\tilde{q}(\boldsymbol{x})}\right)^{-s} = \sum_{n \geq 0} \binom{-s}{n} \left(\sum_{|\boldsymbol{m}| \geq \alpha+1} b(\boldsymbol{m}) \frac{\boldsymbol{x}^{\boldsymbol{m}}}{\tilde{q}(\boldsymbol{x})} \right)^n$$

converges on $B_{\varepsilon/(2M)} \cap K$, and multiplying through by $\tilde{q}^{-s}$ yields (11.12). Convergence on any neighborhood of the origin implies the estimate (11.13). □

Exercise 11.7. Let $q(x, y) = y - x - x^2$. Find $\tilde{q}$ and R, the expansion (11.12) to second order, and the exponent in the remainder term in (11.13).

Recall from Definition 6.46 that $\deg(Q, w)$ denotes the order of vanishing of an analytic function Q at a point w, equal to zero if $Q(z) \neq 0$. We define the order of vanishing of a real power $Q(z)^s$ to be $s \cdot \deg(Q, z)$, and remark that if Q is a Laurent polynomial then a branch of Q^s may be defined on the domain of convergence of any Laurent series for F. Finally, we also define the degree of a product $Q(z) = \prod_{j=1}^k Q_j(z)^{s_j}$ of Laurent polynomials by

$$\deg(Q, w) = \sum_{j=1}^{k} s_j \deg(Q_j, w)$$

and abuse notation slightly by writing

$$\texttt{amoeba}(Q) = \texttt{amoeba}\left(\prod_{j=1}^{k} Q_j\right) = \bigcap_{j=1}^{k} \texttt{amoeba}(Q_j).$$

Lemma 11.37 (Big-O Lemma). *Let $Q(z) = \prod_{j=1}^k Q_j^{s_j}(z)$ for Laurent polynomials $Q_1, \ldots, Q_k$ and real numbers $s_1, \ldots, s_k$ that are not negative integers, let $F = P(z)/Q(z)$ for a Laurent polynomial P coprime to Q, and let $f = F \circ \exp$. Fix a component B of $\texttt{amoeba}(Q)^c$ corresponding to the convergent expansion $F(z) = \sum_r a_r z^r$ on B, and fix a direction $\hat{r}_*$ such that $-\hat{r}_* \cdot x$ is minimized on $\overline{B}$ at a unique point $x_* \in \partial B$. Let $z = \exp(x_* + iw)$ for some $w \in \texttt{localarg}(\hat{r}_*)$. Assume the last conclusion of Lemma 11.27 and let $\overline{C}_\delta(z) = \exp[\overline{c}_\delta(w)]$ be a chain satisfying the conclusion of Lemma 11.32. If $\hat{r}_*$ is a non-obstructed vector in the interior to the dual cone $\texttt{N}(z)$ then the following estimates hold uniformly as $r \to \infty$ while $\hat{r} = r/|r|$ varies over a neighborhood of $\hat{r}_*$.*

(i) If $\phi(z)$ is any function that is $O(|z|^\beta)$ at the origin with $\beta + d > 0$ then

$$|z^r| \int_{\overline{c}(w)} \exp(-r \cdot z')\phi(z')\, dz' = O(|r|)^{-d-\beta}. \tag{11.14}$$

(ii) The same estimate holds for the chain $\overline{c}_\delta(w)$ in place of $\overline{c}(w)$.

(iii) Let $D = \deg(F, z) = \deg(P, z) - \sum_{j=1}^k s_j \deg(Q_j, z)$. Then for any bounded function ψ,

$$|z^r| \int_{\overline{c}_\delta(w)} \psi(z')q(z') \exp(-r \cdot z')\, dz' = O\left(|r|^{-d-D}\right).$$

(iv) The estimate $|z|^r a_r = O(|r|^{-d-D_*})$ holds where $D_* = \min_{w \in \texttt{local}} \deg(F, w)$.

Proof The cone $\overline{c}(w)$ is a subset of an infinite cone $\bigcup_{\lambda \geq 0} \lambda S$, and we may decompose $dz = t^{d-1}\, dt \wedge dS$ where dS is a finite measure on S. It follows from conclusion (*iv*) of Lemma 11.32 that there is a $\theta > 0$ for which $\text{Re}\{\boldsymbol{r} \cdot \boldsymbol{y}\} \geq \theta|\boldsymbol{r}|$ on S. Thus,

$$
\begin{aligned}
|\boldsymbol{z}^{\boldsymbol{r}}| \left| \int_{\overline{c}_\delta(\boldsymbol{w})} \exp(-\boldsymbol{r} \cdot \boldsymbol{z}')\phi(\boldsymbol{z}')\, d\boldsymbol{z}' \right| &\leq \int_0^\infty \left(\int_S C\, e^{-\theta t|\boldsymbol{r}|} t^\beta \, dS \right) t^{d-1}\, dt \\
&\leq \int_0^\infty C' e^{-\theta t|\boldsymbol{r}|} t^{\beta+d-1}\, dt \\
&= O(|\boldsymbol{r}|)^{-d-\beta},
\end{aligned}
$$

proving (*i*).

The chains $\overline{c}_\delta(\boldsymbol{w})$ are all homotopic in $\mathcal{M}$. Thus, for any fixed $\boldsymbol{r}$ the integral in (*ii*) is independent of δ. We have seen that $\beta + d > 0$ implies absolute integrability on $\overline{c}(\boldsymbol{w})$. The same estimates imply that the integral over the intersection of $\overline{c}_\delta(\boldsymbol{w})$ with an ε-neighborhood of $\boldsymbol{w}$ goes to zero as $\varepsilon \to 0$ uniformly in δ. This implies convergence of the integrals in (*ii*) to the integral in (*i*), and because the integrals in (*ii*) are all the same, they are all equal to the integral in (*i*), proving the second conclusion. The third conclusion follows from the first with $\phi = \psi q$ and from the estimate $q(\boldsymbol{z}') = O(|\boldsymbol{z}'|)^D$ on $\overline{c}_\delta(\boldsymbol{w})$, which is a consequence of Lemma 11.36 with $q = f$. The fourth conclusion follows from the second and Corollary 11.35. □

11.3.1 Fourier transforms

Our results up to this point hold for an arbitrary product $Q(\boldsymbol{z}) = \prod_{j=1}^k Q_j(\boldsymbol{z})^{s_j}$ of powers of Laurent polynomials. We now specialize to $k = 1$ and let Q be a polynomial whose leading term is a Lorentzian quadratic. The summands in Corollary 11.35 are evidently Fourier transforms which, as we will shortly see, are classically known. Recall that the standard Lorentzian quadratic $S(\boldsymbol{y}) = y_1^2 - \sum_{j=2}^d y_j^2$ (discussed in Example 11.22 above) is in fact equivalent to any other Lorentzian quadratic q in the sense that there is a real linear map M such that $q = S \circ M^{-1}$. The Fourier transform of a Lorentzian quadratic is known [Rie49; ABG70; BP11], as is the transform of any power S^{-s} as long as s is not 0 or $d/2-1$. In the formula that follows, S^* denotes the dual quadratic which has an identical formula $r_1^2 - \sum_{j=2}^d r_j^2$, and $q^*(\boldsymbol{r}) = S^*(M^*\boldsymbol{r})$ where M^* is the adjoint of the linear map M. Note that what we call the *Fourier transform*, while standard, is called the *inverse Fourier transform* in [ABG70; BP11].

Proposition 11.38 (Fourier transform of a Lorentzian quadratic). *Let s be any real number for which neither s nor $s + 1 - d/2$ is a non-positive integer. The* *generalized Fourier transform of S^{-s} is given by*

$$\widehat{S^{-s}}(\boldsymbol{r}) = e^{i\pi s} \frac{S^*(\boldsymbol{r})^{s-d/2}}{2^{2s-1}\pi^{(d-2)/2}\Gamma(s)\Gamma(s+1-d/2)}.$$

More generally, for any monomial $\boldsymbol{x}^{\boldsymbol{m}}$ and any Lorentzian quadratic q, the Fourier transform of $\boldsymbol{x}^{\boldsymbol{m}}\, q^{-s}$ is given by

$$\widehat{\boldsymbol{x}^{\boldsymbol{m}} q^{-s}}(\boldsymbol{r}) = e^{i\pi s}\, i^{|\boldsymbol{m}|}\, \frac{|M|(\partial/\partial\boldsymbol{r})^{\boldsymbol{m}} q^*(\boldsymbol{r})^{s-d/2}}{2^{2s-1}\pi^{(d-2)/2}\Gamma(s)\Gamma(s+1-d/2)}. \tag{11.15}$$

□

There is a catch to the statements in Proposition 11.38: the function $\boldsymbol{x}^{\boldsymbol{m}} q^{-s}$ will fail to be integrable at infinity if the homogeneous degree $|\boldsymbol{m}|-2s$ is $-d$ or more, and the integral defining the Fourier transform blows up at the origin if $|\boldsymbol{m}|-2s \le -d$. Proposition 11.38 is thus properly stated in terms of *generalized functions*. These generalized functions are defined as limits of actual functions on $\boldsymbol{u} + i\mathbb{R}^d$ as $\boldsymbol{u} \to 0$ in a cone of hyperbolicity of the quadratic, their integrals over noncompact sets are defined by weak limits of compact integrals, and their Fourier transforms are defined not by direct integration against $e^{i\boldsymbol{r}\cdot\boldsymbol{x}}$ but by their integrals against (classical) Fourier transforms of smooth, compactly supported functions. When s or $s+1-d/2$ is a non-positive integer, a Fourier transform can be constructed that is itself a generalized function (in particular, a sort of delta function supported on the hypersurface $S = 0$). For further details of generalized functions, we refer to [GS16] or the summary in [BP11].

We do not worry about these subtleties here, and use only the following result from [BP11, Lemma 6.3]. The proof is not trivial, involving the right choice of insertions of compactly supported functions and truncation estimates. The statement in [BP11] assumes that Q is the product of quadratic and linear factors, but in fact relies only on the conclusion of Theorem 11.33, and thus holds whenever conclusion *(iii)* of Lemma 11.27 holds.

Lemma 11.39. *Fix $\boldsymbol{x}_* \in \partial B$, let $\boldsymbol{w}$ be one of finitely many points of* `localarg`, *and let* `N` *denote the negative dual of* $\tan_{\boldsymbol{x}_*}(B)$. *Assume the last conclusion of Lemma 11.27. As $\boldsymbol{r} \to \infty$ through a compact set of non-obstructed directions in* `N`, *the generalized Fourier transform $\widehat{S^{-s}}(\boldsymbol{r})$ correctly computes the integral of* $\exp(-\boldsymbol{r}\cdot\boldsymbol{x})S^{-s}(\boldsymbol{x})$ *over the chain $\overline{c}_p^{\delta}(\boldsymbol{w})$. More generally, the same is true of $\boldsymbol{x}^{\boldsymbol{m}} q^{-s}$ and of $F \circ \exp$ when F is a Laurent polynomial.* □

11.3.2 Main result on coefficient asymptotics

Let $Q(\boldsymbol{z}) = \prod_{j=1}^{k} Q_j(\boldsymbol{z})^{s_j}$ be a product of powers of Laurent polynomials satisfying the conclusions of Theorem 11.33, and let B be a component of

$\mathtt{amoeba}(Q)^c$ corresponding to a convergent Laurent expansion of the form $F(z) = P(z)/Q(z) = \sum_{r} a_{r} z^{r}$. Pick $x_* \in \partial B$ and suppose that there is a non-obstructed direction $\hat{r}_*$ in the interior of $\mathtt{N}_{x_*}(B)$, such that the set $\mathtt{local}(\hat{r}_*)$ is a singleton $\{z\} = \{\exp(x_* + iy)\}$. Writing $q_j = Q_j \circ \exp$ and $\tilde{q}_j = \mathtt{hom}(q_j, x_* + iy)$, Lemma 11.36 allows us to develop each $q_j^{-s_j}$ as a sum of terms of the form $c(m, n, j) x^{m} \tilde{q}_j^{-s-n}$. Multiplying these series and then multiplying by P gives an expansion of the form

$$f(x) = F(\exp(x)) = \sum_{m} c(n, m) x^{m} \tilde{q}(x)^{-s-n}, \tag{11.16}$$

where $\tilde{q}(x)^{-s-n} = \prod_{j=1}^{k} \tilde{q}_j(x)^{-s_j-n_j}$, the vector n has nonnegative integer coordinates, the sum contains only terms whose degrees are the degree of $\tilde{q}$ or greater, and the sum contains only finitely many terms of any fixed degree.

Theorem 11.40. *Under the setup above, let $b = (b_1, \ldots, b_k)$ be the sequence of degrees of homogeneous polynomials $q_1, \ldots, q_k$ and let $\chi_{n,m}$ be the generalized Fourier transform of $x^{m} Q^{-s-n}$. Then there is an asymptotic development*

$$a_{r} \approx z^{-r} (2\pi)^{-d} \sum_{n,m} c(n, m) \chi_{n,m}(r) \tag{11.17}$$

valid when $r \to \infty$ with $\hat{r}$ restricted to some neighborhood of $\hat{r}_*$. For any N, there are only finitely many terms with $|m| + d - (s + n) \cdot b < N$. When $N > 0$, truncating the sum to the finitely many terms satisfying this inequality yields a remainder of $O\left(|r|^{-N}\right)$.

If $\mathtt{local}(\hat{r}_*)$ has cardinality greater than 1 and $\hat{r}_*$ is interior to $\mathtt{N}(z)$ and non-obstructed for every $z \in \mathtt{local}(\hat{r}_*)$ then the series on the right-hand side of (11.17) *can be summed over $z \in \mathtt{local}(\hat{r}_*)$ to give an asymptotic series for a_{r}.*

Proof We have seen in Corollary 11.35 that, up to a term of lower exponential order, a_{r} is computed by a sum of integrals over chains $\overline{c}_\rho^\delta(w)$ of Fourier integrands $f(z) e^{-r \cdot z}\, dz$. We assume without loss of generality that $\mathtt{local}(\hat{r}_*) = \{z\} = \{\exp(x_* + iy)\}$; the case of cardinality greater than one follows by a similar argument. Expand f via the series (11.16), ordered by increasing homogeneous degree. Equation (11.13) of Lemma 11.36 shows that the series is a true asymptotic development, in the sense that the remainders beginning with a term of a given homogeneous degree β are $O(|x|^{\beta})$ near $x + iy$ on any closed cone avoiding $\log \mathcal{V}$.

Lemma 11.37 tells us we can integrate $f(z) e^{-r \cdot z}\, dz$ term by term over $\overline{c}_\rho^\delta(y)$ for all terms of homogeneous degree less than β, and as long as $\beta > -d$ the remainder of the integral will be $O(|r|^{-\beta-d})$. By Lemma 11.39 each integral over

$\bar{c}$ is given by its generalized Fourier transform $\chi_{n,m}$, proving the theorem in the case that $\beta > -d$. Finally, if $\beta \le -d$ we observe that the generalized Fourier transform of a homogeneous function of degree α is always homogeneous of degree $-\alpha - d$. Therefore, letting α run over all degrees of terms that are in the interval $[\beta, -d]$, the remainder is expressed as the sum of finitely many terms of type $O(|\boldsymbol{r}|^{-\alpha-d})$ together with a remainder that is at most $O(1)$. This establishes that the remainder is $O(|\boldsymbol{r}|^{-\beta-d})$ and finishes the proof. □

11.4 Examples and consequences

We now give several examples computing asymptotics determined by a minimal point that is a Lorentzian quadratic.

Example 11.41 (power of a cone)**.** Let $F(x, y, z) = 1/Q(x, y, z)^\beta$ where

$$Q(x, y, z) = (1 - x)(1 - y) + (1 - x)(1 - z) + (1 - y)(1 - z).$$

Friedrichs and Lewy [FL28] were interested in nonnegativity of the power series coefficients of F when $\beta = 1$ while they studied a discretized time-dependent wave equation in two spatial dimensions; such nonnegativity was proven for $\beta > 1/2$ in classical work of Szegő [Sze33]. A vast generalization of this result, applying to the reciprocal of the Tutte polynomial of a graph with all variables e replaced by $1 - e$, was later studied by Scott and Sokal [SS14].

In logarithmic coordinates $(u, v, w) = \exp(x, y, z)$, the leading homogeneous term of Q at the singular point $(0, 0, 0)$ of $\log \mathcal{V}$ is the second elementary symmetric function $q(u, v, w) = uv + uw + vw$. The power series expansion of F corresponds to the component of $\mathtt{amoeba}(Q)^c$ containing points $(-N, -N, -N)$ with N sufficiently large, and the normal cone at this singular point is the cone of all directions (r, s, t) such that $r^2 + s^2 + t^2 < 2(rs + rt + st)$. This cone is contained in the positive orthant and its boundary is tangent to the coordinate hyperplanes along the directions $(1, 1, 0), (1, 0, 1)$, and $(0, 1, 1)$. Thus, $\mathtt{N} = \mathtt{N}_{(0,0,0)}(B)$ is a symmetric cone inscribed in the positive orthant. Because nonnegativity of the coefficients is known for some β, the boundary of the amoeba is contained in the real variety; the real variety does not intersect the projective dual cone in log space, so this example satisfies Assumption (11.7).

Substituting the result of Proposition 11.38 into Theorem 11.40, taking β to be greater than $1/2$, we compute that the dual to q is $2(rs+st+rt)-(r^2+s^2+t^2)$ and deduce

$$a_{\boldsymbol{r}} \sim \frac{4^{1-\beta}}{\sqrt{\pi}\Gamma(\beta)\Gamma(\beta - 1/2)}(2rs + 2rt + 2st - r^2 - s^2 - t^2)^{\beta-3/2}$$

as $\boldsymbol{r} \to \infty$, uniformly as $\hat{\boldsymbol{r}}$ varies over compact subsets of the interior of $\mathtt{N}$. ◁

Stringing together several facts we have accumulated concerning amoebas, tangent cones, and hyperbolicity leads to a useful one-sided bound. Given a Laurent polynomial Q and a component B of the complement of $\mathtt{amoeba}(Q)$, we know from Proposition 11.26 that any $\boldsymbol{z} = \exp(\boldsymbol{x} + i\boldsymbol{y}) \in \partial B \cap \mathcal{V}$ has a cone of hyperbolicity K containing $\tan_{\boldsymbol{x}}(B)$. From the remarks following Definition 11.12, we know that K is convex and is a component of the complement of the zero set of $A = \mathtt{hom}(Q \circ \exp, \boldsymbol{x})$ in $\mathbb{R}^d$. The homogeneous polynomial A vanishes on the boundary of the cone K. Dualizing, we see that the algebraic dual A^* to A vanishes on the boundary of the dual cone $\mathtt{N}(\boldsymbol{z})$. It follows that $\mathtt{N}(\boldsymbol{z})$ is a subset of the largest subset L of the halfspace dual to $\boldsymbol{u}$ that is bounded by the algebraic dual A^* to A. From this, it follows that for any $\boldsymbol{r} \notin L$ the set $\mathtt{local}(\boldsymbol{r})$ is empty, whence by Theorem 11.4 the exponential rate $\overline{\beta}(\hat{\boldsymbol{r}})$ is strictly less than $\beta^*(\hat{\boldsymbol{r}}) = -\hat{\boldsymbol{r}} \cdot \boldsymbol{x}$. This proves an ACSV analogue of the *Paley–Wiener Theorem* for Fourier transforms, which asserts that the generalized Fourier transform of a homogeneous function vanishes outside the dual cone.

Theorem 11.42 (Paley–Wiener Theorem for ACSV). *Let B be a component of the complement of* $\mathtt{amoeba}(Q)$, *let Q vanish at $\boldsymbol{z} = \exp(\boldsymbol{x} + i\boldsymbol{y})$, and let $\boldsymbol{u}$ be any element of* $\tan_{\boldsymbol{x}}(B)$. *Then $\overline{\beta}(\hat{\boldsymbol{r}}) < -\hat{\boldsymbol{r}} \cdot \boldsymbol{x}$ for any $\hat{\boldsymbol{r}}$ outside the closed dual cone* $\mathtt{normal}_{\boldsymbol{x}}(B) = (-\,\mathtt{tan}_{\boldsymbol{x}}(B))^*$. *In particular, when* $\mathtt{tan}_{\boldsymbol{x}}(B)$ *is bounded by the algebraic tangent cone then* $\mathtt{normal}_{\boldsymbol{x}}(B)$ *is bounded by the algebraic dual cone.* □

When the point $\boldsymbol{x}$ is the origin, as it is in many probabilistic examples, Theorem 11.42 implies that the Laurent coefficients $a_{\boldsymbol{r}}$ decay exponentially as $\boldsymbol{r} \to \infty$ with $\hat{\boldsymbol{r}}$ bounded outside the algebraic dual cone at the origin.

Example 11.43 (cube groves). Let

$$F(x, y, z) = \frac{1}{1 + xyz - (x + y + z + xy + xz + yz)/3}$$

be the cube grove creation generating function, let $\boldsymbol{x} = \mathbf{0}$, and let B be the component of $\mathtt{amoeba}(Q)^c$ containing the negative orthant. Then

$$A = \mathtt{hom}(Q \circ \exp) = 2xy + 2xz + 2yz$$

is twice the second elementary symmetric function. This quadratic form is represented by the matrix

$$M = \begin{bmatrix} 0 & 1 & 1 \\ 1 & 0 & 1 \\ 1 & 1 & 0 \end{bmatrix}$$

while the dual of A is represented by the matrix

$$M^{-1} = \frac{1}{2} \begin{bmatrix} -1 & 1 & 1 \\ 1 & -1 & 1 \\ 1 & 1 & -1 \end{bmatrix}.$$

Thus

$$A^*(r, s, t) = rs + rt + st - \frac{1}{2}\left(r^2 + s^2 + t^2\right)$$

and the zero set of A^* is a circular cone ∂L tangent to the three bounding planes of the positive orthant at the diagonals $\{x = y, z = 0\}$, $\{x = z, y = 0\}$, and $\{y = z, x = 0\}$ bounding a solid cone L. It follows from Theorem 11.42 that $a_{\boldsymbol{r}}$ decays exponentially as $\boldsymbol{r} \to \infty$ in any closed cone disjoint from L. ◁

Example 11.43 derives the "easy" direction for asymptotics, which follows directly from Theorem 11.42 and the computation of the dual cone. Nevertheless, this computation and its counterpart for orientation probabilities (where the denominator has an extra factor of $1 - z$) are the main results in the paper that introduced cube groves [PS05]. The analysis here is simpler because the hyperbolicity results above reduce geometric questions in complex codimension 1 to corresponding analyses in real codimension 1, where one can use connectivity and natural orientations. The machinery of algebraic duals and Theorem 11.42 combine to make it almost automatic to show exponential decay outside a set of directions whose boundary is the algebraic dual.

The "hard" direction for asymptotics is showing that there is no exponential decay when $\boldsymbol{r}$ is in the interior of L. This is harder to do because we typically need to evaluate the integral near the points of $\texttt{local}(\hat{\boldsymbol{r}})$ to show that asymptotics are indeed not decaying exponentially. In this case we can apply Theorem 11.40 and Proposition 11.38 directly to obtain the following result for cube grove creation rates.

Corollary 11.44. *The creation rates $\{a_{r,s,t}\}$ defined by the power series expansion of $F(x, y, z)$ in* (11.43) *satisfy*

$$a_{r,s,t} \sim \frac{1}{\pi}\left[rs + rt + st - \frac{1}{2}(r^2 + s^2 + t^2)\right]^{-1/2}$$

as $(r, s, t) \to \infty$ within a closed subcone in the interior of L. □

Example 11.45 (Aztec Diamond). Let $Q(x,y,z) = 1-(x+x^{-1}+y+y^{-1})z/2+z^2$. The Laurent expansion of

$$F(x,y,z) = \frac{z/2}{(1-yz)Q(x,y,z)}$$

in the domain of convergence containing the point $(1,1,0)$ enumerates information about the *Aztec Diamond*, a combinatorial structure we do not describe here. The singular variety $\mathcal{V}_Q$ is smooth except at $\pm(1,1,1)$, where it is represented by a Lorentzian quadratic. The homogenization $\tilde{q}$ of $q = Q \circ \exp$ at the origin is the circular cone $2z^2 - (x^2 + y^2)$ and the cone of hyperbolicity containing the negative z-axis, which is the one of interest, is the cone $B_- = \{(x,y,z) : z < -\sqrt{(x^2+y^2)/2}\}$. Its dual is given by

$$B_-^* = \left\{(r,s,t) : r^2 + s^2 \leq \frac{1}{2}t^2\right\}.$$

The other factor $1 - yz$ is smooth at $\pm(1,1,1)$ and is in fact already linear when put in logarithmic coordinates: if $(x,y,z) = \exp(u,v,w)$ then $yz = 1$ becomes $v + w = 0$. The cones of hyperbolicity are the two halfspaces $\mathcal{H}_- = \{v+w < 0\}$ and $\mathcal{H}_+ = \{v+w > 0\}$, the former containing the negative z-axis. The amoeba of a product is the intersection of the amoebas of the factors, hence the component corresponding to the chosen Laurent expansion is $B_0 = B_- \cap \mathcal{H}_-$. Dualizing, B_0^* is equal to the convex hull of $B_-^* \cup \mathcal{H}_-^*$. Projectively, B_-^* is the cone over the circle $\{\hat{r}^2 + \hat{s}^2 \leq 1/2\}$ while $\mathcal{H}_-^*$ is the single point $(0,1)$. The convex hull of the union is the teardrop shape shown in Figure 11.4. We thus have the following consequence of Theorem 11.42.

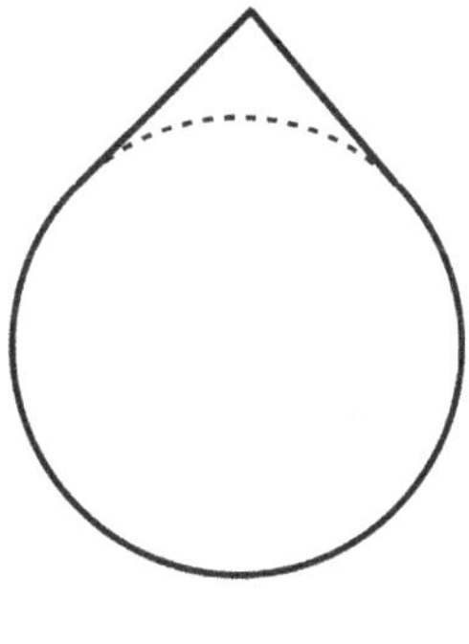

Figure 11.4 The teardrop-shaped region in Example 11.45.

Corollary 11.46. *Outside of the teardrop shaped region given by the convex hull of* $B_-^* \cup \{(0,1)\}$*), the Aztec Diamond placement probabilities decay exponentially. More specifically, the north-going placement probabilities decay* *exponentially unless either $\hat{r}^2 + \hat{s}^2 \le 1/2$, in which case these converge to a quantity in $(0,1)$ for fixed $\hat{\mathbf{r}}$ as $\mathbf{r} \to \infty$, or $\hat{r}^2 + \hat{s}^2 > 1/2$ and $\hat{s} > |\hat{r}|$, in which case the north-going placement probability converges to 1.*

◁

Example 11.47 (fortress tiling). The *fortress tiling ensemble* is a combinatorial structure enumerated by a generating function of the form $G/(Q_1 \cdots Q_k \cdot Q)$, where the Q_i are all smooth at the point $(1,1,1)$ and Q is a nondegenerate quartic and the homogeneous part of $Q \circ \exp$ at $(0,0,0)$ is given by

$$A(x,y,z) = 200z^2\left(2z^2 - x^2 - y^2\right) + 9\left(x^2 - y^2\right)^2$$

(see, for example, [Du11]). The zero set of A is a cone over the curve $400 - 200x^2 - 200y^2 + 9\left(x^2 - y^2\right)^2 = 0$ depicted in Figure 11.5.

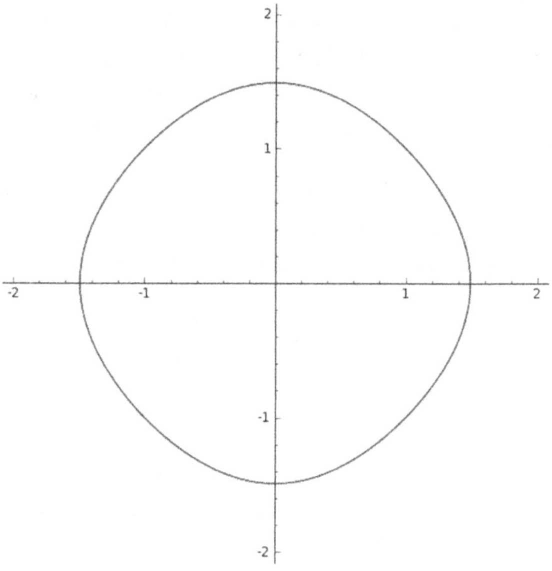

Figure 11.5 Cross-section of the homogeneous part of the fortress generating function.

The Fourier transform of A will be computed in forthcoming work of Baryshnikov and Pemantle. Without this, however, we can still prove that the coefficients a_{rst} decay exponentially outside the algebraic dual curve. Computing a Gröbner basis of the ideal $\langle r - A_x, s - A_y, t - A_z, A\rangle$ produces a basis whose first entry is the algebraic dual

$$\begin{aligned} A^*(r,s,t) = {} & 729\,t^8 - 13608\,t^6 s^2 - 22896\,s^4 t^4 + 64000\,s^6 t^2 + 102400\,s^8 \\ & - 13608\,r^2 t^6 + 412992\,s^2 t^4 r^2 - 1104000\,s^4 r^2 t^2 + 870400\,s^6 r^2 \\ & - 22896\,r^4 t^4 - 1104000\,r^4 s^2 t^2 + 2054400\,s^4 r^4 + 64000\,r^6 t^2 \\ & + 870400\,r^6 s^2 + 102400\,r^8\,. \end{aligned}$$

This projective curve is a cone over the octic affine curve

$$\begin{aligned} q^*(r,s,t) = {} & 729 - 13608\,s^2 - 22896\,s^4 + 64000\,s^6 + 102400\,s^8 - 13608\,r^2 \\ & + 412992\,s^2 r^2 - 1104000\,s^4 r^2 + 870400\,s^6 r^2 - 22896\,r^4 \\ & - 1104000\,r^4 s^2 + 2054400\,s^4 r^4 + 64000\,r^6 + 870400\,r^6 s^2 + 102400\,r^8 \end{aligned}$$

whose zero set is shown in Figure 11.6. The real part of this octic curve has two

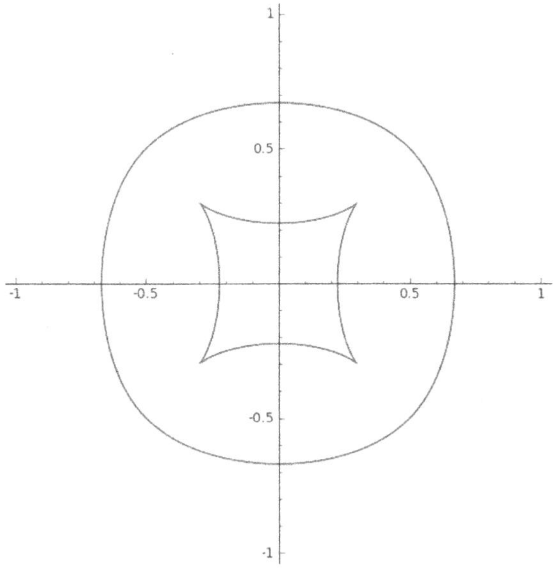

Figure 11.6 The fortress dual curve.

components, consisting of a concave aster-shaped region inside a nearly circular region. The dual cone $\mathtt{N}(1,1,1)$ must be contained within the outer curve, leading to the following "octic circle" result, conjectured by Cohn and Pemantle in 1998 and proved when Kenyon and Okounkov obtained asymptotics for this ensemble [KO07].

Theorem 11.48. *Let K be the cone over the region bounded by the outer dual curve q^*. Then, uniformly over closed cones disjoint from the closure of K, the coefficients of the fortress generating function decay exponentially. A normalized vector $(r/t, s/t, 1)$ is in the region of exponential decay if q^* is positive there and it is not in the inner area which is separated from the outer circle by any circle lying between them (such as $\hat{r}^2 + \hat{s}^2 = 1/4$).* □

◁

Product of a cone and a plane

The Fourier transforms for the Aztec Diamond and the cube grove probability generating functions, which involve the product of a function with a cone point singularity and a smooth hyperplane, are explicitly computable. The computation, contained in [BP11], is too long to include here, but the result can be understood intuitively as follows.

The Fourier transform of a linear function is a *Heaviside function*, the delta function of a ray $\{t\hat{\boldsymbol{u}} : t \geq 0\}$. The Fourier transform of a product is the convolution of the transforms of the factors, whence the Fourier transform of a product $\mathcal{L} \cdot L$, where L is linear, is given by

$$\int_0^\infty \hat{\mathcal{L}}(\boldsymbol{r} - t\hat{\boldsymbol{u}})\, dt\,.$$

When $\hat{\mathcal{L}}$ is supported on a cone (e.g., by the Paley–Wiener Theorem) and for any $\boldsymbol{r}$ the integrand vanishes for sufficiently large t, this leads to a simple integral. Formalizing this intuition involves a lot of checking that certain limits commute and appears daunting. Instead, the integral was computed in a roundabout but rigorous manner in [BP11, Section 4.1–4.2], leading to inverse trigonometric functions.

Theorem 11.49 ([BP11, Theorem 4.1–4.2])**.** *Let $\{a_{r,s,t}\}$ be the series coefficients for*

$$A(x, y, z) = \frac{z/2}{(1 - yz)(1 - (x + x^{-1} + y + y^{-1})/2 + z^2)}$$

corresponding to the amoeba complement component B containing the negative z-axis. These coefficients satisfy

$$a_{rst} \sim (1 + (-1)^{i+j+n+1})\frac{1}{2\pi} \arctan\left(\frac{\sqrt{t^2 - 2r^2 - 2s^2}}{t - 2s}\right)$$

uniformly over compact subsets of the interior of the disconnected set which is the teardrop-shaped region in Figure 11.4, taken to exclude the dashed line.

Here, the arctangent is taken in $[0, \pi]$ *so that it varies continuously as* s/t *increases through* $1/2$.

Let $\{b_{r,s,t}\}$ *be the series coefficients of*

$$B(x, y, z) = \frac{2z^2}{(1 - z)(1 + xyz - (x + y + z + xy + xz + yz)/3)}$$

corresponding to the amoeba complement component B *containing the negative diagonal. These coefficients satisfy*

$$b_{rst} \sim \frac{1}{\pi} \arctan\left(\frac{\sqrt{2(rs + rt + st) - (r^2 + s^2 + t^2)}}{r + s - t} \right)$$

uniformly over compact subsets of the interior of the region shown in Figure 11.7 in symmetrized coordinates. Again, the arctangent is taken in $[0, \pi]$ *and the solid vertical line is excluded from the region.* □

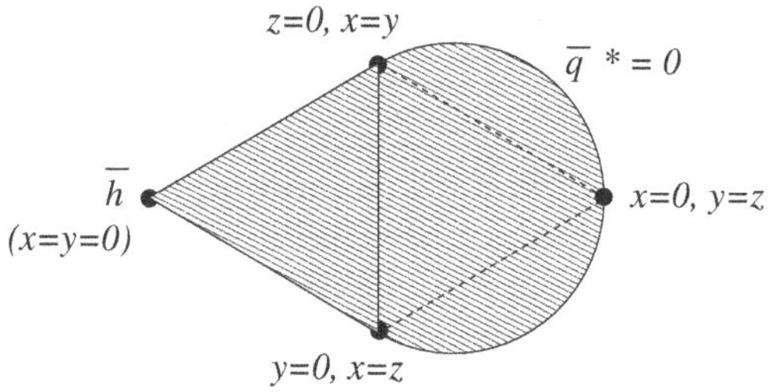

Figure 11.7 The dual cone in symmetrized coordinates.

Exercise 11.8. Use the same intuition to recompute asymptotics for the power series expansion of the multiple point at $(1, 1, 1)$ in the generating function $\frac{1}{(1-x)(1-y)(1-xy)}$. The Fourier transforms of the factors are delta functions in the directions of the rays in the respective directions $(1, 0), (0, 1)$, and $(1, 1)$. The convolution of the first two is 1 on the positive quadrant and 0 elsewhere. Up to a constant multiple, what do you get when you convolve this with the third delta function? Check your answer against Example 10.47.

Presence of a lacuna

Given the dimension d of the space and the power s of the Lorentzian quadratic S, the Fourier transforms $\chi_{n,m}(r)$ of S^{-s-n} in the expansion of (11.17) are well known; see for example [BP11, Theorem 6.4]. When $s = d/2 - 1$, this Fourier transform is not a function but rather a distribution (*generalized function* in the terminology of [GS16]) which vanishes on the *inside* of the normal cone $\mathtt{N}$ as well as the outside, being supported entirely on $\partial\mathtt{N}$. In the optics literature, this phenomenon is called a ***lacuna***. For directions $\hat{\boldsymbol{r}}$ interior to N the asymptotic development in Theorem 11.40, which has decreasing scales $|\boldsymbol{r}|^{-k}\boldsymbol{z}^{-\boldsymbol{r}}$ for integers k going to infinity, is everywhere zero, and the result only states that the asymptotics inside the cone are smaller in magnitude than any such term.

While Theorem 11.40 relies on Assumption (11.7) which does not hold for this example, topological analysis [BMP19, Theorem 2.3] gives us everything we need. It shows that for $\hat{\boldsymbol{r}} \in \mathtt{N}^{\circ}$ and $\boldsymbol{x} \in B$, where B is the component of the amoeba complement corresponding to ordinary power series, the torus $\mathbf{T}_{e}(\boldsymbol{x})$ is homologous to a cycle whose maximum height is less than the height of the cone point. This, together with a computation ruling out CPAI and a Morse theoretic analysis along the lines of Theorem 7.35, shows that asymptotics are characterized by critical points of lower height. We include such an example here, pulling together many of the methods developed thus far to exhibit a minimal point $\boldsymbol{p}$ for which $n_{\boldsymbol{p}} = 0$ even when $\hat{\boldsymbol{r}}$ is interior to the normal cone $\mathtt{N}$ at $\boldsymbol{p}$, and illustrating non-zero coefficients not in $\{\pm 1, 0\}$.

Example 11.50. Gillis, Reznick, and Zeilberger [GRZ83] consider a family of 4-variable generating functions $F_{\lambda}(\boldsymbol{z}) = 1/(1 - z_1 - z_2 - z_3 - z_4 + \lambda z_1 z_2 z_3 z_4)$ with λ a real parameter, having origins in the earlier work [AG72; Sze33]. The behavior of the function and its coefficient array differs in the cases $\lambda < 27$ and $\lambda > 27$. At the critical parameter value $\lambda = 27$ the denominator Q fails to be smooth, having instead a singularity at the diagonal point $\boldsymbol{p} = (1/3, 1/3, 1/3, 1/3)$ diffeomorphic to a Lorentzian quadratic. For a direction $\hat{\boldsymbol{r}}$ in the interior of $\mathtt{N}(\boldsymbol{p})$, one would normally expect $\overline{\beta}(\hat{\boldsymbol{r}}) = -\sum_{j=1}^{4} r_j \log(1/3)$. In other words, if $n_{\boldsymbol{p}} \neq 0$, then $\log |a_{r,s,t,u}| \sim (r + s + t + u) \log 3$ inside the normal cone and $n^{-1} \log a_{n,n,n,n} \to \log 81$.

However, the number of variables d is 4, the minimum for $d/2 - 1$ to be a positive integer. The coefficients of $F_{27}(\boldsymbol{z}) = Q(\boldsymbol{z})^{1-d/2} = Q(\boldsymbol{z})^{-1}$ exhibit a lacuna, and the exponential rate is in fact given by $\log |a_{r,s,t,u}| \sim (r + s + t + u)(\log 3)/2$ so that $n^{-1} \log a_{n,n,n,n} \to \log 9$. The bulk of [BMP19] is devoted to showing that the contour for the Cauchy integral for the main diagonal of the ordinary power series expansion is homologous to a cycle Γ supported at height less than $\log 81$. This is not achieved by a deformation but rather by an explicit cobordism which is the limit of deformations in perturbed varieties. Informally, one might say there is a deformation that passes through the singularity at the cone point. Assuming this as a black box result, we complete the analysis as follows.

A quick Gröbner basis computation shows that the critical points in the diagonal direction are precisely $\boldsymbol{p} = (1/3, 1/3, 1/3, 1/3)$, $\boldsymbol{w} = (\zeta, \zeta, \zeta, \zeta)$, and

$\overline{\boldsymbol{w}} = (\overline{\zeta}, \overline{\zeta}, \overline{\zeta}, \overline{\zeta})$, where $\zeta = (-1 + i\sqrt{2})/3$ has modulus $\sqrt{3}$. Further computer algebra shows $\mathcal{V}$ to have the geometry of a Lorentzian cone near $\boldsymbol{p}$, and Algorithm 1 from Chapter 8 verifies that there are no CPAI in the main diagonal direction. Applying Corollary 7.17 to the black box result, we see that the cycle Γ may be pushed down further to the level of the next critical points $\boldsymbol{w}$ and $\overline{\boldsymbol{w}}$. As these are the only two remaining critical points, and the asymptotics are not eventually zero, we know that

$$a_{\boldsymbol{r}} \approx \kappa\,(\Phi_{\boldsymbol{w}}(\boldsymbol{r}) + \Phi_{\overline{\boldsymbol{w}}}(\boldsymbol{r})) = 2\kappa\,\mathrm{Re}\{\Phi_{\boldsymbol{w}}(\boldsymbol{r})\} \tag{11.18}$$

for some non-zero integer κ and $\Phi_{\boldsymbol{w}}$ given by (9.4).

We compute κ for the main diagonal direction, which extends automatically to a neighborhood of the diagonal, by comparing (11.18) to asymptotics computed from an analysis of the ODE obtained by representing the diagonal as a D-finite function using the techniques described in Section 8.4.2 of Chapter 8. In particular, creative telescoping methods compute that the generating function $f(z)$ of the main diagonal satisfies

$$z^2(81z^2 + 14z + 1)f^{(3)}(z) + 3z(162z^2 + 21z + 1)f^{(2)}(z) + (21z + 1)(27z + 1)f'(z) + 3(27z + 1)f(z) = 0\,. \tag{11.19}$$

This ODE is Fuchsian: all singular points of solutions are regular and are contained in the set of roots $\{0, \zeta^4, \overline{\zeta^4}\}$ of the leading polynomial coefficient $z^2(81z^2+14z+1)$. The techniques described in Chapter 8 for D-finite functions compute a basis for the three-dimensional vector space of solutions to (11.19), whose expansions at the origin begin

$$\begin{aligned}
a_1(z) &= \log(z)^2\left(\frac{1}{2} - \frac{3z}{2} + \frac{9z^2}{2} + \cdots\right) + \log(z)\big(-4z + 18z^2 + \cdots\big)\\
&\quad + \big(8z^2 - 48z^3 + \cdots\big)\\
a_2(z) &= \log(z)\big(1 - 3z + 9z^2 + \cdots\big) + \big(-4z + 18z^2 + \cdots\big)\\
a_3(z) &= 1 - 3z + 9z^2 + \cdots,
\end{aligned}$$

and another basis of solutions whose expansions around ζ^4 begin

$$b_1(z) = 1 + \left(\frac{13}{2} + \frac{43\sqrt{2}}{4}i\right)(z-\zeta^4)^2 + \left(\frac{8165}{48} + \frac{943\sqrt{2}}{30}i\right)(z-\zeta^4)^3 + \cdots$$

$$b_2(z) = \sqrt{z-\zeta^4} + \left(\frac{13}{3} - \frac{365}{48\sqrt{2}}i\right)(z-\zeta^4)^{\frac{3}{2}} - \left(\frac{7071}{1024} - \frac{1041}{16\sqrt{2}}i\right)(z-\zeta^4)^{\frac{5}{2}} + \cdots$$

$$b_3(z) = (z-\zeta^4) + \left(\frac{17}{3} - \frac{31\sqrt{2}}{6}i\right)(z-\zeta^4)^2 - \left(\frac{1013}{72} + \frac{1805\sqrt{2}}{36}i\right)(z-\zeta^4)^3 + \cdots .$$

In the $\{a_j\}$ basis, the diagonal generating function $f(z)$ is $a_3(z)$, because this is the only element of the span that is continuous in a neighborhood of the origin and has constant coefficient 1. A change of basis matrix between the $\{a_j\}$ and $\{b_j\}$ bases can be computed numerically, in this case giving

$$f(z) = a_3(z) = C_1 b_1(z) + C_2 b_2(z) + C_3 b_3(z)$$

where C_1, C_2, and C_3 are constants which can be rigorously approximated to arbitrary precision. As $b_2(z)$ is the only element of the $\{b_j\}$ basis that is singular at $z = \mathbf{w}$, the dominant singular term in the expansion of $f(z)$ near $z = \mathbf{w}$ is $C_2\sqrt{z-\zeta^4}$ where

$$C_2 = -\Big(\big(3.5933098558743233\ldots\big) + i\big(0.38132214909311386\ldots\big)\Big) .$$

Thus, $f(z)$ has a singularity at $z = \zeta^4$ and the asymptotic contribution of this singularity to $a_{n,n,n,n}$ is

$$\Psi_1(n) = \frac{\left(4i\sqrt{2}-7\right)^n}{n^{3/2}} \frac{\Big(\big(0.543449606382202\ldots\big) + i\big(0.259547320313100\ldots\big)\Big)}{\sqrt{\pi}} + O(9^n n^{-5/2}) .$$

Repeating the same analysis at the point $z = \overline{\zeta}^4$ gives an asymptotic contribution

$$\Psi_2(n) = \frac{\left(4i\sqrt{2}+7\right)^n}{n^{3/2}} \frac{\Big(\big(0.543449606382202\ldots\big) - i\big(0.259547320313100\ldots\big)\Big)}{\sqrt{\pi}} + O(9^n n^{-5/2}) ,$$

so that $a_{n,n,n,n}$ has the asymptotic expansion $a_{n,n,n,n} = \Psi_1(n) + \Psi_2(n)$.

Comparing this expansion to (11.18) gives two expressions for asymptotics: one with complex coefficients numerically determined to arbitrary precision and another with undetermined coefficients that are known to be integers. Since one needs only 1 decimal digit to identify an unknown integer, combining these expansions proves that the unknown integer $\kappa = 3$. ◁

Notes

Hyperbolicity, as defined here, arose first in the context of PDEs. If f is a complex polynomial, let $f(D)$ denote the corresponding linear partial differential operator obtained by replacing each x_j by $\partial/\partial x_j$. For example, if f is the standard Lorentzian quadratic $x_1^2 - \sum_{j=2}^d x_j^2$, then $f(D)$ is the wave operator. Gårding set out to investigate when the equation

$$f(D)\, u = g$$

with g supported on a halfspace has a solution supported in the same halfspace. When $f(D)$ is the wave operator, this is true and the solution is unique. It turns out that the class of homogeneous polynomials f for which this is true is precisely characterized as the hyperbolic homogeneous polynomials.

The concept of hyperbolicity was used in the study of lacunas by [ABG70] to construct the deformations we have borrowed in this chapter. Later, the property of hyperbolicity turned up in algebraic combinatorics under the name of the *real root property*. Polynomials with this property are called *real stable polynomials*. Hyperbolicity and real stability are linked to a wide range of theorems and conjectures, including the van der Waerden conjecture, the Strong Rayleigh property, and the Kadison–Singer problem, solved in 2013 by Marcus, Spielman, and Srivastava. The importance of these concepts seems to stem from the closure of the class of real stable polynomials under a great many algebraic and combinatorial operations, surveyed in [Wag11]. For polynomials which are generating functions of joint distributions of binary variables, the property implies a number of negative dependence properties, and this has resulted in the solution of a number of outstanding conjectures in the theory of negative dependence [BBL09]. Although Proposition 11.15 follows from the conclusion (3.45) of [ABG70, Lemma 3.42], the self-contained proof provided above is satisfying, and was told to the authors of [BP11] by J. Borcea (personal communication).

In this chapter, while we have mostly excerpted and condensed [BP11], we have also taken the opportunity to correct some errors. The second implication of Proposition 11.25 is proved in [BP11, Proposition 2.11], however the last sentence of the cited result mistakenly claims a reverse implication (see Exercise 11.9). The implication in the wrong direction *(ii)* $\Rightarrow$ *(iii)* is then used to prove Proposition 11.26, one of the key results original to [BP11, Proposition 2.12]. Our statement here of Proposition 11.25 corrects [BP11] by eliminating the mistaken reverse implication in [BP11, Proposition 2.11], and our statement and proof of Proposition 11.26 derives [BP11, Proposition 2.12] without using the mistaken converse in the previous result. Beyond this, the first sentence of the last paragraph of [BP11, page 3177], "We piece these together ... with a partition of unity argument as before," while not known to be wrong, requires considerable justification not provided there. While this will be corrected in a forthcoming erratum, we have opted here to provide a more direct fix, Lemma 11.27 part *(iii)*, whose hypotheses cover all the examples we discuss in Section 11.4 except for one covered by a direct topological analysis. Part *(iii)* of Lemma 11.27 is used to amplify the problematic sentence from [BP11] into an argument by cases consuming three paragraphs at the end of the proof of Theorem 11.33; the lemma is used crucially in case *(iii)*. Most other results in Section 11.2 correspond to [BP11] as follows. Theorems 11.4, 11.5, 11.33, and 11.42 above correspond to Lemma 5.1, Lemma 5.3, Theorem 5.8, and Lemma 6.8 of [BP11], respectively. Lemma 11.32 corresponds to [BP11, Theorem 5.6].

There is another possible approach to asymptotics governed by points other than smooth and multiple points, namely resolution of singularities. A resolution at a singular point $\mathbf{z}$ is a change of variables which is one-to-one away from $\mathbf{z}$ and after which the local geometry at $\mathbf{z}$ is a *normal crossing*, that is, one or more smooth, transversely intersecting sheets. Resolution of singularities is effective [BM97], however the phase function becomes highly degenerate, complicating the integral substantially.

Additional exercises

Exercise 11.9. (weak hyperbolicity versus hyperbolicity of the homogeneous part) Let $f(x, y) = x^2 + y^3$. Show that f is weakly hyperbolic in direction $(0, 1)$ at the origin but that $\mathtt{hom}(f, 0) = x^2$ is not hyperbolic in direction $(0, 1)$.

Exercise 11.10. (tangent cones at a cubic point) Let $f(x, y, z) = xy + z^3$. First, compute a stratification of the zero set $\mathcal{V}$ of f. The tangent cones to $\mathcal{V}$ at (x, y, z) vary continuously as (x, y, z) moves within a stratum. Describe these, then prove or disprove that there exists a lower semi-continuous choice of tangent cone $K(x, y, z)$ in some neighborhood of the origin.

Exercise 11.11. (Explicitly constructing the vector field) Let $f(x, y, z) = z^2 - x^2 - y^2$ be the standard Lorentzian quadratic and let $B = \{(x, y, z) : z < -\sqrt{x^2 + y^2}\}$ be the cone of hyperbolicity for f containing the downward direction. Let $\hat{\mathbf{r}}_* = e_3$, the elementary vector in the positive z-direction. Find a projective vector field $\boldsymbol{v}(\boldsymbol{y})$ such that for all $\boldsymbol{y} \neq \mathbf{0}$, the function $\boldsymbol{w} \mapsto f(i\boldsymbol{y}+\boldsymbol{w})$ does not vanish on $t\boldsymbol{u} + (1 - t)\boldsymbol{v}(\boldsymbol{y})$ for $t \in [0, 1]$. You may use the proof of Lemma 11.32 or provide a sketch.

12

Combinatorial applications

The utility of any mathematical theory is ultimately determined by the breadth of problems it can solve. In this chapter we illustrate the techniques of ACSV developed in previous chapters on a large selection of combinatorial examples. These examples are arranged taxonomically in Section 12.1, helping readers identify a template for their work when trying to apply ACSV to new problems. Sections 12.2, 12.3, and 12.4 give detailed applications of our basic theory to the study of Riordan arrays, Lagrange inversion, and the transfer matrix method, respectively. Section 12.5 discusses the use of higher order asymptotics, and Section 12.6 studies algebraic generating functions by encoding them as subseries of higher-dimensional rational generating functions. Section 12.7 presents miscellaneous examples chosen to illustrate particular aspects of the theory. Combinatorics and discrete probability are closely related, and Section 12.8 applies the results of Chapter 9 to prove probabilistic limit laws for asymptotics governed by smooth points, leading to a local central limit theorem in Theorem 12.36.

12.1 Some classifications

We begin with a guide to help users of ACSV find examples similar to their intended application, with some of the examples occurring in earlier chapters and some in this chapter. We classify the examples by local geometry, by form of generating function, and by intended application. We also point to some examples where our standard hypotheses fail to hold. The website for this book contains links to Sage worksheets computing many of the examples listed here.

Local geometry of GF	Examples
Smooth nondegenerate; explicit critical points	9.10, 9.11, 9.13, 9.14, 12.11, 12.13, 12.15, 12.27, 12.30
Smooth nondegenerate; implicit critical points	12.5, 12.6, 12.9, 12.10, 12.17, 12.18, 12.20, 12.22, 12.23, 12.24, 12.25, 12.26
Smooth nondegenerate; periodicity	9.14, 12.11, 12.13
Smooth nondegenerate; torality	12.11, 12.13, 11.11
Smooth degenerate	9.32, 9.39
Multiple transverse $n < d$	12.29
Multiple transverse $n = d$	10.28, 10.27, 10.66
Multiple arrangement $n > d$	10.35, 10.47, 12.28
Multiple not arrangement	10.69, 13.3
Cone point	11.41, 11.43, 11.45, 11.47, 11.50

Table 12.1 *Guide by local geometry (dimension d, number of local sheets n).*

Classification by geometry of contributing points

Table 12.1 collates examples arranged by local geometry. Because smoothness is a generic property, smooth singular critical points dictate asymptotics in many applications. Although the coordinates of critical points can be solved in radicals for simple examples, such as Examples 9.10 and 9.11, this is usually not the case. Thus, many examples use the algebraic techniques discussed in Chapter 8 to work with critical points implicitly. A generating function with more than one contributing point for a given direction leads to periodicity in coefficients – for instance, the rational function

$$F(x,y) = \frac{1}{1 - x^2 - y^2}$$

has four contributing singularities in the main diagonal direction, reflecting the fact that the only terms that appear in any series expansion of F are those with even exponents. We may even have a continuum of critical points, such as when $F(x,y) = 1/(1 - xy)$, which can be handled under the strong torality hypothesis discussed in Section 9.1 of Chapter 9. Although the vast majority of our results require nondegenerate critical points, an example with cubic degeneracy was studied in Example 9.39 of Chapter 9. Degeneracies of any order can be handled in two dimensions using Theorem 9.38.

The difficulty of analyzing a multiple arrangement point $\mathbf{w}$ depends both on the dimension of the problem and on the number of smooth sheets intersecting at $\mathbf{w}$. The simplest cases occur when there is a single sheet (which is the smooth case) or when the number of sheets equals the dimension (where there is a complete intersection). Arrangement points with more sheets than factors are handled through an algebraic decomposition.

Form of GF	Examples
Denominator linear in a variable	9.10, 9.11, 9.13, 9.14, 9.39, 12.5, 12.6, 12.9, 12.10, 12.11, 12.13, 12.15, 12.17
Bivariate	9.10, 9.11, 9.13, 9.14, 9.32, 9.39, 10.28, 10.27, 10.35, 10.47, 10.66, 10.69, 11.11, 12.5, 12.6, 12.9, 12.15, 12.18, 12.22, 12.23, 12.25
Trivariate	11.41, 11.43, 11.45, 11.47, 11.50, 12.20, 12.26, 12.29
Higher/arbitrary dimension	12.10, 12.17, 12.27
Repeated factors, $m < d$	12.29
Repeated factors, $m = d$	10.28, 10.27, 10.66
Repeated factors, $m > d$	10.32, 10.35, 10.47, 12.28
Meromorphic, not rational	12.24, 12.30
Algebraic	12.18, 12.20
Non-combinatorial	9.32

Table 12.2 *Guide by form of GF (dimension d, number of denominator factors m).*

Multiple points that are not arrangement points are tricky, and the general theory has not been worked out. Two sheets that are tangent behave basically like a single repeated sheet, as seen in Proposition 10.68 of Chapter 10, although more complicated singularities can arise, as in Example 10.9. Chapter 11 contains essentially all that we know for more complicated singularities, with explicit results for cone points.

Classification by form of generating function

Table 12.2 classifies our examples by the algebraic form of the generating function $F(\mathbf{z}) = P(\mathbf{z})/Q(\mathbf{z})$. The simplest case occurs when Q is linear in one of its variables, and the (perhaps surprising) ubiquity of examples of this form is a reflection of the fact that the sequence construction on combinatorial classes (described in Section 2.2 of Chapter 2) corresponds to the quasi-inverse map $f \mapsto 1/(1 - f)$ on generating functions. The technique of Lagrange inversion can also be incorporated into this framework. Sections 12.2 and 12.3 cover applications to Riordan arrays and Lagrange inversion, respectively. Section 12.4 discusses the transfer matrix method.

Our formulae are flexible enough to work in any dimension, and even for families with arbitrary dimension as a parameter, although computations are often simpler in lower dimensions. Repeated denominator factors correspond to higher order poles, and thus change asymptotic behavior, while multiple distinct factors lead to multiple points.

Structure/application	Typical type of GF	Examples
regular languages, words, strings	Rational, smooth	12.6, 12.10, 12.17
lattice walks	Rational, smooth/multiple	12.26
trees	Algebraic or Riordan	12.9
quantum walks	Rational, toral	12.11, 12.13
tilings	Rational, cone/nasty	11.43, 11.45, 11.47
sums of independent random variables	Riordan	12.34
number triangles	Riordan	9.10, 9.11, 9.14, 12.15, 12.20, 12.24
constant coefficient linear recurrences	Rational, smooth/multiple	12.22
partitions	Infinite product	12.30

Table 12.3 *Guide by application area.*

Most of our generating functions are rational, but our asymptotic results hold more generally for meromorphic functions. Some examples, such as $F(x, y) = 1/(1 - e^x - e^y)$, require solving transcendental equations for critical points, while others, such as

$$F(x, y) = \prod_{i=1}^{\infty} \frac{1}{1 - x^i - y^i}$$

can be reduced to cases with polynomial denominators. We can also find asymptotics of algebraic generating functions by embedding them in rational series of higher dimension.

Classification by application

Many combinatorial families gives rise to multivariate generating functions with the same type of behavior, which can be analyzed together, and Table 12.3 gives a rough guide for readers seeking to quickly find a relevant application. Note that some problems fall into multiple areas, due to bijections between various combinatorial objects.

Examples where our standard hypotheses fail to hold

Most asymptotic expansions derived in this text hold for nondegenerate contributing points, require intersections of multiple denominator factors to be transverse, and occur in directions in the interior of cones where asymptotic behavior transitions smoothly with direction. Furthermore, the first-order terms

Exception	Examples
Degenerate contributing points	9.39
Non-transverse intersection	10.69
Boundary directions	10.66
Vanishing numerator	12.18, 12.20

Table 12.4 *Guide to non-generic examples.*

in our asymptotic expansions typically do not vanish. Table 12.4 collects examples where the above assumptions fail to hold.

Non-transverse intersections may be arbitrarily complicated, and we cover only a simple example. Similarly, we discuss one example with a degenerate contributing point. From the point of view of Fourier–Laplace integrals, asymptotics in directions on boundaries of the cones dictating uniform behavior are half of what they would be if the direction was interior to the cone. The first-order term in our asymptotic expansions of a sequence with generating function F vanishes when the numerator of F is zero at its contributing singularities. In this case, we can usually determine dominant asymptotic behavior by computing higher order coefficients in the expansion.

12.2 Powers, quasi-powers, and Riordan arrays

Let $v(\mathbf{z})$ be a power series (or polynomial) and suppose that we want to estimate the coefficient $[\mathbf{z}^{\mathbf{r}}]v(\mathbf{z})^k$ of a large power of v. This coefficient equals the coefficient of $\mathbf{z}^{\mathbf{r}}w^k$ in the power series expansion

$$F(\mathbf{z}, w) = \frac{1}{1 - wv(\mathbf{z})}, \tag{12.1}$$

so we can determine asymptotics using the tools of ACSV. The combinatorial constructions discussed in Section 2.2 of Chapter 2 show some ways in which generating functions of this form arise. Another common application comes from probability: if $v(\mathbf{z}) = \sum_{\mathbf{r}} a_{\mathbf{r}}\mathbf{z}^{\mathbf{r}}$, where $a_{\mathbf{r}} = \Pr(X_j = \mathbf{r})$ for a family $\{X_j\}$ of independent, identically distributed random variables taking values in $\mathbb{N}^d$ then $v(\mathbf{z})^n$ is the probability generating function for the partial sum $S_n = \sum_{j=1}^n X_j$, and hence

$$\mathbb{P}(S_n = \mathbf{r}) = [\mathbf{z}^{\mathbf{r}}]v(\mathbf{z})^n .$$

It has long been known that, under suitable hypotheses, such *large powers* lead to Gaussian behavior. An early work on multivariate analytic combinatorics [BR83] observed that this behavior is robust enough to hold not only for exact powers, but also for ***quasi-powers***, meaning sequences of functions $\{f_n(\mathbf{z})\}$ satisfying

$$f_n(\mathbf{z}) \sim C_n g(\mathbf{z}) \cdot h(\mathbf{z})^n \tag{12.2}$$

uniformly as $\mathbf{z}$ ranges over certain polydisks. Gaussian behavior of coefficients of quasi-powers is the basis for the *GF-sequence method* developed by Bender, Richmond, and collaborators in a series of papers including [Ben73; BR83; GR92; BR99]; see also the work of Hwang extending this to algebraic-logarithmic singularities [Hwa96; Hwa98a; Hwa98b]. These papers give conditions under which a multivariate generating function

$$F(z_1, \ldots, z_d, w) = \sum_{n=0}^{\infty} f_n(\mathbf{z}) w^n \tag{12.3}$$

is a quasi-power in the sense of (12.2). They then show that if g and h are analytic in a Δ-domain (recall Figure 3.1), if h has a unique dominant singularity where the boundary of the region intersects the positive real axis, and if the quadratic part of h is nondegenerate there, then, after a rescaling, the coefficients of $f_n(\mathbf{z})$ have a Gaussian limit distribution as $n \to \infty$.

Riordan arrays

An important combinatorial family of quasi-powers is the set of ***Riordan arrays*** $\{a_{nk} : n, k \geq 0\}$ whose generating functions $F(x, y) = \sum_{n,k\geq 0} a_{nk} x^n y^k$ satisfy

$$F(x, y) = \frac{\phi(x)}{1 - yv(x)} \tag{12.4}$$

for some analytic functions ϕ and v with $\phi(0) \neq 0$. Just as (12.1) represents sums of independent, identically distributed random variables when v is a probability generating function, the function (12.4) is a *delayed renewal sum* [Dur04, Section 3.4], where an initial random variable X_0 may be added that is distributed differently from the others. Riordan arrays cover an enormous number of examples arising in applications, including many lattice path problems, and are useful for simplifying sums because of the Pascal-like recurrences the terms satisfy – see [Mer+97; Spr94].

Remark 12.1. Some authors require that $v(0) = 0$ in (12.4), but we do not make this assumption.

Define the functions

$$\mu(v;x) := \frac{xv'(x)}{v(x)} \tag{12.5}$$

$$\sigma^2(v;x) := \frac{x^2 v''(x)}{v(x)} + \mu(v;x) - \mu(v;x)^2 = x\frac{d\mu(v;x)}{dx}, \tag{12.6}$$

whose symbols are motivated by probability theory (see Section 12.8 below).

Theorem 12.2. *Let $(v(x), \phi(x))$ determine a Riordan array with generating function* (12.4). *Suppose that $v(x)$ has radius of convergence $R \in (0, \infty]$ and is aperiodic with nonnegative coefficients. If ϕ has radius of convergence at least R then*

(i) the function $\mu(v;x)$ is strictly increasing for $x \in (0, R)$, and its range contains the interval $J = (A, B)$, where $A = \mu(v;0)$ and $B = \mu(v;R)$ are defined as one-sided limits;

(ii) there is an asymptotic expansion of the form

$$a_{rs} \approx x^{-r} v(x)^s s^{-1/2} \sum_{k=0}^{\infty} b_k(r/s) s^{-k} \tag{12.7}$$

uniformly as r/s varies over compact subsets of J, where x is the unique positive real solution to $\mu(v;x) = r/s$. The leading term in this expansion is

$$a_{rs} \sim x^{-r} v(x)^s \frac{\phi(x)}{\sqrt{2\pi s \sigma^2(v;x)}}. \tag{12.8}$$

Proof Writing $P(x,y) = \phi(x)$ and $Q(x,y) = 1 - yv(x)$, we see that $\mathcal{V}$ is smooth because Q and Q_y never simultaneously vanish. Furthermore, the smooth critical point equations simplify to show that $(x, 1/v(x))$ is critical in the direction (r, s) if and only if $\mu(v;x) = r/s$. Lemma 6.41 shows that all points of the form $(x, 1/v(x))$ for $x \in (0, R)$ are strictly minimal points. The Hessian determinant appearing in the asymptotic expansion (9.2) simplifies to $\sigma^2(v;x)$, so the result follows from an application of Theorem 9.4 after we show that μ is strictly increasing for $x \in (0, R)$. The latter result follows from (12.6). □

Remark 12.3. If v has coefficients of mixed sign, more complicated behavior can occur – see Exercise 12.8.

Remark 12.4. Riordan arrays are often specified by a recursion of the form

$$a_{n+1,k+1} = \sum_{j=0}^{\infty} c_j a_{n,k+j},$$

where the generating function $C(t) = \sum_{j=1}^{\infty} c_j t^j$ is known explicitly but $v(x)$ is known only implicitly through the equation $v(x) = xC(v(x))$. Subtleties that arise in computations when dealing with such implicitly defined v are discussed in [Wil05].

Example 12.5 (Packing paths in paths). Došlić [Doš19] derives the bivariate generating function

$$F(x,y) = \sum_{n,k} a_{nk} x^n y^k = \frac{1 - x^m}{1 - x - x^m(1 - x^m)y}$$

for the number a_{nk} of ways to maximally pack a path of length m in a path of length n, using exactly k copies of the smaller path. This is a Riordan array with $\phi(x) = (1 - x^m)/(1 - x)$ and $v(x) = x^m \phi(x)$, and ϕ and v have infinite radii of convergence since they are both polynomials.

Taking derivatives shows that

$$\mu(v; x) = m + \frac{x + 2x^2 + \cdots + (m-1)x^{m-1}}{1 + x + x^2 + \cdots + x^{m-1}},$$

and μ increases from m to $2m - 1$ as x increases from zero to infinity, covering all directions of combinatorial interest [Doš19, Proposition 2.1]. Thus, when $\lambda = n/k$ remains in a compact sub-interval of $(m, 2m - 1)$ the Gaussian asymptotic expansion (12.7) holds.

To determine the leading term in this expansion with respect to λ, one can compute a lexicographic Gröbner basis of the ideal $\langle \mu(v; x) - \lambda, \sigma^2(v; x) - S, \phi(x) - T \rangle$ in $\mathbb{Q}(\lambda)[x, S, T]$ to eliminate x and write σ^2 and ϕ in terms of λ. For instance, when $m = 3$ we obtain the irreducible elimination polynomials

$$p(S; \lambda) = S^2 + \left(2\lambda^2 - 16\lambda + \frac{88}{3}\right) S + \lambda^4 - 16\lambda^3 + \frac{281}{3}\lambda^2 - \frac{712}{3}\lambda + 220$$

$$q(T; \lambda) = (\lambda - 5)^2 T^2 + (3\lambda - 16)T + 3.$$

In this case we can use the quadratic formula to express σ^2 and ϕ explicitly in terms of λ, but for general m one must identify the correct branch of the elimination polynomials implicitly. Note that the system $p(S, \lambda) = p_S(S, \lambda) = p_\lambda(S, \lambda) = 0$ has no solutions, so the two branches given by solving $p(S, \lambda)$ for S do not meet. Because σ^2 is a continuous function of λ on the interval $[m, 2m - 1]$, and $\lim_{\lambda \to m^+} \sigma^2(v; x) = \lim_{x \to 0^+} \sigma^2(v; x) = 0$, the value of σ^2 as a function of λ is the branch of $p(S, \lambda) = 0$ passing through the point $(S, \lambda) = (0, m)$. The correct branch of $q(S, \lambda) = 0$ for ϕ with respect to λ is determined analogously. ◁

The condition in Theorem 12.2 that ϕ has at least as large a radius of convergence as v is satisfied in many, but not all, applications.

Example 12.6 (Maximum number of distinct subsequences)**.** Flaxman, Harrow, and Sorkin [FHS04] study strings of length n over the alphabet $\{1, 2, \ldots, d\}$ that contain as many distinct (not necessarily contiguous) subsequences of length k as possible. Let a_{nk} denote the maximum number of distinct subsequences of length k that can be found in a single string of length n. Initial segments $S|_n$ of the infinite string S consisting of repeated blocks of the string $12 \cdots d$ turn out always to be maximizers, meaning $S|_n$ has exactly a_{nk} distinct subsequences of length k. The generating function for $\{a_{nk}\}$ is then computed to be

$$F(x, y) = \sum_{n,k} a_{nk} x^n y^k = \frac{1}{1 - x - xy(1 - x^d)},$$

meaning a_{nk} is a Riordan array with $\phi(x) = (1-x)^{-1}$ and $v(x) = x + x^2 + \cdots + x^d$.

Assume for non-triviality that $d \geq 2$. The singular variety $\mathcal{V}$ is the union of the line $x = 1$ and the smooth curve $y = 1/v(x)$, which meet transversely at the double point $(1, 1/d)$; see Figure 12.1 for an illustration with $d = 3$.

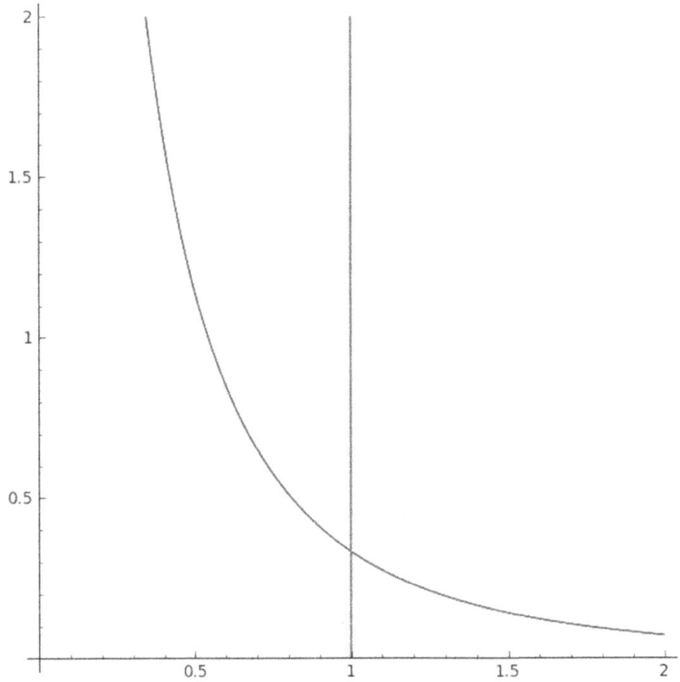

Figure 12.1 $\mathcal{V}$ in the case $d = 3$.

The radius of convergence of ϕ, namely 1, is now less than the radius of convergence of v, which is infinity. Taking derivatives shows that

$$\mu(v;x) = \frac{1}{1-x} - \frac{dx^d}{1-x^d} = \frac{1+2x+3x^2+\cdots+dx^{d-1}}{1+x+x^2+\cdots+x^{d-1}},$$

so that μ increases from 1 to $(d+1)/2$ as x increases from 0 to 1. The Gaussian asymptotics of (12.7) still hold, but only when $\lambda = n/k$ remains in a compact sub-interval of $(1, \frac{d+1}{2})$, and the computations to compute terms in this asymptotic expansion are analogous to those for Example 12.5 above.

Proposition 6.41 implies that there is a strictly minimal point in the positive quadrant controlling asymptotics in any direction. When $\lambda \geq (d+1)/2$ this turns out to be the non-smooth point $(1, 1/d)$, and Corollary 10.14 from Chapter 10 implies that $a_{\lambda k,k} \sim d^k$ in this case. Note that this is trivial when $\lambda \geq d$ because any prefix of the infinite string S with length at least dk will allow all possible k-subsequences to occur, meaning $a_{nk} = d^k$ in this regime. ◁

12.3 Lagrange inversion

On many families of recursively defined combinatorial classes, such as tree enumeration problems, the combinatorial constructions discussed in Chapter 2 yield generating functions satisfying functional equations of the form $f(z) = zv(f(z))$ for some function v that is analytic at the origin and does not vanish there. Although this equation can be solved exactly for f in some simple cases, the method of *Lagrange inversion* gives an invaluable tool for computing the coefficients of f, and their asymptotic behavior, directly from this functional equation.

Proposition 12.7 (*Lagrange inversion formula*). *If $f(z) = zv(f(z))$ with v analytic and $v(0) \neq 0$ then*

$$[z^n]f(z) = \frac{1}{n}\left[y^{n-1}\right]v(y)^n. \tag{12.9}$$

Proof Change variables to $y = f(z)$ so that the implicit equation for f implies $z = y/v(y)$ and $dz = dy[1/v(y) - yv'(y)/v(y)^2]$. The Cauchy integral representation

$$[z^n]f(z) = \frac{1}{2\pi i}\int z^{-n-1}f(z)\,dz$$

over a circle sufficiently close to the origin becomes the integral

$$\frac{1}{2\pi i}\int\left[\left(\frac{v(y)}{y}\right)^n - \left(\frac{v(y)}{y}\right)^{n-1}v'(y)\right]dy$$

around the origin in the y-plane. The difference between this integral and

$$\frac{1}{n}\left[y^{n-1}\right]v(y)^n = \frac{1}{2\pi i}\int \frac{1}{n}\left(\frac{v(y)}{y}\right)^n dy$$

equals

$$\frac{1}{2\pi i}\int \left[\left(\frac{v(y)}{y}\right)^{n-1} - \frac{n-1}{n}\left(\frac{v(y)}{y}\right)^n\right] dy = \frac{1}{2\pi i}\int d\left[\frac{y}{n}\left(\frac{v(y)}{y}\right)^n\right] dy = 0,$$

as claimed. □

To estimate the right-hand side of (12.9) via multivariate asymptotic analysis, we consider the generating function

$$\frac{1}{1 - xv(y)} = \sum_{n=0}^{\infty} x^n v(y)^n$$

which generates the powers of v, so that

$$[z^n]f(z) = \frac{1}{n}\left[x^n y^n\right]\frac{y}{1 - xv(y)}. \tag{12.10}$$

This formula holds at the level of formal power series and, if v has a non-zero radius of convergence, at the level of analytic functions.

Asymptotics for $[z^n]f(z)$ can be derived in terms of the power series coefficients of v using (12.10). For example, it follows from Theorem VI.6 of [FS09] that

$$[z^n]f(z) \sim \frac{1}{\sqrt{2\pi v''(y_0)/v(y_0)}}\, n^{-3/2} v'(y_0)^n, \tag{12.11}$$

where y_0 is the least $y > 0$ such that the tangent line to v at $(y, v(y))$ passes through the origin. In our notation y_0 is the smallest positive solution to $\mu(v; y) = 1$.

For a fixed power k, the generalization

$$[z^n]f(z)^k \sim \frac{k}{n}\frac{y_0^{k-1}}{\sqrt{2\pi n v''(y_0)/v(y_0)}} v'(y_0)^n$$

can also be obtained with univariate methods. Using multivariate methods, however, we may go further and derive bivariate asymptotics for $[z^n]f(z)^k$ as $k, n \to \infty$, holding uniformly as $\lambda = k/n$ varies over compact subsets of $(0, 1)$.

Proposition 12.8. *Let v be analytic and nonvanishing at the origin, where its power series expansion is aperiodic with nonnegative coefficients, and of order at least 2 at infinity. Let f be the nonnegative series satisfying $f(z) = zv(f(z))$ and define μ and σ^2 by equations (12.5) and* (12.6) *above, respectively. Let*

$\lambda = \lambda(k,n) = k/n$ and let x_λ be the positive real solution of the equation $\mu(v;x) = 1 - \lambda$. Then

$$[z^n]f(z)^k \sim v(x_\lambda)^n x_\lambda^{k-n} \frac{\lambda}{\sqrt{2\pi n \sigma^2(v;x_\lambda)}} = (1-\lambda)^{-n} v'(x_\lambda)^n \frac{\lambda x_\lambda^k}{\sqrt{2\pi n \sigma^2(v;x_\lambda)}}, \tag{12.12}$$

uniformly as λ varies over any compact subset of (0, 1).

Proof Exercise 12.5 below asks the reader to prove that if ψ is analytic at the origin then

$$[z^n]\psi(f(z)) = \frac{1}{n}\left[y^{n-1}\right]\psi'(y)v(y)^n, \tag{12.13}$$

a classic extension of Proposition 12.7. Assuming this result and taking $\psi(y) = y^k$, we see that

$$\begin{aligned} [z^n]f(z)^k &= \frac{k}{n}\left[y^{n-k}\right]v(y)^n \\ &= \frac{k}{n}[x^n y^{n-k}]\frac{1}{1 - xv(y)}, \end{aligned} \tag{12.14}$$

representing coefficients in the powers of f as a Riordan array determined by the (known) function v. We thus apply Theorem 12.2 with $\phi \equiv 1$, after reversing the roles of x and y to obtain the Riordan array $\{a_{rs}\}$ with generating function $1/(1 - yv(x))$, yielding

$$\begin{aligned} [z^n]f(z)^k &= \frac{k}{n}a_{n-k,k} \\ &\sim \lambda v(x_\lambda)^n x_\lambda^{k-n} n^{-1/2} \frac{1}{\sqrt{2\pi\sigma^2(v;x_\lambda)}}. \end{aligned}$$

The final equality in (12.12) follows from the defining equation $\mu(v;x_\lambda) = 1 - \lambda$. □

Examples seen in previous chapters, including binomial coefficients and Delannoy numbers, fit into this framework, but we present a more interesting example here.

Example 12.9 (Forests of trees with restricted offspring sizes)**.** Consider the class of unlabeled plane trees with the restriction that the number of children of each node must lie in a prescribed finite subset $\Omega \subseteq \mathbb{N}$. The generating function $f(z)$ counting such trees by their number of vertices satisfies $f(z) = zv(f(z))$ where $v(z) = \sum_{j\in\Omega} y^j$ (see [FS09, Section VII.3]) and asymptotic behavior follows from Proposition 12.8. For instance, unary-binary trees are defined by

$\Omega = \{0, 1, 2\}$, giving $\nu(z) = 1 + z + z^2$ and

$$\mu(\nu; z) = \frac{2z^2}{z^2 + z + 1}$$
$$\sigma^2(\nu; z) = \frac{z\left(z^2 + 4z + 1\right)}{(z^2 + z + 1)^2}.$$

When $k/n = 1/4$, so that the average tree size in a forest is 4, Proposition 12.8 implies that the number $[z^n]f(z)^k$ of forests of unary-binary trees with n nodes and k trees is asymptotically given by $C\alpha^n n^{-1/2}$ where $C \approx 0.12666642608296$ and $\alpha \approx 2.8610046903287$. ◁

12.4 Transfer matrices

The univariate transfer matrix method, discussed in Chapter 2, is easily extended to multivariate generating functions that enumerate multiplicatively weighted paths. If M is a matrix indexed by a finite set S, the ***weight*** $w(\boldsymbol{x})$ of the path $\boldsymbol{x} = (x_0, \ldots, x_n) \in S^n$ under M is $\prod_{r=1}^{n} M_{x_{r-1}, x_r}$. The sum of weights of all paths of length n from i to j under M is given by $(M^n)_{i,j}$, yielding the generating function $F_{ij}(z) = (I - zM)^{-1}_{ij}$ enumerating weighted walks from i to j by length. Similarly, the multivariate generating function enumerating weighted walks by length while also tracking d additive integer valued functions $\nu_1, \ldots, \nu_d$ defined by their values $\boldsymbol{v}(i, j)$ on single steps (i, j) is

$$F_{\boldsymbol{v}}(\boldsymbol{y}, z)_{i,j} = (I - zM_{\boldsymbol{v}})^{-1}_{ij},$$

where $M_{\boldsymbol{v}}$ is the matrix whose (i, j)-coefficient is $M_{i,j}\boldsymbol{y}^{\boldsymbol{v}(i,j)}$. We consider two applications of this observation, message passing and quantum random walks.

Example 12.10 (Message passing)**.** Let G be the graph on $K + L + 2$ vertices which is the union of two complete graphs of sizes $K + 1$ and $L + 1$, with a loop at every vertex and one edge between them. Paths on this graph correspond to a message or task being passed around two workgroups, with communication between the workgroups not allowed except between the bosses. If we sample uniformly among paths of length n, how much time does the message spend among the common (non-boss) members?

To analyze this problem we build a new graph, with vertices $\{\nu_1, \nu_2, \nu_3, \nu_4\}$ where ν_1 represents the common Group 1 members, ν_2 represents the Group 1 boss, ν_3 represents the Group 2 boss, and ν_4 represents the common Group 2 members. Every time the message moves to ν_1 it can do so in K ways, and every move to ν_4 can be done in L ways. The generating function counting paths by time spent among the common members of each workgroup and by total length is $(I - zA)^{-1}$, where

$$A = \begin{bmatrix} Ku & Ku & 0 & 0 \\ 1 & 1 & 1 & 0 \\ 0 & 1 & 1 & 1 \\ 0 & 0 & Lv & Lv \end{bmatrix}$$

with u tracking time among common Group 1 members, v tracking time among common Group 2 members, and z tracking the total time. The entries of $(I - zA)^{-1}$ are rational functions with common denominator $Q(Ku, Lv, z)$, where

$$Q(u, v, z) = uz^2 + uz^2v - uz - uz^4v + z^2v - 2z - zv + 1 + z^3v + uz^3 ,$$

and the coefficient of z^n in the power series expansion of any entry gives a probability distribution for the amount of time spent among the common members of each group after time n. Using Gröbner bases (or simply by computing a resultant) it can be shown that the system $Q(u, v, z) = Q_z(u, v, z) = 0$ has no solutions with $u, v > 0$. Thus, by the limit theorem discussed in Theorem 12.33 below, the times spent in Group 1 and Group 2 as a portion of the length n converge to $\mathbf{m} = -\nabla_{\log} Q(K, L, z_0)$, where z_0 is the minimal modulus root of $f(z) = Q(K, L, z)$.

The portion of time spent among the common members of Group 1 is given by $KQ_u(K, L, z)/(z_0 Q_z(K, L, z_0))$. Plugging in $K = L = 1$, for example, we see that $Q(1, 1, z) = 1 - 4z + 3z^2 + 2z^3 - z^4$ so $z_0 \cong 0.382$ and a proportion of approximately 0.154 of the time is spent among the common members of Group 1. If bosses and employees had equal access to communication then, by symmetry, this portion would be $1/4$, so the effect of communicating only through bosses reduced the time each message spends with each non-boss by nearly 40%. This effect is more marked when the workgroups have different sizes: increasing the size of the second group to 2, we plug in $K = 1$ and $L = 2$ to find that $z_0 \cong 0.311$ and the fraction of time spent among the common members in Group 1 plummets to just under 0.038. ◁

Example 12.11 (One-dimensional quantum walk). In Example 9.47 the notion of a quantum random walk, and its associated spacetime generating function F, was introduced. Letting $p(r, n)$ denote the amplitude for the random walk to be at location r at time n, we have

$$F(\mathbf{z}) = \sum_{r,n \geq 0} p(r, n) x^r y^n = (I - yMU)^{-1},$$

where U is a $k \times k$ unitary matrix and M is a $k \times k$ diagonal matrix whose entries

x^a run through the k possible steps (a, b) of the walk. It is shown in [BP07] that when $k = 2$ there is no loss of generality in taking U to be the real matrix

$$U_c = \begin{bmatrix} c & \sqrt{1-c^2} \\ \sqrt{1-c^2} & -c \end{bmatrix}$$

and taking the entries of M to be 1 and x, meaning that the walk either stays where it is or moves one to the right. The universal spacetime generating function for a two-dimensional quantum walk is therefore given by

$$F_c(x, y) = \frac{P_c(x, y)}{Q_c(x, y)} = \frac{P_c(x, y)}{1 - cy + cxy - xy^2},$$

where the numerator P_c depends on initial chiralities (one of k hidden states of the walk) and plays no special role. For example, if $k = 2$ and starting and ending chiralities are both in state 2 then $P_c(x, y) = 1 - cy$. ◁

Theorem 12.12 (Spacetime asymptotics for one-dimensional quantum walk). *There is a real phase function $\rho(r, s)$ such that*

$$p(r, s) = \frac{2}{\pi} \frac{\lambda \sqrt{1-c^2}}{(1-\lambda)s\sqrt{-((1-c^2) - 4\lambda + 4\lambda^2)}} \cos^2(\rho(r, s)) + O\left(s^{-3/2}\right)$$

uniformly as $\lambda = \frac{r}{s}$ varies over any compact subset of the interior of $J_c = [(1-c)/2, (1+c)/2]$. Conversely, if λ varies over a compact subset of the complement of J_c then $p(r, s) \to 0$ exponentially.

The variation of probabilities in the feasible region for $c = 1/2$ is illustrated in Figure 12.2. Qualitatively similar results hold for the other starting and ending chiralities, and for combinations of chiralities.

Proof The denominator $Q = 1 - c(1 - x)y - xy^2$ is quadratic in y and linear in x, so the critical point equations can be solved explicitly in radicals. Furthermore, Q satisfies the strong torality hypothesis (Definition 9.20) so that Corollary 9.21 applies. The intersection $\mathcal{V}_1$ of $\mathcal{V}$ with the unit torus is a topological circle $x = (cy - 1)/(cy - y^2)$ winding twice around the torus in the x direction and once in the y direction. The logarithmic Gauss map is a smooth map on this circle with two extreme values, $(1-c)/2$ and $(1+c)/2$, and no other critical points. Therefore, for each λ in the interior of J_c there are precisely two points (x_1, y_1) and (x_2, y_2) in $\mathcal{V}_1$ with $\nabla_{\log} Q(x_j, y_j) \parallel (\lambda, 1)$ and an application of (9.7) implies

$$p(r, s) = \sum_{j=1}^{2} \frac{1 - cy_j}{\sqrt{2\pi}} x_j^{-r} y_j^{-s} \sqrt{\frac{-y_j Q_y(x_j, y_j)}{s\underline{Q}(x_j, y_j)}} + O\left(s^{-3/2}\right),$$

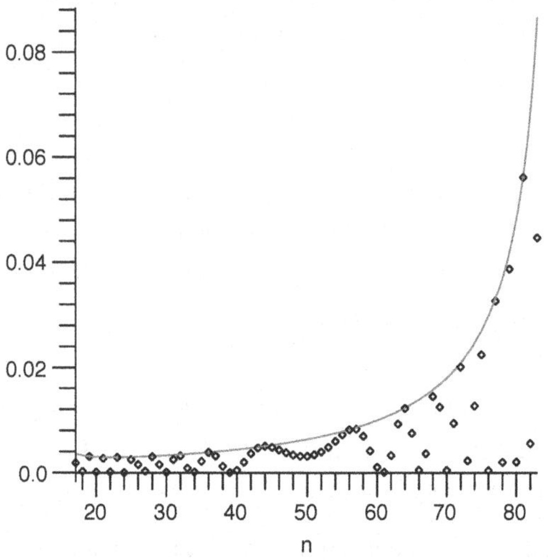

Figure 12.2 The time $n = 100$ probabilities starting and ending in state 2 when $c = 1/2$, and their upper envelope obtained by dropping the $\cos^2(\rho)$ term in Theorem 12.12.

where $\underline{Q}$ is given by (9.6). The two critical points (x_1, y_1) and (x_2, y_2) are conjugate, meaning

$$p(r,s) = 2\,\mathrm{Re}\left\{\frac{1-cy_j}{\sqrt{2\pi}} x_j^{-r} y_j^{-s} \sqrt{\frac{-y_j Q_y(x_j, y_j)}{s\,\underline{Q}(x_j, y_j)}}\right\} + O\left(s^{-3/2}\right).$$

Defining ρ to be the argument of the expression in braces, writing (x, y) for either one of the two points, and taking the square modulus, we obtain

$$p(r,s) = \frac{2}{\pi} \cos^2(\rho) \left|(1-cy)^2 \frac{-yQ_y(x,y)}{s\,\underline{Q}(x,y)}\right| + O\left(s^{-3/2}\right).$$

Using the techniques of Chapter 8, we compute that at the critical point (x, y) the quantity

$$w = (1-cy)^2 \frac{-yQ_y(x,y)}{n\,\underline{Q}(x,y)}$$

satisfies

$$\lambda^2(1-c^2) + 4\left(\frac{(1+c)}{2} - \lambda\right)\left(\lambda - \frac{(1-c)}{2}\right)(1-\lambda)^2 w^2 = 0,$$

where λ and c are parameters. Solving, we find

$$|w| = \frac{\sqrt{1-c^2\lambda}}{(1-\lambda)s\sqrt{-((1-c^2)-4\lambda+4\lambda^2)}},$$

proving the claimed asymptotics when $\lambda \in J_c$.

If $\lambda \notin J_c$ then either $(0,0)$ does not minimize the height function in the direction $(\lambda, 1)$ on the amoeba complement component corresponding to the power series expansion of F_c, meaning $\overline{\beta} \leq \beta^* < 0$, or $(0,0)$ is the minimizing point but there are no critical points on $T(1,1)$, so Theorem 11.4 implies $\overline{\beta} < 0$. In either case, $p(r,s)$ decays exponentially. □

Example 12.13 (QRWs in higher dimensions)**.** Example 9.47 of Chapter 9 shows that, in general, the spacetime generating function for a quantum walk with steps $\boldsymbol{v}^{(1)}, \ldots, \boldsymbol{v}^{(k)}$ has the form

$$F(\boldsymbol{z}) = (I - z_{d+1}MU)^{-1},$$

where $\boldsymbol{z}^\circ = (z_1, \ldots, z_d)$ are d space variables, the variable z_{d+1} tracks time, and M is the diagonal matrix whose (j,j)-entry is $(\boldsymbol{z}^\circ)^{\boldsymbol{v}^{(j)}}$. The common denominator of the F_{ij} is $Q(\boldsymbol{z}) = \det(I - z_{d+1}MU)$. As noted in Example 9.47, the feasible velocity region R is the image of the logarithmic Gauss map and the limit law for the amplitudes can be written, up to an oscillatory term, in terms of the Gaussian curvature.

Consider the two-dimensional family of walks with the nearest neighbor unit vectors $\boldsymbol{v}^{(1)}, \ldots, \boldsymbol{v}^{(4)}$ as steps and the unitary matrix

$$U = S(p) = \begin{pmatrix} \frac{\sqrt{p}}{\sqrt{2}} & \frac{\sqrt{p}}{\sqrt{2}} & \frac{\sqrt{1-p}}{\sqrt{2}} & \frac{\sqrt{1-p}}{\sqrt{2}} \\ -\frac{\sqrt{p}}{\sqrt{2}} & \frac{\sqrt{p}}{\sqrt{2}} & -\frac{\sqrt{1-p}}{\sqrt{2}} & \frac{\sqrt{1-p}}{\sqrt{2}} \\ \frac{\sqrt{1-p}}{\sqrt{2}} & -\frac{\sqrt{1-p}}{\sqrt{2}} & -\frac{\sqrt{p}}{\sqrt{2}} & \frac{\sqrt{p}}{\sqrt{2}} \\ -\frac{\sqrt{1-p}}{\sqrt{2}} & -\frac{\sqrt{1-p}}{\sqrt{2}} & \frac{\sqrt{p}}{\sqrt{2}} & \frac{\sqrt{p}}{\sqrt{2}} \end{pmatrix}.$$

Setting $\alpha = \sqrt{2p}$ to simplify notation, we have

$$\begin{aligned} Q(x,y,z) &= (x^2y^2 + y^2 - x^2 - 1 + 2xyz^2)z^2 - 2xy \\ &\quad - \alpha z(xy^2 - y - x + z^2y - z^2x + z^2xy^2 + z^2x^2y - x^2y) \end{aligned}$$

and a lexicographic Gröbner Basis for the ideal generated by Q and its partial derivatives includes the polynomial

$$z\alpha^2(\alpha^2 - 1)(\alpha^2 - 2) = 2zp(2p-1)(2p-2).$$

The only root of this polynomial with $x, y, z \neq 0$ and $0 < p < 1$ occurs when

$p = 1/2$, and back substitution of this value implies that $-z+z^5 = z^3+2y-z = -z-z^3+2x = 0$. The first of these polynomials vanishes on the unit circle when $z \in \{\pm 1, \pm i\}$, however when $z = \pm 1$ the second polynomial only vanishes when $y = 0$, and when $z = \pm i$ the third polynomial only vanishes when $x = 0$, so Q and its partial derivatives do not simultaneously vanish on $T(\mathbf{1})$.

Thus, if Q_z vanishes at some point (a, b, c) of $\mathcal{V}_1 = \mathcal{V} \cap T(\mathbf{1})$ then (a, b, c) contributes to nonvanishing asymptotics in a direction $(r, s, 0)$ for some $(r, s) \neq (0, 0)$. This is ruled out from our knowledge of the generating function, because the velocity of QRW is at most the longest step, so Q_z does not vanish on $\mathcal{V}_1$ and the projection $(x, y, z) \to (x, y)$ is a smooth four-fold cover of the unit torus in $\mathbb{C}^2$. There are many four-fold covers of the 2-torus, but in this case some trigonometry [Bar+10, Proposition 4.6] shows that $\mathcal{V}_1$ is in fact the union of four 2-tori, each mapping diffeomorphically to the 2-torus under the logarithmic Gauss map. Figure 12.13 shows the four components for the parameter value $p = 1/2$ by graphing z as a function of x and y with the torus depicted as the unit cube with wraparound boundary conditions.

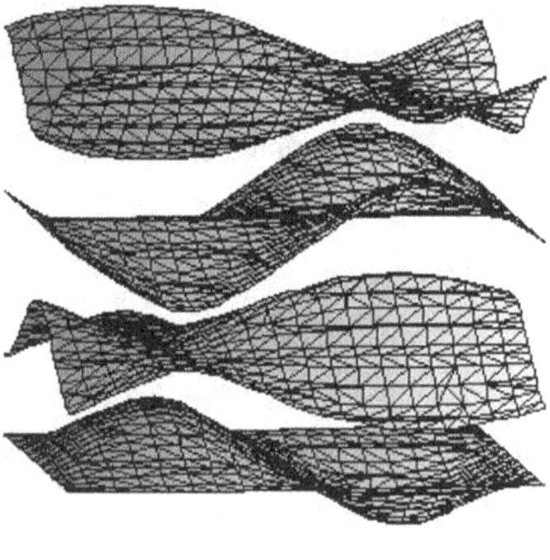

Figure 12.3 The four tori comprising $\mathcal{V}_1$ for $S(1/2)$.

Asymptotic behavior follows from Corollary 9.46 in Chapter 9.

Theorem 12.14. *For each $\hat{\mathbf{r}}$ in the image of the logarithmic Gauss map on $\mathcal{V}_1$*

let $\mathbf{W}$ *be the set of four preimages of* $\mathbf{r}$ *in* $\mathcal{V}_1$ *for the* $S(p)$ *walk with* $0 < p < 1$ *and fixed states* i *and* j. *Then, as* $\mathbf{r} \to \infty$ *with* $\hat{\mathbf{r}}$ *in a compact subset of the feasible velocity region* R, *the amplitude* $p(\mathbf{r})$ *satisfies*

$$p(\mathbf{r}) = (-1)^{\delta} \frac{1}{2\pi|\mathbf{r}|} \sum_{\mathbf{z}\in\mathbf{W}} \mathbf{z}^{-\mathbf{r}} \frac{P(\mathbf{z})}{||\nabla_{\log} Q(\mathbf{z})||_2^2} \frac{1}{\sqrt{|\mathcal{K}(\mathbf{z})|}} e^{-i\pi\tau(\mathbf{z})/4} + O\left(|\mathbf{r}|^{-3/2}\right), \tag{12.15}$$

where $\delta = 1$ *if* $\nabla_{\log} Q$ *is a negative multiple of* $\hat{\mathbf{r}}$ *(to account for the absolute value in Corollary 9.46) and zero otherwise.*

◁

12.5 Higher order asymptotics

Our results give effective methods for computing asymptotic expansions of multivariate generating functions. Although the first term in such an expansion generically dictates behavior of the sequence under consideration, there are several reasons for wanting to compute higher order terms. Most obviously, computing more terms in such expansions gives better approximations.

Example 12.15. Consider the $(4,3)$-diagonal of the Delannoy numbers with generating function $F(x,y) = 1/(1-x-y-xy)$. The critical points in this direction are $(-2,-3)$ and $(1/2,1/3)$, both of which are smooth, and the point $\mathbf{w} = (1/2,1/3)$ is strictly minimal. Corollary 5.17 implies

$$[x^{4n}y^{3n}]F(x,y) = 432^n \left(a_1 n^{-1/2} + a_2 n^{-3/2} + O(n^{-5/2})\right)$$

as $n \to \infty$, where

$$a_1 = \frac{\sqrt{30}}{10\sqrt{\pi}} \approx 0.3090193616$$

$$a_2 = -\frac{\sqrt{30}}{288\sqrt{\pi}} \approx -0.01072983895\,.$$

Comparing this approximation with the actual Delannoy numbers for small n gives (after scaling out the exponential growth) the results in Table 12.5. The error in the 1-term approximation clearly decays as $1/n$ while the 2-term error decays as $1/n^2$ – note the extreme accuracy even for $n = 1$. The question of the optimal order to which to truncate an asymptotic series in a given application goes beyond our scope here.

n	1	2	4	8	16
Exact	0.2986	0.2148	0.1532	0.1088	0.07709
1-term	0.3090	0.2185	0.1545	0.1093	0.07726
Rel. error	-0.03486	-0.01742	-0.008698	-0.004345	-0.002171
2-term	0.2983	0.2147	0.1532	0.1088	0.07709
Rel. error	0.001077	0.0002450	0.00005820	0.00001417	0.000003496

Table 12.5 *Approximations to scaled Delannoy numbers* $432^n F_{4n,3n}$.

Higher order terms in an asymptotic expansion for any direction $\hat{r}$ with positive coordinates can be computed symbolically with the coordinates of $\hat{r}$ as parameters. ◁

Although the first terms in our asymptotic expansions are typically non-zero, and thus determine dominant asymptotic behavior, they can vanish. In this case, higher order terms need to be computed in order to determine how the sequence behaves.

Example 12.16. The rational functions

$$F_1(x,y) = \frac{1}{1-x-y}, \quad F_2(x,y) = \frac{y(1-2y)}{1-x-y}, \quad \text{and} \quad F_3(x,y) = \frac{x-y}{1-x-y}$$

all admit $(x,y) = (1/2, 1/2)$ as a minimal contributing point in the main diagonal direction. Applying the results of smooth ACSV gives an asymptotic expansion (9.4) of the coefficients of F_1 whose leading term is non-zero. The leading term in the asymptotic expansion of the coefficients of F_2 vanishes, as the numerator of F_2 vanishes at the contributing point, but the second order term in this expansion is non-zero (in fact, the main diagonal of F_2 is the generating function of the Catalan numbers – see Example 12.18 below). On the other hand, it can be shown that the main diagonal of F_3 is identically zero, so all coefficients in the asymptotic expansion must vanish. It is not obvious how to detect automatically the vanishing of all terms in the asymptotic expansion for a general rational function. ◁

Another reason why we may need higher order terms is because of cancellation of terms when we combine asymptotic expansions of related functions.

Example 12.17. Consider the $(d+1)$-variate function

$$W(x_1, \ldots, x_d, y) = \frac{A(\boldsymbol{x})}{1 - yB(\boldsymbol{x})},$$

where

$$A(\boldsymbol{x}) = \frac{1}{1 - \sum_{j=1}^{d} \frac{x_j}{1+x_j}}$$

$$B(\boldsymbol{x}) = 1 - (1 - e_1(\boldsymbol{x}))A(\boldsymbol{x})$$

for the elementary symmetric function $e_1(\boldsymbol{x}) = \sum_{i=1}^{d} x_j$. The combinatorial constructions discussed in Chapter 2 imply that W enumerates words over the alphabet $\{1, \ldots, d\}$, where x_j marks occurrences of the letter j and y marks occurrences of ***snaps***, which are nonoverlapping pairs of duplicate letters. The factor $A(\boldsymbol{x})$ counts snapless words over X, which are the Smirnov words described in Example 2.11. If ψ denotes the random variable counting snaps among words with n occurrences of each letter then the expected value and variance of ψ satisfy

$$\mathbb{E}(\psi) = \frac{[\boldsymbol{x}^{n\mathbf{1}}]W_y(\boldsymbol{x}, 1)}{[\boldsymbol{x}^{n\mathbf{1}}]W(\boldsymbol{x}, 1)} = \frac{[\boldsymbol{x}^{n\mathbf{1}}]A(\boldsymbol{x})^{-1}B(\boldsymbol{x})(1 - e_1(\boldsymbol{x}))^{-2}}{[\boldsymbol{x}^{n\mathbf{1}}](1 - e_1(\boldsymbol{x}))^{-1}}$$

and $\mathbb{V}(\psi) = \mathbb{E}(\psi^2) - \mathbb{E}(\psi)^2$ with

$$\begin{aligned}\mathbb{E}(\psi^2) &= \frac{[x^{n\mathbf{1}}]\left(W_{yy}(\boldsymbol{x}, 1) + W_y(\boldsymbol{x}, 1)\right)}{[\boldsymbol{x}^{n\mathbf{1}}]W(\boldsymbol{x}, 1)} \\ &= \frac{[\boldsymbol{x}^{n\mathbf{1}}]A(\boldsymbol{x})^{-2}B(\boldsymbol{x})(B(\boldsymbol{x}) + 1)(1 - e_1(\boldsymbol{x}))^{-3}}{[\boldsymbol{x}^{n\mathbf{1}}](1 - e_1(\boldsymbol{x}))^{-1}}.\end{aligned}$$

Each of the above coefficient extractions applies to a rational function whose denominator is a power of $Q(\boldsymbol{x}) = 1 - e_1(\boldsymbol{x})$. By Lemma 6.41, asymptotics can be determined by applying the smooth ACSV results of Chapter 9 to the strictly minimal critical point $(1/d, \ldots, 1/d)$ of each function, giving

$$\begin{aligned}\mathbb{E}(\psi) &= \frac{\frac{3\sqrt{3}}{8\pi} - \frac{61\sqrt{3}}{192\pi}n^{-1} + O(n^{-2})}{\frac{\sqrt{3}}{2\pi}n^{-1} - \frac{\sqrt{3}}{9\pi}n^{-2} + O(n^{-3})} \\ &= (3/4)n - 15/32 + O(n^{-1})\end{aligned}$$

and

$$\begin{aligned}\mathbb{E}(\psi^2) &= \frac{\frac{9\sqrt{3}}{32\pi}n - \frac{35\sqrt{3}}{128\pi} + O(n^{-1})}{\frac{\sqrt{3}}{2\pi}n^{-1} - \frac{\sqrt{3}}{9\pi}n^{-2} + O(n^{-3})} \\ &= (9/16)n^2 - (27/64)n + O(1)\end{aligned}$$

so that $\mathbb{V}(\psi) = (9/32)n + O(1)$. Note that determining the leading term of $\mathbb{V}(\psi)$ requires the second-order terms in the expansions of $\mathbb{E}(\psi)$ and $\mathbb{E}(\psi^2)$.

Comparing these approximations when $d = 3$ with the actual values for small n, we obtain the following table.

n	1	2	4	8
$\mathbb{E}(\psi)$	0	1.00000	2.50909	5.52056
$(3/4)n$	0.75	1.5	3	6
$(3/4)n - 15/32$	0.28125	1.03125	2.53125	5.53125
One-term relative error	Undefined	0.50000	0.19565	0.086846
Two-term relative error	Undefined	0.031250	0.0088315	0.0019363
$\mathcal{V}(\psi)$	0	0.8	1.20057	2.31961
$(9/32)n$	0.28125	0.5625	1.125	2.25
Relative error	Undefined	0.29688	0.062942	0.030013

◁

12.6 Algebraic generating functions

Algebraic generating functions in one variable can be analyzed asymptotically by the transfer theorems of Section 3.4 in Chapter 3, and often exactly by Lagrange inversion as described in Section 12.3 above. Of course, the situation in several variables is more complicated. For the purposes of asymptotics, we can sometimes ignore the fact that the generating function is algebraic, because the contributing points determining asymptotics might occur in regions where the generating function is meromorphic (this happens in several Riordan array examples). The more difficult cases occur when asymptotics are determined by an algebraic singularity. Assuming that we have an analytic branch at the origin, we aim to compute asymptotics of the coefficients of its power series expansion as done for rational and meromorphic functions in previous chapters.

Section 2.4 gives an approach to asymptotics by embedding algebraic generating functions as subseries of higher-dimensional rational functions. Unfortunately, there are several difficulties with this approach. First, the results of Chapters 9–11 cannot capture asymptotics of all algebraic generating functions: for any $\beta \in \mathbb{Z}$ there exists a univariate sequence a_n with an algebraic generating function such that $a_n \sim C\alpha^n n^\beta$, but our results for nondegenerate critical points only give asymptotics of this form with $\beta = k/2$ for $k \in \mathbb{Z}$. This asymptotic behavior also shows that the leading terms in our asymptotic expansions often vanish when applied to rational functions constructed by such lifting procedures (Exercise 12.11 asks you to prove that this vanishing always occurs for an embedding given by the method presented in Proposition 2.36).

Algebraic generating functions are often well-behaved as they are built from nice combinatorial constructions, but the rational functions obtained from a lifting procedure are built from algebraic machinery and are thus often worse behaved. For instance, the lifted functions usually have negative power series coefficients, making the determination of minimal points harder to verify, and may have critical points at infinity. Finally, there is no guarantee that the rational generating function will have a contributing singularity of a type that we can deal with using current technology.

Despite these qualifications, the rational embedding approach does work in many situations.

Example 12.18. Consider the shifted Catalan number GF $f(x) = \sum_n a_n x^n = (1 - \sqrt{1-4x})/2$ with minimal polynomial $P(x, y) = y^2 - y + x$. Proposition 2.34 implies that $f(x)$ is the main diagonal of the bivariate rational function $F(y, z) = y(1 - 2y)/(1 - x - y)$. The singular variety $\mathcal{V}$ of F is globally smooth and the critical point equations yield the single solution $(1/2, 1/2)$ which is strictly minimal. The first term in the expansion (9.4) vanishes, as expected, while the second term does not vanish and implies $a_n \sim 4^{n-1}/\sqrt{\pi n^3}$. ◁

Remark 12.19. For bivariate algebraic singularities it is also possible to determine asymptotics through a transfer theorem [Gre18].

Example 12.20. As shown in Example 2.37, the generating function for the Narayana numbers a_{rs} is the subseries of

$$F(w, x, y) = \frac{w(1 - 2w - wx(1 - y))}{1 - w - xy - wx(1 - y)}$$

consisting of terms whose powers of w and x are equal. Since F can be written

$$F(w, x, y) = \frac{(1 - w)^{-1}}{1 - xy - \frac{wx}{1-w}},$$

its power series expansion is combinatorial. To determine asymptotics of $a_{\alpha n, \beta n}$ as $n \to \infty$ with $\alpha > \beta > 0$, we compute asymptotics of the power series coefficients of F in the direction (α, α, β). The critical point equations yield the unique solution

$$(w, x, y) = \left(\frac{\beta}{\alpha}, \frac{(\alpha - \beta)^2}{\alpha\beta}, \frac{\beta^2}{(\alpha - \beta)^2}\right),$$

which can be shown to be minimal as F is combinatorial. An asymptotic expansion can thus be computed using Theorem 9.5 in Chapter 9. In particular, the exponential growth of $a_{\alpha n, \beta n}$ is $\left(\frac{\alpha^\alpha}{\beta^\beta(\alpha-\beta)^{\alpha-\beta}}\right)^2$, with the maximum exponential growth of 4 achieved when $\alpha/\beta = 2$. ◁

One positive feature of the embedding approach is that there are many possible embeddings of a given algebraic generating function, so if a rational function obtained by one embedding procedure is too difficult to analyze, it may be possible to find one better suited to ACSV.

Example 12.21. The combinatorial univariate algebraic generating function $\sqrt{1-x} - 1 = \sum_n a_n x^n$ is the main diagonal of

$$\frac{2\,(Yx-1)(Y+1)Y}{Y^2x + 2\,Yx - Y + x - 2}.$$

It is easily checked that among affine points of the singular variety, there are no non-smooth points and no critical points for the main diagonal direction, so asymptotics are determined by a critical point at infinity. The generating function $2xY/(2-x-Y)$ has the same main diagonal but is much easier to analyze, giving $a_n \sim (\pi n)^{-1/2}$, as we expect from Newton's binomial theorem and Stirling's approximation. ◁

See Section 13.5 for further discussion on ACSV and algebraic generating functions.

12.7 Additional worked examples

We now run through several other examples.

Example 12.22 (A constant coefficient bivariate recurrence)**.** Consider the constant coefficient linear recurrence

$$f(m,n) = f(m-1,n-1) + f(m-1,n-2) + 2f(m-2,n-1) \qquad (m,n \geq 2)$$

with boundary conditions $f(m,n) = 0$ if $m < 0$ or $n < 0$ and $f(m,n) = 1$ when $0 \leq m \leq 1$ or $0 \leq n \leq 1$. Grau Ribas [Gra] asked for the limit of $f(n+1,n+1)/f(n,n)$ which, based on numerical computations such as $f(100,100)/f(99,99) = 2.70265\ldots$, appears to exist and be close to e.

Introducing the generating function $F(x,y) = \sum_{m,n} f(m,n)x^m y^n$, the defining recurrence of f implies

$$F(x,y) = \frac{1}{Q(x,y)}$$

with $Q(x,y) = 1 - xy - xy^2 - 2x^2y$. Gröbner basis computations verify that $\mathcal{V}$ is smooth, and the critical points in the main diagonal direction are the points $(x,2x)$ where x is a root of $8x^3 + 2x^2 - 1$. If $w = 0.85\ldots$ denotes the unique real root of $8x^3 + 2x^2 - 1$ then Lemma 6.41 implies that $(w,2w)$ is minimal, and it is the only critical point on the torus $T(w,2w)$. The limit

$$\frac{f(n+1,n+1)}{f(n,n)} \to \frac{1}{2w^2}$$

is algebraic, and therefore not equal to e. Reducing $R = 1/(2w^2)$ modulo the minimal polynomial for w implies R is the unique positive root of the polynomial $8 - R + 2R^2 - R^3$. ◁

Example 12.23 (Horizontally Convex Polyominoes)**.** The generating function

$$F(x, y) = \sum_{r,s \geq 0} a_{rs} x^r y^s = \frac{xy(1-x)^3}{(1-x)^4 - xy(1 - x - x^2 + x^3 + x^2 y)} \tag{12.16}$$

enumerates ***horizontally convex polyominoes*** (HCPs) [Pól69; Odl95; Wil06; Sta97] by total size r and number of rows s. Letting Q denote the denominator of F in (12.16), we know from Example 8.4 that $\mathcal{V}$ is smooth except at the point $(1, 0)$. By Proposition 6.38, there is a part of the graph of Q in the first quadrant consisting of minimal points, which are the points shown in Figure 12.4 with $x \leq 1$.

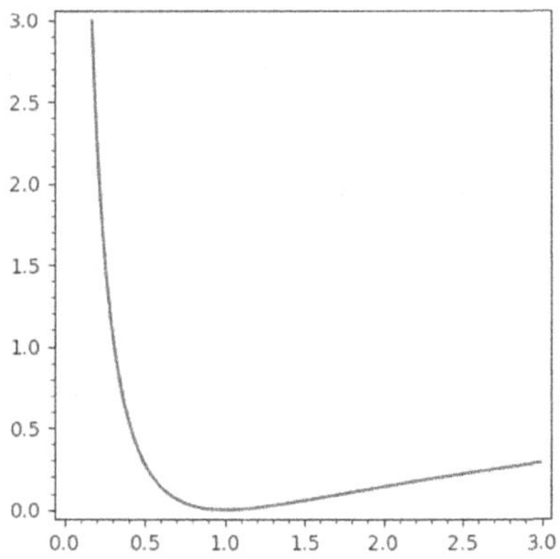

Figure 12.4 Minimal points of $\mathcal{V}$ in the positive real quadrant.

The only combinatorially interesting directions occur in $\Xi = \{(\hat{r}, \hat{s}) : 0 < \hat{s} < 1/2\}$, because an HCP cannot have more rows than its size. As $\hat{\mathbf{r}}$ varies over Ξ from the horizontal to the diagonal, the unique contributing point in the direction $\hat{\mathbf{r}}$ moves along this graph from $(1, 0)$ to $(0, \infty)$. The numerator $P = xy(1-x)^3$ of F in (12.16) is nonvanishing on this component and, using Gröbner bases, we find that the quantity $\underline{Q}$ in the asymptotic expansion (9.7) does not vanish at any of these contributing points.

Since for each direction there are only finitely many critical points, and no others lie on the same torus as the one we have identified, it follows from Theorem 9.12 that asymptotics for a_{rs} are uniform as s/r varies over a compact subset of the interval $(0, 1)$ and have the form $a_{rs} \sim C x^{-r} y^{-s} s^{-1/2}$. Algebraic methods may then be used to determine x, y, and C as explicit functions of $\lambda = s/r$, giving asymptotics for the number of HCPs that are uniform as long as s/r remains in a compact subinterval of $(0, 1)$. For instance, when $\lambda = 1/2$ the smooth critical points (x, y) satisfy $3x^2 + 18x - 5$ and $75y^2 - 288y + 256$, with the contributing singularity occurring at

$$(x_0, y_0) = \left(\sqrt{\frac{32}{3}} - 3, \frac{48 - \sqrt{512}}{25} \right) \approx (0.265986, 1.397442),$$

and a floating point computation gives $a_{n,n/2} \sim (0.2373\ldots)(3.1803\ldots)^n n^{-1/2}$. Note that the exponential growth of $3.1803\ldots$ in this case is only slightly less than the exponential growth $3.2055\ldots$ for all HCPs – a reflection of the fact that the exponential growth varies quadratically around its maximum.

We can also derive limit laws. For instance, let $q(x) = Q(x, 1) = 1-5x+7x^2-4x^3$ and let $a \approx 0.3120$ be the root of q with minimum modulus. Theorem 12.33 below, with the roles of x and y switched, proves that the number of rows k of a uniformly chosen HCP of total size n satisfies $k/n \to m$ in probability, where

$$m = \frac{yQ_y(a, y)}{aQ_x(a, y)} = \frac{5 - 9a + 11a^2}{4(5 - 14a + 12a^2)} = \frac{1}{2.207\ldots}.$$

In other words, the average row size converges to a little over 2.2.

◁

Example 12.24 (Symmetric Eulerian numbers). The symmetric ***Eulerian numbers*** $A(r, s)$ count the number of permutations of the set $[r+s+1] = \{1, 2, \ldots, r+s+1\}$ with precisely r descents, and admit the exponential generating function

$$F(x, y) = \frac{e^x - e^y}{xe^y - ye^x} = \sum_{r,s \geq 0} \frac{A(r, s)}{r!\, s!}, \tag{12.17}$$

see [Com74, page 246] and [GJ04, p. 2.4.21]. To represent the numerator and denominator of F as analytic functions with no common divisor we factor out a term $x-y$ from each, writing $F(x, y) = P(x, y)/Q(x, y)$ for $P = (e^x - e^y)/(x-y)$ and $Q = (xe^y - ye^x)/(x - y)$. We know by its combinatorial definition that the power series coefficients of F are nonnegative, and the power series expansion of Q is aperiodic, so the results of Section 6.4 imply that the minimal points of F have positive coordinates. The graph of Q in this quadrant of $\mathbb{R}^2$ is monotone decreasing with x and asymptotic to both axes; see Figure 12.5.

It can be verified by a direct computation that the singular variety of F is smooth, and the quantity $\underline{Q}$ in the asymptotic expansion (9.7) does not vanish at positive real points – in fact, $\underline{Q}(x, y)$ reaches its minimum value of $e^3/12$ at the point $(1, 1)$. The logarithmic gradient of Q at any point (x, y) in the first

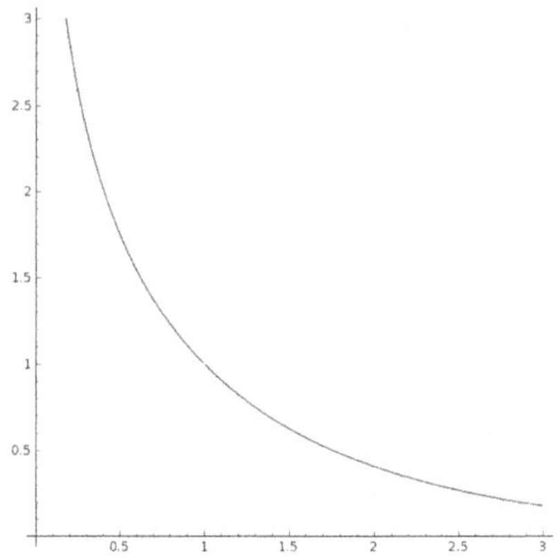

Figure 12.5 The zero set of $(xe^y - ye^x)/(x - y)$ in the first quadrant.

quadrant is a non-zero multiple of $(\alpha, 1 - \alpha)$, where

$$\alpha = \frac{xQ_x(x,y)}{xQ_x(x,y) + yQ_y(x,y)},$$

which simplifies to

$$\alpha = \frac{1-x}{y-x}$$

for points on $\mathcal{V}$, except when $(x, y) = (1, 1)$ and $\alpha = 1/2$. Thus, Theorem 9.5 implies that asymptotics in the direction $(\alpha, 1 - \alpha)$ can be computed by solving the system

$$\alpha = \frac{1-x}{y-x}, \qquad xe^y = ye^x$$

for positive real values (x, y) and using the expansion (9.7). The result, recalling that F is an exponential generating function, is that $A(r, s)$ is asymptotically estimated by

$$A(r,s) \sim C_\alpha (r+s)^{-1/2} \gamma^{r+s} r!s!$$

where $\gamma = x^{-\alpha} y^{\alpha-1}$ and C_α is a messy constant determined by (9.6) and (9.7). ◁

Example 12.25 (Number of successes in a coin-flipping game)**.** Consider a single player game with biased coins, so that heads appears with probability $p = 2/3$ for the first n flips and $p = 1/3$ thereafter. The player is told to get r heads and s tails, and is allowed to choose n. On average, how many choices of $n \leq r + s$ will be winning choices?

Decomposing the probability that n is a winning choice for the player by how many heads are rolled in the first n throws shows that the probability is

$$\sum_{a+b=n} \binom{n}{a}(2/3)^a(1/3)^b\binom{r+s-n}{r-a}(1/3)^{r-a}(2/3)^{s-b}.$$

If a_{rs} is the sum of the winning probability over all n, then the array $\{a_{rs}\}$ is the convolution of the arrays $\binom{r+s}{r}(2/3)^r(1/3)^s$ and $\binom{r+s}{r}(1/3)^r(2/3)^s$, so the generating function $F(x,y) = \sum_{rs} a_{rs}x^ry^s$ is the product

$$F(x,y) = \frac{P(x,y)}{Q(x,y)} = \frac{1}{\left(1 - \frac{1}{3}x - \frac{2}{3}y\right)\left(1 - \frac{2}{3}x - \frac{1}{3}y\right)}.$$

The complete intersection asymptotic results of Chapter 10 imply that $a_{rs} = 3$ plus an error term which is exponentially small as $r, s \to \infty$ provided that $r/(r+s)$ stays in a compact subinterval of $(1/3, 2/3)$. A purely combinatorial analysis of the sum may be carried out to yield the leading term 3, but says nothing about the error terms. ◁

Example 12.26 (Lattice paths constrained to a quadrant)**.** Consider the class of lattice paths on $\mathbb{Z}^2$ that start at the origin and must remain in the nonnegative quadrant $\mathbb{N}^2$ while using the set $\{(1,0),(1,-1),(-1,0),(-1,1)\}$ of allowable steps. The *kernel method* [Mel21, Chapter 4] allows one to prove that the generating function enumerating such walks by number of steps is the main diagonal of

$$F(x,y,t) = \frac{(x+1)(x^{-2} - y^{-1})(x-y)(x+y)}{1 - xyt(x + xy^{-1} + yx^{-1} + x^{-1})}.$$

This is a rational function with smooth singular variety, and minimal critical points $\mathbf{p}_1 = (1,1,1/4)$ and $\mathbf{p}_2 = (-1,1,1/4)$. As the numerator of F has a zero of order 2 at $\mathbf{p}_1$ and a zero of order 3 at $\mathbf{p}_2$, only $\mathbf{p}_1$ contributes to dominant asymptotics. Computation with a computer algebra system shows that the counting sequence for the number of walks on these steps satisfies

$$s_n = \frac{4^n}{n^2} \cdot \frac{8}{\pi} + O\left(\frac{4^n}{n^3}\right).$$

ACSV provides a powerful framework to determine asymptotics of such lattice path models [Mel21, Chapters 6 and 9]. ◁

Example 12.27 (alignments)**.** Fix a positive integer d. The *$(d,\mathbf{n})$-alignments* are $d \times m$ binary matrices for some m, such that no columns are identically zero and the ith row sum is n_i. These structures have relevance to bioinformatics [RT05; Wat95], and the generating function enumerating such alignments by multi-index $\boldsymbol{n}$ is

$$F(\boldsymbol{z}) = \frac{1}{2 - \prod_{i=1}^{d}(1+z_i)} = \sum_{\boldsymbol{n}\in\mathbb{N}^d} a_{\boldsymbol{n}} \boldsymbol{z}^{\boldsymbol{n}}.$$

In the main diagonal direction, the symmetry of F and the combinatorial aperiodic nature of the problem combine with Lemma 6.41 to imply that there is a single, strictly minimal, contributing point $(z, \ldots, z)$ in the positive orthant, where $z = 2^{1/d} - 1$. Theorem 9.5 then yields

$$a_{n,\ldots,n} \sim \frac{2^{(1-d^2)/d}}{(2^{1/d}-1)\sqrt{d\pi^{d-1}}}(2^{1/d}-1)^{-kn} n^{-\frac{d-1}{2}},$$

recovering a result of [Gri+90]. ◁

Example 12.28 (integer solutions to linear equations). Let A be a $d \times m$ matrix of nonnegative integers and, for $\boldsymbol{r} \in \mathbb{N}^d$, let $a_{\boldsymbol{r}}$ denote the number of nonnegative integer solutions to $A\boldsymbol{x} = \boldsymbol{r}$. The generating function for the array $\{a_{\boldsymbol{r}}\}$ is given by

$$F(\boldsymbol{z}) = \sum_{\boldsymbol{r}\in\mathbb{N}^d} a_{\boldsymbol{r}} \boldsymbol{z}^{\boldsymbol{r}} = \prod_{j=1}^{m} \frac{1}{1 - \boldsymbol{z}^{Ae_j^T}},$$

where e_j is the jth elementary vector. This enumeration problem is discussed at length in [DS03] (see also [Sta97, Section 4.6]) which uses the running example

$$A = \begin{bmatrix} 1 & 0 & 0 & 1 & 1 \\ 0 & 1 & 0 & 1 & 0 \\ 0 & 0 & 1 & 0 & 1 \end{bmatrix}$$

with generating function

$$F(\boldsymbol{z}) = \frac{1}{Q_1 Q_2 Q_3 Q_4 Q_5} = \frac{1}{(1-x)(1-y)(1-z)(1-xy)(1-xz)}.$$

The divisors of F are all binomials of the form $1 - \boldsymbol{z}^{\boldsymbol{\alpha}}$, and their zero sets all intersect at $(1, 1, 1)$. The logarithmic gradients of the divisors at $(1, 1, 1)$ are the columns of A.

Every triple of columns of A except for $(1, 2, 4)$ and $(1, 3, 5)$ forms a linearly independent set. The circuits of the matroid defined by A are therefore these triples and the only quadruple $(2, 3, 4, 5)$ not containing either triple, so the broken circuits are $(1, 2), (1, 3)$, and $(2, 3, 4)$, and the bases containing no broken circuit are $(1, 4, 5), (2, 3, 5), (2, 4, 5)$, and $(3, 4, 5)$. By Theorem 10.33, a basis for rational functions with simple poles on $\mathcal{V}_{Q_1}, \ldots, \mathcal{V}_{Q_5}$ is given by

$$\left\{ \frac{1}{Q_1 Q_4 Q_5}, \frac{1}{Q_2 Q_3 Q_5}, \frac{1}{Q_2 Q_4 Q_5}, \frac{1}{Q_3 Q_4 Q_5} \right\}.$$

To reduce F to the sum of terms whose support is in $\mathsf{BC}(F)$ we use relations expressing Q_i in terms of $\{Q_j : j \in C \setminus \{i\}\}$, where i is the greatest element in a circuit. A little scratch work uncovers these relations for the respective circuits $(1,2,4), (1,3,5)$, and $(2,3,4,5)$, giving

$$Q_4 = Q_1 + Q_2 - Q_1 Q_2 \tag{12.18a}$$

$$Q_5 = Q_1 + Q_3 - Q_1 Q_3 \tag{12.18b}$$

$$Q_5 = -y^{-1} Q_2 + y^{-1} Q_3 + x Q_4 . \tag{12.18c}$$

The first relation (12.18a) divided by QQ_4 yields

$$\frac{1}{Q} = \frac{1}{Q_2 Q_3 Q_4^2 Q_5} + \frac{1}{Q_1 Q_3 Q_4^2 Q_5} - \frac{1}{Q_3 Q_4^2 Q_5} . \tag{12.19}$$

We are finished manipulating the third term of (12.19), as its support $\{3, 4, 5\}$ is in $\mathsf{BC}(F)$. The second term of (12.19), after an application of (12.18b), becomes

$$\frac{1}{Q_3 Q_4^2 Q_5^2} + \frac{1}{Q_1 Q_4^2 Q_5^2} - \frac{1}{Q_4^2 Q_5^2} .$$

The first term of (12.19), after an application of (12.18c), yields

$$\frac{-1/y}{Q_3 Q_4^2 Q_5^2} + \frac{1/y}{Q_2 Q_4^2 Q_5^2} + \frac{x}{Q_2 Q_3 Q_4 Q_5^2}$$

and, using (12.18c) once again on the last of these three terms, replaces that term by

$$\frac{-x/y}{Q_3 Q_4 Q_5^3} + \frac{x/y}{Q_2 Q_4 Q_5^3} + \frac{x^2}{Q_2 Q_3 Q_5^3} .$$

Putting this all together gives the decomposition

$$\begin{aligned} F = & -\frac{1}{Q_3 Q_4^2 Q_5} + \frac{1}{Q_3 Q_4^2 Q_5^2} + \frac{1}{Q_1 Q_4^2 Q_5^2} - \frac{1}{Q_4^2 Q_5^2} \\ & + \frac{-1/y}{Q_3 Q_4^2 Q_5^2} + \frac{1/y}{Q_2 Q_4^2 Q_5^2} \\ & + \frac{-x/y}{Q_3 Q_4 Q_5^3} + \frac{x/y}{Q_2 Q_4 Q_5^3} + \frac{x^2}{Q_2 Q_3 Q_5^3} , \end{aligned}$$

and the techniques of Chapter 10 can now be applied to each term to determine asymptotic behavior. ◁

Example 12.29 (Serial Dictatorship). The trivariate sequence

$$a_{njr} = \begin{cases} \frac{\binom{r-s}{t-s}}{\binom{r}{t-1}} & \text{if } 1 \le s \le t \le r \\ 0 & \text{otherwise} \end{cases}$$

arises in the study of the Serial Dictatorship algorithm for allocating indivisible goods (it gives the probability, under IID uniform strict preferences of r agents over r items, that the sth agent receives its tth most preferred item).

The generating function for the numerators $b_{rst} = \binom{r-s}{t-s}$ is

$$F(x, y, z) = \sum_{r,s,t \ge 0} b_{rst} x^r y^s z^t = \frac{xyz}{(1 - x - xz)(1 - xyz)}.$$

Let $Q_1(x, y, z) = 1 - z - xz$ and $Q_2(x, y, z) = 1 - xyz$. Because Q_1 is independent of y, points on the stratum where Q_1 vanishes but Q_2 is non-zero can only be critical points for directions (r, s, t) with $s = 0$. Similarly, the stratum where Q_1 is non-zero but Q_2 vanishes contains only critical points in the main diagonal direction (which results in a trivial sequence). Thus, all interesting directions correspond to points where both factors vanish, which is the curve parametrized by $(x, 1/(1 - x), (1 - x)/x)$ for $x \ne 0$.

A point on this curve is a contributing point for asymptotics in the direction (r, s, t) if and only if the direction lies in the lognormal cone at the point. This means that $(r, s, t) = \lambda(1, 0, 1 - x) + \mu(1, 1, 1)$ for some $\lambda, \mu \ge 0$. Solving this system gives the unique contributing point

$$\left(\frac{r-t}{r-s}, \frac{r-s}{t-s}, \frac{t-s}{r-t} \right)$$

for any such direction, when $r > t > s > 0$. Each such point is minimal by Corollary 6.39. Theorem 10.38 applies and yields the dominant asymptotic behavior. We obtain the first-order approximation for the numerator

$$b_{rst} \sim \frac{(r-s)^{r-s+\frac{1}{2}}}{\sqrt{2\pi}(r-t)^{r-t+\frac{1}{2}}(t-s)^{t-s+\frac{1}{2}}}.$$

The relative error is less than 2% even for $(r, s, t) = (20, 5, 10)$. Note that the approximation is uniform in the union of the cones spanned by $(1, 1, 1)$ and $(1, 0, 1 - x)$ as x ranges over any compact subset of $(0, 1)$. The cones cover the entire range of interest $r > t > s > 0$. ◁

Example 12.30 (Infinite products: Quivers and Littlewood–Richardson coefficients). Consider the infinite product generating function

$$F(x, y) = \prod_{i=1}^{\infty} \left(1 - x^i - y^i\right)^{-1}$$

arising both in the study of chiral operators in four-dimensional quiver gauge theories [RWZ20] and in the enumeration of *Littlewood–Richardson coefficients* $c^{\lambda}_{\mu\nu}$, since

$$F(x,y) = \sum_{\lambda,\mu,\nu} \left(c^{\lambda}_{\mu\nu}\right)^2 x^{|\mu|} y^{|\nu|}$$

(see Harris and Willenbring [HW14] for a derivation and more on Littlewood–Richardson coefficients). Although F is meromorphic and not rational, its analysis is surprisingly simple. Write

$$F(x,y) = \frac{P(x,y)}{1-x-y},$$

where $P(x,y) = \prod_{i=2}^{\infty}\left(1 - x^i - y^i\right)^{-1}$. It is easy to compute that for a direction (r,s) with positive coordinates the unique critical point on $\mathcal{V}_{1-x-y}$ occurs at $\boldsymbol{w} = \left(\frac{r}{r+s}, \frac{s}{r+s}\right)$. Because the coordinates of $\boldsymbol{w}$ lie in $(0,1)$, the polynomial $1 - x^i - y^i$ does not vanish at $(x,y) = \boldsymbol{w}$ when $i > 1$ and $\boldsymbol{w}$ is a smooth strictly minimal singularity of F where P does not vanish. Theorem 9.5 and Equation (9.7) thus imply that

$$a_{r,s} \sim \frac{P\left(\frac{r}{r+s}, \frac{s}{r+s}\right)}{\sqrt{2\pi}} \frac{(r+s)^{r+s+\frac{1}{2}}}{r^r s^s \sqrt{rs}}.$$

◁

12.8 Limit laws from probability theory

We now show how to use the techniques of ACSV to derive limit laws. Although probabilistic interpretations and limit laws can hold for series with negative coefficients, to simplify our presentation we restrict to the most common case of a series F with nonnegative coefficients. Suppose the array $\{a_{\boldsymbol{r}}\}$ encodes some combinatorial structure, and that the size of an object is given by a map $\gamma(\boldsymbol{r})$ into the integers – most commonly $\gamma(\boldsymbol{r}) = r_d$ or $\gamma(\boldsymbol{r}) = |\boldsymbol{r}|$.

To discuss typical behavior, we define a ***grand measure*** μ on $\mathbb{Z}^d$ as a sum of point mass measures

$$\mu = \sum_{\boldsymbol{r}\in\mathbb{Z}^d} a_{\boldsymbol{r}} \delta_{\boldsymbol{r}},$$

where $\delta_{\boldsymbol{r}}(\boldsymbol{s}) = 1$ if $\boldsymbol{s} = \boldsymbol{r}$ and 0 otherwise, and normalize its slices

$$\mu_k = \frac{\sum_{\gamma(\boldsymbol{r})=k} a_{\boldsymbol{r}} \delta_{\boldsymbol{r}}}{\sum_{\gamma(\boldsymbol{r})=k} a_{\boldsymbol{r}}}$$

on the objects of size k to be probability measures. One can ask for limit laws with varying degrees of subtlety. A *weak law* tells us that μ_k is concentrated on a region of diameter $o(k)$. More precisely, a weak law with limit $\boldsymbol{m}$ holds for the sequence $\{\mu_k\}$ if, for any $\varepsilon > 0$,

$$\mu_k \left\{ \boldsymbol{r} : \left| \frac{\boldsymbol{r}}{k} - \boldsymbol{m} \right| > \varepsilon \right\} \to 0 \tag{12.20}$$

as $k \to \infty$. More delicate is a *central limit theorem*, which states that the distribution of μ_k in a region of diameter $O(k^{1/2})$ around $k\boldsymbol{m}$ has Gaussian behavior:

$$\mu_k \left\{ \boldsymbol{r} : \frac{\boldsymbol{r} - k\boldsymbol{m}}{k^{1/2}} \in A \right\} \to \Phi(A) \tag{12.21}$$

as $k \to \infty$, where A is any "nice" region and Φ is a multivariate normal distribution.

Exercise 12.1. When a central limit theorem holds, what (if anything) can you conclude from (12.21) about the probabilities $\mu_k \{\lfloor k\boldsymbol{m} \rfloor\}$, where the greatest integer function is applied coordinatewise? What about the probabilities

$$\mu_k \left\{ \boldsymbol{r} : \frac{|\boldsymbol{r} - k\boldsymbol{m}|}{k} > c \right\}$$

for fixed $c > 0$ as $k \to \infty$?

Even better is a *local central limit theorem* (LCLT) which estimates μ_k at individual points. A LCLT tells us that $\mu_k(\boldsymbol{r}) \sim \mathfrak{n}(\boldsymbol{r})$, where $\mathfrak{n}$ is the density of a multivariate normal distribution. In the case where the size parameter is r_d, for example, the normal density looks like

$$\mathfrak{n}(\boldsymbol{r}) = (2\pi\, r_d)^{(1-d)/2} \sqrt{\det N} \exp\left[\frac{-1}{2} (\boldsymbol{r}^\circ - r_d \boldsymbol{v})^T N (\boldsymbol{r}^\circ - r_d \boldsymbol{v}) \right], \tag{12.22}$$

where $\boldsymbol{r}^\circ$ denotes $(r_1, \ldots, r_{d-1})$, the vector $\boldsymbol{v}$ is the mean of the limit Gaussian distribution, and the matrix N is the inverse covariance matrix. As we will see, so-called *Ornstein–Zernike* behavior $a_{\boldsymbol{r}} \sim C(\hat{\boldsymbol{r}})|\boldsymbol{r}|^{(1-d)/2} \boldsymbol{z}^{-\boldsymbol{r}}$ leads to Gaussian estimates for individual probabilities if the coordinates of $\boldsymbol{z}$ are nonnegative real numbers. Ornstein–Zernike asymptotics are precisely the conclusion of our asymptotic estimates in the case where there is a single smooth point and the Hessian matrix is nondegenerate. Thus, all we require for a weak law and LCLT is that asymptotics are governed by a single smooth point with nondegenerate Hessian matrix. The remainder of this section is devoted to the statement and proof of a weak law and local central limit theorem, holding under smoothness and nondegeneracy assumptions.

Exercise 12.2. Suppose $a_{\boldsymbol{r}} = 0$ unless $\sum_{j=1}^d r_j$ is even.

(a) How would you expect (12.22) to be altered?
(b) How would you expect this to be reflected in the generating function?

Weak laws

Unless otherwise specified, our results in this section apply to the power series expansion of a generating function $F(\mathbf{z}) = P(\mathbf{z})/Q(\mathbf{z})$ with nonnegative coefficients, where Q is a polynomial containing all variables $z_1, \ldots, z_d$. Confining our statements to power series rather than Laurent series simplifies matters by ensuring finite support of the cross-sectional measures μ_k.

Theorem 12.31 (weak law for diagonal slices). *Let $q(z) = Q(z, \ldots, z)$ and let $p > 0$ be the smallest value such that $q(p) = 0$. If $z = p$ is a strictly minimal simple zero of $q(z)$ and $P(p, \ldots, p) \neq 0$ then the sequence $\{\mu_k\}$ of probability measures defined above satisfies a weak law* (12.20) *with limit $\boldsymbol{m}$, where*

$$\boldsymbol{m} = -\nabla_{\log} Q(p, \ldots, p).$$

Proof The domain of convergence of the power series defined by F corresponds to the component B of the complement of $\texttt{amoeba}(Q)^c$ that contains all points of the form $-(N, \ldots, N)$ for all sufficiently large $N > 0$. In fact, because B is convex the set of points $(x, \ldots, x)$ for $x \in \mathbb{R}$ must intersect ∂B (or else B would be all of $\mathbb{R}^d$). Proposition 6.38 then implies that $q(e^x) = Q(e^x, \ldots, e^x) = 0$ has a real solution, so the polynomial q has positive roots and $p > 0$ is well defined.

Due to nonnegativity of the coefficients, every point of ∂B is a zero of $Q \circ \exp$. Write $\boldsymbol{p} = (p, \ldots, p)$ and define $\log \boldsymbol{p}$ coordinate-wise. If $\nabla Q(\boldsymbol{p}) = \mathbf{0}$ then $q(z)$ would have a zero of order at least two at $z = p$, so our assumptions imply $\nabla Q(\boldsymbol{p}) \neq \mathbf{0}$, hence $\nabla(Q \circ \exp)(\log \boldsymbol{p}) \neq \mathbf{0}$ and the zero set of $Q \circ \exp$ in a real neighborhood of $\log \boldsymbol{p}$ is a smooth hypersurface normal to $\boldsymbol{m}$.

Letting Z denote the real points of $\mathcal{V}$, we claim that the logarithm maps a neighborhood of $\boldsymbol{p}$ in Z into ∂B. Denoting the homeomorphism type of a $(d-1)$-ball with a distinguished interior point (N, x), the intersection I of $\overline{B}$ with a closed ball around $\log \boldsymbol{p}$ is compact and, via projection from the center of a sphere in the interior, has homeomorphism type (N, x). By the implicit function theorem, a neighborhood of $\boldsymbol{p}$ in Z also has type (N, x). Any homeomorphism from one pair of type (N, x) to another covers a neighborhood of the distinguished point. Therefore, the logarithm maps some neighborhood of $\boldsymbol{p}$ in Z into a neighborhood of $\log \boldsymbol{p}$ in ∂B, and we see that Z coincides locally with ∂B.

Thus, for any $\boldsymbol{r}$ not parallel to $\boldsymbol{m}$ the maximum value of $\boldsymbol{r} \cdot \boldsymbol{x}$ over $\boldsymbol{x} \in B$

is strictly greater than $\boldsymbol{r} \cdot \log \boldsymbol{p}$, so $a_{\boldsymbol{r}} = O((p+\varepsilon)^{-|\boldsymbol{r}|})$ for some $\varepsilon > 0$ whose choice is uniform as $\hat{\boldsymbol{r}}$ varies over any neighborhood not containing $\hat{\boldsymbol{m}}$. The generating function $\sum_{k \geq 0} C_k z^k$ of the sequence $C_k = \sum_{\gamma(\boldsymbol{r})=k} a_{\boldsymbol{r}}$ is $F(z, \ldots, z)$, which has a strictly minimal simple pole at $z = p$. Proposition 3.1 then implies that $C_k \sim cp^{-k}$ for some constant c. In particular, $\sum_{\boldsymbol{r} \in \mathcal{R}} a_{\boldsymbol{r}} = o(C_k)$ as $k \to \infty$ where $\mathcal{R}$ consists of all indices whose directions are in a compact set not containing $\hat{\boldsymbol{m}}$, so a weak law holds with limit $\boldsymbol{m}$. □

Example 12.32 (multinomial distribution)**.** Let $c_1, \ldots, c_d > 0$ sum to 1 and let $F(\boldsymbol{z}) = 1/(1 - \mathbf{c} \cdot \boldsymbol{z})$ be the generating function for the multinomial distribution

$$a_{\boldsymbol{r}} = \binom{|\boldsymbol{r}|}{r_1, \ldots, r_d} c_1^{r_1} \cdots c_d^{r_d}$$

with parameters $\mathbf{c}$. The denominator of F is $Q(\boldsymbol{z}) = 1 - \mathbf{c} \cdot \boldsymbol{z}$. It follows that $Q(z, \ldots, z) = 1 - z$ regardless of $\mathbf{c}$ and the hypotheses of Theorem 12.31 are satisfied with $p = 1$. Thus, there is a weak law with limit $\boldsymbol{m} = -\nabla_{\log} Q(p, \ldots, p) = \mathbf{c}$, recovering the well-known weak law for repeated rolls of a die with weights $\mathbf{c}$.

◁

Often combinatorial classes are enumerated by a size parameter that is not the sum $|\boldsymbol{r}|$ of the indices but is just one of the indices, say r_d. A similar weak law holds in this case, except that the probability measures $\{\mu_k\}$ no longer have finite support and an added hypothesis is required. Adding this hypothesis allows us to work with Laurent series, instead of only power series.

Theorem 12.33 (weak law for coordinate slices)**.** *Let $F(\boldsymbol{z}) = P(\boldsymbol{z})/Q(\boldsymbol{z}) = \sum_{\boldsymbol{r}} a_{\boldsymbol{r}} \boldsymbol{z}^{\boldsymbol{r}}$ be a Laurent series with nonnegative coefficients converging on a component B of* amoeba$(Q)^c$*. Suppose there is some $p > 0$ such that $Q(\mathbf{1}, p) = 0$ and $(\mathbf{0}, \log x) \in B$ if and only if $0 < x < p$. Then $C_k = \sum_{r_d=k} a_{\boldsymbol{r}}$ is finite for all k and if $z = p$ is also a strictly minimal simple zero of $q(z) = Q(\mathbf{1}, z)$ and $P(\mathbf{1}, p) \neq 0$ then the sequence $\{\mu_k\}$ of probability measures defined by*

$$\mu_k = \frac{1}{C_k} \sum_{r_d=k} a_{\boldsymbol{r}} \delta_{\boldsymbol{r}}$$

satisfies a weak law with limit

$$\boldsymbol{m} = -\nabla_{\log} Q(\mathbf{1}, p).$$

Proof Arguing as in the proof of Theorem 12.31, we see again that $\exp(\boldsymbol{u}) \in \mathcal{V}$ for every $\boldsymbol{u} \in \partial B$ and that $\nabla(Q \circ \exp)$ is nonvanishing at $(\mathbf{0}, \log p)$. Because $(\mathbf{1}, \log x) \in B$ for $0 < x < p$, we have convergence of the sum

$$F(\mathbf{1}, z_d) = \sum_{r_d=k} a_{\boldsymbol{r}} z_d^{r_d} = \sum_{k \geq 0} C_k z_d^k$$

whenever $0 < z_d < p$, so C_k takes on finite values and its univariate generating function has radius of convergence p. Just as in the proof of Theorem 12.31, our hypothesis that $q(z)$ has a simple strictly minimal zero at $z = p$ implies $C_k \sim cp^{-k}$ for some constant c, and the total weight of μ_k is $o(p^{-k})$ on sets for which $\hat{r}$ is bounded away from $\hat{m}$, since any hyperplane through $(\mathbf{0}, \log p)$ other than the hyperplane normal to the dth coordinate plane intersects the interior of B. □

Example 12.34 (IID sums). Let μ be a probability measure on a finite subset $E \subseteq \mathbb{Z}^{d-1}$. The spacetime generating function F for convolutions of μ is given by

$$F(\boldsymbol{z}) = \sum_{k\geq 0} \sum_{\boldsymbol{r}\in\mathbb{Z}^{d-1}} \mu^{(k)}(\boldsymbol{r})\boldsymbol{z}^{(\boldsymbol{r},k)} = \frac{1}{1 - z_d\phi_\mu(z_1, \dots, z_{d-1})},$$

where ϕ_μ is the $(d-1)$-variable generating function for μ. Then $\mathbf{1}$ is a simple pole of Q and is strictly minimal as long as ϕ is aperiodic. Directly, $\nabla_{\log} Q(\mathbf{1}) = (\boldsymbol{m}, 1)$, where $\boldsymbol{m}$ is the mean vector of μ. Theorem 12.33 then recovers the weak law of large numbers for sums of IID samples from μ. ◁

Remark. More generally, we may allow μ to be any measure on $\mathbb{Z}^{d-1}$ whose moment generating function is finite everywhere. This takes us out of the theory of amoebas of polynomials. However, all the facts that are required concerning logarithmic domains of convergence still hold. Because we have not developed the theory of analytic amoebas, we do not include a statement or proof of this result. The greatest generality for weak laws via this type of argument is achieved by weakening the hypothesis to finiteness of the moment generating function in a neighborhood of the origin.

Exercise 12.3. Find a weak law for the binomial coefficients $a_{rs} = \binom{r+s}{s}$ with generating function $1/(1-x-y)$ under the size $\gamma(r,s) = r + 2s$. Rather than using Stirling's formula, maximize $m^{-1} \log a_{rs}$ over $r + 2s = m$ using that the exponential growth of a_{rs} is $x^{-r}y^{-s}$ where $(x, y) = (r/(r+s), s/(r+s))$ is the critical point in the direction $\boldsymbol{r} = (r, s)$ computed in Example 9.10.

Central limits

We now derive a local central limit theorem for the profile $\{a_{\boldsymbol{r}} : r_d = k\}$ as $k \to \infty$. Similar limit theorems for profiles such as $\{a_{\boldsymbol{r}} : |\boldsymbol{r}| = k\}$ also hold, but the arguments are similar and we find it simplest to stick to the case where the size parameter is the last coordinate, mirroring classical limit theorems for the spacetime generating function of a stochastic process on $\mathbb{Z}^{d-1}$. We do, however, weaken our hypotheses to allow F to be a (non-rational) meromorphic function in a suitable domain.

Before giving a limit theorem we need a lemma describing Ornstein–Zernike behavior in this setting. We use the notation e_j for the jth elementary unit vector and recall the notation $\mathbf{T}_e(\boldsymbol{x})$ for the set of complex vectors $\boldsymbol{z}$ with $(\log|z_1|, \ldots, \log|z_d|) = \boldsymbol{x}$.

Lemma 12.35. *Let $F(\boldsymbol{z}) = \sum_{\boldsymbol{r} \in \mathbb{Z}^{d-1} \times \mathbb{N}} a_{\boldsymbol{r}} \boldsymbol{z}^{\boldsymbol{r}}$ be a d-variate series with logarithmic domain of convergence of convergence $B \subseteq \mathbb{R}^d$. Suppose that $\overline{B}$ intersects the negative e_d-axis in the ray $(-\infty, t]$ for some real number t, that F is meromorphic on a neighborhood of the torus $\mathbf{T}_e(\mathbf{0}, t)$ with $F(\boldsymbol{z}) = P(\boldsymbol{z})/Q(\boldsymbol{z})$ for analytic functions P and Q in a neighborhood of $\boldsymbol{w} = (\mathbf{1}, e^t)$, and that $\boldsymbol{w}$ is a strictly minimal pole of F where $Q_{z_d}(\boldsymbol{w}) \neq 0$. Then the logarithmic pole variety $\log \mathcal{V}$ of F is a smooth complex analytic hypersurface in a neighborhood of $\log \boldsymbol{w} = (\mathbf{0}, t)$.*

Let $\boldsymbol{m}$ denote the vector $(\nabla_{\log} Q)(\mathbf{1}, e^t)$ scaled so that $m_d = 1$, and let g be the function parametrizing $(\boldsymbol{x}^\circ, x_d) \in \log \mathcal{V}$ by $x_d = g(\boldsymbol{x}^\circ)$ near $\log \boldsymbol{w}$. If the Hessian matrix $\mathcal{H}$ for g is nonsingular at the origin then as $\boldsymbol{r}$ varies over a neighborhood of $\boldsymbol{m}$ in $S = \mathbb{R}^{d-1} \times \{1\}$ the point $\boldsymbol{w}(\boldsymbol{r}) \in \mathcal{V}$ near $\boldsymbol{w}$ with $(\nabla_{\log} Q)(\boldsymbol{w}(\boldsymbol{r})) = \boldsymbol{r}$ varies smoothly. In this case there is an Ornstein–Zernike estimate

$$a_{\boldsymbol{r}} \sim \frac{(2\pi |r_d|)^{(1-d)/2}}{\sqrt{\operatorname{sgn}(r_d) \det \mathcal{H}}} \cdot \frac{-\operatorname{sgn}(r_d) P(\boldsymbol{w}(\boldsymbol{r}))}{z_d Q_{z_d}(\boldsymbol{w}(\boldsymbol{r}))} \cdot \exp(-\boldsymbol{r} \cdot \boldsymbol{w}(\boldsymbol{r})) . \tag{12.23}$$

Proof The first conclusion, that $\log \mathcal{V}$ is a smooth complex analytic hypersurface near $\log \boldsymbol{w}$, follows from the implicit function theorem, since $\nabla(Q \circ \exp)$ is nonvanishing at $\log \boldsymbol{w}$ under our assumptions. Proposition 9.44 implies that nonsingularity of the Jacobian of the Gauss map is equivalent to nondegeneracy of the critical point $\boldsymbol{w}$, and to nonvanishing of the Gaussian curvature of $\mathcal{V}$. Thus, the inverse of the map $\boldsymbol{z} \mapsto (\nabla_{\log} Q)(\boldsymbol{z})$ is smooth near $\boldsymbol{w}$ and $\boldsymbol{w}(\boldsymbol{r})$ varies smoothly with $\boldsymbol{r}$. Theorem 9.4 and Remark 9.19 from Chapter 9 then yield (12.23). □

Theorem 12.36 (LCLT). *Let F be a d-variate generating function satisfying the hypotheses of Lemma 12.35 and let $M = \mathcal{H}(\mathbf{0})$ denote the Hessian matrix of g at the origin. If M is nonsingular then there is a constant c such that*

$$e^{tk} a_{\boldsymbol{r}^\circ, k} \sim c \, \mathfrak{n}_k(\boldsymbol{r}^\circ)$$

as $k \to \infty$ with $|\boldsymbol{r}^\circ - k\boldsymbol{m}| = o(k^{2/3})$, where

$$\mathfrak{n}_k(\boldsymbol{r}^\circ) = \frac{(2\pi k)^{(1-d)/2}}{\sqrt{\det M}} \exp\left[-\frac{1}{2k} (\boldsymbol{r}^\circ - k\boldsymbol{m})^T M^{-1} (\boldsymbol{r}^\circ - k\boldsymbol{m}) \right] \tag{12.24}$$

denotes the $(d-1)$*-variate* ***normal density*** *with mean* $k\boldsymbol{m}$ *and covariance* M. *It follows that*

$$\sup_{\boldsymbol{r}:r_d=k} k^{(d-1)/2} \left| e^{tk} a_{\boldsymbol{r}} - c\,\mathfrak{n}(\boldsymbol{r}^\circ) \right| \to 0$$

as $k \to \infty$.

Remark. We can compute the constant c by comparing (12.23) and (12.24), although a LCLT traditionally does not require knowledge of the normalizing constant.

Proof Let $\boldsymbol{x}(\boldsymbol{r}) = \log \boldsymbol{w}(\boldsymbol{r})$. Comparing (12.24) to (12.23) in Lemma 12.35, it is sufficient to prove that the rate function $\beta(\boldsymbol{r}) = -\boldsymbol{r} \cdot \boldsymbol{x}(\boldsymbol{r})$ satisfies

$$\beta(k\boldsymbol{m} + \boldsymbol{y}) = -\frac{1}{2k} \boldsymbol{y}^T M^{-1} \boldsymbol{y} + C_k + o(1)$$

as $k \to \infty$ with $|\boldsymbol{y}| = o(k^{2/3})$. The rate function is homogeneous of degree one, so this is the same as

$$\beta(\boldsymbol{m} + \boldsymbol{y}) = -\frac{1}{2} \boldsymbol{y}^T M^{-1} \boldsymbol{y} + C'_k + o(k^{-1}), \tag{12.25}$$

where we have scaled $\boldsymbol{y}$ by $1/k$ so it is now restricted to be $o(k^{-1/3})$.

The hyperplane with normal $\boldsymbol{r}$ going through $\boldsymbol{x}(\boldsymbol{r})$ is a support hyperplane to B, so $\boldsymbol{x}(\boldsymbol{r})$ is a minimizing point for $-\boldsymbol{r} \cdot \boldsymbol{x}$ on $\overline{B}$. When $\boldsymbol{r} \in S$ we may write $\boldsymbol{r} = (\boldsymbol{r}^\circ, 1)$, and we write $\boldsymbol{x} = (\boldsymbol{x}^\circ, g(\boldsymbol{x}^\circ))$ for points $\boldsymbol{x}$ in a neighborhood of $\boldsymbol{x}(\boldsymbol{m})$ in $\log \mathcal{V} \cap \partial B$. The function g is concave, because locally the logarithmic domain of convergence $\overline{B}$ is described by $\{(\boldsymbol{x}, u) : u \le g(\boldsymbol{x})\}$ and logarithmic domains of convergence are convex. Thus,

$$\beta(\boldsymbol{r}^\circ, 1) = \inf_{\boldsymbol{x}^\circ \in \mathbb{R}^{d-1}} \{-g(\boldsymbol{x}^\circ) - \boldsymbol{r}^\circ \cdot \boldsymbol{x}^\circ\}$$

which is the negative of the convex dual of the convex function $-g$. As discussed in Chapter 6, the convex dual of the quadratic form $\boldsymbol{x} \mapsto \boldsymbol{x}^T A \boldsymbol{x}$ represented by a positive definite matrix A is represented in the dual basis by the inverse matrix $\boldsymbol{r} \mapsto \boldsymbol{r}^T A^{-1} \boldsymbol{r}$. The quadratic Taylor expansion of the convex dual at a point $\boldsymbol{r}$ is determined by the quadratic Taylor expansion of the function (assuming this is nondegenerate) at the point where the minimum occurs.

The minimizing point for $\boldsymbol{r} = \boldsymbol{m}$ is at the origin and the quadratic term in the expansion of $-g$ at the origin is the matrix $-M$ representing the Hessian of $-g$ at the origin. Therefore the Taylor expansion of β about $\boldsymbol{m}$ on S is given by

$$\beta(\boldsymbol{m} + \boldsymbol{y}) = \beta(\boldsymbol{m}) - \frac{1}{2} \boldsymbol{y}^T M^{-1} \boldsymbol{y} + O(|\boldsymbol{y}|^3).$$

The condition $|\boldsymbol{y}| = o(k^{-1/3})$ is exactly what is needed for $O(|\boldsymbol{y}|^3)$ to be $o(k^{-1})$

and taking C'_k to be $\beta(\boldsymbol{m})$ (not depending on k after all) establishes (12.25) and the first conclusion.

Pick ν with $1/2 < \nu < 2/3$. When $|\boldsymbol{r} - k\boldsymbol{m}| \leq k^\nu$, the first conclusion implies that

$$|e^{tk} a_{\boldsymbol{r}} - c\, \mathfrak{n}(\boldsymbol{r})| = o(\mathfrak{n}(\boldsymbol{r})) = o(k^{(1-d)/2}).$$

It remains to establish the second conclusion when $|\boldsymbol{r} - k\boldsymbol{m}| \geq k^\nu$, which we do by showing that both terms being compared are small separately. The term $\mathfrak{n}(\boldsymbol{r})$ is in fact bounded above by $\exp(-ck^{\nu-1/2})$ for some $c > 0$. On the other hand, when $\boldsymbol{r} \in S$ the quantity $\boldsymbol{r} \cdot \boldsymbol{x}(\boldsymbol{r})$ differs from its value at $\boldsymbol{r} = \boldsymbol{m}$ by at least a constant multiple of $|\boldsymbol{r} - \boldsymbol{m}|^2$. In general, when $r_d = k$ the value of $\boldsymbol{r} \cdot \boldsymbol{x}(\boldsymbol{r})$ differs from its value at $\boldsymbol{r} = k\boldsymbol{m}$ by at least a constant multiple of $k^{-1}|\boldsymbol{r} - k\boldsymbol{m}|^2$. When $|\boldsymbol{r} - k\boldsymbol{m}| \geq k^\nu$ this is of order at least $k^{2\nu-1}$, which is a positive power of k. Plugging this into (12.23) shows that $a_{\boldsymbol{r}}$ is also at most $\exp(-ck^{2\nu-1})$, completing the proof. □

Classical LCLT

We end by deriving the classical LCLT for sums of independent lattice random variables whose moment generating functions are everywhere finite using Theorem 12.36. Let μ be an aperiodic probability distribution on $\mathbb{Z}^d$, let $\mu^{(k)}$ denote the k-fold convolution of μ, and let

$$F(\boldsymbol{z}, z_{d+1}) = \sum_{(\boldsymbol{r},k)\in\mathbb{Z}^d\times\mathbb{N}} \mu^{(k)}(\boldsymbol{r})\boldsymbol{z}^{\boldsymbol{r}} z_{d+1}^k$$

denote the spacetime generating function for the random walk with increments distributed as μ. The d-dimensional probability generating function ϕ for μ is defined by

$$\phi(\boldsymbol{z}) = \sum_{\boldsymbol{r}\in\mathbb{Z}^d} \mu(\boldsymbol{r})\boldsymbol{z}^{\boldsymbol{r}}$$

and finiteness of the moment generating function implies ϕ is an entire function. The spacetime generating function F is related to the moment generating function ϕ for the distribution μ via

$$F(\boldsymbol{z}) = \frac{1}{1 - z_{d+1}\,\phi(\boldsymbol{z})}.$$

The singular variety of F is globally defined by $z_{d+1} = 1/\phi(\boldsymbol{z})$ so the logarithmic singular variety is the graph of the function

$$g(\boldsymbol{x}) = -\log\phi(\exp(\boldsymbol{x})).$$

Nonnegativity of series coefficients implies that $(\boldsymbol{x}, t)$ is in the interior of the domain of convergence when $t < g(\boldsymbol{x})$, while no point $(\boldsymbol{x}, g(\boldsymbol{x}))$ lies in the logarithmic domain because it is on $\log \mathcal{V}$. When $\boldsymbol{z}$ is on the torus $\mathbf{T}_e(\boldsymbol{x}, g(\boldsymbol{x}))$, aperiodicity implies that $\phi(\boldsymbol{z}) \neq 0$ unless $\boldsymbol{z}$ is real. Therefore, each point of $\mathcal{V}_{\mathbb{R}} = \mathcal{V} \cap \mathbb{R}^d$ is the only point of $\mathcal{V}$ on its torus, and such a minimal point is strictly minimal.

As μ is a probability measure, $g(\mathbf{0}) = \log(1) = 0$. The chain rule further implies

$$\frac{\partial}{\partial x_j} g(\mathbf{0}) = -\frac{\partial}{\partial x_j}(\log \circ \phi \circ \exp)(\mathbf{0}) = -\frac{e^{\mathbf{0}} \phi_{z_j}(\exp(\mathbf{0}))}{\phi(\exp(\mathbf{0}))}$$
$$= -\phi_{z_j}(\mathbf{1}),$$

and this partial derivative evaluates to $\sum_{\boldsymbol{r} \in \mathbb{Z}^d} r_j \mu(\boldsymbol{r})$, so

$$\boldsymbol{m} = -\nabla(\phi \circ \exp)(\mathbf{0}) = \sum_{\boldsymbol{r} \in \mathbb{Z}^d} \boldsymbol{r} \mu(\boldsymbol{r})$$

is the mean of the distribution μ. Differentiating again, if $i \neq j$ we find that

$$\frac{\partial}{\partial x_i x_j} g(\mathbf{0}) = -\left[\frac{e^{x_i} e^{x_j} \phi_{z_i z_j} \circ \exp}{\phi \circ \exp} - \frac{e^{x_i} e^{x_j} (\phi_{z_i} \circ \exp)(\phi_{z_j} \circ \exp)}{(\phi \circ \exp)^2}\right](\mathbf{0})$$
$$= \phi_{z_i z_j}(\mathbf{1}) - m_i m_j$$

so that the (i, j) entry of the Hessian of g at the origin is indeed the covariance of the i and j coordinates under μ. A similar computation works for $i = j$ and establishes that the Hessian matrix of g at the origin is the covariance matrix for μ. Applying Theorem 12.36, we see that $\mu^{(k)}(\boldsymbol{r})$ is asymptotically equal to $c\, \mathfrak{n}_k(\boldsymbol{r})$. There is no need to compute c because we know $\sum_{\boldsymbol{r}^\circ} \mu^{(k)}(\boldsymbol{r}^\circ) = 1$. Thus $c = 1$ and we recover the classical LCLT.

Theorem 12.37. *If μ is an irreducible aperiodic probability measure on $\mathbb{Z}^n$ with moment generating function ϕ everywhere finite, then*

$$\mu^{(k)}(\boldsymbol{r}) \sim \mathfrak{n}_k(\boldsymbol{r})$$

as $k \to \infty$ with $|\boldsymbol{r} - k\boldsymbol{m}| = o(k^{2/3})$, where $\mathfrak{n}_k$ is defined by (12.24) with $d = n+1$, the vector $\boldsymbol{m}$ equal to the mean of μ, and M equal to the covariance matrix of μ. It follows that

$$\sup_{\boldsymbol{r} \in \mathbb{Z}^d} k^{d/2} |a_{\boldsymbol{r}} - \mathfrak{n}(\boldsymbol{r})| \to 0$$

as $k \to \infty$. □

Example 12.38. Recalling nonnegative Riordan arrays from Section 12.2, observe that setting $x = 1$ gives

$$\mu(v; 1) = \frac{v'(1)}{v(1)}$$

and

$$\sigma^2(v; 1) = \frac{v(1)v''(1) - v'(1)^2 + v(1)v'(1)}{v(1)^2}.$$

Thus, when the hypotheses of Theorem 12.33 are satisfied, a WLLN holds with mean $\boldsymbol{m} = \mu(v; 1)$. Here $\mu(v; 1)$ is simply the mean of the renormalized distribution on the nonnegative integers with probability generating function v, and $\sigma^2(v; 1)$ is the variance of the renormalized distribution. The quadratic form in the exponent of (12.24) in Theorem 12.36 is given by $(s-\mu(v; 1)r)^2/(2k\sigma^2(v; 1))$, meaning a local central limit theorem holds with variance $\sigma^2(v; 1)$. ◁

Exercise 12.4. Large deviation theory is used to provide bounds on $\mu^{(k)}\{|\boldsymbol{r} - k\boldsymbol{m}| > Ck\}$, which are exponentially small in k. What bounds on this kind of event follow from Theorem 12.37?

Notes

The material in Sections 12.2 and 12.3 is largely taken from [PW08, Section 4.3], as is the message passing example in Section 12.4. The idea for Exercise 12.6 comes from [Nob10], and Exercise 12.8 comes from [PW02, Example 3.4]. Exercise 12.9 is suggested by a line of work on rook walks, see for example [KZ11]. The results of Exercise 12.10 are generalized in [Wil15] to diagonal asymptotics of products of combinatorial classes. Elementary derivations of some of the limit theorems presented here are given in Melczer [Mel21, Section 5.3.3].

Riordan arrays have been widely studied. In addition to enumerating a great number of combinatorial classes, Riordan arrays also behave in an interesting way under matrix multiplication (note that the condition $v(0) = 0$ implies $a_{nk} = 0$ for $k < n$, and, by triangularity of the infinite array, that multiplication in the Riordan group is well defined). Surveys of the Riordan group and its combinatorial applications may be found in [Spr94; Sha+91].

There are many multivariate generalizations of the Lagrange Inversion Formula, but we know of none that are useful for our purposes. Proposition 12.8 was given in [Wil05]. The asymptotic behavior of univariate QRWs is derived in several papers, of which [CIR03] is perhaps the most complete. Some of our presentation comes from [BP07], and our second QRW example comes from [Bar+10, Section 4.1].

It is an interesting question to pick how many terms to compute in an asymptotic expansion when the goal is to numerically approximate a fixed coefficient. The books [PK01; Par11] give a good introduction, and applications to integrals arising from coefficient extraction are treated in [DH02].

Additional exercises

Exercise 12.5. (general Lagrange inversion) Use the change of variables described in the proof of Proposition 12.7 and the exact differential $d\left[\frac{\psi(y)}{n}\left(\frac{v(y)}{y}\right)^n\right]$ to prove the more general Lagrange inversion formula (12.13).

Exercise 12.6. Compute dominant asymptotics for $\mu(n,n)$, where $\mu(m,n) = \sum_{k=0}^{n}(-1)^k\binom{n}{k}\binom{2m}{k}$. *Hint:* Replace -1 by z, multiply by $x^m y^n$ and sum over k, m, n to obtain a trivariate generating function.

Exercise 12.7. Using the results in Section 12.2, compute asymptotics for the generalized Dyck paths described in Section 2.3.

Exercise 12.8. Let $F(x,y) = 1/(3 - 3x + x^2 - y)$ be the generating function of a generalized Riordan array $\{a_{rs}\}$. Compute asymptotics for directions (r,s) when $r/s > 1$, then do the same thing when $0 < r/s < 1$. What happens when $r = s$?

Exercise 12.9. Let $a_{\boldsymbol{r}}$ count the number of ways in which a chess rook can move from the origin to $\boldsymbol{r}$ by moves that increase one coordinate and do not decrease any other. The methods of Chapter 2 yield the generating function

$$F(\boldsymbol{z}) = \sum_{\boldsymbol{r}\in\mathbb{N}^d} a_{\boldsymbol{r}}\boldsymbol{z}^{\boldsymbol{r}} = \frac{1}{1 - \sum_{i=1}^{d}\frac{z_i}{1-z_i}} = \frac{\prod_{i=1}^{d}(1-z_i)}{\sum_{j=0}^{d}(-1)^j(j+1)e_j(\boldsymbol{z})},$$

where e_j is the jth elementary symmetric polynomial.

a) Use Theorem 2.32 in Chapter 2 to find the generating function of the main diagonal of F when $d = 2$. What happens when you try this for $d = 3$?
b) Compute the first-order asymptotic approximation to $a_{\boldsymbol{r}}$ for $d = 3$.
c) Use a computer algebra system, or write your own program, to compute $a_{\boldsymbol{r}}$ exactly, for values of d up to 10. Compare with the first-order asymptotic when $\boldsymbol{r} = (100, \ldots, 100)$.

d) Compute the next term in the expansion and determine how much better is the accuracy of the 2-term asymptotic approximation when $d = 3$ and $\mathbf{r} = (100, 100, 100)$.

Exercise 12.10. Derive the bivariate generating function $\sum_{r,s\geq 0} a_{rs}x^r y^s$ for the number of ordered pairs of ordered sequences of integers with parts in a fixed set $A \subset \mathbb{N}$, the first summing to r and the second to s, each having the same number of parts. Compute the asymptotics of the coefficients on the main diagonal. Compare your results and methods with those in [BH12].

Exercise 12.11. Prove that embedding an algebraic generating function A into a rational one R using the method of Lemma 2.36 always makes the numerator of R vanish at the contributing points of R.

13

Challenges and extensions

In this final chapter we look to the future of ACSV, discussing the most important challenges and extensions of current results. Work attacking several of these problems is ongoing. The breadth of behavior exhibited by multivariate generating functions is vast, and new applications arise constantly that require additional techniques.

13.1 Contributing singularities and diagonals

Let $F(\mathbf{z})$ be the generating function of a sequence $(a_{\mathbf{r}})$. Theorems 7.20 and 7.35 represent $a_{\mathbf{r}}$ as an integer sum of saddle point integrals near critical points of F, which can be analyzed to determine asymptotics of $a_{\mathbf{r}}$. Unfortunately, identifying the integer coefficients in this sum seems to be extremely difficult, if not undecidable. Even identifying the contributing singularities of F, which are the critical points of highest height with non-zero coefficients, is currently only possible in general for minimal critical points, in two dimensions, or when F is the product of linear factors.

Being able to identify the contributing singularities of a general rational function would be an important theoretical breakthrough for ACSV. One (topological) approach is to generalize the two-dimensional algorithm discussed in Section 9.3 to higher dimensions. Another (computational) approach is to use software for D-finite functions. Recall from Section 8.4.2 that for any fixed $\mathbf{r} \in \mathbb{Z}^d$ the $\mathbf{r}$-diagonal of a multivariate rational function is D-finite, and the methods of creative telescoping produce a D-finite equation satisfied by the diagonal. Thus, it is possible to study rational diagonals using both ACSV and techniques for D-finite functions. In fact, these methods are complementary: ACSV often determines asymptotics up to *unknown integers that are difficult to determine*, while D-finite techniques often determine asymptotics up to *unknown complex numbers that can be rigorously approximated.* The combination of these two representations was used in Example 11.50 to determine asymptotics for a multivariate generating function with cone singularities, and it is natural to ask how far this combined method can be pushed – both to characterize the behavior of multivariate rational diagonals, and to attack the connection problem for the asymptotics of sequences with D-finite generating functions.

Problem 13.1. Classify the types of rational functions for which this hybrid ACSV and D-finite numeric method applies.

Remark 13.1. The diagonal of any bivariate rational function is algebraic, so asymptotics in any fixed direction can always be decided by computing the minimal polynomial for the diagonal and applying univariate techniques. However, the complexity of the computation to reduce to the one-dimensional case can increase with the size of the integers representing a direction of interest, and does not work for general direction. Thus, it is interesting to study even bivariate rational diagonals using multivariate methods.

Example 13.2. The (a, b)-diagonal of the bivariate generating function

$$F(x, y) = \frac{1}{1 - x - y - xy}$$

for the Delannoy numbers has the representation

$$\begin{aligned} G(x) &= \left[t^0\right] F\left(x^{1/a}/t^b, t^a\right) \\ &= \frac{1}{2\pi i}\int_{\gamma_x} \frac{F\left(x^{1/a}/t^b, t^a\right)}{t}dt \\ &= \frac{1}{2\pi i}\int_{\gamma_x} \frac{t^{b-1}}{t^b - x^{1/a} - t^{a+b} - x^{1/a}t^a}dt\,, \end{aligned}$$

for x sufficiently close to the origin and γ_x a circle around the origin that approaches the origin as $x \to 0$. This integrand has a single pole $t = s(x)$ satisfying $\lim_{x\to 0} s(x) = 0$, which is a pole of order b, so

$$\begin{aligned} G(x) &= \operatorname*{Res}_{t=s(x)} F\left(x^{1/a}/t^b, t^a\right)/t \\ &= \lim_{t\to s(x)} \frac{1}{(b-1)!}\partial_t^{b-1}\left((t - s(x))^b F\left(x^{1/a}/t^b, t^a\right)/t\right). \end{aligned}$$

The product rule gives this limit as an algebraic expression in $s(x)$, which can be combined with the defining algebraic equation for $s(x)$ to give an algebraic equation satisfied by G, however this expression is extremely unwieldy for large a and b, and the complexity of the operations grows with a and b. In contrast, asymptotics of the Delannoy numbers in all diagonal directions was given in Example 9.11 using ACSV. ◁

13.2 Phase transitions

Our asymptotic approximations typically hold uniformly as $\boldsymbol{r} \to \infty$ with $\hat{\boldsymbol{r}}$ staying in certain cones of directions, corresponding to contributing points at which the local geometry of $\mathcal{V}$ does not change. When the local geometry of $\mathcal{V}$ does change, asymptotic behavior is no longer uniform. For instance, consider the situation of Example 9.39 in Chapter 9: asymptotic behavior grows like a constant times $r^{-1/2}$ and is uniform in any direction bounded away from the main diagonal, while asymptotic behavior on the main diagonal grows like a constant times $r^{-1/3}$. Without some kind of result to bridge the gap we cannot, for instance, conclude that

$$\limsup \frac{\log a_{\boldsymbol{r}}}{\log |\boldsymbol{r}|} = -1/3 . \tag{13.1}$$

A similar issue for trivariate functions arises in the analysis of spacetime generating functions for two-dimensional quantum random walks, where the logarithmic Gauss map maps a 2-torus to a simply connected subset Ξ of the plane. Such a map must have entire curves on which it folds over itself, and some points of greater degeneracy where such curves meet or fold on themselves.

There is some work in this area. A combinatorial generating function with the behavior of Example 9.39 was discussed in [Ban+01] under the name *Airy phenomena* (in the rescaled window $s = \lambda r + O(r^{1/3})$, the leading term converges to an Airy function). A start on a general formulation of such asymptotics in dimension $d = 2$ was made by Lladser [Lla03], and (13.1) follows from [Lla03, Corollary 6.12]. Lladser [Lla06] also shows that if there is a change of degree of the amplitude and the phase does not change degree, then we can derive a uniform formula for the coefficients in the expansion.

Problem 13.2. Characterize asymptotic transitions in more than two variables.

13.3 Degenerate phase

Most of our results in previous chapters have relied on reduction to a stationary phase integral for which the phase is quadratically nondegenerate at an isolated critical point, and hence amenable to a Complex Morse Lemma argument, but more complicated situations can arise.

Example 13.3. Recall from Example 10.69 the generating function

$$F(x,y) = \sum_{r,s} a_{rs} x^r y^s = \frac{1}{(1-xy)(1-x/2-y/2)}$$

with main diagonal terms $a_{rr} = \sum_{j=0}^{r} 4^{-j}\binom{2j}{j}$.

The critical point equations for the factor $H_1 = 1 - x/2 - y/2$ in the direction $(1,1)$ have a unique solution $(1,1)$, while every point on the torus $|x| = |y| = 1$ satisfies the critical point equations for the factor $H_2 = 1 - xy$. The point $(1,1)$ is therefore a minimal, but not strictly minimal, critical point which is a double point of the singular variety. In addition to a torus of singularities with the same coordinate-wise modulus, the varieties $\mathcal{V}(H_1)$ and $\mathcal{V}(H_2)$ intersect tangentially at $(1,1)$.

A systematic application of the surgery approach reduces the problem of finding asymptotics of a_{rr} as $r \to \infty$ to finding asymptotics of

$$\frac{1}{2\pi}\int_D A(\theta,t)\exp\left(-\lambda\phi(\theta,t)\right)\,d\mu$$

as $\lambda \to \infty$, where

$$A(\theta,t) = \frac{2}{2-e^{i\theta}}$$
$$\phi(\theta,t) = -\log\left(1 - t\left(1 - \frac{e^{-i\theta}}{2-e^{i\theta}}\right)\right),$$

the domain of integration $D = [-\pi,\pi]\times[0,1] \subset \mathbb{R}^2$, and μ is the Lebesgue measure. Note that $\operatorname{Re}\phi$ is nonnegative on D, with minimum value 0. The stationary points of ϕ on D are $(0,t)$ for $0 \le t \le 1$, and $(\theta,0)$ for $-\pi \le \theta \le \pi$. Not only is the phase not equivalent via a smooth change of variables to $t^2+\theta^2$ (it looks more like $t\theta^2$), the stationary phase set consists of more than a single point, being a 1-dimensional T-shaped subset of the rectangle. ◁

Example 10.69 derived asymptotics for Example 13.3 using an *ad hoc* approach.

Example 13.4. Develop a systematic theory for such degenerate integrals. ◁

Example 13.5. If S denotes the set of $n\times n$ Hermitian matrices with Frobenius norm 1, and E denotes the subset of S containing matrices with repeated eigenvalues, then E has positive codimension in S but has a volume with respect to the Riemannian metric inherited from S. It turns out[1] that this volume is, up

[1] K. Kozhasov (personal communication)

to an easily computed constant, the coefficient of $(z_1z_2z_3z_4)^n$ in the generating function

$$M(z_1, z_2, z_3, z_4) = \frac{\prod_{1 \le i < j \le 4} (z_i - z_j)}{1 - e_1(\mathbf{z})^2 + 4e_2(\mathbf{z})},$$

where e_j is the jth elementary symmetric function in the variables z_1, z_2, z_3, z_4. While symmetry initially makes the problem tractable, the stationary phase set is a union of curves rather than points and, to make things worse, the numerator vanishes to different orders on these curves and their intersections. By computing a D-finite equation satisfied by the main diagonal of M, and using the numeric methods discussed in Section 8.4.2 and Section 13.1 above, it can be shown that the coefficient of interest has asymptotic behavior $Cn^{-5/2}64^n$ as $n \to \infty$ for a constant $C \approx 0.4527$. It would be interesting to derive this result using multivariate techniques. ◁

In the case of real phase, it is possible to compute a degenerate integral as a Laplace integral by determining volumes of level sets. Suppose we wish to compute an integral of the form

$$\int_D \exp(-\lambda\phi(\boldsymbol{x}))A(\boldsymbol{x})\, d\boldsymbol{x},$$

where $D = [0, 1]^d \subset \mathbb{R}^d$ in some dimension $d \ge 1$, the parameter λ is large, and ϕ and A are analytic functions on D. Fubini's Theorem tells us that for a nonnegative measurable function f defined on a measure space (X, μ) we have

$$\int_X f(\boldsymbol{x})\, d\mu(\boldsymbol{x}) = \int_0^\infty \mu\left(\{\boldsymbol{x} : f(\boldsymbol{x}) \ge z\}\right) dz,$$

and the change of variable $z = \exp(-u)$ converts this integral to

$$\int_{-\infty}^\infty e^{-u}\mu\left(\{\boldsymbol{x} : -\log f(\boldsymbol{x}) \le u\}\right) du.$$

Let V_u denote the measure of the *sub-level set* $\mu\left(\{\boldsymbol{x} : -\log f(\boldsymbol{x}) \le u\}\right)$. Letting $f(\boldsymbol{x}) = e^{-\lambda\phi(\boldsymbol{x})}$, we obtain

$$V_u = \int_D \mathbf{1}_{\phi(\boldsymbol{x}) \le u/\lambda} d\mu(\boldsymbol{x})$$

and, for simple enough ϕ, it may be possible to compute V_u explicitly. Proposition 4.7 can then be applied to determine asymptotics.

Exercise 13.1. Compute asymptotics as $\lambda \to \infty$ for

$$\int_0^1 \int_0^1 e^{-\lambda xy^2}\, xy\, dx\, dy.$$

Degeneracies of phase can also, in principle, be handled by resolving singularities to obtain a *normal crossing*, using a change of coordinates mapping the phase function into a monomial, expanding the resulting amplitude function into a power series, and then applying known exact asymptotics to each term. Resolution of singularities, together with the methods of [Var77], implies that the possible asymptotic behaviors for any rational asymptotics fall in a limited set of leading terms.

13.4 Critical points at infinity

Let $T = T(\boldsymbol{x})$, where $\boldsymbol{x}$ is in some component B of the complement of the amoeba of the denominator Q of some rational function. As seen in Chapter 7, non-existence of CPAI in the direction $\hat{\boldsymbol{r}}$ is a sufficient condition for $[T]$ to be representable as the sum of cycles corresponding to attachments at affine critical points in direction $\hat{\boldsymbol{r}}$. However, this non-existence is by no means necessary. For instance,

- The trajectories flowing to a CPAI may not be trajectories of any stratified gradient-like flow.
- A trajectory flowing to a CPAI may be a gradient-like flow, but the particular torus $[T]$ may flow down to an affine critical point and not be pulled down a path leading to this CPAI.
- $[T]$ may not flow down without being pulled to infinity, but there may be a cobordism between T and a cycle lower than any CVAI.
- Even if $[T]$ is pulled to infinity by some gradient-like vector field, the CPAI there may not alter the topology and it might be possible to deform $[T]$ to "come back from infinity."

Problem 13.3. a) Can there be unreachable CPAI, meaning that there is a downward gradient-like field that has no flows reaching a CPAI?

b) If so, how can we compute which CPAI are like this?

c) Do all CPAI alter the topology of $\mathcal{V}_*$? Is there an attachment theory for CPAI, giving a way to compute the topological effect of each CPAI?

d) If there is a local attachment, can we find a cycle α representing it and determine $\int_\alpha \boldsymbol{z}^{-\boldsymbol{r}-\boldsymbol{1}} F(\boldsymbol{z})\, d\boldsymbol{z}$?

Another useful tool would be the capability to rule out the existence of CPAI in some direction without an overly long computer algebra computation. An early conjecture, which turned out not to be true, was that all CPAI must be parallel to some face of the Newton polytope of Q. Some ongoing work proves that this conjecture is true when the polytope is *schön*, a property defined in terms of compactifications via toric varieties [Huh13, Definition 3.6]. The set of directions parallel to a face of the Newton polytope is a small set (it has positive codimension) so it is useful to know when CPAI are restricted to these directions. Generally, we would like to find computable restrictions on the possible directions of CPAI.

13.5 Algebraic GFs

Section 12.6 showed how to study algebraic generating functions by embedding them as diagonals of higher-dimensional rational functions. The simplest embedding method, due to Furstenberg, is easy to apply when it works, but does not apply to any algebraic generating function intersected by one of its algebraic conjugates at the origin. There are several known methods for resolving singularities in such cases, for instance the algorithm of Safonov [Saf00] mentioned in Theorem 2.41 and the less constructive procedure of Denef and Lipshitz [DL87].

Example 13.6. Let $f(x) = x/\sqrt{1-x}$ be an algebraic function with minimal polynomial $P(x,y) = (1-x)y^2 - x^2$. Because $f(x) = x + O(x^2)$ and its algebraic conjugate $-f(x) = -x + O(x^2)$ intersect at the origin, the embedding method of Furstenberg does not apply. Safonov's procedure subtracts some initial terms via the substitution $y = xz+x$, yielding a minimal polynomial $(1-x)(z+1)^2-1$ to which Proposition 2.34 now applies. Converting back to the original variables then gives f as the main diagonal of

$$F(x,y) = \frac{\left(2\,y^3x + 3\,y^2x - 2\,y^2 + 2\,yx - 3\,y + x - 2\right)yx}{y^2x + 2\,yx - y + x - 2}.$$

We remark that F has no affine critical points in the main diagonal direction, so it admits critical points at infinity which determine asymptotic behavior, making the analysis difficult. An alternative embedding, following the method of [DL87], is obtained through the much less obvious substitution $y = x/(1-z)$, expressing $f(x)$ as the main diagonal of

$$G(x,y) = \frac{2xy}{1-x-y}.$$

In contrast to the difficult behavior of F, the function G is combinatorial and has a smooth contributing critical point at $(1/2, 1/2)$, allowing for an easy asymptotic analysis. ◁

Problem 13.4. Give a complete analysis of algebraic generating functions $f(x)$ with quadratic minimal polynomials.

Problem 13.5. Can an algebraic series $f(x)$ with nonnegative coefficients always be embedded as the diagonal of a combinatorial rational function? Find an efficient algorithm that converts coefficient extraction for algebraic functions to coefficient extraction for rational functions in a way that preserves the combinatorial nature of the problem.

Recent work [Gre+22] develops software to analyze a variety of algebraic generating functions, ultimately cataloguing 20 combinatorial examples. An alternative approach being developed [BJP23] integrates algebraic generating functions directly. If f is an algebraic function defined by $P(\mathbf{z}, f) = 0$ and $f(\mathbf{0}) = c$, with $y = c$ a simple zero of $P(\mathbf{0}, y)$, coefficients of the power series expansion of f at the origin are given by

$$a_{\mathbf{r}} = \frac{1}{(2\pi i)^d} \int_T f(\mathbf{z})\mathbf{z}^{-\mathbf{r}-\mathbf{1}}\, d\mathbf{z}, \tag{13.2}$$

where T is a torus about the origin, sufficiently small so the polydisk with the same radii contains no singularities of f. Because $(\mathbf{0}, c)$ is a simple zero of P, there is a neighborhood $\mathcal{N}$ of $(\mathbf{0}, c)$ in $\mathbb{C}^{d+1}$ such that projection π onto the first d coordinates of the hypersurface $\mathcal{V}_P$ in $\mathbb{C}^{d+1}$ restricted to this neighborhood is a bi-analytic map to a neighborhood of the origin in $\mathbb{C}^d$. Choosing T smaller if necessary so as to be contained in this neighborhood of the origin, the set $C = \pi^{-1}(T)$ is a small torus in $\mathcal{V}_P$ and (13.2) becomes

$$a_{\mathbf{r}} = \frac{1}{(2\pi i)^d} \int_C y\mathbf{z}^{-\mathbf{r}-\mathbf{1}}\, d\mathbf{z}. \tag{13.3}$$

Aside from the high negative powers of $z_1, \ldots, z_d$, the integrand y has no denominator, however the coefficients $a_{\mathbf{r}}$ may be recovered the same way as one recovers coefficients of rational functions. In the absence of critical points at infinity, the d-dimensional complex variety $\mathcal{V}_* = \mathcal{V}_P \cap \mathbb{C}_*^d$ has a Morse theoretic decomposition into cycles attaining their maximum height near critical points of the height function $h_{(\hat{\mathbf{r}},0)}(\mathbf{z}) = \sum_{j=1}^d r_j \log |z_j|$. One then resolves the chain C in this basis. For instance, if the surface $\mathcal{V}_P$ is smooth and there are no CPAI for P in the direction $(\hat{\mathbf{r}}, 0)$ then the asymptotics of $a_{\mathbf{r}}$ are given by some linear combination of $\Phi_{\mathbf{w}}$ from (9.4) over critical points $\mathbf{w}$ of P in the direction $(\hat{\mathbf{r}}, 0)$.

Example 13.7. Let $f(x) = \frac{1-\sqrt{1-4x}}{2x}$ be the Catalan generating function, with minimal polynomial $P(x, y) = xy^2 - y + 1 = 0$. The point $(0, f(0)) = (0, 1)$ is a smooth point of $\mathcal{V}_P$, and a small circle about $(0, 1)$ in $\mathcal{V}_*$ projects by π to a small circle about the origin in $\mathbb{C}^1$. The smooth surface $\mathcal{V}_P$ has precisely one critical point $\boldsymbol{p} = (1/4, 2)$ in the direction $(1, 0)$, defined by the simultaneous vanishing of P and P_y. There is a critical point at infinity because the value of P_y goes to zero as $x \to 0$, however the height function goes to infinity as $x \to 0$ so Morse theory tells us that the initial curve C can be deformed in $\mathcal{V}_*$ to a smooth contour γ in P passing through $\boldsymbol{p}$ so that the minimum of $\log|\boldsymbol{x}|$ on γ occurs at $\boldsymbol{p}$. A standard stationary phase integral for $\int_\gamma x^{-n-1} y dx$ leads to an asymptotic series for the Catalan numbers. Because the amplitude y is stationary at $\boldsymbol{p}$, there will be one more negative power of n than the usual $n^{-1/2}$ obtained from a univariate saddle point integral. This recovers without computation the fact that the nth Catalan number is $\Theta\left(n^{-3/2} 4^n\right)$. ◁

Exercise 13.6 below explores what happens when $(0, c)$ is not a simple pole of P. We conclude with an exercise illustrating a multivariate algebraic function.

Exercise 13.2. Let

$$f(x, y) = \frac{1 + x(y-1) - \sqrt{1 - 2x(y+1) + x^2(y-1)^2}}{2}$$

be the Narayana bivariate generating function from Example 2.37, defined by the minimal polynomial

$$P(x, y, w) = w^2 - w\left[1 + x(y-1)\right] + xy$$

and the fact that $f(0, 0) = 0$.

(a) Show that P is smooth.
(b) Find $c = f(0, 0)$ and determine whether $(0, 0, c)$ is a simple zero of P.
(c) Find the critical points for P in the direction $(2, 1, 0)$.
(d) Among the critical points, which have finite height?
(e) Are there critical points at infinity in the direction $(2, 1, 0)$?
(f) What asymptotics for $a_{2n,n}$ do you get from integrating $\int_\gamma w x^{-2n-1} y^{-n-1}\, dxdy$ over an appropriate contour γ?

13.6 Asymptotic formulae

In Section 9.4.2 we presented a geometric interpretation, in terms of Gaussian curvature, of the first term in our basic smooth point formula.

Problem 13.6. Give a coordinate-free formula for the next term in the basic smooth point asymptotic expansion. Give similar formulae for arrangement points.

Theorem 10.38 and Corollary 10.41 in Chapter 10 imply that the asymptotic contribution of a minimal arrangement point is unchanged when the factors in the denominator are replaced by their first-order terms.

Problem 13.7. Let $Q(\mathbf{z})$ be any polynomial in d variables and let $\mathbf{p}$ be a zero of Q such that the homogeneous part $H(\mathbf{z})$ of Q at $\mathbf{p}$ (in the sense of Definition 6.46) factors into $k < d$ linearly independent factors. Under what conditions is the dominant asymptotic contribution of $\mathbf{p}$ to the series coefficients of $1/Q(\mathbf{z})$ the same as the dominant asymptotic contribution of $\mathbf{p}$ to the series coefficients of $1/H(\mathbf{z})$?

An approach to Problem 13.7 is provided by a series of results in [BP11]. Lemma 2.24 of that paper shows that in the interior of the normal cone at $\mathbf{p}$, the function $1/Q(\mathbf{z})$ can be expanded in negative powers of $H(\mathbf{z})$, while Lemma 6.3 there shows that the Cauchy integral for the leading negative power is the inverse Fourier transform. These results are stated for points with specific types of local factorizations, but in fact appear to be much more general. As summarized in Chapter 11 of this text, the construction of a conical contour, and the error estimates that follow, rely only on the direction $\hat{\mathbf{r}}$ being non-obstructed. In fact the types of local divisors allowed are Lorentzian quadratics and smooth divisors, which as a degenerate case (having no quadratics) include arrangement points. Thus, solving Problem 13.7 for a large class of functions should be possible by an application of the results in [BP11].

Some caution is indicated due to the asymptotics for two smooth tangential divisors, worked out in a special case in Proposition 10.68, and the fact that asymptotics for the square of a single smooth divisor are a special case but do not capture the results of Proposition 10.68 in general. The difficulty here may be traced back to the fact that the expansion in [BP11, Lemma 2.24] only holds in the interior of a cone where the homogeneous part does not vanish; because two tangential curves cannot be separated by a cone, the expansion does not hold near where Q vanishes.

13.7 Symmetric functions

Multivariate generating functions often possess some degree of symmetry. For example, the denominators in the Delannoy generating function, the cube grove generating function, the Friedrichs–Lewy–Szegő generating function, and the Gillis–Reznick–Zeilberger generating functions are all symmetric polynomials. The denominator in the Aztec Diamond generating function is symmetric in two of its variables.

A symmetric function Q always has critical points in the main diagonal direction, since $\nabla_{\log} Q(\mathbf{z}) \parallel \mathbf{1}$ whenever $\mathbf{z} = (w, \ldots, w)$ and w is a root of the univariate polynomial $Q(z, \ldots, z)$. When Q is symmetric and *multi-affine*, meaning Q has degree 1 in each variable, then there must be a minimal critical point.

Theorem 13.8 ([Bar+18, Lemma 15]). *Let Q be a multi-affine elementary symmetric function and let δ^Q denote the univariate diagonalization $\delta^Q(z) := Q(z, \ldots, z)$. If w is a root of δ^Q of minimal modulus then $(w, \ldots, w)$ is a minimal point for Q in the main diagonal direction.*

Proof Denote the roots of δ^Q by $\alpha_1, \ldots, \alpha_k$, where $|\alpha_1|$ is a root of least modulus, and let

$$M(\mathbf{z}) = \prod_{j=1}^{k} (z_j - \alpha_j).$$

For any $\varepsilon > 0$, the polynomial M has no zeros in the polydisk $\mathcal{D}$ centered at the origin whose radii are all $|\alpha_1| - \varepsilon$. For any d-variable polynomial P, denote its *symmetrization* by

$$P_*(\mathbf{z}) := \frac{1}{d!} \sum_{\pi \in S_d} P\left(z_{\pi(1)}, \ldots, z_{\pi(d)}\right).$$

Then $M_* = Q$ and the Borcea–Brändén symmetrization lemma [BB09, Theorem 2.1] implies that Q has no zeros in the polydisk $\mathcal{D}$. Because $\varepsilon > 0$ was arbitrary, we conclude that $(\alpha_1, \ldots, \alpha_1)$ is a minimal point of Q. □

Exercise 13.3. In d variables, the jth *elementary symmetric function* is the polynomial

$$e_j(\mathbf{z}) = \sum_{S \in \mathcal{E}_j} \prod_{i \in S} z_i,$$

where $\mathcal{E}_j$ consists of all subsets of $\{1, \ldots, d\}$ with j elements. Use Theorem 13.8 to find minimal points for the following denominators without resorting to geometric arguments.

(a) In 2 variables, $Q(x, y) = 1 - e_1(x, y) - e_2(x, y)$ (the Delannoy generating function).
(b) In 3 variables, $Q(x, y, z) = 3 - e_1(x, y, z) - e_2(x, y, z) + 3e_3(x, y, z)$ (the cube grove generating function).
(c) In 4 variables, $Q(\mathbf{z}) = 1 - e_1(\mathbf{z}) + 27e_4(\mathbf{z})$ (the GRZ generating function with critical parameter).

Problem 13.8. Find a way analogous to Theorem 13.8 to conclude minimality in some off-diagonal direction for some class of multi-affine polynomials. Replace the multi-affine hypothesis in Theorem 13.8 by something weaker so that the conclusion still holds.

Naively applying Gröbner basis methods typically breaks symmetry, but recent research in computer algebra has given effective methods for solving polynomial systems with symmetric polynomials, including critical point systems [HL16; Fau+23].

Problem 13.9. Incorporate software for symmetric polynomial solving into packages for ACSV computations.

Exercise 13.4. In four variables, let $G = 1 - e_1 + 27e_4$ be the Gillis–Reznick–Zeilberger denominator, let $K = 1 - e_1 + 2e_3 + 4e_4$ be the Kauers–Zeilberger denominator, let $S = e_3(1 - x, 1 - y, 1 - z, 1 - w)$ be the Szegő denominator, and let $L = e_2(1 - x, 1 - y, 1 - z, 1 - w)$ be the Lewy–Askey denominator.

(a) Express G, K, S, and L as polynomials $P_1, \dots, P_4$ in the elementary symmetric functions $e_1, \dots, e_4$.
(b) Compute a Gröbner basis for $\langle P_1, \dots, P_4 \rangle$ as polynomials in the variables $e_1, \dots, e_d$ and describe the variety $\mathcal{V}_e$ defined by the points $(e_1, \dots, e_d)$ where the P_j simultaneously vanish.
(c) Use this computation to find the elements of $\mathcal{V}(G, K, S, L)$.

13.8 Conclusion

This book aims to illustrate effective methods for computing asymptotic approximations to the coefficients of multivariate generating functions. Such methods have many applications, and have already been used to study problems arising in, among other areas, dynamical systems, bioinformatics, number theory, statistical physics, algebraic statistics, string theory, information theory, and queueing theory. We expect the number of applications to grow steadily. While many (most?) applied problems can be tackled by a smooth point analysis, there are many interesting problems that involve much more complicated local geometry, such as the tiling models discussed in Section 11.4.

From the standpoint of mathematical analysis, many of the tools required to extend the basic ACSV results already exist. Problems for which minimal points control asymptotics usually sidestep complicated topology, and the Morse-theoretic intuition behind our results can often be ignored in such cases by casual users seeking to solve a specific problem. However, substantial topological difficulties can arise when dealing with contributing points that are not necessarily minimal. We believe that to make further progress in this area, substantial work in the Morse-theoretic framework will be required.

This book is certainly not the last word on the subject, but rather an invitation to the reader to join in further development of this research area, which combines beauty, utility, and tractability, and which has given the current authors considerable challenge and enjoyment over many years.

Notes

ACSV was the subject of an AMS-sponsored Mathematical Research Community in 2020–2022, and a 2022 workshop at the American Institute of Mathematics. Among the topics discussed at these events, which still have active research collaborations, are characterizations of CPAI [Gil22], software for ACSV [LMS22], rational embeddings of algebraic functions [Gre+22], and work in progress by Drmota and Pak on multivariate characterizations of so-called $\mathbb{N}$-algebraic functions. The methodology for algebraic functions given at the end of Section 13.5 is contemporaneous to this edition and appears in the preprint [BJP23], along with a new formula for coefficient asymptotics in terms of the defining algebraic function.

Proving minimality by conventional means in Exercise 13.3 (c) above is quite challenging; it is the basis of Problem B5 on the 2020 Putnam examination. The approach of Exercise 13.4 was suggested by Brendan Rhoades, and Example 13.2 is adapted from [Sta99, Section 6.3].

Additional exercises

Exercise 13.5. (binomial transition) Consider the binomial coefficient generating function $(1 - x - y)^{-1}$, and compute first-order asymptotics for the coefficient a_{rs}, where $s/r \to 0$ as $r, s \to \infty$. How many different cases are there in the analysis?

Exercise 13.6. Let $P(x, y) = (1 - x)y^2 - x^2$ as in Example 13.6.

(a) Show that for sufficiently small $\varepsilon > 0$ there are two liftings by π^{-1} of the centered circle of radius ε in $\mathbb{C}_*$ to a contour in $\mathcal{V}_*$ and that one of them describes the positive square root $y = x/\sqrt{1-x}$.
(b) Find all affine critical points of $\mathcal{V}_*$ in direction $(1, 0)$.
(c) Find all critical points at infinity of $\mathcal{V}_*$ in direction $(1, 0)$.

(d) Which of these critical points are at finite height?

(e) Which is more of a problem for computation, the double zero of P or the existence of a critical point at infinity?

Exercise 13.7. Find a general formula for $\det \mathbf{\Gamma}_\Psi$ in terms of the partial derivatives of Q when Q vanishes to degree 3 and is locally the product of three transversely intersecting smooth divisors.

Appendix A

Integration on manifolds

Our first two appendices develop the theoretical background necessary to understand what it means to represent the series coefficients $a_{\mathbf{r}}$ of an analytic function $F(\mathbf{z})$ by an integral

$$a_{\mathbf{r}} = \left(\frac{1}{2\pi i}\right)^d \int_C \omega$$

defined up to homologous cycles C and cohomologous forms ω in the domain of holomorphicity $\mathcal{M}$ of ω. To make sense of this statement we define differential forms and their integrals, state the multivariate Cauchy Integral Formula, construct singular homology and cohomology over the domain $\mathcal{M}$, and connect the singular homology of $\mathcal{M}$ to the integration of exact forms over cycles in $\mathcal{M}$.

We begin in this appendix by formally constructing the apparatus to integrate differential forms on real and complex manifolds.

- In **Section A.1** embedded and abstract manifolds are defined, an important example of the latter being projective spaces.
- In **Section A.2** vector fields and differential forms are introduced.
- In **Section A.3** integration of forms over chains is defined, leading to Stokes's Theorem (Theorem A.24).
- In **Section A.4** complex holomorphic forms are defined and Cauchy's integral formula in a polydisk and polyannulus (Propositions A.28 and A.29) are stated.

A.1 Manifolds

The notion of a manifold may already be familiar to some readers, but there are several different formalizations so we start with basic definitions. Except

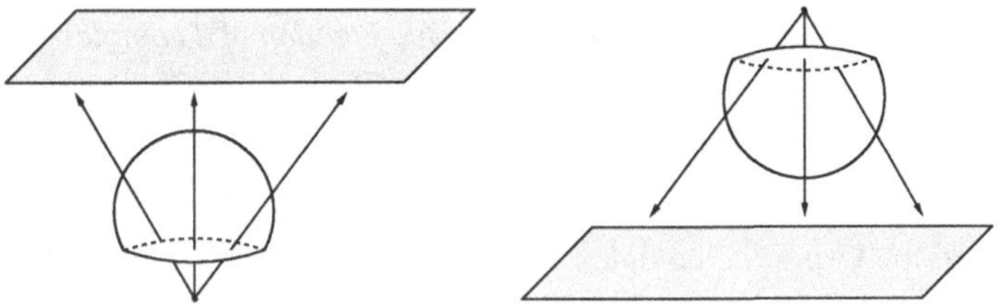

Figure A.1 Covering the sphere with two charts.

for projective spaces, the manifolds most relevant to us will be submanifolds of Euclidean space, and we present this more concrete case first.

A.1.1 Embedded manifolds

An ***embedded real d-manifold*** is a subset $\mathcal{M} \subset \mathbb{R}^n$ such that every point $\boldsymbol{p} \in \mathcal{M}$ has a neighborhood in $\mathbb{R}^n$ whose intersection U with $\mathcal{M}$ is diffeomorphic to the open unit ball in $\mathbb{R}^d$. Here, diffeomorphic means there is a map $\phi : U \to \mathbb{R}^d$ such that the coordinates of ϕ are *smooth* (that is, members of the class C^∞ of infinitely differentiable functions) and the rank of its Jacobian matrix at each point is d (which implies that ϕ has a smooth inverse). If (U, ϕ) is a pair of such a set and diffeomorphism then we call U a ***chart*** and ϕ a ***chart map***.

Example A.1. The implicit function theorem implies that if Q is a smooth function and $\boldsymbol{w}$ is a zero of Q where the gradient of Q does not vanish, then the zero set of Q is locally diffeomorphically parametrized near $\boldsymbol{w}$ by any $(d-1)$ coordinates $x_{k_1}, \ldots, x_{k_{d-1}}$ such that the span of the gradients $\nabla x_{k_1}, \ldots, \nabla x_{k_{d-1}}$ at $\boldsymbol{w}$ does not contain $(\nabla Q)(\boldsymbol{w})$. For instance, if $S^d = \{\boldsymbol{x} \in \mathbb{R}^{d+1} : \sum_{j=1}^d x_j^2 = 1\}$ is the ***unit d-sphere*** in $\mathbb{R}^{d+1}$ then taking $Q(\boldsymbol{x}) = \sum_{j=1}^d x_j^2 - 1$ establishes that $S^d = \{\boldsymbol{x} \in \mathbb{R}^{d+1} : Q(\boldsymbol{x}) = 0\}$ is a manifold. The unit d-sphere can be parametrized with as few as two charts, by projecting the points above and below certain z_d values; see Figure A.1. ◁

The definition of an ***embedded complex d-manifold*** is analogous to that of an embedded real manifold, except we require the stricter condition that the chart maps be *holomorphic* (or, equivalently, analytic) instead of smooth.

Remark A.2. Because $\mathbb{C}^d$ can be identified with $\mathbb{R}^{2d}$ by writing $z_j = x_j + iy_j$ for real coordinates $(x_1, y_1, \ldots, x_d, y_d)$, subsets of $\mathbb{C}^d$ can also be viewed as real manifolds. This identification is crucial to many of our arguments.

The implicit function theorem directly implies the following result.

Proposition A.3. *Suppose Q is a holomorphic function of d complex variables and let $\mathcal{V}_Q$ be the affine holomorphic variety $\mathcal{V}_Q = \{z \in \mathbb{C}^d : Q(z) = 0\}$. If $\nabla Q(z)$ is nonvanishing for every $z \in \mathcal{V}_Q$ then $\mathcal{V}_Q$ is an embedded real $(2d-2)$-manifold.* □

Remarks. (i) Using the complex implicit function theorem, instead of mapping to $\mathbb{R}^{2d}$ and using the real implicit function theorem, shows that $\mathcal{V}_Q$ will be a complex $(d-1)$-manifold. We do not need this extra structure at present. (ii) When Q is the product of distinct irreducible polynomials then this result is sharp: if ∇Q vanishes at some $z \in \mathcal{V}_Q$ then no neighborhood of z in Q is diffeomorphic to $\mathbb{C}^{d-1}$.

Exercise A.1. For which polynomials Q is $\mathcal{V}_Q$ a manifold?

(1) $Q(x, y) = x^2 + y^2 + z^2 - c$, where c is a constant. Does the value of c matter?
(2) $Q(x, y) = y^3 - x^2$
(3) $Q(x, y) = z^2 - (x + y)^2$
(4) $Q(x, y) = 1 - x - y - xy$

Example A.4. The ***unit torus*** in $\mathbb{C}^d$ is the set $\{z \in \mathbb{C}^d : |z_j| = 1 \text{ for all } 1 \le j \le d\}$. Identifying $\mathbb{C}^d$ with $\mathbb{R}^{2d}$, this set is defined by $x_j^2 + y_j^2 = 1$ for all $1 \le j \le d$, and is thus always a real d-manifold. On the other hand, the unit torus is never a complex manifold. When d is odd this follows immediately from the fact that it has the wrong dimension to be a complex manifold (when d is even then the concept of tangent spaces, described below, can be used to prove the unit torus is not a complex manifold). See Exercise A.14 for another example of lack of complex structure. ◁

A.1.2 Abstract manifolds and projective spaces

Instead of working directly with subsets of $\mathbb{R}^d$ or $\mathbb{C}^d$, it is possible to define abstract manifolds whose ground sets can be anything. Although the *Whitney embedding theorem* states that an abstract d-manifold can be embedded in $\mathbb{R}^{2d}$, in some cases it is more natural to consider a definition requiring only a local embedding into some real or complex space. In particular, expanding to abstract manifolds helps us discuss projective spaces, which arise in the formulation of critical points at infinity and the singular integrals of Chapter 11.

Let $\mathcal{M}$ be a second-countable Hausdorff space. A ***chart of dimension*** d on $\mathcal{M}$ is an open subset $U \subset \mathcal{M}$ and a homeomorphism ϕ from U onto an open subset of $\mathbb{R}^d$. We say that the chart (U, ϕ) ***contains*** $\boldsymbol{p}$ if $p \in U$. Two charts (U, ϕ) and (V, ψ) are ***compatible*** if the ***transition maps***

$$\phi \circ \psi^{-1} : \psi(U \cap V) \to \phi(U \cap V) \qquad \text{and} \qquad \psi \circ \phi^{-1} : \phi(U \cap V) \to \psi(U \cap V)$$

are smooth as functions on $\mathbb{R}^d$ (this holds trivially when $U \cap V = \emptyset$), and we call $\mathcal{M}$ a ***real d-manifold*** if there exists a collection of pairwise-compatible charts of dimension d that cover $\mathcal{M}$. Any additional charts that are compatible with the charts defining a manifold must also be compatible with each other, so any collection of charts covering a manifold is contained in a uniquely defined maximal set of compatible charts, known as an ***atlas***. This makes some of the notation easier by allowing us to assume that for any given $\mathbf{p} \in \mathcal{M}$ there is a chart map taking $\mathbf{p}$ to the origin in $\mathbb{R}^d$, since we can always subtract a constant from a chart map to obtain another chart map.

Example A.5 (real projective d-space). Let $\mathcal{M} = \left(\mathbb{R}^{d+1} \setminus \mathbf{0}\right) / (\boldsymbol{x} \mapsto \lambda \boldsymbol{x})$ denote the set of non-zero vectors modulo non-zero scalar multiples, which can be identified with the set of lines through the origin in $\mathbb{R}^{d+1}$. It is conventional to write an element of $\mathcal{M}$ as $(x_0 : x_1 : x_2 : \cdots : x_d)$, with the colons in place of commas indicating that multiplying every coordinate by a non-zero real number gives a different representation for the same element. For each $0 \leq \alpha \leq d$ let $U_\alpha = \{\boldsymbol{x} \in \mathcal{M} : x_\alpha \neq 0\}$, and define chart maps $\phi_\alpha : U_\alpha \to \mathbb{R}^d$ by $\phi_\alpha(\boldsymbol{x}) := (x_j/x_\alpha)_{j \neq \alpha}$. This defines the manifold $\mathbb{RP}^d$ of ***real projective d-space***. ◁

Recall from Section 8.2 that if Q is a polynomial in $x_1, \ldots, x_d$ of total degree m then the homogenization of Q is the polynomial $\overline{Q}(x_0, x_1, \ldots, x_d) = x_0^m Q(x_1/x_0, \ldots, x_d/x_0)$ in $x_0, x_1, \ldots, x_d$. Because every monomial appearing in $\overline{Q}$ has degree m by construction, $\overline{Q}(\lambda \boldsymbol{x}) = \lambda^m \overline{Q}(\boldsymbol{x})$ and the zero set of $\overline{Q}$ in $\mathbb{R}^d$ is closed under scalar multiples. Thus, $\overline{Q}$ defines a ***real projective variety*** $\overline{\mathcal{V}}_Q = \{\boldsymbol{x} \in \mathbb{RP}^{d-1} : \overline{Q}(\boldsymbol{x}) = 0\}$.

Exercise A.2. Let $Q(\boldsymbol{x}) = 1 - x^2 - y^2$. (1) What subset of $\mathbb{R}^3$ is the union of equivalence classes making up the real projective variety $\overline{\mathcal{V}}_Q$? (2) Is the projective variety $\overline{\mathcal{V}}_Q$ a manifold? (3) Intuitively, what simple shape is $\overline{\mathcal{V}}_Q$ equivalent to?

A ***complex d-manifold*** is defined analogously to a real d-manifold, except that the chart maps go to $\mathbb{C}^d$ and the transition maps between charts need to be holomorphic. Analytic functions from $\mathbb{C}^d$ to $\mathbb{C}^d$ are smooth when viewed as maps from $\mathbb{R}^{2d}$ to $\mathbb{R}^{2d}$, so all complex d-manifolds may be viewed as real $(2d)$-manifolds. We also define the ***complex projective d-space*** $\mathbb{CP}^d$ analogously to real projective space and, overloading notation, use $\overline{\mathcal{V}}_Q$ to denote the complex projective variety defined by a polynomial Q. When writing $\overline{\mathcal{V}}_Q$, we make sure it is understood whether we are referring to the real or complex projectivization.

Exercise A.3. Give a description of the complex projective variety $\overline{\mathcal{V}}_Q$ when $Q(\boldsymbol{x}) = 1 - x^2 - y^2$.

A real m-manifold is called an ***orientable manifold*** if it can be covered by charts (U_α, ϕ_α) such that the real transition maps $\phi_\alpha \circ \phi_\beta^{-1} : \phi_\beta(U_\alpha \cap U_\beta) \to \phi_\alpha(U_\alpha \cap U_\beta)$ have Jacobian matrices with positive determinants. Such a cover yields an ***orientation***, which is the set of all chart maps ***consistent*** with the chosen orientation (meaning that all transition maps still have positive Jacobian determinants).

Exercise A.4. Show that the standard 2-sphere $S^2 \subseteq \mathbb{R}^3$ is orientable.

In fact, the notion of orientability is interesting only for real manifolds. The following classical result follows from the results of Section A.4 below.

Proposition A.6. *Every complex manifold is orientable when considered as a real manifold.*

A ***smooth map*** between two real manifolds $\mathcal{M}$ and $\mathcal{M}'$ of dimensions m and n is a function $f : \mathcal{M} \to \mathcal{M}'$ such that for any $\boldsymbol{p} \in \mathcal{M}$ and any chart maps ϕ on a chart of $\mathcal{M}$ containing $\boldsymbol{p}$ and ψ on a chart of $\mathcal{M}'$ containing $f(\boldsymbol{p})$, the function $\psi \circ f \circ \phi^{-1}$ from a subset of $\mathbb{R}^m$ to $\mathbb{R}^n$ is smooth. Holomorphic maps between complex manifolds are defined analogously, requiring $\psi \circ f \circ \phi^{-1}$ to be holomorphic instead of smooth.

In this appendix we construct various objects associated with manifolds: tangent bundles, differential forms, chain complexes, homology, etc. Without introducing large amounts of category theory for its own sake, some categorical terminology is useful because it guides us to what should be true and what arguments are predictable symbol manipulation – so-called *diagram chases* – that can be skipped. In particular, our constructions are ***functorial***. In the ***covariant*** case this means that, in addition to associating an object $\mathcal{G}(\mathcal{M})$ with each manifold $\mathcal{M}$, we associate a map $\mathcal{G}_*(f) : \mathcal{G}(\mathcal{M}) \to \mathcal{G}(\mathcal{M}')$ with every map $f : \mathcal{M} \to \mathcal{M}'$ between manifolds, having the property that $\mathcal{G}_*(g \circ f) = \mathcal{G}_*(g) \circ \mathcal{G}_*(f)$. The map $\mathcal{G}_*(f)$ between the objects $\mathcal{G}(\mathcal{M})$ and $\mathcal{G}(\mathcal{M}')$ is said to be the ***induced map*** associated to f. In the ***contravariant*** case, the induced map $\mathcal{G}^*(f) : \mathcal{G}(\mathcal{M}') \to \mathcal{G}(\mathcal{M})$ goes in the other direction (as indicated by the raised *) and $\mathcal{G}^*(g \circ f) = \mathcal{G}^*(f) \circ \mathcal{G}^*(g)$.

Exercise A.5. Let $\mathcal{G}(\mathcal{M})$ denote the space of smooth real-valued functions on $\mathcal{M}$. Determine whether this should be viewed as a covariant or contravariant functor, and define the map $\mathcal{G}_*$ or $\mathcal{G}^*$ as the case may be.

Given a manifold $\mathcal{M}$, an $\mathbb{R}^k$-***bundle*** E over $\mathcal{M}$ is a map of manifolds $\pi : E \to$

$\mathcal{M}$ such that E is locally a product $\mathcal{M}\times\mathbb{R}^k$. More precisely, this means that there is a collection of open sets $U \subseteq \mathcal{M}$ covering $\mathcal{M}$ and maps $\phi_U : U \times \mathbb{R}^k \to E$ satisfying the following conditions.

1. Each ϕ_U is a diffeomorphism between $U \times \mathbb{R}^k$ and $\pi^{-1}(U)$.
2. $\pi(\phi_U(\mathbf{p}, \mathbf{v})) = \mathbf{p}$ for all U and $\mathbf{p} \in U$.
3. If U and V are two of the open sets covering $\mathcal{M}$ and $\mathbf{p} \in U \cap V$ then there is an invertible linear transformation $g = g_{U,V,\mathbf{p}} : \mathbb{R}^k \to \mathbb{R}^k$ such that $\phi_U(\mathbf{p}, \mathbf{v}) = \phi_V(\mathbf{p}, g(\mathbf{v}))$. In other words, the vector space structure on each fiber $\pi^{-1}(\mathbf{p})$ is well defined.

The point $\mathbf{p}$ is said to be the ***basepoint*** of any element of $\pi^{-1}(\mathbf{p})$. A ***section of a bundle*** E is a smooth map $s : \mathcal{M} \to E$ with $s(\mathbf{p}) \in \pi^{-1}(\mathbf{p})$, which is equivalent to the statement that $\pi \circ s$ is the identity on $\mathcal{M}$.

Example A.7 (tangent bundle of S^2, intuitively). Before we define tangent vectors formally, here is an example to help with intuition. Let $\mathcal{M}$ be the sphere $S^2 \subseteq \mathbb{R}^3$. The *tangent bundle* of $\mathcal{M}$ is the set of vectors tangent to S^2 beginning at basepoints $\mathbf{p} \in S^2$, with the projection operator π taking any such vector to its basepoint. The Hairy Ball Theorem, first proved by Poincaré in 1885 [Poi85], states that this tangent bundle has no section that is everywhere non-zero. Thus, although the tangent bundle is locally diffeomorphic to $S^2\times\mathbb{R}^2$ there is no such global diffeomorphism. ◁

We will see that the tangent bundle is functorial: a map $f : \mathcal{M} \to \mathcal{M}'$ induces a map f_* between tangent bundles. As you might expect, f_* maps tangent vectors over the basepoint $\mathbf{p}$ to those over $f(\mathbf{p})$ and preserves the notion of tangency to a curve. Conversely, we will see that the *cotangent bundle* of a manifold is a contravariant functor: if $f : \mathcal{M} \to \mathcal{M}'$ then the induced map f^* sends differential forms on $\mathcal{M}'$ to differential forms on $\mathcal{M}$. The formal definitions of such objects are full of symbols, but building intuition on key examples helps greatly with understanding why the constructions are necessary and useful. The reader is encouraged to have a favorite running example that they use to illustrate the definitions and constructions as they go through this appendix.

A.2 Vector fields and differential forms

In this section we construct the objects necessary to do calculus on manifolds, including the tools to integrate functions and residues over contours that are piecewise submanifolds of algebraic varieties. This involves the use of tangent and cotangent vectors, and the exterior algebra of differential forms, so that integrals are well defined independent of specific parametrizations.

The constructions in this section formalize the following intuitive notion: integrating $f(x, y, z)\, dx\, dy$ on a surface S involves projecting the surface element onto the xy-direction, multiplying by f, and organizing the sum of the infinitesimal contributions into a two-parameter integral. Describing the surface element without natural coordinates requires the introduction of tangent vectors, while describing the role of $dx\, dy$ in projecting the surface element requires the introduction of cotangent vectors. Together these concepts allow us to define integrals along curves, with the framework of exterior algebra extending this to surface integrals of arbitrary dimension. We know how to integrate in $\mathbb{R}^d$, and within a chart we can map from an abstract manifold to $\mathbb{R}^d$ using a chart map, so we can generalize standard definitions from $\mathbb{R}^d$ provided that using different charts for the same object provides the same answer.

A.2.1 Tangent vectors and vector fields

In an embedded manifold $\mathcal{M} \subset \mathbb{R}^d$ the tangent space of $\mathcal{M}$ at $\boldsymbol{p}$ is naturally defined as the vector space containing vectors in $\mathbb{R}^d$ tangent to $\mathcal{M}$ at $\boldsymbol{p}$. Although thinking about tangent vectors in this manner is concrete, it can be messy to naturally identify tangent vectors under different charts, and it can be difficult to generalize certain concepts to abstract manifolds. For this reason, we reinterpret the tangent space as a collection of (directional) derivatives.

Let $\mathcal{M}$ be a real m-manifold. A ***smooth germ*** in $\mathcal{M}$ at $\boldsymbol{p}$ is an equivalence class of smooth functions defined in neighborhoods of $\boldsymbol{p}$ such that two functions f on U and g on V are equivalent if there is some neighborhood $W \subseteq U \cap V$ of $\boldsymbol{p}$ for which $f|_W = g|_W$. Letting $C^\infty_{\boldsymbol{p}}(\mathcal{M})$ denote the set of smooth germs of real-valued functions on $\mathcal{M}$ at $\boldsymbol{p}$, we define a ***derivation*** on $\mathcal{M}$ at $\boldsymbol{p}$ to be any linear map $D : C^\infty_{\boldsymbol{p}}(\mathcal{M}) \to \mathbb{R}$ that satisfies the *Leibniz property*,

$$D(fg) = (Df)g(\boldsymbol{p}) + f(\boldsymbol{p})(Dg).$$

The ***tangent space*** of $\mathcal{M}$ at $\boldsymbol{p}$, denoted $T_{\boldsymbol{p}}(\mathcal{M})$, is the real vector space of derivations on $\mathcal{M}$ at $\boldsymbol{p}$. By analogy with embedded manifolds, we call each element of $T_{\boldsymbol{p}}(\mathcal{M})$ a ***tangent vector***.

Example A.8. If $\mathcal{M} = \mathbb{R}^m$ then $T_{\boldsymbol{p}}(\mathcal{M}) = T_{\boldsymbol{p}}(\mathbb{R}^m)$ is the vector space of directional derivatives at $\boldsymbol{p}$. Let $x_1, \ldots, x_m$ denote the usual coordinates of $\mathbb{R}^m$ and let $\frac{\partial}{\partial x_1}\Big|_{\boldsymbol{p}}, \ldots, \frac{\partial}{\partial x_m}\Big|_{\boldsymbol{p}}$ be the maps that take a smooth function $g(\boldsymbol{x})$ from $\mathbb{R}^m$

to $\mathbb{R}$ and return its partial derivatives evaluated at $\boldsymbol{x} = \boldsymbol{p}$. Then

$$T_{\boldsymbol{p}}(\mathbb{R}^m) = \left\{ \sum_{k=1}^{m} c_k \left. \frac{\partial}{\partial x_k} \right|_{\boldsymbol{p}} : c_k \in \mathbb{R} \right\}$$

and we can identify $T_{\boldsymbol{p}}(\mathbb{R}^m) = \mathbb{R}^m$ by the mapping

$$\sum_{k=1}^{m} c_k \left. \frac{\partial}{\partial x_k} \right|_{\boldsymbol{p}} \quad \leftrightarrow \quad \mathbf{c} \in \mathbb{R}^m.$$

◁

If $f : \mathcal{M} \to \mathcal{M}'$ is a smooth map of manifolds then the ***differential of f at $\boldsymbol{p}$*** is the map $f_* : T_{\boldsymbol{p}}\mathcal{M} \to T_{f(\boldsymbol{p})}\mathcal{M}'$ that takes the tangent vector $X \in T_{\boldsymbol{p}}\mathcal{M}$ and returns the tangent vector $f_*X \in T_{f(\boldsymbol{p})}\mathcal{M}'$ such that

$$(f_*X)(g) = X(g \circ f) \in \mathbb{R} \quad \text{for all smooth } g : \mathcal{M}' \to \mathbb{R} \text{ at } f(\boldsymbol{p}).$$

Exercise A.6. Prove that f_* is an isomorphism of vector spaces when f is a diffeomorphism.

If (U, ϕ) is a chart of $\mathcal{M}$ containing $\boldsymbol{p}$ then ϕ is a diffeomorphism from U to $\mathbb{R}^m$, so ϕ_* is an isomorphism from $T_{\boldsymbol{p}}\mathcal{M}$ to $T_{\phi(\boldsymbol{p})}\mathbb{R}^m = \mathbb{R}^m$. This proves that the dimension of the tangent space at any point of a manifold is the same as the dimension of the manifold.

If $x_1, \ldots, x_m$ are the usual coordinates of $\mathbb{R}^m$ then, as seen in Example A.8, the maps $\left. \frac{\partial}{\partial x_1} \right|_{\phi(\boldsymbol{p})}, \ldots, \left. \frac{\partial}{\partial x_m} \right|_{\phi(\boldsymbol{p})}$ form a natural basis for $T_{\phi(\boldsymbol{p})}\mathbb{R}^m$. Thus, the tangent vectors $\partial_1, \ldots, \partial_m$ defined by

$$\partial_k = \phi_*^{-1} \left(\left. \frac{\partial}{\partial x_k} \right|_{\phi(\boldsymbol{p})} \right)$$

form a basis for $T_{\boldsymbol{p}}\mathcal{M}$. Because ϕ maps into $\mathbb{R}^m$, we can write $\phi = (\phi_1, \ldots, \phi_m)$ for smooth $\phi_k : U \to \mathbb{R}$. To express the dependence of this basis on $\boldsymbol{p}$ and ϕ, we write $\partial_k = \left. \frac{\partial}{\partial \phi_k} \right|_{\boldsymbol{p}}$ and call $\left. \frac{\partial}{\partial \phi_1} \right|_{\boldsymbol{p}}, \ldots, \left. \frac{\partial}{\partial \phi_m} \right|_{\boldsymbol{p}}$ the ***standard basis for $T_{\boldsymbol{p}}\mathcal{M}$*** with respect to the chart (U, ϕ). Explicitly, for any smooth function $g \in C^\infty_{\boldsymbol{p}}(\mathcal{M})$ we have $g \circ \phi^{-1} : \mathbb{R}^m \to \mathbb{R}$ and

$$\left. \frac{\partial}{\partial \phi_k} \right|_{\boldsymbol{p}} (g) = \left. \frac{\partial}{\partial x_k} \left(g \circ \phi^{-1} \right) \right|_{\boldsymbol{x} = \phi(\boldsymbol{p})} \in \mathbb{R}.$$

To ease notation we sometimes suppress the basepoint $\boldsymbol{p}$ when it is understood, writing the standard basis as $\partial/\partial\phi_k$, and also use the shorthand $\left. \frac{\partial}{\partial \phi_k} \right|_{\boldsymbol{p}} (g) = \frac{\partial g}{\partial \phi_k}(\boldsymbol{p})$.

Example A.9. Let $\mathcal{M} = S^1 = \{(x,y) \in \mathbb{R}^2 : x^2 + y^2 = 1\}$ be the one-dimensional circle in $\mathbb{R}^2$. Suppose U is the half-circle of points in S^1 with $y > 0$ and ϕ is the chart map $\phi(x,y) = x$ with inverse $\phi^{-1}(x) = (x, \sqrt{1-x^2})$. If $(a,b) \in U$ and g is a smooth, real-valued function on a neighborhood of S^1 in $\mathbb{R}^2$, then the chain rule applied to g and ϕ^{-1} at (a,b) implies

$$\begin{aligned}\frac{\partial g}{\partial \phi}(a,b) &= \frac{d}{dx}\left[g\left(x, \sqrt{1-x^2}\right)\right]\Big|_{x=a} \\ &= g_x\left(a, \sqrt{1-a^2}\right) - \frac{a}{\sqrt{1-a^2}} g_y\left(a, \sqrt{1-a^2}\right) \\ &= g_x(a,b) - \frac{a}{b} g_y(a,b).\end{aligned}$$

◁

If $f : \mathcal{M} \to \mathcal{M}'$ is a smooth map of manifolds then the differential of f has an explicit representation using local coordinates. Let (U,ϕ) be a chart on $\mathcal{M}$ containing $\boldsymbol{p}$ and (V,ψ) be a chart on $\mathcal{M}'$ containing $f(\boldsymbol{p})$. Working through the definitions above shows that with respect to the bases $\frac{\partial}{\partial \phi_k}\Big|_{\boldsymbol{p}}$ and $\frac{\partial}{\partial \psi_j}\Big|_{f(\boldsymbol{p})}$ the linear map $f_* : T_{\boldsymbol{p}}\mathcal{M} \to T_{f(\boldsymbol{p})}\mathcal{M}'$ given by the differential is represented by the $m \times n$ ***Jacobian*** matrix

$$J(f) = J_{\boldsymbol{p},\phi,\psi}(f) = \left(\frac{\partial(\psi_j \circ f)}{\partial \phi_i}(\boldsymbol{p})\right)_{i,j}.$$

Example A.10. If $f : \mathbb{R}^m \to \mathbb{R}^n$ and we take standard coordinates $\boldsymbol{x} = (x_1, \ldots, x_m)$ and $\boldsymbol{y} = (y_1, \ldots, y_n)$ for $\mathbb{R}^m$ and $\mathbb{R}^n$ respectively, then we can write $f = (f_1(\boldsymbol{x}), \ldots, f_n(\boldsymbol{x}))$ for functions $f_k : \mathbb{R}^m \to \mathbb{R}$. The Jacobian matrix has entries $\frac{\partial(y_i \circ f)}{\partial x_j}(\boldsymbol{p}) = \frac{\partial f_i}{\partial x_j}(\boldsymbol{p})$, so our definition of the Jacobian generalizes the definition from a first calculus course. ◁

Proposition A.11 (tangent vectors map functorially). *If $f : \mathcal{M} \to \mathcal{M}'$ and $g : \mathcal{M}' \to \mathcal{M}''$ are smooth maps between manifolds then $(g \circ f)_* = g_* \circ f_*$.*

Exercise A.7. Prove Proposition A.11 using the definitions above.

Having developed the necessary machinery for the tangent space at a point, we now define the ***tangent bundle*** of a d-manifold $\mathcal{M}$ whose ground set is the disjoint union of the tangent spaces at each point,

$$T\mathcal{M} = \bigcup_{\boldsymbol{p} \in \mathcal{M}} T_{\boldsymbol{p}}\mathcal{M},$$

and whose projection map $\pi : T\mathcal{M} \to \mathcal{M}$ sends all of $T_{\boldsymbol{p}}(\mathcal{M})$ to $\boldsymbol{p}$. For any open $U \subset \mathcal{M}$ we write $TU = \bigcup_{p \in U} T_p\mathcal{M}$ and, if (U, ϕ) is a chart on $\mathcal{M}$, we define $f_{\phi,U} : TU \to \mathbb{R}^{2d}$ by

$$f_{\phi,U}(v) = \big((\phi \circ \pi)(v), \phi_*(v)\big) \in \phi(U) \times T_{\phi(\pi(v))}\mathbb{R}^d \subset \mathbb{R}^{2d}.$$

It can be shown that as (U, ϕ) ranges over an atlas of $\mathcal{M}$ then the collection of $(TU, f_{\phi,U})$ gives an atlas for $T\mathcal{M}$ as a smooth manifold (the main consideration is proving compatibility of chart maps, which follows straightforwardly from the definition of the differential using coordinates; see [Tu11, Ch. 12] for a full derivation). If (U, ϕ) is a chart on $\mathcal{M}$ then under this structure the map $v \mapsto (\pi(v), \phi_*(v))$ is a diffeomorphism from TU to $U \times \mathbb{R}^d$. The maps ϕ_* are vector space isomorphisms, so that $T\mathcal{M}$ is a bundle.

Example A.12. Suppose $\mathcal{M} = \mathbb{R}^d$, with the canonical global chart $U = \mathbb{R}^d$ and chart map $\phi(x) = x$. Unsurprisingly, if we let v_p denote the tangent vector at p that takes a smooth map g and returns the directional derivative $\frac{d}{dt}\big|_{t=0} g(p + tv)$ then $f_{\phi,U}(v_p) = (p, v)$. Thus, $T\mathbb{R}^d$ has a global product structure and (in this trivial case) there is a canonical isomorphism between two different tangent spaces $T_p(\mathbb{R}^d)$ and $T_{p'}(\mathbb{R}^d)$. ◃

Example A.13. Let $\mathcal{M} = S^2$. If $T\mathcal{M}$ had a global product structure, i.e., if there existed an isomorphism $f : \mathcal{M} \times \mathbb{R}^2 \to T\mathcal{M}$, then the map $s(p) = f(p, \mathbf{e}_1)$ would define a smooth global non-zero section of $T\mathcal{M}$, contradicting the Hairy Ball Theorem (see Example A.7). Thus, the tangent bundle of the sphere S^2 does not admit a global product structure. ◃

A ***smooth vector field*** on $\mathcal{M}$ is a section of $T\mathcal{M}$. In other words, a vector field is a smooth map X that takes $p \in \mathcal{M}$ and returns a tangent vector $X_p \in T_p\mathcal{M}$. If (U, ϕ) is a chart on $\mathcal{M}$ then for any $p \in U$ the tangent vectors $\frac{\partial}{\partial \phi_1}\big|_p, \ldots, \frac{\partial}{\partial \phi_m}\big|_p$ form a basis for $T_p\mathcal{M}$. Thus, for any map $X : \mathcal{M} \to T\mathcal{M}$ with $X_p \in T_p\mathcal{M}$ we can write

$$X_p = \sum_{k=1}^{m} c_k(p) \left.\frac{\partial}{\partial \phi_k}\right|_p \qquad \text{for all } p \in U,$$

which defines coefficient maps $c_k : U \to \mathbb{R}$. The mapping X is a vector field (i.e., is smooth) if and only if the coefficient maps c_k constructed this way are all smooth as (U, ϕ) ranges over the charts of $\mathcal{M}$.

A.2.2 Cotangent vectors and 1-forms

Having defined what we need for tangent vectors, we now define cotangent vectors by dualizing everything. It can be difficult to build intuition as to why calculus on manifolds requires the extra complexity introduced by dualizing. Roughly speaking, when there is no canonical parametrization for a manifold then the infinitesimal elements such as dx, dA, and dV that arise in integration must be defined in a way that is independent of parametrization. Cotangent vectors act functorially on tangent vectors, which in turn act functorially on smooth functions, to produce numerical results invariant under changes of parametrization. We recommend the reader work through the examples in the section on integration below, when they come to it, to help understand how to do computations.

Definition A.14 (Cotangent vectors). Let $\mathcal{M}$ be an m-manifold. The ***cotangent space*** to $\mathcal{M}$ at $\boldsymbol{p}$ is the *dual space* $T^{\boldsymbol{p}}\mathcal{M}$ to $T_{\boldsymbol{p}}\mathcal{M}$. In other words, $T^{\boldsymbol{p}}\mathcal{M}$ is the space of linear maps from $T_{\boldsymbol{p}}\mathcal{M}$ to $\mathbb{R}$. If (U, ϕ) is a chart of $\mathcal{M}$ containing $\boldsymbol{p}$ then the ***standard basis for*** $T^{\boldsymbol{p}}\mathcal{M}$ with respect to (U, ϕ) is the basis $\{d\phi_1|_{\boldsymbol{p}}, \ldots, d\phi_m|_{\boldsymbol{p}}\}$ dual to the standard basis for $T_{\boldsymbol{p}}\mathcal{M}$ with respect to (U, ϕ). In other words, the standard basis for the dual is defined by setting

$$(d\phi_i|_{\boldsymbol{p}})\left(\left.\frac{\partial}{\partial\phi_j}\right|_{\boldsymbol{p}}\right) = \begin{cases} 1 & \text{if } i = j \\ 0 & \text{otherwise} \end{cases}$$

and extending linearly. As we did for tangent vectors, we often drop the basepoint $\boldsymbol{p}$ from our notation when it is understood.

Analogously to our construction of the tangent bundle of an m-manifold $\mathcal{M}$, we define the ***cotangent bundle*** $T^*\mathcal{M}$ whose ground set is the disjoint union of the cotangent spaces at each point,

$$T^*\mathcal{M} = \bigcup_{\boldsymbol{p}\in\mathcal{M}} T^{\boldsymbol{p}}\mathcal{M},$$

and whose projection map $\pi : T\mathcal{M} \to \mathcal{M}$ sends all of $T^{\boldsymbol{p}}(\mathcal{M})$ to $\boldsymbol{p}$. If (U, ϕ) is a chart on $\mathcal{M}$ then any cotangent vector $\boldsymbol{v}$ with $\pi(\boldsymbol{v}) \in U$ is an $\mathbb{R}$-linear combination $\boldsymbol{v} = \sum_k c_k(\boldsymbol{v}) d\phi_k|_{\pi(\boldsymbol{v})}$. Analogously to the tangent bundle $T\mathcal{M}$, a smooth structure can be put on $T^*\mathcal{M}$ using maps of the form

$$\boldsymbol{v} \mapsto \big((\phi\circ\pi)(\boldsymbol{v}), c_1(\boldsymbol{v}), \ldots, c_m(\boldsymbol{v})\big) \in \phi(U)\times\mathbb{R}^m \subset \mathbb{R}^{2m},$$

where $\boldsymbol{v}$ ranges over $\boldsymbol{v} \in T^*\mathcal{M}$ with $\pi(\boldsymbol{v}) \in U$. Under this structure the map $\boldsymbol{v} \mapsto (\pi(\boldsymbol{v}), c_1(\boldsymbol{v}), \ldots, c_m(\boldsymbol{v}))$ is a local diffeomorphism when restricted to the bundle T^*U over chart neighborhoods. Changing charts preserves the vector space structure, so $T^*\mathcal{M}$ is a bundle.

We can use the differential on tangent spaces to define a dual map on cotangent spaces. Let $\mathcal{M}$ and $\mathcal{M}'$ be manifolds with $\boldsymbol{p} \in \mathcal{M}$, let $f : \mathcal{M} \to \mathcal{M}'$ be a smooth map, and let $\xi \in T^{f(\boldsymbol{p})}\mathcal{M}'$ be a cotangent vector. Then ξ maps tangent vectors in $T_{f(\mathbf{p})}\mathcal{M}'$ to $\mathbb{R}$, so the composition $f^*(\xi) = \xi \circ f_*$ of ξ with the differential of f is a map from $T_{\mathbf{p}}\mathcal{M}$ to $\mathbb{R}$, which is a cotangent vector in $T^{\mathbf{p}}\mathcal{M}$. In other words, we have constructed a contravariant map $f^* : T^{f(\mathbf{p})}\mathcal{M}' \to T^{\mathbf{p}}\mathcal{M}$, known as the ***pullback*** by f.

Exercise A.8. Let $f : \mathcal{M} \to \mathcal{M}'$ and $g : \mathcal{M}' \to \mathcal{M}''$ be smooth maps. Prove that $(g \circ f)^* = f^* \circ g^*$.

The pullback can be computed explicitly in local coordinates. Suppose (U, ϕ) is a chart of $\mathcal{M}$ containing $\mathbf{p}$ and let (V, ψ) be a chart on $\mathcal{M}'$ containing $f(\mathbf{p})$. We have seen above that the differential of f_* with respect to the standard bases $\{\partial/\partial\phi_i\}$ and $\{\partial/\partial\psi_j\}$ of $T_{\mathbf{p}}\mathcal{M}$ and $T_{f(\mathbf{p})}\mathcal{M}'$ in these charts is given by left-multiplication of the Jacobian matrix $J(f)$. Plugging in the definition of f^*,

$$\begin{aligned} f^*(d\psi_i)\left(\frac{\partial}{\partial\phi_j}\right) &= (d\psi_i)\left(f_*\left(\frac{\partial}{\partial\phi_j}\right)\right) \\ &= (d\psi_i)\left(\sum_k \frac{\partial(\psi_k \circ f)}{\partial\phi_j}(\mathbf{p})\frac{\partial}{\partial\psi_k}\right) \\ &= \frac{\partial(\psi_i \circ f)}{\partial\phi_j}(\mathbf{p}). \end{aligned}$$

In other words, with respect to the dual bases $\{d\phi_j\}$ and $\{d\psi_i\}$ the linear map f^* is defined through right-multiplication by the Jacobian matrix.

If $\mathcal{M}$ is a manifold then a ***differential 1-form*** on $\mathcal{M}$ is a section on $T^*\mathcal{M}$. In other words, a 1-form is a smooth map ω that takes $\mathbf{p} \in \mathcal{M}$ and returns a cotangent vector $\omega|_{\mathbf{p}} \in T^{\mathbf{p}}\mathcal{M}$. If $f : \mathcal{M} \to \mathbb{R}$ is a smooth function on $\mathcal{M}$ then the ***differential of*** f is the 1-form $df : \mathcal{M} \to T^*\mathcal{M}$ defined by

$$(df|_{\mathbf{p}})(X_{\mathbf{p}}) = X_{\mathbf{p}}(f) \quad \text{for all } \mathbf{p} \in \mathcal{M} \text{ and } X_{\mathbf{p}} \in T_{\mathbf{p}}\mathcal{M}.$$

Remark A.15. When x is the standard coordinate on $\mathbb{R}$ and we take $\mathbf{p} \in \mathcal{M}$ and $X_{\mathbf{p}} \in T_{\mathbf{p}}\mathcal{M}$ then

$$f_*(X_{\mathbf{p}}) = (df|_{\mathbf{p}})(X_{\mathbf{p}}) \left.\frac{d}{dx}\right|_{f(\mathbf{p})},$$

where $f_* : T_{\mathbf{p}}\mathcal{M} \to T_{f(\mathbf{p})}\mathbb{R}$ is the differential of f defined previously. Thus, both of our definitions of the differential agree under the identification $T_{f(\mathbf{p})}\mathbb{R} = \mathbb{R}$ given by the standard basis of the tangent space (see Example A.12).

If (U, ϕ) is a chart on $\mathcal{M}$ then for any $\mathbf{p} \in U$ the cotangent vectors $d\phi_1|_{\mathbf{p}}, \ldots, d\phi_m|_{\mathbf{p}}$ form a basis for $T^{\mathbf{p}}\mathcal{M}$. Thus, for any map $\omega : \mathcal{M} \to T^*\mathcal{M}$ with

$\omega|_p \in T^{\mathbf{p}}\mathcal{M}$ we can write

$$\omega|_p = \sum_{k=1}^{m} c_k(p)\, d\phi_k|_p \qquad \text{for all } p \in U,$$

which defines coefficient maps $c_k : U \to \mathbb{R}$. The mapping ω is a 1-form (i.e., is smooth) if and only if the coefficient maps constructed in this way are all smooth as (U, ϕ) ranges over the charts of $\mathcal{M}$.

A.2.3 Differential k-forms

The concept of a differential form will allow us to integrate on manifolds, however 1-forms will only allow us to integrate on one-dimensional manifolds (i.e., curves). To integrate on higher-dimensional surfaces, we need to generalize 1-forms to higher-dimensional objects. The necessary algebra can be developed for any vector space V, although we use it only in the case when $V = T^{\mathbf{p}}\mathcal{M}$ is the cotangent space of a manifold.

Let V be a d-dimensional real vector space. A ***k-form*** on V is a multi-linear map $\alpha : V^k \to \mathbb{R}$ which is *antisymmetric*: swapping any two arguments of α negates the value of the map. Note that linearity and antisymmetry imply that the only k-form when $k > d$ is the zero map. The vector space of k-forms over V is denoted $\Lambda_k(V)$, and the direct sum $\Lambda(V) = \bigoplus_{k=1}^{d} \Lambda_k(V)$ is called the ***exterior algebra*** of V. To give $\Lambda(V)$ a ring structure, we define the ***wedge product*** of k 1-forms $\alpha_1, \ldots, \alpha_k \in \Lambda^1(V)$ to be the k-form $\alpha_1 \wedge \cdots \wedge \alpha_k : V^k \to \mathbb{R}$ determined by

$$(\alpha_1 \wedge \cdots \wedge \alpha_k)(v_1, \ldots, v_k) = \det\big(\alpha_i(v_j)\big) \;\; \text{for all } \boldsymbol{v} \in V^k.$$

If $\alpha_1, \ldots, \alpha_d$ are 1-forms and $I \in \{1, \ldots, d\}^r$ then we write $\alpha_I = \alpha_{i_1} \wedge \cdots \wedge \alpha_{i_r}$. When $\alpha_1, \ldots, \alpha_d$ form a basis for $\Lambda^1(V) = V$ then the k-forms α_I, where $I = (i_1, \ldots, i_k)$ with $i_1 < i_2 < \cdots < i_k$, form a basis for $\Lambda^k(V)$. Thus, the vector space of k-forms has dimension $\binom{d}{k}$ and $\wedge$ defines a bilinear associative product on $\Lambda(V)$ which is *anticommutative*: if α is an a-form and β is a b-form then $\alpha \wedge \beta = (-1)^{ab} \beta \wedge \alpha$.

Example A.16. The space of d-forms has dimension one, and is thus generated by the determinant. ◁

Example A.17. Let x and y be the usual coordinates on $\mathcal{M} = \mathbb{R}^2$ and $V = T^{\mathbf{0}}\mathcal{M}$ be the cotangent space of $\mathbb{R}^2$ at the origin. Since the differentials dx and dy span V, any 1-form on V can be written $a\,dx + b\,dy$ for $a, b \in \mathbb{R}$. The wedge product of two 1-forms is computed by

$$\begin{aligned}(a\,dx + b\,dy) \wedge (p\,dx + q\,dy) &= ap(dx \wedge dx) + aq(dx \wedge dy)\\ &\quad + bp(dy \wedge dx) + bq(dy \wedge dy)\\ &= aq(dx \wedge dy) + bp(dy \wedge dx)\\ &= (aq - bp)(dx \wedge dy),\end{aligned}$$

where $dx \wedge dx = dy \wedge dy = 0$ by anticommutativity. ◁

Bringing our discussion back to differential geometry, let $\mathcal{M}$ be a smooth m-manifold. For any $1 \leq k \leq m$ the ***exterior k-algebra bundle*** of $\mathcal{M}$ whose ground set is the disjoint union

$$\Lambda^k(\mathcal{M}) = \bigcup_{\boldsymbol{p} \in \mathcal{M}} \Lambda^k(T^{\boldsymbol{p}}(\mathcal{M}))$$

is defined by the projection map π that takes $v \in \Lambda^k(T^{\boldsymbol{p}}(\mathcal{M}))$ and returns $\pi(v) = \boldsymbol{p}$. If (U, ϕ) is a chart on $\mathcal{M}$ then for any $\boldsymbol{p} \in U$ the cotangent vectors $d\phi_1|_{\boldsymbol{p}}, \ldots, d\phi_m|_{\boldsymbol{p}}$ form a basis for $T^{\boldsymbol{p}}(\mathcal{M})$. Thus, the k-forms $d\phi_I|_{\boldsymbol{p}} = d\phi_{i_1}|_{\boldsymbol{p}} \wedge \cdots \wedge d\phi_{i_k}|_{\boldsymbol{p}}$ form a basis for $\Lambda^k(T^{\boldsymbol{p}}(\mathcal{M}))$ as I ranges over strictly increasing tuples in $\{1, \ldots, m\}^k$. For any map $\omega : \mathcal{M} \to \Lambda^k(\mathcal{M})$ with $\omega|_{\boldsymbol{p}} \in T^{\boldsymbol{p}}(\mathcal{M})$ this means we can write ω as a sum over $d\Phi_I$ as I ranges over subsets of $[d]$ of cardinality k,

$$\omega|_{\boldsymbol{p}} = \sum_{|I|=k} c_I(\boldsymbol{p})\, d\phi_I|_{\boldsymbol{p}} \quad \text{for all } \boldsymbol{p} \in U\,.$$

As for the tangent and cotangent bundles, we can give $\Lambda^k(\mathcal{M})$ a smooth structure using charts (U, ϕ) and the coefficient functions c_I. A ***differential k-form*** on $\mathcal{M}$ is a section of $\Lambda^k(\mathcal{M})$ under this structure, and a mapping from $\mathcal{M}$ to $\Lambda^k(\mathcal{M})$ is a differential k-form if and only if the coefficient maps c_I are all smooth as (U, ϕ) ranges over the charts of $\mathcal{M}$. The set of differential k-forms is denoted $E^k(\mathcal{M})$ and the union of the $E^k(\mathcal{M})$ over all k is denoted $E^*(\mathcal{M})$. From now on we restrict to this differential setting and use the terms *k-form* and *differential k-form* interchangeably. A ***top level form*** on a manifold of dimension m is any m-form.

Example A.18. Once again, let $\mathcal{M} = S^1$ be the one-dimensional circle in the plane, let U be the halfcircle of points in S^1 with $y > 0$, and let ϕ be the chart map $\phi(x, y) = x$ with inverse $\phi^{-1}(x) = (x, \sqrt{1 - x^2})$. If ω is a 1-form on S^1 then for $(a, b) \in U$ we can write $\omega|_{a,b} = f(a, b) d\phi|_{a,b}$, where $f(x, y)$ is smooth at (a, b). The criterion of smoothness means the real function $(f \circ \phi^{-1})(x) = f(x, \sqrt{1 - x^2})$ is smooth at $x = a$. ◁

Exercise A.9. Let (U, ϕ) be a chart on $\mathcal{M}$ and $f_1, \ldots, f_k : U \to \mathbb{R}$. Show that the wedge product of differentials satisfies

$$(df_1) \wedge \cdots \wedge (df_k) = \sum_{|I|=k} \det J_I(\mathbf{f}, \phi)\, d\phi_I,$$

where $J_I(\mathbf{f}, \phi)$ is the Jacobian matrix $J_I(\mathbf{f}, \phi) = \left(\frac{\partial f_a}{\partial \phi_{i_b}}\right)_{1 \le a,b \le k}$ when $I = (i_1, \ldots, i_k)$ with $i_1 < \cdots < i_k$.

Let $f : \mathcal{M} \to \mathcal{M}'$ be a smooth map of manifolds. If $\boldsymbol{p} \in \mathcal{M}$ then we saw above how to define the pullback $f^* : T^{f(\boldsymbol{p})}\mathcal{M}' \to T^{\boldsymbol{p}}\mathcal{M}$ of cotangent vectors from $\mathcal{M}'$ to $\mathcal{M}$. Working pointwise, we can also define a pullback $f^* : E^k(\mathcal{M}') \to E^k(\mathcal{M})$ on forms. Concretely, if ω is a k-form on $\mathcal{M}'$ then the ***pullback of*** ω by f is the k-form $f^*\omega$ on $\mathcal{M}$ defined for each $\boldsymbol{p} \in \mathcal{M}$ by

$$(f^*\omega)|_{\boldsymbol{p}}(X_1, \ldots, X_k) = \omega|_{f(\boldsymbol{p})}(f_*X_1, \ldots, f_*X_k) \quad \text{for all } X_1, \ldots, X_k \in T_{\boldsymbol{p}}\mathcal{M}.$$

We also define the pullback of a smooth function (which can be considered a 0-form) $g : \mathcal{M}' \to \mathbb{R}$ by $f^*g = g \circ f$. Basic properties of the pullback include

- distributivity over addition: $f^*(\omega + \tau) = f^*\omega + f^*\tau$,
- distributivity over wedge product: $f^*(\omega \wedge \tau) = (f^*\omega) \wedge (f^*\tau)$,
- commutativity with the differential: $f^*(dg) = d(f^*g)$.

Example A.19. Let $\mathcal{M} = \mathcal{M}' = \mathbb{R}^d$ with standard coordinates $x_1, \ldots, x_d$ and let $f : \mathbb{R}^d \to \mathbb{R}^d$ be smooth, so $f = (f_1, \ldots, f_d)$ with each $f_j : \mathbb{R}^d \to \mathbb{R}$ smooth. If ω is a top level form on $\mathbb{R}^d$ (i.e., a d-form) then $\omega = a(\boldsymbol{x})\, dx_1 \wedge \cdots \wedge dx_d$ for some smooth $a : \mathbb{R}^d \to \mathbb{R}$. Since $f^*x_k = x_k \circ f = f_k$, the pullback of ω under f is given by the Jacobian determinant of f,

$$\begin{aligned} f^*\omega = f^*(a(\boldsymbol{x})\, dx_1 \wedge \cdots \wedge dx_d) &= a(f(\boldsymbol{x}))\, d(f^*x_1) \wedge \cdots \wedge d(f^*x_d) \\ &= a(f(\boldsymbol{x}))\, (df_1) \wedge \cdots \wedge (df_d) \\ &= a(f(\boldsymbol{x})) \det\left(\frac{\partial f_j}{\partial x_i}\right) (dx_1 \wedge \cdots \wedge dx_d), \end{aligned}$$

where the final equality follows from Exercise A.9. ◁

A.3 Integration of forms

The only thing remaining before defining integration on manifolds is to discuss domains of integration. For each $p \ge 1$, let Δ^p denote the ***standard p-simplex***

in $\mathbb{R}^p$ defined by

$$\Delta^p = \left\{(x_1, \ldots, x_p) \in \mathbb{R}^p : x_i \geq 0 \text{ for all } i \text{ and } \sum_{i=1}^{p} x_i \leq 1\right\}.$$

By convention, if $p = 0$ then Δ^p is defined to be a single point. Associated with the standard simplices are $p+2$ ways of embedding Δ^p as a face of Δ^{p+1}. To be explicit, for each $1 \leq i \leq p+1$ let $\kappa_i^p : \Delta^p \to \Delta^{p+1}$ be the embedding obtained by inserting a zero in the ith position,

$$\kappa_i^p(x_1, \ldots, x_p) = (x_1, \ldots, x_{i-1}, 0, x_{i+1}, \ldots, x_p),$$

and let $\kappa_0^p : \Delta^p \to \Delta^{p+1}$ embed into the diagonal face,

$$\kappa_0^p(x_1, \ldots, x_p) = \left(1 - \sum_{i=1}^{p} x_i\ , x_1, \ldots, x_p\right).$$

We now fix an m-manifold $\mathcal{M}$ in $\mathbb{R}^n$. For each $0 \leq p \leq m$, a ***smooth p-simplex*** (or simply p-simplex) in $\mathcal{M}$ is defined to be a map $\sigma : \Delta^p \to \mathcal{M}$ which extends to a smooth map of a neighborhood of Δ^p in $\mathbb{R}^p$ into $\mathcal{M}$. Define the ***space*** $\mathbb{C}^p(\mathcal{M})$ ***of p-chains*** on $\mathcal{M}$ to be the space of finite formal linear combinations $\sum c_i\sigma_i$ of p-simplices in $\mathcal{M}$ with coefficients $c_i \in \mathbb{C}$. A ***chain*** $\sum c_i\sigma_i$ is said to be ***supported on a set*** $X \subseteq \mathcal{M}$ whenever all σ_i in the sum map into X.

Definition A.20 (boundary operator). For any $1 \leq p \leq m$, the ***boundary operator*** on p-chains is the linear map $\partial : \mathbb{C}^p \to \mathbb{C}^{p-1}$ which takes a p-simplex $\sigma : \Delta^p \to \mathcal{M}$ and returns the $(p-1)$-chain $\partial\sigma : \Delta^{p-1} \to \mathcal{M}$ defined by

$$\partial\sigma := \sum_{i=0}^{p}(-1)^i\, \sigma \circ \kappa_i^{p-1}. \tag{A.3.1}$$

Figure A.2 illustrates the boundary operator on a singular 2-simplex.

A crucial property of the boundary operator is that it gives the zero map when applied twice.

Proposition A.21. *For any manifold $\mathcal{M}$ and $1 \leq p \leq d$ the map $\partial^2 = 0$.*

Exercise A.10. Verify $\partial^2 = 0$ by proving that $\kappa_i^{p+1} \circ \kappa_j^p = \kappa_{j+1}^{p+1} \circ \kappa_i^p$ for all $p \geq 0$ and $i \leq j$ (where κ_a^b is identically zero if a or b are negative or $a > p$).

For a domain $A \subseteq \mathbb{R}^n$ with standard coordinates $x_1, \ldots, x_n$, we define the integral of a top level form $\omega = f(\boldsymbol{x})dx_1 \wedge \cdots \wedge dx_n$ on A to be the usual Riemann–Lebesgue integral

$$\int_A \omega = \int_A f(\boldsymbol{x})\, dx_1 \wedge \cdots \wedge dx_n = \int_A f(\boldsymbol{x})\, dV,$$

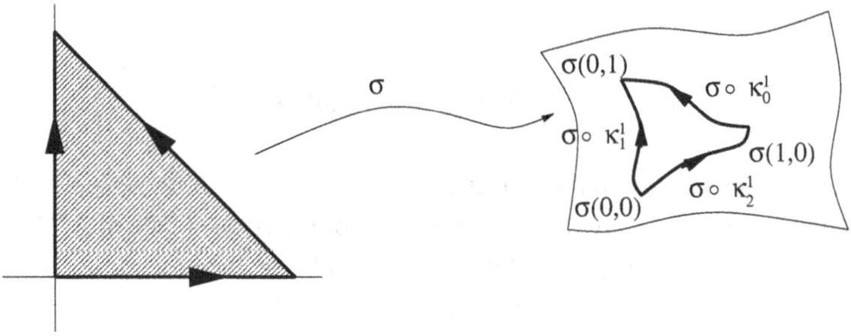

Figure A.2 A 2-simplex and its boundary.

where dV is Lebesgue measure in $\mathbb{R}^n$.

Remark A.22. In order for this definition of integration to be consistent, we must write the wedge product in increasing order. For instance, in two dimensions $dy \wedge dx = -dx \wedge dy$, so

$$\int_A f(x,y)dy \wedge dx = -\int_A f(x,y)dx \wedge dy = -\int_A f(x,y)dxdy$$

even though

$$\int_A f(x,y)dxdy = \int_A f(x,y)dydx$$

as Riemann integrals (since f is smooth).

If $\mathcal{M}$ is an oriented manifold then for any p-simplex $\sigma : \Delta^p \to \mathcal{M}$, integration of a p-form ω over σ is *defined* by

$$\int_\sigma \omega := \int_{\Delta^p} \sigma^*(\omega),$$

where σ^* is the pullback of ω by σ from a form on $\mathcal{M}$ to a form on $\Delta^p \subset \mathbb{R}^p$. It is important to note that we have defined integration not over a *set* (the range of σ) but over a *parametrization* (σ itself). Integration over a chain is defined by linearity,

$$\int_{\sum_{i=1}^r a_i\sigma_i} \omega := \sum_{i=1}^r a_i \int_{\sigma_i} \omega .$$

Example A.23. If $\mathcal{M} = \mathbb{R}^d$ then we can write any d-simplex $\sigma = (\sigma_1, \ldots, \sigma_d)$, where each σ_i is a smooth map from Δ^d to $\mathbb{R}$. Example A.19 implies that for any d-form $\omega = a(\boldsymbol{x})dx_1 \wedge \cdots \wedge dx_d$,

$$\int_\sigma \omega = \int_{\Delta^p} \sigma^*(\omega) = \int_{\Delta^p} a(\sigma(\boldsymbol{x}))(\det J)dx_1 \cdots dx_d ,$$

where J is the Jacobian matrix of $\sigma : \mathbb{R}^d \to \mathbb{R}^d$ and the final integral is the Riemann integral of a smooth function on $\mathbb{R}^d$. This mirrors the usual change of variables formula seen in a first calculus course. ◁

We often want to integrate over manifolds, but our formal definition of integration holds over chains. Integration over a manifold can be defined by dividing the manifold into a finite collection of chains, provided (*i*) how we divide up the manifold does not affect the result, and (*ii*) we can orient the chains in a consistent way so that the integral is uniquely determined up to a sign depending on a global orientation. Recall from above that a manifold $\mathcal{M}$ is called oriented if for any charts (U, ϕ) and (V, ψ) on $\mathcal{M}$ with nonempty intersection $W = U \cap V$ the smooth function $\phi \circ \psi^{-1}$ from $\psi(W) \subset \mathbb{R}^m$ to $\phi(W) \subset \mathbb{R}^m$ has positive Jacobian determinant. We do not attempt to define integration over non-orientable manifolds.

Assume that the p-manifold $\mathcal{M}$ is orientable and fix one of its two orientations. We say that a p-simplex $\sigma : \Delta^p \to \mathcal{M}$ is a ***positively oriented simplex*** with respect to this orientation of $\mathcal{M}$ if for any oriented chart (U, ϕ) of $\mathcal{M}$ the differential $\sigma_* : T_{\mathbf{p}}\mathbb{R}^m \to T_{\sigma(\mathbf{p})}\mathcal{M}$ with respect to the bases $\partial/\partial x_k$ and $\partial/\partial \phi_k$ of $T_{\mathbf{p}}\mathbb{R}^m$ and $T_{\sigma(\mathbf{p})}\mathcal{M}$ has positive determinant. If $\mathcal{M}$ is also compact then a ***triangulation*** of $\mathcal{M}$ is a finite set $\sigma_1, \ldots, \sigma_r$ of positively oriented p-simplices satisfying certain conditions set out at greater length in the definition of a simplicial complex in Section B.1 of Appendix B. These conditions imply that the simplices *divide $\mathcal{M}$ into regions*, meaning that each map is one-to-one on the interior of Δ^m, the images of the interiors are disjoint, and the union of the images is all of $\mathcal{M}$. Triangulation represents $\mathcal{M}$ as a p-chain $\sum_{j=1}^r \sigma_j$, so the integral of a p-form ω over $\mathcal{M}$ is defined by linearity,

$$\int_{\mathcal{M}} \omega := \sum_{j=1}^{r} \int_{\sigma_j} \omega \,. \tag{A.3.2}$$

The integral does not depend on which oriented triangulation we chose [War83, Section 4.8], so (A.3.2) defines the integral of a p-form over an orientable p-manifold (up to a sign that depends on the orientation chosen).

Exercise A.11. Give an explicit oriented triangulation of S^2 and use it to integrate $f(x, y) = xy^2$ on S^2.

Above we defined the differential map, which takes a smooth function on a manifold $\mathcal{M}$ – in other words, a 0-form $f \in E^0(\mathcal{M})$ – and returns a 1-form $df \in E^1(\mathcal{M})$. We now extend this map to the ***differential operator*** (or ***exterior***

derivative operator) $d : E^*(\mathcal{M}) \to E^*(\mathcal{M})$ defined locally in a chart (U, ϕ) by

$$d(f\, d\phi_I) := \sum_{1 \le j \le m} \frac{\partial f}{\partial \phi_j} d\phi_j \wedge d\phi_I .$$

The differential maps a p-form to a $(p+1)$-form, applying d twice gives the zero map (this is usually written $d^2 = 0$), and if $\omega \in E^p(\mathcal{M})$ and $\eta \in E^q(\mathcal{M})$ then

$$d(\omega \wedge \eta) = d(\omega) \wedge \eta + (-1)^p \omega \wedge d(\eta) . \tag{A.3.3}$$

The differential still commutes with pullbacks, meaning $d(\psi^*(\omega)) = \psi^*(d\omega)$ for any smooth map ψ.

Exercise A.12. Prove that (A.3.3) holds.

We introduce the differential to state Stokes's Theorem, our last result on the integration of forms.

Theorem A.24 (Stokes's Theorem). *Let $p \geq 1$ and suppose that ω is a $(p-1)$-form on a manifold $\mathcal{M}$ of dimension at least p. If C is a p-chain on $\mathcal{M}$ then*

$$\int_{\partial C} \omega = \int_C d\omega .$$

Stokes's Theorem follows from our definitions, integration by parts, and some elementary computations; see [War83, Theorem 4.7] for details.

A.4 Complex manifolds and differential forms in $\mathbb{C}^n$

Having reviewed the basics of differential geometry for real manifolds, we now return to the complex case. As noted above, we may identify $\mathbb{C}^n$ with $\mathbb{R}^{2n}$ by mapping $(x_1 + iy_1, \dots, x_n + iy_n) \in \mathbb{C}^n$ to $(x_1, \dots, x_n, y_1, \dots y_n) \in \mathbb{R}^{2n}$ so any complex manifold is a real smooth manifold. However, this elides much of the extra structure of a complex manifold. In particular, we want to understand how complex analytic behavior and the multiplicative structure of the complex numbers can be captured in a setting where complex numbers are represented as real two-vectors. For the remainder of this section, points of $\mathbb{R}^{2n}$ will be referred to by $(x_1, \dots, x_n, y_1, \dots, y_n)$ rather than $(x_1, \dots, x_{2n})$.

For any smooth map u from $\mathbb{C}^n = \mathbb{R}^{2n}$ to $\mathbb{C}$, define

$$\frac{\partial u}{\partial z_j} := \frac{1}{2}\left(\frac{\partial u}{\partial x_j} - i\frac{\partial u}{\partial y_j}\right) \qquad \frac{\partial u}{\partial \overline{z}_j} := \frac{1}{2}\left(\frac{\partial u}{\partial x_j} + i\frac{\partial u}{\partial y_j}\right)$$

$$dz := dx + i\,dy \qquad d\overline{z} := dx - i\,dy$$

$$\partial u := \sum_{j=1}^{d} \frac{\partial u}{\partial z_j}\, dz_j \qquad \overline{\partial} u := \sum_{j=1}^{d} \frac{\partial u}{\partial \overline{z}_j}\, d\overline{z}_j,$$

and $du := \partial u + \overline{\partial} u$. The function u is said to be a ***holomorphic function*** if $\overline{\partial} u = 0$. Holomorphicity of u, under this definition, is equivalent to the vanishing of each $\partial u/\partial \overline{z}_j$, which in turn is equivalent to the Cauchy–Riemann equations holding in each variable. Since the Cauchy–Riemann equations hold precisely when u is complex analytic, holomorphicity is invariant under complex analytic coordinate changes. In fact, the decomposition $du = \partial u + \overline{\partial} u$ is well defined when u is any smooth complex-valued function on a complex manifold: such a manifold is defined by an atlas of chart maps whose transitions are complex analytic and thus preserve holomorphicity. It also follows that for a holomorphic function u, the above formal definition of the derivation $\partial/\partial z_j$ in terms of the tangent vectors $\partial/\partial x_j$ and $\partial/\partial y_j$ agrees with the limit definition

$$\frac{\partial f}{\partial z_j}(\boldsymbol{z}) = \lim_{t\to 0} \frac{u(\boldsymbol{z} + t\mathbf{e}_j) - u(\boldsymbol{z})}{t},$$

where $\boldsymbol{e}_j$ is the elementary basis vector with a 1 in position j and 0 elsewhere.

Exercise A.13. Use the definition of d/dz above to compute $(d/dz)u(z)$ when $u(z) = u(x, y) = (x + iy)^2$ and verify that it is equal to the definition via calculus over the complex numbers.

The computational apparatus representing complex numbers as real two-vectors allows us to prove an important fact about complex manifolds.

Proposition A.25. *Every complex manifold is orientable.*

Proof By definition, overlapping chart maps are related by a complex analytic coordinate change f. It therefore suffices to prove that a bi-holomorphic map $f : U \subseteq \mathbb{C}^m \to \mathbb{C}^m$ always has a positive Jacobian determinant when viewed as a map on $\mathbb{R}^{2m}$. On the complex tangent space, f_* is complex linear represented by an $m \times m$ complex Jacobian matrix $J_{\mathbb{C}} = J(f)$, so that

$$\left(\sum_{j=1}^{m} \mathbf{e}_j z_j\right) = \sum_{j=1}^{m} (J\boldsymbol{v})_j \mathbf{e}_j.$$

Recalling that z is represented as $[\mathrm{Re}(z), \mathrm{Im}(z)]$, the real $2m \times 2m$ Jacobian matrix is given by

$$J_{\mathbb{R}} = \left[\begin{array}{c|c} A & -B \\ \hline B & A \end{array}\right],$$

where $A = \mathrm{Re}(J_{\mathbb{C}})$ and $B = \mathrm{Im}(J_{\mathbb{C}})$. Positivity of $\det J_{\mathbb{R}}$ then follows from nonvanishing of $\det J_{\mathbb{C}}$ and the identity

$$\det\left[\begin{array}{c|c} A & -B \\ \hline B & A \end{array}\right] = \det\left[\begin{array}{c|c} A+iB & -B+iA \\ \hline B & A \end{array}\right] = \det\left[\begin{array}{c|c} A+iB & 0 \\ \hline B & A-iB \end{array}\right]$$
$$= \det(A+iB) \cdot \det(A-iB)$$
$$= |\det(A+iB)|^2 .$$

□

Next, we extend the definitions of $\partial, \overline{\partial}$, and d from functions to differential forms as follows. For any coordinate system $t_1, \ldots, t_{2n}$ and $J \subseteq [2n]$, let dt_J denote the wedge product of dt_j for $j \in J$ in increasing order. For a smooth function u define $d(u\, dt_J) = du \wedge dt_J$ and extend linearly to all forms. The form ω is said to be a ***holomorphic form*** if it is the sum of terms $u_J dz_J$ with no $d\overline{z}_i$ terms and each u_J is holomorphic. The space of holomorphic functions form a subring of $C^\infty(\mathbb{R}^{2n})$, while the space of holomorphic forms have a (noncommutative) ring structure under wedge product and are generated over the ring of holomorphic functions by $\{dz_1, \ldots, dz_n\}$.

Remarks A.26. (i) Functoriality and the Cauchy–Riemann equations imply that these definitions extend to functions on any complex manifold $\mathcal{M}$.

(ii) While the same symbol is used for the boundary operator $\partial\sigma$ and the holomorphic differential ∂u, it is always clear from context which we refer to.

Tangent and cotangent bundles, differential forms, and integration on a complex manifold of dimension m can all be defined via its structure as a real manifold of dimension $2m$. The operator d preserves holomorphicity: if $\overline{\partial}\omega = 0$ then $d\omega = \partial\omega$, which is evidently holomorphic. We also define the ***holomorphic volume form*** $dz := dz_1 \wedge \cdots \wedge dz_n$ in $\mathbb{C}^n$.

Exercise A.14. Show that the conjugation function $\overline{z}$ is not holomorphic on $\mathbb{C}$ by computing $\partial\overline{z}$ and $\overline{\partial}\overline{z}$. Find a continuous, piecewise smooth function $f_\varepsilon : \mathbb{C} \to \mathbb{C}$ that agrees with $\overline{z}$ on the disk $\{|z| \le \varepsilon\}$ and is holomorphic outside of this disk.

The holomorphic volume form is an n-form in $\mathbb{R}^{2n}$, thus *middle-dimensional*, but is the highest dimensional holomorphic form in $\mathbb{R}^{2n}$. This leads to the following useful result.

Theorem A.27. *If C is any $(n+1)$-chain on a domain $U \subseteq \mathbb{C}^n$ and ω is any holomorphic n-form on U then*

$$\int_{\partial C} \omega = 0\,.$$

Proof The d operator preserves holomorphicity, so $d\omega$ is holomorphic. There are no holomorphic forms above rank n and $\partial\omega$ is a $(n+1)$-form, hence $\partial\omega = 0$. Stokes's Theorem then gives

$$\int_{\partial C} \omega = \int_C d\omega = 0\,.$$

□

Exercise A.15. Let C be an n-chain supported on a complex submanifold $\mathcal{M}$ of $\mathbb{C}^n$, where the dimension of $\mathcal{M}$ is strictly less than n. Show that $\int_C \omega$ vanishes for any holomorphic n-chain ω.

The ***polydisk*** with center $\boldsymbol{w} \in \mathbb{C}^d$ and radius $\boldsymbol{r} \in \mathbb{R}^d_{>0}$ is the set $\{\boldsymbol{z} \in \mathbb{C}^d : |z_j - w_j| < r_j \text{ for } 1 \le j \le d\}$, while the ***torus*** with center $\boldsymbol{w}$ and radius $\boldsymbol{r}$ is the set $\{\boldsymbol{z} \in \mathbb{C}^d : |z_j - w_j| = r_j \text{ for } 1 \le j \le d\}$. Our asymptotic arguments always begin by representing the power series coefficients of an analytic function by a complex integral over a polytorus, after which we apply the integration techniques discussed in these appendices. We thus conclude this first appendix by stating and proving the multivariate Cauchy Integral Formula, which is used to derive such representations.

Proposition A.28 (multivariate ***Cauchy Integral Formula***)**.** *Let $U \subset \mathbb{C}^d$ be an open set and $D, T \subset \mathbb{C}^d$ be a polydisk and torus with radii $\boldsymbol{r} \in \mathbb{R}^d_{>0}$, centered around the same point. If $\overline{D} \subset U$ and f is an analytic function on U then, for all $\boldsymbol{w} \in D$,*

$$f(\boldsymbol{w}) = \left(\frac{1}{2\pi i}\right)^d \int_T \frac{f(\boldsymbol{z})}{(z_1 - w_1)\cdots(z_d - w_d)} d\boldsymbol{z}\,. \tag{A.4.1}$$

□

Proof Without loss of generality, we may assume that D and T are centered at the origin. When $d = 1$ this is the usual (univariate) Cauchy Integral Formula. The general case follows by induction. In particular, for any fixed $(a_1, \ldots, a_{d-1}) \in \mathbb{C}^{d-1}$ with each $|a_j| \le r_j$ we know from the univariate case that

$$f(a_1, \ldots, a_{d-1}, w_d) = \frac{1}{2\pi i} \int_{|z_d| = r_d} \frac{f(a_1, \ldots, a_{d-1}, w_d)}{z_d - w_d} dz_d\,. \tag{A.4.2}$$

If the Cauchy Integral Formula holds in $d - 1$ variables then

$$f(\boldsymbol{w}) = \left(\frac{1}{2\pi i}\right)^{d-1} \int_{|z_1|=r_1} \cdots \int_{|z_{d-1}|=r_{d-1}} \frac{f(z_1, \ldots, z_{d-1}, w_d)}{(z_1 - w_1) \cdots (z_{d-1} - w_{d-1})} dz_1 \cdots dz_{d-1} ,$$

and substitution with (A.4.2) implies

$$f(\boldsymbol{w}) = \left(\frac{1}{2\pi i}\right)^{d} \int_T \frac{f(\boldsymbol{z})}{(z_1 - w_1) \cdots (z_d - w_d)} d\boldsymbol{z} ,$$

as desired. □

Given vectors $\boldsymbol{a}, \boldsymbol{b} \in \mathbb{R}^d_{>0}$, the ***polyannulus*** with center $\boldsymbol{w} \in \mathbb{C}^d$, inner radius $\boldsymbol{a}$ and outer radius $\boldsymbol{b}$ is the set $\{\boldsymbol{z} \in \mathbb{C}^d : a_k < |z_k - w_k| < b_k \text{ for all } k\}$. Some of our arguments require a version of the Cauchy Integral Formula in a polyannulus.

Proposition A.29 (Cauchy Integral Formula in a polyannulus). *Let $U \subset \mathbb{C}^d$ be an open set and let $D \subset \mathbb{C}^d$ be a polyannulus of inner radius $\mathbf{a}$ and outer radius $\mathbf{b}$. For $\eta \in \{a, b\}^d$ let T_η denote the torus with radius $\boldsymbol{r}$ where*

$$r_i = \begin{cases} a_i & \text{if } \eta_i = a \\ b_i & \text{if } \eta_i = b \end{cases} ,$$

and let $\operatorname{sgn}(\eta) = (-1)^{\#\{j \,:\, \eta_j = a\}}$. *If $\overline{D} \subset U$ and f is an analytic function on U, then for all $\boldsymbol{w} \in D$*

$$f(\boldsymbol{w}) = \left(\frac{1}{2\pi i}\right)^{d} \sum_{\eta \in \{a,b\}^d} \operatorname{sgn}(\eta) \int_{T_\eta} \frac{f(\boldsymbol{z})}{(z_1 - w_1) \cdots (z_d - w_d)} d\boldsymbol{z} . \tag{A.4.3}$$

Proof When $d = 1$ the conclusion states that

$$f(w) = \frac{1}{2\pi i} \left[\int_{|z|=b} \frac{f(z)}{z - w} dz - \int_{|z|=a} \frac{f(z)}{z - w} dz \right] ,$$

which is the classical univariate Cauchy formula in an annulus [Hör90, Theorem 1.2.1]. Induction now applies exactly as in the previous proposition, noting that as chains

$$\sum_{\eta \in \{a,b\}^d} \operatorname{sgn}(\eta) T_\eta = \sum_{\eta \in \{a,b\}^{d-1}} \operatorname{sgn}(\eta) T_\eta \times (\gamma_b - \gamma_a) ,$$

where γ_c is the circle of radius c oriented counterclockwise. □

Notes

The material in the first three sections of this chapter is standard graduate-level calculus, and our presentation of the material owes a debt to the texts of Tu [Tu11] and Warner [War83]. Additional details on complex manifolds can be found in Griffiths and Harris [GH94].

Additional exercises

Exercise A.16. Is there a difference between $\mathbb{RP}^2$ and $\mathbb{CP}^1$ as an abstract set? As a topological space?

Exercise A.17. Let $f : X \to \mathbb{R}$ be a smooth map on a d-manifold X for which df is everywhere nonvanishing. Let $\mathcal{M}$ be the zero set of f and let $\iota : \mathcal{M} \to X$ denote the inclusion map. Prove that for any $(d-1)$-form η the equality $\iota^*(\eta) = 0$ holds if and only if $\eta \wedge df$ vanishes on $\mathcal{M}$. Repeat this for $k \leq d$ functions $f_1, \ldots, f_k$ whose intersection defines a smooth surface $\mathcal{M}$ of codimension k. *Hint:* Use the implicit function theorem to coordinatize X with first coordinate f and use functoriality of $\wedge$ to reduce to the case $f = x_1$.

Exercise A.18. Do Exercise A.17 when X is a complex d-manifold and $f : X \to \mathbb{C}$ is analytic. *Hint: You can copy the proof, only you need the complex form of the implicit function theorem [Hör90, Theorem 2.1.2] in order to be sure your coordinates are holomorphic.*

Exercise A.19. Define a 2-form ω in $\mathbb{R}^3$ by $\omega = x\,dy \wedge dz + y\,dz \wedge dx + z\,dx \wedge dy$ and define a 2-chain C representing the unit sphere. Compute $\int_C \omega$ directly from the definitions, then figure out a shortcut to the same result using Stokes's Theorem.

Appendix B

Algebraic topology

Having established the necessary background for integration on real and complex manifolds in Appendix A, we now move on to the topological results that allow us to manipulate these integrals in order to derive asymptotics. A differential form ω is said to be a ***closed form*** if $d\omega = 0$ and an ***exact form*** if $\omega = d\tau$ for some form τ. Many of the forms we care about are closed, for instance if ω is any holomorphic n-form in $\mathbb{C}^n$ then $\overline{\partial}\omega$ vanishes by holomorphicity and $\partial\omega$ vanishes because there are no holomorphic $(n+1)$-forms, hence ω is closed. A chain C is called a ***cycle*** if $\partial C = 0$ and a ***boundary*** if $C = \partial \mathcal{D}$ for some chain $\mathcal{D}$. The boundaries form a subset (in fact, a sub-vector space) of the cycles because $\partial^2 = 0$.

By the same reasoning as our proof of Theorem A.27 in the last appendix, the integral of any closed form over a boundary is zero. Thus, by linearity of the integral, if C is a cycle, then $\int_C \omega$ depends only on the equivalence class of C in the quotient space of cycles modulo boundaries. Homology theory is the study of this quotient space, which may be thought of simultaneously as a topological invariant and as classifying contours of integration for closed forms. After studying various forms of homology, we dualize our constructions and define cohomology of differential forms. Just as $\int_C \omega$ depends only on the homology class of the chain C, it also depends only on the cohomology class of the form ω.

B.1 Chain complexes and homology theory

Instead of working only with cycles of integration, we develop homology in a more general setting. This approach better illustrates underlying structure, and allows us to reuse results in different contexts. We therefore introduce the following abstract definitions, which generalize some of the properties discussed above.

(i) A ***chain complex*** is a collection $C = \{C_n : n = 0, 1, 2, \ldots\}$ of complex vector spaces, not necessarily finite dimensional, together with a ***boundary operator*** ∂, which for all n is a linear map ∂_n from the space of ***n-chains*** C_n to the space of $(n-1)$-chains C_{n-1} that satisfies $\partial^2 = 0$ (meaning $\partial_n \circ \partial_{n+1} = 0$ for all n, so that "a boundary has no boundary""). By definition, $\partial = 0$ on C_0.

(ii) The vector space of ***n-cycles*** $Z_n \subseteq C_n$ of a chain complex C is the kernel of ∂_n, and the group $B_n \subset C_n$ of ***n-boundaries*** of C is the image of ∂_{n+1}.

(iii) The nth ***homology group*** of C is the vector space quotient

$$H_n(C) := Z_n / B_n\,.$$

The notation $H_*(C)$ is used to refer collectively to $H_n(C)$ for all n. Cycles in the same equivalence class are called ***homologous***.

Remark. Because we work with complex vector spaces, $H_*(C)$ is sometimes called ***homology with coefficients in*** $\mathbb{C}$ to distinguish it from the analogous construction with integer coefficients. While the theory with integer coefficients is richer, taking coefficients in a field better suits the purposes of computing integrals. With integer coefficients, the spaces of chains, cycles, and boundaries are $\mathbb{Z}$-modules, and their quotients are abelian groups, hence "homology group" rather than "homology vector space."

To discuss the homology of a manifold, we must define an appropriate chain complex. One natural candidate is the chain complex defined by smooth chains together with the boundary map discussed in Section A.3 of Appendix A. For many purposes, however, it is convenient to relax our smoothness condition. If $\mathcal{M}$ is any Hausdorff topological space then a ***singular n-simplex*** in $\mathcal{M}$ is a *continuous* (not necessarily smooth) map $\sigma : \Delta^n \to \mathcal{M}$ from the standard n-simplex Δ^n to $\mathcal{M}$, and a ***singular n-chain*** in $\mathcal{M}$ is a complex linear combination of singular n-simplices in $\mathcal{M}$. Just as for smooth chains, we may use the natural ordering of the faces of Δ^n to define a canonical boundary map through (A.3.1), taking a singular n-simplex to a singular $(n-1)$-simplex and extending linearly to singular chains. It is a foundational result in homology theory that the homology of a manifold is unchanged whether one studies smooth or singular chains.

Proposition B.1. *Let $\mathcal{M}$ be a differentiable manifold, let C be the chain complex whose chains are linear combinations of singular simplices on $\mathcal{M}$, and let C' be the chain complex whose chains are linear combinations of smooth* *simplices on $\mathcal{M}$. Then for any $n \in \mathbb{N}$ the homology groups $H_n(C)$ and $H_n(C')$ are isomorphic.*

Proof See [Eil47]. □

We write $C(\mathcal{M})$ for the chain complex defined by the singular n-chains in $\mathcal{M}$. The homology group $H_n(C(\mathcal{M}))$ is written $H_n(\mathcal{M})$ and called the ***nth singular homology group*** of $\mathcal{M}$. One can think of the rank of the homology group $H_n(\mathcal{M})$ – i.e., the minimum size of a generating set of the group – as indicating how many unique cycles in $\mathcal{M}$ don't bound anything.

Example B.2. The rank of $H_1(\mathcal{M})$ represents the number of nonequivalent circles that can be drawn on $\mathcal{M}$ without bounding something in $\mathcal{M}$. The rank of $H_1(\mathcal{M})$ for a connected space $\mathcal{M}$ is thus zero if $\mathcal{M}$ is simply connected, however the converse does not hold. Homology with coefficients in $\mathbb{C}$ cannot detect the presence of a cycle γ such that γ does not bound anything but k times γ does (homology with integer coefficients is more discerning but typically more complicated to compute). ◁

Topological invariance of homology

The first crucial property of homology is that it is a topological invariant. A ***topological map*** (or simply ***map***) between two topological spaces is a continuous function between them, while a ***chain map*** between two chain complexes is a function between them that commutes with their boundary operators. More precisely, if $(\mathcal{A}, \partial^A)$ and $(\mathcal{B}, \partial^B)$ are two chain complexes then a chain map between them can be considered to be a collection of functions $f = (f_0, f_1, \ldots)$ with $f_n : \mathcal{A}_n \to \mathcal{B}_n$ mapping from the n-chains of $\mathcal{A}$ to the n-chains of $\mathcal{B}$, such that $\partial_n^B \circ f_n = f_{n-1} \circ \partial_n^A$ for all n.

A topological map from X to Y induces a chain map from the singular chain complex of X to the singular chain complex of Y. A map $f : \mathcal{A} \to \mathcal{B}$ between chain complexes in turn induces a homomorphism f_* on homology groups by applying the map to representatives for the equivalence classes of cycles modulo boundaries. Both of these induced maps are functorial.

Proposition B.3. *If the topological spaces X and Y are homeomorphic then the singular homology groups $H_n(X)$ and $H_n(Y)$ are isomorphic for all n.*

Proof A homeomorphism between topological spaces is a topological map whose inverse is also a topological map. Hence, a homeomorphism between two spaces induces an isomorphism between the homology groups of the spaces. □

Homology groups are preserved under more than just homeomorphism. If X and Y are topological spaces then two maps $f, g : X \to Y$ are said to be ***homotopic maps*** if there is a continuous map $\mathbf{H} : X \times [0, 1] \to Y$, called a ***homotopy***, such that $\boldsymbol{H}(x, 0) = f(x)$ and $\boldsymbol{H}(x, 1) = g(x)$ for all x. A topological map $f : X \to Y$ is called a ***homotopy equivalence*** if there is a topological map $g : Y \to X$ such that $f \circ g$ is homotopic to the identity on Y and $g \circ f$ is homotopic to the identity on X. If $f : X \to Y$ is a homotopy equivalence then we say that X and Y are ***homotopic spaces***.

We claim that homotopic maps induce equal maps on homology, hence homotopy equivalent spaces have naturally identical homology. To see this one proves, on the chain level, that a homotopy equivalence between topological spaces induces a ***chain homotopy equivalence*** between the singular chain complexes, which in turn induces an isomorphism between the homology groups.

Proposition B.4. *If a topological map $f : X \to Y$ is a homotopy equivalence then the singular homology groups $H_n(X)$ and $H_n(Y)$ are isomorphic for all n, with f_* inducing one such isomorphism.*

Proof See Theorem 2.10 and Corollary 2.11 of [Hat02]. □

Exercise B.1. Explain why a homeomorphism is always a homotopy equivalence.

Suppose $H : X \times [0, 1] \to X$ is a homotopy with $H(x, 0) = x$ for all X, and $Y \subseteq X$ is a subspace such that $H(x, 1) \in Y$ for all x. If $H(y, t) \in Y$ for all y and t (so that X ends in Y and Y stays in Y) then we call H a ***weak deformation retract*** and say Y is a weak deformation retract of X. If $H(y, t) = y$ for all y and t then we call H a ***strong deformation retract*** (or simply ***deformation retract***) and say that Y is a strong deformation retract of X.

Exercise B.2. Prove that if Y is a weak deformation retract of X then X and Y are homotopy equivalent.

Remark. Imagine looking at a space X that starts deforming in a continuous manner, with points allowed to collide and pass through each other, ultimately ending up in a different space Y. The space Y can be smaller, even a single point, but Exercise B.2 implies that as long as every point that starts in Y stays in Y then the homology of X is isomorphic to the homology of Y.

Various homologies and their equivalence

Working with the homology of smooth chains is nice because they are what we integrate over, but this approach has some disadvantages. Most prominently, the large amount of freedom involved means the collection of chains on a manifold is huge (not having a countable basis) even though the homology groups are finite dimensional. Because of this we introduce new types of spaces with more structure.

- A ***cell complex*** or ***CW-complex*** X is a Hausdorff space defined using a specific inductive procedure. Let X^0 be a discrete collection of points and for any $n \geq 0$ let X^n be the quotient space, with quotient topology, defined by the disjoint union $X^{n-1} \sqcup_\alpha \Delta_\alpha^n$ of X^{n-1} with a collection of standard n-simplices Δ_α^n, where we identify each $x \in \partial\Delta_\alpha^n$ with some $\phi_\alpha(x) \in X^{n-1}$ using ***gluing maps*** $\phi_\alpha : \partial\Delta_\alpha^n \to X^{n-1}$. The set X^n is the ***n-skeleton*** of X, and contains its ***k-cells*** Δ_α^k for $k \leq n$. We consider *finite dimensional* cell complexes, meaning $X = X^n$ for some natural number n and the smallest such n is called the ***dimension of the cell complex*** X. Each simplex Δ_α^n corresponds to a map $\sigma_\alpha^n : \Delta_\alpha^n \to X$ defined by embedding Δ_α^n into $X^{n-1} \sqcup_\alpha \Delta_\alpha^n$ and then taking the quotient by the gluing maps.
- A ***Δ-complex*** is a cell complex X where the gluing map ϕ_α for any simplex Δ_α^n maps each $(n-1)$-dimensional face F of Δ_α^n homeomorphically to one of the $(n-1)$-simplices in X, preserving the ordering of vertices and agreeing with the previously defined gluing map ϕ_F on ∂F. A Δ-complex X may be viewed as a collection of maps $\sigma_\alpha : \Delta^{n_\alpha} \to X$ such that the restriction of σ_α to the interior of Δ^{n_α} is injective, each point of X lies in exactly one such restriction, and the restriction of σ_α to a face of Δ^{n_α} is another one of the maps $\sigma_\beta : \Delta^{n_\alpha - 1} \to X$ in the collection defining X.
- A ***simplicial complex*** is a Δ-complex where each gluing map is injective (so that distinct faces in the boundary of each simplex are glued to distinct lower-dimensional simplices) and each n-simplex is uniquely determined by its vertices. A simplicial complex X may be viewed as a set of simplices such that every face of a simplex in X is also in X and the nonempty intersection of any two simplices $\Delta_1, \Delta_2 \in X$ is a face of both Δ_1 and Δ_2.

We say that a space S is ***represented*** by a cell complex, Δ-complex, or simplicial complex X if S and X are homeomorphic. The representation of a space by a simplicial complex is called a ***triangulation*** of the space. Figure B.1 shows some examples representing a sphere and a circle.

The ***CW approximation theorem*** [Hat02, Proposition 4.13] states that any Hausdorff space X can be *approximated* by a cell complex $\tilde{X}$ meaning, among other things, that X and $\tilde{X}$ have the same singular homology groups. We are most interested in algebraic varieties, or their complements, which are examples of semi-algebraic sets and can therefore be triangulated [BPR03, Theorem 5.43].

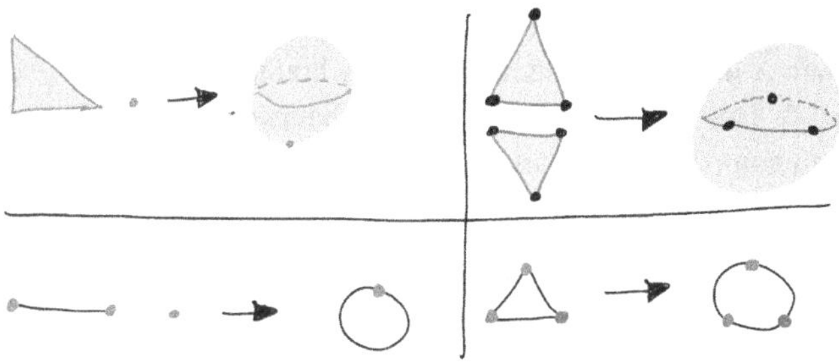

Figure B.1 *Top left:* The sphere S^2 can be represented as a cell complex with one 0-cell and one 2-cell whose boundary is mapped to the 0-cell, but this is not a Δ-complex. *Top right:* A representation of S^2 as a Δ-complex containing three 0-cells, three 1-cells, and two 2-cells. *Bottom left:* The circle S^1 can be represented as a Δ-complex with one 0-cell and one 1-cell whose boundary is identified with the 0-cell, but this is not a triangulation. *Bottom right:* A triangulation of S^1 with three 0-cells and three 1-cells.

The difficulties with singular homology for general topological spaces still arise for Δ and simplicial complexes. However, the additional structure present allows us to define a more rigid notion of homology. If X is a simplicial complex then the ***simplicial homology*** of X is the homology defined by the chain complex C whose n-chains are complex linear combinations of the n-simplices in X, with the boundary map again defined by (A.3.1). Simplicial homology is crucial for calculation, due to the following result.

Proposition B.5. *If X is a simplicial complex then the nth singular homology group of X is isomorphic to the nth simplicial homology group of X. The simplicial homology groups of X are algorithmically computable.*

Proof See [Hat02, Theorem 2.27] for the statement on the equivalence of the homologies, and [Mun84, Theorem 11.5] for an algorithm to determine simplicial homology by representing the linear boundary maps of X as matrices and computing their Smith normal forms. □

Among other corollaries, Proposition B.5 implies that a space which can be represented by a Δ-complex with a finite number of simplices has finitely generated homology groups. Although simplicial complexes have more structure than Δ-complexes, it can often be more efficient to represent a space as a Δ-complex compared to representing it as a simplicial complex.

Example B.6. We compute the homology of the circle S^1 by representing it as the Δ-complex X with one point $X^0 = \{p\}$ and one line segment $X^1 = \{\ell\}$ with both endpoints of ℓ glued at p. The 0-chains in this representation are the complex multiples of p, while the 1-chains are the complex multiples of ℓ. Any 0-chain is trivially a cycle, and the only boundary is $\partial\ell = p - p = 0$, so $H_0(S^1) = \mathbb{C}$. This computation also shows that every 1-chain is a cycle, and there are no non-trivial boundaries because there are no 2-cycles, so $H_1(S^1) = \mathbb{C}$ while $H_k(S^1) = 0$ for $k \geq 2$. ◁

Exercise B.3. Compute the homology of the circle by triangulating it, verifying you get the same result as Example B.6. Compute the homology of the sphere S^2 using similar techniques.

B.2 Tools for homology

Although we can compute the homology of a variety, or the complement of a variety, by triangulating the space and computing simplicial homology, this can be very expensive (such algorithms are generally considered efficient if they run in single-exponential instead of doubly-exponential time; see [Bas08] for a survey of complexity results in this area). Furthermore, instead of computing the entire set of homology groups of a singular variety, we often only need some partial information in our integral manipulations, for instance throwing away topological information that does not affect dominant asymptotics.

Because of such considerations, it is useful to have additional tools to compute homology. One of the most effective approaches is to work recursively, studying a space X using a subspace $A \subset X$ and the quotient space X/A. The relationship between the homology groups of X, X/A, and A is explicit but intricate, with the homologies fitting into a type of nesting structure. We thus require some additional algebraic constructions to describe precisely what is going on.

An ***exact sequence*** of abelian groups (and in particular, complex vector spaces) is a sequence of maps

$$\cdots \to X_{n+1} \xrightarrow{f_{n+1}} X_n \xrightarrow{f_n} X_{n-1} \to \ldots,$$

where the image of each map is equal to the kernel of the next. For instance, an exact sequence of the form $0 \xrightarrow{\epsilon} A \xrightarrow{\alpha} B$ says $0 = \text{Image}(\epsilon) = \text{Kernel}(\alpha)$, so α is injective, while an exact sequence of the form $A \xrightarrow{\alpha} B \xrightarrow{\epsilon} 0$ says $\text{Image}(\alpha) = \text{Kernel}(\epsilon) = B$, so α is surjective. A ***short exact sequence*** is an exact sequence of the form

$$0 \to X \to Y \to Z \to 0,$$

meaning the map from X to Y is injective and the map from Y to Z is surjective.

Remark B.7. When X, Y, and Z are finite dimensional complex vector spaces of dimensions k, ℓ, and m, respectively, then a short exact sequence $0 \to X \to Y \to Z \to 0$ implies $\ell = k + m$ and $Y = X \oplus Z$ as a direct sum. However, this splitting is not natural: X embeds naturally into Y but there is no canonical choice of coset representatives for Y/Z.

A ***short exact sequence of chain complexes*** is a map of chain complexes which is a short exact sequence on the n-chains for all n. One very useful fact about short exact sequences of chain complexes is that they give rise to long exact homology sequences.

Theorem B.8 (Zig-Zag Lemma). *Let* $0 \to A \xrightarrow{\alpha} B \xrightarrow{\beta} C \to 0$ *be a short exact sequence of chain complexes. Then there is an exact sequence*

$$\cdots \to H_{n+1}(C) \xrightarrow{\partial_*} H_n(A) \xrightarrow{\alpha_*} H_n(B) \xrightarrow{\beta_*} H_n(C) \xrightarrow{\partial_*} H_{n-1}(A) \to \dots, \tag{B.2.1}$$

where α_* *and* β_* *are the homology maps induced by the chain maps* α *and* β.

Proof See [Mun84, Lemma 24.1]. □

The exact sequence (B.2.1) in Theorem B.8 is known as the ***long exact sequence*** on homology. The homology map ∂_* has a natural but unwieldy definition; instead of defining it in general we describe it explicitly in the situation most relevant to us in Corollary B.11 below.

Relative homology and excision

Our goal is to apply Theorem B.8 to a short exact sequence of chain complexes related to embedding a subspace A into a space X and then taking the quotient to map into X/A. In fact, we consider a slightly more general setting which is also useful for our asymptotic calculations.

A ***pair of spaces*** (X, A) is any pair of topological spaces with A a subspace of X. A ***pair map*** $f : (X, A) \to (X', A')$ between pairs of spaces is any (topological) map $f : X \to X'$ such that $f(A) \subset A'$. The inclusion $A \hookrightarrow X$ induces an inclusion of chain complexes $C(A) \hookrightarrow C(X)$, and we let $C(X, A)$ denote the ***pair complex*** whose n-chains are the quotient group $C_n(X)/C_n(A)$. The ***relative homology*** of the pair (X, A) is the homology $H_*(X, A) = H_*(C(X, A))$ of the pair complex.

One may think of relative homology roughly as the homology of X if the subspace A were to be shrunk to a point: $H_n(X, A)$ contains ***relative cycles*** $\gamma \in C_n(X)$ with $\partial\gamma \in C_{n-1}(A)$ modulo ***relative boundaries*** $\beta = \partial\zeta + \alpha$, where $\zeta \in C_{n+1}(X)$ and $\alpha \in C_n(A)$. We thus search for cycles that do not bound, but are willing to count a chain as a cycle if its boundary is in A; see Figure B.2.

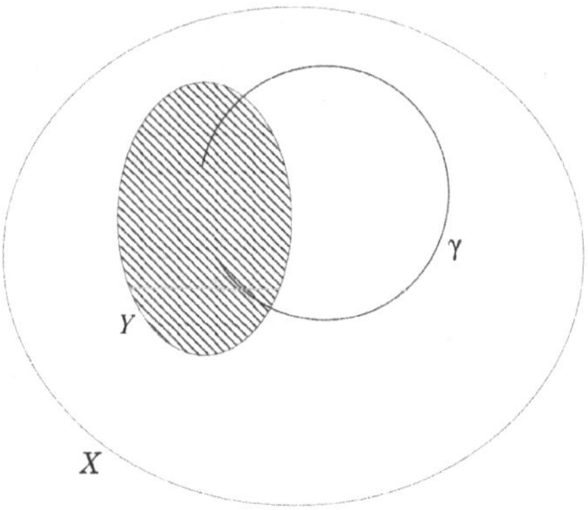

Figure B.2 γ is a relative cycle in $C(X, Y)$.

We often use relative homology to "ignore" regions that do not contribute to dominant asymptotics in our integral manipulations.

The concept of homotopy equivalence is defined for pairs of spaces similarly to the definition for spaces, and homotopy equivalent pairs have the same relative homology.

Example B.9. Let X be the unit ball in $\mathbb{R}^n$ and A be the unit sphere S^{n-1}. To compute the homology of the pair (X, A) we consider the unit n-simplex Δ^n and its boundary: topologically $(\Delta^n, \partial\Delta^n)$ and (B_n, S^{n-1}) are homeomorphic, so their relative homologies are the same. The simplicial chain complex for Δ^n contains all faces of Δ^n, hence it has simplices in dimensions 0 through n. The complex of the pair is non-trivial only in dimension n because every face of dimension less than n is supported on $\partial\Delta^n$. Proposition B.5 (that simplicial and singular homology coincide) extends to pairs composed of a simplicial complex and a subcomplex. Therefore, the singular homology of the pair (B^n, S^{n-1}) is computed by this rather small chain complex, leading to $H_k(B^n, S^{n-1}) \cong \mathbb{C}$ when $k = n$ and 0 otherwise. ◁

Exercise B.4. What is the relative homology of (S^2, S^2_-), where S^2 is the 2-sphere and S^2_- is the closed southern hemisphere?

Relative homology is useful for approximating integrals due to the following result.

Proposition B.10 (asymptotics depend only on relative homology class). *Let X be a manifold of dimension n with submanifold Y also of dimension n, and let ϕ be a smooth complex-valued function on X satisfying $\Re\{\phi\} \leq \beta$ on Y for some $\beta \in \mathbb{R}$. Suppose that $\omega = \omega_\lambda = \exp(\lambda\,\phi(z))\,\eta$ for some closed k-form η on*

X with $k \le n$. If C and C' are k-chains on X with $C \equiv C'$ in $H_k(X, Y)$ then, as $\lambda \to \infty$,

$$\int_C \omega_\lambda = \int_{C'} \omega_\lambda + O\left(e^{\lambda\beta}\right).$$

Proof By definition, the difference between C and C' is a relative boundary $C - C' = \partial\mathcal{D} + C''$ with C'' supported on Y. Using Stokes's Theorem (Theorem A.24),

$$\left|\int_C \omega - \int_{C'} \omega\right| = \left|\int_{\partial\mathcal{D}} \omega + \int_{C''} \omega\right| = \left|\int_{\mathcal{D}} d\omega + \int_{C''} \omega\right|$$
$$= \left|\int_{C''} \omega\right| \le \int_{C''} e^{\lambda\beta}|\eta| \le K e^{\lambda c'},$$

where $K = \int_{C'} |\eta|$ and the appearance of $|\eta|$ in an integral means that when η is pulled back to integrate, we integrate the modulus of the result. □

Let (X, A) be a pair for which A is a strong deformation retract of an open neighborhood in X. It can be shown [Mun84, Ex. 39.2] that for all $n \ge 1$ the relative homology group $H_n(X, A)$ is isomorphic to the singular homology group $H_n(X/A)$ of the quotient X/A, obtained from X by *shrinking A to a point*. This gives a way of computing the homology of a quotient X/A: applying Theorem B.8 to the short exact sequence of chain complexes $0 \to C(A) \to C(X) \to C(X, A) \to 0$ gives a long exact sequence computing $H_*(X, A)$, and hence $H_*(X/A)$, from $H_*(X)$, $H_*(A)$ and some knowledge of the maps in the long exact sequence.

Corollary B.11. *Let A be a subspace of X. Then there is a* ***long exact sequence of the pair*** *(X, Y),*

$$\cdots \to H_{n+1}(X, A) \xrightarrow{\partial_*} H_n(A) \xrightarrow{i_*} H_n(X) \xrightarrow{j_*} H_n(X, A) \xrightarrow{\partial_*} H_{n-1}(A) \to \cdots,$$

where i_* and j_* are the maps induced by the inclusions of A into X, and X into (X, A), respectively, and ∂_* is the map induced by taking a relative cycle $\gamma \in C_n(X)$ to its boundary $\partial\gamma \in C_{n-1}(A)$. When A is a deformation retract of an open neighborhood in X then this long exact sequence holds with $H_n(X, A)$ replaced by $\tilde{H}_n(X/A)$, where the ***reduced homology group*** *$\tilde{H}_n$ is the same H_n when $n > 0$, and has dimension one less when $n = 0$.*

Exercise B.5 (computing homology of S^{n-1})**.** Let $X = \Delta^n$ and let Y be the subcomplex of cells with dimension strictly less than n. Use the long exact sequence for the pair (X, Y) to determine $H_{n-1}(Y)$.

One important feature of relative homology is the ***excision property***.

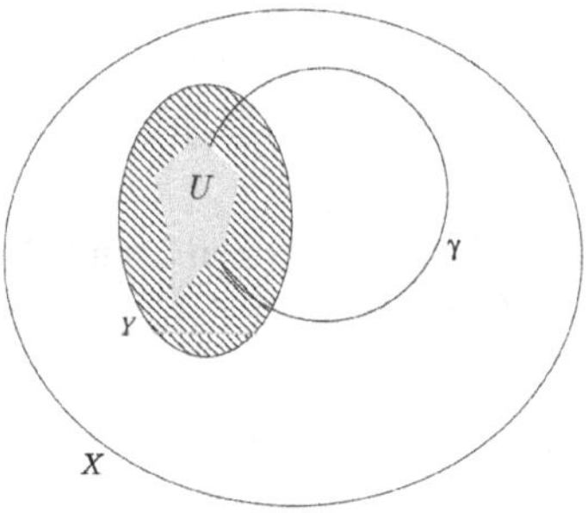

Figure B.3 $(X \setminus U, Y \setminus U) \hookrightarrow (X, Y)$ is a homology isomorphism.

Proposition B.12 (excision theorem). *Given subspaces $U \subset Y \subset X$ such that $\overline{U} \subset \operatorname{int}(Y)$, the inclusion $(X \setminus U, Y \setminus U) \hookrightarrow (X, Y)$ induces an isomorphism $H_*(X \setminus U, Y \setminus U) \cong H_*(X, Y)$.*

Proof See [Hat02, Theorem 2.20]. □

Informally, Proposition B.12 says that the relative homology of (X, Y) *cannot see* the interior of U; Figure B.3 gives an illustration. Relative homology can also be used to define another important homology theory for cell complexes. If $X = X^n$ is a cell complex then the ***cellular chain complex*** of X is the complex

$$\cdots \to H_{n+1}(X^{n+1}, X^n) \xrightarrow{\partial_{n+1}} H_n(X^n, X^{n-1}) \xrightarrow{\partial_n} H_{n-1}(X^{n-1}, X^{n-2}) \to \cdots,$$

where we define $X_{-1} = \varnothing$ and ∂_n is the composition of the boundary operator mapping $H_n(X^n, X^{n-1})$ to $H_{n-1}(X^{n-1})$ with the inclusion of $H_{n-1}(X^{n-1})$ into $H_{n-1}(X^{n-1}, X^{n-2})$. The homology groups of this cellular chain complex form the ***cellular homology groups*** of X. Cellular homology is *homology of (relative) homology*, and can be surprisingly useful. Although its definition might seem technical and contrived, $H_n(X^n, X^{n-1})$ can be interpreted easily as linear combinations of the n-cells in X. The following result makes cellular homology a tool for computation.

Proposition B.13. *If X is a cell complex then the nth singular homology group of X is isomorphic to the nth cellular homology group of X.*

Among other things, Proposition B.13 implies that if X can be represented by a cell complex with no n-cells then $H_n(X) = 0$. See [Hat02, Section 2.2] for a full discussion of cellular homology, its implications, and a proof of Proposition B.13.

The homology of a union

Instead of decomposing a space X using a subspace and the corresponding quotient space, we can instead represent X as a union $X = A \cup B$ for subspaces $A, B \subset X$ such that $X = \text{int}(A) \cup \text{int}(B)$ and describe the homology of X in terms of the homologies of A, B, and $A \cap B$. In particular, there is a short exact sequence of chain complexes defined by the short exact sequence in each dimension,

$$0 \to C_n(A \cap B) \xrightarrow{p} C_n(A) \oplus C_n(B) \xrightarrow{s} C_n(A) + C_n(B) \to 0,$$

where $p(a) = (-a, a)$ and $s(a, b) = a + b$, with $C_n(A) + C_n(B) \subseteq C_n(X)$ denoting a sum inside the space of chains on X. The corresponding long exact sequence is known as the ***Mayer–Vietoris sequence***.

Theorem B.14 (Mayer–Vietoris sequence). *Let $A, B \subset X$ be such that X is the union of the interiors of A and B. Then the inclusion of the chain complex $C_*(A) + C_*(B) \hookrightarrow C_*(X)$ induces an isomorphism in homology. It follows that there is a long exact homology sequence*

$$\cdots \to H_k(A \cap B) \to H_k(A) \oplus H_k(B) \to H_k(X) \to H_{k-1}(A \cap B) \to \cdots .$$

Proof See [Hat02, Proposition 2.21] or [Mun84, Theorem 33.1]. □

Exercise B.6. Use the Mayer–Vietoris sequence to re-compute the homology of S^n by decomposing it as the union of two hemispheres, expanded a little so their interiors cover S^n. How does the result of this computation relate to a geometric understanding of two balls glued along their boundary? Try visualizing in dimensions one and two for greatest intuition.

Attachments and the homology of a product

Relative homology is useful for our asymptotic computations because it allows us to "ignore" points that are asymptotically negligible. In practice, to study complex integrals whose domains of integration are allowed to vary in some set $\mathcal{M}$ we describe $\mathcal{M}$ by attaching different spaces together as needed. In essence, we express $\mathcal{M}$ on-the-fly as a cell complex until we have enough information to perform the necessary asymptotic computations (much more information on this approach is given in the next appendix on Morse theory). This section, our final on the basics of homology, describes product complexes, attachments, and products of pairs.

Let C' and C'' be chain complexes with boundary maps ∂' and ∂''. The ***tensor product complex*** (or simply ***product complex***) of C' and C'' is the chain complex $C = C' \otimes C''$ whose n-chains C_n are defined by the direct sum $C_n = \bigoplus_{k=0}^{n} C'_k \otimes C''_{n-k}$, where a basis for the tensor product $C'_k \otimes C''_{n-k}$ is given by elements $\sigma_k \otimes \tau_{n-k}$ as σ_k ranges over a basis for C'_k and τ_{n-k} ranges over a basis for C''_{n-k}. The boundary operator ∂ of C is defined by $\partial(\sigma_k \otimes \tau_{n-k}) = (\partial'\sigma_k) \otimes \tau_{n-k} + (-1)^k \sigma_k \otimes (\partial''\tau_{n-k})$. The product $Z = X \times Y$ of two simplicial complexes X and Y is naturally a cell complex, and this definition is constructed so that the product chain complex $C(X) \otimes C(Y)$ is isomorphic [Mun84, Theorem 57.1] to the cellular chain complex of Z.

Example B.15. Consider $C = \Delta_1$ as a cell complex with two 0-cells $\{[0], [1]\}$ and one 1-cell $\sigma_1 = [0, 1]$ oriented from 0 to 1, meaning that $\partial\sigma_1 = [1] - [0]$. Then $C \times C$ has four 0-cells $\{(0, 0), (1, 0), (0, 1), (1, 1)\}$, four 1-cells $\{\sigma_1 \times [0], \sigma_1 \times [1], [0] \times \sigma_1, [1] \times \sigma_1\}$, and one 2-cell $\sigma_1 \times \sigma_1$. ◁

Exercise B.7. Write the circle S^1 as a cell complex, then describe the product cell complex $S^1 \times S^1$. Compute the homology of $S^1 \times S^1$ from this cell complex.

The homology of a product is given by the *Künneth formula*. Because we study homology with complex coefficients, the formula is relatively simple.

Theorem B.16 (Künneth product formula). *If C' and C'' are the singular chain complexes for two cell complexes then there is a natural isomorphism*

$$\bigoplus_{p+q=n} H_p(C') \otimes H_q(C'') \to H_n(C' \otimes C'') .$$

Proof See [Mun84, Theorem 58.5]. □

Exercise B.8. Use the Künneth formula to compute the homology of $S^1 \times S^1$ and verify it is the same as you computed in Exercise B.7.

Generalizing the attaching maps for cell complexes, the ***attachment*** of a space Y to a space X along a closed subset $Y_0 \subseteq Y$ by the map $\phi : Y_0 \to X$ is the topological quotient $(X \sqcup Y)/\phi$ obtained from the disjoint union of X and Y by identifying each $y \in Y_0$ with $\phi(y) \in X$. The triple (Y, Y_0, ϕ) is known as ***attachment data***.

Remark. When ϕ is one-to-one, attachments are more or less the same as unions; that is, both X and Y naturally embed in the attachment and their union covers the attachment. In general, an attachment can be thought of as a union of a space X with a quotient Y/ϕ, where points $y_1, y_2 \in Y_0$ are identified if $\phi(y_1) = \phi(y_2)$. While the two spaces in a union play symmetric roles, the attachment of a space Y to a space X is described asymmetrically. This coincides with the framework of *filtered spaces* built up by successive attachments, described in the next appendix. Thus attachments, while not entirely new, provide a useful way to build up a space.

Finally, we define a product on pairs by

$$(X', Y') \times (X'', Y'') = (X' \times X'', X' \times Y'' \cup Y' \times X'') .$$

One nice property of this definition is that the singular chain complex $C(X, Y)$ for a pair $(X, Y) = (X', Y') \times (X'', Y'')$ turns out to be isomorphic to a tensor product of the singular chain complexes for the pairs (X', Y') and (X'', Y''). The Künneth product formula for chain complexes results in a homology formula for products of pairs.

Corollary B.17 (Künneth formula for pairs). *For pairs* (X', X'') *and* (Y', Y''),

$$\begin{aligned} H_n(X' \times X'', X' \times Y'' \cup Y' \times X'') &= H_n((X', Y') \times (X'', Y'')) \\ &\cong \bigoplus_{p+q=n} H_p(X', Y') \otimes H_q(X'', Y'') . \end{aligned}$$

B.3 Cohomology

Given a chain complex

$$\cdots \to C_{n+1} \xrightarrow{\partial_{n+1}} C_n \xrightarrow{\partial_n} C_{n-1} \to \cdots$$

we may replace each vector space C_n by its dual C^n consisting of linear maps from C_n to $\mathbb{C}$. As for homology, we consider cohomology with complex coefficients. The elements of C^n are called ***n-cochains*** and the boundary map $\partial_n : C_n \to C_{n-1}$ on n-chains induces a dual map $\delta^n : C^{n-1} \to C^n$ on n-cochains: if $f \in C^{n-1}$ is a linear map from C_{n-1} to $\mathbb{C}$ then $\delta^n(f) \in C^n$ is the linear map $f \circ \partial_n$ from C_n to $\mathbb{C}$. It can be verified that $\delta^n \circ \delta^{n-1} = 0$ for all n, so we have a ***cochain complex***

$$\cdots \leftarrow C^{n+1} \xleftarrow{\delta^n} C^n \xleftarrow{\delta^{n-1}} C^{n-1} \leftarrow \cdots .$$

The quotient of the kernel of δ^n (the ***n-cocycles***) by the image of δ^{n-1} (the ***n-coboundaries***) is called the ***nth cohomology group*** of C and is denoted $H^n(C)$. The value of an n-cocycle ν evaluated at an n-cycle σ depends only on the cohomology class $[\nu]$ of ν and the homology class $[\sigma]$ of σ, so this evaluation defines a product $\langle \omega, \eta \rangle$ for $\omega \in H^n(C)$ and $\eta \in H_n(C)$. If $C = C(X)$ is the singular chain complex (or smooth chain complex) of a topological space X, then we use the notation $C^n(X) = C^n$ for the ***singular n-cochains*** of X and $H^n(X) = H^n(C)$ for the ***nth singular cohomology group*** of X. If X is a cell complex then $H^n(X)$ is the dual space of $H_n(X)$.

The functor taking a topological space to its singular or smooth chain complex is covariant, as is the functor from a chain complex to its homology groups. Hence, as noted above, any map $f : X \to Y$ of topological spaces induces maps $f_* : C(X) \to C(Y)$ and $f_* : H_*(X) \to H_*(Y)$. Conversely, the singular or smooth cochain complex of a space is a contravariant functor, so a map $f : X \to Y$ induces a map $f^* : H^*(Y) \to H^*(X)$.

On a manifold we may identify a p-form ω with the smooth p-cochain defined by $\alpha \mapsto \int_\alpha \omega$. Using the definition of the coboundary δ and Stokes's Theorem,

$$\delta\omega(C) = \omega(\partial C) = \int_{\partial C} \omega = \int_C d\omega = d\omega(C)\,.$$

In other words, $\delta\omega = d\omega$, so cocycles correspond to closed forms and we have the following.

Theorem B.18 (integral depends only on homology class)**.** *Let ω be a closed p-form holomorphic on an embedded complex manifold $\mathcal{M} \subseteq \mathbb{C}^n$ (if $p = n$ then ω is always closed). Let C be a singular p-cycle on $\mathcal{M}$. Then $\int_C \omega$ depends on C only via the homology class $[C]$ of C in $H_p(\mathcal{M})$ and on ω only via the cohomology class $[\omega]$ of ω in $H^p(\mathcal{M})$.* □

This theorem is one reason for our detour into topology. Another is the de Rham Theorem. Let ι be the map that takes the smooth p-form ω and maps it to the p-cochain $C \mapsto \int_C \omega$ as C varies over p-chains. This map is in general not a bijection: there may be linear maps on chains that are not represented by integrals of smooth forms. Nevertheless, the induced map ι_* will be an isomorphism from the singular cohomology of $\mathcal{M}$ with coefficients in $\mathbb{R}$ or $\mathbb{C}$ to the cohomology H^*_{DR} of the ***de Rham complex*** of smooth p-forms with coboundary given by the differential operator d.

Theorem B.19 (de Rham Theorem)**.** *Let X be a real manifold. The identification of p-forms with cochains induces an isomorphism $H^*_{\mathrm{DR}}(X) \cong H^*(X)$.*

Proof See [Lee03, Theorem 18.7]. □

Remark. A product called the ***cup product*** may be defined on cochains, satisfying a product rule with respect to the d operator. The cup product endows cohomology with the structure of a graded $\mathbb{C}$-algebra, and the isomorphism in the de Rham Theorem is in fact a ring isomorphism, mapping the wedge product to the cup product.

B.4 Topology of complex manifolds

For us, complex manifolds usually arise as (subsets of) varieties or the complements of varieties in $\mathbb{C}_*^d$, with ACSV requiring the integration of holomorphic forms over chains of real dimension p contained in complex p-manifolds. This final section shows how the complex structure of a complex p-manifold makes it behave in several ways like a real p-manifold, even though it actually has real dimension $2p$.

Proposition B.20 (Andreotti–Frankel Theorem). *If X is a complex p-manifold embedded in $\mathbb{C}^n$ for some $n \geq p$ then X is homotopy equivalent to a CW complex of dimension at most p. It follows that the singular homology groups $H_k(X)$ and singular cohomology groups $H^k(X)$ vanish for all $k > p$.*

Proof Andreotti and Frankel [AF59] proved this for smooth algebraic (and analytic) varieties using Morse theory. A sketch is given at the end of Appendix C. □

Remark B.21. The complex projective space $\mathbb{CP}^k$ is a complex manifold having nonvanishing homology in all even dimensions up to $2k$. Therefore it violates the conclusions of the Andreotti–Frankel Theorem, and cannot be embedded in $\mathbb{C}^n$ for any n. This contrasts to the real case, where the Whitney embedding theorem states that any real k-manifold can be embedded into $\mathbb{R}^{2k}$.

In fact, we can compute homology by considering only holomorphic forms. If $\mathcal{M}$ is a complex p-manifold then the operator $\omega \mapsto d\omega$ preserves holomorphicity, so the holomorphic forms on $\mathcal{M}$ define a sub-cochain complex $C^{n,\text{holo}}$ of the de Rham complex C^n, called the ***holomorphic de Rham complex***. The inclusion $C^{n,\text{holo}} \hookrightarrow C^n$ does not, in general, induce an isomorphism on cohomology, but once again this difficulty can be overcome by restricting to manifolds embedded in complex space.

Proposition B.22 (holomorphic de Rham cohomology). *Let $\mathcal{M}$ be a complex p-manifold embedded in $\mathbb{C}^n$. Then the inclusions $C^{n,holo}(\mathcal{M}) \hookrightarrow C^n(\mathcal{M})$ induce an isomorphism of cohomology rings. In particular, $H^{holo,k}(\mathcal{M}) \cong H^k(\mathcal{M})$ for all $k \geq 0$.* □

Proof See Voisin [Voi02] and the notes at the end of this appendix. □

We finish this appendix by observing a corollary of Proposition B.10 in the complex setting.

Corollary B.23 (asymptotics depend only on relative homology class). *Let X be a complex manifold of dimension n with submanifold Y also of dimension* *n, and define* $\omega = \omega_\lambda = \exp(\lambda\,\phi(\mathbf{z}))\,\eta$ *for some holomorphic n-form* η *and holomorphic function* ϕ *on X. When* $C \equiv C'$ *in* $H_n(X, Y)$ *with* $\mathrm{Re}\{\phi\} \leq \beta$ *on Y then, as* $\lambda \to \infty$,

$$\int_C \omega_\lambda = \int_{C'} \omega_\lambda + O\left(e^{\lambda\beta}\right).$$

Proof This follows immediately from Proposition B.10, because $d\omega = 0$ for a holomorphic n-form on a complex n-manifold. □

Notes

From their origins near the end of the nineteenth century, homology and cohomology have become crucial tools in many areas of mathematics. Much of our presentation of the material in this appendix follows Hatcher [Hat02] and Munkres [Mun84], and further details can be found in those sources.

The Andreotti–Frankel Theorem is true in much greater generality than Proposition B.20: for instance, it holds for all algebraic varieties in complex affine space, regardless of whether they are smooth or singular. This was first proved in [Kar79] via stratified Morse theory. The complement of a variety $\mathcal{V}_Q$ is biholomorphically equivalent to the variety $\mathcal{V}_{1-z_{d+1}Q}$ in one greater dimension, hence complements of d-dimensional affine algebraic varieties are also homotopy equivalent to d-dimensional cell complexes.

Voisin [Voi02] proves Proposition B.22 by showing that holomorphic de Rham *hypercohomology* (cohomology with coefficients in a sheaf resolution) is the same as the ordinary de Rham cohomology, hence the same as smooth cohomology and singular cohomology. For *Stein spaces*, such as embedded complex manifolds, this resolution is flat and holomorphic de Rham hypercohomology boils down to the cohomology of the holomorphic de Rham complex itself. Special cases were known earlier; for example, if $\mathcal{A}$ is a complex hyperplane arrangement then Brieskorn [Bri73] showed that the forms df/f as f varies over annihilators of hyperplanes in $\mathcal{A}$ generate the cohomology ring of the complement $\mathcal{M}$ of $\mathcal{A}$.

Additional exercises

Exercise B.9. Define the ***Möbius strip*** as the quotient of the cell complex representing the unit square as $\Delta_1 \times \Delta_1$ via the three identifications $(0,0) \sim (1,1)$, $(0,1) \sim (1,0)$ and $(0) \times \Delta_1 \sim -(1) \times \Delta_1$.

(1) What is the dimension of this cell complex?

(2) Give a basis for each space Z_0, B_0, Z_1, B_1, Z_2, and B_2.
(3) Compute the homology of the Möbius strip with coefficients in $\mathbb{C}$ from this cell complex.
(4) What changes, if anything, if you use coefficients in $\mathbb{Z}$ instead of coefficients in $\mathbb{C}$?

Exercise B.10. Let $X = \mathbb{C}_* = \mathbb{C} \setminus \{0\}$ be the punctured plane, the simplest case of the complement of a hyperplane arrangement. To establish the complex de Rham Theorem for $H^1(X)$ we need to show that holomorphic 1-forms ω and θ map to the same element of the dual of $H_1(X)$ if and only if they differ by a coboundary df.

(1) Use Theorem B.18 to prove the forward implication.
(2) Compute the homology of X by verifying that the embedding of S^1 into X is a homotopy equivalence.
(3) Let ω be any holomorphic 1-form on X. Use Stokes's Theorem to show that $\int_C \omega = 0$ for any C homologous to zero in $H_1(X)$.
(4) Let γ be the unit circle oriented, say, counterclockwise, and let $\eta = \omega - c\,dz/z$, where $c = (2\pi i)^{-1} \int_\gamma \omega$. Show that $\int_C \eta = 0$ for every $C \in Z_1(X)$. *Hint:* Use part 2.
(5) Show that $\int_\gamma \omega = \int_\gamma \theta$ implies $\omega - \theta = df$ for some holomorphic function f. *Hint:* You can construct f by integrating from an arbitrary fixed basepoint.

Exercise B.11. Let X be the complex curve $\{(x, y) \in \mathbb{C}^2 : x^2 + y^2 = 1\}$. By the Andreotti–Frankel Theorem, it is homotopy equivalent to a cell complex of (real) dimension 1. Demonstrate this by finding a deformation retract of X onto a one-dimensional manifold.

Appendix C

Residue forms and classical Morse theory

In our first two appendices, we developed the mathematics needed to prove that the Cauchy integral representation

$$a_{\boldsymbol{r}} = \left(\frac{1}{2\pi i}\right)^{s} \int_{C} F(\boldsymbol{z})\, \boldsymbol{z}^{-\boldsymbol{r}-\boldsymbol{1}} d\boldsymbol{z}$$

for the coefficients of a convergent Laurent series $F(\boldsymbol{z}) = \frac{P(\boldsymbol{z})}{Q(\boldsymbol{z})} = \sum_{\boldsymbol{r}} a_{\boldsymbol{r}} \boldsymbol{z}^{\boldsymbol{r}}$ depends only on the singular homology class of the chain C and the de Rham cohomology class of the form $F(\boldsymbol{z})\, \boldsymbol{z}^{-\boldsymbol{r}-\boldsymbol{1}} d\boldsymbol{z}$ in the domain of holomorphicity

$$\mathcal{M} = \left\{ \boldsymbol{z} \in \mathbb{C}^{d} : Q(\boldsymbol{z}) \prod_{j=1}^{d} z_j \neq 0 \right\}$$

of the integrand.

In this appendix, we begin to discuss how such a representation allows us to manipulate Cauchy integrals into a form where we can derive asymptotic information. First, we discuss intersection classes and residue forms, which illustrate how to convert the Cauchy integral into an integral lying "on" the singular set $\mathcal{V} \subset \mathbb{C}^d$ of F. After introducing these concepts, we discuss how to use Morse theory to manipulate integrals over chains in $\mathcal{V}$ into representations that will ultimately allow us to use saddle point approximations. Morse theory is a large subject, and our treatment is restricted to the core topics we need: height functions, attachments, homology groups, and homotopy type. In this appendix we focus on the case where $\mathcal{V}_* := \mathcal{V} \cap \mathbb{C}^d_*$ is a complex manifold. Appendix D, our final appendix, describes extensions of this material to general algebraic sets (and their complements) using *stratified* Morse theory.

C.1 Intersection classes

Before describing how to generalize residues from the classical univariate setting to several variables, we first need to describe the domains of integration over which we can take multidimensional residues of differential forms with singularities on $\mathcal{V}$. These domains of integration will be *intersection classes*.

The intuition behind intersection classes is captured in Figure C.1. A torus T on one side of $\mathcal{V}$ expands to a torus T' on the other side of $\mathcal{V}$. Mathematically, this expansion could be obtained by expanding each coordinate at a constant rate, or by a more complicated deformation, or perhaps not by a deformation at all but through a *cobordism*, meaning some $(d+1)$-chain whose boundary is $T' - T$. In getting from T to T' this expansion crosses $\mathcal{V}$; if the crossing is transverse, as it will be generically, it sweeps out a $(d-1)$-chain $\gamma \subseteq \mathcal{V}$. For the intersection class to be well defined for our purposes, the homology class of γ in $H_{d-1}(\mathcal{V}_*)$ should depend only on the homology classes of T and T' in $\mathcal{M}$.

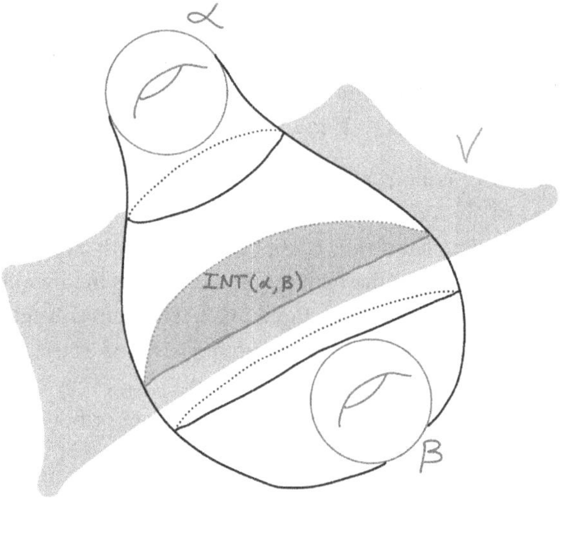

Figure C.1 The intersection class of a cobordism from α to β.

The concepts involved in defining an intersection class are analytic in nature, so we work with analytic functions instead of restricting ourselves to polynomials. Let $\mathcal{V} = \mathcal{V}_Q$ be the complex manifold in $\mathbb{C}^d$ defined by the vanishing of an analytic function $Q(\mathbf{z})$ on $\mathbb{C}^d$ *whose gradient does not vanish* on $\mathcal{V}$. Our first step in constructing the intersection class is to derive a diffeomorphism between a neighborhood of $\mathcal{V}$ in $\mathbb{C}^d$ and a product $A \times B$, where A is a connected open set in $\mathcal{V}$ and B is a neighborhood of the origin in $\mathbb{C}$. This is accomplished by considering the embedded complex manifold $\mathcal{V}$ in its ambient space $\mathbb{C}^d$. The tangent bundle of $\mathcal{V}$ may be identified with a sub-bundle of the tangent space to $\mathbb{C}^d$ by sending $\mathbf{v}$ to $e_*(\mathbf{v})$, where $e : \mathcal{V} \to \mathbb{C}^d$ is the embedding of $\mathcal{V}$ into $\mathbb{C}^d$.

Recall from Appendix A that for any $\mathbf{w} \in \mathbb{C}^d$ there is a natural identification ϕ of the tangent space $T_{\mathbf{w}}\mathbb{C}^d$ with $\mathbb{C}^d$ using the standard basis $\partial/\partial z_1, \ldots, \partial/\partial z_d$ for the holomorphic tangent space. Intuitively, we decompose $\mathbb{C}^d$ near $\mathbf{w}$ by taking the tangent plane to $\mathcal{V}$ at $\mathbf{w}$ and its orthogonal complement. Formally, the ***embedded tangent space*** and the ***embedded normal space*** of $\mathcal{V}$ at $\mathbf{w}$ are the subsets $S_{\mathbf{w}} := \{\mathbf{w}+\phi(e_*(X)) : X \in T_{\mathbf{w}}\mathcal{V}\}$ and $S'_{\mathbf{w}} := \{\mathbf{w}+\mathbf{v} : \mathbf{v} \in N_{\mathbf{w}}\mathcal{V}\}$ of $\mathbb{C}^d$, respectively, where $N_{\mathbf{w}}\mathcal{V} \subseteq \mathbb{C}^d$ is the orthogonal complement to $\phi(e_*(T_{\mathbf{w}}\mathcal{V}))$.

Under our assumptions, $N_{\mathbf{w}}\mathcal{V}$ is the one-dimensional complex vector space (or two-dimensional real vector space) described in local coordinates as the span of the vector $(\nabla Q)(\mathbf{w})$. The ***total space of the normal bundle*** to $\mathcal{V}$ is the set $\{(\mathbf{w}, \mathbf{v}) \in \mathcal{V} \times \mathbb{C}^d : \mathbf{v} \in N_{\mathbf{w}}\mathcal{V}\}$ pairing elements of $\mathcal{V}$ and normal vectors.

Lemma C.1 (Collar Lemma)**.** *Under our running assumption that the gradient of Q is nonvanishing on $\mathcal{V}$, there is an open neighborhood of $\mathcal{V}$ in $\mathbb{C}^d$ that is diffeomorphic to the total space of the normal bundle to $\mathcal{V}$ under a diffeomorphism that maps $\mathbf{w} \in \mathcal{V}$ to the vector $(\mathbf{w}, \mathbf{0})$.*

Proof Because ∇Q is nonvanishing on $\mathcal{V}$, the gradient ∇Q is non-zero in a neighborhood of $\mathcal{V}$ and thus defines a complex line bundle whose integral surfaces have real dimension two. If U is any sufficiently small neighborhood of $\mathcal{V}$, we let $a : U \to \mathcal{V}$ be the map sending $\mathbf{z} \in U$ to the unique point of $\mathcal{V}$ on whose integral curve it lies; see Figure C.2. The map ψ sending $\mathbf{z} \in U$ to $\psi(\mathbf{z}) = (a(\mathbf{z}), \rho(\mathbf{z}))$ is the desired diffeomorphism, where $\rho(\mathbf{z})$ is the projection of $\mathbf{z} - a(\mathbf{z})$ onto the affine set $S'_{a(\mathbf{z})}$, because $\rho(\mathbf{z}) \in N_{a(\mathbf{z})}\mathcal{V}$ by construction and the kernels of da and $d\rho$ are transverse on $\mathcal{V}$ (they are orthogonal subspaces), hence also transverse in a sufficiently small neighborhood of $\mathcal{V}$. □

Lemma C.1 implies that for any k-chain γ in $\mathcal{V}$ we can define a $(k+1)$-chain $\circ\gamma$, which we call a ***tube around*** γ, by taking the union of small circles in the fibers of the normal bundle with centers in γ. The radii of these disks should be positive and small enough to fit into the domain of the collar map, but can (continuously) vary with the point on the base. Different choices of the radii of these circles lead to homologous tubes. Similarly, we let $\bullet\gamma$ denote the union of solid disks in the fibers of the normal bundle with centers in γ. The elementary rules for boundaries of products imply

$$\partial(\circ\gamma) = \circ(\partial\gamma) \quad \text{and} \quad \partial(\bullet\gamma) = \circ\gamma \cup \bullet(\partial\gamma)\,.$$

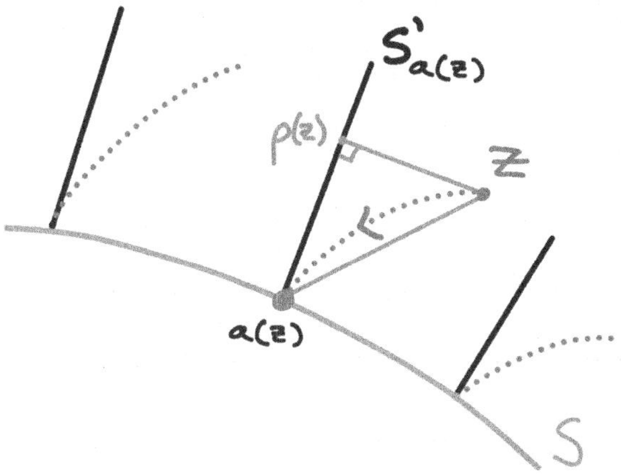

Figure C.2 Integral curves (dotted) making up the normal bundle, and a decomposition of $\boldsymbol{z}$ into $(a(\boldsymbol{z}), \rho(\boldsymbol{z}))$.

Because $\circ$ commutes with ∂, cycles map to cycles, boundaries map to boundaries, and the map $\circ$ on the singular chain complex of $\mathcal{V}$ induces a map from $H_{k-1}(\mathcal{V}_*)$ to $H_k(\mathcal{M})$, where $\mathcal{M} = \mathbb{C}_*^d \setminus \mathcal{V}$. To simplify notation, we also denote this map on homology by $\circ$.

We are now ready to define intersection classes after recalling a few constructions from differential geometry. Two submanifolds $A, B \subset \mathbb{C}^d$ are said to ***intersect transversely*** if for all $\boldsymbol{w} \in A \cap B$ the tangent spaces of A and B at $\boldsymbol{w}$ jointly span $\mathbb{C}^d$. Two classic results of differential geometry state that if A and B intersect transversely then $A \cap B$ is a manifold, and that if B is fixed and A is any manifold then A can be slightly perturbed into a manifold A' that intersects B transversely (i.e., transversality is a *generic* property) – see [Hir76, Chapter 3], for instance.

Theorem C.2 (intersection classes). *Define* $\circ : H_{d-1}(\mathcal{V}) \to H_d(\mathcal{M})$ *as above. Under our running assumption that* $\mathcal{V}$ *is a manifold,*

(i) $\circ$ *is injective and its image is the kernel of the map* ι_* *induced by the inclusion* $\mathcal{M} \xrightarrow{\iota} \mathbb{C}_*^d$.

(ii) Given $\alpha \in \ker(\iota_*)$ *one may compute the inverse* $\mathcal{I}(\alpha) = \circ^{-1}(\alpha)$ *by intersecting* $\mathcal{V}_*$ *with any* $(d+1)$*-chain in* $\mathbb{C}_*^{d+1}$ *whose boundary is* α *and for which the intersection with* $\mathcal{V}_*$ *is transverse.*

When $\alpha = C - C'$ in Theorem C.2, where C and C' are two d-cycles in $\mathcal{M}$ homologous in $\mathbb{C}_*^d$, we call $\mathbf{INT}[C, C'; \mathcal{V}] = \mathcal{I}(C - C')$ the ***intersection class*** of C and C'. We usually use the intersection class when $C = T$ and $C' = T'$ are tori and T' can be deformed to points where the Cauchy integral representing a multivariate sequence is asymptotically negligible.

Proof of Theorem C.2 The *Thom–Gysin long exact sequence* implies exactness of a sequence

$$0 \to H_{d-1}(\mathcal{V}_*) \xrightarrow{\circ} H_d(\mathcal{M}) \to H_d(\mathbb{C}_*^d). \tag{C.1.1}$$

This may be found in [Gor75, page 127] (where, in the notation of that source, $W = \mathbb{C}_*^d$) though in the particular situation at hand it goes back to Leray [Ler59]. Injectivity of $\circ$ follows from exactness at $H_{d-1}(\mathcal{V}_*)$ while the rest of part (*i*) follows from exactness at $H_d(\mathcal{M})$.

For part (*ii*), we begin by showing that the map I which takes a subset S of $\mathbb{C}_*^d$ transverse to $\mathcal{V}_*$ and returns $I(S) = S \cap \mathcal{V}_*$ induces a well-defined map from $\ker(\iota_*)$ to $H_{d-1}(\mathcal{V}_*)$. Because transversality is generic, given any $\alpha \in \ker(\iota_*)$ there exist $(d+1)$-chains intersecting $\mathcal{V}_*$ transversely whose boundary is α. If $\mathcal{D}$ is such a chain and $C = I(\mathcal{D})$ then C is a cycle, since

$$\partial C = \partial(\mathcal{D} \cap \mathcal{V}_*) = (\partial \mathcal{D}) \cap \mathcal{V}_* = \alpha \cap \mathcal{V}_* = \emptyset .$$

Let $\mathcal{D}_1$ and $\mathcal{D}_2$ be two such chains, and define $C_j = \mathcal{D}_j \cap \mathcal{V}_*$. Observe that $\mathcal{D}_1 - \mathcal{D}_2$ is null homologous because there is no $(d+1)$-homology in $\mathbb{C}_*^d$. Thus, $\mathcal{D}_1 - \mathcal{D}_2 = \partial \mathbf{H}$ for some $(d+1)$-chain $\mathbf{H}$ in $\mathbb{C}_*^d$. Choosing $\mathbf{H}$ transverse to $\mathcal{V}_*$,

$$C_1 - C_2 = I(\mathcal{D}_1 - \mathcal{D}_2) = \partial(\mathbf{H} \cap \mathcal{V}_*)$$

is a boundary in $\mathcal{V}_*$. Thus, the class $[I(\mathcal{D})]$ in $H_{d-1}(\mathcal{V}_*)$ is the same for any $\mathcal{D}$ with $\partial \mathcal{D} = \alpha$. If $\alpha = \circ(\gamma)$ then taking $\mathcal{D} = \bullet(\gamma)$ gives $I(\mathcal{D}) = \gamma$, showing that I does in fact invert $\circ$ and thus computes $\mathcal{I}$. □

Remark C.3. Because $\mathbb{C}_*$ is topologically a circle, the homology group $H_d(\mathbb{C}_*^d)$ is cyclic and generated by a product of small circles about the coordinate axes, and $H_k(\mathbb{C}_*^d)$ vanishes for $k > d$. The kernel of ι_* consists of the classes that don't *link* the origin in $\mathbb{C}_*^d$, i.e., the classes σ for which the integer invariant $\ell(\sigma) = (2\pi i)^{-d} \int_\sigma dz_1/z_1 \wedge \cdots \wedge dz_d/z_d$ vanishes. This holds, for example, if $\sigma = T - T'$, where T and T' are standard oriented tori around the origin, since $\ell(T) = \ell(T') = 1$.

There is a version of the intersection class in relative homology as well. This will be useful to us when we integrate over a difference of tori, one being the starting domain of integration in the Cauchy integral and the other being a "large" torus, because it helps us ignore whether we have chosen a large enough torus to avoid $\mathcal{V}_*$ at points that are asymptotically negligible. We omit the proof of this construction, which is similar to the proof of Theorem C.2.

Corollary C.4. *Let Y be a closed subspace of $\mathbb{C}_*^d$ and let α and β be relative cycles in the pair $(\mathcal{M}, \mathcal{M} \cap Y)$ that are homologous in $(\mathbb{C}_*^d, Y)$. There is a well-defined intersection class* $\mathbf{INT}[\alpha, \beta; \mathcal{V}]_Y \in H_{d-1}(\mathcal{V}_*, Y)$ *such that if* $\mathbf{H}$ *is any* $(d+1)$*-chain in* $\mathbb{C}_*^d$ *with* $\partial \mathbf{H} = \alpha' - \beta' + \gamma$*, where* $[\alpha'] = [\alpha]$ *and* $[\beta'] = [\beta]$ *in* $H_d(\mathcal{M}, \mathcal{M} \cap Y)$ *and* $\gamma \in Y$*, and if* $\mathbf{H}$ *intersects* $\mathcal{V}$ *transversely, then* $\mathbf{H} \cap \mathcal{V}$ *is a relative cycle in the class* $\mathbf{INT}[\alpha, \beta; \mathcal{V}]_Y$. □

By the excision property of homology, the pair $(\mathcal{M}, \mathcal{M} \cap Y)$ is homotopy equivalent to the pair $(\mathcal{M} \setminus Y^\circ, \partial Y)$. This allows us to extend Corollary C.4 to the case where α and β can intersect $\mathcal{V}$, but only in the interior of Y.

Corollary C.5. *Let Y be a closed subspace of $\mathbb{C}_*^d$ and let α and β be relative cycles homologous in $(\mathbb{C}^d, Y)$ intersecting $\mathcal{V}$ only in the interior of Y. There is a well-defined intersection class* $\mathbf{INT}[\alpha, \beta; \mathcal{V}]_Y \in H_{d-1}(\mathcal{V}_*, \mathcal{V}_* \cap Y)$*, depending only on the class of* $\alpha - \beta$ *in* $H_d(\mathcal{M}, \mathcal{M} \cap Y)$*, such that if* $\mathbf{H}$ *is a* $(d+1)$*-chain in* $\mathbb{C}_*^d$ *with* $\partial \mathbf{H} = \alpha - \beta + \gamma$*, where* γ *is supported on the interior of* Y*, and if the intersection of* $\mathbf{H}$ *with* $\mathcal{V}$ *is transverse away from the interior of* Y*, then* $\mathbf{H} \cap \mathcal{V}$ *is a relative cycle representing the class* $\mathbf{INT}[\alpha, \beta; \mathcal{V}]_Y$. □

In the special case of Corollary C.5, where $\beta = 0$ and Y is the set of points at height c or less, we denote the ***relative intersection class*** by $\mathbf{INT}[\alpha; \mathcal{V}]_{\leq c}$.

C.2 Residue forms and the residue integral theorem

Integrating a differential form over a difference of chains can often be reduced to integrating a *residue form* over an intersection cycle. Because residues depend on local behavior, we work with subsets of $\mathbb{C}^d$ that are *locally* defined by analytic functions. An ***analytic hypersurface*** (or simply hypersurface) is a set $\mathcal{V} \subset \mathbb{C}^d$ such that for any $w \in \mathcal{V}$ and any sufficiently small neighborhood $\mathcal{D}$ of w in $\mathbb{C}^d$ there is an analytic function $Q_\mathcal{D}$ on $\mathcal{V} \cap \mathcal{D}$ such that $\mathcal{V} \cap \mathcal{D} = \{z \in \mathcal{D} : Q_\mathcal{D}(z) = 0\}$. If the function $Q_\mathcal{D}$ can be chosen to have nonvanishing gradient on $\mathcal{V} \cap \mathcal{D}$ then we say $\mathcal{V}$ is a ***smooth analytic hypersurface at*** w, and we call $\mathcal{V}$ a ***smooth analytic hypersurface*** if it is a smooth analytic hypersurface at every point.

Although the theory of multivariate residues is much more involved than its univariate counterpart, we require it only for differential forms whose singularities lie on unions of smooth analytic hypersurfaces. We build our results in four degrees of generality, starting with forms having smooth simple poles, then forms with smooth higher order poles, followed by forms with transversely intersecting smooth sheets of simple poles, and concluding with forms having transversely intersecting smooth sheets with higher order poles.

C.2.1 Residue forms for smooth simple poles

Fix a smooth analytic hypersurface $\mathcal{V}$ defined locally by analytic functions $Q_{\mathcal{D}}$ as above, and recall our notation $\mathcal{M} = \mathbb{C}_*^d \setminus \mathcal{V}$ and $d\mathbf{z} = dz_1 \wedge \cdots \wedge dz_d$. We say that a d-form $\omega \in E^d(\mathcal{M})$ has ***smooth poles of order*** k on $\mathcal{V}$ if for any $\mathbf{a} \in \mathcal{V}$ there is a sufficiently small neighborhood $\mathcal{D}$ of $\mathbf{a}$ in $\mathbb{C}_*^d$ such that $Q_{\mathcal{D}}(\mathbf{z})^k \omega$ extends to a holomorphic form on $\mathcal{D}$ but $Q_{\mathcal{D}}(\mathbf{z})^j \omega$ does not extend to such a holomorphic form for any $0 \le j < k$. A form with smooth poles of order one is said to have *simple poles*.

Proposition C.6. *Let* ω *be a holomorphic d-form with smooth simple poles on* $\mathcal{V}$*, represented as a quotient* $\omega = P(\mathbf{z})/Q(\mathbf{z})\, d\mathbf{z}$ *of analytic functions on* $\mathcal{M} \cap \mathcal{D}$ *for some domain* $\mathcal{D} \subset \mathbb{C}^d$ *on which the gradient of* Q *does not vanish. If* $\mathcal{W} = \mathcal{V} \cap \mathcal{D}$ *and* $\iota : \mathcal{W} \hookrightarrow \mathcal{D}$ *is the inclusion of* $\mathcal{W}$ *into* $\mathbb{C}^d$ *then there is a* $(d-1)$*-form* θ *on* $\mathcal{D}$ *solving* $dQ \wedge \theta = P\, d\mathbf{z}$ *and any such solution restricts to a unique* $(d-1)$*-form* $\operatorname{Res}(\omega) = \iota^* \theta$ *on* $\mathcal{W}$ *called the* ***residue of*** ω *on* $\mathcal{W}$.

Remark C.7. Our definition of the residue is both ***natural***, meaning it does not depend on the particular polynomials P and Q used to represent $\omega = P(\mathbf{z})/Q(\mathbf{z})\, d\mathbf{z}$ on $\mathcal{D}$, and ***functorial***, meaning $\operatorname{Res}(f^*\omega) = f^* \operatorname{Res}(\omega)$ for smooth functions f. Because the residue is natural, we can define $\operatorname{Res}(\omega)$ on all of $\mathcal{V}$ by defining it locally over the elements of a cover of $\mathcal{V}$ by sufficiently small domains using Proposition C.6.

Proof We must show both that $dQ \wedge \theta = P\, d\mathbf{z}$ always has a holomorphic solution, and that the restriction of any such solution to $\mathcal{V}$ is unique. Uniqueness follows from Exercise A.18 at the end of Appendix A: if θ_1 and θ_2 are two solutions then $dQ \wedge (\theta_1 - \theta_2) = 0$, hence Exercise A.18 implies $\iota^*\theta_1 = \iota^*\theta_2$. The existence of a solution follows from the following proposition, which expresses the residue explicitly in local coordinates in sufficiently small neighborhoods, by combining residues in local neighborhoods as discussed in Remark C.7. □

Proposition C.8. *Under the hypotheses of Proposition C.6, if the partial derivative* $\partial Q / \partial z_k$ *is nonvanishing on* $\mathcal{D}$ *for some fixed* $k \le d$ *and* $\mathbf{r} \in \mathbb{Z}^d$ *then*

$$\operatorname{Res}(\mathbf{z}^{-\mathbf{r}}\omega) = (-1)^{k-1} \frac{\mathbf{z}^{-\mathbf{r}} P(\mathbf{z})}{Q_{z_k}(\mathbf{z})} d\mathbf{z}_{\hat{k}}\,, \tag{C.2.1}$$

where $d\mathbf{z}_{\hat{k}} = dz_1 \wedge \cdots \wedge dz_{k-1} \wedge dz_{k+1} \wedge \cdots \wedge dz_d$.

Proof If $k = 1$ and θ is the right-hand side of (C.2.1), then

$$dQ \wedge \theta = \left(\sum_{j=1}^{d} Q_{z_j}(\mathbf{z}) dz_j \right) \wedge \left(\frac{\mathbf{z}^{-\mathbf{r}} P}{Q_{z_1}(\mathbf{z})} dz_2 \wedge \cdots \wedge dz_d \right) = \mathbf{z}^{-\mathbf{r}} P\, d\mathbf{z}\,,$$

as desired. In the general case, the sign $(-1)^{k-1}$ comes from the position of dz_k in the wedge product. □

Exercise C.1. Let $\omega = 1/Q(x,y)\,dxdy$, where $Q(x,y) = 1 - x - xy + y^2$. Find a formula for $\text{Res}(\omega)$ on $\mathcal{V} = \mathcal{V}_Q$ in terms of dx only, and another in terms of dy only. Prove that the restrictions of these forms to $\mathcal{V}$ are equal.

Theorem C.9 (residue integral theorem). *Suppose $\mathcal{V} = \mathcal{V}_Q = \{z \in \mathbb{C}^d : Q(z) = 0\}$ is defined globally by a function Q that is analytic on a neighborhood of $\mathcal{V}$ and has nonvanishing gradient on $\mathcal{V}$, and let ω be a holomorphic d-form on $\mathcal{M}$ with smooth simple poles on $\mathcal{V}$. If α and β are d-cycles in $\mathcal{M}$ whose projections to $H_d(\mathbb{C}_*^d)$ are equal then*

$$\int_\alpha \omega - \int_\beta \omega = 2\pi i \int_{\mathbf{INT}[\alpha,\beta;\mathcal{V}]} \text{Res}(\omega)\,.$$

Proof Vanishing of $[\alpha - \beta]$ in $H_d(\mathbb{C}_*^d)$ by definition implies the existence of a $(d+1)$-chain $\mathbf{H}$ on $\mathbb{C}_*^d$ with boundary $\alpha - \beta$. Perturbing slightly if necessary, we can assume without loss of generality that $\mathbf{H}$ intersects $\mathcal{V}$ transversely. Letting N denote the intersection of $\mathbf{H}$ with a small neighborhood of $\mathcal{V}$ and $\Theta = \mathbf{H} - N$, the vanishing of holomorphic integrals of d-forms over boundaries (Theorem A.27) implies that the integral of the holomorphic d-form ω over $\partial\Theta$ vanishes. In other words,

$$\int_\alpha \omega - \int_\beta \omega = \int_{\partial N} \omega\,.$$

The Collar Lemma (Lemma C.1) implies that N is homotopic to a product $\sigma \times B_\varepsilon$, where $\sigma = \mathbf{H} \cap \mathcal{V}$. Thus ∂N is homotopic to $\partial(\sigma \times B_\varepsilon)$, which is equal to $\sigma \times \partial B_\varepsilon$ because σ is a cycle, giving

$$\int_{\sigma\times\partial B_\varepsilon} \omega = \int_\sigma \left(\int_{\partial B_\varepsilon} \omega \right).$$

Using functoriality of the residue, we may change coordinates so that $\mathcal{V}$ is the complex hyperplane defined by $z_1 = 0$. Thus we need only prove our claim in the case where $Q(z) = z_1$. Writing $\omega = (P/z_1)dz_1 \wedge (dz_2 \wedge \cdots \wedge dz_d)$, the iterated integral is

$$\int_\sigma \left[\int_{\partial B_\varepsilon} \frac{P(z)}{z_1}\, dz_1 \right] dz_2 \wedge \cdots \wedge dz_d\,.$$

By standard univariate complex analysis, the inner integral at a point $(z_2, \ldots, z_d)$ is the residue with respect to t of the meromorphic function $P(t, z_2, \ldots, z_d)/t$ at the pole $(0, z_2, \ldots, z_d)$. This is equal to

$$(2\pi i) \int_\sigma P(0, z_2, \ldots, z_d)\,,$$

which in this special case is precisely $\int_\sigma \mathrm{Res}(\omega)$. □

There is also a relative version of this result.

Theorem C.10 (relative residue integral theorem). *Let $\mathcal{V}$ and ω be as in Theorem C.9. If Y is any closed subspace of $\mathbb{C}_*^d$ such that $H_d(\mathbb{C}_*^d, Y)$ vanishes, and if α is a d-cycle in $\mathcal{M}$, then*

$$\int_\alpha \omega = 2\pi i \int_{\mathbf{INT}[\alpha,0;\mathcal{V}]} \mathrm{Res}(\omega) + \int_{C'} \omega \tag{C.2.2}$$

for some chain C' supported on the interior of Y. In particular, if $\omega = \boldsymbol{z}^{-\boldsymbol{r}}\eta$ for some holomorphic form η on $\mathcal{M}$ and if Y is the set where the real part of $h_{\hat{\boldsymbol{r}}} = -\hat{\boldsymbol{r}} \cdot \log \boldsymbol{z}$ is at most c then, as $\lambda \to \infty$,

$$\int_\alpha \omega = 2\pi i \int_{\mathbf{INT}[\alpha,0;\mathcal{V}]} \mathrm{Res}(\omega) + O\left(e^{\lambda c'}\right) \tag{C.2.3}$$

for any $c' > c$.

Proof By the vanishing of $H_d(\mathbb{C}_*^d, Y)$ there is a $(d+1)$-chain $\mathbf{H}$ with $\partial \mathbf{H} = \alpha + \gamma$ and γ supported on the interior of Y. Let N denote the intersection of $\mathbf{H}$ with a neighborhood of $\mathcal{V}$. As before,

$$\int_\alpha \omega = \int_\gamma \omega + \int_{\partial N} \omega\,.$$

Letting $\sigma = \mathbf{H} \cap \mathcal{V}$, we recall that ∂N is homotopic to $\sigma \times B_\varepsilon$ plus a piece γ' in the interior of Y. Taking $C' = \gamma + \gamma'$, the rest of the proof of (C.2.3) is the same as that of Theorem C.9. The asymptotic estimate follows because $\left|\int_{C'} \omega\right| \le e^{\lambda c} \int_{C'} |\eta|$, as in the proof of Proposition B.10. □

C.2.2 Residue forms on smooth higher order poles

Let ω and ω' be holomorphic d-forms on $\mathcal{M}$ with simple poles on the smooth variety $\mathcal{V} = \mathcal{V}_Q$. If ω and ω' are cohomologous in $H^d(\mathcal{M})$ then $[\mathrm{Res}(\omega)] = [\mathrm{Res}(\omega')]$ in $H^{d-1}(\mathcal{V}_*)$, and the study of residue classes of forms with smooth higher order poles can be reduced to those with smooth simple poles using this property.

Lemma C.11 (Gelfand–Shilov reduction). *If ω is a holomorphic d-form on $\mathcal{M}$ with smooth poles of order $k \geq 2$ on $\mathcal{V}$ and the representation $\omega = P(z)/Q(z)^k \, dz$ holds on a domain $\mathcal{D}$ then*

$$\begin{aligned}\omega &= \frac{dQ}{Q^k} \wedge \psi + \frac{\theta}{Q^{k-1}} \\ &= d\left(\frac{-\psi}{(k-1)Q^{k-1}}\right) + \frac{\theta_1}{Q^{k-1}}\end{aligned}$$

for some holomorphic forms ψ and θ, where $\theta_1 = \theta + d\psi/(k-1)$. Thus, any d-form on $\mathcal{M}$ with smooth poles of order k is cohomologous to a d-form on $\mathcal{M}$ with smooth poles of order $k-1$.

Proof See [AY83, Lemma 17.1]. □

If ω is any d-form on $\mathcal{M}$ with smooth poles then Lemma C.11 implies ω is cohomologous to a d-form ω' on $\mathcal{M}$ with smooth simple poles, and we define the ***residue class*** $[\mathrm{Res}(\omega)]$ of ω to be the class $[\mathrm{Res}(\omega')] \in H^{d-1}(\mathcal{V}_*)$. To simplify notation we usually write $\mathrm{Res}(\omega)$ for the class $[\mathrm{Res}(\omega)]$. As our integrals of residues depend only on their cohomology classes, there is no harm in this abuse of notation. This inductive definition gives the following corollary of Theorem C.9.

Corollary C.12. *Suppose the assumptions of Theorem C.9 hold, except that ω can have smooth poles of any order on $\mathcal{V}$. If α and β are d-cycles in $\mathcal{M}$ whose projections to $H_d(\mathbb{C}^d_*)$ are equal then the identity*

$$\int_\alpha \omega - \int_\beta \omega = 2\pi i \int_{\mathbf{INT}[\alpha,\beta;\mathcal{V}]} \mathrm{Res}(\omega)$$

still holds.

Just as for smooth poles, there is an explicit formula for the residue of a form with higher order poles. We state the following theorem for the types of integrands that arise in our asymptotic analyses.

Lemma C.13. *Let $dz_{\hat{k}}$ denote the $(d-1)$-form $dz_1 \wedge \cdots \wedge dz_{k-1} \wedge dz_{k+1} \wedge \cdots \wedge dz_d$. Wherever the functions $P(z)z^{-r}$ and $Q(z)$ are analytic and the partial derivative $Q_{z_k}(z)$ does not vanish,*

$$\mathrm{Res}\left(z^{-r} \frac{P(z)}{Q(z)^\ell} \, dz\right) = z^{-r} \Phi_{r_k}(z) \tag{C.2.4}$$

for a polynomial

$$\Phi_{r_k}(z) = \left[(-1)^{k-1} \binom{-r_k}{\ell-1} z_k^{-(\ell-1)} \frac{P(z)}{Q_{z_k}(z)^\ell} + O\left(r_k^{\ell-2}\right)\right] dz_{\hat{k}}$$

in r_k of degree $\ell - 1$ whose coefficients are analytic functions of $\mathbf{z}$ explicitly given in terms of derivatives of P and Q.

Proof We induct on ℓ, with the case $\ell = 1$ handled by Proposition C.8. Assume for an induction that the lemma holds for $\ell - 1$. Because the residue of an exact form is zero, we let

$$\eta = (-1)^{k-1} \mathbf{z}^{-\mathbf{r}} \frac{P(\mathbf{z})}{(\ell - 1)Q(\mathbf{z})^{\ell-1} Q_{z_k}(\mathbf{z})} d\mathbf{z}_{\hat{k}}$$

and examine

$$\begin{aligned} 0 &= \operatorname{Res}(d\eta) \\ &= \operatorname{Res}\left[\mathbf{z}^{-\mathbf{r}} \frac{P_{z_k}(\mathbf{z})}{(\ell - 1)Q(\mathbf{z})^{\ell-1} Q_{z_k}(\mathbf{z})} d\mathbf{z} + \mathbf{z}^{-\mathbf{r}} \frac{-r_k P(\mathbf{z}) z_k^{-1}}{(\ell - 1)Q(\mathbf{z})^{\ell-1} Q_{z_k}(\mathbf{z})} d\mathbf{z} \right. \\ &\qquad \left. - \mathbf{z}^{-\mathbf{r}} \frac{P(\mathbf{z})}{Q(\mathbf{z})^{\ell}} d\mathbf{z} - \mathbf{z}^{-\mathbf{r}} \frac{P(\mathbf{z}) Q_{z_k, z_k}(\mathbf{z})}{(\ell - 1)Q(\mathbf{z})^{\ell-1} Q_{z_k}(\mathbf{z})} d\mathbf{z} \right]. \end{aligned}$$

Isolating the third term on the right yields

$$\begin{aligned} \operatorname{Res}\left(\mathbf{z}^{-\mathbf{r}} \frac{P(\mathbf{z})}{Q(\mathbf{z})^{\ell}} d\mathbf{z} \right) &= \operatorname{Res}\left(\mathbf{z}^{-\mathbf{r}} \frac{-r_k P(\mathbf{z}) z_k^{-1}}{(\ell - 1)Q(\mathbf{z})^{\ell-1} Q_{z_k}(\mathbf{z})} d\mathbf{z} \right) \\ &\quad + \operatorname{Res}\left(\mathbf{z}^{-\mathbf{r}} \frac{A(\mathbf{z})}{Q(\mathbf{z})^{\ell-1}} d\mathbf{z} \right), \end{aligned} \tag{C.2.5}$$

for an analytic function A independent of r_k. Applying the induction hypothesis to the first residue on the right-hand side of (C.2.5) shows that it equals

$$(-1)^{k-1} \left[\frac{-r_k}{\ell - 1} \binom{-r_k - 1}{\ell - 2} \mathbf{z}^{-\mathbf{r}} z_k^{-(\ell-2)} \frac{P(\mathbf{z}) z_k^{-1} / Q_{z_k}(\mathbf{z})}{Q_{z_k}(\mathbf{z})^{\ell-1}} + O\left(r_k^{\ell-3}\right) \right] d\mathbf{z}_{\hat{k}},$$

while applying the induction hypothesis to the second residue on the right-hand side of (C.2.5) proves that it is $O\left(r_k^{\ell-2}\right)$. Combining powers of $Q_{z_k}(\mathbf{z})$ and powers of z_k, and simplifying $\frac{-r_k}{\ell - 1} \binom{-r_k - 1}{\ell - 2} = \binom{-r_k}{\ell - 1}$, then gives the stated result. □

C.2.3 Iterated residue forms for simple poles on transverse sheets

In this section we summarize a generalization of residue forms to the case where $\mathcal{V}$ is the union of a finite number of smooth analytic hypersurfaces that intersect transversely. A full treatment of residues for forms with transverse poles can be found in [AY83, Section 16.5].

Definition C.14. If $\mathcal{V} \subset \mathbb{C}^d$ is an analytic hypersurface then we call $\boldsymbol{w} \in \mathcal{V}$ a ***transverse multiple point*** of $\mathcal{V}$ if there exists a neighborhood $\mathcal{D}$ of $\boldsymbol{w}$ in $\mathbb{C}^d$ such that $\mathcal{D} \cap \mathcal{V} = \mathcal{D} \cap (\mathcal{V}_{Q_1} \cup \cdots \cup \mathcal{V}_{Q_k})$ for smooth analytic hypersurfaces $\{\mathcal{V}_{Q_j} : 1 \leq j \leq k\}$ defined by analytic functions $Q_j(\boldsymbol{z})$ whose gradients at $\boldsymbol{z} = \boldsymbol{w}$ are linearly independent. When this collection of analytic functions is understood and $\boldsymbol{m} \in \mathbb{N}^k$ then we write $\boldsymbol{Q}(\boldsymbol{z})^{\boldsymbol{m}} = Q_1(\boldsymbol{z})^{m_1} \cdots Q_k(\boldsymbol{z})^{m_k}$. If every point of $\mathcal{V}$ is a transverse multiple point then we call $\mathcal{V}$ a ***transverse analytic hypersurface***.

Example C.15. Every smooth analytic hypersurface is a transverse analytic hypersurface. ◁

Fix a transverse analytic hypersurface $\mathcal{V}$ and let ω be a d-form on $\mathcal{M} = \mathbb{C}^d \setminus \mathcal{V}$. We say that ω has a ***transverse pole*** (or ***transverse multiple point***) of order $\boldsymbol{m} \in \mathbb{N}^k$ at $\boldsymbol{w} \in \mathcal{V}$ if

- there exists a neighborhood $\mathcal{D}$ of $\boldsymbol{w}$ in $\mathbb{C}^d$ and analytic functions $Q_1, \ldots, Q_k$ on $\mathcal{D}$ such that $\mathcal{D} \cap \mathcal{V} = \mathcal{D} \cap (\mathcal{V}_{Q_1} \cup \cdots \cup \mathcal{V}_{Q_k})$ and the gradients of the Q_i are linearly independent at $\boldsymbol{w}$ (in particular, they are all non-zero),
- there exists an analytic function P on $\mathcal{D}$ such that $\omega = P(\boldsymbol{z})/\boldsymbol{Q}(\boldsymbol{z})^{\boldsymbol{m}} d\boldsymbol{z}$ when $\boldsymbol{z} \in \mathcal{D} \cap \mathcal{M}$, and
- there is no possible choice of $\boldsymbol{Q}$ and P such that these properties hold with any coordinate of $\boldsymbol{m}$ decreased.

A transverse multiple point of order $\mathbf{1}$ is called a ***transverse simple pole*** (or ***transverse simple point***). The final item in this definition implies that the numerator P and denominator factors Q_k in the local representation of ω are coprime in the ring of germs of analytic functions, ensuring that $\boldsymbol{m}$ is the correct notion of order (no unwanted cancellation can occur).

Let $\boldsymbol{p}$ be a transverse simple pole of the d-form ω, with the local representation

$$\omega = \frac{P(\boldsymbol{z})}{Q_1(\boldsymbol{z}) \cdots Q_k(\boldsymbol{z})} d\boldsymbol{z}$$

in some neighborhood of $\boldsymbol{p}$. To simplify notation we write $\mathcal{V}_i = \mathcal{V}_{Q_i}$ and let $\mathcal{S} = \bigcap_{i=1}^k \mathcal{V}_i$ be the ***stratum*** of $\mathcal{V}$ containing $\boldsymbol{p}$. Because the gradients of the Q_i are linearly independent at $\boldsymbol{p}$, there exist coordinates $\boldsymbol{\pi} = \{\pi_1, \ldots, \pi_{d-k}\}$ that locally analytically parametrize $\mathcal{S}$ near $\boldsymbol{p}$. In particular, writing $\boldsymbol{z}_{\boldsymbol{\pi}} = (z_{\pi_1}, \ldots, z_{\pi_{d-k}})$ there exists a neighborhood $\mathcal{D}$ of $\boldsymbol{p}$ in $\mathbb{C}^d$ and analytic functions $\zeta_i(\boldsymbol{z}_{\boldsymbol{\pi}})$ on $\mathcal{D}$ for $i \notin \boldsymbol{\pi}$ such that $\boldsymbol{z} \in \mathcal{D}$ lies in $\mathcal{S}$ if and only if $z_i = \zeta_i(\boldsymbol{z}_{\boldsymbol{\pi}})$ for all $i \notin \boldsymbol{\pi}$.

As detailed later in these appendices, if $\mathcal{D}$ is sufficiently small then $\mathcal{M} \cap \mathcal{D}$ has a local product structure $\tilde{N} \times \mathcal{S}$. Because $\boldsymbol{p}$ is a transverse multiple point with k sheets, the factor $\tilde{N}$ is homotopy equivalent to a k-torus and we can represent the homology of $\tilde{N}$ using a product of k circles around $\mathbf{p}$. To make this explicit, we note that the map $\Psi : \mathcal{D} \to \mathbb{C}^d$ defined by

$$\Psi(\mathbf{z}) = (Q_1(\mathbf{z}), \ldots, Q_k(\mathbf{z}), z_{\pi_1} - p_{\pi_1}, \ldots, z_{\pi_{d-k}} - p_{\pi_{d-k}}) \tag{C.2.6}$$

is a bi-analytic change of coordinates taking $\mathcal{D} \cap \mathcal{S}$ to a neighborhood of the origin in $\{\mathbf{0}\} \times \mathbb{C}^{d-k}$. Let $T_\varepsilon \subseteq \mathbb{C}^k \times \{\mathbf{0}\}$ denote the product of circles of radius ε in each of the first k coordinates. If ε is sufficiently small then $T_\varepsilon \subset \Psi(\mathcal{D})$ and the cycle $T = \Psi^{-1}(T_\varepsilon)$ will be a generator for $H_k(\tilde{N})$. We give $\mathcal{D}$ the local product structure that Ψ^{-1} induces from the product structure on $\mathbb{C}^d$.

Definition C.16. If f is a differentiable function then the ***logarithmic gradient*** of f at $\mathbf{z}$ is

$$(\nabla_{\log} f)(\mathbf{z}) := (z_1 f_{z_1}(\mathbf{z}), \ldots, z_d f_{z_d}(\mathbf{z})) .$$

For each $\mathbf{z} \in \mathcal{S}$, the ***augmented lognormal matrix*** is the $d \times d$ matrix

$$\mathbf{\Gamma}_\Psi(\mathbf{z}) = \begin{pmatrix} (\nabla_{\log} Q_1)(\mathbf{z}) \\ \vdots \\ (\nabla_{\log} Q_k)(\mathbf{z}) \\ z_{\pi_1} \mathbf{e}_{\pi_1} \\ \vdots \\ z_{\pi_{d-k}} \mathbf{e}_{\pi_{d-k}} \end{pmatrix},$$

where $\mathbf{e}_j$ denotes the jth elementary basis vector. Equivalently, $\mathbf{\Gamma}_\Psi = J_\Psi D$ where D is the diagonal matrix with entries $z_1, \ldots, z_d$ and J_Ψ is the Jacobian matrix of the map Ψ.

Remark. The definition of $\mathbf{\Gamma}_\Psi$ depends on the choice of factorization, each factor being determined only up to a complex multiple. Suitable normalizations are assumed later in the definition of the torus $\mathcal{T}$ following (10.3) and the determination of the orientation in the proof of Theorem 10.25.

Theorem C.17 (iterated residues). *Under the setup discussed above, let ω be the holomorphic d-form $\omega = \frac{P(\mathbf{z})}{\prod_{j=1}^k Q_j(\mathbf{z})} d\mathbf{z}$ on $\mathcal{M} \cap \mathcal{D}$ and write $\mathcal{S}_\mathcal{D} = \mathcal{S} \cap \mathcal{D} = \mathcal{V}_1 \cap \cdots \cap \mathcal{V}_k \cap \mathcal{D}$.*

(i) **Iterated residue is well defined.** *The restriction to $\mathcal{S}_\mathcal{D}$ of any d-form θ on $\mathcal{D}$ satisfying*

$$dQ_1 \wedge \cdots \wedge dQ_k \wedge \theta = P\, d\mathbf{z} \tag{C.2.7}$$

is independent of the particular solution θ.

(ii) **Formula for the iterated residue.** *Denoting the* **iterated residue** *defined by this restriction by* $\text{Res}(\omega; \mathcal{S}_{\mathcal{D}})$*, there is a formula*

$$\text{Res}\left(\frac{P(\boldsymbol{z})}{\prod_{j=1}^{k} Q_j(\boldsymbol{z})} d\boldsymbol{z}\,;\, \mathcal{S}_{\mathcal{D}}\right) = \left.\frac{P(\boldsymbol{z})}{\det J_\Psi(\boldsymbol{z})}\right|_{\substack{z_i=\zeta_i(\boldsymbol{z}_{\boldsymbol{\pi}}) \\ \textit{for all } i\notin\boldsymbol{\pi}}} dz_{\pi_1} \wedge \cdots \wedge dz_{\pi_{d-k}} . \tag{C.2.8}$$

(iii) **Residue integral identity.** *Let* σ *be any* $(d-k)$*-chain in* $\mathcal{S}_{\mathcal{D}}$ *and* $T = \Psi^{-1}(T_\varepsilon)$ *be as above. Then*

$$\frac{1}{(2\pi i)^k} \int_{T\times\sigma} \frac{P(\boldsymbol{z})\, d\boldsymbol{z}}{\prod_{j=1}^{k} Q_j(\boldsymbol{z})} = \int_\sigma \text{Res}\left(\frac{P(\boldsymbol{z})}{\prod_{j=1}^{k} Q_j(\boldsymbol{z})}\,;\, \mathcal{S}_{\mathcal{D}}\right). \tag{C.2.9}$$

(iv) **Formula for Cauchy integral.** *In particular,*

$$\frac{1}{(2\pi i)^k} \int_{T\times\sigma} \frac{\boldsymbol{z}^{-\boldsymbol{r}-\boldsymbol{1}} P(\boldsymbol{z})}{\prod_{j=1}^{k} Q_j(\boldsymbol{z})} d\boldsymbol{z} = \int_\sigma \left.\frac{\boldsymbol{z}^{-\boldsymbol{r}} P(\boldsymbol{z})}{\det \boldsymbol{\Gamma}_\Psi(\boldsymbol{z})}\right|_{\substack{z_i=\zeta_i(\boldsymbol{z}_{\boldsymbol{\pi}}) \\ \textit{for all } i\notin\boldsymbol{\pi}}} dz_{\pi_1} \wedge\cdots\wedge dz_{\pi_{d-k}} . \tag{C.2.10}$$

Proof We first prove all four parts under the assumption that $Q_j(\boldsymbol{z}) = z_j$ for all $1 \le j \le k$. Setting $\pi_i = k+i$ for all $1 \le i \le d-k$, the form $\theta = P(\boldsymbol{z})\, dz_{k+1} \wedge \cdots \wedge dz_d$ satisfies (C.2.7). As in the proof of Proposition C.6 above, the result of Exercise A.17 implies that $\iota^*\theta$ is well defined, yielding (*i*). The formula (C.2.8) is also evident in this case: J_Ψ is the identity matrix, hence (C.2.8) agrees with our choice of θ after setting $z_i = 0$ for $1 \le i \le k$, proving (*ii*). For (*iii*), we write the left-hand side as an iterated integral

$$\frac{1}{(2\pi i)^k} \int_\sigma \int_{\gamma_1} \cdots \int_{\gamma_k} \frac{P(\boldsymbol{z})\, d\boldsymbol{z}}{\prod_{j=1}^{k} Q_j(\boldsymbol{z})},$$

where γ_j is the circle of radius ε about the origin in the jth coordinate. Applying the univariate residue theorem to each of the inner k integrals leaves

$$\int_\sigma P(\boldsymbol{z})\, dz_{k+1} \wedge \cdots \wedge dz_d ,$$

proving (*iii*). Finally, (*iv*) follows from (*iii*) by replacing $P(\boldsymbol{z})$ with $\boldsymbol{z}^{-\boldsymbol{r}-\boldsymbol{1}} P(\boldsymbol{z})$ in (C.2.8) and absorbing one factor of each z_j in the denominator when going from $\det J_\Psi$ to $\det \boldsymbol{\Gamma}_\Psi$.

For the general case, map by Ψ and use functoriality. The fact that Res is well defined and functorial follows from the same argument as in the proof of Proposition C.6. Applying the case already proved to the image space and pulling back by Ψ^{-1}, it remains only to observe that $dz_{k+1} \wedge \cdots \wedge dz_d$ pulls back to $\left.\frac{1}{\det J_\Psi(\boldsymbol{z})}\right|_{z_i=\zeta_i(\boldsymbol{z}_{\boldsymbol{\pi}})\,:\, i\notin\boldsymbol{\pi}} dz_{\pi_1} \wedge\cdots\wedge dz_{\pi_{d-k}}$, and $P(\boldsymbol{0})$ pulls back to $P(\boldsymbol{z})|_{z_i=\zeta_i(\boldsymbol{z}_{\boldsymbol{\pi}})\,:\, i\notin\boldsymbol{\pi}}$.

□

Remark. The residue depends on Q_j only via its gradient. The sign of the residue form depends on the order of the factors in the denominator, and we account for this when using residue forms to determine asymptotics.

When the stratum $\mathcal{S}$ is a single point (meaning $k = d$) the residue at $\mathbf{p}$ is just a number, simplifying the conclusions of Theorem C.17 as follows.

Corollary C.18. *Suppose the hypotheses of Theorem C.17 hold in the special case where $k = d$, so that the residue of ω at $\mathbf{p}$ is a number θ_0. Then*

(i) $P(\mathbf{p})\, d\mathbf{z} = \theta_0 \, (dQ_1 \wedge \cdots \wedge dQ_d)\,(\mathbf{p})$,

$$(ii)\ \operatorname{Res}\left(\frac{P(\mathbf{z})}{\prod_{j=1}^{d} Q_j(\mathbf{z})} d\mathbf{z}\,;\, \mathbf{p}\right) = \frac{P(\mathbf{p})}{\det J_\Psi(\mathbf{p})},$$

$$(iii)\ \frac{1}{(2\pi i)^d}\int_{\Psi^{-1}(T_\varepsilon)} \frac{P(\mathbf{z})\, d\mathbf{z}}{\prod_{j=1}^{d} Q_j(\mathbf{z})} = \operatorname{Res}\left(\frac{P(\mathbf{z})}{\prod_{j=1}^{d} Q_j(\mathbf{z})}\,;\, \mathbf{p}\right),$$

$$(iv)\ \frac{1}{(2\pi i)^d}\int_{\Psi^{-1}(T_\varepsilon)} \frac{\mathbf{z}^{-\mathbf{r}-\mathbf{1}} P(\mathbf{z})}{\prod_{j=1}^{s} Q_j(\mathbf{z})} d\mathbf{z} = \frac{\mathbf{p}^{-\mathbf{r}-\mathbf{1}} P(\mathbf{p})}{\det J_\Psi(\mathbf{p})} = \frac{\mathbf{p}^{-\mathbf{r}} P(\mathbf{p})}{\det \mathbf{\Gamma}_\Psi(\mathbf{p})}.$$

Example C.19 (two lines in $\mathbb{C}^2$)**.** Let

$$Q(x, y) = \left(1 - \frac{1}{3}x - \frac{2}{3}y\right)\left(1 - \frac{2}{3}x - \frac{1}{3}y\right)$$

so that $\mathcal{V}_Q$ has a transverse multiple point at $(x, y) = (1, 1)$. The gradients of the factors of Q are $(1/3, 2/3)$ and $(2/3, 1/3)$, which are also their logarithmic gradients when $x = y = 1$. The determinant of $\mathbf{\Gamma}_\Psi$ is therefore one of $\pm 1/3$, the sign choice depending on the order in which we choose the factors. Up to sign, the iterated residue of $Q(x, y)^{-1} dx \wedge dy$ at $(1, 1)$ is thus the number 3. ◁

Example C.20 (dimension three with two factors)**.** Consider the generating function

$$F(x, y, z) = \frac{16}{(4 - 2x - y - z)(4 - x - 2y - z)},$$

whose singular set consists of two planes meeting at the complex line $\mathcal{S} = \{(1, 1, 1) + \lambda(-1, -1, 3) : \lambda \in \mathbb{C}\}$. In this case we can parametrize $\mathcal{S}$ globally by any of its three coordinates (i.e., we can take $\mathcal{D} = \mathbb{C}^3$). Choosing the third coordinate, making $\pi_1 = 3$, we obtain

$$J_\Psi(x, y, z) = \begin{bmatrix} -2 & -1 & -1 \\ -1 & -2 & -1 \\ 0 & 0 & 1 \end{bmatrix},$$

whence $\det \mathbf{\Gamma}_\Psi = 3$ and

$$\text{Res}(F(x, y, z)\, dx \wedge dy \wedge dz; \mathcal{S}) = \frac{16}{3}\, dz\,.$$

Choosing one of the first two coordinates leads to an equivalent answer: the first two rows of J_Ψ are unchanged while the third row becomes either $(1, 0, 0)$ or $(0, 1, 0)$, ultimately giving the representations $-16dx$ and $16dy$. These are all equal, up to sign, as 1-forms on $\mathcal{S}$. ◁

C.2.4 Iterated residue forms for higher order poles on transverse divisors

In Section C.2.2 above we used Gelfand–Shilov reduction (Lemma C.11) to define a residue for higher order smooth poles in terms of the residue for smooth simple poles. A version of Gelfand–Shilov reduction also works for iterated residues leading, through a computation analogous to the ones used to establish Theorem C.17, except messier, to the following result.

Proposition C.21. *Let $\mathcal{S}$ be a smooth codimension k variety in $\mathbb{C}_*^d$ defined by the vanishing of k analytic functions $Q_1, \dots, Q_k$, let U denote the module over holomorphic functions of all meromorphic forms that can be written as $\psi / \prod_{j=1}^k Q_j^{n_j}$, where ψ is holomorphic in a neighborhood of $\mathcal{S}$ in $\mathbb{C}^d$, and let $R = U/E$, where E is generated by the forms $\{\text{Res}(d\eta) : \eta \in R\}$. Then every class in R has a representative in which each power n_j is equal to 1.* □

The rest of this section is devoted to the statement and proof of Theorem C.24, an explicit formula for the residue in the specific case we use in this text. We begin with a lemma indicating what form the answer will take.

Lemma C.22. *Let $f, f_1, \dots, f_d$ be smooth functions of $\boldsymbol{u} \in \mathbb{C}^k$. Then*

$$\left(\frac{\partial}{\partial \boldsymbol{u}}\right)^{\boldsymbol{n}} f(\boldsymbol{u}) f_1(\boldsymbol{u})^{r_1} \cdots f_d(\boldsymbol{u})^{r_d} = f(\boldsymbol{u}) f_1(\boldsymbol{u})^{r_1} \cdots f_d(\boldsymbol{u})^{r_d} \Phi(\boldsymbol{r}, \boldsymbol{u}),$$

where Φ is a polynomial in $\boldsymbol{r}$ of degree $|\boldsymbol{n}| = n_1 + \cdots + n_k$. The leading term of Φ is $\mathcal{K}(\boldsymbol{r}, \boldsymbol{u})^{\boldsymbol{n}} = \prod_{j=1}^k \mathcal{K}_j(\boldsymbol{r}, \boldsymbol{u})^{n_j}$, where

$$\mathcal{K}_j(\boldsymbol{r}, \boldsymbol{u}) = \sum_{i=1}^d r_i \frac{\partial \log f_i}{\partial u_j}\,.$$

Proof We show by induction that

$$\left(\frac{\partial}{\partial \boldsymbol{u}}\right)^{\boldsymbol{n}} f(\boldsymbol{u})f_1(\boldsymbol{u})^{r_1}\cdots f_d(\boldsymbol{u})^{r_d} = f(\boldsymbol{u})f_1(\boldsymbol{u})^{r_1}\cdots f_d(\boldsymbol{u})^{r_d}\left[\mathcal{K}(\boldsymbol{r},\boldsymbol{u})^{\boldsymbol{n}} + Q(\boldsymbol{r},\boldsymbol{u})\right] \tag{C.2.11}$$

for all $\boldsymbol{n}$, where Q is a polynomial in $\boldsymbol{r}$ of degree less than $|\boldsymbol{n}|$. When $\boldsymbol{n} = \boldsymbol{0}$ this holds with $Q = 0$. Assuming this holds for $\boldsymbol{n}$, taking the logarithm and differentiating with respect to u_j gives, after some algebraic simplification, that $\left(\frac{\partial}{\partial \boldsymbol{u}}\right)^{\boldsymbol{n}+\delta_j} f(\boldsymbol{u})f_1(\boldsymbol{u})^{r_1}\cdots f_d(\boldsymbol{u})^{r_d}$ equals the right-hand side of (C.2.11) when $\mathcal{K}(\boldsymbol{r},\boldsymbol{u})^{\boldsymbol{n}} + Q(\boldsymbol{r},\boldsymbol{u})$ is replaced by

$$\frac{\partial \log f}{\partial u_j} + \mathcal{K}_j\cdot(\mathcal{K}^{\boldsymbol{n}} + Q) + \frac{\partial}{\partial u_j}(\mathcal{K}^{\boldsymbol{n}} + Q)\,.$$

The terms in this expression other than $\mathcal{K}_j\cdot\mathcal{K}^{\boldsymbol{n}} = \mathcal{K}^{\boldsymbol{n}+\delta_j}$ are polynomials in $\boldsymbol{r}$ of degree at most $|\boldsymbol{n}|$, completing the induction. □

We now specialize to our context.

Corollary C.23. *Let Ψ be the parametrization defined in* (C.2.6) *with Jacobian matrix $J_\Psi(\boldsymbol{z})$. If we parametrize $\boldsymbol{z} = \boldsymbol{z}(\boldsymbol{u})$ for variables $\boldsymbol{u} \in \mathbb{C}^k$ and $\boldsymbol{m}$ is a vector of positive integers then*

$$\left(\frac{\partial}{\partial \boldsymbol{u}}\right)^{\boldsymbol{m}-\boldsymbol{1}}\left(\frac{\boldsymbol{z}(\boldsymbol{u})^{-\boldsymbol{r}}P\left(\Psi^{-1}(\boldsymbol{u})\right)}{\prod_{j=1}^d z_j(\boldsymbol{u})\ \det J_\Psi\left(\Psi^{-1}(\boldsymbol{u})\right)}\right) = \frac{\boldsymbol{z}(\boldsymbol{u})^{-\boldsymbol{r}}P\left(\Psi^{-1}(\boldsymbol{u})\right)\mathcal{P}(\boldsymbol{r},\boldsymbol{u})}{\prod_{j=1}^d z_j(\boldsymbol{u})\ \det J_\Psi\left(\Psi^{-1}(\boldsymbol{u})\right)}\,, \tag{C.2.12}$$

where

$$\mathcal{P}(\boldsymbol{r},\boldsymbol{u}) = \left[\prod_{j=1}^k\left(\sum_{i=1}^d r_i\frac{\partial \log z_i(\boldsymbol{u})}{\partial u_j}\right)^{m_j-1} + R(\boldsymbol{r},\boldsymbol{u})\right] \tag{C.2.13}$$

for some polynomial R in $\boldsymbol{r}$ of degree less than $|\boldsymbol{m}| - k$.

Theorem C.24. *Under our running assumptions, the iterated residue has a computable expression*

$$\operatorname{Res}\left(\boldsymbol{z}^{-\boldsymbol{r}-\boldsymbol{1}}\frac{P(\boldsymbol{z})}{\prod_{j=1}^k Q_j(\boldsymbol{z})^{m_j}}d\boldsymbol{z}\,;\,S_{\mathcal{D}}\right) = \boldsymbol{z}^{-\boldsymbol{r}}\frac{\mathcal{P}(\boldsymbol{r},\boldsymbol{z})}{\prod_{j\in\pi} z_j}\Bigg|_{z_i=\zeta_i(\boldsymbol{z}_\pi)\,:\,i\notin\pi} d\boldsymbol{z}_\pi\,, \tag{C.2.14}$$

where $\mathcal{P}(\boldsymbol{r},\boldsymbol{z})$ is a polynomial in $\boldsymbol{r}$ of degree $|\boldsymbol{m}| - k$. The leading term of $\mathcal{P}(\boldsymbol{r},\boldsymbol{z})$ is

$$\mathcal{P}(\boldsymbol{r},\boldsymbol{z}) \sim \frac{(-1)^{|\boldsymbol{m}-\boldsymbol{1}|}}{(\boldsymbol{m}-\boldsymbol{1})!}\ \frac{P(\boldsymbol{z})}{\det \boldsymbol{\Gamma}_\Psi(\boldsymbol{z})}\ (\boldsymbol{r}\boldsymbol{\Gamma}_\Psi^{-1})^{\boldsymbol{m}-\boldsymbol{1}}\,, \tag{C.2.15}$$

where $\boldsymbol{\Gamma}_\Psi$ is the matrix from Definition C.16, the notation $(\boldsymbol{r}\boldsymbol{\Gamma}_\Psi^{-1})^{\boldsymbol{m}-\boldsymbol{1}}$ stands *for* $\prod_{i=1}^k (\boldsymbol{r}\boldsymbol{\Gamma}_\Psi^{-1})_i^{m_i-1}$, *and* $(\boldsymbol{m}-\boldsymbol{1})! = \prod_{i=1}^k (m_i - 1)!$. *When* $k = d$, *the formula* (C.2.14) *simplifies slightly to*

$$\operatorname{Res}\left(\boldsymbol{z}^{-\boldsymbol{r}-\boldsymbol{1}} \frac{P(\boldsymbol{z})}{\prod_{j=1}^d Q_j(\boldsymbol{z})^{m_j}} d\boldsymbol{z}\,;\, \boldsymbol{p} \right) = \boldsymbol{p}^{-\boldsymbol{r}} \mathcal{P}(\boldsymbol{r}, \boldsymbol{p}). \tag{C.2.16}$$

Remark. Recall that the factors $\{Q_i : 1 \le i \le k\}$ are defined only up to transformations multiplying each Q_i by a complex number λ_i, satisfying $\prod_{i=1}^k \lambda_i^{m_1} = 1$. This multiplies $\det \boldsymbol{\Gamma}_\Psi$ by $\prod_{i=1}^k \lambda_i$ and divides $(\boldsymbol{r}\boldsymbol{\Gamma}_\Psi^{-1})^{\boldsymbol{m}-\boldsymbol{1}}$ by $\prod_{i-1}^k \lambda_i^{m_i-1}$, thus leaving the ratio $(\boldsymbol{r}\boldsymbol{\Gamma}_\Psi^{-1})^{\boldsymbol{m}-\boldsymbol{1}} / \det \boldsymbol{\Gamma}_\Psi$ which appears in (C.2.15) invariant. Later, when we need to compute orientations, it will be convenient to normalize each Q_i to have constant term 1, simultaneously normalizing P to have constant term a_0.

Example C.25. Let a and b be positive integers and consider the function

$$F(x, y, z) = \frac{16}{(4 - 2x - y - z)^a \, (4 - x - 2y - z)^b},$$

generalizing the function in Example C.20. Choosing to parametrize the line $\mathcal{S}$ defined by the common zero sets of the denominator factors of F by the coordinate z, we have the matrix

$$\boldsymbol{\Gamma}_\Psi(x, y, z) = \begin{bmatrix} -2x & -y & -z \\ -x & -2y & -z \\ 0 & 0 & 1 \end{bmatrix},$$

whence $\det \boldsymbol{\Gamma}_\Psi = 3xy$ and, writing $\boldsymbol{r} = (r, s, t)$,

$$\boldsymbol{r}\boldsymbol{\Gamma}_\Psi^{-1} = \left(\frac{sx - 2ry}{3xy}, \frac{ry - 2sx}{3xy}, \frac{3txy - ryz - sxz}{3xy} \right).$$

Since we can parametrize $x = y = g(z)$ on $\mathcal{S}$ where $g(z) = (4 - z)/3$, we have

$$\det \boldsymbol{\Gamma}_\Psi |_{x=y=g(z)} = \frac{(4 - z)^2}{3}$$

and

$$(\boldsymbol{r}\boldsymbol{\Gamma}_\Psi^{-1})^{\boldsymbol{m}-\boldsymbol{1}} \big|_{x=y=g(z)} = \left(\frac{2r - s}{z - 4} \right)^{a-1} \left(\frac{2s - r}{z - 4} \right)^{b-1},$$

where $\boldsymbol{m} = (a, b)$. Thus

$$\operatorname{Res}\left[\boldsymbol{z}^{-\boldsymbol{r}-\boldsymbol{1}} F(\boldsymbol{z})\, d\boldsymbol{z}; \mathcal{S} \right] = x^{-r} y^{-s} z^{-t-1} \left[\mathcal{P}_0(z) + O\left((r + s)^{a+b-3} \right) \right] dz, \tag{C.2.17}$$

where, taking into account $(-1)^{|\boldsymbol{m}-\boldsymbol{1}|} = (-1)^{(a-1)+(b-1)}$ to change factors of $z - 4$

into $4 - z$, we see

$$\mathcal{P}_0(z) = \frac{48}{(4-z)^2\,(a-1)!\,(b-1)!}\left(\frac{2r-s}{4-z}\right)^{a-1}\left(\frac{2s-r}{4-z}\right)^{b-1}.$$

◁

Proof of Theorem C.24. Fix any index t with $1 \le t \le k$ and let η be the $(k-1)$-form defined by

$$\eta = \frac{\tilde{P}(\boldsymbol{u})}{\boldsymbol{u}^{\boldsymbol{m}-\delta_t}} d\boldsymbol{u}_{\hat{t}}\,,$$

where $d\boldsymbol{u}_{\hat{t}}$ denotes the form $du_1 \wedge \cdots \wedge du_{t-1} \wedge du_{t+1} \wedge \cdots \wedge du_k$ and $\tilde{P}$ is an analytic function to be chosen later. Direct computation shows

$$d\eta = \frac{(\partial/\partial u_t)\tilde{P}(\boldsymbol{u})}{\boldsymbol{u}^{\boldsymbol{m}-\delta_{jt}}} d\boldsymbol{u} - \frac{(m_t-1)\tilde{P}(\boldsymbol{u})}{\boldsymbol{u}^{\boldsymbol{m}}} d\boldsymbol{u}\,,$$

and the fact that $\text{Res}[d\eta] = 0$ implies

$$\text{Res}\left[\frac{\tilde{P}(\boldsymbol{u})}{\boldsymbol{u}^{\boldsymbol{m}}} d\boldsymbol{u}\right] = \frac{1}{m_t - 1}\,\text{Res}\left[\frac{(\partial/\partial u_t)\tilde{P}(\boldsymbol{u})}{\boldsymbol{u}^{\boldsymbol{m}-\delta_{jt}}} d\boldsymbol{u}\right]$$

(all residues with respect to forms in $\boldsymbol{u}$ are taken around the origin, which we suppress for readability). Applying this maneuver $m_t - 1$ times for each $1 \le t \le k$ then yields

$$\text{Res}\left[\frac{\tilde{P}(\boldsymbol{u})}{\boldsymbol{u}^{\boldsymbol{m}}} d\boldsymbol{u}\right] = \frac{1}{(\boldsymbol{m}-\mathbf{1})!}\,\text{Res}\left[\frac{(\partial/\partial \boldsymbol{u})^{\boldsymbol{m}-\mathbf{1}}\tilde{P}(\boldsymbol{u})}{u_1 \cdots u_k} d\boldsymbol{u}\right], \tag{C.2.18}$$

and using Theorem C.17 on the right-hand side of (C.2.18) implies

$$\text{Res}\left[\frac{\tilde{P}(\boldsymbol{u})}{\boldsymbol{u}^{\boldsymbol{m}}} d\boldsymbol{u}\right] = \frac{1}{(\boldsymbol{m}-\mathbf{1})!}\left(\frac{\partial}{\partial \boldsymbol{u}}\right)^{\boldsymbol{m}-\mathbf{1}} \tilde{P}(\boldsymbol{u})\,. \tag{C.2.19}$$

By the definition of the map Ψ we can parametrize $\boldsymbol{z}$ on $\mathcal{S}_{\mathcal{D}}$ by $\boldsymbol{z}(\boldsymbol{u}) = \Psi^{-1}(\mathbf{0}, \boldsymbol{u})$ for $\boldsymbol{u}$ in a neighborhood of the origin in $\mathbb{C}^{d-k}$. To simplify notation we write $\Psi^{-1}(\mathbf{0}, \boldsymbol{u})$ as $\Psi^{-1}(\boldsymbol{u})$, understanding the first k coordinates are implicitly zero. We now select

$$\tilde{P}(\boldsymbol{u}) = \frac{\boldsymbol{z}(\boldsymbol{u})^{-\boldsymbol{r}-\mathbf{1}} P(\Psi^{-1}(\boldsymbol{u}))}{J_\Psi(\Psi^{-1}(\boldsymbol{u}))}\,,$$

which is chosen so that

$$\Psi^*\left(\frac{\tilde{P}(\boldsymbol{u})}{\boldsymbol{u}^{\boldsymbol{m}}} d\boldsymbol{u}\right) = \boldsymbol{z}^{-\boldsymbol{r}-\mathbf{1}} \frac{P(\boldsymbol{z})}{\boldsymbol{Q}(\boldsymbol{z})^{\boldsymbol{m}}} d\boldsymbol{z}\,.$$

Functoriality of the residue, combined with (C.2.19), now implies

$$\operatorname{Res}\left[z^{-r-1}\frac{P(z)}{Q(z)^m}dz; S_{\mathcal{D}}\right]$$
$$= \frac{1}{(m-1)!}\left(\frac{\partial}{\partial u}\right)^{m-1}\left(\frac{z(u)^{-r-1}P(\Psi^{-1}(u))}{J_\Psi(\Psi^{-1}(u))}\right)\Bigg|_{(0,u)=\Psi(z)} dz_\pi . \qquad \text{(C.2.20)}$$

Applying Corollary C.23 to the right-hand side of (C.2.20) and noting that

$$\Gamma_\Psi = \left(\frac{\partial u_i}{\partial \log z_j(u)}\right) \quad \text{implies} \quad \frac{\partial \log z_i(u)}{\partial u_j} = \left(\Gamma_\Psi^{-1}\right)_{ij}$$

we obtain an expression

$$\left(\frac{\partial}{\partial u}\right)^{m-1}\left(\frac{z(u)^{-r}P(\Psi^{-1}(u))}{\prod_{j=1}^d z_j(u)J_\Psi(\Psi^{-1}(u))}\right)\Bigg|_{(0,u)=\Psi(z)} = z^{-r}\frac{P(z)}{\det \Gamma_\Psi(z)}\frac{\tilde{\mathcal{P}}(r,z)}{\prod_{j\in\pi} z_j}$$

whose leading term is as stated. □

Remarks. The leading term (C.2.15) depends on the divisors Q_j only through their gradients. When the stratum $\mathcal{S}$ is a single point ($k = d$), the residue at p is a 0-form – i.e., a polynomial $\mathcal{P}(r)$ in r.

C.3 Classical Morse theory

After using residues to replace our starting Cauchy integral with a residue integral over an intersection class σ in $\mathcal{V}$, we need to understand how to deform σ in $\mathcal{V}$. The possible deformations we can make, and which deformations will allow us to compute asymptotic behavior, depend on the topological properties of $\mathcal{V}$. Morse theory attempts to describe the topology of a space X by means of the geometry of X near critical points of a smooth, proper function $h : X \to \mathbb{R}$.

Our destination in this appendix is Theorems C.38 and C.39, which state that we may find a basis for each homology group $H_k(X)$ consisting of *quasi-local cycles* at the critical points of h: for each critical point p there will be a cycle with height bounded by $h(p) - \varepsilon$ except in an arbitrarily small neighborhood of p. We establish this result by studying the sublevel sets $X_{\le a} := \{x \in X : h(x) \le a\}$ as a increases and showing that the homotopy type of X does not change (the Morse Lemma C.27) except at critical points, where a cell is attached (Theorem C.28). Along the way, a description of X as a cell complex is given in Theorem C.32. A description of the attachments in terms of relative homology is also given in the last section.

Our material here covers classical (smooth) Morse theory, which assumes that the space X under consideration is a manifold. More general spaces are handled in Appendix D.

Homotopy equivalence except at critical points

Let X be a smooth manifold and let $h : X \to \mathbb{R}$ be a smooth function; we think of h as giving the points on X a *height* (see Figure C.4 below). The ***critical points of the height function*** h are the points $\mathbf{p} \in X$ for which the differential $dh|_{\mathbf{p}}$ is zero on the tangent space $T_{\mathbf{p}}(X)$. The values $h(\mathbf{p})$ of h at its critical points $\mathbf{p}$ are called the ***critical values of the height function*** h. A critical point $\mathbf{p}$ is a ***nondegenerate critical point for*** h if the quadratic form given by the quadratic terms in the Taylor approximation for h at $\mathbf{p}$ has no zero eigenvalues. In coordinates, this means that the determinant of the Hessian matrix $\left[\frac{\partial^2 h}{\partial x_i \partial x_j}(\mathbf{p})\right]$ is non-zero when X is locally coordinatized by $x_1, \ldots, x_d$ near $\mathbf{p}$. While the Hessian matrix itself depends on the coordinates, its (non)singularity does not; see [Mil63, Section 2.1]. While it is traditional to require Morse functions to be proper and have distinct critical values, we will not require this.

Definition C.26 (Morse function)**.** A smooth function $h : X \to \mathbb{R}$ is called a ***Morse function*** if the critical points of h are nondegenerate. If h is a proper map (meaning the inverse image of any closed and bounded interval is compact) then we call h a ***proper Morse function***. If the critical values of h are distinct, then h is a ***Morse function with distinct critical values***.

Exercise C.2. In which of the following cases is h a proper Morse function on X?

(1) X is the surface of a doughnut lying on a table and h is height.
(2) X is the infinite cylinder $\{(x, y, z) : x^2 + y^2 = 1\}$ and h is the z coordinate.
(3) X is the unit sphere and h is the distance to the point $(-2, 0, 0)$.

Let X be a smooth manifold with proper Morse function h. If a is a real number, we let $X_{\leq a}$ denote the topological subspace $\{\mathbf{x} \in X : h(\mathbf{x}) \leq a\}$. The fundamental *Morse Lemma* states that the topology of $X_{\leq a}$ changes only when a is a critical value of h.

Lemma C.27 (Morse Lemma)**.** *Let $a < b$ be real numbers, suppose that the interval $[a, b]$ contains no critical values of h, and assume that $h^{-1}([a, b])$ is compact. Then the inclusion $X_{\leq a} \hookrightarrow X_{\leq b}$ is a homotopy equivalence.*

Proof The Morse Lemma is proven in [Mil63, Theorem 3.1] by constructing a homotopy on $X_{\leq b}$ that follows the orthogonal trajectories of the level-sets

$h = c$ for constants $c \in [a, b]$. This is accomplished using a *downward gradient flow* constructed locally using the gradient of $h(\boldsymbol{x})$ (which never vanishes when $h(\boldsymbol{x}) \in [a, b]$ due to the absence of critical points). □

Exercise C.3. Let X be the torus embedded in $\mathbb{R}^3$ and let f be the distance from points on X to a fixed point not on X. Use the Morse Lemma to prove that f has a critical point on X that is either degenerate or is neither a maximum nor a minimum.

Attachment at critical points

Suppose now that there is precisely one critical point $\boldsymbol{p}$ with $h(\boldsymbol{p}) \in [a, b]$. The Hessian of h at $\boldsymbol{p}$ is a real symmetric matrix and therefore has real eigenvalues. We define the ***Morse index of** h **at** $\boldsymbol{p}$* to be the number of negative eigenvalues of the Hessian. The Morse index can range from 0 at a local minimum to the dimension d of X at a local maximum. We now describe the topology of $X_{\leq b}$ as an attachment of a space Y to $X_{\leq a}$, where Y and the attaching map depend on the Morse index of h at $\boldsymbol{p}$. Following standard terminology in Morse theory, a *k-cell* (more properly a ***topological** k**-cell***) is a ball of dimension k (in Appendix A we used this term for k-simplices, but topologically a k-ball and k-simplex are equivalent).

Theorem C.28. *Suppose that $h^{-1}([a, b])$ is compact and contains precisely one critical point $\boldsymbol{p}$, with critical value $h(\boldsymbol{p})$ strictly between a and b. Then the space $X_{\leq b}$ has the homotopy type of $X_{\leq a}$ with a λ-cell attached along its boundary, where λ is the Morse index of the critical point $\boldsymbol{p}$ (a 0-cell is a point with empty boundary).*

Proof See [Mil63, Theorem 3.2]. □

Example C.29. Suppose X is the unit sphere in $\mathbb{R}^3$ and consider the height function $h(x, y, z) = z$ (when working in $\mathbb{R}^3$, we often set $h(x, y, z) = z$ so that the “height” function measures actual height). There are only two critical points of h, namely its minimum $(0, 0, -1)$ at height -1 and maximum $(0, 0, 1)$ at height 1.

Let us follow $X_{\leq a}$ as a increases from $-\infty$ to $+\infty$. For $a < -1$, the set $X_{\leq a}$ is empty. As a passes -1, Theorem C.28 states that a 0-cell is added with no identification, making $X_{\leq a}$, homotopically, a point. Geometrically, $X_{\leq a}$ with $a \in (-1, 1)$ is a small dish, which is contractible to a point. The only other attachment occurs at the top of the sphere. For $a < 1 \leq b$, the set $X_{\leq b} \setminus X_{\leq a}$ is a polar cap. Thus, geometrically as well as homotopically, a 2-cell is attached

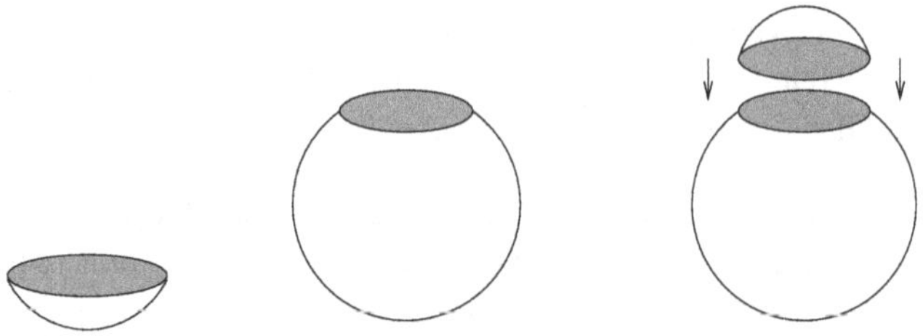

Figure C.3 Sublevel sets of a sphere, which form a contractible subset of the sphere until reaching its maximum, when a cap is attached to complete the sphere.

along its bounding circle. All spaces resulting from attaching a k-cell to a contractible space are homotopy equivalent to attaching a k-cell to a point. In the present case $k = 2$ and the resulting space is homotopy equivalent to a 2-sphere, recovering the homotopy class of X; see Figure C.3. We remark that analyzing the attachments recovers only the homotopy type, not the homeomorphism class. ◁

Example C.30. Let X be the torus in $\mathbb{R}^3$ obtained by rotating the circle $(x - 5)^2 + (y - 5)^2 = 1$ about the y-axis and let $h(x, y, z) = z$. The function h has four critical points, all on the z-axis: a maximum (Morse index 2) at $(0, 0, 6)$, a minimum (Morse index 0) at $(0, 0, -6)$, and saddle points (Morse index 1) at $(0, 0, 4)$ and $(0, 0, -4)$; see Figure C.4.

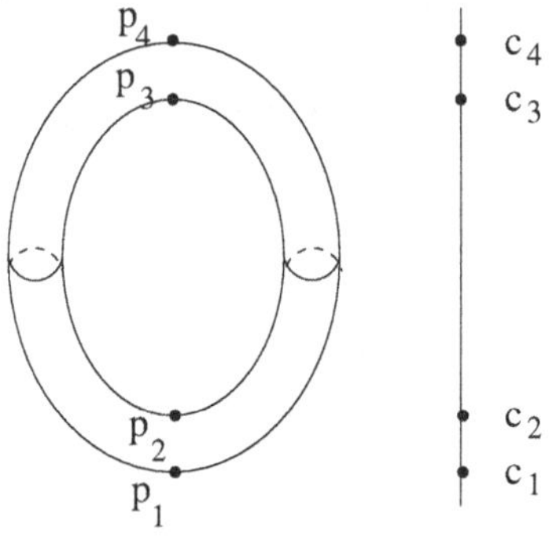

Figure C.4 The critical points on a torus for the standard height function.

For $-6 \leq a < -4$ we see that, as in Example C.29, $X_{\leq a}$ is homotopic to a point (geometrically, it is a dish). As a passes -4, the theorem tells us to add a 1-cell along its boundary. The only way of attaching a 1-cell to a point is to map both endpoints to the point, leaving a circle. Geometrically, if $-4 < a < 4$

then the set $X_{\leq a}$ is a patch in the shape of a 2-cell with two disjoint segments of its boundary attached to two disjoint segments of the geometric boundary of $X_{\leq a}$; see Figure C.5.

Figure C.5 Crossing the critical value c_2.

The critical point at height 4 adds another 1-cell modulo its boundary, making the homotopy type of $X_{\leq a}$ for $4 \leq a < 6$ the union of two circles touching at a point. Finally, crossing the top of the sphere a 2-cell B is added modulo its boundary. There is more than one choice for the homotopy type of the attachment, and keeping track only of the homotopy type throughout the process of attaching cannot resolve this choice – one must look at the geometry of the attachment. In this case, because the attachment is to a topological circle (the $6-\varepsilon$ level set) and results in a nonintersecting surface in $\mathbb{R}^3$, there are only two possibilities for the attaching map in homology: ∂B is mapped to the circle in one of two orientations. Either choice results in a torus.

We remark that knowing the mapping of the last attachment in homology is not sufficient to compute the homotopy type of the space. For example, attaching a 2-cell to two circles joined at a point by mapping the boundary of the 2-cell to the common point produces a sphere with two circular handles, which is not homotopy equivalent to a torus. ◁

Although Theorem C.28 specifies the attachment pair when the topology of X changes, Example C.30 shows that the computation of the attaching map is not automatic. It will help to have some results that narrow down this computation to certain constructions local to the critical point: the homotopy in the Morse Lemma may be improved so that outside of a neighborhood of $\boldsymbol{p}$, every point is pushed down at least to a level $c - \varepsilon$.

Let $\boldsymbol{p}$ be a critical point with height $c = h(\boldsymbol{p})$ and suppose $a < c < b$ are such that $\boldsymbol{p}$ is the only critical point with height in $[a, b]$. Given $\varepsilon > 0$, let $B_\varepsilon(\boldsymbol{p})$ denote the ε-neighborhood of $\boldsymbol{p}$. We will see a formal version of the local Morse Lemma, in a more general context, in Appendix D. For now, we note that the local Morse Lemma implies that the homotopy type of $X_{\leq b}$ is the same as the homotopy type of $X_{\leq c-\varepsilon} \cup B_\varepsilon(\boldsymbol{p})$ for sufficiently small ε.

Definition C.31. Let $X_{c^+} := X_{\leq c-\varepsilon} \cup B_\varepsilon(\boldsymbol{p})$ for any sufficiently small $\varepsilon > 0$, and let $X^{\boldsymbol{p},\text{loc}}$ denote the pair $(X_{c^+}, X_{\leq c-\varepsilon})$ depicted in Figure C.6.

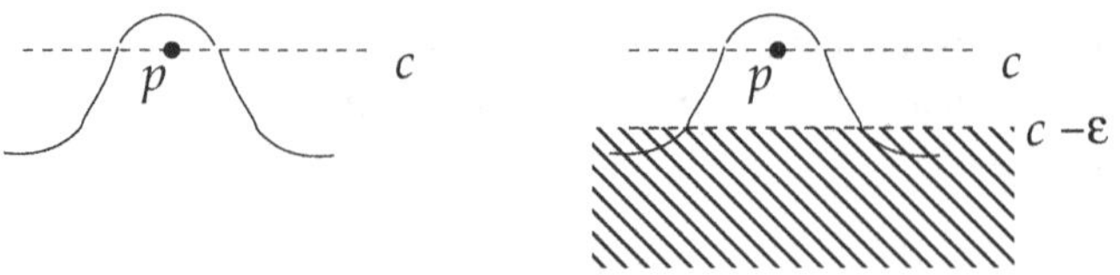

Figure C.6 The space X_{c^+} with points of height c represented by a dotted line.

The discussion above implies that the attachment pair $(X_{\le b}, X_{\le a})$ is homotopy equivalent to $X^{p,\text{loc}}$. Suppose now that h is a Morse function whose critical values need not be distinct. If $[a, b]$ contains the unique critical value $c \in (a, b)$ then the homotopy pushes points down to $X_{\le c-\varepsilon}$ except in a neighborhood of the set of critical points whose value is $\mathbf{p}$. Since this set of critical points is discrete under our assumptions,

$$(X_{\le b}, X_{\le a}) \simeq \widetilde{\bigoplus_{\mathbf{p}:h(\mathbf{p})=c}} X^{\mathbf{p},\text{loc}} \tag{C.3.1}$$

for sufficiently small $\varepsilon > 0$, where the tilde sum denotes the ***wedge*** of spaces, meaning a disjoint union with the second space in each pair identified. This equivalence states that $(X_{\le b}, X_{\le a})$ is homotopy equivalent to the wedge of the local pairs at all critical points $\mathbf{p}$ with value c. The reduced homology of a wedge is the direct sum of the reduced homologies of the individual spaces.

Exercise C.4. Intuitively, why is (C.3.1) a (reduced) *direct* sum? That is, explain why cycles in different summands cannot cancel each other.

The last step for this section is to put all this information together to produce a global topological picture of X. At the level of homotopy type, the result is that X has the topology of a cell complex, about which certain information is known.

Theorem C.32. *Let X be a manifold and $h : X \to \mathbb{R}$ be a differentiable function with no degenerate critical points. Suppose each sublevel set $X_{\le a}$ is compact. Then X has the homotopy type of a cell complex with one cell of dimension λ for each critical point of Morse index λ in $X_{\le a}$.*

Proof A full proof of Theorem C.32 can be found in [Mil63, Theorem 3.5], but we give a sketch here. The proof for the case of finitely many critical points with distinct critical values involves showing inductively that for any critical value c the homotopy equivalence between $X_{\le c-\varepsilon}$ and a cell complex may be extended, via the attachment of a cell, to a homotopy equivalence between $X_{\le c+\varepsilon}$ and a cell complex with one more cell. The restriction on distinct critical values is then removed by homotopically perturbing h so as to satisfy the conditions, and a limiting argument removes the assumption of a finite number of critical points. □

Example C.33. The 2-sphere from Example C.29 is a cell complex with one 2-cell and one 0-cell; as noted above, up to homotopy equivalence, there is only one choice for the attachment map. The 2-torus from Example C.30 is a cell complex with one 0-cell, two 1-cells, and one 2-cell. Up to homotopy equivalence, the one skeleton must be the wedge of two circles. There are a number of ways to attach a 2-cell to a wedge of two circles, and the right attaching map can be worked out by knowing what the boundary (i.e., the level set ε below the maximum height) looks like. ◁

Remark C.34. Let X be a complex d-manifold in $\mathbb{C}^n$ and, for $\mathbf{p} \in \mathbb{C}^n$, let $h_{\mathbf{p}}$ denote the function mapping $\mathbf{z} \in \mathbb{C}^n$ to the complex distance $\|\mathbf{z} - \mathbf{p}\| = (\sum_{j=1}^n |z_j - p_j|^2)^{1/2}$. Andreotti and Frankel's original proof of Proposition B.20 from Appendix B proved that $\mathbf{p}$ can be chosen to make $h_{\mathbf{p}}$ a Morse function by establishing that the set of $\mathbf{p}$ for which it is not a Morse function has positive codimension, then showing that the Morse index of any critical point on X for $h_{\mathbf{p}}$ is at most d.

C.4 Description at the level of homology

For our purposes it is useful to consider the successive attachments from the last section on the level of homology. Suppose that c is a critical value and (B, A) is any pair with the same homotopy type of the attachment $(X_{\leq c+\varepsilon}, X_{\leq c-\varepsilon})$. The long exact sequence has a portion

$$H_{n+1}(B, A) \xrightarrow{\partial_{n+1}} H_n(A) \xrightarrow{\iota_*} H_n(B) \xrightarrow{\pi_*} H_n(B, A) \xrightarrow{\partial_n} H_{n-1}(A),$$

which implies

$$\frac{H_n(A)}{\text{Image}(\partial_{n+1})} = \frac{H_n(A)}{\ker(\iota_*)} \cong \text{Image}(\iota_*) = \ker(\pi_*).$$

In particular, there is a short exact sequence

$$0 \to \frac{H_n(A)}{\text{Image}(\partial_{n+1})} \to H_n(B) \to \ker(\partial_n) \to 0$$

and, by Remark B.7, $H_n(B)$ decomposes as a direct sum of the kernel of ∂_n and the cokernel of ∂_{n+1}.

This decomposition allows us to construct a basis for the homology groups

$H_n(B)$ from knowledge of the homology of A and the boundary map ∂_*: starting with a basis for $H_n(A)$ we identify basis elements differing by elements in the image of ∂_{n+1} and then add new basis elements indexed by a basis for the kernel of ∂_n. These new basis elements have an explicit geometric description. The group $H_n(B, A)$ consists of equivalence classes of chains in B whose boundaries lie in A. If C is a chain in the kernel of ∂_n then the image $\partial_*([C])$ is the class of $\partial C \in H_{n-1}(A)$, which bounds some n-chain D in A. The inverse image of the class $[C]$ by π_* is the class of the chain $C - D$, which is a cycle because $\partial C = \partial D$. Heuristically, we write

$$\pi_*^{-1}([C]) = C - \partial_A^{-1}(\partial C) \tag{C.4.1}$$

and view $\pi_*^{-1}([C])$ as the relative cycle C in $Z_n(B, A)$, completed to an actual cycle in a way that stays within A.

Remark C.35. The choice of D in this construction is not natural (see Remark B.7). A particular composition of a space B as a subspace A attached to $C = \overline{B \setminus A}$ comes with an explicit inclusion map from ∂C to A, and this induces the ∂_* operator. There may, however, be more than one way to reassemble A, C, and ∂_* into B, giving homotopy equivalent spaces with different homology bases.

One further remark on notation: when attaching a space Y along Y_0, the pair (Y, Y_0) is commonly referred to as the ***attachment data*** or, in the case of Morse theory, the ***Morse data*** for the attachment. This data should really include the homotopy type of the attachment map, or else the homotopy type of X and the attachment data do not determine the homotopy type of the new space. On the level of homology what we need to know is the relative homology of the pair (Y, Y_0), which is the homology of the new space relative to the old space, *together with the ∂_* map.*

Filtered spaces

A ***filtered space*** X_n is the end of a nested sequence $X_0 \subseteq X_1 \subseteq \cdots \subseteq X_n$ of topological spaces. We use the terminology of filtered spaces to describe how homology changes among sublevel sets $X_j = X_{\leq a_j}$, and our first result concerns the homology of a chain that is successively pushed toward lower heights.

Lemma C.36 (Pushing Down Lemma). *Let $X_0 \subseteq \cdots \subseteq X_n$ be a filtered space and let C be a non-zero homology class in $H_k(X_n, X_0)$ for some k. Then there is a unique positive $j \leq n$ such that for some $C_* \in H_k(X_j, X_{j-1})$,*

$$\iota(C_*) = \pi(C) \neq 0 \text{ in } H_k(X_n, X_{j-1}), \tag{C.4.2}$$

where ι is the map induced by the inclusion of pairs $(X_j, X_{j-1}) \to (X_n, X_{j-1})$ and π is the map induced by the projection of pairs $(X_n, X_0) \to (X_n, X_{j-1})$. If ι is an injection then C_* is unique.

Proof To prove uniqueness of j, suppose that (C.4.2) is satisfied for some minimal j with a chain C_*, and let $j < r \leq n$. The composition of the two maps

$$(X_j, X_{j-1}) \to (X_n, X_{j-1}) \to (X_n, X_{r-1})$$

induces the zero mapping on homology because any class in the image of the first map has a cycle representative in X_j. Letting π' denote projection of (X_n, X_0) to (X_n, X_{r-1}), we have $\pi'(C) = \pi'(\pi(C)) = \pi'(\iota(C_*)) = 0$ and therefore (C.4.2) cannot hold for $r > j$.

For existence we argue by induction on n. The case $n = 1$ is trivial because then $j = 1$ and $C_* = C$. Assume the result for $n-1$ and let C be a non-zero class in $H_k(X_n, X_0)$. If the image of C under the projection of (X_n, X_0) to (X_n, X_{n-1}) is non-zero then we may take C_* to be this image and j to be n. Assume therefore that C projects to zero. The short exact sequence of chain complexes for the pairs

$$0 \to (X_{n-1}, X_0) \to (X_n, X_0) \to (X_n, X_{n-1}) \to 0$$

induces the exact sequence

$$H_k(X_{n-1}, X_0) \to H_k(X_n, X_0) \to H_k(X_n, X_{n-1}).$$

By assumption C is in the kernel of the second map, hence is the image under the first map of some non-zero class C'. Applying the inductive hypothesis to C' yields some $j \leq n-1$ and a cycle $C_* \in H_k(X_j, X_{j-1})$ satisfying (C.4.2) with C' in place of C. The commuting diagram

$$\begin{array}{ccccc} & & C' \in (X_{n-1}, X_0) & \xrightarrow{\iota_3} & C \in (X_n, X_0) \\ & & \downarrow \pi_1 & & \downarrow \pi \\ C_* \in (X_j, X_{j-1}) & \xrightarrow{\iota_1} & (X_{n-1}, X_{j-1}) & \xrightarrow{\iota_2} & (X_n, X_{j-1}) \end{array}$$

allows us to conclude $\pi(C) = \pi(\iota_3(C')) = \iota_2(\pi_1(C')) = \iota_2(\iota_1(C_*)) = \iota(C_*)$, verifying (C.4.2). □

Building up by successive attachments

If we understand the topology of each pair (X_{k+1}, X_k) of consecutive elements in a filtration $X_0 \subseteq \cdots \subseteq X_n$ and we understand the homology groups of X_0 then, using the argument above and induction, we understand the homology groups of all X_k. Furthermore, if X is a smooth manifold with a proper height function h then the Morse Lemma implies that the topology of the continuum of spaces $\{X_{\le t}\}$ is captured by a filtration of sets described by the critical values $c_0 < c_1 < \cdots < c_{n-1}$ of h. The ***Morse filtration*** of X with respect to h is the filtration defined by $X_j = X_{\le c_j - \varepsilon}$ for $0 \le j \le n-1$ and $X_n = X_{\le c_{n-1}+\varepsilon}$, where ε is any sufficiently small positive number.

When h has distinct critical values the pairs (X_{i+1}, X_i) are homotopy equivalent to $X^{\mathbf{p}_i,\mathrm{loc}}$, where $\mathbf{p}_i$ are the critical points listed in order of increasing height. In general, the successive pairs are homotopy equivalent to $\widetilde{\bigoplus_{h(\mathbf{p})=c_j}} X^{\mathbf{p},\mathrm{loc}}$ as c_j increases through all critical values. We could describe how to keep track of generators and relations for the homologies of X_j inductively on j in the general case, however what we will need is both more specialized (our spaces are complex algebraic or analytic varieties) and more general (our spaces may not be manifolds). Accordingly, we restrict the discussion here to one illustration, continuing our example of the torus to show what can happen.

Example C.37. In Example C.30 we examined a height function on the torus X with four critical points: one of Morse index 0, two of Morse index 1, and one of Morse index 2. All ∂_* maps vanish so the homology groups $H_0(X), H_1(X)$, and $H_2(X)$ are cyclic groups of rank $1, 2$, and 1, respectively. The filtration consists of $X_0 = \emptyset$, X_1 which is contractible to a point, X_2 which is homeomorphic to a cylinder, X_3 which is homeomorphic to a punctured torus, and X_4 which is the whole torus.

As an illustration of the non-naturality of the homology basis in (C.4.1), consider the second 1-cell to be added. Let α be the homology class in $H_1(X_2)$ of the first 1-cell. Then the second 1-cell, which is a well-defined relative homology class β in $H_1(X_3, X_2)$, may be completed to an absolute class in $H_1(X_3)$ in many different ways, resulting in cycles differing by multiples of α. Geometrically, one may for example complete β to the circle defined by $x^2 + z^2 = 1$, or instead wrap around the torus any integer number of times. ◁

Exercise C.5. Let γ be the cycle pictured in Figure C.7, going around the torus between the critical points p_4 and p_3. Applying the Pushing Down Lemma with respect to the Morse filtration for the pictured height function, what is j and what cycle represents $\iota(C_*) = \pi(C)$?

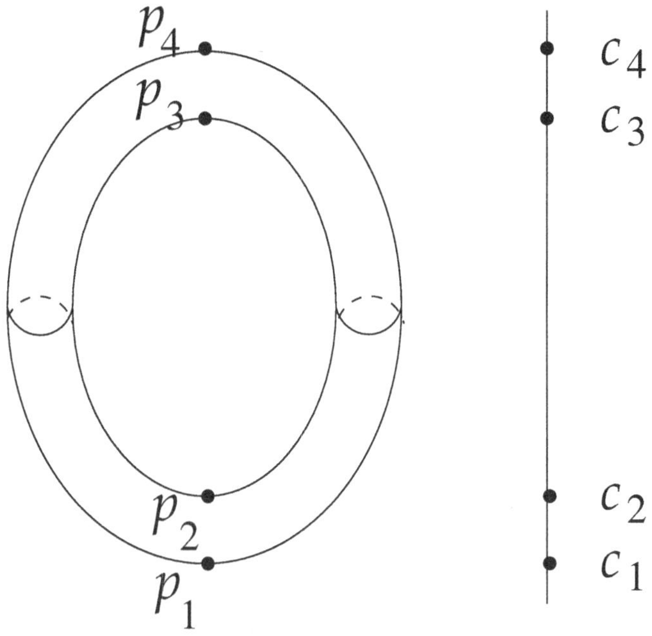

Figure C.7 A cycle on a torus.

Assume now that $X \subseteq \mathbb{C}_*^d$ is a smooth algebraic hypersurface (a real manifold of dimension $2d-2$) and that the specific height function $h(\boldsymbol{z}) = h_{\boldsymbol{r}}(\boldsymbol{z}) = -r_1 \log|z_1| - \cdots - r_d \log|z_d|$ is a proper Morse function on X. The purpose of introducing critical points at infinity in Chapter 7 is to remove the strong assumption that h is proper, however, to see how everything works, we now derive the results of Chapter 7 in this setting. The height function h is the real part of (a branch of) a holomorphic function $-\boldsymbol{r} \cdot \log \boldsymbol{z}$, and is thus harmonic, so all critical points for h on X have middle Morse index $d-1$. Let $\boldsymbol{p}^{(j)}$ for $1 \le j \le m$ enumerate the critical points, ordered so that the critical values $c_j = h(\boldsymbol{p}_j)$ are nondecreasing.

By the Morse Lemma, any cycle supported on $X_{<c_1}$ can be deformed via gradient flow to a cycle in $X_{\le t}$ for t arbitrarily small, therefore integrals of $\boldsymbol{z}^{-\boldsymbol{r}} F(\boldsymbol{z}) d\boldsymbol{z}$ over such cycles decay faster than any exponential. For this reason, it suffices to describe the homology of $(X, -\infty)$, where $-\infty$ stands for $X_{\le t}$ for any $t < c_1$. The final set of results in this appendix describes this homology.

Theorem C.38. *Suppose $h = h_{\boldsymbol{r}}$ is a proper Morse function on a smooth algebraic hypersurface X in $\mathbb{C}_*^d$. Then $H_k(X, -\infty)$ vanishes in dimensions $k \neq$*

$d-1$. The $(d-1)$-homology is given by $H_{d-1}(X, -\infty) \cong \mathbb{C}^m$, where m is the number of critical points. A basis $\{\gamma_{\mathbf{p}}\}$ may be chosen, indexed by the critical points $\mathbf{p}$ of h on X, with the property that h is maximized on $\gamma_{\mathbf{p}}$ at $\mathbf{p}$. By the isomorphism of smooth homology and singular homology, each $\gamma_{\mathbf{p}}$ may be chosen to be smooth.

Proof First assume distinct critical values $c_1 < \cdots < c_m$ with corresponding critical points $\mathbf{p}^{(1)}, \ldots, \mathbf{p}^{(m)}$, and let X_j denote the space $X_{\leq c_j+\varepsilon}$. Inducting on j, we show that the conclusion of the theorem holds for X_j in place of X. First, we note that each pair (X_j, X_{j-1}) is homotopy equivalent to a $(d-1)$-ball B^{d-1} modulo its boundary, whose homology has rank 1 in dimension $d-1$ and zero in every other dimension. For the base step $j = 1$, where we take $X_0 = X_{\leq t}$ for any $t < c_1$, the conclusion is immediate.

Now assume the conclusion holds with $X = X_j$ for some $j < m$, and consider the short exact sequence of chain complexes of pairs

$$0 \to C_*(X_j, -\infty) \to C_*(X_{j+1}, -\infty) \to C_*(X_{j+1}, X_j) \to 0 .$$

None of these pairs has any homology in dimensions higher than $d-1$, therefore the long exact sequence is as follows, with an arrow from the rightmost element of each row other than the bottom row to the leftmost element of the next row down.

$$\begin{array}{ccccccc}
0 & \to & H_{d-1}(X_j, -\infty) & \to & H_{d-1}(X_{j+1}, -\infty) & \to & H_{d-1}(X_{j+1}, X_j) \\
 & & H_{d-2}(X_j, -\infty) & \to & H_{d-2}(X_{j+1}, -\infty) & \to & H_{d-2}(X_{j+1}, X_j) \\
 & & H_{d-3}(X_j, -\infty) & \to & H_{d-3}(X_{j+1}, -\infty) & \to & H_{d-3}(X_{j+1}, X_j) \\
 & & \vdots & & \vdots & & \vdots \\
 & & H_0(X_j, -\infty) & \to & H_0(X_{j+1}, -\infty) & \to & H_0(X_{j+1}, X_j) \to 0.
\end{array}$$

Identifying each $H_k(X_{j+1}, X_j)$ as $\mathbb{C}$ if $k = d-1$ and 0 otherwise, and using the induction hypothesis, fills in most of this sequence:

$$\begin{array}{ccccccc}
0 & \to & \mathbb{C}^j & \to & H_{d-1}(X_{j+1}, -\infty) & \to & \mathbb{C} \\
 & & 0 & \to & H_{d-2}(X_{j+1}, -\infty) & \to & 0 \\
 & & 0 & \to & H_{d-3}(X_{j+1}, -\infty) & \to & 0 \\
 & & \vdots & & \vdots & & \vdots \\
 & & 0 & \to & H_0(X_{j+1}, -\infty) & \to & 0,
\end{array}$$

from which we deduce that $H_k(X_{j+1}, -\infty)$ has rank one more than that of $H_k(X_j, -\infty)$ when $k = d-1$ and rank zero otherwise. The extra generator comes from the attachment of B^{d-1} modulo its boundary, for which the generator γ_{j+1} may be chosen to maximize h at $\mathbf{p}^{(j)}$ (see Exercise C.10), thereby completing the induction. Finally, the assumption of distinct critical values may be removed via (C.3.1). □

Exercise C.6. Let X be a smooth complex algebraic variety of complex dimension d in $\mathbb{C}_*^m$ for $m > d$ and let h be a Morse function on X which is the real part of a complex analytic function. Suppose X has five critical points. Can you determine the homotopy type of the pair $(X_{\le b}, X_{\le a})$ when $a \to -\infty$ and $b \to +\infty$?

We now have the tools to state a result analogous to what Theorem C.38 tells us about X for $\mathcal{M} = \mathbb{C}_*^d \setminus X$. Recall the tube operator $\mathsf{o} : H_{d-1}(\mathcal{V}_*) \to H_d(\mathcal{M})$ from Theorem C.2 – this is not only an injection, but (as we will see in the next appendix) is an isomorphism on (X_{j+1}, X_j). An induction then gives the following result.

Theorem C.39. *Suppose the height function $h_{\mathbf{r}}$ is a proper Morse function on a smooth complex algebraic hypersurface X in $\mathbb{C}_*^d$ and let $\mathcal{M} = \mathbb{C}_*^d \setminus X$. Then there is a basis for $H_d(\mathcal{M}, -\infty)$ consisting of a single generator $\gamma_{\mathbf{p}}$ for each critical point $\mathbf{p}$ of $h_{\mathbf{r}}$, which is tube around a cycle reaching maximum height at $\mathbf{p}$ and is homotopy equivalent to $S^1 \times (B^{d-1}, \partial B^{d-1})$.* □

We cover Theorem C.39, and generalizations, in Appendix D.

Notes

Detailed treatments of multivariate residues are given in [Pha11; AY83], while the classic text on Morse theory, which we have based our presentation around, is [Mil63].

Additional exercises

Exercise C.7 (univariate residues via Stokes's Theorem)**.** Let f be a meromorphic function inside and on a closed contour γ such that f has no singularities on γ. The familiar residue theorem in one variable states that

$$\frac{1}{2\pi i}\int_\gamma f = \sum_a \operatorname{Res}(f; a),$$

where the sum is over the poles a of f inside γ. Derive this from Theorem C.9. What are $\alpha, \beta, \omega, \mathcal{V}$, and $\mathbf{INT}[\alpha, \beta; \mathcal{V}]$?

Exercise C.8. Let P be a polynomial in two complex variables such that $P(0, y)$ has only simple, non-zero roots. Let α be a small torus of polyradius $(\varepsilon, \varepsilon)$ and let β be a torus of polyradius (ε, M), where M is much larger than any root of $P(0, y)$. Compute the intersection class $\mathbf{INT}[\alpha, \beta; \mathcal{V}]$. *Hint:* Use the obvious homotopy and parametrize $\mathcal{V}_*$ as $y = f_j(x)$ near each root y_j of $P(0, y)$.

Exercise C.9 (lumpy sphere)**.** Let X be a sphere with a lump, that is, a patch on the northern hemisphere where the surface is raised to produce a local, but not global, maximum of the height function. List the critical points of the lumpy sphere and determine the homotopy types of the attachments. This gives a description of the lumpy sphere as a cell complex different from the complex with just two cells. Use this to compute the homology and verify it is the same as for the non-lumpy sphere.

Exercise C.10. Let $\mathcal{M}$ be a manifold with Morse function h having distinct critical values and let $\boldsymbol{x}$ be a critical point of Morse index k. Let P be any submanifold of $\mathcal{M}$ diffeomorphic to an open k-ball about $\boldsymbol{x}$ such that h is strictly maximized on P at $\boldsymbol{x}$. Prove that P is a homology generator for the local homology group $H_k(\mathcal{M}^{h(\boldsymbol{x})+\varepsilon}, \mathcal{M}^{h(\boldsymbol{x})-\varepsilon})$. *Hint:* This is true of any embedded k-disk through $\boldsymbol{x}$ in $\mathcal{M}^{h(\boldsymbol{x})}$ that intersects the ascending $(n-k)$-disk transversely.

Exercise C.11. Let $\iota_d : \mathbb{CP}^{d-1} \hookrightarrow \mathbb{CP}^d$ denote the embedding $\iota_d(z_0 : \cdots : z_d) = (z_0 : \cdots : z_d : 0)$.

(i) Show that $\mathbb{CP}^d \setminus \mathrm{Image}(\iota)$ is homeomorphic to a $(2d)$-ball.
(ii) Describe $\mathbb{CP}^d$ as $\mathbb{CP}^{d-1}$ with a $(2d)$-cell attached. What is the attachment map on homology?
(iii) Use induction on d to compute the homology of $\mathbb{CP}^d$.

Appendix D

Stratification and stratified Morse theory

In this final appendix we extend the Morse-theoretic decompositions of Appendix C to handle general algebraic varieties and their complements. More specifically, we cover results from *stratified* Morse theory [GM88] that characterize the topology of a stratified space X through changes in topology in the sublevel sets $X_{\leq c}$ as c passes through critical values (in a stratified sense) of a height function. We develop from scratch the notion of a Whitney stratified space, Morse functions, and stratified critical points. We discuss non-proper extensions of this material and then summarize a number of basic results of [GM88], including specific properties enjoyed by complex algebraic varieties.

D.1 Whitney stratified spaces

Ideally one would use the apparatus of manifolds, developed in the previous appendices, to do calculus on complex algebraic varieties, however many varieties that appear in interesting combinatorial problems are not manifolds. The right generalization for our purposes is the notion of a *stratified space*, which can contain *non-manifold points* whose neighborhoods in the variety are not diffeomorphic to any Euclidean space $\mathbb{R}^d$. One well-known example of such spaces are *manifolds with boundary*, and we begin with a recap of these objects. The following material on manifolds with boundary can be skipped if desired, as it is subsumed by our discussion of stratified spaces, however we include it because it is likely familiar to many readers.

A ***d-manifold with boundary*** is a subset $\mathcal{M} \subset \mathbb{R}^n$ such that every point $x \in \mathcal{M}$ has a neighborhood in $\mathbb{R}^n$ whose intersection $\mathcal{N}$ with $\mathcal{M}$ is either diffeomorphic to $\mathbb{R}^d$ or diffeomorphic to the closed halfspace $\mathbb{R}^{d-1} \times [0, \infty)$. In the former case $\boldsymbol{x}$ is called a ***manifold point*** or ***interior point*** of $\mathcal{M}$, while in the latter case $\boldsymbol{x}$ is called a ***boundary point*** of $\mathcal{M}$.

Example D.1. A closed ball in any dimension d is a manifold with boundary, while a cube $[0, 1]^d$ and a simplex $\{\boldsymbol{x} \in \mathbb{R}^d : x_j \geq 0 \text{ for all } j \text{ and } \sum_{j=1}^d x_j = 1\}$ are not. ◁

Example D.2. If $H = \{(x, y) \in \mathbb{R}^2 : y \geq 0\}$ is the upper half-plane then H is a manifold with boundary, the boundary points being the x-axis. Now let $K = \{(x, y) \in \mathbb{R}^2 : x, y \geq 0\}$ denote the positive quarter plane. The map $\phi(x, y) = (x^2 - y^2, 2xy)$ from K to H, constructed by taking the real and imaginary parts of $(x + iy)^2$, is analytic and one to one, so it may seem that K is diffeomorphic to H and thus a manifold with boundary. However, ϕ^{-1} is not differentiable at the origin, and in fact no neighborhood of the origin in K is diffeomorphic to a neighborhood of the origin in a half-plane. Thus, K is not a manifold with boundary. ◁

Exercise D.1. Give an example of a manifold $\mathcal{M} \subseteq \mathbb{R}^d$ with closure $\overline{\mathcal{M}}$ whose boundary $\overline{\mathcal{M}} \setminus \mathcal{M}$ is also a manifold, such that $\mathcal{M}$ is not a manifold with boundary.

A generalization of manifolds with boundary is the notion of a ***d-manifold with corners***, where every point has a neighborhood diffeomorphic to some orthant $\mathbb{R}^{d-k} \times \mathbb{R}^k_{\geq 0}$ for some $k \leq d$; see Figure D.1. Cubes and simplices are manifolds with corners, however complex algebraic varieties that are not smooth are also typically not manifolds with corners. This difficulty is why we introduce the generality of stratified spaces.

Figure D.1 The closed quadrant $\mathcal{M}$ is a manifold with corners, but not a manifold with boundary.

Stratifications

As a first attempt to perform calculus on an algebraic variety $\mathcal{V}$, one might partition $\mathcal{V}$ into a finite disjoint union of smooth sets and then work on each piece of the partition. Although such ***smooth partitions*** can be easily understood, and easily computed using standard algebro-geometric techniques, they are not sufficient for our (and many other) purposes. The problem is that the pieces in an arbitrary smooth partition may not *fit together nicely*; among other difficulties, this means the local behavior of $\mathcal{V}$ near points in the same piece of the partition can be very different.

The issue of determining the "right" type of smooth partition to use for topological arguments was taken up by Whitney [Whi65b], who introduced what we now call (Whitney) stratifications. An ***I-decomposition*** of a space $X \subseteq \mathbb{R}^n$ is a finite disjoint union $\bigcup_{\alpha \in I} S_\alpha$ of smooth manifolds of various dimensions, indexed by a partially ordered set I, such that for every $\alpha, \beta \in I$,

$$S_\alpha \cap \overline{S_\beta} \neq \emptyset \iff S_\alpha \subset \overline{S_\beta} \iff \alpha \leq \beta. \tag{D.1.1}$$

Definition D.3 (Whitney stratification). Let Z be a closed subset of $\mathbb{R}^n$. A ***Whitney stratification*** of Z is an I-decomposition of Z with the additional property that whenever

- $\alpha < \beta$, and
- the sequences $\{x_i \in S_\beta\}$ and $\{y_i \in S_\alpha\}$ both converge to some $y \in S_\alpha$, and
- the lines $\ell_i = \overline{x_i y_i}$ converge to a line ℓ, and
- the tangent planes $T_{x_i}(S_\beta)$ converge to a plane T,

then $\ell \subseteq T$. We call Z a ***Whitney stratified space***.

Remark. In the original definition, in addition to $\ell \subseteq T$ (the so-called ***second Whitney condition***), it was required that $T_y(S_\alpha) \subseteq T$ (the so-called ***first Whitney condition***). The second condition turns out to imply the first, so the first condition is usually omitted.

This definition is well crafted: the conditions are easy to fulfill – for example, every algebraic variety admits a Whitney stratification, see [Whi65b, Theorem 18.11] or [Hir73, Theorem 4.8] – and the conditions have strong consequences (for instance, they are strong enough for stratified Morse theorems to hold). Stratifications of algebraic varieties are also effectively computable. A classic approach to algorithmic stratification through quantifier elimination and real algebraic geometry, relying on cylindrical algebraic decomposition, is discussed in [Ran98; MR91]. Recently, [DJ21] and [HN22] have given more practical algorithms[1] for the stratification of algebraic varieties using Gröbner basis computations.

Proposition D.4. *Every algebraic variety in $\mathbb{R}^d$ or $\mathbb{C}^d$ admits a Whitney stratification.* □

In examples arising from combinatorial applications, it is often possible to deduce a stratification directly from the form of the polynomials under consideration.

Example D.5. A smooth manifold is a stratified space with a single stratum. ◁

Example D.6. If X is a finite union of affine subspaces of $\mathbb{R}^n$ then a Whitney stratification of X is obtained by taking the set $\mathcal{A}$ of all intersections of the affine subspaces, and choosing the elements of $\{A \setminus B : A, B \in \mathcal{A} \text{ with } A \supsetneq B\}$ as strata. ◁

Example D.7. Let Z be a real algebraic curve $\{(x, y) \in \mathbb{R}^2 : f(x, y) = 0\}$ with f irreducible and let $Y = \{(x, y) : \nabla f(x, y) = 0\}$ be the finite set of singular points of Z. Taking $Z \setminus Y$ to be one stratum and each singleton $\{(x, y)\}$ for $(x, y) \in Y$ to be another produces a Whitney stratification of Z. The following figure shows two examples of this, the first curve $x^2 - y^3$ having a cusp at the origin and the second curve $19 - 20x - 20y + 5x^2 + 14xy + 5y^2 - 2x^2y - 2xy^2 + x^2y^2$ having a self-intersection at $(1, 1)$. ◁

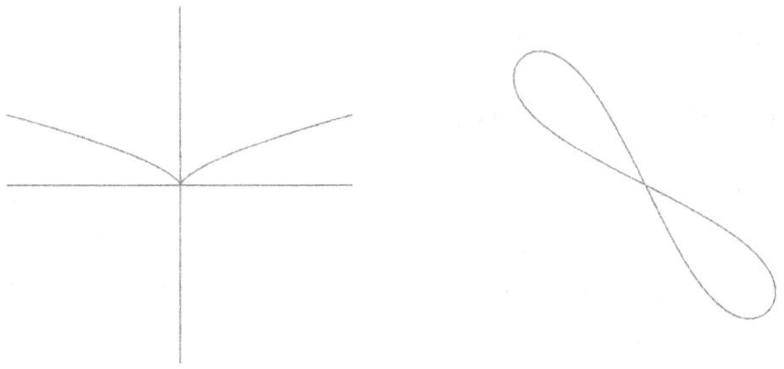

Figure D.2 Two curves, each stratified by taking one stratum consisting of a singular point and another stratum consisting of the rest of the curve.

Let $\mathcal{V}$ be any complex variety. As discussed in Chapter 8, it is possible to decompose $\mathcal{V}$ into smooth sets by determining algebraic equations for the set

[1] Helmer and Nanda [HN22] give an implementation of both of these algorithms in Macaulay2, available at `http://martin-helmer.com/Software/WhitStrat/`.

Σ_0 of its singular points, letting $\mathcal{S}_0 = \mathcal{V} \setminus \Sigma_0$ encode the smooth points of $\mathcal{V}$, then recursively computing the sets Σ_{n+1} and $\mathcal{S}_{n+1}$ of smooth and singular points of Σ_n until arriving at some $\Sigma_N = \emptyset$. From the previous two examples, one might get the idea that this decomposition is always a Whitney stratification, but Exercise D.6 below shows this not to be the case. It is true, however, that any stratification must be at least this coarse.

Example D.8. Let Z be a complex algebraic hypersurface in $\mathbb{C}^3$ defined by $f(x, y, z) = 0$ and suppose ∇f vanishes along an algebraic curve γ. It is possible that $\{\gamma, Z \setminus \gamma\}$ is a Whitney stratification for Z. On the other hand, if γ is not smooth then a Whitney stratification of Z will have at least three strata, one containing singularities of γ, one containing the rest of γ, and one containing $Z \setminus \gamma$. ◁

Exercise D.2. Compute a Whitney stratification of the real variety $\mathcal{V}_Q$ where $Q(x, y, z) = z^2 - x^2 - y^2$.

The following exercise implies, with a little more work, that any manifold with corners (including any manifold with boundary) is a Whitney stratified space, with strata $\{S_j : 0 \le j \le d\}$ defined by the union of the open j-dimensional faces.

Exercise D.3. Let $H = \mathbb{R}^d_{\geq 0}$ be the positive orthant, let F be a (open) face of H and let $\boldsymbol{x}$ be a point of F. Prove directly that the interior $S_\beta = H^\circ$ and face $S_\alpha = F$ satisfy the Whitney condition (Definition D.3) at $\boldsymbol{x}$.

One fundamental result of stratified spaces concerns their local product structure, implying that the local behavior of a stratified space "looks the same" in neighborhoods of different points on the same stratum. The proof of this fact is long and difficult, but we sketch some of it in the next section.

Theorem D.9 (local product structure). *Let $\mathbf{p}$ be a point in a k-dimensional stratum $\mathcal{S}$ of a stratified space Z. There is a topological space* $\mathtt{N}$, *called the* **normal slice**, *depending only on $\mathcal{S}$ and not the choice of $\mathbf{p} \in \mathcal{S}$, such that some neighborhood of $\mathbf{p}$ in Z is homeomorphic to $B^k \times N$, where B^k is a k-dimensional ball.* □

We end this section with the following concept.

Definition D.10 (stratification of a pair)**.** If $Y \subseteq X$ are closed subsets of real space then a ***stratification of the pair*** (X, Y) is defined to be a stratification of X such that intersecting each stratum with Y gives a stratification of Y and intersecting each stratum with $X \setminus Y$ gives a stratification of $X \setminus Y$.

A result of Whitney implies that if $(X \setminus Y, Y)$ is a decomposition of X into two smooth manifolds satisfying (D.1.1) then some Whitney stratification of X refines this, and is a stratification of the pair (X, Y); see, for instance, [LT10, Proposition 2.1]. Proposition D.4 extends to the following.

Proposition D.11. *If $\mathcal{V}_*$ is a complex algebraic variety in $\mathbb{C}_*^d$ with stratification $\{S_\alpha : \alpha \in I\}$ then adding the stratum $\mathcal{M} = \mathbb{C}_*^d \backslash \mathcal{V}_*$ produces a stratification of the pair $(\mathbb{C}_d^*, \mathcal{V}_*)$.* □

D.2 Critical points and the fundamental lemma

We now extend the geometric concepts discussed in previous appendices to stratified spaces.

Critical points for stratified spaces

Fix a Whitney stratification $\{S_\alpha : \alpha \in I\}$ of a closed subset X of a smooth manifold $M \subseteq \mathbb{R}^n$ and let $f = h|_X$ be the restriction to X of a smooth function $h : M \to \mathbb{R}$.

Definition D.12 (stratified critical points and Morse functions). Any point $\mathbf{p} \in X$ is contained in a unique stratum $S = S(\mathbf{p})$, and we say that $\mathbf{p}$ is a ***critical point of the height function*** h ***on the stratified space*** X if $\mathbf{p}$ is a critical point of $h|_{S(\mathbf{p})}$ (in other words, if the restriction of the differential of h to the tangent space $T_{\mathbf{p}}S(\mathbf{p})$ is zero). We call h a ***Morse function*** if

(1) the restriction $h|_{S_\alpha}$ is a Morse function for each $\alpha \in I$, meaning that its critical points are nondegenerate (i.e., its Hessian is nonsingular at each critical point), and
(2) whenever $\mathbf{p} \in S_\alpha$ is a critical point of $h|_{S_\alpha}$ and T is a limit of tangent planes $T_{\mathbf{p}_i}(S_\beta)$ as $\mathbf{p}_i \to \mathbf{p}$ in a stratum S_β with $\beta > \alpha$, then either $T = T_{\mathbf{p}}(S_\alpha)$ or T contains a tangent vector on which $dh(\mathbf{p})$ does not vanish.

This generalization of Morse functions to stratified spaces appears in [Pig79, Section 3]; see also [Laz73]. In many contexts it is assumed that Morse functions have distinct critical values – in which case we say we have a ***Morse function with distinct critical values*** – or are proper – in which case we say we have a ***proper Morse function***. Figure D.3 shows two height functions, one failing condition (2) in Definition D.12 and one satisfying it: on the left, the limit of tangent lines at the cusp is horizontal, and is therefore annihilated by

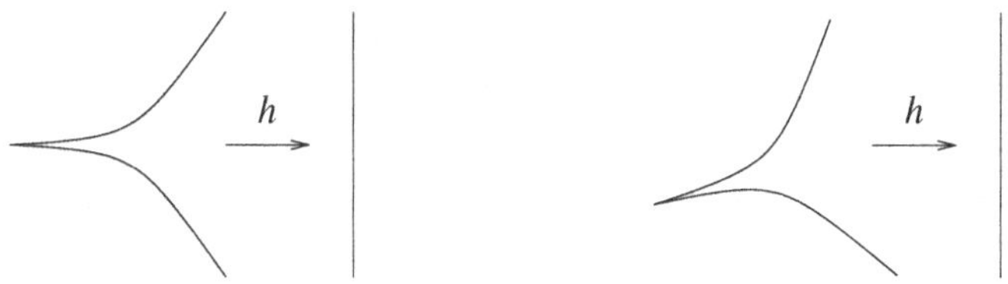

Figure D.3 A non-Morse function (left) and a Morse function (right).

dh. A standard perturbation argument shows that coinciding critical values do not affect topology.

The stratified version of the Fundamental Morse Lemma (Lemma C.27) is the following.

Theorem D.13 (Stratified Morse Lemma [GM88, Theorem SMT part A]). *Let $X \subseteq \mathbb{C}_*^d$ be a stratified space with proper Morse function h and let $a < b$ be real numbers such that the interval $[a, b]$ contains no critical values of h. If $h^{-1}([a, b])$ is compact, then the inclusion $X_{\le a} \hookrightarrow X_{\le b}$ is a homotopy equivalence.*

□

Tangent vector fields

The argument behind Theorem D.13 is worth understanding for readers who have made it this far into the appendices. Both Theorem D.13 and Theorem D.9 will be derived from Thom's Isotopy Lemma, stated as Lemma D.16 below. For non-experts, the geometric intuition behind Theorem D.13 is not apparent, and it can be instructive to pursue a line of reasoning that sometimes fails but more closely parallels classical Morse theory.

Proposition D.14. *In the following cases, the local product structure in Theorem D.9 is induced by a diffeomorphism.*

1. *When Z is a smooth algebraic hypersurface.*
2. *When Z is the simplex or the complexification of a simplex.*
3. *When Z is a hyperplane arrangement.*
4. *When Z is the product of two spaces on which the local product is induced by a diffeomorphism.*

Proof In Cases 2 and 3, diffeomorphisms can be explicit constructed. Case 1 follows from the smooth implicit function theorem, while Case 4 follows from taking a product diffeomorphism. □

Proof sketch of Theorem D.13 (assuming diffeomorphic product structure)
Step 1: Each stratum $\mathcal{S}$ is a smooth manifold. The nonvanishing of the gradient of $h|_{\mathcal{S}}$ implies the existence of a nonvanishing downward gradient vector field $\boldsymbol{v}_{\mathcal{S}}$ parallel to $\mathcal{S}$. More specifically, there is a smooth nonvanishing section of the tangent bundle (i.e., a map $\boldsymbol{v}_{\mathcal{S}} : \mathcal{S} \to T\mathcal{S}$) such that $dh(\boldsymbol{v}_{\mathcal{S}}) < 0$.

Step 2: By assumption of diffeomorphic local product structure, for each point $\boldsymbol{p}$ in each k-dimensional stratum $\mathcal{S}$ of X there is a $\mathbb{C}^d$-neighborhood $\mathcal{N}$ of $\boldsymbol{p}$ and a smooth change of coordinates in $\mathcal{N}$ under which $\mathcal{S} \cap \mathcal{N} = \{\boldsymbol{z} \in \mathcal{N} : z_j = 0 \text{ for } j > k\}$ and $X \cap \mathcal{N} = \{\boldsymbol{z} \in \mathcal{N} : (z_{j+1}, \ldots, z_d) \in N'\}$, where N' is the normal slice consisting of all $(d-j)$-tuples $(z_{j+1}, \ldots, z_d)$ such that $(0, \ldots, 0, z_{j+1}, \ldots, z_d) \in X$. Strata in this neighborhood are the products of strata of N' in the first k coordinates with $\mathbb{R}^{d-k}$. Vectors $\boldsymbol{v}$ tangent to $\mathcal{S}$ in this neighborhood have $v_j = 0$ for $j > k$ and are therefore tangent to all strata in the neighborhood.

The within-stratum downward gradient flows $\boldsymbol{v}_{\mathcal{S}}$ can be stitched together via a partition of unity to form a single gradient-like flow $\boldsymbol{v}$ with Lipschitz constant 1. More specifically, each point $\boldsymbol{p}$ in a stratum has a neighborhood $U_{\boldsymbol{p}}$ in $\mathbb{C}^d$ that intersects only strata whose closure contains $\mathcal{S}(\boldsymbol{p})$, the stratum containing $\boldsymbol{p}$, and on which $dh(\boldsymbol{v}_{\mathcal{S}(\boldsymbol{p})}) < 0$. If $\{\psi_{U_{\boldsymbol{p}}} : \boldsymbol{p} \in E\}$ is a partition of unity subordinate to a finite subcover of $h^{-1}[a, b]$ by these neighborhoods, then

$$\boldsymbol{v} = \sum_{\boldsymbol{p} \in E} \psi_{U_{\boldsymbol{p}}} \boldsymbol{v}_{\mathcal{S}(\boldsymbol{p})} \tag{D.2.1}$$

defines the required flow. It is gradient-like because $dh(\boldsymbol{v})$ is a convex combination of values $dh(\boldsymbol{v}_{\mathcal{S}}(\boldsymbol{p}))$, which are all negative. It is tangent to each stratum because $\boldsymbol{v}(\boldsymbol{p}')$ is a convex combination of vectors $\boldsymbol{v}(\mathcal{S}(\boldsymbol{p}))$ tangent to strata $\mathcal{S}(\boldsymbol{p})$ whose tangent spaces are contained in the tangent space to $\boldsymbol{p}'$. Choosing $U_{\boldsymbol{p}}$ small enough that some constant multiple of each $\boldsymbol{v}_{\mathcal{S}(\boldsymbol{p})}$ can be chosen to have Lipschitz constant 1 on $U_{\boldsymbol{p}}$, convexity implies that $\boldsymbol{v}$ globally has Lipschitz constant 1.

Figure D.4 shows a picture of this. The left-hand picture shows that the vector field $\boldsymbol{w}(\boldsymbol{p}) = \boldsymbol{v}_{\mathcal{S}(\boldsymbol{p})}(\boldsymbol{p})$ is gradient-like but not continuous. It changes direction sharply when approaching a substratum, because $\mathcal{S}(\boldsymbol{p})$ changes discontinuously from one stratum to a substratum. The right-hand picture shows these blended by a partition of unity, so as to become smooth while remaining gradient-like.

Step 3: Let $c > 0$ be the infimum value of $|dh(\boldsymbol{v})|$ on X, and let $\Psi : X \times [0, \infty] \to X$ be the flow defined by $(d/dt)\Psi(x, t) = \boldsymbol{v}(\Psi(x, t))$, stopped when it hits $h^{-1}(c)$. Such a flow exists and is unique because $\boldsymbol{v}$ is Lipschitz, being a convex combination of locally constant vector fields (in the natural

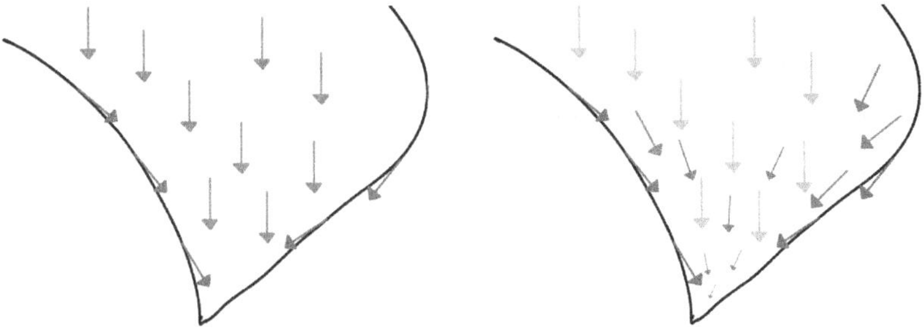

Figure D.4 *Left:* A flow in a 2D stratum that turns sharply when reaching a boundary. *Right:* A partition of unity blends the flow smoothly between strata (note that the flow smoothly becomes zero in a neighborhood of the zero-dimensional stratum).

identification of tangent spaces with subspaces of the tangent space to the ambient space $\mathbb{R}^d$). Fixing any $T \geq (b-a)/c$ the time T map defines a deformation retract of $X_{\leq b}$ onto $X_{\leq a}$, proving homotopy equivalence. □

The problem with this sketched proof is that, in general, the local product structure is not witnessed by a diffeomorphism. This is shown by Whitney's counterexample [Whi65a], reproduced in Goresky's introduction [Gor12] to Mather's cleaned up notes [Mat12] as motivation for the work that follows. Figure D.5 shows three planes and a ruled surface in $\mathbb{R}^3$, whose common intersection is the x-axis. Intersecting with a plane parallel to the yz-plane moving down the x-axis results in a configuration of four lines, the first three constant and the fourth becoming more sloped. Any coordinate system in which the first three lines remain fixed as the slice moves down the x-axis also fixes the slope at the origin of the fourth line, and therefore cannot represent the figure as a product of the x-axis with a four-line configuration.

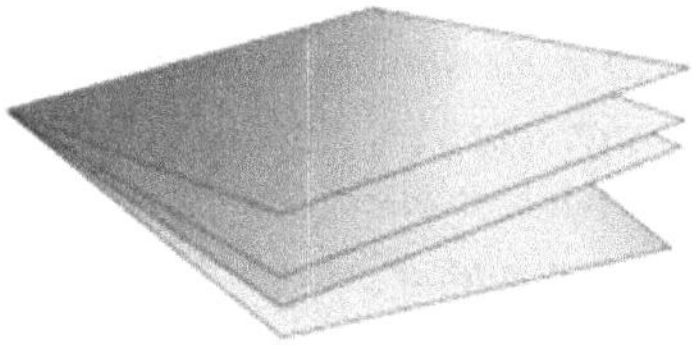

Figure D.5 Whitney's counterexample to smooth isotopy.

The trouble is that the category in which one most naturally deals with stratified spaces is *smooth within strata and continuous across strata*. Whitney's conditions do not guarantee the existence of a differential structure that is smooth across strata, even for algebraic hypersurfaces in Euclidean space. Nevertheless, it is true that there is a continuous isotopy moving the yz-plane to the right while continuously deforming a sector so that the line of intersection with the ruled surface in each slice remains identified. Working in the smooth within strata continuous across strata category, one can obtain a vector field but it will generally not be Lipschitz. The flow in Step 3 will not necessarily exist, and the argument falls apart.

Remark D.15. In the neighborhood of a hyperbolic point of a complex algebraic hypersurface, a Lipschitz vector field can be constructed explicitly from a lower-semicontinuously varying family of cones. This is carried out in [BP11] (see Lemma 5.1 there) and is based on the lengthier development in [ABG70]; the construction is summarized in Section 11.2 of this book. Thus, the three steps above prove Theorem D.13 when X is a complex algebraic hypersurface with all critical points hyperbolic, even though Proposition D.14 will not necessarily hold.

Isotopy

To repair the stratified gradient flow argument, one needs a statement of Thom's Isotopy Lemma strong enough to imply the deformation retract in Step 3 directly, as well as implying Theorem D.9. This lemma is proved by giving up on the idea that the desired vector field can be continuous, providing instead a ***controlled*** vector field satisfying a set of axioms allowing one to infer that the vector field defines a continuous flow with the desired properties. We will not go into the theory of controlled vector fields, being content to quote where they are used and referring the reader to [Mat12, Proposition 11.1] for the proof of the following results and full details of controlled vector fields for stratified spaces.

Lemma D.16 (Thom's Isotopy Lemma)**.** *Let Z be a Whitney stratified space Z that is a closed subset of some smooth manifold $\mathcal{M}$, and suppose that $\pi : \mathcal{M} \to P$ is a smooth proper mapping to a connected manifold P such that the restriction $\pi|_S$ of π to each stratum $\mathcal{S}$ of Z is a submersion (surjective on tangent spaces). Then any smooth vector field V on P has a lift $\tilde{V}$ to a controlled vector field on Z. By a lift, we mean that V is a (not necessarily continuous) section of the tangent bundle of each stratum of Z such that $\pi_* \circ \tilde{V} = V \circ \pi$. Although $\tilde{V}$ is not necessarily continuous, it has a continuous flow $\tilde{\Psi}$ that projects under π to the flow Ψ defined by V on P. The fact that there is a continuous flow lifting the flow of V implies that $\pi|_Z : Z \to P$ is a locally trivial fiber bundle.* □

Proof of Theorem D.13 Apply Thom's Isotopy Lemma with manifolds $P = \mathbb{R}$ and $\mathcal{M} = \mathbb{C}^*_d$, stratified space $Z = X \cap h^{-1}(a - \varepsilon, b + \varepsilon)$, and mapping $\pi = h$. If $h : X \to \mathbb{R}$ has no critical values in $[a, b]$ then it has no critical values in $[a - \varepsilon, b + \varepsilon]$, hence h is a submersion on each stratum of X. The conclusion of the lemma is that the level surfaces of h are fibers of a local product bundle, hence the flow $\tilde{V}$ witnesses a strong deformation retraction of $X_{\leq b}$ onto $X_{\leq a}$. □

To conclude this section, we show how Thom's Isotopy Lemma can be used to derive the local product topological structure of stratified spaces.

Proof of Theorem D.9 Let Z be a stratified space in $\mathbb{R}^d$ and let $\mathcal{S}$ be a stratum of dimension k, with $\bullet\mathcal{S}$ denoting a closed tubular neighborhood of $\mathcal{S}$ in $\mathbb{R}^d$ and $\pi : \bullet\mathcal{S} \to \mathcal{S}$ denoting the projection map. Then $\bullet\mathcal{S}$ is a manifold with boundary $\circ\mathcal{S}$ and an interior which we denote $(\bullet\mathcal{S})^\circ$. If the tubular neighborhood was chosen sufficiently small, then $X = Z \cap \bullet\mathcal{S}$ is naturally stratified with strata of the form $W \cap (\bullet\mathcal{S})^\circ$ and $W \cap \circ\mathcal{S}$, where W runs over strata whose closure contains $\mathcal{S}$.

The mapping π on X satisfies the conditions of Thom's isotopy lemma. Consequently, its normal slice $\mathtt{N} = \pi^{-1}(\mathbf{p}) \cap Z$ is stratified by its intersection with the strata of X. Taking $U_{\mathbf{p}}$ to be a small ball around $\mathbf{p}$ in the stratum $\mathcal{S}$ that contains $\mathbf{p}$, there is a stratum-preserving homeomorphism, smooth in each stratum, given by $\pi^{-1}(U_p) \cap Z \cong U_{\mathbf{p}} \times N$. Since $\pi^{-1}(U_{\mathbf{p}})$ is a neighborhood of $\mathbf{p}$ in Z, we have shown that each stratum has a neighborhood that is locally a topological product of a k-ball $U_{\mathbf{p}}$ with the normal slice. □

D.3 Description of the attachments

Let $\mathcal{V}_*$ denote the intersection $\mathcal{V} \cap \mathbb{C}^d_*$ of an affine algebraic hypersurface $\mathcal{V}$ with $\mathbb{C}^d_*$, and let $\mathcal{M} = \mathbb{C}^d_* \setminus \mathcal{V}$. We return to our plan to use Morse theory to find generators for $H_d(\mathcal{M})$. Because we may want to describe either $\mathcal{M}$ or $\mathcal{V}_*$, depending on the situation, results in the literature are often stated in two parts, so as to cover both cases, and we continue to adhere to this. For what follows we fix a Whitney stratification $\{\mathcal{S}_\alpha : \alpha \in I\}$ of the pair $(\mathbb{C}^d_*, \mathcal{V}_*)$ as in Proposition D.11, so that $\mathcal{M}$ will be the unique stratum of dimension $2d$. The function $h = h_{\hat{\mathbf{r}}}$ is assumed to be a Morse function and the space X may denote either $\mathcal{V}_*$ or $\mathcal{M}$. The point $\mathbf{p}$ denotes a stratified critical point for h in the stratum $\mathcal{S}$, and we let $N = N_{\mathbf{p}}(\mathcal{V})$ denote the complex normal space to $\mathcal{V}_*$ at $\mathbf{p}$.

The *tangential Morse data* is defined in terms of p and $\mathcal{S}$, regardless of whether $X = \mathcal{V}_*$ or $X = \mathcal{M}$.

Definition D.17 (tangential Morse data). The ***tangential Morse data*** at p is the homotopy type of the pair $(B^\lambda, \partial B^\lambda)$, where λ is the Morse index of $h|_S$ at p and B^λ denotes the ball of dimension λ. By Theorem C.28, this is the Morse data at p for the height function $h|_S$ on the smooth manifold S.

The *normal Morse data* is defined in terms of the intersection of X with a slice normal to the stratum $\mathcal{S}$, localized to the point p. If D is an arbitrarily small disk in $N_p(\mathcal{V})$ centered at p then the ***normal slice*** at p is $\mathtt{N}(X) := X \cap D$. To visualize this, it sometimes helps to picture the ***normal link*** $\mathcal{L}(X)$ at p, defined by $\mathcal{L}(X) := X \cap \partial D$. When $X = \mathcal{V}_*$ the normal slice $\mathtt{N}(X)$ is homeomorphic to a cone over $\mathcal{L}(X)$ from the point p. In particular, $\mathtt{N}(X)$ is contractible. When $X = \mathcal{M}$ the point p is absent from the normal slice, which then retracts onto $\mathcal{L}(X)$, hence $\mathtt{N}(X) \simeq \mathcal{L}(X)$.

Example D.18. Let $\mathcal{V}$ be the union of two complex planes in complex 3-space meeting at the line $\mathcal{S}$ and let p be a point on $\mathcal{S}$. This line is the stratum containing p, and the tangent space at p or any other point on $\mathcal{S}$ is the translation of $\mathcal{S}$ to the origin. The normal space $N_p(\mathcal{V})$ at p (or any other point on $\mathcal{S}$) is the complex two-space orthogonal to $\mathcal{S}$.

First consider the case $X = \mathcal{V}_*$. The intersection of X with a normal plane to $\mathcal{S}$ at p is two complex lines meeting at p. The normal slice $\mathtt{N}(X)$ is the intersection of this with a ball around p, and thus is two disks joined by identifying their centers. The link $\mathcal{L}(X) = X \cap \partial D$ is the union of two disjoint circles, each on one of the complex lines, and the normal slice $\mathtt{N}(X)$ is the cone over these circles.

Alternatively, if $X = \mathcal{M}$ then $\mathcal{L}(X) = X \cap \partial D$ is the complement of two intersecting lines in a small bi-disk, which is the product of two punctured disks. Each punctured disk retracts to its boundary, so the four-dimensional space $\mathtt{N}(X)$ retracts to the three-dimensional space $\mathcal{L}(X)$, which retracts to a two-dimensional torus $S^1 \times S^1$. ◁

Definition D.19 (normal Morse data). Let X be $\mathcal{V}_*$ or $\mathcal{M}$. The ***normal Morse data*** for X at p is defined to be the homotopy type of the pair

$$\left(\mathtt{N}(X) \cap h^{-1}([c-\varepsilon, c+\varepsilon]), \mathtt{N}(X) \cap h^{-1}(c+\varepsilon)\right), \tag{D.3.1}$$

where the disk D in the definition of $\mathtt{N}(X)$ is sufficiently small, and ε is a sufficiently smaller positive number. It is proved in [GM88] that these homotopy types are the same for all D and ε sufficiently small.

Example D.20. Suppose that $\mathcal{V}_*$ is a smooth algebraic hypersurface near one of its points $\mathbf{p}$.

(1) If $X = \mathcal{V}$ then $\mathtt{N}(X)$ is the single point $\mathbf{p}$. Formally, the homotopy type is that of $(\{\mathbf{p}\}, \emptyset)$.
(2) If $X = \mathcal{M}$ then $\mathtt{N}(X)$ has the homotopy type of $(D \setminus \mathbf{0}, \mathbf{q})$, where D is a small disk and $\mathbf{q}$ is a point on the boundary of D. This is the reduced homotopy type of a circle, cyclic in dimension 1 and null in every other dimension. ◁

The following theorem is stated for the case $X = \mathcal{V}_*$ in [GM88, Theorem SMT B on page 8] and for the case $X = \mathcal{M}$ in [GM88, unnamed theorem on page 12]; the equivalent characterizations of the homotopy type are stated in [GM88, pages 7, 66–67, 120–122].

Theorem D.21 (attachments are determined by Morse data). *Let X be either $\mathcal{V}_*$ or $\mathcal{M}$ with a Whitney stratification as above, and let $\mathbf{p}$ be a critical point for h in a stratum $\mathcal{S}$ with critical value $c = h(\mathbf{p})$.*

1. *The homotopy type of the attachment at $\mathbf{p}$ is the product, in the category of pairs, of the normal and tangential Morse data as given in Definitions D.19 and D.17.*
2. *The tangential data for a stratum of codimension k is always the reduced homology of a $(d-k)$-sphere: rank 1 in dimension $(d-k)$ and vanishing otherwise.*
3. *The normal data has the following characterizations.*
   (i) *When $X = \mathcal{V}_*$, the normal data is homotopy equivalent to the pair $(\mathrm{Cone}(\ell^-), \ell^-)$, where $\mathrm{Cone}(Y)$ is the topological quotient $Y \times [0,1] \,/\, Y \times \{1\}$ and ℓ^- is the* ***lower halflink*** *defined as the level set of $\mathtt{N}(X)$ at height $c - \varepsilon$ for sufficiently small $\varepsilon > 0$.*
   (ii) *When $X = \mathcal{M}$, the normal data is homotopy equivalent to the pair $(\mathcal{L}^+(X), \partial\mathcal{L}^+(X))$, where $\mathcal{L}^+(X)$ is the part of $\mathcal{L}(X)$ at height at least c.*
   (iii) *When $X = \mathcal{M}$, the normal data is also homotopy equivalent to the pair $(\mathcal{L}^+(X), \mathcal{L}^0(X))$, where $\mathcal{L}^0(X)$ is the intersection of $\mathcal{L}(X)$ with the level set $\{\mathbf{z} \in X : h(\mathbf{z}) = c\}$.*

□

Remark. Goresky and MacPherson have this to say [GM88, page 9]: "Theorem SMT Part B, although very natural and geometrically evident in examples, takes 100 pages to prove rigorously in this book."

Example D.22 (complement of S^2 in $\mathbb{R}^3$). Let X be the complement of the unit

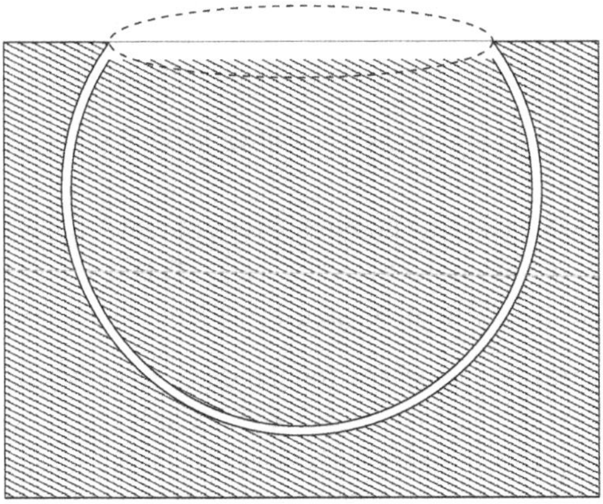

Figure D.6 The complement of the unit sphere up to height $+1/2$.

sphere $S \subseteq \mathbb{R}^3$. The function $h(x, y, z) = z$ extends to a proper height function on $\mathbb{R}^3$, which is Morse with respect to the stratification $\{S, X\}$.

There are no critical points in X but there are two in S: the South pole and the North pole. In each case the normal slice is an interval minus a point, so the normal data is homotopy equivalent to (S^0, S^0_-), where S^0 is two points, one higher than the other, and S^0_- is the lower of the two points. For the South pole, which has Morse index 0, the tangential data is a point, so the attachment is (S^0, S^0_-), which is the addition of a disconnected point. Figure D.6 illustrates that for $-1 < a < 1$, the space $X_{\leq a}$ is in fact the union of two contractible components. The North pole has Morse index 2, so the tangential data at the North pole is $(D^2, \partial D^2)$, a polar cap modulo its boundary. Taking the product with the normal data gives two polar caps modulo all of the lower one and the boundary of the upper one. This is the upper polar cap sewn down along its boundary, the boundary being a point in one of the components. Thus, one component becomes a sphere and the other remains contractible. ◁

Suppose we have a closed space $Y \subset \mathbb{R}^d$ whose complement X we view as a stratified space with Morse function h. If $\boldsymbol{p}$ is a critical point for h in some stratum S then there is a local coordinatization of Y as $S \times B_{\boldsymbol{p}}$, where $B_{\boldsymbol{p}}$ is a small ball of dimension $d - k$ and k is the dimension of S. The set $B_{\boldsymbol{p}} \setminus Y$ is this ball minus the origin, so it is a cone over $L(\boldsymbol{p})$ with vertex $\boldsymbol{p}$. Any chain in $B_{\boldsymbol{p}} \setminus Y$ may be brought arbitrarily close to $\boldsymbol{p}$.

D.4 Stratified Morse theory for complex manifolds

If X is a complex variety then the Morse data has an alternate description obeying the complex structure of X. Let S be a stratum containing a critical point p, let $N(p)$ be a small ball in the normal space to S at p, and define the ***complex link*** $\mathcal{L}(S)$ to be the intersection of X with a generic hyperplane $A \subseteq N(p)$ that comes sufficiently close to p but does not contain it. It is shown in [GM88, page 16] that the normal Morse data at $p \in X$ is given in terms of $\mathcal{L}(S)$ by the pair

$$(\mathrm{Cone}^{\mathbb{R}}(\mathcal{L}(S)), \mathcal{L}(S)), \tag{D.4.1}$$

where $\mathrm{Cone}^{\mathbb{R}}(\mathcal{L}(S))$ denotes the real cone over $\mathcal{L}(S)$. In other words, the normal link has the homotopy type of the pair $(\mathcal{L}(S)\times[0,1] \,/\, \mathcal{L}(S)\times\{1\}\,,\ \mathcal{L}(S)\times\{0\})$, where the real cone (the first space of this pair) is defined as a quotient.

Suppose that X has dimension d, the stratum S has dimension k, and the ambient space has dimension n (all dimensions are complex). Then $N(p)$ is a complex space of dimension $n-k$, its intersection with a generic hyperplane has dimension $n-k-1$, and thus

$$\dim_{\mathbb{C}} \mathcal{L}(S) = d - k - 1\,.$$

In fact the homeomorphism type of the complex link depends on X and S but not on the individual choice of $p \in S$, nor the ambient space, nor the choice of proper Morse function h on the stratified space X (see [GM88, Section II:2.3]).

Suppose next that X is the complement of a d-dimensional variety in $\mathbb{C}^{d+1}$. A formula for the Morse data at a point $p \notin X$ in a stratum S is given [GM88, page 18] by

$$(\mathcal{L}(S), \partial\mathcal{L}(S)) \times (B^1, \partial B^1)\,, \tag{D.4.2}$$

where B^1 is a real interval (which can be interpreted as a 1-ball).

Theorem D.23. (i) *If X is a complex analytic variety of dimension d then X has the homotopy type of a cell complex of dimension at most d.* (ii) *If X is the complement in a domain of $\mathbb{C}^n$ of a complex variety of dimension d then X has the homotopy type of a cell complex of dimension at most* $2n-d-1$.

Remark. The proof of this result in [GM88] is somewhat difficult, mostly due to the necessity of establishing the invariance properties of the complex link. The result, however, is very useful. For example, suppose that X is the complement of the zero set of a polynomial in n variables. Then $d = n-1$ and the homotopy dimension of X is at most n. Note that X may have strata of any complex dimension $j \le d$, and that the complement of a j-dimensional complex space in $\mathbb{C}^n$ has homotopy dimension $2n-2j$. The theorem asserts that the complex structure prevents the dimensions of contributions at strata of dimensions $j < d$ from exceeding the dimension of the contributions from d-dimensional strata.

Proof sketch (*i*) Assume that the variety is embedded in $\mathbb{C}^n$ and that the height function h has been chosen to be the square of the distance from a generic point. We examine the homotopy type of the attachment at a point $\mathbf{p}$ in a stratum of dimension k. It suffices, as in the proof of Theorem C.39, to show that each attachment has the homotopy type of a cell complex of dimension at most d.

First, if $k = d$ ($\mathbf{p}$ is a smooth point) then, as was observed prior to stating Theorem C.39, the Morse index of h is at most d. The attachment is $(B^i, \partial B^i)$ where i is the Morse index of h, so in this case the homotopy type of the attachment is at most d.

When $k < d$, we proceed by induction on d. The tangential Morse data has the homotopy type of a cell complex of dimension at most k. The space $\mathcal{L}(S)$ is a complex analytic space, with complex dimension one less than the dimension of the normal slice, meaning it has dimension $d-k-1$. The induction hypothesis shows that the homotopy dimension of $\mathcal{L}(S)$ is at most $d - k - 1$. Taking the cone brings the dimension to at most $d - k$ and adding the dimension of the tangential data brings this up to at most d, completing the induction.

(*ii*) When X is the complement of a variety $\mathcal{V}$, still assuming h to be the square of the distance to a generic point, all critical points with respect to the pair $(X, \mathcal{V})$ are contained in $\mathcal{V}$, not in X. Again it suffices to show that the attachments all have homotopy dimension at most $2n - d - 1$, and again we start with the case $k = d$. Here $\mathbf{p}$ is a smooth point of $\mathcal{V}$, so the normal data is the same as for the complement of a point in $\mathbb{C}^{n-d}$, which is $S^{2(n-d)-1}$. The tangential data has homotopy dimension at most d, so the attachment has dimension at most $2n - d - 1$.

When $k < d$, we again proceed by induction on d. The link $\mathcal{L}(S)$ is the complement of $\mathcal{V} \cap A$ in a generic hyperplane A. We have directly $\dim_{\mathbb{C}} N(\mathbf{p}) = n - k$ and $\dim_{\mathbb{C}}(A) = n - k - 1$, and $\dim_{\mathbb{C}}(\mathcal{V} \cap A) = d - k - 1$ because $\mathcal{V}$ has codimension $n - d$, intersects A generically, and $k \leq d - 1$. The induction hypothesis applied to the complement of $\mathcal{V} \cap A$ in A shows that $\mathcal{L}(S)$ has the homotopy type of a cell complex of dimension at most $2(n - k - 1) - (d - k - 1) - 1 = 2n - d - k - 2$. The normal Morse data is the product of this with a 1-complex, hence it has homotopy dimension at most $2n - d - k - 1$, and taking the product with the tangential Morse data brings the dimension up to at most $2n - d - 1$, completing the induction. □

It is useful for the main part of this book to summarize the results from this section for complements of manifolds, applying the Künneth formula to obtain a description of the attachments in terms of specific relative cycles.

Definition D.24 (quasi-local cycles). A (relative or absolute) ***local cycle*** at a point $\mathbf{p}$ is a cycle which may be deformed so as to be in an arbitrarily small neighborhood of $\mathbf{p}$. Given a stratified space with Morse function h, a ***quasi-local cycle*** at a critical point $\mathbf{p}$ of the stratification is a cycle $C_\perp \times C_\|$, where $C_\|$ is a disk in S on which h is strictly maximized at $\mathbf{p}$, $B_\mathbf{p}$ is a small ball around $\mathbf{p}$ in the normal slice, $C^\perp$ is a local cycle in $(B_\mathbf{p} \setminus Y, (B_\mathbf{p} \setminus Y)_{\leq h(\mathbf{p})-\varepsilon})$, and the product is taken in any local coordinatization of a neighborhood of $\mathbf{p}$ by $B_\mathbf{p} \times S$.

Theorem D.25. *Let X be the complement of a complex variety of dimension d in $\mathbb{C}^{d+1}$. Then X may be built by attaching spaces that are homotopy equivalent to cell complexes of dimension at most $d+1$. Consequently, $H_d(X)$ has a basis of quasi-local cycles which may be described as $\mathcal{B} = \{\sigma_{\mathbf{p},i}\}_{\mathbf{p},i}$, where $\mathbf{p}$ ranges over critical points in different strata, and each $\sigma_{\mathbf{p},i} \in X^{c,\mathbf{p}}$. For each fixed $\mathbf{p}$, the projection $\pi_* : X^{c,\mathbf{p}} \to (X^{c,\mathbf{p}}, X_{\leq c-\varepsilon}) = X^{\mathbf{p},\mathrm{loc}}$ maps the set $\{\sigma_{\mathbf{p},i}\}$ to a basis for the relative homology group $H_d(X^{\mathbf{p},\mathrm{loc}})$.* □

Notes

The idea to use Morse theory to evaluate integrals was not one of the original purposes of Morse theory. Nevertheless, the utility of Morse theory for this purpose has been known for over 50 years. Much of the history appears difficult to trace: the present authors learned it from Yuliy Baryshnikov, who related it as mathematical folklore from Arnold's seminar. The smooth Morse theory in this chapter (and some of the pictures) is borrowed from Milnor's classic text [Mil63]. Stratified Morse theory is a relatively new field, in which the seminal text is [GM88]; most of our understanding came from this text.

The result usually quoted as the description of the attachment in the stratified case (a stratified version of Theorem C.28) is an unnumbered result named "Theorem" in [GM88, Section 3.12]. This computes the change in topology of a stratified space X on which the function h is proper. When h is a continuous function on $\mathbb{C}_*^d$, this requires the subset X to be closed. We are chiefly interested in the space $X = \mathcal{V}^c$ which is not closed. Dealing with nonproper height functions requires two extra developmental steps. The first is to develop a system for keeping track of the change in topology of the complement of a closed space up to a varying height cutoff. This computation is similar to the one for the space itself. Goresky and MacPherson state the two results together in a later version of the "Main Theorem" of [GM88], and we have followed their example, stating the results together in Theorem D.21. The second way h can fail to be proper occurs at infinity. The results of [GM88] across the height interval $[a, b]$ can be extended to unbounded spaces when there are no critical points at infinity with heights in $[a, b]$. This was the motivation for the results on CPAI derived in [BMP22], which we use in Chapter 7.

Additional exercises

Exercise D.4 (Whitney umbrella)**.** Let $f(x, y, z) = x^2 + y^2 z$ be the polynomial whose real variety $\mathcal{V}_f$ forms the ***Whitney umbrella***. Decompose $\mathcal{V}_f$ into the union of smooth sets by computing algebraic equations for its singularities, the singularities of its singularities, and so on until no singularities remain. Either prove that this decomposition is a Whitney stratification of $\mathcal{V}_f$, or prove that it is not and find a refinement that is.

Exercise D.5. Let X be the complement in $\mathbb{C}^2$ of the smooth curve $x^2 + y^2 = 1$. Define a Morse function and use it to compute the homology of X.

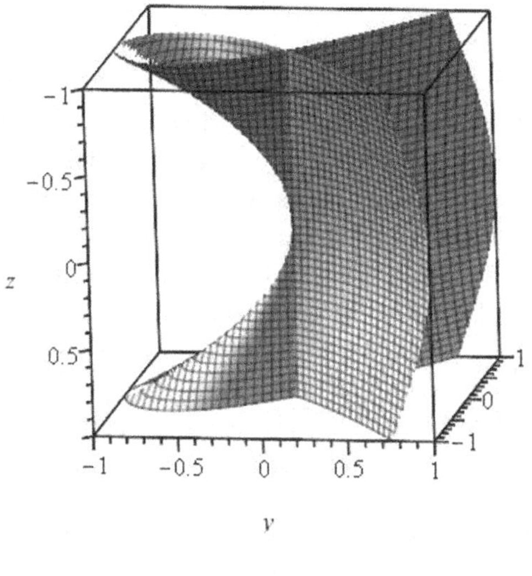

Figure D.7 The real variety in Exercise D.6.

Exercise D.6. Let $Q = x^2 - y^3 - z^2 y^2$ and let $\mathcal{V}_Q$ denote the corresponding real affine variety shown in Figure D.7. Compute the set S of singularities of $\mathcal{V}_Q$, and then determine whether $\{\mathcal{V}_Q \setminus S, S\}$ is a Whitney stratification of $\mathcal{V}_Q$. *Hint:* Consider points $x_n = (0, -t^2, t) \in \mathcal{V}_Q \setminus S$ and $y_n = (0, t, 0) \in S$.

References

[AB05] D. Aldous and A. Bandyopadhyay. "A survey of max-type recursive distributional equations". *Ann. Appl. Prob.* **15** (2005), 1047–1110 (cit. on p. 75).

[AB13] B. Adamczewski and J. P. Bell. "Diagonalization and rationalization of algebraic Laurent series". *Ann. Sci. Éc. Norm. Supér. (4)* **46** (2013), 963–1004 (cit. on p. 56).

[ABG70] M. F. Atiyah, R. Bott, and L. Gårding. "Lacunas for hyperbolic differential operators with constant coefficients I". *Acta Mathematica* **124** (1970), 109–189 (cit. on pp. 15, 126, 211, 217, 340, 353, 355, 365, 379, 522).

[AF59] A. Andreotti and T. Frankel. "The Lefschetz theorem on hyperplane sections". *Ann. of Math. (2)* **69** (1959), 713–717 (cit. on p. 477).

[AG72] R. Askey and G. Gasper. "Certain rational functions whose power series have positive coefficients". *Amer. Math. Monthly* **79** (1972), 327–341 (cit. on p. 376).

[Ald98] D. Aldous. "A Metropolis-Type Optimization Algorithm on the Infinite Tree". *Algorithmica* **22** (1998), 388–412 (cit. on p. 75).

[Amb+01] A. Ambainis et al. "One-dimensional quantum walks". In: *Proceedings of the 33rd Annual ACM Symposium on Theory of Computing*. New York: ACM Press, 2001, 37–49 (cit. on p. 286).

[And89] Y. André. *G-functions and geometry*. Aspects of Mathematics, E13. Friedr. Vieweg & Sohn, Braunschweig, 1989 (cit. on p. 241).

[AY83] I. A. Aĭzenberg and A. P. Yuzhakov. *Integral representations and residues in multidimensional complex analysis*. Vol. 58. Translations of Mathematical Monographs. Providence, RI:

American Mathematical Society, 1983, x+283 (cit. on pp. 9, 328, 337, 489, 490, 511).

[Ban+01] C. Banderier et al. "Random maps, coalescing saddles, singularity analysis, and Airy phenomena". *Random Structures Algorithms* **19** (2001), 194–246 (cit. on p. 426).

[Bar+10] Y. Baryshnikov et al. "Two-dimensional quantum random walk". *J. Stat. Phys.* **142** (2010), 78–107 (cit. on pp. 219, 281, 286, 289, 398, 422).

[Bar+18] Y. Baryshnikov et al. "Diagonal asymptotics for symmetric rational functions via ACSV". In: *29th International Conference on Probabilistic, Combinatorial and Asymptotic Methods for the Analysis of Algorithms (AofA 2018)*. Vol. 110. Dagstuhl, 2018, 12 (cit. on p. 434).

[Bar94] A. Barvinok. "A polynomial time algorithm for counting integral points in polyhedra when the dimension is fixed." *Math. of Operations Res.* **19** (1994), 769–779 (cit. on p. 338).

[Bas08] S. Basu. "Algorithmic semi-algebraic geometry and topology—recent progress and open problems". In: *Surveys on discrete and computational geometry*. Vol. 453. Contemp. Math. Amer. Math. Soc., Providence, RI, 2008, 139–212 (cit. on p. 468).

[BB09] J. Borcea and P. Brändén. "The Lee-Yang and Pólya-Schur programs, II: Theory of stable polynomials and applications". *Comm. Pure Appl. Math.* **62** (2009), 1595–1631 (cit. on pp. 159, 434).

[BBL09] J. Borcea, P. Brändén, and T. Liggett. "Negative dependence and the geometry of polynomials". *J. AMS* **22** (2009), 521–567 (cit. on p. 379).

[Ben+12] I. Bena et al. "Scaling BPS solutions and pure-Higgs states". *J. High Energy Phys.* (2012), 171, front matter + 36 (cit. on p. 289).

[Ben73] E. A. Bender. "Central and local limit theorems applied to asymptotic enumeration". *J. Combinatorial Theory Ser. A* **15** (1973), 91–111 (cit. on pp. 8, 386).

[Ben74] E. A. Bender. "Asymptotic methods in enumeration". *SIAM Rev.* **16** (1974), 485–515 (cit. on p. 8).

[Ber71] G. M. Bergman. "The logarithmic limit-set of an algebraic variety". *Trans. Amer. Math. Soc.* **157** (1971), 459–469 (cit. on p. 164).

[BER99] S. Baouendi, P. Ebenfelt, and J. Rothschild. *Real Submanifolds in Complex Space and Their Mappings*. Princeton: Princeton University Press, 1999, xi1+404 (cit. on p. 126).

[BG91] C. A. Berenstein and R. Gay. *Complex variables*. Vol. 125. Graduate Texts in Mathematics. New York: Springer-Verlag, 1991, xii+650 (cit. on p. 8).

[BH12] C. Banderier and P. Hitczenko. "Enumeration and asymptotics of restricted compositions having the same number of parts". *Discrete Appl. Math.* **160** (2012), 2542–2554 (cit. on p. 423).

[BH86] N. Bleistein and R. A. Handelsman. *Asymptotic expansions of integrals*. Second edition. New York: Dover Publications Inc., 1986, xvi+425 (cit. on pp. 13, 112, 131).

[BJ82] T. Bröcker and K. Jänich. *Introduction to Differential Topology*. New York: Cambridge University Press, 1982, vii+160 (cit. on p. 272).

[BJP23] Y. Baryshnikov, K. Jin, and R. Pemantle. "Coefficient asymptotics of algebraic multivariable generating functions". *Preprint* (2023), 30 (cit. on pp. 431, 436).

[BLS13] A. Bostan, P. Lairez, and B. Salvy. "Creative telescoping for rational functions using the Griffiths-Dwork method". In: *ISSAC 2013—Proceedings of the 38th International Symposium on Symbolic and Algebraic Computation*. ACM, New York, 2013, 93–100 (cit. on p. 241).

[BM93] A. Bertozzi and J. McKenna. "Multidimensional residues, generating functions, and their application to queueing networks". *SIAM Rev.* **35** (1993), 239–268 (cit. on pp. 10, 310, 337).

[BM97] E. Bierstone and P. Millman. "Canonical desingularization in characteristic zero by blowing up the maximum strata of a local invariant". *Inventiones Mathematicae* **128** (1997), 207–302 (cit. on p. 380).

[BMP19] Y. Baryshnikov, S. Melczer, and R. Pemantle. "Asymptotics of multivariate sequences in the presence of a lacuna". *arXiv preprint arXiv:1905.04174* (2019) (cit. on p. 376).

[BMP22] Y. Baryshnikov, S. Melczer, and R. Pemantle. "Stationary points at infinity for analytic combinatorics". *Found. Comput. Math.* **22** (2022), 1631–1664 (cit. on pp. 200, 215–217, 219, 234, 235, 289, 530).

[BMP23] Y. Baryshnikov, S. Melczer, and R. Pemantle. "Asymptotics of multivariate sequences IV: generating functions with poles on a hyperplane arrangement". *Accepted to Annals of Combinatorics* (2023) (cit. on pp. 302, 307, 312, 335, 337).

[Bos+07] A. Bostan et al. "Differential equations for algebraic functions". In: *ISSAC 2007*. New York: ACM, 2007, 25–32 (cit. on p. 56).

[BP00] M. Bousquet-Mélou and M. Petkovšek. "Linear recurrences with constant coefficients: the multivariate case". *Discrete Math.* **225** (2000), 51–75 (cit. on pp. 32, 35, 36, 56).

[BP07] A. Bressler and R. Pemantle. "Quantum random walks in one dimension via generating functions". In: *2007 Conference on Analysis of Algorithms, AofA 07*. Discrete Math. Theor. Comput. Sci. Proc., AH. Assoc. Discrete Math. Theor. Comput. Sci., Nancy, 2007, 403–412 (cit. on pp. 259, 395, 422).

[BP11] Y. Baryshnikov and R. Pemantle. "Asymptotics of multivariate sequences, part III: quadratic points". *Advances in Mathematics* **228** (2011), 3127–3206 (cit. on pp. 15, 212, 219, 289, 340, 341, 352–354, 359, 361, 365, 366, 374, 375, 379, 380, 433, 522).

[BP21] Y. Baryshnikov and R. Pemantle. "Elliptic and hyper-elliptic asymptotics of trivariate generating functions with singularities of degree 3 and 4". *Manuscript in progress* (2021) (cit. on p. 212).

[BPR03] S. Basu, R. Pollack, and M.-F. Roy. *Algorithms in real algebraic geometry*. Vol. 10. Algorithms and Computation in Mathematics. Springer-Verlag, Berlin, 2003 (cit. on pp. 184, 227, 466).

[BR83] E. A. Bender and L. B. Richmond. "Central and local limit theorems applied to asymptotic enumeration. II. Multivariate generating functions". *J. Combin. Theory Ser. A* **34** (1983), 255–265 (cit. on pp. xiv, 8, 9, 288, 385, 386).

[BR99] E. A. Bender and L. B. Richmond. "Multivariate asymptotics for products of large powers with applications to Lagrange inversion". *Electron. J. Combin.* **6** (1999), Research Paper 8, 21 pp. (electronic) (cit. on pp. 9, 386).

[Bri10] E. Briand. "Covariants vanishing on totally decomposable forms". In: *Liaison, Schottky problem and invariant theory*. Vol. 280. Progr. Math. Basel: Birkhäuser Verlag, 2010, 237–256 (cit. on p. 325).

[Bri73] E. Brieskorn. "Sur les groupes de tresses". In: *Séminaire Boubaki*. Vol. 317. Berlin: Springer-Verlag, 1973, 21–44 (cit. on p. 478).

[BRW83] E. A. Bender, L. B. Richmond, and S. G. Williamson. "Central and local limit theorems applied to asymptotic enumeration. III.

Matrix recursions". *J. Combin. Theory Ser. A* **35** (1983), 263–278 (cit. on p. 9).

[Bry77] T. Brylawski. "The broken-circuit complex". *Trans. AMS* **234** (1977), 417–433 (cit. on p. 311).

[Buc65] B. Buchberger. "Ein Algorithmus zum Auffinden der Basiselemente des Restklassenringes nach einem nulldimensionalen Polynomideal". PhD thesis. University of Innsbruck, 1965 (cit. on p. 243).

[BW59] F. Bruhat and H. Whitney. "Quelques propriétés fondamentales des ensembles analytiques-réels". *Comment. Math. Helvetici* **33** (1959), 132–160 (cit. on p. 126).

[BW93] T. Becker and V. Weispfenning. *Gröbner bases*. Vol. 141. Graduate Texts in Mathematics. Springer-Verlag, New York, 1993 (cit. on p. 226).

[CDE06] W. Y. C. Chen, E. Deutsch, and S. Elizalde. "Old and young leaves on plane trees". *European Journal of Combinatorics* **27** (2006), 414–427 (cit. on p. 59).

[Chr15] G. Christol. "Diagonals of rational fractions". *Eur. Math. Soc. Newsl* **97** (2015), 37–43 (cit. on p. 57).

[CIR03] H. A. Carteret, M. E. H. Ismail, and L. B. Richmond. "Three routes to the exact asymptotics for the one-dimensional quantum walk". *J. Phys. A* **36** (33 2003), 8775–8795 (cit. on p. 421).

[CLO05] D. Cox, J. Little, and D. O'Shea. *Using algebraic geometry*. Second edition. Vol. 185. Graduate Texts in Mathematics. New York: Springer, 2005, xii+572 (cit. on p. 244).

[CLO07] D. Cox, J. Little, and D. O'Shea. *Ideals, varieties, and algorithms*. Third edition. Undergraduate Texts in Mathematics. New York: Springer, 2007, xvi+551 (cit. on pp. 222, 224, 233, 243).

[CLP04] S. Corteel, G. Louchard, and R. Pemantle. "Common intervals of permutations". In: *Mathematics and computer science. III*. Trends Math. Basel: Birkhäuser, 2004, 3–14 (cit. on p. 24).

[CM09] E. R. Canfield and B. McKay. "The asymptotic volume of the Birkhoff polytope". *Online J. Anal. Comb.* **4** (2009), 4 (cit. on p. 338).

[Com64] L. Comtet. "Calcul pratique des coefficients de Taylor d'une fonction algébrique". *Enseignement Math. (2)* **10** (1964), 267–270 (cit. on p. 56).

[Com74] L. Comtet. *Advanced combinatorics*. Enlarged edition. Dordrecht: D. Reidel Publishing Co., 1974, xi+343 (cit. on pp. 24, 251, 406).

[Con78a] C. Conley. *Isolated invariant sets and the Morse index*. Vol. 38. CBMS Regional Conference Series in Mathematics. Springer-Verlag, 1978 (cit. on p. 204).

[Con78b] J. B. Conway. *Functions of one complex variable*. Second edition. Vol. 11. Graduate Texts in Mathematics. New York: Springer-Verlag, 1978, xiii+317 (cit. on pp. 8, 62, 164, 349).

[Cox20] D. Cox. "Reflections on elimination theory". In: *ISSAC'20—Proceedings of the 45th International Symposium on Symbolic and Algebraic Computation*. ACM, New York, 2020, 1–4 (cit. on p. 244).

[CS98] F. Chyzak and B. Salvy. "Non-commutative elimination in Ore algebras proves multivariate identities". *J. Symbolic Comput.* **26** (1998), 187–227 (cit. on p. 47).

[dALN15] R. F. de Andrade, E. Lundberg, and B. Nagle. "Asymptotics of the extremal excedance set statistic". *European J. Combin.* **46** (2015), 75–88 (cit. on p. xi).

[dBru81] N. G. de Bruijn. *Asymptotic methods in analysis*. Third edition. New York: Dover Publications Inc., 1981, xii+200 (cit. on pp. 6, 12, 112).

[DeV10] T. DeVries. "A case study in bivariate singularity analysis". In: *Algorithmic probability and combinatorics*. Vol. 520. Contemp. Math. Providence, RI: Amer. Math. Soc., 2010, 61–81 (cit. on pp. 10, 270).

[DeV11] T. DeVries. "Algorithms for bivariate singularity analysis". PhD thesis. University of Pennsylvania, 2011 (cit. on pp. 262, 269).

[DH02] E. Delabaere and C. J. Howls. "Global asymptotics for multiple integrals with boundaries". *Duke Math. J.* **112** (2002), 199–264 (cit. on p. 422).

[DJ21] S. T. Dinh and Z. Jelonek. "Thom isotopy theorem for nonproper maps and computation of sets of stratified generalized critical values". *Discrete Comput. Geom.* **65** (2021), 279–304 (cit. on pp. 227, 515).

[DL87] J. Denef and L. Lipshitz. "Algebraic power series and diagonals". *J. Number Theory* **26** (1987), 46–67 (cit. on pp. 50, 56, 430).

[DL93] R. A. DeVore and G. G. Lorentz. *Constructive approximation*. Vol. 303. Grundlehren der Mathematischen Wissenschaften. Berlin: Springer-Verlag, 1993, x+449 (cit. on p. 331).

[Doš19] T. Došlić. "Block allocation of a sequential resource". *Ars Mathematica Contemporanea* **17** (2019), 79–88 (cit. on p. 388).

[DS03] J. A. De Loera and B. Sturmfels. "Algebraic unimodular counting". *Math. Program.* **96** (2003), 183–203 (cit. on pp. 338, 409).

[Du11] P. Du. *The Aztec Diamond edge-probability generating function.* Masters thesis, Department of Mathematics, University of Pennsylvania. 2011 (cit. on p. 372).

[Dur04] R. Durrett. *Probability: theory and examples.* Third edition. Belmont, CA: Duxbury Press, 2004, 497 (cit. on p. 386).

[DvdHP11] T. DeVries, J. van der Hoeven, and R. Pemantle. "Effective asymptotics for smooth bivariate generating functions". *Online J. Anal. Comb.* **6** (2011) (cit. on p. 262).

[dWol13] T. de Wolff. "On the Geometry, Topology and Approximation of Amoebas". PhD thesis. Frankfurt: Johann Wolfgang Goethe-Universität, 2013 (cit. on p. 165).

[dWol17] T. de Wolff. "Amoebas and their tropicalizations—a survey". In: *Analysis meets geometry.* Trends Math. Birkhäuser/Springer, Cham, 2017, 157–190 (cit. on p. 165).

[Eil47] S. Eilenberg. "Singular homology in differentiable manifolds". *Ann. of Math. (2)* **48** (1947), 670–681 (cit. on p. 464).

[Eis95] D. Eisenbud. *Commutative algebra.* Vol. 150. Graduate Texts in Mathematics. New York: Springer-Verlag, 1995, xvi+785 (cit. on p. 227).

[Fau+23] J.-C. Faugère et al. "Computing critical points for invariant algebraic systems". *J. Symbolic Comput.* **116** (2023), 365–399 (cit. on p. 435).

[FHS04] A. Flaxman, A. W. Harrow, and G. B. Sorkin. "Strings with maximally many distinct subsequences and substrings". *Electron. J. Combin.* **11** (2004), Research Paper 8, 10 pp. (electronic) (cit. on p. 389).

[FIM99] G. Fayolle, R. Iasnogorodski, and V. Malyshev. *Random walks in the quarter-plane.* Vol. 40. Applications of Mathematics (New York). Berlin: Springer-Verlag, 1999, xvi+156 (cit. on p. xiv).

[FL28] K. Friedrichs and H. Lewy. "Das Anfangswertproblem einer beliebigen nichtlinearen hyperbolischen Differentialgleichung beliebiger Ordnung in zwei Variablen. Existenz, Eindeutigkeit und Abhängigkeitsbereich der Lösung." *Math. Annalen* **99** (1928), 200–221 (cit. on p. 368).

[FO90] P. Flajolet and A. M. Odlyzko. "Singularity analysis of generating functions". *SIAM J. Discrete Math.* **3** (1990), 216–240 (cit. on pp. 71, 75, 84).

[For+19] J. Forsgård et al. “Lopsided approximation of amoebas”. *Math. Comp.* **88** (2019), 485–500 (cit. on p. 165).

[FPT00] M. Forsberg, M. Passare, and A. Tsikh. “Laurent determinants and arrangements of hyperplane amoebas”. *Adv. Math.* **151** (2000), 45–70 (cit. on pp. 143, 165).

[FS09] P. Flajolet and R. Sedgewick. *Analytic combinatorics*. Cambridge University Press, 2009, 824 (cit. on pp. xiv, 12, 15, 17, 49, 56, 68, 76, 84, 112, 156, 270, 391, 392).

[Fur67] H. Furstenberg. “Algebraic functions over finite fields”. *J. Algebra* **7** (1967), 271–277 (cit. on pp. 47, 48).

[Går50] L. Gårding. “Linear hyperbolic partial differential equations with constant coefficients”. *Acta Math.* **85** (1950), 1–62 (cit. on pp. 348, 353).

[GE20] E. Granet and F. H. L. Essler. “A systematic $1/c$-expansion of form factor sums for dynamical correlations in the Lieb-Liniger model”. *SciPost Phys.* **9** (2020), Paper No. 082, 76 (cit. on p. xi).

[Geo21] T. George. “Grove arctic curves from periodic cluster modular transformations”. *Int. Math. Res. Not. IMRN* (2021), 15301–15336 (cit. on p. xi).

[Ges81] I. M. Gessel. “Two theorems of rational power series”. *Utilitas Math.* **19** (1981), 247–254 (cit. on p. 56).

[GFS21] O. Gordon, Y. Filmus, and O. Salzman. “Revisiting the Complexity Analysis of Conflict-Based Search: New Computational Techniques and Improved Bounds”. *Proceedings of the Fourteenth International Symposium on Combinatorial Search (SoCS 2021)* (2021) (cit. on p. xi).

[GH94] P. Griffiths and J. Harris. *Principles of algebraic geometry*. Wiley Classics Library. New York: John Wiley & Sons Inc., 1994, xiv+813 (cit. on p. 461).

[Gil22] S. Gillen. “Critical Points at Infinity for Hyperplanes of Directions”. *arXiv preprint arXiv:2210.05748* (2022) (cit. on p. 436).

[GJ04] I. P. Goulden and D. M. Jackson. *Combinatorial enumeration*. Mineola, NY: Dover Publications Inc., 2004, xxvi+569 (cit. on pp. 17, 27, 56, 406).

[GKZ08] I. M. Gel'fand, M. M. Kapranov, and A. V. Zelevinsky. *Discriminants, resultants and multidimensional determinants*. Modern Birkhäuser Classics. Birkhäuser Boston, Inc., Boston, MA, 2008 (cit. on pp. xv, 164, 165, 243).

[GM88] M. Goresky and R. MacPherson. *Stratified Morse theory*. Vol. 14. Berlin: Springer-Verlag, 1988, xiv + 272 (cit. on pp. 14, 185, 204, 217, 219, 513, 519, 524, 525, 527, 529, 530).

[Gor12] M. Goresky. "Introduction to the papers of R. Thom and J. Mather". *Bull. AMS* **49** (2012), 469–474 (cit. on p. 521).

[Gor75] G. Gordon. "The residue calculus in several complex variables". *Trans. AMS* **213** (1975), 127–176 (cit. on p. 484).

[GP74] V. Guillemin and A. Pollack. *Differential Topology*. Englewood Cliffs, NJ: Prentice-Hall, Inc., 1974, xvi+222 (cit. on p. 272).

[GR92] Z. Gao and L. B. Richmond. "Central and local limit theorems applied to asymptotic enumeration. IV. Multivariate generating functions". *J. Comput. Appl. Math.* **41** (1992), 177–186 (cit. on pp. 9, 386).

[Gra] J. M. Grau Ribas. *What is the limit of* $a(n+1)/a(n)$? MathOverflow. eprint: `https://mathoverflow.net/q/389034` (cit. on p. 404).

[Gre+22] T. Greenwood et al. "Asymptotics of coefficients of algebraic series via embedding into rational series (extended abstract)". *Sém. Lothar. Combin.* **86B** (2022), Art. 30, 12 (cit. on pp. 431, 436).

[Gre18] T. Greenwood. "Asymptotics of bivariate analytic functions with algebraic singularities". *J. Comb. Theory A* **153** (2018), 1–30 (cit. on p. 403).

[Gri+90] J. R. Griggs et al. "On the number of alignments of k sequences". *Graphs Combin.* **6** (1990), 133–146 (cit. on p. 409).

[GRZ83] J. Gillis, B. Reznick, and D. Zeilberger. "On elementary methods in positivity theory". *SIAM J. Math. Anal.* **14** (1983), 396–398 (cit. on p. 376).

[GS16] I. M. Gel'fand and G. E. Shilov. *Generalized functions. Vol. 1: Properties and operations. Translated from the Russian by E. Saletan. Reprint of the 1964 original published by Academic Press*. Providence, RI: AMS Chelsea Publishing, 2016, xvii + 423 (cit. on pp. 366, 375).

[GS96] X. Gourdon and B. Salvy. "Effective asymptotics of linear recurrences with rational coefficients". In: *Proceedings of the 5th Conference on Formal Power Series and Algebraic Combinatorics (Florence, 1993)*. Vol. 153. 1996, 145–163 (cit. on p. 62).

[Gül97] O. Gülen. "Hyperbolic polynomials and interior point methods for convex programming". *Math. Oper. Res.* **22** (1997), 350–377 (cit. on p. 348).

[GWW21] J. S. Geronimo, H. J. Woerdeman, and C. Y. Wong. "Spectral density functions of bivariable stable polynomials". *Ramanujan J.* **56** (2021), 265–295 (cit. on p. xi).

[Hat02] A. Hatcher. *Algebraic topology*. Cambridge University Press, Cambridge, 2002 (cit. on pp. 465–467, 472, 473, 478).

[Hay56] W. K. Hayman. "A generalisation of Stirling's formula". *J. Reine Angew. Math.* **196** (1956), 67–95 (cit. on pp. xiv, 78, 84).

[Hen88] P. Henrici. *Applied and computational complex analysis. Vol. 1*. Wiley Classics Library. New York: John Wiley & Sons Inc., 1988, xviii+682 (cit. on p. 8).

[Hen91] P. Henrici. *Applied and computational complex analysis. Vol. 2*. Wiley Classics Library. New York: John Wiley & Sons Inc., 1991, x+662 (cit. on pp. 8, 84, 112).

[Hir73] H. Hironaka. "Subanalytic sets". In: *Number theory, algebraic geometry and commutative algebra, in honor of Yasuo Akizuki*. Tokyo: Kinokuniya, 1973, 453–493 (cit. on p. 515).

[Hir76] M. W. Hirsch. *Differential topology*. Graduate Texts in Mathematics, No. 33. Springer-Verlag, New York-Heidelberg, 1976 (cit. on p. 483).

[HK71] M. L. J. Hautus and D. A. Klarner. "The diagonal of a double power series". *Duke Math. J.* **38** (1971), 229–235 (cit. on p. 47).

[HL16] E. Hubert and G. Labahn. "Computation of invariants of finite abelian groups". *Mathematics of Computation* **85** (2016), 3029–3050 (cit. on p. 435).

[HL17] J. D. Hauenstein and V. Levandovskyy. "Certifying solutions to square systems of polynomial-exponential equations". *J. Symbolic Comput.* **79** (2017), 575–593 (cit. on p. 237).

[HN22] M. Helmer and V. Nanda. "Conormal Spaces and Whitney Stratifications". *Foundations of Computational Mathematics* (2022) (cit. on pp. 227, 515, 516).

[Hör83] L. Hörmander. *The analysis of linear partial differential operators. I*. Vol. 256. Grundlehren der Mathematischen Wissenschaften. Berlin: Springer-Verlag, 1983, ix+391 (cit. on pp. 129, 353).

[Hör90] L. Hörmander. *An introduction to complex analysis in several variables*. Third edition. Vol. 7. North-Holland Mathematical Library. Amsterdam: North-Holland Publishing Co., 1990, xii+254 (cit. on pp. 323, 460, 461).

[HPS77] M. Hirsch, C. Pugh, and M. Shub. "Invariant manifolds". Lecture Notes in Mathematics **583** (1977) (cit. on p. 204).

[HR00a] G. H. Hardy and S. Ramanujan. "Asymptotic formulæ for the distribution of integers of various types [Proc. London Math. Soc. (2) **16** (1917), 112–132]". In: *Collected papers of Srinivasa Ramanujan*. AMS Chelsea Publ., Providence, RI, 2000, 245–261 (cit. on p. 84).

[HR00b] G. H. Hardy and S. Ramanujan. "Une formule asymptotique pour le nombre des partitions de n [Comptes Rendus, 2 Jan. 1917]". In: *Collected papers of Srinivasa Ramanujan*. AMS Chelsea Publ., Providence, RI, 2000, 239–241 (cit. on p. 84).

[HRS18] J. D. Hauenstein, J. I. Rodriguez, and F. Sottile. "Numerical computation of Galois groups". *Foundations of Computational Mathematics* **18** (2018), 867–890 (cit. on p. 327).

[Huh13] J. Huh. "The maximum likelihood degree of a very affine variety". *Compositio Math.* **149** (2013), 1245–1266 (cit. on p. 430).

[HW14] P. E. Harris and J. F. Willenbring. "Sums of squares of Littlewood–Richardson coefficients and GL_n-harmonic polynomials". In: *Symmetry: representation theory and its applications*. Springer, 2014, 305–326 (cit. on p. 412).

[Hwa96] H.-K. Hwang. "Large deviations for combinatorial distributions. I: Central limit theorems". *Ann. Appl. Probab.* **6** (1996), 297–319 (cit. on p. 386).

[Hwa98a] H.-K. Hwang. "Large deviations of combinatorial distributions. II: Local limit theorems". *Ann. Appl. Probab.* **8** (1998), 163–181 (cit. on p. 386).

[Hwa98b] H.-K. Hwang. "On convergence rates in the central limit theorems for combinatorial structures". *Eur. J. Comb.* **19** (1998), 329–343 (cit. on p. 386).

[IK94] E. Isaacson and H. B. Keller. *Analysis of numerical methods*. New York: Dover Publications Inc., 1994, xvi+541 (cit. on p. 37).

[JPS98] W. Jockusch, J. Propp, and P. Shor. "Random Domino Tilings and the Arctic Circle Theorem". *ArXiv Mathematics e-prints* (Jan. 1998). eprint: `arXiv:math/9801068` (cit. on p. 15).

[Kar79] K. Karchyauskas. *Homotopy properties of complex algebraic sets*. Leningrad: Steklov Institute, 1979 (cit. on p. 478).

[Kes78] H. Kesten. "Branching Brownian motion with absorption". *Stochastic Processes Appl.* **7** (1978), 9–47 (cit. on p. 29).

[KLM21] J. Khera, E. Lundberg, and S. Melczer. "Asymptotic enumeration of lonesum matrices". *Adv. in Appl. Math.* **123** (2021), 102–118 (cit. on p. xi).

[Knu06] D. Knuth. *The Art of Computer Programming*. Vol. I–IV. Upper Saddle River, NJ: Addison-Wesley, 2006 (cit. on p. xiv).

[KO07] R. Kenyon and A. Okounkov. "Limit shapes and the complex Burgers equation". *Acta Math.* **199** (2007), 263–302 (cit. on p. 373).

[Kog02] Y. Kogan. "Asymptotic expansions for large closed and loss queueing networks". *Math. Probl. Eng.* **8** (2002), 323–348 (cit. on p. 310).

[Kov19] M. Kovačević. "Runlength-limited sequences and shift-correcting codes: asymptotic analysis". *IEEE Trans. Inform. Theory* **65** (2019), 4804–4814 (cit. on p. xi).

[KP16] R. Kenyon and R. Pemantle. "Double-dimers, the Ising model and the hexahedron recurrence". *J. Comb. Theory, ser. A* **137** (2016), 27–63 (cit. on p. 212).

[Kra01] S. G. Krantz. *Function theory of several complex variables*. AMS Chelsea Publishing, Providence, RI, 2001, xvi+564 (cit. on p. 165).

[KY96] Y. Kogan and A. Yakovlev. "Asymptotic analysis for closed multichain queueing networks with bottlenecks". *Queueing Systems Theory Appl.* **23** (1996), 235–258 (cit. on p. 10).

[KZ11] M. Kauers and D. Zeilberger. "The computational challenge of enumerating high-dimensional rook walks". *Advances in Applied Mathematics* **47** (2011), 813–819 (cit. on p. 421).

[Lai16] P. Lairez. "Computing periods of rational integrals". *Math. Comp.* **85** (2016), 1719–1752 (cit. on p. 241).

[Las15] J. B. Lasserre. *An introduction to polynomial and semi-algebraic optimization*. Cambridge Texts in Applied Mathematics. Cambridge University Press, Cambridge, 2015 (cit. on p. 149).

[Laz73] F. Lazzeri. "Morse theory on singular spaces". *Astérisque* **7–8** (1973), 263–268 (cit. on p. 518).

[Lee03] J. M. Lee. *Introduction to smooth manifolds*. Vol. 218. Graduate Texts in Mathematics. New York: Springer-Verlag, 2003, xviii+628 (cit. on pp. 127, 476).

[Len+23] A. Lenz et al. "Exact Asymptotics for Discrete Noiseless Channels". *Proceedings of the 2023 IEEE International Symposium on Information Theory (ISIT)* (2023) (cit. on p. xi).

[Ler59] J. Leray. "Le calcul différentiel et intégral sur une variété analytique complexe. (Problème de Cauchy. III)". *Bull. Soc. Math. France* **87** (1959), 81–180 (cit. on pp. 9, 484).

[Lic91] B. Lichtin. "The asymptotics of a lattice point problem associated to a finite number of polynomials. I". *Duke Math. J.* **63** (1991), 139–192 (cit. on p. 10).

[Lip88] L. Lipshitz. "The diagonal of a D-finite power series is D-finite". *J. Algebra* **113** (1988), 373–378 (cit. on p. 46).

[Lip89] L. Lipshitz. "D-finite power series". *J. Algebra* **122** (1989), 353–373 (cit. on pp. 44, 45, 56).

[LL99] M. Larsen and R. Lyons. "Coalescing particles on an interval". *J. Theoret. Probab.* **12** (1999), 201–205 (cit. on pp. 15, 33).

[Lla03] M. Lladser. "Asymptotic enumeration via singularity analysis". PhD thesis. The Ohio State University, 2003 (cit. on pp. 10, 426).

[Lla06] M. Lladser. "Uniform formulae for coefficients of meromorphic functions in two variables. I". *SIAM J. Discrete Math.* **20** (2006), 811–828 (electronic) (cit. on pp. 10, 426).

[LMS22] K. Lee, S. Melczer, and J. Smolčić. "Homotopy Techniques for Analytic Combinatorics in Several Variables". In: *Proceedings of the 24th International Symposium on Symbolic and Numeric Algorithms for Scientific Computing (SYNASC)*. 2022, 27–34 (cit. on pp. 236, 237, 436).

[LP04] V. Limic and R. Pemantle. "More rigorous results on the Kauffman-Levin model of evolution". *Ann. Probab.* **32** (2004), 2149–2178 (cit. on p. 58).

[LT10] D. T. Lê and B. Teissier. "Geometry of characteristic varieties". In: *Algebraic approach to differential equations*. World Sci. Publ., Hackensack, NJ, 2010, 119–135 (cit. on p. 518).

[Mat12] J. Mather. "Notes on topological stability". *Bull. AMS* **49** (2012), 475–506 (cit. on pp. 521, 522).

[Mat70] J. Mather. "Notes on topological stability". *Mimeographed notes* (1970) (cit. on pp. 126, 216, 217).

[Mel21] S. Melczer. *An Invitation to Analytic Combinatorics: From One to Several Variables*. Texts & Monographs in Symbolic Computation. Springer International Publishing, 2021 (cit. on pp. xi, 10, 41, 46, 56, 57, 62, 153, 184, 227, 242–244, 408, 421).

[Mer+97] D. Merlini et al. "On some alternative characterizations of Riordan arrays". *Canad. J. Math.* **49** (1997), 301–320 (cit. on p. 386).

[Mez16] M. Mezzarobba. "Rigorous multiple-precision evaluation of D-finite functions in SageMath". *arXiv preprint arXiv:1607.01967* (2016) (cit. on p. 242).

[Mez19] M. Mezzarobba. "Truncation bounds for differentially finite series". *Ann. H. Lebesgue* **2** (2019), 99–148 (cit. on p. 237).

[Mik00] G. Mikhalkin. “Real algebraic curves, the moment map and amoebas”. *Ann. of Math. (2)* **151** (2000), 309–326 (cit. on p. 165).

[Mik04] G. Mikhalkin. “Amoebas of algebraic varieties and tropical geometry”. In: *Different faces of geometry*. Vol. 3. Int. Math. Ser. (N. Y.) Kluwer/Plenum, New York, 2004, 257–300 (cit. on p. 165).

[Mil63] J. Milnor. *Morse theory*. Based on lecture notes by M. Spivak and R. Wells. Annals of Mathematics Studies, No. 51. Princeton, N.J.: Princeton University Press, 1963, vi+153 (cit. on pp. 14, 185, 500, 501, 504, 511, 529).

[Mis19] M. Mishna. *Analytic Combinatorics: A Multidimensional Approach*. CRC Press, 2019 (cit. on p. xi).

[MM08] Ž. Mijajlović and B. Malešević. “Differentially transcendental functions”. *Bull. Belg. Math. Soc. Simon Stevin* **15** (2008), 193–201 (cit. on p. 57).

[MM16] S. Melczer and M. Mishna. “Asymptotic lattice path enumeration using diagonals”. *Algorithmica* **75** (2016), 782–811 (cit. on p. xi).

[MR91] T. Mostowski and E. Rannou. “Complexity of the computation of the canonical Whitney stratification of an algebraic set in $\mathbf{C}^n$”. In: *Applied algebra, algebraic algorithms and error-correcting codes (New Orleans, LA, 1991)*. Vol. 539. Lecture Notes in Comput. Sci. Berlin: Springer, 1991, 281–291 (cit. on pp. 227, 515).

[MS21] S. Melczer and B. Salvy. “Effective Coefficient Asymptotics of Multivariate Rational Functions via Semi-Numerical Algorithms for Polynomial Systems”. *Journal of Symbolic Computation* **103** (2021), 234–279 (cit. on pp. 62, 229, 236, 237).

[MS22] S. Melczer and J. Smolčić. “Rigorous two dimensional analytic combinatorics in Sage”. 2022 (cit. on pp. 269, 270).

[Mum95] D. Mumford. *Algebraic geometry. I*. Classics in Mathematics. Berlin: Springer-Verlag, 1995, x+186 (cit. on p. 228).

[Mun84] J. R. Munkres. *Elements of algebraic topology*. Menlo Park, CA: Addison-Wesley Publishing Company, 1984, ix+454 (cit. on pp. 467, 469, 471, 473, 474, 478).

[MW19] S. Melczer and M. C. Wilson. “Higher dimensional lattice walks: connecting combinatorial and analytic behavior”. *SIAM J. Disc. Math.* **33** (2019), 2140–2174 (cit. on pp. xi, 318).

[Nob10] R. Noble. “Asymptotics of a family of binomial sums”. *J. Number Theory* **130** (2010), 2561–2585 (cit. on p. 421).

[Odl95] A. M. Odlyzko. “Asymptotic enumeration methods”. In: *Handbook of combinatorics, Vol. 1, 2.* Amsterdam: Elsevier, 1995, 1063–1229 (cit. on pp. xiv, 9, 405).

[OT92] P. Orlik and H. Terao. *Arrangements of hyperplanes.* Vol. 300. Grundlehren der Mathematischen Wissenschaften [Fundamental Principles of Mathematical Sciences]. Berlin: Springer-Verlag, 1992, xviii+325 (cit. on pp. 297, 312).

[Pan17] J. Pantone. “The Asymptotic Number of Simple Singular Vector Tuples of a Cubical Tensor”. *Online Journal of Analytic Combinatorics* **12** (2017) (cit. on p. xi).

[Par11] R. B. Paris. *Hadamard expansions and hyperasymptotic evaluation.* Vol. 141. Encyclopedia of Mathematics and its Applications. Cambridge: Cambridge University Press, 2011, viii+243 (cit. on p. 422).

[Pem00] R. Pemantle. “Generating functions with high-order poles are nearly polynomial”. In: *Mathematics and computer science (Versailles, 2000).* Trends Math. Basel: Birkhäuser, 2000, 305–321 (cit. on p. 338).

[Pem10] R. Pemantle. “Analytic combinatorics in d variables: an overview”. In: *Algorithmic probability and combinatorics.* Vol. 520. Contemp. Math. Providence, RI: Amer. Math. Soc., 2010, 195–220 (cit. on p. 219).

[Pha11] F. Pham. *Singularities of integrals: homology, hyperfunctions and microlocal analysis.* Universitext. New York: Springer, 2011, xxii+217 (cit. on pp. 337, 511).

[Pig79] R. Pignoni. “Density and stability of Morse functions on a stratified space”. *Ann. Scuola Nor. Sup.. Pisa, ser. IV* **6** (1979), 593–608 (cit. on p. 518).

[PK01] R. B. Paris and D. Kaminski. *Asymptotics and Mellin-Barnes integrals.* Vol. 85. Encyclopedia of Mathematics and its Applications. Cambridge: Cambridge University Press, 2001, xvi+422 (cit. on p. 422).

[Poi85] H. Poincaré. “Sure les courbes définies par les équations différentielles”. *J. Math. Pure et Appliquées* **4** (1885), 167–244 (cit. on p. 443).

[Pól69] G. Pólya. “On the number of certain lattice polygons”. *J. Combinatorial Theory* **6** (1969), 102–105 (cit. on p. 405).

[Poo60] E. G. C. Poole. *Introduction to the theory of linear differential equations*. Dover Publications, Inc., New York, 1960 (cit. on p. 241).

[PPT13] M. Passare, D. Pochekutov, and A. Tsikh. "Amoebas of complex hypersurfaces in statistical thermodynamics". *Math. Phys. Anal. Geom.* **16** (2013), 89–108 (cit. on p. 165).

[PS05] T. K. Petersen and D. Speyer. "An arctic circle theorem for Groves". *J. Combin. Theory Ser. A* **111** (2005), 137–164 (cit. on p. 370).

[PS98] G. Pólya and G. Szegő. *Problems and theorems in analysis. II.* Classics in Mathematics. Berlin: Springer-Verlag, 1998, xii+392 (cit. on p. 37).

[PV19] F. Pinna and C. Viola. "The saddle-point method in $\mathbb{C}^N$ and the generalized Airy functions". *Bull. Math. Soc. France* **147** (2019), 211–257 (cit. on p. 131).

[PW02] R. Pemantle and M. C. Wilson. "Asymptotics of multivariate sequences. I. Smooth points of the singular variety". *J. Combin. Theory Ser. A* **97** (2002), 129–161 (cit. on pp. 10, 15, 219, 289, 337, 421).

[PW04] R. Pemantle and M. C. Wilson. "Asymptotics of multivariate sequences. II. Multiple points of the singular variety". *Combin. Probab. Comput.* **13** (2004), 735–761 (cit. on pp. 10, 15, 219, 317, 337).

[PW08] R. Pemantle and M. C. Wilson. "Twenty combinatorial examples of asymptotics derived from multivariate generating functions". *SIAM Rev.* **50** (2008), 199–272 (cit. on pp. 10, 421).

[PW10] R. Pemantle and M. C. Wilson. "Asymptotic expansions of oscillatory integrals with complex phase". In: *Algorithmic probability and combinatorics*. Vol. 520. Contemp. Math. Providence, RI: Amer. Math. Soc., 2010, 221–240 (cit. on pp. 10, 130).

[PWZ96] M. Petkovšek, H. S. Wilf, and D. Zeilberger. $A = B$. Wellesley, MA: A K Peters Ltd., 1996, xii+212 (cit. on p. 57).

[Ran98] E. Rannou. "The complexity of stratification computation". *Discrete Comput. Geom.* **19** (1998), 47–78 (cit. on pp. 227, 515).

[Rie49] M. Riesz. "L'intégrale de Riemann-Liouville et le problème de Cauchy". *Acta Mathematica* **81** (1949), 1–223 (cit. on p. 365).

[Roc66] R. T. Rockefellar. *Convex analysis*. Princeton: Princeton University Press, 1966, xiii+451 (cit. on p. 165).

[RT05] M. Régnier and F. Tahi. *Generating functions in computational biology*. Preprint available at http://algo.inria.fr/regnier/publis/ReTa04.ps. (2005) (cit. on p. 408).

[Rub83] L. A. Rubel. "Some research problems about algebraic differential equations". *Trans. Amer. Math. Soc.* **280** (1983), 43–52 (cit. on p. 57).

[Rub92] L. A. Rubel. "Some research problems about algebraic differential equations. II". *Illinois J. Math.* **36** (1992), 659–680 (cit. on p. 57).

[Rud69] W. Rudin. *Function theory in polydiscs.* W. A. Benjamin, Inc., New York-Amsterdam, 1969 (cit. on p. 165).

[RW08] A. Raichev and M. C. Wilson. "Asymptotics of coefficients of multivariate generating functions: improvements for smooth points". *Electron. J. Combin.* **15** (2008), Research Paper 89, 17 (cit. on pp. 10, 289).

[RW11] A. Raichev and M. C. Wilson. "Asymptotics of coefficients of multivariate generating functions: improvements for multiple points". *Online J. Anal. Comb.* **6** (2011), 21 (cit. on p. 10).

[RWZ20] S. Ramgoolam, M. C. Wilson, and A. Zahabi. "Quiver asymptotics: free chiral ring". *Journal of Physics A: Mathematical and Theoretical* **53** (2020), 105401 (cit. on pp. xi, 412).

[Saf00] K. V. Safonov. "On power series of algebraic and rational functions in $\mathbf{C}^n$". *J. Math. Anal. Appl.* **243** (2000), 261–277 (cit. on pp. 49–51, 430).

[Sha+91] L. W. Shapiro et al. "The Riordan group". *Discrete Appl. Math.* **34** (1991), 229–239 (cit. on p. 421).

[Sha13] I. Shafarevich. *Basic Algebraic Geometry 1: Varieties in Projective Space.* Third edition. New York: Springer, 2013, xviii+310 (cit. on p. 182).

[Spr94] R. Sprugnoli. "Riordan arrays and combinatorial sums". *Discrete Math.* **132** (1994), 267–290 (cit. on pp. 386, 421).

[SS14] A. D. Scott and A. Sokal. "Complete monotonicity for inverse powers of some combinatorially defined polynomials". *Acta Mathematica* **213** (2014), 323–392 (cit. on p. 368).

[Sta15] R. P. Stanley. *Catalan numbers.* Cambridge University Press, 2015 (cit. on p. 31).

[Sta80] R. P. Stanley. "Differentiably finite power series". *European J. Combin.* **1** (1980), 175–188 (cit. on p. 56).

[Sta97] R. P. Stanley. *Enumerative combinatorics. Vol. 1.* Vol. 49. Cambridge Studies in Advanced Mathematics. Cambridge: Cambridge University Press, 1997, xii+325 (cit. on pp. 3, 15, 17, 56, 405, 409).

[Sta99] R. P. Stanley. *Enumerative combinatorics. Vol. 2.* Vol. 62. Cambridge Studies in Advanced Mathematics. Cambridge: Cambridge University Press, 1999, xii+581 (cit. on pp. 41, 42, 47, 56, 436).

[Ste93] E. M. Stein. *Harmonic analysis: real-variable methods, orthogonality, and oscillatory integrals.* Vol. 43. Princeton Mathematical Series. Princeton, NJ: Princeton University Press, 1993, xiv+695 (cit. on pp. 105, 112, 121, 131).

[Stu02] B. Sturmfels. *Solving systems of polynomial equations.* Vol. 97. CBMS regional conference series in mathematics. Providence: American Mathematical Society, 2002, viii+152 (cit. on p. 244).

[Sze33] G. Szegő. "Über gewisse Potenzreihen mit lauter positiven Koeffizienten". *Math. Z.* **37** (1933), 674–688 (cit. on pp. 368, 376).

[Tei82] B. Teissier. "Variétés polaires. II. Multiplicités polaires, sections planes, et conditions de Whitney". In: *Algebraic geometry (La Rábida, 1981).* Vol. 961. Lecture Notes in Math. Springer, Berlin, 1982, 314–491 (cit. on p. 227).

[The02] T. Theobald. "Computing amoebas". *Experiment. Math.* **11** (2002), 513–526 (cit. on p. 165).

[Tim18] S. Timme. "Fast Computation of Amoebas, Coamoebas and Imaginary Projections in Low Dimensions". MA thesis. Technische Universität Berlin, 2018 (cit. on p. 165).

[Tu11] L. W. Tu. *An introduction to manifolds.* Second edition. Universitext. Springer, New York, 2011 (cit. on pp. 447, 461).

[Var77] A. N. Varchenko. "Newton polyhedra and estimation of oscillating integrals". *Functional Anal. Appl.* **10** (1977), 175–196 (cit. on p. 429).

[VG87] A. Varchenko and I. Gelfand. "Combinatorics and topology of configuration of affine hyperplanes in real space". *Funk. Analiz i ego Prilozh.* **21** (1987), 11–22 (cit. on p. 304).

[Vid17] R. Vidunas. "Counting derangements and Nash equilibria". *Ann. Comb.* **21** (2017), 131–152 (cit. on p. xi).

[vLW01] J. H. van Lint and R. M. Wilson. *A Course in Combinatorics.* Second edition. Cambridge: Cambridge University Press, 2001, xiv+602 (cit. on p. 17).

[Voi02] C. Voisin. *Hodge theory and complex algebraic geometry. I.* Vol. 76. Cambridge Studies in Advanced Mathematics. Cambridge University Press, Cambridge, 2002 (cit. on pp. 477, 478).

[Wag11] D. G. Wagner. “Multivariate Stable Polynomials: theory and application”. *Bull. AMS* **48** (2011), 53–84 (cit. on p. 379).

[Wan22] H.-Y. Wang. “A bivariate rational Laurent series of interest”. *Personal communication* (2022) (cit. on pp. 274, 276).

[War10] M. Ward. “Asymptotic rational approximation to Pi: Solution of an unsolved problem posed by Herbert Wilf”. *Disc. Math. Theor. Comp. Sci.* **AM** (2010), 591–602 (cit. on p. 85).

[War83] F. W. Warner. *Foundations of differentiable manifolds and Lie groups*. Vol. 94. Graduate Texts in Mathematics. New York: Springer-Verlag, 1983, ix+272 (cit. on pp. 455, 456, 461).

[Wat95] M. S. Waterman. “Applications of combinatorics to molecular biology”. In: *Handbook of combinatorics, Vol. 1, 2*. Amsterdam: Elsevier, 1995, 1983–2001 (cit. on p. 408).

[Whi65a] H. Whitney. “Local properties of analytic varieties”. In: *Differential and Combinatorial Topology*. Princeton, NJ: Princeton University Press, 1965 (cit. on p. 521).

[Whi65b] H. Whitney. “Tangents to an analytic variety”. *Annals Math.* **81** (1965), 496–549 (cit. on p. 515).

[Wil05] M. C. Wilson. “Asymptotics for generalized Riordan arrays”. In: *2005 International Conference on Analysis of Algorithms*. Discrete Math. Theor. Comput. Sci. Proc., AD. Assoc. Discrete Math. Theor. Comput. Sci., Nancy, 2005, 323–333 (cit. on pp. 10, 388, 421).

[Wil06] H. S. Wilf. *generatingfunctionology*. Third edition. Wellesley, MA: A. K. Peters, 2006, x+245 (cit. on pp. 12, 17, 405).

[Wil15] M. C. Wilson. “Diagonal Asymptotics for Products of Combinatorial Classes”. *Combinatorics, Probability and Computing* **24** (2015), 354–372 (cit. on pp. xi, 421).

[Won01] R. Wong. *Asymptotic approximations of integrals*. Vol. 34. Classics in Applied Mathematics. Philadelphia, PA: Society for Industrial and Applied Mathematics (SIAM), 2001, xviii+543 (cit. on pp. 112, 131).

[WZ85] J. Wimp and D. Zeilberger. “Resurrecting the asymptotics of linear recurrences”. *J. Math. Anal. Appl.* **111** (1985), 162–176 (cit. on p. 57).

[Zei82] D. Zeilberger. “Sister Celine’s technique and its generalizations”. *J. Math. Anal. Appl.* **85** (1982), 114–145 (cit. on p. 56).

Author Index

Subject Index

A **boldface** page number indicates the page on which the term is defined.

Printed in the United States
by Baker & Taylor Publisher Services